T0178160

Stream Ecology

J. David Allan • María M. Castillo •
Krista A. Capps

Stream Ecology

Structure and Function of Running Waters

Third Edition

 Springer

J. David Allan
School for Environment
and Sustainability
University of Michigan
Ann Arbor, MI, USA

María M. Castillo
Departamento de Ciencias
de la Sustentabilidad, El Colegio
de la Frontera Sur
Villahermosa, Tabasco, México

Krista A. Capps
Odum School of Ecology
and the Savannah River Ecology Laboratory
University of Georgia, Athens, GA, USA

ISBN 978-3-030-61288-7   ISBN 978-3-030-61286-3   (eBook)
https://doi.org/10.1007/978-3-030-61286-3

1st edition: © J. David Allan 1995
2nd edition: © Springer 2007
3rd edition: © Springer Nature Switzerland AG 2021
This work is subject to copyright. All rights are reserved by the Publisher, whether the whole or part of the material is concerned, specifically the rights of translation, reprinting, reuse of illustrations, recitation, broadcasting, reproduction on microfilms or in any other physical way, and transmission or information storage and retrieval, electronic adaptation, computer software, or by similar or dissimilar methodology now known or hereafter developed.
The use of general descriptive names, registered names, trademarks, service marks, etc. in this publication does not imply, even in the absence of a specific statement, that such names are exempt from the relevant protective laws and regulations and therefore free for general use.
The publisher, the authors and the editors are safe to assume that the advice and information in this book are believed to be true and accurate at the date of publication. Neither the publisher nor the authors or the editors give a warranty, expressed or implied, with respect to the material contained herein or for any errors or omissions that may have been made. The publisher remains neutral with regard to jurisdictional claims in published maps and institutional affiliations.

Cover photo: Students conducting macroinvertebrate sampling in the Guare River, Venezuela

This Springer imprint is published by the registered company Springer Nature Switzerland AG
The registered company address is: Gewerbestrasse 11, 6330 Cham, Switzerland

Dedicated to our families, and to colleagues, students, and all those whose efforts advance our understanding of flowing waters to protect them for future generations

Preface to the Third Edition

The underlying science of stream ecology has undergone dramatic advances since the first edition over 20 years ago. The subject matter of other disciplines increasingly must be addressed, including hydrology, geomorphology, an array of topics in the earth sciences and biogeochemistry, along with advances in socioeconomic studies. The published literature is enormous, spread across many different professional journals. We cannot do justice to all of these advances, but hopefully, we have provided an entry point for readers new to the field. Some of the material will be challenging to some readers, depending on background. Each chapter has a summary that clarifies the main ideas covered, and it may be helpful to view that first, to keep the main ideas in focus.

This Third Edition of *Stream Ecology* will be used primarily in electronic form, as an e-book, and viewers will have the opportunity to acquire individual chapters rather than the entire book. Recognizing that readers may not have the opportunity to refer back and forth among different chapters, we have made an effort to ensure that each chapter can stand alone. The result is a moderate amount of repetition, which we hope will serve either as a useful explanation or reminder, ensuring that each chapter's narrative is accessible without depending upon significant cross-referencing to other chapters.

We are grateful to many individuals who have discussed ideas and shared references with us in preparation of this Third Edition of *Stream Ecology*. We especially thank a number of colleagues who read and improved earlier versions of chapters, including Sebastian Birk, Alex Flecker, Hal Halvorson, Nick Hudson, Susan Jackson, Peter McIntyre, Thomas Parr, LeRoy Poff, Amber Ulseth, and Ellen Wohl. We also appreciate the generosity of colleagues who shared or helped us acquire photographs, including Francesco Comiti, Carolyn Cummins, Chris Dutton, Jeff Duda, John Gussman, Angus McIntosh, Jeremy Monroe, Luca Messina, Jim O'Connor, Julian Olden, Amanda Subalusky, and John Warrick. We are especially grateful to Jesús Montoya for his excellent work in figure preparation. We appreciate the support of our editors Judith Terpos and Nel van der Werf during the preparation of this edition, and the efficient work of their production team.

Although the material in this third edition is substantially modified and updated, it nonetheless includes material from previous editions. We acknowledge our debt to many wonderful colleagues, who provided helpful guidance to previous editions of *Stream Ecology*.

Ann Arbor, USA — J. David Allan
Villahermosa, México — María M. Castillo
Athens, USA — Krista A. Capps

Preface to the Second Edition

The diversity of running water environments is enormous. When one considers torrential mountain brooks, large rivers of lowlands, and great rivers whose basins occupy subcontinents, it is apparent how location-specific environmental factors contribute to the sense of uniqueness and diversity of running waters. At the same time, however, our improved understanding of ecological, biogeochemical, hydrological, and geomorphological processes provides insight into the structural and functional characteristics of river systems that brings a unifying framework to this field of study. Inputs and transformations of energy and materials are important in all river systems, regional species richness and local species interactions influence the structure of all riverine communities, and the interaction of physical and biological forces is important to virtually every question that is asked. It seems that the processes acting in running waters are general, but the settings often are unique.

We believe that it helps the reader, when some pattern or result is described, to have some image of what kind of stream or river is under investigation, and also where it is located. Stream ecology, like all ecology, depends greatly on context: place, environmental conditions, season, species. The text includes frequent use of descriptors like "small woodland stream", "open pastureland stream", or "large lowland river", and we believe that readers will find these useful clues to the patterns and processes that are reported. For most studies within North America we have included further regional description, but have done so less frequently for studies from elsewhere around the globe. We apologize to our international readers for this pragmatic choice, and we have made every effort to include examples and literature from outside of North America.

Some locations have established themselves as leading centers of study due to the work of many researchers carried out over decades. The Hubbard Brook Experimental Forest in New Hampshire, Coweeta Hydrologic Laboratory in North Carolina, and some individual streams including Walker Branch in Tennessee, Sycamore Creek in Arizona, Río Las Marías in Venezuela, and the Taieri and Whatawhata in New Zealand, are locations that appear frequently in the pages that follow. Knowing what these places are like, and how they may or may not be typical, in our view justifies the frequent use of place names and brief descriptions. The names of organisms also appear frequently and may at first overwhelm the reader. It may be easiest to pay them little attention until they gradually become familiar. Ultimately it is difficult to really comprehend the outcome of a study without some appreciation for the organisms that were present.

As is true for every area of ecology in present time, the study of streams and rivers cannot be addressed exclusive of the role of human activities, nor can we ignore the urgency of the need for conservation. This is a two-way street. Ecologists who study streams without considering how past or present human modifications of the stream or its valley might have contributed to their observations do so at the risk of incomplete understanding. Conservation efforts that lack an adequate scientific basis are less likely to succeed. One trend that seems safe to forecast in stream ecology is toward a greater emphasis on understanding human impacts. Fortunately, signs of this trend are already apparent.

We have organized the flow of topics in a way that is most logical to us, but no doubt some readers will prefer to cover topics in whatever order they find most useful. For this reason, we have strived to explain enough in each chapter that it is comprehensible on its own. This leads to a certain amount of intentional repetition, which we hope will provide clarification or a reminder that will benefit the reader's understanding.

We are extremely grateful to the many colleagues who shared ideas, provided references, and reviewed chapters in draft form. Space doesn't permit us to thank everyone who answered a query with a helpful explanation and suggestions for source material; however we do wish to acknowledge the individuals who carefully read and improved our chapters. Any remaining shortcomings or errors are the authors' responsibility, but hopefully these are few, thanks to the efforts of: Robin Abell, Brian Allan, Fred Benfield, Barb Downes, David Dudgeon, Kurt Fausch, Stuart Findlay, Alex Flecker, Art Gold, Sujay Kaushal, Matt Kondolf, Angus McIntosh, Peter McIntyre, Rich Merritt, Judy Meyer, Pat Mulholland, Bobbi Peckarsky, LeRoy Poff, Brian Roberts, Doug Shields, Al Steinman, Jan Stevenson, Jen Tank, Paul Webb, Jack Webster, Kevin Wehrly, and Kirk Winemiller. All were generous with their time and their knowledge, and we are in their debt.

We also wish to thank those who provided helpful reviews of chapters in the first edition of this book, including Fred Benfield, Art Benke, Art Brown, Scott Cooper, Stuart Findlay, Alex Flecker, Nancy Grimm, David Hart, Chuck Hawkins, Bob Hughes, Steve Kohler, Gary Lamberti, Rex Lowe, Rich Merritt, Diane McKnight, Judy Meyer, Bobbi Peckarsky, Pete Ode, Walt Osterkamp, M.L. Ostrofsky, Margaret Palmer, LeRoy Poff, Karen Prestergaard, Ike Schlosser, Len Smock, Al Steinman, Scott Wissinger, and Jack Webster.

Other individuals provided invaluable assistance with important aspects of manuscript production. Mary Henja and Jamie Steffes did extensive proof reading and arranged all the figure permissions. Haymara Alvarez, Susana Martinez and Dana Infante assisted with production of figures, and Jesus Montoya did a superb job of taking figures made in many different styles and re-drafting them to a common style and high quality. Funding for MMC release time and travel to Michigan was provided by Dirección de Desarrollo Profesoral of Universidad Simon Bolivar, and the Horace H Rackham School of Graduate Studies of the University of Michigan. We also wish to thank our editors at Springer, Suzanne Mekking and Martine van Bezooijen, and our prior editor Anna Besse-Lototskaya, for their support, encouragement, and patience. It has been a pleasure to work with them all.

Lastly, our deepest thanks go to our families for their love and support, and we must admit for a good deal of tolerance as well, during the writing of this book. It has been an enjoyable experience for both of us, and we hope that the current edition will serve as a useful guide for the next generation of stream ecologists.

Contents

About the Authors

J. David Allan is Professor Emeritus in the School for Environment and Sustainability at the University of Michigan. His work emphasizes the application of ecological knowledge to species conservation and ecosystem management. Research interests center on the influence of human activities on the condition of rivers and their watersheds, including the effects of land use on stream health, assessment of variation in flow regime, and estimation of nutrient loads and budgets. Additional, collaborative activities are directed at the translation of aquatic science into useful products for management, conservation, and restoration of running waters.

María M. Castillo is a Research Scientist in the Departmento de Ciencias de la Sustentabilidad at El Colegio de la Frontera Sur, México. The aim of her research is to better understand the influence of natural and anthropogenic drivers on tropical stream and river ecosystems. Her work emphasizes the influence of watershed and riparian processes, hydrological seasonality, and river-floodplain interactions on fluvial ecosystems. Her major research interests are the effects of land use change on water quality and aquatic communities, nutrient transport by rivers, and ecological functioning of floodplain ecosystems.

Krista A. Capps is an Assistant Professor in the Odum School of Ecology and the Savannah River Ecology Laboratory at the University of Georgia. Her research is dedicated to understanding how anthropogenic activities alter community structure and ecosystem processes in temperate and tropical freshwater ecosystems. She attempts to view her research through a social-ecological lens, acknowledging the powerful impacts that public policy and economic considerations can have on the quality and quantity of freshwater resources, the abundance and diversity of aquatic organisms, and the function of freshwater ecosystems.

Streams and rivers occur in almost bewildering variety. In common usage, rivers are larger than streams, and some 30 or so are referred to as big or great rivers based on discharge, basin area, and length (Best 2019). Partly because the vast majority of river length is in the smaller headwater streams, and partly because these smaller systems have received considerably more study, many researchers consider themselves 'stream ecologists'. Attempting to understand how the principles of fluvial systems are manifested across scale is one of the primary themes of this book. Fluvial ecosystems vary in many additional features, of course. Some are the color of tea due to high concentrations of dissolved plant matter, while others have fewer chemical constituents and so remain clear; these are known as blackwater and clearwater rivers, respectively. Rivers can tumble and cascade down steep slopes over large boulders, meander through gentle valleys, or flow majestically across broad flats as they approach the sea. Food webs in temperate forested streams derive much of their food base from autumn leaf fall, whereas streams that are open, shallow, and stony typically develop a rich film of algae and microbes. Rivers that still have an intact floodplain exchange organic matter and nutrients with the adjacent land, and all fluvial ecosystems exhibit high connectivity laterally, longitudinally, and vertically. Rivers transport vast quantities of sediments, nutrients, and carbon from uplands to the sea. Along a river's length, sediment mobilization and deposition shape channels, and nutrients are incorporated into the biota and then released in cycles that are more like spirals as water flows downstream.

River science attempts to catalog this diversity, reveal the underlying processes that are responsible for the variety of patterns that we observe, and understand how those processes interact with different environmental settings and across scale from the smallest headwater streams to great rivers. Most rivers today, except those in remote regions, flow through human-dominated landscapes and are managed to varying degrees to meet societal and environmental goals.

River science thus is very much an applied field as it seeks to understand pressures on ecosystem condition and identify management actions to meet human needs and maintain healthy ecosystems. This introductory chapter sets the stage for the rest of this book: first, we describe key elements of a healthy river; second, we provide an overview of the many threats that rivers face; lastly we provide an outline of forthcoming chapters, as they build toward an understanding of the fundamental science that underpins our best efforts to repair, restore, and protect streams and rivers wherever possible.

1.1 Structure and Function of River Systems

The diversity of running water environments is enormous. When one considers torrential mountain brooks, large rivers of lowlands, and great rivers whose basins occupy subcontinents, it is apparent how location-specific environmental factors contribute to the sense of uniqueness and diversity of running waters. At the same time, however, our improved understanding of ecological, biogeochemical, hydrological, and geomorphological processes provides insight into the structural and functional characteristics of river systems that bring a unifying framework to this field of study. Inputs and transformations of energy and materials are important in all river systems, regional species richness and local species interactions influence the structure of all riverine communities, and the interaction of physical and biological forces is important to virtually every question that is asked. It seems that the processes acting in running waters are general, but the settings often are unique.

1.1.1 Physical Setting

Each stream or river drains an area of land referred to as its catchment or watershed, defined topographically by ridge

© Springer Nature Switzerland AG 2021
J. D. Allan et al., *Stream Ecology*,
https://doi.org/10.1007/978-3-030-61286-3_1

lines that divide one catchment from another. Precipitation that falls within this area is routed downslope, usually by subsurface pathways, replenishing groundwater and reaching stream channels after a lag time that varies with slope, soil composition, vegetation, and many other factors. The volume of streamflow tracks precipitation and snowmelt, which accounts for flow variability, but perennial streams flow even during long periods without rain, as groundwater supply maintains baseflow throughout the year. At larger spatial scales, one sees the joining of individual small streams, so the combined catchment area becomes greater, as does the volume of water carried and the size of the river. Thus, streams and the landscape units they drain form nested hierarchies. The smallest permanently flowing stream is referred to as first order. The union of two first-order streams results in a second-order stream, the union of two streams of order two results in a stream of order three, and so on. Stream order is an approximate measure of stream size and correlates with a number of other, more precise size measures including the area drained, volume of water discharged, and channel dimensions. Each tributary drains a sub-catchment (sub-watershed), and so multiple sub-catchments are nested within a higher-order catchment or watershed. To describe the drainage area of large rivers, the term river basin is usually preferred.

Identifying where a stream or river falls within the fluvial hierarchy provides useful context, but simply visiting any stream at multiple points along its length quickly reveals a great deal of variation from place to place. Fast-flowing, shallow areas, termed riffles, may alternate with slower, deeper locations, termed pools, and with areas of more uniform flows, termed runs. The shape of the channel varies as it shifts first one way and then the other; substrate on the stream bed may be composed of sand, gravel, or boulders, and fallen trees may create additional complexity. All of this affects what habitats are present and what organisms are likely to be found there. Within even a small stream one can usually identify river segments, typically extending between an upstream and downstream tributary juncture, and from one to tens of kilometers in length. Individual reaches are recognizably homogeneous units within a valley segment. In practice, they often are defined as a repeating sequence of channel units (such as a riffle-pool-run sequence) or by a sampling convention, such as a distance equal to 25 stream widths. A reach can be 100 m or less in length in a small stream, and several km in a larger river. Macrohabitats such as a pool or riffle occur within a reach, and smaller units such as a gravel patch or a leaf accumulation along the stream margin constitute microhabitats. The reach is the scale at which much fieldwork is done and where features important to the biota are apparent.

As one proceeds along the length of a large river, beginning with its many small tributaries originating at the highest points in the catchment, and coalescing into a larger and larger river system, a characteristic longitudinal profile can be seen. Rivers typically are steeper in the uplands where they originate, and have a more gradual slope in the lowlands near their terminus. This longitudinal profile can be divided into roughly three zones from the perspective of sediment transport: an upper zone of erosion, a middle zone of transfer, and a lower zone of sediment deposition. In addition to their steeper slopes, headwaters often have deep, v-shaped valleys, contain rapids and waterfalls, and export sediments. The mid-elevation transfer zone is characterized by broader valleys and gentler slopes. Tributaries merge, and some meandering develops. Sediments are received from the headwaters and delivered to lower sections of the river system. In the lower elevation depositional zone, the river meanders across a broad, nearly flat valley, and may become divided into multiple channels as it flows across its own deposited sediments. Because the river's power to transport sediment is a function of gradient and volume of flow, and more power is required to move large versus small particles, the river also is a sediment sorting machine. Indeed, many of the channel types and features that contribute to the variety of rivers, such as boulder cascades, rapids, riffle-pool sequences, and so on can be seen to exhibit a longitudinal progression determined by sediment supply and stream power.

Stream ecologists have long recognized the profound influence that surrounding lands have on the stream ecosystem. Rain and snow that fall within the catchment reach the stream by myriad flow paths. Some, notably surface and shallow sub-surface flows, reach the stream rapidly and so high flows quickly follow storms. Other inputs, primarily of groundwater, are so gradual that streamflow barely responds to rain events. The valley slope is the source of much of the sediment input to the headwaters, which exports all but the largest particles downstream, while the river's channel walls become increasingly important as a sediment source in middle and lower sections as the river's meandering and flooding drives endless cycles of erosion and deposition. Thus, key aspects of the river's hydrology, its sediment load, channel shape, and chemistry are the consequence of the geology, topography, and vegetation of the valley, and their interaction with climate.

Vegetation bordering the stream is especially important, affecting multiple stream functions. Called the riparian zone, and including the floodplain in locations where the river frequently overflows its banks, the influence of the stream margin and its vegetation cannot be overstated. Roots stabilize banks and prevent slumping, branches and trunks of trees create habitat diversity wherever they fall into streams, canopy shade prevents excessive warming, and the infall of vegetation and invertebrates are major sources of energy to stream food webs. As Hynes (1975) so succinctly expressed it, "in every respect, the valley rules the stream". For these reasons, streams and rivers are highly vulnerable to human actions within their catchments, including altered hydrologic flowpaths, channel erosion, loss of habitat, and other impairments to ecosystem function.

1.1.2 The Fluvial Ecosystem

The fluvial ecosystem integrates organisms and biological interactions with all of the interacting physical and chemical processes to determine key aspects of system function: the diversity of energy sources, the number of species and feeding roles represented, how efficiently nutrients are used, and overall production and metabolism. In fluvial food webs, all energy available to consumers ultimately originates as primary production, but not necessarily from aquatic plants or within the stream itself. The primary producers of greatest importance within the wetted channel include algae, diatoms, and other microscopic producers. These are found on stones, wood, and other surfaces and occur where light, nutrients, and other conditions are suitable for their growth. Organic matter that enters the stream from the surrounding land, such as leaf fall and other plant and animal detritus, is an important energy source in most streams, and is of primary importance in many. Bacteria and fungi are the immediate consumers of organic matter, and in doing so create a microbe-rich and nutritious food supply for consumers, including biofilms on both inorganic and organic surfaces, and autumn-shed leaves riddled with fungal mycelia.

Rivers typically receive organic matter from upstream and also laterally, depending upon the nature of the riparian vegetation and the river's connectivity with a floodplain. In forested headwater streams and large floodplain rivers, much of the energy supply to food webs is received as external inputs, termed allochthonous sources. Streams flowing over a stony bottom in an open meadow often develop a rich algal turf on the substrate, and so most of the energy is produced internally, that is, from an autochthonous source. Typically, the food webs of streams and rivers are fueled by a complex mixture of energy sources that originate within the stream as well as from upstream and laterally. Consumers both small and large exhibit a wide variety of food-gathering abilities and feeding roles, which are shaped by the food supplies available to them and the habitats in which they forage. The macroinvertebrates of streams, including insects, crustaceans, mollusks, and other taxa, can be grouped based on similarities in how food is gathered as well as the food type. Grazers and scrapers consume algae from substrate surfaces, stones in particular; shredders consume autumn-shed leaves enriched with microbes; predators consume other animals; and collector-gatherers feed on the abundant and amorphous fine organic particles originating from the breakdown of leaves and everything once living. Because functional feeding groups place primary emphasis on how the food is obtained rather than where it originated, they may imply greater distinctiveness in trophic pathways than actually is the case. This is especially true in the case of biofilms, which appear to be ubiquitous and likely make a direct contribution to the trophic needs of all consumers.

The vertebrates of fluvial ecosystems likewise exhibit considerable diversification in their feeding roles and their adaptations to exploit available resources. The various trophic categories include algivore, detritivore, omnivore, invertivore, and piscivore; and feeding location (stream bed versus water surface, for example) may also be distinguished. Many fishes consume primarily an invertebrate diet, but so also do salamanders, and some birds and mammals. Algae are the primary diet of a number of fish species, especially in the tropics, and of some larval anurans. Other fishes with elongated guts are able to digest detritus, including leaf matter and ooze. In addition to omnivores, a term generally used to describe species whose diet includes plant (or detritus) and animal matter, and piscivores, which are invertivores early in their life histories, many species feed more broadly than these categories imply.

The biological communities of fluvial ecosystems are assembled from the organisms that are adapted to regional conditions, including the physical environment and available food resources, and are further refined through interactions with other species. Persistence at a location requires that species are adapted to physical habitat, food resources, and specific temperature and flow conditions. Competition, predation, and even more subtle types of interactions, such as when risk of predation affects the timing or location of species foraging, are important filters on community assembly. Diversity of habitat promotes diversity of feeding roles and of species present at the local scale, while the overall diversity of habitat types at some larger scale, such as amongst river reaches or different tributaries, largely determines the biological diversity of the river system.

In any ecosystem, nutrient cycling describes the uptake of some nutrient, usually from a dissolved inorganic phase, and its subsequent incorporation into biological tissue. That material resides for a time in organic form within the plant or microbe, and likely passes through other consumers, but eventually is mineralized or released by excretion, egestion, or respiration, thus completing the cycle. In running waters, downstream transport of nutrients occurs in both the inorganic and organic phases, but especially in the former, stretching the cycle into a spiral. Thus, uptake distance, rather than time, becomes a useful measure of biological availability and demand. Access to nutrients, especially nitrogen and phosphorus, can limit important ecological processes including photosynthesis and the decomposition of organic

Fig. 1.1 The river continuum concept summarizes expected longitudinal changes in energy inputs and consumer feeding roles as one proceeds from a first-order stream to a large river. A low ratio of primary production to ecosystem respiration (P/R) indicates that the majority of the energy supplied to the food web derives from organic matter and microbial activity, and mostly originates as terrestrial production outside the stream channel. A P/R approaching 1 indicates that much more energy to the food web is supplied by primary production within the stream channel. An important upstream–downstream linkage is the export of fine particulate organic matter (FPOM) from the headwaters, where inputs of autumn-shed leaves and other coarse particles (CPOM) predominate, to locations downstream (Reproduced from Vannote et al. 1980.)

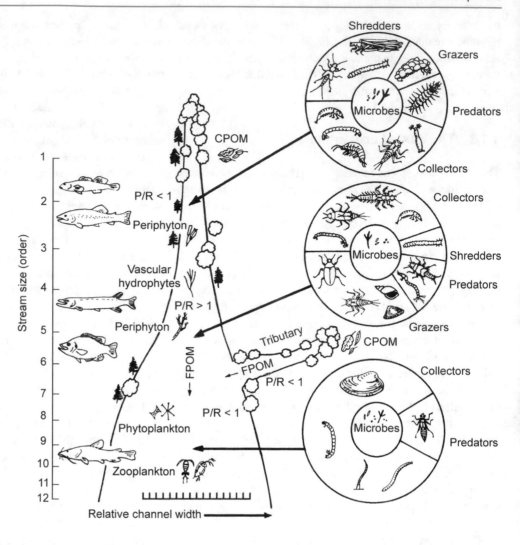

matter. Nutrient poor systems tend to cycle nutrients rapidly and are often characterized by lower productivity. The addition of nutrients to a river or stream through anthropogenic activities, such as agricultural development, wastewater discharge, and surface runoff from urban and suburban landscapes, can enhance primary production in rivers and streams and lead to eutrophication, often seen in excessive algal growth. The quantity of nutrients available in a system is not the only factor limiting biological activities in flowing waters. The ratio, or stoichiometry, of nutrients available to aquatic organisms is important in regulating biological structure and ecosystem function, as the nutrient whose availability is least relative to biological demand will ultimately control system productivity.

Patterns in productivity and respiration, collectively termed ecosystem metabolism, have long been used to characterize rivers and streams. Influenced by physicochemical factors, including but not limited to, temperature, flow, light, and nutrient availability; by biological factors such as species composition; and by socio-political factors including

watershed land use and water infrastructure, patterns of productivity and respiration can provide important insights into river condition.

The river continuum concept integrates stream order, energy sources, food webs, and to a lesser degree nutrients into a longitudinal model of stream ecosystems (Fig. 1.1). As originally conceived for a river system flowing through a forested region, the headwaters (order 1–3) are heavily shaded and receive abundant leaf litter, but algal growth often will be light-limited. Streams of order four through six are expected to support more plant life because they are wider and less shaded, and in addition receive organic particles from upstream. The headwaters have more allochthonous inputs, indicated by a ratio of primary production to respiration well below unity, whereas the mid reaches have more autochthonous production and a higher P/R ratio. Higher-order rivers are too wide to be dominated by riparian leaf fall and too deep for algal production on the bed to be important. Instead, organic inputs from upstream and the floodplain, along with river plankton, play a greater role.

1.1.3   Vision of a Healthy River

A healthy river is defined by its intertwined physical, chemical, and biological properties (Fig. 1.2). It need not be pristine—healthy rivers can co-exist with human activities, even in regions where societal drivers are strong—provided that core functions are not compromised greatly. Thus, it is important that flows retain much of their natural quantities and variability, sediments are eroded and re-deposited in natural cycles, and substrate, water temperature, and water quality remain within normal ranges. To ensure health of the entire ecosystem, nutrients should remain within normal limits and continue to cycle between the biota and the environment, supporting biological productivity and

biological diversity characteristic of regional conditions. Because rivers are connected longitudinally from source to mouth, laterally with their riparian zones, floodplains, and landscapes, and vertically with groundwater sources (Fig. 1.3), conditions at a specific location are highly vulnerable to threats originating at considerable distance. Ultimately, healthy rivers are unlikely to be found other than in healthy landscapes.

Environmental monitoring and assessment can assess a river's condition on numerical or qualitative scales, indicating that not all rivers are healthy, or as healthy as they could be. In the most recent nationwide survey in the United States, which sampled 1,924 river and stream sites across the country representing nearly 2 million river km, biological

Fig. 1.2 A mountain stream in the Colorado Rockies of the United States. Picturesque, popular with trout fishers and hikers, and the location of much ecological study by researchers at the nearby Rocky Mountain Biological Laboratory, this stream has experienced its share of human impacts, as have so many others. A now abandoned mining boomtown was located just outside this frame at the end of the 19th Century. Introduced species of trout have largely displaced the native cutthroat, and a non-native diatom known colloquially as 'rock snot' recently colonized substantial areas of the stream bed. The impact of climate change is beginning to be recorded in surrounding ecosystems. Photo by Angus McIntosh

Fig. 1.3 The fluvial ecosystem
with its three major axes:
upstream/downstream,
channel/margins, and
surficial/underground
environments (Reproduced from
Piégay and Schumm 2003.)

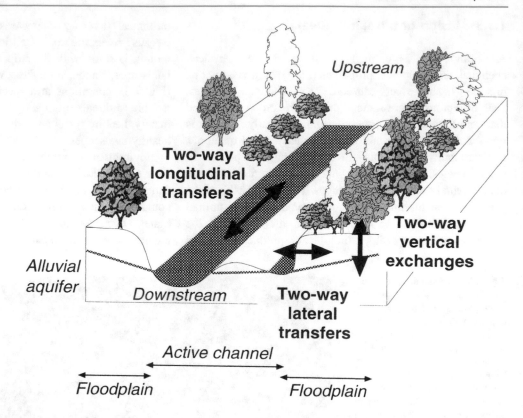

surveys revealed that 28% of river and stream length was in
good condition, 25% was rated fair, and 46% was in poor
condition (USEPA 2016). Widespread stressors included
excessive nutrients and sediments, and poor riparian conditions. This is important, motivating information. To address
the challenge of ensuring healthy rivers, it is important to
know what characteristics define them, and understand the
threats that they face.

1.2 Rivers in the Anthropocene

"Anthropocene" is a new construct, in wide use only since
the early 21st century (Steffen et al. 2007). It posits that
modern human impact on the Earth merits the designation of
a new geologic epoch. The beginning date is contested: some
would select the mid-20th century; other would go back three
centuries to the great expansion of technology and commerce
following the early industrial revolution, and categorize the
more recent date as the "great acceleration". As scholars
endeavor to determine whether modern human impact on the
Earth merits the designation of a new geologic epoch,
changes in the state of river systems may provide some of the
strongest support. Dams and other waterworks, both large
and small, have proliferated to the point that today, nearly
half of the global river volume is moderately to severely
impacted (Grill et al. 2015; Lange et al. 2018). Humans now

capture >50% of available freshwater runoff (Jackson et al.
2001), reservoirs trap 25% of the global sediment load before
it reaches the oceans (Vörösmarty and Sahagian 2000), and
several of the world's great rivers, including the Ganges–
Brahmaputra, Huang He, Nile, and Colorado, have stopped
flowing to the sea during dry periods (Postel 2000). Dams are
amongst the most obvious of human impacts on river systems, but altered land use and a changing climate also have
profound effects. Outcomes include significant gains in
human well-being and grand engineering accomplishments,
but often at great environmental and social cost.

Although human influence on river systems has a more
than 4000-year history in basins that have given birth to
major civilizations, monitoring data spanning more than the
past 30–50 years are scarce. Long duration studies of the
Seine River near Paris provide an exceptional case study of
trends in organic pollution, eutrophication, nitrate pollution,
and metal contamination from as early as 1880 to the present
(Meybeck et al. 2016). Such a long view supports the concept of a trajectory in river quality, with several distinct
timelines: first, for scientific recognition and compilation of
evidence as human impacts gradually become more apparent; second, for social recognition of an issue; and third, for
societal responses such as environmental advocacy and
political actions. For the Lower Seine, the entire sequence
spanned 50–150 years or has yet to adequately address all
impacts, depending upon variable considered.

1.2.1 State of the World's River Systems

Today, many of the world's rivers are moderately to severely impacted by human activities (Best 2019). Occupying less than 1% of the Earth's surface, rivers supply approximately 80% of renewable fresh water to society (Vörösmarty et al. 2015); protecting this resource to meet human needs and ensure healthy freshwater ecosystems is one of the grand challenges for a sustainable future (Bunn 2016). River systems today face a number of threats from human activities that degrade water quality and ecosystem condition, such that a large fraction of the world's population faces challenges to both water security and freshwater biodiversity (Vörösmarty et al. 2010).

The many threats to rivers are most easily discussed within some classification of human pressures, and a framework that distinguishes human actions from system response. Figure 1.4 presents the DPSIR framework (Drivers, Pressures, State, Impact, Response), a causal model describing the interactions between society and the environment. Driving forces reflect societal developments, such as population and economic growth. Pressures include pollution, hydrologic alteration, habitat loss, and more that result from drivers. State refers to observable system condition, such as water quality or animal population abundances, and usually their negative changes due to pressures. Impacts refer to environmental harm due to changes in system conditions, including biodiversity loss, impaired ecosystem function, and loss of human benefits. Finally,

responses are societal efforts to un-do or reduce the harm, including via laws, regulations, and cooperative initiatives. Although portrayed as a linear sequence, because of the interdependence of the components it is evident that humans have impacts on the environment and vice versa. It is worth noting that many variants of the DPSIR language are used. The term 'stressor' is often used instead of pressure, changes in system state are often called responses or impacts, and 'threats' is a useful catch-all for all human influences. Here we identify six major categories of pressures influencing river systems, and a number of the most important changes in system state that result from these pressures. The six categories are pollution, flow modification, habitat degradation, over-exploitation, species invasions, and climate change (Table 1.1). Various authors surveying threats to rivers have converged on similar categories (Malmqvist and Rundle 2002; Meybeck 2003; Dudgeon et al. 2006; Carpenter et al. 2011). And as Strayer and Dudgeon (2010) have documented, many of these threats have been increasing over time, and show no sign of abating.

Pollution refers to the altered physical or chemical state of fresh water in human-dominated river basins resulting from the release of urban, industrial, and agricultural chemicals and waste products. It includes agricultural chemicals (nutrients, herbicides, pesticides), urban and industrial wastes (sewage, metals, and other industrial by-products), as well as pollutants of emerging concern such as pharmaceuticals and personal care products, and plastic wastes including microscopic fragments that enter food chains and may serve as

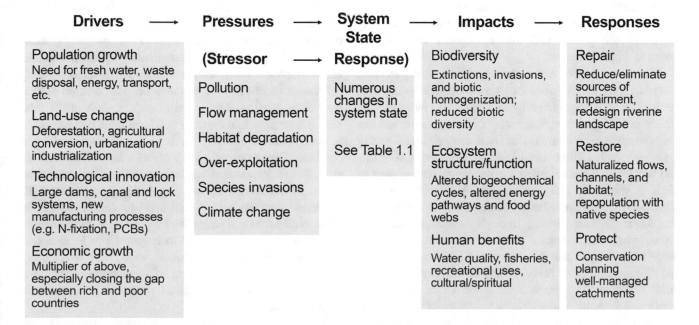

Fig. 1.4 A generalized DPSIR framework depicting the causal relationships between human societies and river systems. In reality the components are inter-related and difficult to cleanly separate. The term 'stressor' is often used instead of pressure, changes in system state are often called responses or impacts, and 'threats' is a useful catch-all for all human influences

Table 1.1 Major categories of pressures influencing river systems, and common symptoms observed in the altered state of rivers experiencing these pressures, drawing upon analyses by Malmqvist and Rundle (2002), Meybeck (2003), Dudgeon et al. (2006), and Carpenter et al. (2011), among others. Many of these pressures interact: for example, flow modification can affect habitat quality or favor species invasions. The list is not exhaustive

Pressure	Symptoms (Change in state of system)
Pollution	Organic waste decomposition lowers oxygen levels; metals and industrial byproducts may be toxic and long-lived; excess nutrients cause eutrophic conditions and biotic shifts
Flow modification	Altered frequency, magnitude, duration, timing etc. of flows disrupt ecosystems; withdrawals and diversions reduce water levels and flows; further impacts on water and habitat quality frequently occur. Groundwater withdrawals reduce streamflows during summer or year-round, may cause cessation of flow
Habitat degradation	Channel modifications including straightening, dredging, levees, often reducing floodplain connectivity; river fragmentation by impoundments limits movements by biota and alters sediment and thermal regimes
Over-exploitation	Primarily of fish populations but also birds, reptiles, amphibians and some invertebrates; population declines may significantly impact species composition and ecosystem function; gear may target largest individuals and species in tropical multi-species assemblages
Species invasions	Non-native species spread via canals, ship ballast water, and deliberate introductions for sports fisheries, pest control, etc.; typically a fraction become abundant and harmful, significantly impacting species composition and ecosystem function; results in homogenization of biota by successful colonizers
Climate change	Direct influence on temperatures, precipitation, and sea levels; many follow-on changes occur in species distributions due to altered thermal and flow conditions; incursion of salt water affects tidal rivers

carriers of toxic substances. Pollutants enter waterways by direct discharge, diffuse runoff following rain events, and as atmospheric deposition. Examples of the later include mercury and acid-forming gases that are released into the air by coal-fired power plants and enter waterways as wet and dry deposition. Pollution includes not only man-made contaminants but also physical changes due to human actions. Important examples include intensified land use causing excess sediment inputs that can smother streambed habitat, thermal pollution due to return of water used in cooling thermoelectric plants, and the temperature changes that occur in impoundments.

The consequences of excessive organic waste entering rivers of great European cities became evident by the 19th century onward. Organic pollution is indicated by an oxygen deficit in rivers resulting from microbial respiration, and can be inferred from minimum dissolved oxygen levels during summer months. Pronounced summer episodes of very low oxygen levels were recorded for the Seine downstream of Paris in the 1880s (Meybeck et al. 2016), and likely had developed well before first measurements were made. For the River Thames flowing through London, human and animal wastes have been a documented source of foulness since at least 1620, and the summer of 1858 was known as the "Year of the Great Stink" owing to the odors emanating from the river. The problem of untreated human sewage continued into the mid-1950s, when parts of the Thames around London became anaerobic from microbial respiration driven by organic waste, and sparked pollution control efforts that have led to substantial recovery.

To make water accessible for human use where or when it is not readily available, human efforts to store or divert water when it is present in excess, or procure it when it is lacking, are global in scope. These are accomplished through dams and impoundments, and by irrigation withdrawal, drainage, groundwater pumping, impoundment, levee construction, and interbasin transfer. Human demand for freshwater approaches or exceeds current supply in many regions, and approximately 2.4 billion people live in water-stressed environments (Oki and Kanae 2006), prompting wide use of surface water storage and withdrawals of deep groundwater. Worldwide, today's approximately 50,000 large dams create impoundments with a cumulative storage capacity equivalent to nearly 10% of the water stored in all natural freshwater lakes on Earth, and about one-sixth of the total annual river flow into the oceans. The number of small dams is poorly known but certainly is in the millions, and has a combined storage capacity on a par with large dams (Lehner et al. 2011). Impounded sections resemble lakes more than rivers, downstream river sections have altered flow, sediment, and thermal regimes, and loss of natural connectivity restricts dispersal and migration by organisms. Groundwater extraction for human use and to irrigate crops in semi-arid areas can dramatically reduce streamflow, especially during summer, resulting in collapses of large-stream fishes and expansion of small-stream fishes where hydrologic conditions were altered most (Perkin et al. 2017).

Habitat degradation refers to any and all changes in river systems that reduce the amount or quality of habitat for organisms. Some habitat degradation is associated with pressures just mentioned. Organic pollution can reduce oxygen levels to life-threatening levels, excess sediment can render the stream bed uninhabitable to invertebrates that feed on the algae and microbes found on the surfaces of stones, and altered thermal regimes will shift thermal niches towards different species. Flow modification can profoundly change habitat, as when the reduced flows of an impounded river favor a different biota than occurred in its original, fast-flowing state. The molluscan fauna of the southeastern US, once the most diverse molluscan fauna in the world, has lost one-third its species due to habitat and flow alteration (Haag and Williams 2014). In addition to these causes, habitat commonly is degraded due to land use and river engineering. Deforestation and the intensification of agriculture affect sediment, flow, and thermal regimes of rivers, and often cause channel erosion. River engineering includes dredging for ship transport, construction of levees to prevent flooding, channel straightening to speed floodwaters downstream, bank stabilization to prevent erosion, and more. Severing the relationship between a river and its floodplain prevents access to critical rearing habitat for many fishes and invertebrates. Indeed, natural floodplains are highly imperiled ecosystems worldwide, with over 90% under cultivation in Europe and North America (Tockner and Stanford 2002). Collectively, these actions have profoundly simplified instream habitat. River bends, backwaters, and shallows frequently are replaced with a straightened and deepened channel. Submerged wood and trees are removed, eliminating the structural diversity important to all levels of the food chain. Because the diversity of species is closely related to diversity of habitats, habitat loss profoundly affects biological diversity.

Overexploitation is of primary concern for fishes and some other vertebrates including amphibians, turtles, crocodilians, and certain water-associated birds and mammals. Few freshwater invertebrates are imperiled due to direct harvest, although mussels, valued for their shells, pearls, and meat since prehistoric times, have been extensively harvested in Europe and North America (Strayer 2006). Some reports describe harvest quantities that would be unimaginable today, such as the more than 13 million kg of shells taken from the waters of Illinois in 1913 (Anthony and Downing 2001). Overexploitation is the single most important threat to the freshwater turtles of Asia, where the majority of individuals are sold as food but most species are also sold for traditional medicine and as pets (Cheung and Dudgeon 2006). Commercial and recreational fisheries have each been responsible for serious declines of highly valued freshwater fishes. Worldwide, the total freshwater commercial catch has been recorded since the 1950s, and these data show a steady increase with a possible plateau since the late 1990s (Allan et al. 2005). Changes in fishing gear are one indicator of overfishing, as nets of finer mesh are more

time-consuming and expensive to make than coarser mesh sizes, and will be adopted by fishers only out of necessity. In foothill rivers of the Venezuelan Andes, local fishers use cast-nets to catch migrating *Prochilodus*, a member of the most widely harvested freshwater fish family in South America. Net mesh is measured by the width of several fingers of the hand, and over the past 20 years net mesh has declined from "four fingers" to "two fingers" to accommodate the decreasing size of the target (Taylor et al. 2006). As global fish stocks decline, aquaculture production has expanded in recent decades to meet increasing demand for fish, and nearly half of the global aquaculture production comes from freshwater ecosystems. While clearly providing benefits to humans, the expansion of aquaculture has negative consequences, including increases in the harvesting of wild fish for feed, water pollution, altered hydrologic flows, and the accidental release of non-native species.

Non-native species that become abundant in new environments are referred to as invasive, implying harm to the recipient community. Some introductions are purposeful, to enhance sports fishing, for aquaculture, and as agents of biological control. Others are unintentional, including escapees from fish farms, the release of aquarium pets, unnoticed species that "hitch-hike" with a planned introduction, and those carried in ship ballast water or dispersed through canal systems. Although only a fraction of introduced species establish self-sustaining populations, and a fraction of those become a nuisance, spread, and cause harm, species invasion is nevertheless an important pressure on river ecosystems. Invasive species bring about declines in native species and changes in food web structure via species interactions, habitat alterations, introductions of diseases or parasites, trophic alterations, and hybridization. In some cases, a successful invasion effectively swaps a non-native for a native species, with loss of biodiversity but little change in system functioning. In other cases the successful invader has no ecological analogue and may have system-wide impacts, as seen with the common carp in North America. The worldwide spread of invasive species coupled with the decline or extirpation of native species results in a fauna that is becoming progressively more similar across regions, which in turn lessens the uniqueness of local faunas. This is the phenomenon known as biological homogenization, and freshwater fishes are a prime example (Rahel 2002).

Climate change directly affects river systems via increased temperatures, greater variability of precipitation among locations and over time, and rising sea levels. Warming decreases the amount of oxygen that can dissolve in water and affects organisms' thermal niches, causing range expansions and contractions and thus shifts in species composition. Climate change will surely interact with river flows as precipitation and evaporation both change, with often unclear consequences for the amount and timing of runoff. With rising sea levels, salt water extends further upstream in tidal rivers. In addition, climate change will interact with other pressures, enhancing some regional water shortages, favoring species invasions, and acting as an additional stressor on the biota. Most freshwater organisms are adapted to a particular temperature range, and so are projected to undergo distributional shifts to higher latitudes or altitudes under warmer climate scenarios. Because dispersal ability varies among taxa and few rivers are barrier-free, some and perhaps many species will face limited opportunities. While changes in species composition may be complex and unpredictable, an overall increase in system productivity is likely to be a common response to climate warming. Streamside vegetation will almost certainly change under future climates, affecting the nature, timing, and supply of organic matter inputs, which commonly are a critical carbon source to stream food webs.

1.2.2 What is at Stake

Threats to the health of freshwater ecosystems have serious consequences for human well-being. A joint analysis of stressors threatening human water security and biodiversity found that nearly 80% (4.8 billion) of the world's population live in areas where threats to human water security or biodiversity are high (Vörösmarty et al. 2010). Regions of intensive agriculture and dense settlement, and water-scarce dryland areas, exhibit high threat (Fig. 1.5), while remote areas of the world including the high north and unsettled parts of the tropics show the lowest threat levels. The world's rivers have indeed entered the Anthropocene, sparing only remote rivers in regions relatively untouched by human drivers. What are the impacts of all of these changes to the state of rivers, and what is at stake from both an ecological and a human perspective? Lost biodiversity, impaired ecosystem function, and compromised human benefits are the likely outcomes, and each describes an impoverishment that we should strive to avoid.

Freshwater biodiversity is highly threatened. Occupying <1% of the Earth's surface, freshwater habitats support more than 125,000 species of freshwater animals, comprising nearly 10% of all known species and approximately one-third of vertebrate species (Balian et al. 2008; Strayer and Dudgeon 2010). The total diversity of fresh waters is far from fully catalogued, especially for invertebrates and microbes, and in tropical latitudes where the diversity of freshwater fishes is known to be much greater than at temperate latitudes (LéVêque et al. 2008). Strayer (2006) notes that the best studied invertebrate groups (insects, crustaceans, and mollusks) have roughly the same number of described species as freshwater fish, but have received only about 1/10th of the attention from scientists. Even less is

known of the "freshwater animals that have been living on the dark side of the scientific moon", including many of the smallest taxa and those dwelling interstitially in river beds. The modest level of scientific study of many species means that we are simply unaware of the extent and status of much of life's diversity.

Nonetheless, the extent of imperilment can be approximated for well-studied groups, and evidence to date tells us that declines in biodiversity are far greater in fresh waters than in the most affected terrestrial ecosystems (Sala et al. 2000). Using the Red List Categories and Criteria maintained by the International Union for the Conservation of Nature, Collen et al. (2014) generated conservation assessments for freshwater species in six groups: mammals, reptiles, amphibians, fishes, crabs, and crayfish. Results showed that almost one in three freshwater species is threatened with extinction world-wide, and all groups evaluated in this analysis exhibited a higher risk of extinction than their terrestrial counterparts. Extinction risk also is higher in lotic habitats in comparison with wetlands and lakes. In addition, the extent of endemism (denoting a unique and possibly limited geographic range) also is poorly known, a further concern because ecological degradation poses the risk that for endemic species of limited range, local extirpation becomes extinction (Strayer 2006). Superimposing fish diversity on a map of over 400 freshwater ecosystems of the world, Abell et al. (2008) found that roughly half of freshwater fishes may be restricted to just one region.

Humans have modified global biogeochemical cycles, and this has had profound, and increasingly serious, effects on rivers and streams. Nutrient pollution and subsequent eutrophication can alter the structure of aquatic communities and ecosystem processes, including primary productivity and ecosystem respiration. Eutrophic systems can often be dominated by algal and bacterial species that release toxins and compromise the quality of water for animal and human use. Aquatic species that are sensitive to environmental conditions often are lost as systems transition from oligotrophic to eutrophic in response to nutrient pollution.

River systems provide numerous human benefits. These are known as ecosystem goods and services, and can be grouped into provisioning, regulating, supporting, and cultural services. Provisioning services include harvestable fish, other animals, and plants; and water for drinking, irrigation, industry, and hydropower. Examples of regulating services include water purification, nourishment of floodplains, flood control, and climate mitigation. Supporting services include organic matter processing, nutrient cycling, and energy production. Cultural services include many recreational activities, such as fishing and boating, as well as the spiritual and aesthetic values that are part of our enjoyment of nature. The end products of the services are human benefits, some of which are part of the market economy but many are not,

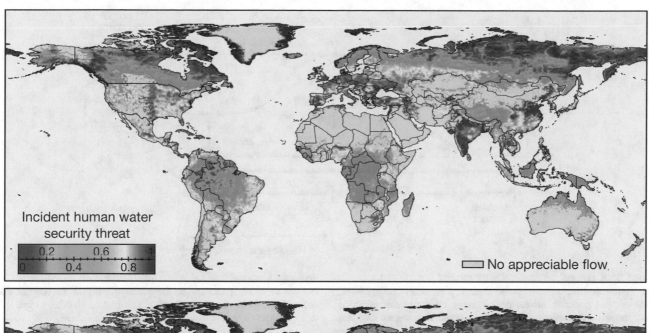

Fig. 1.5 Global geography of incident threat to human water security (top panel) and biodiversity (bottom panel). Maps combine multiple stressors organized under four themes (catchment disturbance, pollution, water resource development and biotic factors) to derive cumulative threat indices. Expert assessment of stressor impacts on human water security and biodiversity produced two distinct weighting sets, which in turn yielded separate maps of incident threat reflecting each perspective (Modified from Vörösmarty et al. 2010.)

making estimation of their value more challenging. For example, commercial fish harvest and hydropower can be directly quantified, but many services are estimated indirectly by surveys that ask respondents to value an experience, compare property values with indicators of water quality, or value a nature experience based on distance travelled. Regardless of the difficulty of placing a dollar value on ecosystem services, they likely contribute a sizable but poorly quantified fraction of local economies. Most importantly, people rely on these services, fresh water is a non-substitutable resource, and humans are the beneficiaries of measures to protect freshwater ecosystems (Brauman et al. 2007).

1.2.3 What is to Be Done?

In river systems with a long history of degradation from human activities, and where societal benefits also have been compromised, a trajectory termed "impair-then-repair" is generally observed (Vörösmarty et al. 2015). It can be viewed as a time series of human-water interactions in a particular river basin or region, or as a contemporary snapshot of rivers distributed along a global gradient of impairment. In the earliest stage (O–A of Fig. 1.6), rivers are minimally impacted, if at all. They provide basic goods (such as food) and services (such as floodplain agriculture), and societies adapt well to their dynamics. Impacts from

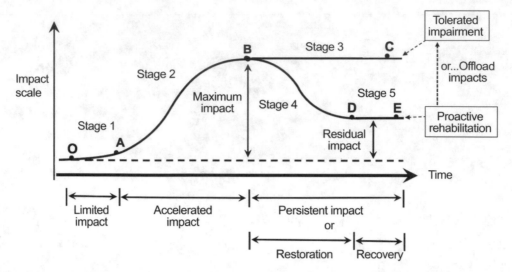

Fig. 1.6 A generalized time series for societal impacts and responses for some freshwater ecosystem. Rising population and economic development generate increasing pressures on freshwater resources, until impacts are sufficient to motivate societal responses that limit further increase or ameliorate to varying degrees. This schematic can depict the historical time series of a highly developed region such as for Europe or North America, or the status of regions or countries at different levels of economic development (poor countries to the left, rich countries to the right) (Reproduced from Vörösmarty et al. 2015.)

human and agricultural pollution are absorbed, but increase with population growth. Over time, land-use change, mill dams, and other activities contribute to greater degradation. Today, only remote rivers remain at stage 1, as environmental degradation has accelerated in most areas, typically linked to urbanization and agricultural intensification, and with pollution increasing faster than population. Much of this acceleration occurred in countries with richer economies after about 1800, and by the mid-twentieth century rich countries began to respond with sewage collection and waste purification systems. What follows next—stage 3 or stage 4 of Fig. 1.6—depends on a number of factors, including a region's wealth, environmental awareness, and political commitment, and also the specific impact to be repaired. Organic contamination and fecal contamination can be successfully remediated in a few decades, as in the Seine below Paris, whereas metal and radioactive wastes are much longer-term challenges. Source reduction through regulatory bans, such as phasing out the manufacture of a harmful chemical, can also produce marked improvements in a few decades. Stage 5 represents sustained recovery when river systems are managed to maintain their rehabilitated state. This is not a return to natural or pre-impact conditions, but to a modern river that provides an array of benefits to society and aquatic biota alike through well-designed, co-use strategies. In human-dominated river basins in richer countries, contaminant levels are likely to remain well above historic background, albeit stabilized below peak values, illustrating the "impair-then-repair" sequence often seen where environmental awareness and financial resources are sufficient. However, adequate response will be difficult for many of the growing megacities of the world, especially if located on relatively small rivers, with faster rates of economic development and perhaps fewer financial resources.

River restoration often targets systems where human domination is less severe, one or a few pressures dominate, and the technological knowledge to end or ameliorate the stress is relatively well developed. Restoration projects undertaken by government agencies and environmental groups have grown rapidly in number and ambition since the 1990s (Bernhardt et al. 2005; Friberg et al. 2016), generating demand for consultants trained in engineering, hydrology, and biology. Restoration projects vary widely in scope and expense. More modest endeavors include the addition of habitat elements such as submerged logs and spawning gravel, fencing to keep livestock out of stream channels, and stabilization of eroding banks with vegetation. Others are larger in scope and budget, as when small dams are completely removed, meanders are restored to previously channelized river sections, and water release from dams is managed to mimic natural variation. Still a relatively young field, there is active debate over the choice of terms, with both restoration and rehabilitation in common use. At the heart of this discussion is whether the desired goal is framed as returning the system to its historic and pristine state, restoring its natural functioning, or simply improvement. That in turn depends largely on the limitations imposed by human presence in the catchment, and existing pressures. Whether the action is viewed as restoration, rehabilitation, or simply as river management, the described activities are aimed at restoring lost or reduced functionality to river systems, and the desired outcomes are articulated not only in terms of water quality, but also in enhanced biodiversity and ecosystem functioning.

Protected area designation and conservation planning are motivated by concerns for the reduction and loss of biodiversity, ecosystem function, and diversity of river types (Moilanen et al. 2008; Nel et al. 2009). Identifying river systems and catchments deserving of conservation efforts is in essence a prioritization process, and various planning and computational approaches have been employed to target conservation action in terrestrial and marine ecosystems. Adapting these methods to freshwater systems can be challenging owing to the importance of upstream–downstream connectivity and the need to incorporate the landscape by means of catchment or sub-catchment-based planning units. The desired outcomes of conservation planning are to conserve the full variety of biodiversity features in a planning region, maintain the natural processes that support and generate biodiversity, and set quantitative targets such as the number of occurrences of a particular river type or species. Locations can be prioritized in a number of ways: overall biodiversity, species of concern, extent of human disturbance, and river system connectivity are suitable criteria, as are considerations to minimize cost or required land area.

Effective implementation of protected areas will require a management plan that, in addition to on-going scientific study, evaluates current and long-term threats, identifies and involves key stakeholders, and works within the social, economic, and institutional context of the area (Nel et al. 2009). Authors agree that the catchment scale is appropriate for freshwater conservation (Dudgeon et al. 2006), but problematic in practice because the area required can be impracticably large and the exclusion of people rarely is feasible. Although small areas may be set aside with freshwater conservation as their sole priority, and even some larger-scale river systems may be protected in their entirety in remote regions, human use of freshwater resources will need to be accommodated in most instances. Abell et al. (2007) argue that the solution requires looking beyond the protection of individual sites, and instead developing a spatially distributed set of conservation strategies intended to protect specific populations or target areas.

Where societal response falls along the continuum from impair-then-repair, to restore, to protect and conserve, is a reflection of both the condition of the river system and how societies view its uses and values. Regardless, the goal should be to ensure sustainable water for human use and ecosystem protection. This requires a collaborative approach that includes citizens and government agencies, is informed by best science, and seeks to manage rivers within their catchments to meet shared goals of human and ecosystem well-being. Important corollaries include conceptualization of rivers and their catchments as complex, adaptive social-ecological systems that are inherently unpredictable and difficult to control, requiring integrative approaches focused on learning, governance, and the human dimensions of management. Known by a variety of terms, including integrated river basin, catchment, or watershed management, their common basis is a shift in emphasis from a limited focus on meeting human needs while attempting to mitigate environmental costs, to an emphasis on sustained human benefits, including water for direct human use and water to support healthy ecosystems and the services they provide. This new paradigm is laudatory, ambitious, and remains a work in progress, and we shall return to consider it in detail in the concluding chapter.

1.3 What to Expect in This Book

The ecology of rivers and streams is a fascinating, complex subject that draws on diverse specialty areas. Life exists within limits set by the physical and chemical environment. Lotic (running water) systems are, by definition, characterized by flowing water, and it is the variability in the amount and timing of flows that give rise to the many types of running waters. In Chap. 2, Streamflow, we explore the tremendous variability that fluvial ecosystems exhibit in the quantity, timing, and temporal patterns of flow, giving rivers and streams their distinctive character. Because rivers have enormous value to society as a source of drinking water and means of transport, as well as for hydropower, waste removal, and irrigation, humans have extracted, diverted, and impounded river flows since earliest civilizations. By linking hydrologic analysis to ecological responses, scientists are developing a process-based understanding of flow-ecology relationships that can identify the range of flows needed to sustain resilient, socially valued ecological characteristics in a changing world. Of course, water does not flow in fixed channels (unless they are man-made), and so flow interacts with sediment supply to shape channel features at small and large scales. Chapter 3, Fluvial Geomorphology, explores the dynamic interplay between rivers and landscapes in the shaping of river channels and drainage networks. A central theme in fluvial geomorphology is that the development of stream channels and entire drainage networks, and the existence of various regular patterns in the shape of channels, indicate that rivers are in dynamic equilibrium between erosion and deposition, and governed by common hydraulic processes. And wherever human actions have altered the balance between river flow and sediment supply, such as trapping sediments behind dams or mining sediments from streambeds, an ecologically degraded river is the frequent outcome. The growing field of river restoration relies heavily on principles of geomorphology in efforts designed to improve the natural state and functioning of river systems in support of ecological and social goals.

In lotic ecosystems, organisms exist in a physical and chemical environment that can be characterized by the chemical constituents of water, and the influence of current, substrate, and temperature. Chapter 4, Streamwater Chemistry, describes the chemical composition of river water, and the many factors that are responsible for its variation from place to place. Chemical variation among streams is primarily governed by the type and composition of rocks and by the amount and chemical composition of precipitation in a given region. However, human activities can significantly impact the chemical composition of rivers and streams, indirectly by changing land use and by altering the chemical composition of precipitation, and directly by the input of agricultural, industrial, and domestic waste. These direct inputs may be addressed in richer countries by following the "impair-then-repair" sequence, but such actions may often be infeasible in poorer regions. Chapter 5, The Abiotic Environment, focuses on current, substrate, and temperature as three critical aspects of the physical environment that strongly influence which species occur in a particular habitat. Furthermore, the diversity of habitats within a region has a profound influence on which and how many species are found there. Environments that are structurally simple or extreme tend to support fewer species, whereas more moderate and heterogeneous habitats support more species. A high frequency of disturbance tends to diminish biological richness, although a moderate level of disturbance potentially may enhance diversity by maintaining an ever-changing spatial mosaic of conditions. Unfortunately, anthropogenic degradation and homogenization of habitat due to human alteration of streamflow, geomorphic processes, and other activities poses significant risks to both biodiversity and ecosystem function, prompting many recent efforts to reverse habitat loss through ecological restoration.

Lotic ecosystems are rich in life, including microbes, plants, and animals linked through complex food webs. Functionally, these can be reduced to two classes: autotrophs, the primary producers that acquire energy from sunlight and materials from non-living sources; and heterotrophs, organisms that obtain their energy and materials by consuming living organisms and dead organic matter. The autotrophs include algae, higher plants, and some bacteria and protists capable of photosynthesis. The heterotrophs include all animals, as well as fungi and many bacteria and protists. Chapter 6, Primary Producers, discusses the benthic algae (periphyton) found on almost any surface receiving light in streams and rivers, the phytoplankton (suspended algae) occurring in the water column of larger and slower rivers, and the macrophytes or larger plants most commonly found along the margins of larger rivers and in backwaters. Periphyton are especially important in fluvial food webs, especially in headwater and midsized streams, and they also influence the benthic habitat and nutrient cycling. Proximate factors that may limit benthic algal communities include light and nutrients, which, along with temperature, influence biomass accrual; and disturbance and grazing, which are the factors that lead to algal dislodgement and biomass loss. Macrophytes are important to fish and invertebrates as habitat and as refuge from predators, and they can increase habitat heterogeneity for aquatic organisms by modifying water velocity and trapping sediments and organic matter. They are directly consumed by some herbivorous animals, and also can be an important source of dead organic matter, entering food webs via consumers of detritus. Phytoplankton can become abundant in slowly moving rivers and backwaters where their doubling rates exceed downstream losses due to current. However, downstream transport as well as vertical mixing to depths below the level of adequate light are often limiting. Anthropogenic stressors such as acidification, increased water temperature, eutrophication, pollution, and invasive species can have major impacts on autotroph biomass and diversity in running waters. Because algae and macrophytes are sensitive to changes in environmental conditions, they can be a useful tool in stream bioassessments to determine human impacts on a river ecosystem.

Chapter 7, Detrital Energy and the Decomposition of Organic Matter, explores the enormous importance of nonliving organic matter to lotic food webs. Detritus includes all forms of nonliving organic carbon, including fallen leaves, the waste products and carcasses of animals, fragments of organic matter of unknown origin, and organic compounds. It occurs as particles of varying size, and as dissolved compounds of varying molecular size and assimilative quality. In small streams shaded by terrestrial vegetation but receiving an abundance of autumn-shed leaves, and large, turbid rivers receiving fine particulate organic matter from upstream and floodplain sources, detritus can provide much of the energy to the food web. Because most detritus enters the stream food web from its banks, in the form of terrestrial litter, human alteration of riparian vegetation and floodplain connectivity can have important consequences for the amount and quality of this basal food supply to lotic ecosystems.

Chapter 8, Stream Microbial Ecology, describes the important roles of microbes, both bacteria and fungi, and very small invertebrates in processing available carbon from both detritus and autotrophs. Occurring primarily as biofilms on benthic substrates but also found in suspension in large rivers, microbes are heterotrophs that obtain energy from detrital material as well as from organic compounds in the surrounding water. Biofilms are diverse assemblages of microorganisms growing on the streambed and other surfaces, often in symbiotic association with algae. Organic molecules exuded by actively growing algae can be used by microbes, while microbial decomposition of detritus releases inorganic compounds beneficial to algal production. Microbial production forms the base of many stream food webs, and the flow of carbon and elements through microbial communities and into higher trophic levels involves complex linkages. Still relatively under-studied, a food web involving protists and very small metazoans, termed meiofauna (<0.5 mm in length) exists within biofilms and in plankton assemblages, possibly serving as a link to higher trophic levels, and possibly dissipating most of the energy available within a "microbial loop".

Chapter 9, Trophic Relationships, introduces the animal consumers that populate the food web and subsist on the algae, plants, and diverse sources of dead organic matter that constitute the basal energy supply within stream ecosystems. These larger consumers, mainly invertebrates and fishes, are classified into feeding groups referred to as functional feeding groups and trophic guilds, respectively. These classifications rely only partly on what foods are eaten, but also on the where and how of food acquisition. Among invertebrates, grazers ingest periphyton, shredders feed on large particles of organic matter such as leaves, collectors feed on small organic particles either from suspension or the streambed, and predators feed on other animals. Fishes are often categorized by their preferred feeding habitat, such as midwater versus benthic. In addition, fishes can exhibit a great deal of functional specialization in mouthparts, digestive capabilities, and sensory modalities in detecting prey. Other vertebrates, including anurans, reptiles and turtles, birds, and mammals also can play important roles in lotic food webs.

In Chap. 10, Species Interactions, we go beyond recognition of trophic roles, and explore how linked pairs of species affect one another's behaviors, abundance, and fitness. Grazing on benthic algae by invertebrates, some fishes, and a few amphibian larvae is the most important pathway of herbivory in streams, and has received by far the most study. Grazers can have a number of impacts on algae, reducing their abundance, altering assemblage composition, and even stimulating algal growth and overall productivity through the removal of senescent cells and the recycling of nutrients. However, extremes of flow can alter the grazer-algal dynamic by reducing grazer abundance, and heavy grazing pressure can reduce algal biomass to a level where it is less vulnerable to scouring during high flows. Predation affects all organisms at some stage of the life cycle, and many species encounter predation risk throughout their lives. It affects individuals and populations directly through consumption and mortality, and can result in behavioral and morphological adaptations that may entail some fitness cost to the prey. Prey species depart from risky environments, restrict the time of day and location of foraging, and evolve defensive morphologies that may exact a cost in growth or subsequent reproduction. Competition between consumers for a shared resource either through its mutual exploitation or by aggressive interference depends on the extent of niche overlap versus niche segregation. The large literature on resource partitioning among stream-dwelling invertebrates and fishes provides much insight into the specialization of individual species, but because the extent to which resources actually are limiting often is unknown, this is weak evidence for the importance of competition. Experimental studies with invertebrates have documented numerous cases of aggressive interference, mainly involving space limitation, and in some cases the interaction is as much predation as competition. Freshwater ecosystems harbor pathogens and parasites responsible for a number of well-known animal diseases including salmonid whirling disease, amphibian chytridiomycosis, and crayfish plague, as well as diseases affecting humans, such as malaria, cholera, and river blindness. The arrival of a novel pathogen can result in such a rapid decline that the host population virtually disappears, as evidenced by many species of neotropical amphibians infected by a virulent fungal pathogen. Severe infections may arise when brought to a new region by an invasive species that hosts but is relatively resistant to the parasite or pathogen, or when some change to environmental conditions favor its emergence.

Chapter 11, Lotic Communities, explains how assemblages of species are formed from species within the region that are potential members of a community, and the matching of individual species to available resources, habitat, and one another. The number and relative abundance of species, their traits and functional roles, and energy pathways through food webs are useful descriptors of community structure. The processes that influence the assembly and maintenance of local communities include niche-based models that focus on the interplay between biotic interactions and abiotic forces, the habitat template model based on the association of individual species with habitat features, and disturbance models that emphasize the interplay between species interactions and variation in environmental factors that periodically reduce the abundance of some or all species in the community. Chapter 12, Nutrient and Energy Flow in Aquatic Communities, focuses on the network of interactions portrayed in a food web, and their dependence on energy pathways that vary longitudinally, with season, and among ecosystems. Food web networks ideally describe all interactions within a community of species, but also recognize that much of the energy flows through a subset of species, and basal resources include not only energy produced within the stream, but important energy subsidies from beyond its banks. Functional classifications of species' roles in nutrient and energy flows can be based on a variety of species attributes. Often, organisms will have complementary or overlapping roles for a given functional attribute, but at least in some cases, species appear to be functionally irreplaceable.

Chapter 13, Nutrients, emphasizes the importance of various inorganic materials to the productivity of lotic ecosystems. Whether nutrients are acquired from the surrounding environment, as is typically the case for autotrophs, or from ingesting food and water, as is typical of heterotrophs, a variety of major nutrients, especially nitrogen and phosphorus, and other elements, such as iron, manganese, and silica, are necessary for life. Following their incorporation into tissue, egestion, excretion, and death and decomposition return these materials to the inorganic state, referred to as mineralization. The cycling of nutrients via uptake and release is a critical ecosystem process that in rivers is affected by longitudinal transport, especially in the dissolved inorganic state, and so is called nutrient spiraling. Rates of uptake and release are important measures of ecosystem dynamics and productivity. The ratio of elements, particularly of C:N:P, varies in stream water, in algal tissue, and in herbivore tissue, such that whichever element is in least supply and greatest demand will limit the growth of both producer and consumer.

Chapter 14, Stream Ecosystem Metabolism, integrates all energy producing and energy consuming processes, providing a whole-ecosystem perspective on lotic ecosystems. Studies of stream ecosystem metabolism address two central questions: the relative magnitude of internal versus external energy sources, including their variation along a river's length and with landscape setting; and the efficiency of the stream ecosystem in metabolizing those energy supplies versus export to downstream ecosystems and possibly to the oceans. Principal approaches include the comparison of gross primary production to ecosystem respiration, mass balance estimation of all inputs and exports, and measures of the efficiency with which organic carbon is utilized. The relationship between gross primary production and ecosystem respiration can indicate whether an ecosystem is reliant on internal production or requires external subsidies of organic matter to sustain whole-system respiration. Wherever gross primary production is low relative to ecosystem respiration, a stream clearly is dependent upon external energy inputs, either from the adjacent terrestrial ecosystem or from upstream sources. Stream ecosystems where productivity exceeds respiration are likely to export organic matter to downstream locations.

In Chap. 15, How We manage Rivers and Why, we return to the issues raised in this chapter, of human alteration of running waters. Armed with a more detailed understanding of the abiotic forces, biological processes, and their integration into ecosystem-level fluxes in nutrients and metabolic processes, we can ask what can be done to repair, restore, and protect streams and rivers to maintain their integrity and benefit society. The motivation and urgency for these efforts derives from the recognition that rivers and their biota are highly threatened, yet rivers and stream provide many human benefits, and have intrinsic values beyond their material values. Successful management actions to repair, restore, and protect rivers will require expertise from many sectors, confidence that actions taken are likely to produce desired results, and support from the public and institutions. We conclude by describing the three pillars of river management: fundamental science, measurement of progress, and societal support. Despite justifiable concern that efforts to date have not always delivered the hoped-for improvements, the acceleration of knowledge, concern, and effort is relatively recent. Lotic ecosystems are complex entities with many interacting parts, and there is still much to learn about the timeline and pathway of recovery. Continued efforts are called for.

References

Abell R, Allan JD, Lehner B (2007) Unlocking the potential of protected areas for freshwaters. Biol Conserv 134:48–63. https://doi.org/10.1016/j.biocon.2006.08.017

Abell R, Thieme ML, Revenga C et al (2008) Freshwater ecosystems of the World: A new map of biogeographic units for freshwater biodiversity conservation. Bioscience 58:403–414

Allan JD, Abell R, Hogan Z, et al (2005) Overfishing of inland waters. Bioscience 55:1041–1051

Anthony JL, Downing JA (2001) Exploitation trajectory of a declining fauna: A century of freshwater mussel fisheries in North America. Can J Fish Aquat Sci 58:2071–2090. https://doi.org/10.1139/f01-130

Balian EV, LeVeque C, Segers H, Martens K (2008) Freshwater animal diversity assessment. Hydrobiologia 595:1–637. https://doi.org/10.1007/978-1-4020-8259-7

Bernhardt ES, Palmer MA, Allan JD, et al (2005) Synthesizing U.S. river restoration efforts. Science (80-) 308:636–638

Best J (2019) Anthropogenic stresses on the world's big rivers. Nat Geosci 12:7–21. https://doi.org/10.1038/s41561-018-0262-x

Brauman KA, Daily GC, Duarte TK, Mooney HA (2007) The nature and value of ecosystem services: an overview highlighting hydrologic services. Annu Rev Environ Resour 32:67–98. https://doi.org/10.1146/annurev.energy.32.031306.102758

Bunn SE (2016) Grand challenge for the future of freshwater ecosystems. Front Environ Sci 4:1–4. https://doi.org/10.3389/fenvs.2016.00021

Carpenter SR, Stanley EH, Vander Zanden MJ (2011) State of the World's freshwater ecosystems: physical, chemical, and biological changes. Annu Rev Environ Resour 36:75–99. https://doi.org/10.1146/annurev-environ-021810-094524

Cheung SM, Dudgeon D (2006) Quantifying the Asian turtle crisis: market survey in Southern China, 2000–2003. Aquat Conserv Mar Freshw Ecosyst 16:751–770. https://doi.org/10.1002/aqc

Collen B, Whitton F, Dyer EE et al (2014) Global patterns of freshwater species diversity, threat and endemism. Glob Ecol Biogeogr 23:40–51. https://doi.org/10.1111/geb.12096

Dudgeon D, Arthington AH, Gessner MO et al (2006) Freshwater biodiversity: importance, threats, status and conservation challenges. Biol Rev 81:163–182

Friberg N, Angelopoulos NV, Buijse AD et al (2016) Effective river restoration in the 21st century: from trial and rrror to novel evidence-based approaches. Adv Ecol Res 55:535–611. https://doi.org/10.1016/bs.aecr.2016.08.010

Grill G, Lehner B, Lumsdon AE et al (2015) An index-based framework for assessing patterns and trends in river fragmentation and flow regulation by global dams at multiple scales. Environ Res Lett Lett 10:1–15. https://doi.org/10.1088/1748-9326/10/1/015001

Haag WR, Williams JD (2014) Biodiversity on the brink: an assessment of conservation strategies for North American freshwater mussels. Hydrobiologia 735:45–60. https://doi.org/10.1007/s10750-013-1524-7

Hynes HBN (1975) The stream and its valley. SIL Proc 19:1–15. https://doi.org/10.1080/03680770.1974.11896033

Jackson RB, Carpenter SR, Dahm CN et al (2001) Water in a changing world. Ecol Appl 11:1027–1045. https://doi.org/10.1890/0012-9623(2008)89[341:iie]2.0.co;2

Lange K, Meier P, Trautwein C et al (2018) Basin-scale effects of small hydropower on biodiversity dynamics. Front Ecol Environ 16:397–404. https://doi.org/10.1002/fee.1823

Lehner B, Liermann CR, Revenga C et al (2011) High-resolution mapping of the world's reservoirs and dams for sustainable river-flow management. Front Ecol Environ 9:494–502. https://doi.org/10.1890/100125

LeVeque C, Oberdorff T, Paugy D et al (2008) Global diversity of fish (Pisces) in freshwater. Hydrobiol 595:545–567

Malmqvist B, Rundle S (2002) Threats to the running water ecosystems of the world. Environ Conserv 29:134–153. https://doi.org/10.1017/S0376892902000097

Meybeck M (2003) Global analysis of river systems: from Earth system controls to Anthropocene syndromes, 1935–1955. https://doi.org/10.1098/rstb.2003.1379

Meybeck M, Lestel L, Carré C et al (2016) Trajectories of river chemical quality issues over the Longue Durée: the Seine River (1900S–2010). Environ Sci Pollut Res. https://doi.org/10.1007/s11356-016-7124-0

Moilanen A, Leathwick J, Elith J (2008) A method for spatial freshwater conservation prioritization. Freshw Biol 53:577–592. https://doi.org/10.1111/j.1365-2427.2007.01906.x

Nel JL, Roux DJ, Abell R et al (2009) Progress and challenges in freshwater conservation planning. Aquat Conserv Mar Freshw Ecosyst 19:474–485. https://doi.org/10.1002/aqc.1010

Oki T, Kanae S (2006) Global hydrological cycles and world water resources. Science (80-) 313:1068–1072

Perkin JS, Gido KB, Falke JA et al (2017) Groundwater declines are linked to changes in Great Plains stream fish assemblages. Proc Natl Acad Sci U S a 114:7373–7378. https://doi.org/10.1073/pnas.1618936114

Postel SL (2000) Entering an era of water scarcity: the challenges ahead. Ecol Appl 10:941–948. https://doi.org/10.1890/1051-0761(2000)010[0941:EAEOWS]2.0.CO;2

Rahel FJ (2002) Homogenization of freshwater faunas. Annu Rev Ecol Syst 33:291–315. https://doi.org/10.1146/annurev.ecolsys.33.010802.150429

Sala OE, Chapin FS, Armesto JJ, et al (2000) Global biodiversity scenarios for the year 2100. Science (80-) 287:1770–1774. https://doi.org/10.1126/science.287.5459.1770

Steffen W, Crutzen PJ, McNeill JR (2007) The anthropocene: are humans now overwhelming the great forces of nature? Ambio 36:614–621

Strayer DL (2006) Challenges for freshwater invertebrate conservation. J North Am Benthol Soc 25:271–287. https://doi.org/10.1899/0887-3593(2006)25[271:cffic]2.0.co;2

Strayer DL, Dudgeon D (2010) Freshwater biodiversity conservation: recent progress and future challenges. J North Am Benthol Soc 29:344–358. https://doi.org/10.1899/08-171.1

Taylor BW, Flecker AS, Hall ROJ (2006) Loss of a harvested fish species disrupts carbon flow in a diverse tropical river. Scienc 313:833–836

Tockner K, Stanford JA (2002) Riverine flood plains: present state and future trends. Environ Conserv 29:308–330

USEPA (2016) National national rivers and streams assessment 2008–2009: a collaborative survey. Washington, DC

Vannote RL, Minshall GW, Cummins KW et al (1980) The river continuum concept. Can J Fish Aquat Sci 37:130–137

Vörösmarty CJ, McIntyre PB, Gessner MO et al (2010) Global threats to human water security and river biodiversity. Nature 467:555–561. https://doi.org/10.1038/nature09440

Vörösmarty CJ, Meybeck M, Pastore CL (2015) Impair-then-repair: a brief history & global-scale hypothesis regarding human-water interactions in the anthropocene. Daedalus 144:94–109. https://doi.org/10.1162/DAED_a_00345

Vorosmarty CJ, Sahagian D (2000) Anthropogenic disturbance of the terrestrial water cycle. Bioscience 50:753. https://doi.org/10.1641/0006-3568(2000)050[0753:adottw]2.0.co;2

Streamflow

Fluvial ecosystems exhibit a wide range of natural variability in the quantity, timing, and temporal variability of river flow. How much water does a river need, and what is the importance of natural variation? To answer these questions, an understanding of flow and its far-reaching consequences for the physical, chemical, and biological condition of rivers is essential. In this chapter we show how a basic appreciation of the water cycle is critical to understanding the magnitude and timing of streamflow and discuss how natural variation in flow often has been altered by human actions. Increasingly today, the tools of hydrologic analysis are being combined with other elements of river science to ensure that "environmental flows" are sufficient to protect and restore stream ecosystems.

Human societies extract great quantities of water from rivers, lakes, wetlands, and underground aquifers to meet agricultural, municipal, and industrial demands. Yet freshwater ecosystems also need enough water, of sufficient quality and at the right time, to provide economically valuable commodities and services to society. The benefits of functionally intact and biologically complex freshwater ecosystems include water for human consumption, flood control, transportation, recreation, purification of human and industrial wastes, provision of habitat for plants and animals, and production of fish and other foods and marketable goods (Barron et al. 2002). Unfortunately, however, existing and projected future increases in water demand are resulting in intensifying conflicts between these human uses and the conservation and management of intact, functioning fluvial ecosystems (Postel and Richter 2003).

The human impact on rivers and other surface fresh waters is staggering. Over half of the world's accessible runoff is appropriated for human use, and that fraction is projected to continue to increase (Jackson et al. 2001). Impoundments, surface and groundwater abstractions, interbasin water transfers, and a vast number of small dams, weirs, and diversions alter natural flow patterns, reduce surface flows, and fragment river channels. Increases in the size and affluence of the human population will place even greater demands on surface water supplies in the future, some aquifers will become exhausted, and climate change introduces further uncertainty into water availability. To understand how water appropriation and flow modifications will influence fluvial ecosystems in the future, the analysis of water balance and their influence over streamflow is of central importance.

2.1 The Water Cycle

Until the 16th century, oceans were thought to be the source of rivers and springs via underground seepage. Palissy and others suggested that storage of rainwater was the real source, based on several lines of reasoning. It was noted that springs would not dry up in summer if oceans were the source, since the oceans do not decrease noticeably. Springs should be more common at low elevations if they derive from oceanic water. However, springs often do dry up in summer, they are more common on mountain slopes, and springs are fresh. In 1674, measurements by Perrault showed that precipitation into the Seine basin was six times greater than the river's discharge (Morisawa 1968). This finding changed the focus from whether rainfall is sufficient to provision rivers, to where does the rest of the rainfall go.

The hydrologic cycle describes the continuous cycling of water from atmosphere to earth and oceans, and back again to the atmosphere (Fig. 2.1). Conceptually this cycle can be viewed as a series of storage places and transfer processes, although water in rivers is both a storage place, however temporary, and a transfer between land and sea. The hydrologic cycle is powered by solar energy. This drives evaporation, transferring water from the surface of the land and especially from the oceans, into the atmosphere, and also the water loss that plants experience as a consequence of gas exchange necessary for photosynthesis. Together these are referred to as evapotranspiration (ET).

© Springer Nature Switzerland AG 2021
J. D. Allan et al., *Stream Ecology*,
https://doi.org/10.1007/978-3-030-61286-3_2

Precipitation, primarily as rain and snow, transfers water from the atmosphere to the land surface. These inputs immediately run off as surface water, or follow a number of alternative subsurface pathways, some of which (e.g., groundwater) release to the stream channel much more slowly and so are, in effect, storage places as well.

Despite the enormous significance of rivers in the development of civilizations and the shaping of land forms, the amount of water in rivers at any one time is tiny in comparison to other storages. Only 2.8% of the world's total water occurs on land. Ice caps and glaciers make up the majority (2.24%), and groundwater (0.61%) also is a sizeable percentage. Only 0.009% of the total water is stored in lakes, about 0.001% is stored in the atmosphere, and rivers contain ten times less, 0.0001% of the world's water. Because the volume in the atmosphere and rivers at any instant in time is small, the average water molecule cycles through them rapidly, residing only days to weeks, compared with much longer residence times of water in other compartments.

Estimates of the amount of water discharged annually by rivers to the world's oceans vary, but the value of 40,000 km^3 is widely used. The world's sixteen largest rivers in terms of runoff volume account for nearly one-third of the total, and the Amazon alone contributes nearly 15% (Dingman 2002). Within the United States, the Mississippi contributes some 40% of total discharge, while the Columbia, Mobile, and Susquehanna together contribute an additional 20%. Globally, the greatest runoff occurs in tropical and subtropical areas, because these latitudes also receive the greatest rainfall (Milliman and Farnsworth 2013). By continent, South America is the wettest, Antarctica is the driest, and Australia has the lowest runoff per unit area (Table 2.1).

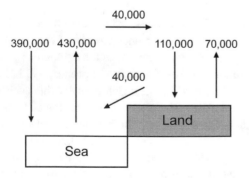

Fig. 2.1 A simplified depiction of the global water cycle. Flows are approximate, in cubic kilometers per year. Downward arrows signify precipitation, upward arrows evapotranspiration. The upper horizontal arrow represents the transfer of moisture from sea to land; the lower arrow represents runoff from land to sea (Reproduced from Postel et al. 1996)

2.1.1 Global Water Cycle

The global water cycle depicted in Fig. 2.1 emphasizes the importance of the transfer of atmospheric moisture from sea to land. Oceans receive 79% of global precipitation and contribute 88% of global evapotranspiration (Dingman 2002). Precipitation on the earth's land surfaces exceeds ET by 40,000 km^3, and this is balanced by an equal amount that is runoff from land to sea. The annual runoff of 40,000 km^3 is the total amount of water potentially available for all human uses (excluding groundwater extractions), including drinking and other municipal needs, for industry, and for irrigated agriculture (non-irrigated agriculture is fed by rain and returns most of this water to the atmosphere as ET over land). However, only about 12,500 km^3 of runoff is truly accessible, because the majority occurs in lightly populated areas or in seasonal floods that are only partly captured (Postel et al. 1996). As of the 1990s, humans appropriated over half of this accessible runoff, and today the fraction no doubt is greater.

2.1.2 Water Balance of a Catchment

For any catchment or region, a water balance equation can be written as follows:

$$P + G_{in} - (Q + ET + G_{out}) = \Delta S \qquad (2.1)$$

where P is precipitation, G_{in} is groundwater inflow to the area, Q is stream outflow, ET is water loss by evapotranspiration, G_{out} is groundwater outflow, and ΔS refers to change in storage (Dingman 2002). When averaged over a period of years with no significant climate trends or anthropogenic influences, changes in storage can be assumed to be zero (but this is not true over short time intervals), and so we can re-write this equation as:

$$P + G_{in} = Q + ET + G_{out} \qquad (2.2)$$

Runoff (R) includes both surface flow (Q) and groundwater outflow, but the latter is usually small and unmeasured, and so the two terms are not always distinguished in common use.

If groundwater inflows and outflows are roughly in balance or small enough to ignore, then precipitation leaves the system as streamflow and ET. This leads to the simplest water balance formulation:

$$Q \,(\text{or } R) = P - ET \qquad (2.3)$$

P and ET vary both spatially and temporally, and thus are primarily responsible for the variability in streamflow that

Table 2.1 Water balances of the continents. From Dingman (2002)

Continent	Area (10^6 km^2)	Precipitation km^3 yr^{-1}	mm yr^{-1}	Evapotranspiration km^3 yr^{-1}	mm yr^{-1}	Runoff km^3 yr^{-1}	mm yr^{-1}
Europe	10.0	6,600	657	3,800	375	2,800	282
Asia	44.1	30,700	696	18,500	420	12,200	276
Africa	29.8	20,700	695	17,300	582	3,400	114
Australia[a]	7.6	3,400	447	3,200	420	200	27
North America	24.1	15,600	645	9,700	403	5,900	242
South America	17.9	28,000	1,564	16,900	946	11,100	618
Antarctica	14.1	2,400	169	400	28	2,000	141
Total land[b]	148.9	111,100	746	71,400	480	39,700	266

[a]Not including New Zealand and adjacent islands; [b]including New Zealand and adjacent islands

we shall discuss at length later in this chapter. First, it will be helpful to describe each term in more detail.

Precipitation includes rain and snow. Its rate varies more on hourly and daily time scales than over months or years, and we are all familiar with patterns in the average values of the latter, recognized as wet and dry months, and from annual mean values that help define the climate of a region and cause us to remark on unusually wet or dry years. Rain infiltrates the land surface or runs off rapidly, but snow is stored on the earth's surface for hours to months before melting. In many areas snow is the main source of surface water supply and groundwater recharge, and melt water is influential in spring flood cycles and in maintaining summer base flows.

Evapotranspiration includes all processes by which water at or near the land surface is returned to the atmosphere. Mainly it includes evaporation from land surfaces, including moisture from soils and standing water from puddles to lakes; and water lost by plants during the exchange of carbon dioxide and oxygen for photosynthesis, referred to as plant transpiration. Globally, about 62% of precipitation that falls on land becomes ET, and ET exceeds runoff for most rivers and for all continents except Antarctica (Table 2.1). Water loss by plant transpiration constitutes a major flux back to the atmosphere. When an experimental forest in New Hampshire was clear–cut and subsequent regrowth was suppressed with herbicides, stream runoff increased 40% on an annual basis, and 400% during summer (Likens and Bormann 1995). This represented water that would have returned to the atmosphere primarily via transpiration in an intact forest. Subsequently, when herbicide treatment was suspended, the forest rapidly regenerated with species with high transpiration rates, and streamflow declined to levels below that of the mature forest.

Seasonal variation in ET, due to the combined effects of temperature and water demand by plants, often is greater than seasonal variation in precipitation. This can be seen in comparisons of monthly averages for precipitation and runoff for a series of North American rivers (Fig. 2.2), in which ET can be inferred as the difference. In cold climates ET is reduced and so a larger fraction of precipitation leaves the catchment as runoff, as Fig. 2.3 illustrates for rivers along the Atlantic Coast from Canada to the southeastern US.

2.1.3 Surface Versus Groundwater Pathways

Precipitation destined to become runoff travels by a number of pathways that are influenced by gradient, vegetation cover, soil properties, and antecedent moisture conditions. Some rainwater evaporates from the surface of vegetation immediately during and after a rainstorm, never reaching the ground, or being absorbed by plants. This is referred to as interception and is included within ET. Some rainfall passes through spaces in the canopy (throughfall), some runs down stems and trunks (stemflow), and some intercepted water later falls to the ground (canopy drip). The latter two pathways may play a role in nutrient transfers, a topic for a later chapter.

Once rain or melt water encounters the ground, it follows several pathways in reaching a stream channel or groundwater (Fig. 2.4). Approximately three-fourths of land-area precipitation infiltrates into the soil. In unsaturated, porous soils, water infiltrates at some maximum rate, termed the infiltration capacity. This capacity declines during a rain event, normally approaching a constant after one to two hours of rain (Dunne and Leopold 1978). The downward percolation of water results in a series of hydrologic horizons. The unsaturated (vadose) zone lies above the saturated (groundwater, phreatic) zone whose upper limit is the water table surface. Soil moisture usually is least in the rooted zone, which is the uppermost horizon of the unsaturated

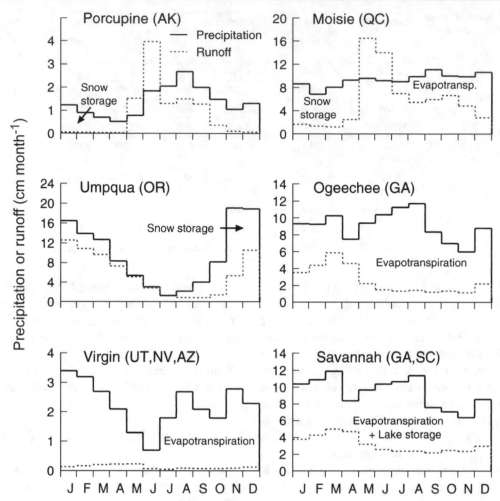

Fig. 2.2 Patterns of precipitation and runoff for rivers from a diversity of regions in North America. Porcupine River, Alaska, showing snowmelt peak of runoff during low precipitation. Moisie River, Quebec, showing snowmelt peak of runoff. Umpqua River, Oregon, showing runoff peak following seasonal precipitation. Ogeechee River, Georgia, showing runoff pattern caused by seasonal changes in evapotranspiration. Virgin River, Utah, Nevada, Arizona, showing very low runoff due to low precipitation and high evapotranspiration. Savannah River, Georgia and South Carolina, showing flattened runoff pattern due to regulation (compare to Ogeechee) (Reproduced from Benke and Cushing 2005)

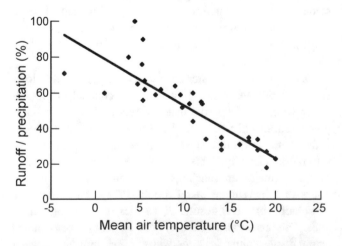

Fig. 2.3 Annual runoff as a percentage of precipitation versus mean annual air temperature for rivers draining into the Atlantic Ocean from Canada to the southeastern US (Reproduced from Allan and Benke 2005)

zone, due to evaporation, plant uptake, and downward infiltration. The water table is the fluctuating upper boundary of the groundwater zone. These horizons fluctuate seasonally depending on rainfall, and generally rise at the end of growing season when ET is low. Soil moisture thus varies with prior rainfall and season, and the degree of soil saturation influences whether new moisture percolates downward to recharge groundwater, moves laterally through the soil, or rises vertically above the soil surface.

Rain that reaches the groundwater will discharge to the stream slowly and over a long period of time. Base flow or dry-weather flow in a river is due to groundwater entering the stream channel from the saturated zone. Above the saturated zone, some infiltration water will move downslope as interflow, which is subsurface runoff in response to a storm event (Fig. 2.4). Interflow is lowest in unsaturated soils and

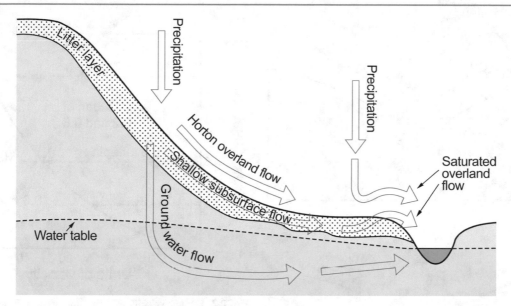

Fig. 2.4 Pathways of water moving downhill. Overland flow occurs when precipitation exceeds the infiltration capacity of the soil. Water that enters the soil adds to groundwater flow and usually reaches streams, lakes, or the oceans. A relatively impermeable layer will cause water to move laterally through the soil as shallow subsurface storm flow. Saturation of the soil can force subsurface water to rise to the surface where, along with direct precipitation, it forms saturation overland flow. The stippled area is relatively permeable topsoil

when grain size (and thus pore size) is small; it can reach 11 m day^{-1} through sandy loam on a steep hill (Linsley et al. 1958). Rainfall in excess of infiltration capacity accumulates on the surface, and any surface water in excess of depression storage capacity will move as an irregular sheet of overland flow. In extreme cases, 50–100% of the rainfall can travel as overland flow (Horton 1945), attaining velocities of 10–500 m hr^{-1}. Overland flow tends to occur in semi-arid to arid regions, wherever human activities have created impervious surfaces or compacted the soil, when the soil surface is frozen, and over smoother surfaces and steeper slopes (Dingman 2002). However, overland flow rarely occurs in undisturbed humid regions because their soils have high infiltration capacities. Lastly, when there is a large enough rainstorm or a shallow enough water table, the water table may rise to the ground surface, causing subsurface water to escape from the saturated soil as saturation overland flow. This is composed of return flow forced up from the soil and direct precipitation onto the saturated soil (Dunne and Leopold 1978). Velocities are similar to the lower range of Horton overland flow.

Most rivers continue to flow during periods of little rainfall. These are called perennial, as opposed to intermittent, and most of the water in the channel comes from groundwater. In humid regions the water table slopes toward the stream channel, with the consequence that groundwater discharges into the channel. Discharge from the water table into the stream accounts for base flow during periods without precipitation, and also explains why base flow increases as one proceeds downstream, even without tributary input.

Such streams are called gaining or effluent (Fig. 2.5a). Streams originating at high elevation sometimes flow into drier areas where the local water table is below the bottom of the stream channel. Depending upon permeability of materials underlying the streambed, the stream may lose water into the ground. This is referred to as a losing or influent stream (Fig. 2.5b). The same stream can shift between gaining and losing conditions along its course due to changes in underlying lithology and local climate, or temporally due to alternation of base flow and storm flow conditions. The exchange of water between the channel and groundwater will turn out to be important to the dynamics of nutrients and the ecology of the biota that dwells within the substrate of the stream bed.

2.2 Streamflow

The volume of flow moving past a point over some time interval is referred to interchangeably as discharge or streamflow. It can be calculated from measurements of width (w, in m), depth (d, in m) and current velocity (v, in m.s^{-1}) and expressed in m^3 s^{-1} (cms or cumecs) or cubic feet per second (cfs) (Fig. 2.6).

$$Q = wdv \qquad (2.4)$$

In practice, discharge is estimated by dividing the stream cross–section into segments, measuring area and velocity for each, and summing the discharge estimates for the segments (Fig. 2.6).

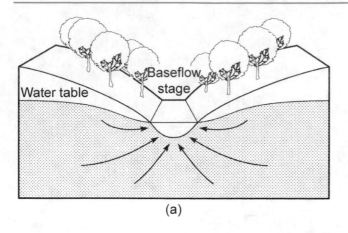

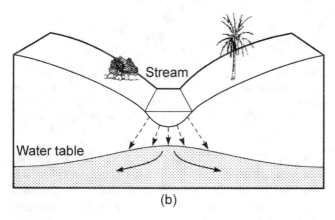

Fig. 2.5 (**a**) Cross-section of a gaining stream, typical of humid regions, where groundwater recharges the stream. (**b**) Cross-section of a losing stream, typical of arid regions, where streams can recharge groundwater. (Reproduced from Fetter 1988)

$$Q = \sum a_i v_i \qquad (2.5)$$

Velocity is measured at the midpoint of the segment and (in shallow streams) at 0.6 depth below the surface. At least ten subsections are recommended, and none should have more than 10% of the total flow (Whiting 2016). Ideally, the flow is uniform and the reach is straight.

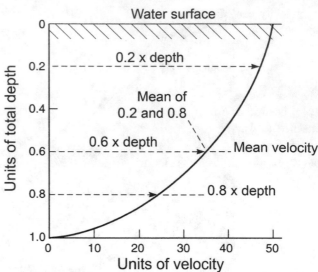

Fig. 2.7 Current velocity as a function of depth in an open channel. Mean velocity is obtained at a depth of 0.6 from the surface when depth is <0.75 m, and from the average of measurements at 0.2 and 0.8 depth in deeper rivers

Current velocity varies considerably within a stream's cross–section owing to friction with the bottom and sides, and to channel sinuosity and obstructions. Highest velocities are found where friction is least, generally at or near the surface and near the center of the channel. In shallow streams, velocity is greatest at the surface due to attenuation of friction with the bed, and in deeper rivers velocity is greatest just below the surface because of friction with the atmosphere (Gordon et al. 2004). Velocity then decreases as a function of the logarithm of depth (Fig. 2.7), approaching zero at the substrate surface. In streams with logarithmic velocity profiles, one can obtain an average value fairly easily by measuring current speed at 0.6 of the depth from the surface to the bottom. At depths greater than 0.75 m, velocities measured at 0.2 and 0.8 beneath the surface can be averaged, and in very turbulent water it may be necessary to measure velocity at 0.1 depth intervals.

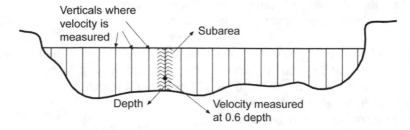

Fig. 2.6 Estimation of discharge from the integration of point measurements of velocity and associated area of flow in sub-sections of the channel cross-section. Velocity is measured at 0.6 depth from the surface in shallow streams (Reproduced from Whiting 2016)

Velocity can be measured in a number of ways. In the absence of an appropriate current meter, the travel of a stick or orange timed over a measured distance gives a rough measure of surface velocity, usually multiplied by 0.85 to adjust for reduced flows along the channel bed and margins. Mechanical current meters with rotating cups or propellers have been the primary measurement tool for decades, attached to wading rods in shallow streams or lowered by cable from boats, bridges, or a small cable car into larger rivers. Electromagnetic current meters with electrodes that measure voltage produced as water moves through their electric field are now widely used, as well as more sophisticated instruments.

Methods of discharge estimation are numerous, but the integration of point measurements of velocity and associated areas of flow is most common, using Eq. 2.5 (Whiting 2016). Many rivers have permanent gauges consisting of a well on the stream bank which is connected by a horizontal pipe to the deepest part of the channel to measure river height, or stage. Discharge is then estimated from Eq. 2.5 from multiple simultaneous measurements of stage and discharge, and a stage–discharge rating curve is constructed for that location. Thereafter, discharge is estimated hourly or continuously by monitoring stage.

Discharge varies over all time scales, from hourly and daily responses to a storm event, through seasonal, annual, and decadal intervals, and over historic and geologic time. It usually increases along a stream network due to inputs from tributaries and groundwater. Over a stream reach of a few hundred meters, assuming that inputs from tributaries and groundwater are negligible, discharge should be the same at any of several transects, even though the channel shape and water velocity vary from transect to transect. This is known as the continuity of flow relationship.

Many countries have extensive networks of stream gages and make data available through the internet, including in real time. For example, the US Geological Service maintains a network of approximately 7,000 active gages, of which roughly half have continuous or near-continuous records of more than 30 years (USGS 2005). A subset of 1,600 stream gages where the discharge is considered to be primarily influenced by climatic variations, known as the USGS Hydro-Climatic Data Network (HCDN) was developed to study such issues as flood frequency, drought severity, and long-term climatic change (Slack and Landwehr 1992). Flow hydrographs including real time data and various summary statistics can be downloaded from government websites such as Waterdata.usgs.gov/nwis in the US, and from similar sites such as the Water Survey (Canada), National River Flow Archive (UK), and Hydrobank (France).

2.2.1 The Hydrograph

A continuous record of discharge plotted against time is called a hydrograph. It can depict in detail the passage of a flood event over several days (Fig. 2.8), or the discharge pattern over a year or more. A hydrograph has several characteristics that reflect the pathways and rapidity with which precipitation inputs reach the stream or river. Base flow represents groundwater input to river flow. Rainstorms result in increases above base flow, called storm flow or quickflow. The rising limb will be steepest when overland and shallow sub-surface flows predominate, and more gradual when water reaches the stream through deeper pathways. The magnitude of the hydrograph peak is influenced by the severity of the storm and the relative importance of various pathways by which rainwater enters the stream (Fig. 2.4). The lag to peak measures the time between the center of mass of rainfall and the peak of the hydrograph. The recession limb describes the return to base flow conditions.

Hydrographs exhibit wide variation over all time scales, from small streams to large rivers, and among geographic regions, influenced by the amount and distribution of precipitation throughout the year, its storage as snow, the size and topography of the basin, and soil and vegetation characteristics. Substantial overland flow causes a rapid and pronounced rising limb to the hydrograph and can result in significant sediment erosion from the land surface. Such streams are called "flashy"; aridland streams often are good examples. Because little or no overland flow occurred in a humid New Hampshire forest even after deforestation (Likens and Bormann 1995), it is clear that soil permeability, enhanced by leaf litter, other organic matter, and root structure, provides an infiltration capacity that rarely is exceeded in forested catchments of humid areas. Because water travels more slowly in subsurface pathways, the rising limb of the hydrograph is less pronounced. Furthermore, the likelihood of sediment transport from the landscape is reduced, while the transport of dissolved materials is enhanced.

Another general pattern is for flood hydrographs to become broader and less sharp as a river gathers tributaries in a downstream direction. This is due to differences among tributary sub-basins in the amount and intensity of precipitation received, causing the sum of tributary discharges to a larger river to be less well defined than the individual events. In addition, a flood peak attenuates as it travels downstream owing to friction with the channel and temporary storage. This attenuation will be greatest when a river is connected to its floodplain and has natural bends and channel roughness.

Fig. 2.8 Streamflow hydrograph resulting from a rainstorm

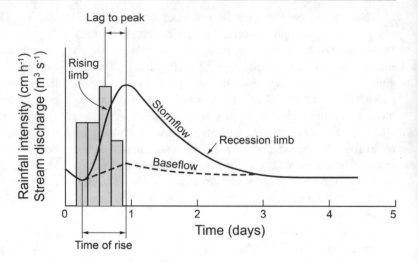

However, when a river is straightened and separated from its flood plain by levees, floods will pass very quickly downstream, where they may cause significant damage. The Upper Mississippi experienced a major flood in 1993 in response to 1-in-200 year rainfall, with an orientation along the river channel that favored the convergence of flood waters, and that occurred during a period of above-normal soil moisture conditions and below-normal ET (Kunkel et al. 1994).

2.3 Flow Variation

The characterization of streamflow has practical application for the design of flood-control structures, evaluation of channel stability, and assessment of whether sufficient water is available at the appropriate time to meet the needs of both people and the ecosystem. Abundant data often are available for gauged sites, in some instances as much as a century of continuous monitoring. Methods also exist that allow extrapolation to ungauged sites, such as obtaining data from a nearby gaged site and adjusting for any differences in drainage area. This has led to a great deal of hydrologic analysis of the spatial and temporal variation in streamflow.

Daily records of streamflow from stream gage networks allow a wide variety of statistical analyses, of averages, extremes, and variability, of regional differences, and of changes over time whether due to natural or human causes. The USGS makes it surface-water data for the nation available online, including with daily, monthly, and annual means (see https://waterdata.usgs.gov/nwis/sw). Total and mean annual discharge are useful measures of the volume of water leaving a catchment. The mean is calculated by dividing the sum of all the individual flows by the number of flows recorded for the period of record. If mean annual flows are available for a number of years, these may be averaged to obtain the long-term mean annual flow. One can compare the relative sizes of rivers of the world by length, drainage area, or discharge. Table 2.2 lists the largest rivers in the world by mean annual discharge in $km^3 \ yr^{-1}$. Ranking by drainage area or length would differ substantially.

Table 2.2 Ten largest rivers of the world, ranked by discharge. Note differences in drainage areas. Data from Milliman and Farnsworth (2011)

River	Country	Average discharge ($km^3 \ yr^{-1}$)	Drainage area (x $10^3 \ km^2$)
Amazon	Brazil	6,300	6,300
Congo	Congo, DR	1,300	3,800
Orinoco	Venezuela	1,100	1,100
Yangtse	China	900	1,800
Brahmaputra	Bangladesh	630	670
Yenisei	Russia	620	2,600
Mekong	Vietnam	550	800
Lena	Russia	520	2,500
Ganges	Bangladesh	490	980
Mississippi	USA	490	3,300
Parana	Argentina	460	2,600

2.3.1 The Likelihood of Extreme Events

Often, we wish to know how frequently a flow of a given magnitude is exceeded in an average year, or the likelihood of an extreme annual flood such as one that we might characterize as a ten-year or 50-year event. Several methods are available for estimating the probability of extreme events, whether flood or drought. The magnitude of events is inversely related to their frequency (or probability) of occurrence, a relationship captured in a flow-duration curve. These typically are constructed using daily streamflows over many years, so they include both seasonal and interannual variability.

Flow duration curves describe the percentage of time flow is greater than a specified value (percent exceedance). A flow-duration curve plots the cumulative frequency of daily records that equal or exceed a given value of average daily discharge against flow magnitude (Fig. 2.9). Low flows are equaled or exceeded on most days, whereas high flows are equaled or exceeded only a small percentage of the time. For small basins with rapid response times, daily data may be too coarse, and estimates as frequent as every 15 min may be needed.

Several useful metrics are easily obtained from a flow duration graph. The median flow ($Q_{.50}$) is that exceeded 50% of the time. Because of the influence of a few large floods on the mean value, in humid regions the mean is exceeded only on 20–30% of the days (Dingman 2002). The $Q_{.05}$ is a streamflow exceeded only 5% of the time (18 days in an average year), so is a reasonable value for high flows that occur infrequently. Similarly, $Q_{.95}$ is a streamflow exceeded 95% of the time. This value indicates how much water is available most of the time, and also provides a threshold below which we can identify extremely low flows.

If one wishes to compare among streams, it is usual to normalize flow to drainage area by plotting flow divided by drainage area. Such comparisons distinguish 'flashy' from stable streams, as Fig. 2.9 illustrates for two locations in Michigan. The relatively flat curve of the Au Sable River indicates stable flow throughout the year compared to the much more variable Black River.

A widely used metric of low flow frequency is the 7Q10, the lowest 7-day average flow that occurs on average once every 10 years, and thus also is exceeded on average in 9 out of every 10 years. It is calculated from an annual series of the smallest values of mean discharge over any 7 consecutive days during the annual period (Riggs 1980). Data usually are for April 1 through March 31 of the following year (referred to as a water year) to avoid artificially separating low-flow periods that often extend from late fall to early spring. A time series typically spans 10 or more years of streamflow, and data are fit to a probability distribution (commonly the log-Pearson Type III distribution) to compute a 10-year recurrence probability. Low-flow frequency metrics also can be computed on a seasonal or monthly basis by limiting the daily data used for the annual series to just the season, such as summer, or to the month of interest.

The 7Q10 is widely used in the US for water quality standards and toxic waste loads relating to chronic effects on aquatic life, but other statistics including the 7Q2 (lowest 7-day flow occurring over two years) are also used. Elsewhere, slightly different methods may be employed such as the mean annual 7-day minimum, which is the average of 7-day low flows over 10 or more years.

Estimating the probability of an extreme annual flood, such as one that might occur on average once in ten or fifty years, also is useful. Typically, one estimates the probability of a "1-in-N-year" event of a flood of a given size or larger. Thus a 1-in-100-year flood has a 1% likelihood of occurring in any year and the average recurrence interval is expected to be 100 years between two floods of that magnitude or larger. Flood probability and average recurrence are inversely related. Given a record of annual maximum flows or other measures of flood events, a number of methods can be used to estimate flood probability (Gordon et al. 2004). One begins with a list of the single highest flow of each year, preferably based on the instantaneous peak of the flood hydrograph rather than average daily discharge. This is

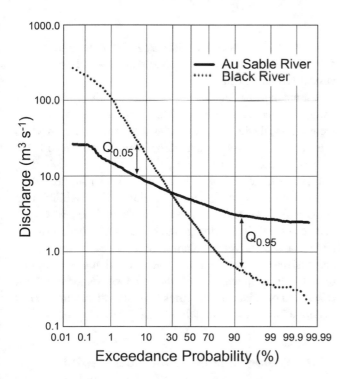

Fig. 2.9 Flow duration curves for two rivers in Michigan, showing the high flow discharge ($Q_{0.05}$) that is exceeded only 5% of the time, and the low flow discharge ($Q_{0.95}$) that is exceeded 95% of the time. Because the two watersheds are of similar area, discharge was not normalized to drainage area. Graphs were constructed from daily records for 1990–2000

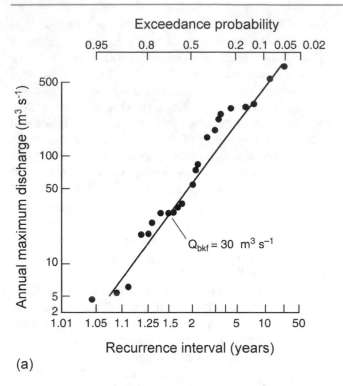

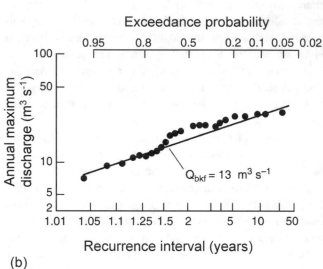

(a)   (b)

Fig. 2.10 Example of a flood–frequency analysis for two rivers, based on annual peak instantaneous flows from a 20-plus year gauge record. The bankful flood is estimated using T = 1.5 years, and the probability or recurrence interval for more extreme events (e.g., 20 and 50-year floods) can be read from the graph. Lines are fitted by eye.

(a) Sycamore Creek, Arizona, is an aridland stream subject to flash floods. (b) The Colorado River in its upper reaches, near Grand Lake, Colorado, has a highly regular snowmelt-driven hydrograph. Note the steeper slope of the graph for Sycamore Creek

especially important in small rivers where the peak flow passes in hours and will be under-estimated by the 24 h average, although this may not be critical for large rivers. By fitting a probability distribution to the data set, it is possible to predict the average recurrence interval for floods of a given magnitude or, conversely, the magnitude of the flood that occurs with a given frequency. The recurrence interval (T) for an individual flood is calculated as:

$$T = \frac{n+1}{m} \qquad (2.6)$$

where n = years of record and m = rank magnitude of that flood. The largest event is scored as m = 1. Figure 2.10 illustrates flood–frequency curves for Sycamore Creek, Arizona, which experiences irregular flash floods, and for the upper reaches of the Colorado River, with a highly repeatable, snowmelt-driven flow regime. The recurrence interval for floods of a given magnitude is read directly from the graph. One also can determine the likelihood (1/T) that the annual maximum flood for a given year will equal or exceed the value of a 10-year, 20-year, or 50-year flood event.

Flood-frequency analysis also is used to estimate the flood magnitude that has a 1–2-year recurrence. This is often considered to be an estimate of the flood that just overtops the banks, the bankfull discharge (Q_{bkf}), which in turn is a surrogate for the effective discharge, the flow that is considered most influential in maintaining channel form (discussed further in Chap. 3).

Estimating the likelihood of rare events obviously is risky, and becomes more so when only a short hydrologic record is available for analysis, as any given string of years can include an individual flood whose true recurrence interval is actually much longer than the record. In addition, standard estimation methods assume that streamflow fluctuates within an unchanging envelope of variability (the "stationarity" assumption), yet changes in land use or climate can result in a heterogeneous data set. Whether due to climate trends, river regulation or changes in hydrologic responsiveness of a catchment over its period of streamflow records, the implications of the "end of stationarity" for estimation of streamflow metrics is an area of current interest (Vogel et al. 2011; Serinaldi and Kilsby 2015).

2.3.2 Ecologically Relevant Flow Metrics

Estimates of the likelihood of a flood or drought of given magnitude, mean and median discharges, and standardized low-flow metrics such as the 7Q10 are widely used in water management, including to address ecological concerns.

However, ecologists have developed their own set of statistical approaches with the aim of characterizing streamflows, and their alteration by human activities, using metrics likely to be especially relevant to the biota and ecosystem processes. The Indicators of Hydrologic Alteration (IHA), developed by Richter et al. (1996) and supported by public-domain software (Mathews and Richter 2007), is widely used to analyze within and between-year variation in daily streamflow, characterize flow regimes, and detect ecologically relevant hydrological differences. The IHA contains 33 hydrologic parameters that characterize five aspects of flow regimes: magnitude of monthly streamflows, magnitude and duration of annual extreme flows, timing of annual extreme flows, frequency and duration of high and low pulses, and rate and frequency of flow changes (Fig. 2.11 and Table 2.3). Some minimal time record, often 20 years, is analyzed to capture the range of variability. The extent of human alteration is inferred by comparing two time series of flows representing natural and altered conditions, such as before and after construction of a dam, or between sets of stream gages that are near-natural compared with human-influenced sites. There are many additional flow metrics in use, especially in the hydrologic literature, and similar approaches have been used for model forecasts that provide estimates based on monthly rather than daily values. However, the conceptual framework of the IHA provides an ecologically relevant framework that captures the five principal aspects of flow variability.

2.3.3 Hydrologic Classification

Hydrologic classification is the process of systematically arranging rivers or catchments into groups that are most similar with respect to the characteristics or determinants of their flow regime (Olden et al. 2012). The classification of flow regimes is important to understand riverine flow variability, its association with geographic features, and to explore the influence of streamflow on biological communities and ecological processes. By assigning rivers or river segments to a particular hydrologic type, relationships between flow alteration and ecological response metrics can be developed for an entire river type and hypothesized to be similar within river types and vary among river types (Poff et al. 2010). For example, Fig. 2.12 shows five river types developed for 420 streams with unmodified flow regimes in the United States (Poff 1996) and defined in terms of 11 flow metrics. Ellipses represent the 90% confidence limits for each river type expressed in terms of two of the flow classification variables (baseflow stability and flood predictability) that are ecologically relevant and amenable to management action. Based on these natural differences in flow regime, it is reasonable to expect different ecological responses to similar types of flow alteration. For example, empirical studies have found differences in fish assemblages between stream types with stable versus more flashy runoff (Poff and Allan 1995), and their ecological response to flow alteration may differ as well.

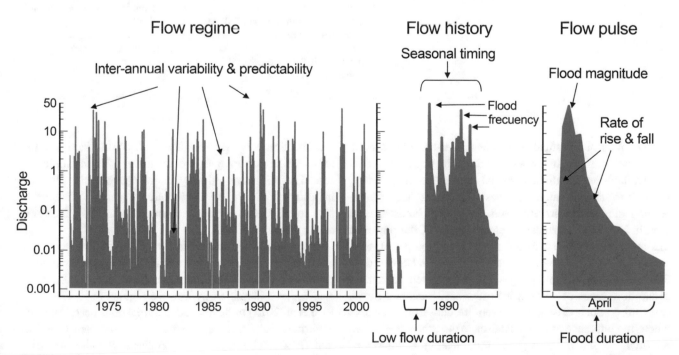

Fig. 2.11 Different components of the flow regime may be characterized over varying temporal scales for use in streamflow classifications (Reproduced from Olden et al. 2012)

Table 2.3 Summary of parameters of the Indicators of Hydrologic Alteration (IHA), and their influences on habitats, organisms and ecosystem processes. (Modified from The Nature Conservancy 2009)

IHA parameter group	Hydrologic parameters	Ecosystem influences
1. Magnitude of monthly flows	Mean or median value for each calendar month *Subtotal 12 parameters*	Habitat availability Reliability of water supply Influences water temperature, oxygen levels, ecosystem processes
2. Magnitude and duration of annual extreme flows	Annual minima, 1-day, 3-day, 7-day, 30 day and 90-day means Annual maxima, 1-day, 3-day, 7-day, 30 day and 90-day means Number of zero-flow days Baseflow Index: 7-day minimum flow/mean flow for year *Subtotal 12 parameters*	Multiple species interactions Importance of abiotic vs biotic factors Duration of stressful low-flow conditions Duration of high flows affecting waste removal, channel properties
3. Timing of annual extreme flow conditions	Julian date of each annual 1-day maximum Julian date of each annual 1-day minimum *Subtotal 2 parameters*	Compatibility with life cycles of organisms Predictability/avoidability of stressful conditions Access to special habitats for reproduction and rearing Spawning cues for migratory fish
4. Frequency and duration of high and low pulses	Number of low pulses within each water year Mean or median duration of low pulses (days) Number of low pulses within each water year Mean or median duration of low pulses (days) *Subtotal 4 parameters*	Frequency and duration of soil moisture stress for plants Frequency and duration of anaerobic stress for plants Availability of floodplain habitats Nutrient and organic matter exchanges Influences channel sediment dynamics and substrate disturbance
5. Rate and frequency of rise and fall of flows	Rise rates: mean or median of all positive differences between consecutive daily values Fall rates; mean or median of all negative differences between consecutive daily values Number of hydrologic reversals *Subtotal 3 parameters* *Grand total 33 parameters*	Drought stress on plants (fall rates) Entrapment of organisms on islands, floodplains (rise rates) Desiccation stress on low-mobility stream-edge organisms

A number of statistical approaches are used in hydrologic classification. In regions where good quality streamflow data are available with sufficient spatial and temporal coverage, streamflow classification involves identifying and characterizing similarities among rivers across the landscape, as seen in Fig. 2.12. In an early demonstration of the power of this approach, Poff (1996) developed a hydrologic classification for streamflows within the conterminous US based on 806 reference stream gauges, identified ten dominant streamflow types of varying intermittency, perennial flows, and timing. Such classifications require data from locations where the influence of dams, landscape disturbance, etc., are minimal, so much effort goes into creating a dataset of 'reference' conditions. A more recent and expanded classification by Mcmanamay et al. (2014) included 1715 reference stream gauges and identified 12 hydrologic classes. These classes display some geographic clustering indicative of differences in climatic and geologic features but also some spatial overlap, so stream flow class and region are an imperfect match.

Because patterns in streamflow exhibit some geographic clustering and are strongly influenced by regional patterns in precipitation and ET, vegetation, topography, etc., various studies have identified hydrologic regions (e.g., Lins 1997; Rice et al. 2015) or developed classifications based on shared environmental features. The River Environment Classification (REC) scheme for New Zealand (Snelder and Biggs 2002) is an example of environmental classification in which

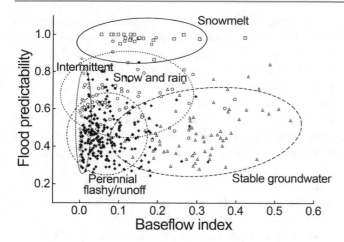

Fig. 2.12 Plot of five river types in the US based on 420 stream gauges and defined in terms of 11 flow classification metrics but plotted here in two-dimensional space defined by flood predictability and baseflow index. Ellipses reflect 90% confidence intervals and show natural range of variability for the two flow metrics for each of five river types: snowmelt (open squares), snow and rain (open circles), stable groundwater (open triangles), perennial flashy/runoff (closed diamonds) and intermittent (open diamonds—combined harsh intermittent, intermittent flashy and intermittent). (Reproduced from Poff et al. 2010)

a mapped hydro-geomorphic topology of rivers is based on a combination of watershed climate and topography, which are assumed to be the dominant causes of hydrologic variation. Similarly, a continental-scale classification of hydrologic regimes for Australia identified 12 classes of distinctive flow-regime types differing in the seasonal pattern of discharge, degree of flow permanence (perennial versus intermittent), variations in flood magnitude and frequency, and other aspects of flow predictability and variability (Kennard et al. 2010). Geographic, climatic, and some catchment topographic factors were generally strong discriminators of flow-regime classes. Importantly, however, geographical distribution of flow-regime classes showed varying degrees of spatial cohesion, as stream gauges from certain flow-regime classes often showed a non-contiguous distribution across the continent. This was particularly pronounced for certain of the Australian flow regime classes including those described as stable baseflow, predictable winter highly intermittent, and variable summer highly intermittent. This is an important caution if extrapolating flow-regime characteristics from individual gauges to ungauged areas, even those within relatively close proximity.

Although the characterization of flow regimes into regionally distinct groups may continue to be refined, the concept that each individual river has a natural flow regime upon which its ecological integrity depends has become firmly established (Poff et al. 1997). Climate, vegetation, geology, and terrain determine the natural flow regime, as discussed earlier; and humans alter flow regimes by

changing any of the terms in Eq. 2.2, and through impacts on flow pathways and response times.

2.4 Human Influence on Streamflow

Human activities modify streamflows at global, regional, and local scales, including by dams and diversions, changing land use, and groundwater withdrawals. By altering the balance of precipitation and ET, climate change can have complex effects on streamflow. Thus it is not surprising that, across the conterminous US, analysis of several thousand stream-gage records for 1980–2007 found flow alteration to be widespread (Carlisle et al. 2011). However, the relative importance of different human activities can be difficult to tease apart.

Reservoirs and water diversions generally reduce annual and seasonal river discharge directly, as well as indirectly through evaporation and plant transpiration. Further, reservoir management profoundly affects the variability of flows. Studies of global water flux generally agree that land conversion to agriculture has increased river runoff. Replacement of natural vegetation with crops has reduced global ET due to shorter growing periods (i.e., intermittent periods with fallow land) or lower rooting depths of crops compared to natural vegetation. At the catchment scale, land conversion to agricultural or urban use generally increases water conveyance, resulting in flashier flows. Groundwater pumping is important in some areas at the local to regional scale, and can substantially lower water tables and thus groundwater inputs. It is difficult to generalize about the effects of climate change on water balance and streamflow, but it is likely to exacerbate other water challenges just described, all with varying degrees of influence depending on river basin. We now examine each of these influences in turn.

2.4.1 Dams and Impoundments

Dams and impoundments provide important human benefits by mitigating floods, securing water supplies, and providing hydropower, thereby contributing to improved human health, expanded food production, and economic growth. There are also negative human impacts. Dams and their reservoirs can displace local populations, cause social disruption, and increase disease incidence. Ecological consequences are numerous and significant. Water storage and release by reservoirs affects the quantity, quality, and timing of downstream flows, and dams serve as physical barriers, disrupting the ecological connectivity of rivers. Dams also trap sediments, alter temperature patterns, and limit movement of biota, all topics to be discussed in later chapters.

The extent of alteration of river flow and loss of river connectivity due to dams and impoundments is staggering. Worldwide, the number of large dams (defined as >15 m in height) is estimated to be about 50,000 (Fig. 2.13), creating impoundments with a cumulative storage capacity in the range of 7,000–8,300 km³ (Vörösmarty et al. 2003; Lehner et al. 2011). For comparison, this is nearly 10% of the water stored in all natural freshwater lakes on Earth, and about one-sixth of the total annual river flow into the oceans. While the number of large dams and impoundments (typically defined by dam height, power generation, or reservoir capacity) is quantified in a world database (WCD 2000), the number of small impoundments, including numerous farm ponds, can only be approximated. An estimated 16.7 million reservoirs larger than 0.01 ha may exist worldwide (Lehner et al. 2011). Their combined storage capacity of approximately 8,000 km³ is on a par with storage by all large dams. For the US, approximately 2.5 million smaller water control structures exist (Poff and Hart 2002).

Alteration of the natural flow regime due to the construction of dams and impoundments is widely documented. Based on 21 sites with adequate long-term hydrologic records, Magilligan and Nislow (2005) found that dams generally caused increases in low-flow and decreases in high-flow statistics, reductions in seasonality, and declines in the mean rate of rise and fall. Comparing 186 sites below dams with 317 undammed reference locations in the US, average maximum flows consistently declined and minimum flows generally increased (Poff et al. 2007). In addition, inter-annual variation in maximum and minimum flows generally declined. By strongly modifying natural flow regimes and reducing natural regional differences, the authors concluded that dams are imposing environmental homogeneity across broad geographic scales.

Although historically a snowmelt-driven hydrograph with most of its runoff generated in its upper basin, the Colorado River today is an extreme example of dam-controlled flows. The Glen Canyon Dam, built over 1956–1966, created Lake Powell, one of the largest man-made reservoirs in the US. Lake Powell helps ensure equitable distribution of water between upper and lower basin states by storing runoff during wet years for release during dry years. It also generates hydropower (second in the southwestern US only to Hoover Dam) and provides recreational uses. The environmental consequences include large evaporative water loss from the reservoir due to its desert environment, and dramatic alteration of the seasonal hydrograph (Fig. 2.14).

In addition to altering the natural variability of a river's flow, dams fragment once-continuous river systems into a series of largely disconnected segments. River fragmentation diminishes the natural connectivity within and among river systems, limiting species migration and dispersal as well as the transport of organic and inorganic matter downstream and into riparian zones and floodplains. The extent of river fragmentation has been quantified in several ways: from the number of free-flowing river segments greater than a certain length, by calculating the number of dams per watershed area, termed dam density, or by more complicated formulae that use dam inventories and river-routing models to estimate river fragmentation indices (or, inversely, connectivity indices) world-wide. In one of the first such estimates, Benke (1990) inventoried rivers of the 48 contiguous US states, finding that only 42 high-quality, free-flowing rivers >200 km in length remained.

There is abundant evidence that most of the largest river systems of North America, Europe, and former Soviet Union are highly or moderately affected by fragmentation of their main channels (Dynesius and Nilsson 1994). An analysis of

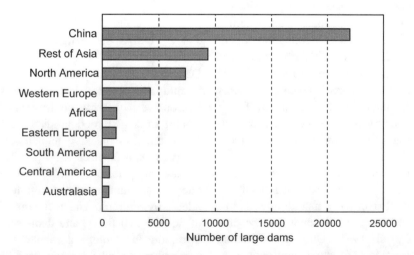

Fig. 2.13 Distribution of large dams by region. China and Australasia (Australia, New Zealand, Papua New Guinea, and Fiji) were treated separately from the rest of Asia, and Central America including Mexico was separated from North America (United States and Canada). Large dams are >15 m in height or, if between 5 and 15 m in height, have a reservoir capacity >3 × 10⁶ m³ (Reproduced from Tharme 2003)

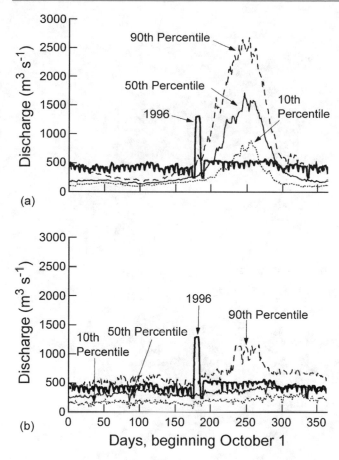

Fig. 2.14 The Glen Canyon Dam (closed in 1963) transformed the lower Colorado River from a spring-pulse system (**a**) to a regulated system (**b**). An experimental flood in March 1996 was a high flow event for the regulated river, but below average for the unregulated Colorado. The 1996 hydrograph (heavy black line in both panels) is compared to long-term average hydrographs for the Colorado River at Lee's Ferry, Arizona, for the (**a**) pre-dam (1922–62) and (**b**) post-dam (1963–1995) periods. The dashed, solid, and dotted lines connect the mean daily discharge values below which 90%, 50%, and 10% of the years, respectively, occur (Reproduced from Schmidt et al. 1998)

channel fragmentation in nearly 300 of the world's largest river systems reported that fragmentation by dams in the main channel occurred in roughly half, and only about one-third were completely unfragmented (Nilsson et al. 2005). Only tundra rivers in the northern hemisphere and some large tropical rivers, particularly in South America, remain predominantly undammed. Grill et al. (2015) conducted a global analysis using a river fragmentation index that estimates either the river length or the volume of water unavailable to fish as potential roaming space, and calculated the cumulative impact of 6374 large existing dams and 3377 planned or proposed dams at basin and sub-basin scales for the period 1930–2030. Their fragmentation analysis found that 43% of the global river volume is moderately to severely impacted today. Extending this analysis to the sub-basin scale showed that highly impacted basins can

include relatively undisturbed sub-basins, due to clustering of dams only in certain tributaries (Fig. 2.15). The Mississippi River in North America, the Parana River in South America, and the Niger, Zambezi, and Nile Rivers in Africa all have sub-basins that are only moderately or weakly affected despite stronger impacts at the basin scale.

Dams vary widely in their size, purpose, and mode of operation, and these differences influence their impact upon river ecosystems (Poff and Hart 2002). Dams also differ in whether water is released from the surface of the dam, near the bottom, or both. Dams constructed for irrigation must store as much water as possible during the rainy season for release during the growing season. Flood control reservoirs maintain only a small permanent pool in order to maximize storage capacity, and they release water as soon as possible after a flood event to restore their capacity. Navigation requires water storage in upper reaches to offset seasonal low flow conditions downstream, which is often accomplished by a system of locks and dams. Hydroelectric dams store water for release to meet regional energy demands, that can vary seasonally or over the course of 24 h. "Run-of-the-river" dams release water at the rate it enters the reservoir, usually are of low height, and are thought to have few adverse effects on hydrology, although they may still disrupt sediment transport and impair longitudinal connectivity. "Peaking" hydropower dams meet daily fluctuations in energy demand by allowing water to flow through turbines only at certain times, usually from mid-morning through early evening as electric demand changes. Fluctuating water levels create unstable habitat conditions that can be especially disruptive to juvenile fishes and limit spawning opportunities for adults (Freeman et al. 2001). The Colorado River within the Grand Canyon experiences fluctuating water levels due to hydro-peaking, and this limits the egg-laying success of aquatic insects (Kennedy et al. 2016). As of 2018, weekend hydropeaking has been halted in an adaptive management experiment to test whether aquatic insect populations can recover. Finally, reservoirs often serve recreational purposes including boating, swimming, and fishing, but typically this is a secondary function of a multiple-purpose facility.

Dam removal, primarily of older, smaller dams that may be hazardous to human recreation and provide few or no benefits, is accelerating but the number of decommissioned dams is small, estimated at about 1200 out of as many as two million or more in the US (Bellmore et al. 2016). Rarely is it feasible to remove a dam of obvious economic value. However, removal of the 33-m high Elwha Dam (Fig. 2.16), completed in 2014 after years of study and debate, reflects the decision that restoring a salmon population was of greater societal benefit than the dam's modest level of hydroelectric output (O'Connor et al. 2015). While the Elwha project followed years of careful research (see

(a)

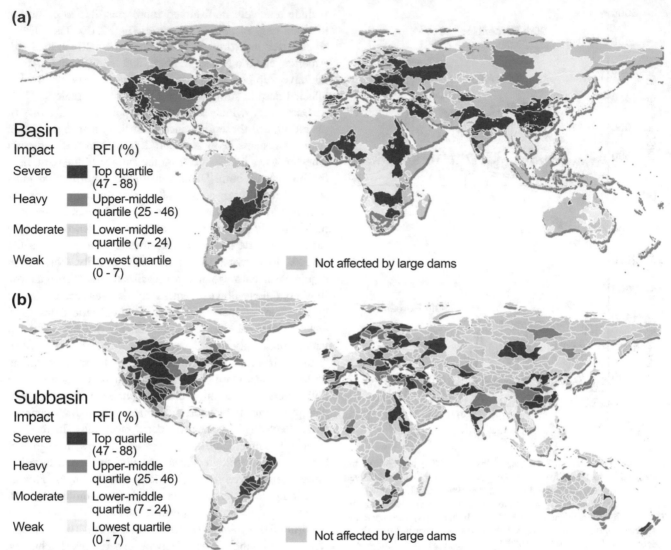

Fig. 2.15 The River Fragmentation Index (RFI) measures river fragmentation by barriers limiting connectivity within basins or subbasins The RFI of an unfragmented river network is 0%, with each subsequent dam increasing the value to a maximum of 100%. The RFI values in quartiles are shown at top panel the basin and bottom panel subbasin scales, circa 2010. Rivers can appear heavily or severely impacted at the overall basin scale, while the subbasin scale may reveal many less impacted areas, e.g., if most dams are clustered only in certain tributaries. Examples are the Mississippi River in North America, the Parana River in South America, or the Niger, Zambezi and Nile Rivers in Africa, which all appear heavily or severely affected at the basin scale but at the subbasin scale larger proportions or even the majority of reaches are only weakly to moderately affected (Reproduced from Grill et al. 2015)

Sect. 3.4.1), the science of dam removal is under-studied (Bellmore et al. 2016), and the success of this project will be followed with much interest.

Regrettably, the rate of dam removal is not remotely close to the pace of dam construction. In response to demand for new sources of renewable energy, major new initiatives in hydropower development are now under way, with at least 3,700 major dams either planned or under construction, primarily in countries with emerging economies (Zarfl et al. 2014). Some of the ecologically most sensitive regions globally are likely to be affected, including the Amazon, Mekong, and Congo basins, and this further construction of hydropower dams will bring fragmentation to 25 of the 120 large river systems currently classified as free-flowing by Nilsson et al. (2005).

2.4.2 Effect of Land Use on Streamflow

Human alteration of land use can have major effects on streamflow by altering the balance between evapotranspiration and runoff and by altering runoff pathways. Most settled regions of the world have seen extensive transformations of their landscapes. In New Zealand, over 80% of the land was forested before agricultural expansion; today, pasture for sheep is the dominant land use in all but the headwaters of

Fig. 2.16 The Elwha River flowing through the remains of the Glines Canyon Dam on 21 February, 2015. The bed of the former reservoir can be seen in the background. (Photo by J. Gussman)

most of New Zealand's streams and rivers (Quinn 2000). Agriculture is the dominant land use in many developed watersheds in the US, comprising more than 40% of the land area of the Lower Mississippi, Upper Mississippi, Southern Plains, Ohio, Missouri, and Colorado River basins (Allan 2004). Urban land use typically makes up a lower percentage of total catchment area, and for large basins urban land is usually less than 5% of catchment area. Because urban stressors have a disproportionate influence on aquatic ecosystems (Paul and Meyer 2001), the influence of urbanization can be important even at low values. When small catchments of low-order streams are the focus, land use can vary from nearly 0–100% coverage of urban, agricultural, or forested land.

Where agriculture replaces forest with crops, average flows, base flows, and peak flows tend to increase for smaller floods but little effect is seen on larger floods (Dingman 2002). Owing to their extensive canopy coverage and deeper roots relative to the majority of shorter vegetation, forest vegetation promotes near maximal interception and transpiration of rainfall. Thus, deforestation usually increases streamflow, especially base flow, as noted earlier for Hubbard Brook. Increased streamflow in the Mississippi River basin since the 1940s correlates with increased precipitation, but changing land use provides an alterative explanation.

The increase is seen mainly in baseflow and can be explained by the conversion of perennial vegetation to seasonal row crops, especially soybeans. This reduction in year-round vegetation cover likely will reduce evapotranspiration, thus routing more precipitation into streams as baseflow than stormflow (Zhang and Schilling 2006).

Extensive land-use change can even alter precipitation, as when deforestation reduces evapotranspiration over a large enough area to lower atmospheric moisture. Numerous studies agree that continental-scale deforestation in Amazonia, Africa, or Southeast Asia leads to a warmer, drier climate over the deforested area (Lawrence and Vandecar 2015). Global circulation models suggest that as deforestation expands, rainfall declines. When deforestation reaches a certain threshold, of perhaps 30–50% of land area, rainfall is substantially reduced, significantly affecting ecosystem structure and function.

Installation of drainage systems, such as tiles buried beneath the soil surface, and channel deepening and straightening for flow conveyance, have the further effect of speeding sub-surface flows and downstream routing of a rain event. Conversion of wetlands into agricultural usage also may contribute to river flooding, since wetlands naturally are locations of surface storage and frequently of groundwater recharge as well. Rivers become much flashier due to

Fig. 2.17 River stage (in meters above mean sea level, msl) at Illinois River mile 137 before water diversions and modern navigation dams (1878–1899) and after many alterations in the catchment and river (1975–1996). Each block shows a year from January to December. The horizontal line indicates a flood at which economic damage occurs (Reproduced from Sparks et al. 1998)

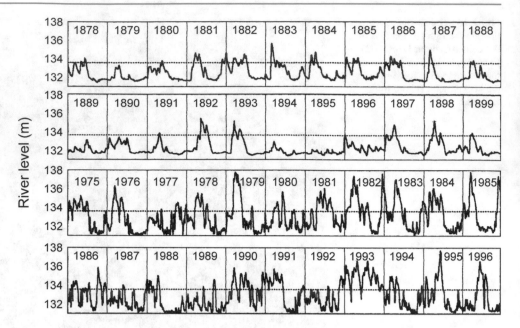

changes to the landscape including drainage tiles, ditches, storm-water conveyances, and impervious surfaces, all of which result in more event-responsive runoff (Fig. 2.17).

Urbanization can have a very strong influence on streamflows. Replacement of vegetation with pavement and buildings reduces transpiration and infiltration, and these impervious surfaces substantially increase the amount of runoff that travels by rapid overland flow. Storm sewers and roadways transport water quickly, and so may require the construction of retention ponds in an effort to retard the flood peak. Runoff approximately doubles when impervious surface area is 10–20% of catchment area and triples at 35–50% impervious surface area (Arnold and Gibbons 1996) (Fig. 2.18). Flood peaks increase, time to peak shortens, and the peak becomes narrower (Paul and Meyer 2001). Because a greater fraction of the water is exported as runoff, less recharge of groundwater occurs, and so base flows are reduced as well.

Groundwater extraction related to land use is important mainly in semi-arid areas with substantial aquifers that support irrigated agriculture. Most of the water used for irrigation is lost into the atmosphere by evapotranspiration, while the remainder either adds to stream run-off or returns to groundwater by infiltration. If groundwater abstraction exceeds the natural recharge for extensive areas and long times, water tables decline with devastating effects on streamflow, groundwater-fed wetlands, and related ecosystems (Wada et al. 2010). From the river's perspective, the primary effect is depletion of flows. Such changes are widely reported for rivers of the US Great Plains, including the Republican and Arkansas River basins (Kustu et al. 2010), where long-term depletion to support irrigated agriculture has caused water tables to drop by more than 50 m in some

locations. Fifty years of streamflow data from 110 gauging stations in eight major river basins throughout Nebraska during 1948–2003 showed decreasing streamflow, mostly in the western region of the state (Wen and Chen 2006). As no significant changes in precipitation and temperature were detected for the same period, groundwater withdrawal for irrigation is strongly implicated as the primary cause. Ecological consequences include collapses of populations of large-stream fishes and expansion of small-stream fishes where hydrologic conditions were altered most (Perkin et al. 2017).

2.4.3   Effect of Climate Change on Streamflow

There is broad consensus that rising levels of CO_2 and other greenhouse gases in the atmosphere have contributed to an approximately 0.85 °C temperature increase over the past century (IPCC 2013). Further increases in greenhouse gases and other atmospheric pollutants are expected to result in significant additional warming by 2100, particularly at latitudes between 40° and 70°, as well as substantial regional and seasonal variation in precipitation. Some future impacts have a high likelihood of occurring, but much uncertainty remains, partly because future climate conditions at regional and local scales are uncertain (Stainforth et al. 2005), and partly because of the complexity that results from climate change acting through multiple pathways. In addition, climate change will interact with other threats to lotic ecosystems, enhancing some regional water shortages, favoring species invasions, and acting as an additional stressor on the biota.

Changes in streamflow in recent decades, most commonly manifested in seasonal timing, indicate that climatic

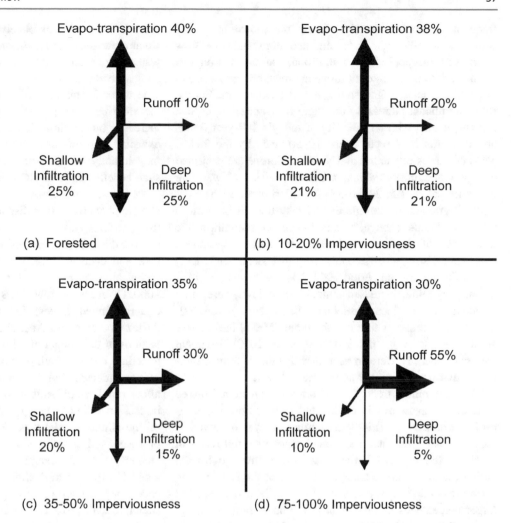

Fig. 2.18 Changes in hydrologic flows with increasing impervious surface cover in urbanizing catchments (Reproduced from Paul and Meyer 2001)

change is already evident in the instrument record. Two caveats apply. First, studies of long-term trends need to be viewed with caution as indicators of climate-driven changes since often they incorporate flow records influenced by river regulation and changing land use, and are derived from time series with variable lengths and start and end dates (Stahl et al. 2010). In addition, patterns may differ between head-water streams and large rivers, and regionally. Second, it is difficult to separate the relative contributions of natural climactic variation and human-induced climate change, especially at the regional and local levels.

Annual flood magnitude assessed at 200 long-term stream gauges (85–127 years of record) in the US showed historical trends in only about one-fourth of long-term gauges analyzed, with both increases and decreases (Hirsch and Ryberg 2012). No universal trend has been observed in the overall extent of drought across the continental United States since 1900 (Dettinger et al. 2015), and globally there is uncertainty regarding the present-day contribution of climate change to droughts. Regionally, however, the worsening severity of recent droughts and streamflow reductions has

been attributed to human-driven climate change (Diffen-baugh et al. 2015). In the Colorado River, the worst 15-year drought on record occurred between 2000 and 2014, but unlike past droughts caused by lack of precipitation, this recent event was the result of high temperatures and greater evapotranspiration (Udall and Overpeck 2017). Future droughts may be even more severe, as climate models predict further temperature increases but neither recent data nor model predictions give confidence that a compensating rise in precipitation will occur. If business-as-usual warming continues, temperature-induced declines in Colorado River flow are estimated conservatively at ~20% by midcentury and 35% by end-century (Udall and Overpeck 2017). For mid-sized streams in arid regions, reduced streamflow is likely to result in more frequent and longer stretches of dry channel fragments, thus limiting the opportunity for native fishes to access spawning habitats and seasonally available refuges (Jaeger et al. 2014).

Although the instrument record does not yet show clear evidence of climate change impact on extreme events, models suggest that floods and droughts may become more frequent in coming decades. Using the global integrated water model WaterGAP to simulate high and low flow regimes in Europe, Lehner et al. (2006) estimate that regions in northern to northeastern Europe are most prone to a rise in flood frequencies, while southern and southeastern Europe show significant increases in drought frequencies. In critical regions, events with an intensity of today's 100-year floods and droughts may recur every 10–50 years by the 2070s. Most countries in Central and Western Europe are projected to experience substantial increase in flood risk at all warming levels modeled (1.5, 2, and 3 °C) (Alfieri et al. 2018).

While trends in extreme events may not yet be attributable to climate change, trends in annual streamflow are indicative of long-term climatic influence. Based on near-natural streamflow records from 441 small catchments across Europe for the period 1962–2004, with some records as early as 1932, increased annual streamflows were more frequent than declines, particularly in northern Europe relative to the south and east, but about one-third of the datasets showed weak or no trends (Stahl et al. 2010). Regional patterns generally were in accord with rainfall trends, indicating that precipitation was the main driver.

One streamflow trend that is both widespread and linked to rising temperatures is earlier and greater flow increases during winter and declines in summer, as was observed in the European study just mentioned. Snowmelt-fed rivers in much of the western United States have trended toward earlier melts and flows since the middle of the last century, owing to declines in spring snowpack, earlier snowmelt, and larger percentages of precipitation falling as rain instead of snow (Fritze et al. 2011). Because winter snowpack is a natural storage reservoir, much of this winter runoff will be lost to the oceans, and some have suggested that new dams will be required to replace the storage previously afforded by snow (Service 2004). It is important to note, however, that considerable regional variation results from geographic differences in climate and topography, and resultant variation in ET and precipitation. Low elevation, rain-dominated coastal basins have consistently seen a shift in greatest runoff from 5 to more than 10 days later, apparently due to wetter conditions in late winter (Fritze et al. 2011).

Rivers in glacier-fed regions are especially likely to be influenced by on-going and future climate change, affecting water supply to ecosystems and for human use (Kaser et al. 2010). More than 80% of the freshwater supply in arid and semiarid areas in the mountain regions of the tropics and the subtropics originates from glaciers (Vuille et al. 2008). Impacts will be dramatic in India, where river systems such as the Indus and Ganges are ultimately fed by glacial meltwater from the Himalayas; and in the western tropical Andes, where climate warming has occurred and is expected to accelerate. Many smaller, low-lying glaciers likely will disappear within a few decades. Even where glaciers do not completely disappear, the change in streamflow seasonality, due to the reduction of the glacial water supply during the dry season, will significantly affect Andean countries that depend on glacial meltwater to provide water for human use and to the aquatic ecosystem. In the short-term, as glaciers retreat, runoff will temporarily increase, but within decades many flowing waters are expected to shrink or entirely disappear. Downstream users may initially benefit from an enhanced water supply, but such benefits are unlikely to be sustained.

A substantial literature explores the likely impacts of climate change on river discharge over the coming century, using various climate projections and hydrology models in combination. Global projections showed an increase in high flow and decrease in low flow for about one-third of the global land surface area for 2071–2100 relative to 1971–2000. Decreases in low flows were projected for the southeastern United States, Europe, eastern China, southern Africa, and southern Australia (van Vliet et al. 2013). Projections of the impact of climate change on streamflows are uncertain because both evapotranspiration and precipitation will be affected, but how the balance between them will affect runoff is difficult to predict. As a consequence, the impact on runoff is geographically variable and dependent on the climate model used. Broadly speaking, as warming proceeds and influences both ET and soil moisture, the amount of runoff generated by a given amount of precipitation is expected to decline (see Fig. 2.3). Regional differences are important, however. For example, annual streamflow is projected to increase in the northwest and decline in the southwest of the US, in accord with projected changes in precipitation (Dettinger et al. 2015).

2.5 Environmental Flows

Historically, river discharge has been managed for human benefit in the context of water resource infrastructure, such as dams, diversion weirs, and water abstraction locations. Over time, flow management has broadened to meet environmental goals. This has resulted in a merging of hydrologic and ecological sciences to identify the environmental flows ("e-flows") needed to sustain healthy ecosystems, based on the understanding that an unaltered river has a natural flow regime, and deviations from that natural flow regime potentially are ecologically damaging (Poff et al. 1997). Broadly speaking, various components of the flow regime (Fig. 2.11) are characterized by metrics such as the IHA (Table 2.3), and changes in flow regime are expected to lead to ecological responses, resulting in an altered and possibly degraded ecosystem. Over the past several decades the science of e-flows has grown in scope and impact, becoming a cornerstone of river management world-wide

(Arthington et al. 2018b; Poff 2018). Methods of hydrologic analysis have evolved considerably since the mid-twentieth century, when the main focus was on minimum flows to sustain valued sports fisheries and to prevent water quality deterioration at very low flows. Advances in hydrologic analysis and ecological understanding resulted in such a proliferation of methods that by 2000, a global review identified over 200 methods in use across 44 countries and six continents (Tharme 2003).

Environmental flows describe the quantity, timing, and quality of freshwater flows necessary to sustain aquatic ecosystems, which, in turn, support human cultures, economies, sustainable livelihoods, and well-being (Arthington et al. 2018a). Its central message is that flows have been compromised or are at risk in most aquatic systems around the world, with serious consequences for biodiversity, aquatic ecosystem health, ecological services, and society. Actions that protect and restore environmental flows serve to protect and restore freshwater-dependent aquatic ecosystems, and to sustain delivery of important and wide-ranging ecological and societal services. Fortunately, this message has resulted in policies to protect river health and national laws to protect and manage environmental water in many countries around the world.

The many methods of flow analysis can be categorized into four main approaches (Tharme 2003; Poff et al. 2017). Hydrologic methods that use daily or monthly flow records and easily calculated indices of flow can serve as a first step in analysis and offer some ecosystem protection, often using a prescribed minimum flow. A percentage of mean annual flow, the 7Q10, and flow duration metrics are examples. The Tennant method (Tennant 1976) uses a percentage of mean annual flow (often 30% of MAF), adjusted seasonally, to recommend minimum flows. Many variants and thresholds are used (Reiser et al. 1989; Tharme 2003), some as low as 5 and 10% of MAF. A more comprehensive statistical characterization is captured using the indicators of hydrologic alteration (IHA) and its five categories of flow variation.

Hydraulic rating methods rely on relationships between a range of flows and general habitat characteristics such as a channel's depth or wetted perimeter. The relationship between discharge and wetted perimeter often is estimated for riffles because they tend to be areas of high macroinvertebrate production and are first to go dry. Streams in Montana showed clear breakpoints such that flows at 10% of the MAF protected about half of the maximum wetted perimeter and flows greater than 30% of MAF protected nearly all of the maximum wetted perimeter (Tennant 1976).

Hydrodynamic habitat modeling is a further extension that simulates the relationship between hydraulic variables and habitat by incorporating detailed measurement of the habitat needs of individual species or biological assemblages, usually to set a minimum flow considered sufficient to sustain target species or the entire biological community. Model output is typically some estimate of the wetted usable area (WUA) of habitat for target species. The instream flow incremental methodology (IFIM, Stalnaker and Arnette 1976) was an early example of this approach, intended to model the quantity and suitability of biological habitat using hydraulic variables such as depth, velocity, and substrate.

Fig. 2.19 Diagrammatic representation of the natural flow regime of a river showing how it influences aquatic biodiversity via several, inter-related mechanisms (Principles 1–4) that operate over different spatial and temporal scales (Reproduced from Bunn and Arthington 2002)

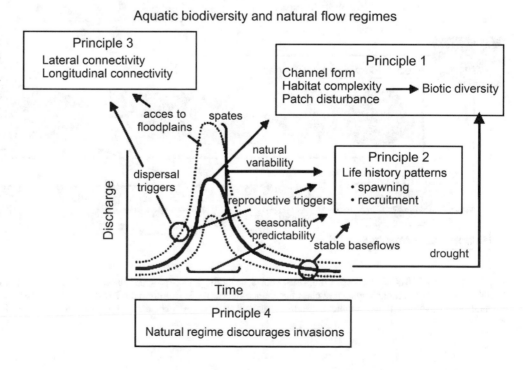

Hydraulic and habitat simulation models then can be combined to predict changes in available habitat in response to changes in flow on a species-specific basis. As of about 2000 this approach was in use in at least 20 countries, despite extensive criticisms of the model approach and the often poor or untested ecological predictive capability of IFIM output (Tharme 2003).

A holistic, ecosystem approach is now at the forefront of environmental flow methodologies, and is the most demanding of combined ecological and hydrologic knowledge. Ideally this will use data on the relationship between flow and ecology over a wide range of flows and species, including life cycle stages and seasonal timing (Fig. 2.19). Such a holistic approach has the potential to recommend a hydrologic regime linked to explicit quantitative or qualitative ecological, geomorphological, and perhaps also social and economic responses.

The ecological limits of hydrologic analysis (ELOHA) provides a framework for determining environmental flows needed to meet ecological and societal needs (Arthington

et al. 2006; Poff et al. 2010). This is a multi-step process that begins with hydrologic analysis and streamflow classification, assigns flow-altered streams to a presumed pre-impact stream type, and uses relationships between flow alteration and ecological responses drawn from existing data and knowledge or new studies (Fig. 2.20). The intent is to establish the extent of ecological change associated with specific flows, and through a social process arrive at recommended environmental flows that address conflicting goals. More than a science framework, ELOHA seeks to incorporate expert and traditional knowledge and differing priorities and social perspectives to provide a decision-making framework to aid planning and address water conflicts.

A number of studies have shown how hydrologic analyses combined with detailed knowledge of flow-ecology relationships can benefit the ecosystem by establishing more natural flows. In a small, regulated stream dominated by non-native fish species in California, US, changes in flow regime designed to mimic the seasonal timing of natural

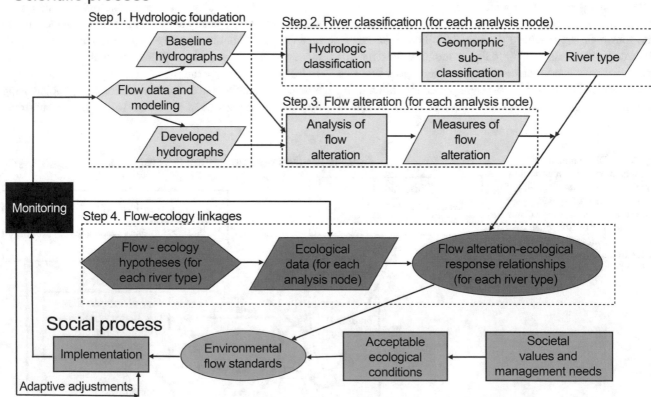

Fig. 2.20 The ELOHA framework comprises both a scientific and social process. Hydrologic analysis and classification are developed in parallel with flow alteration–ecological response relationships, which provide scientific input into a social process that balances this information with societal values and goals to set environmental flow standards (Reproduced from Poff et al. 2010)

increases and decreases in stream flow successfully reestablished native fishes by creating more favorable spawning conditions, water temperatures, and continuous flows, while abundances of non-native fishes were reduced by high-discharge events (Kiernan et al. 2012). The riparian forests of several rivers in western North America have seen extensive recruitment of cottonwoods and willows along previously impoverished reaches as a result of collaborative efforts with dam operators to release high spring flows, followed by a gradual decline to promote seedling survival (Rood et al. 2005). In each case the desired outcome was achieved by manipulating stream flows at biologically important times of the year requiring only a small increase in the total volume of water delivered downstream during most water years, and by taking advantage of high flow years when sufficient water is available for both economic commitments and environmental needs.

The science and practice of e-flows has seen enormous growth in recent years, serving as an excellent example of sound science guiding effective management. This subject also faces a number of challenges, including the need for improved understanding of underlying processes, more effort to quantify flow-ecosystem and social-ecological outcomes at basin and regional scales in diverse river types, and accounting for spatial variability, stochasticity, and uncertainty. In addition, as noted by Poff (2018), this approach has embedded assumptions about climatic and ecological stationarity that call into question the use of historical flow regimes to establish reference conditions and to evaluate outcomes of e-flow interventions in future applications. As a changing climate influences hydrologic regimes, changing land use affects runoff, and species invasions alter biological communities, the past may not adequately foretell the river ecosystem's future response to flow restoration. This challenge may be met in part by research aimed at characterizing time-varying flow dynamics as well as a more holistic view of ecosystem impacts and responses. In addition, a shift in perspective may be called for, toward an approach that uses historical variability for guidance, but relies on a process-based understanding of flow-ecology relationships and emphasizes prescription of a range of flows that can sustain resilient, socially valued ecological characteristics in a changing world.

2.6 Summary

Fluvial ecosystems exhibit tremendous variability in the quantity, timing, and temporal patterns of river flow, and this profoundly influences their physical, chemical, and biological condition. Vast quantities of fresh water are extracted to meet agricultural, municipal, and industrial demands, yet freshwater ecosystems also need enough water, in the right amounts and at the right time, to remain ecologically intact and provide economically valuable commodities and services to society. Increasingly today, the tools of hydrologic analysis are being combined with other elements of river science to address the question, "How much water does a river need?" and ensure that flows are sufficient to protect and restore stream ecosystems.

The hydrologic cycle describes the continuous cycling of water from atmosphere to earth and oceans, and back to the atmosphere. Evaporation from the oceans exceeds that over land, while precipitation on the earth's land surfaces exceeds evaporation and plant water loss. This excess provides the runoff from land to sea that is primarily river discharge, but includes groundwater as well. At the scale of an individual catchment and averaged over years, a water budget consists of inputs from precipitation and groundwater inflow, and outputs due to streamflow, evapotranspiration (ET, water loss from evaporation and by plants), and groundwater outflow. Over short time intervals, imbalances between inputs and outflows result in increases or decreases in storage. Globally, almost two-thirds of precipitation that falls on land becomes ET, which exceeds runoff for most rivers and for all continents except Antarctica. Both precipitation and ET vary with climate and vegetation, resulting in seasonal and regional differences in the amount and timing of streamflow.

The fraction of precipitation that becomes runoff travels by a number of pathways that are influenced by slope, vegetation cover, soil properties, and antecedent moisture conditions. Surface and shallow sub-surface flows reach streams much more quickly than water that percolates to the water table and discharges into the stream as groundwater. Thus, the stream hydrograph, which describes the rise and fall of streamflow over time, will exhibit a strong or a more gradual response to a rain event depending on soils, slopes, and human actions that affect flow paths. Most rivers continue to flow during periods of little rainfall, and this base flow comes from groundwater that discharges into the channel more or less continuously, depending on fluctuation in the level of the water table.

The characterization of streamflow has practical application for the design of flood-control structures, evaluation of channel stability, and in determining whether sufficient water is available at the appropriate time to meet the needs of both people and the ecosystem. One can estimate the frequency with which flows of a given magnitude are expected to occur, or the probability of occurrence of a flow of a given magnitude. The flood magnitude that has a 1–2 year recurrence often is used as an estimate of the flood that just overtops the banks, which is the flow considered most influential in maintaining channel form.

Flow analyses tell us that each individual river has a natural flow regime characterized by the magnitude of flows and their frequencies, as well as duration, timing, and rate of change. Climate, vegetation, geology, and terrain place broad constraints on natural flow regime, conditions at the catchment scale contribute to local variation, and a wide range of human influences including climate change further alter flow regimes by changing flow pathways and response times. Yet, the accumulated scientific evidence reveals that rivers and their floodplains need much of the spatial and temporal variability of their natural flow regime to maintain their ecological integrity.

Unquestionably, human influence on river flows is widespread. Water storage and release by reservoirs affects the quantity, quality, and timing of downstream flows, and dams serve as physical barriers, disrupting the ecological connectivity of rivers. The extent of flow alteration depends on dam configuration and mode of operation, but declines in maximum flows, increases in minimum flows, reduced overall variation, and altered timing are frequent outcomes. Dams also fragment once-continuous river systems into a series of largely disconnected segments, limiting species migration and dispersal as well as the transport of organic and inorganic matter downstream and into riparian zones and floodplains.

Changing land use also has diverse effects on flow regimes. Where agriculture replaces forest with crops, average flow, dry-season flow, and peak flow tend to increase, resulting in flashier hydrographs. In urban areas, impervious surfaces and stormwater conveyances result in very rapid streamflow response to rain events, but may reduce groundwater recharge and thus baseflow. Groundwater extraction, often for irrigated agriculture, has lowered water tables and reduced streamflows in a many semi-arid regions. The effects of climate change are complex, because both precipitation and evapotranspiration are likely to change. A number of studies show increased streamflows over recent decades with a correlation to precipitation trends, indicative of a climate impact, but instrument records do not yet show clear evidence of a climate impact on extreme floods and droughts.

The science of environmental flows, or e-flows, has evolved considerably from its roots in hydrologic analysis of multiple components of flow and the conceptual framework of a river's natural flow regime. A holistic, ecosystem approach is now at the forefront of e-flow methodologies. Ideally using data on the relationship between flow and ecology over a wide range of flows and species, including life cycle stages and seasonal timing, such a holistic approach has the potential to recommend a hydrologic regime that can achieve desired outcomes linked to explicit quantitative or qualitative ecological, geomorphological, and perhaps also social and economic responses. Although increasingly sophisticated and an excellent example of sound science guiding effective management, this approach faces challenges due to its embedded assumptions about climatic and ecological stationarity that call into question the use of historical flow regimes to establish reference conditions and to evaluate outcomes of e-flow interventions. As a changing climate influences hydrologic regimes, changing land use affects runoff, and species invasions alter biological communities, the past may not adequately foretell the river ecosystem's future response to flow restoration. This challenge may be met in part by research but perhaps also by a shift in perspective toward an approach that uses historical variability for guidance, but relies on a process-based understanding of flow-ecology relationships and emphasizes prescription of a range of flows that can sustain resilient, socially valued ecological characteristics in a changing world.

References

Alfieri L, Dottori F, Betts R et al (2018) Multi-model projections of river flood risk in Europe under global warming. Climate 6:16. https://doi.org/10.3390/cli6010016

Allan JD (2004) Landscapes and riverscapes: the influence of land use on stream ecosystems. Annu Rev Ecol Evol Syst 35:257–284

Allan JD, Benke AC (2005) Overview and prospects. In: Benke AC, Cushing CE (eds) Rivers of North America. Elsevier, London, pp 1087–1101

Arnold CL, Gibbons CJ (1996) Impervious surface coverage: the emergence of a key environmental indicator. Am Planners Assoc J 62:243–258

Arthington AH, Bunn SE, Poff NL, Naiman RJ (2006) The challenge of providing environmental flow rules to sustain river ecosystems. Ecol Appl 16:1311–1318

Arthington AH, Bhaduri A, Bunn SE et al (2018a) The Brisbane declaration and global action agenda on environmental flows (2018). Front Environ Sci 6:1–15. https://doi.org/10.3389/fenvs.2018.00045

Arthington AH, Kennen JG, Stein ED, Webb JA (2018b) Recent advances in environmental flows science and water management—innovation in the Anthropocene. Freshw Biol 63:1022–1034. https://doi.org/10.1111/fwb.13108

Barron JS, Poff NL, Angermeier PL et al (2002) Meeting ecological and societal needs for freshwater. Ecol Appl 12:1247–1260. https://doi.org/10.1890/04-0922

Bellmore JR, Duda JJ, Craig LS, et al (2016) Status and trends of dam removal research in the United States. Wires Interdiscip Rev Water 4. 10.1002/wat2.1164

Benke AC (1990) A perspective on America's vanishing streams. J North Am Benthol Soc 9:77–98

Benke AC, Cushing CE (2005) Background and approach. In: Benke AC, Cushing CE (eds) Rivers of North America. Elsevier, London, pp 1-18

Bunn SE, Arthington AH (2002) Basic principles and ecological consequences of altered flow regimes for aquatic biodiversity. Environ Manag 30:492–507

Carlisle DM, Wolock DM, Meador MR (2011) Alteration of streamflow magnitudes and potential ecological consequences: a multiregional assessment. Front Ecol Environ 9:264–270. https://doi.org/10.1890/100053

Dettinger M, Udall B, Georgakakos A (2015) Western water and climate change. Ecol Appl 25:2069–2093

Diffenbaugh NS, Swain DL, Touma D, Lubchenco J (2015) Anthropogenic warming has increased drought risk in California. Proc Natl Acad Sci U S A 112:3931–3936. https://doi.org/10.1073/pnas.1422385112

Dingman SL (2002) Physical hydrology, 2nd edn. Waveland Press Inc, Long Grove, IL

Dunne T, Leopold LB (1978) Water in environmental planning. W H Freeman and Company, San Francisco

Dynesius M, Nilsson C (1994) Fragmentation and flow regulation of river systems in the world. Science (80-) 266:753–762

Fetter CW (1988) Applied Hydrogeology. Merrill, Columbus, Ohio

Freeman MC, Bowen ZH, Bovee KD, Irwin ER (2001) Flow and habitat effects on juvenile fish abundance in natural and altered flow regimes. Ecol Appl 11:179–190

Fritze H, Stewart IT, Pebesma E (2011) Shifts in western North American snowmelt runoff regimes for the recent warm decades. J Hydrometeorol 12:989–1006. https://doi.org/10.1175/2011JHM1360.1

Gordon ND, McMahon TA, Finlayson BL et al (2004) Stream hydrology: an introduction for ecologists. Wiley, Chichester, West Sussex

Grill G, Lehner B, Lumsdon AE et al (2015) An index-based framework for assessing patterns and trends in river fragmentation and flow regulation by global dams at multiple scales. Environ Res Lett Lett 10:1–15. https://doi.org/10.1088/1748-9326/10/1/015001

Hirsch RM, Ryberg KR (2012) Has the magnitude of floods across the USA changed with global CO2 levels? Hydrol Sci J 57:1–9. https://doi.org/10.1080/02626667.2011.621895

Horton RE (1945) Erosional development of streams and their drainage basins: hydrophysical approach to quantitative morphology. Bull Geol Soc Am 56:275–370

IPCC (2013) Climate change 2013: the physical science basis. Cambridge University Press, Cambridge

Jackson RB, Carpenter SR, Dahm CN, et al (2001) Water in a changing world. Ecol Appl 11:1027–1045. https://doi.org/10.1890/0012-9623(2008)89%5b341:iie%5d2.0.co;2

Jaeger KL, Olden JD, Pelland NA (2014) Climate change poised to threaten hydrologic connectivity and endemic fishes in dryland streams. Proc Natl Acad Sci U S A 111:13894–13899. https://doi.org/10.1073/pnas.1320890111

Kaser G, Großhauser M, Marzeion B (2010) Contribution potential of glaciers to water availability in different climate regimes. PNAS 107:20223–20227. https://doi.org/10.1073/pnas.1008162107

Kennard MJ, Pusey BJ, Olden JO et al (2010) Classification of natural flow regimes in Australia to support environmental flow management. Freshw Biol 55:171–193. https://doi.org/10.1111/j.1365-2427.2009.02307.x

Kennedy TA, Muehlbauer JD, Yackulic CB et al (2016) Flow management for hydropower extirpates aquatic insects, undermining river food webs. Bioscience 66:561–575. https://doi.org/10.1093/biosci/biw059

Kiernan JD, Moyle PB, Crain PK (2012) Restoring native fish assemblages to a regulated California stream using the natural flow regime concept. Ecol Appl 22:1472–1482

Kunkel KE, Changnon SA, Angel JR (1994) Climatic aspects of the 1993 Upper Mississippi river basin flood. Bull Am Meteorol Soc 811–822

Kustu MD, Fan Y, Robock A (2010) Large-scale water cycle perturbation due to irrigation pumping in the US high plains: a synthesis of observed streamflow changes. J Hydrol 390:222–244. https://doi.org/10.1016/j.jhydrol.2010.06.045

Lawrence D, Vandecar K (2015) Effects of tropical deforestation on climate and agriculture. Nat Clim Chang 5:27–36

Lehner B, Doll PD, Alcamo J et al (2006) Estimating the impact of global change on flood and drought risks in Europe: a continental, integrated analysis. Clim Change 75:273–299. https://doi.org/10.1007/s10584-006-6338-4

Lehner B, Liermann CR, Revenga C et al (2011) High-resolution mapping of the world's reservoirs and dams for sustainable river-flow management. Front Ecol Environ 9:494–502. https://doi.org/10.1890/100125

Likens GE, Bormann FH (1995) Biogeochemistry of a forested ecosystem, 2 nd ed. Springer, New York

Lins HF (1997) Regional streamflow regimes and hydroclimatology of the United States. Water Resour Res 33:1655–1667. https://doi.org/10.1029/97WR00615

Linsley RK, Kohler MA, Paulhus MH (1958) Hydrology for engineers. McGraw-Hill, New York

Magilligan FJ, Nislow KH (2005) Changes in hydrologic regime by dams. Geomorphology 71:61–78. https://doi.org/10.1016/j.geomorph.2004.08.017

Mathews R, Richter BD (2007) Application of the indicators of hydrologic alteration software in environmental flow setting. J Am Water Resour Assoc 43:1400–1413. https://doi.org/10.1111/j.1752-1688.2007.00099.x

Mcmanamay RA, Bevelhimer MS, Kao SC (2014) Updating the US hydrologic classification: an approach to clustering and stratifying ecohydrologic data. Ecohydrology 7:903–926. https://doi.org/10.1002/eco.1410

Milliman JD, Farnsworth KL (2011) River discharge to the coastal oceans: a global synthesis. Cambridge University Press, Cambridge

Morisawa M (1968) Streams: their dynamics and morphology. McGraw-Hill, New York

Nilsson C, Reidy CA, Dynesius M, Revenga C (2005) Fragmentation and flow regulation of the world's large river systems. Science (80-) 308:405–408. https://doi.org/10.1126/science.1107887

O'Connor JE, Duda J, Grant GE (2015) 1000 dams down and counting. Science (80-) 348:2015–2017

Olden JD, Kennard MJ, Pusey BJ (2012) A framework for hydrologic classification with a review of methodologies and applications in ecohydrology. Ecohydrology 5:503–518. https://doi.org/10.1002/eco.251

Paul MJ, Meyer JL (2001) Streams in the urban landscape. Annu Rev Ecol Syst 32:333–365

Perkin JS, Gido KB, Falke JA et al (2017) Groundwater declines are linked to changes in Great Plains stream fish assemblages. PNAS 114:7373–7378. https://doi.org/10.1073/pnas.1618936114

Poff NL (1996) A hydrogeography of unregulated streams in the United States and an examination of scale-dependence in some hydrological descriptors. Freshw Biol 36:71–79. https://doi.org/10.1046/j.1365-2427.1996.00073.x

Poff NL, Allan JD (1995) Functional organization of stream fish assemblages in relation to hydrological variability. Ecology 76:606–627

Poff NL, Allan JD, Bain MB et al (1997) The natural flow regime. Bioscience 47:769–784. https://doi.org/10.1007/s11277-014-1857-1

Poff NL, Hart DD (2002) How dams vary and why it matters for the emerging science of dam removal. Bioscience 52:659–668

Poff NL, Olden JD, Merritt DM, Pepin DM (2007) Homogenization of regional river dynamics by dams and global biodiversity implications. PNAS 104:5732–5737

Poff NL, Richter B, Arthington AH et al (2010) The ecological limits of hydrologic alteration (ELOHA): a new framework for developing regional environmental flow standard. Freshw Biol 55:147–170. https://doi.org/10.1111/j.1365-2427.2009.02204.x

Poff NL, Tharme R, Arthington A (2017) Evolution of e-flows: principles and methodologies. In: Horne A, Webb A, Stewardson M, et al. (eds) Water for the Environment: Policy, Science, and Integrated Management. Elsevier, London, pp 203–236

Poff NLR (2018) Beyond the natural flow regime? Broadening the hydro-ecological foundation to meet environmental flows challenges in a non-stationary world. Freshw Biol 63:1011–1021. https://doi.org/10.1111/fwb.13038

Postel S, Daily GC, Ehrlich PR (1996) Human appropriation of renewable fresh water. Science (80-.). 271:785–788

Postel S, Richter BD (2003) Rivers for life: managing water for people and nature. Island Press, Washington, DC

Quinn JM (2000) Effects of pastoral development. In: Collier KJ, Winterbourn MJ (eds) New Zealand stream, invertebrates: ecology and implications for management. New Zealand Limnological Society, Christchurch, pp 208–229

Reiser DW, Wesche TA, Estes C (1989) Status of instream flow legislation and practice in North America. Fisheries 14:22–29

Rice JS, Emanuel RE, Vose JM, Nelson SAC (2015) Continental U.S. streamflow trends from 1940 to 2009 and their relationships with watershed spatial characteristics. Water Resour Res 6262–6275. https://doi.org/10.1002/2014WR016367

Richter BD, Baumgartner JV, Powell J, Braun DP (1996) A method for assessing hydrologic alteration within ecosystems. Conserv Biol 10:1163–1174

Riggs HC (1980) Characteristics of low flows. J Hydraul Eng 106: 717–731

Rood SB, Samuelson GM, Braatne JH et al (2005) Managing river flows to restore floodplain forests. Front Ecol Environ 3:193–201

Schmidt JC, Webb RH, Valdez RA, et al. (1998) Science and values in river restoration in the Grand Canyon. Bioscience 48:735–747

Serinaldi F, Kilsby CG (2015) Stationarity is undead: Uncertainty dominates the distribution of extremes. Adv Water Resour 77:17–36. https://doi.org/10.1016/j.advwatres.2014.12.013

Service RF (2004) As the west goes dry. Science (80-) 303:1124–1127

Slack JR, Landwehr JM (1992) Hydro-climatic data network (HCDN); a U.S. Geological Survey streamflow data set for the United States for the study of climate variations, 1874–1988. USGS Water-Resources Investigations Report 93-4076

Snelder TH, Biggs BJF (2002) Multiscale river environment classification for water resources management. J Am Water Resour Assoc 38:1225–1239. https://doi.org/10.1111/j.1752-1688.2002.tb04344.x

Sparks RE, Nelson JC, Yin Y (1998) Naturalization of the flood regime in regulated rivers. Bioscience 48:706-720

Stahl K, Hisdal H, Hannaford J et al (2010) Streamflow trends in Europe: evidence from a dataset of near-natural catchments. Hydrol Earth Syst Sci 14:2367–2382. https://doi.org/10.5194/hess-14-2367-2010

Stainforth DA, Aina T, Christensen C et al (2005) Uncertainty in predictions of the climate response to rising levels of greenhouse gases. Nature 433:403–406. https://doi.org/10.1038/nature03301

Stalnaker CB, Arnette SC (1976) Methodologies for the determination of stream resource flow requirements: An assessment. U S Fish and Wildlife Services, Washington, DC

Tennant DL (1976) Instream flow regimens for fish, wildlife, recreation and related environmental resources. Fisheries 1:6–10

Tharme RE (2003) A global perspective on environmental flow assessment: emerging trends in the development and application of environmental flow methodologies for rivers. River Res Appl 19:397–441. https://doi.org/10.1002/rra.736

The Nature Conservancy. (2009) Indicators of hydrologic alteration, version 7.1. The Nature Conservancy

Udall B, Overpeck J (2017) The twenty-first century Colorado River hot drought and implications for future. Water Resour Res 53:2404–2418

USGS (2005) Streamflow trends in the United States, U.S. Geologic Survey Fact Sheet, FS-2005-3017

van Vliet MTH, Franssen WHP, Yearsley JR et al (2013) Global river discharge and water temperature under climate change. Glob Environ Chang 23:450–464. https://doi.org/10.1016/j.gloenvcha.2012.11.002

Vogel RM, Yaindl C, Walter M (2011) Nonstationarity: Flood magnification and recurrence reduction factors in the United States. J Am Water Resour Assoc 47:464–474. https://doi.org/10.1111/j.1752-1688.2011.00541.x

Vörösmarty CJ, Meybeck M, Fekete B et al (2003) Anthropogenic sediment retention: major global impact from registered river impoundments. Glob Planet Change 39:169–190. https://doi.org/10.1016/S0921-8181(03)00023-7

Vuille M, Francou B, Wagnon P et al (2008) Climate change and tropical Andean glaciers: past, present and future. Earth-Science Rev 89:79–96. https://doi.org/10.1016/j.earscirev.2008.04.002

Wada Y, Van Beek LPH, Van Kempen CM et al (2010) Global depletion of groundwater resources. Geophys Res Lett 37:1–5. https://doi.org/10.1029/2010GL044571

WCD (2000) Dams and development: A new framework for decision-making. The report of the World Commission on Dams. Earthscan Publications, London

Wen F, Chen X (2006) Evaluation of the impact of groundwater irrigation on streamflow in Nebraska. J Hydrol 327:603–617. https://doi.org/10.1016/j.jhydrol.2005.12.016

Whiting P (2016) Flow measurement and characterization. In: Koldolf GM, Piegay H (eds) Tools in fluvial geomorphology, 2nd edn. Wiley, West Sussex, pp 323–346

Zarfl C, Lumsdon AE, Berlekamp J et al (2014) A global boom in hydropower dam construction. Aquat Sci 77:161–170. https://doi.org/10.1007/s00027-014-0377-0

Zhang Y-K, Schilling KE (2006) Increasing streamflow and baseflow in Mississippi River since the 1940s: Effect of land use change. J Hydrol 324:412–422

Fluvial Geomorphology

Before the close of the 18th century, the erosive capabilities of running water were not appreciated. Streams were believed to flow in valleys because the valleys already were there, not because the stream cut the valley. By the late 18th century it began to be appreciated that rivers could cut their own canyons, given sufficient time, and concepts of landscape development and cycles of erosion were developed by the latter part of the 19th century. Fluvial geomorphology advanced rapidly during the 1930s–1950s with the recognition of a number of quantitative relationships among river features, described later in this chapter. Emphasis since has shifted from description to explanation of how rivers change over time and space, linking form with process within a framework defined by physical principles (Wohl 2014a).

Fluvial geomorphology is the study of channel forms and the processes and interactions among channel, floodplain, river network, and catchment. Thus, the fluvial system is more than the river channel, but extends to the entire drainage network from the hillslope sources of sediments and runoff to deltas and alluvial fans, and includes the flow of water and sediments throughout the system (Kondolf and Piegay 2016). Channel adjustments involve linkages among water and sediment supplies that may be manifested locally but are influenced by interactions at large spatial scales and over considerable time. Human activities, climate change, and long-term geological processes can shift a system from one state of equilibrium to another, or to disequilibrium. Methods employed to gain greater understanding of the physical dynamics of river systems are diverse. They include field studies of channel characteristics and sediments, stratigraphic analyses, experimental studies of sediment transport in flumes, statistical analysis and modeling of physical processes, comparisons of landforms, historical records, satellite and aerial imagery, and sophisticated statistical approaches. This chapter summarizes some of the main ideas of this field; readers wishing more depth should consult the books of Charlton (2007), Kondolf and Piegay (2016), and Wohl (2014b).

Fluvial geomorphology has evolved into a multi-disciplinary field with important applications in river engineering and ecology that are key elements of sustainable river basin management. An understanding of its basic principles is important to stream ecologists for several reasons. It helps make sense of the enormous variety of rivers and streams, revealing how stream channels are shaped, and suggesting useful ways to classify stream types. Variability in riverine features over space and time is responsible for the diverse range of habitats in which organisms live. In addition, quantification of the relationships among river features and analysis of the underlying processes contribute to a deeper understanding of how rivers and riverine habitats change in response to environmental influences, both natural and anthropogenic. This knowledge can in turn be used to avoid unwise management choices and inform the design of healthier rivers in human-dominated landscapes.

3.1 Geomorphological Features of a River System

A central theme in fluvial geomorphology is that alluvial rivers (those with unconsolidated and mobile sediments, as opposed to bedrock channels) determine the location and shape of their channels through complex interactions among hydrology, geology, topography, and vegetation. Alluvial rivers are "self-formed", meaning that their channels are shaped by the magnitude and frequency of the floods that they experience, and the ability of these floods to erode, transport, and deposit sediment. The development of stream channels and entire drainage networks, and the existence of various regular patterns in channel shape and dimensions, indicate that rivers are in dynamic equilibrium between erosion and deposition. A river system's discharge, sediment load, and elevational extent from origin to base are controlling or independent variables that the river cannot control, and therefore must adjust to. Adjustable or dependent

© Springer Nature Switzerland AG 2021
J. D. Allan et al., *Stream Ecology*,
https://doi.org/10.1007/978-3-030-61286-3_3

variables include channel pattern, meander wavelength, and local slope, width, and depth. Because channel geometry is three dimensional with a long profile, a cross–section, and a plan view (what one would see from above), and because these mutually adjust over a time scale of years to centuries, a wide range of local channel shapes is possible. For a particular reach, slope exerts significant control over channel characteristics because it adjusts more slowly than other variables. In concept, the mutual adjustment of these variables results in a graded stream, one whose shape exhibits a balance between its transporting capacity and the sediment load available to it. In reality, changing climatic and other factors may prevent a reach from achieving or long remaining in such an ideal state. The usefulness of this concept lies in the fact that streams tend to respond to perturbation by moving in the direction of some equilibrium state.

We begin by describing the shape and form of rivers—along their length, as seen from above, and as viewed across the channel and the valley. Rivers exhibit a number of remarkably predictable relationships among key hydrologic and geometric variables, and we will explore the underlying processes that account for these patterns. Stream channels tend to develop a stable dimension, pattern, and profile, features that are maintained in the face of episodes of high flows, sediment transport, and channel movement, and so it is the relationships that are predictable even though particular channel features may change. We begin by describing the patterns in channel shape, and later consider the processes responsible for these patterns.

3.1.1   The Drainage Network

Water on the ground surface moves downhill, creating small channels or rills that, over time, become persistent channels. The channels join others, forming a tree-like network of increasing drainage area. In reality each additional tributary causes drainage area to increase in steps rather than smoothly, but as an approximation, channel length increases as the 0.6 power of drainage area.[1] Stream networks generally increase in length more than width, and develop particular shapes depending on topography and the erodibility of the land surface. A stream draining a narrow valley often results in a central channel with numerous, short tributaries entering almost at right angles, whereas more gentle terrain can lead to

a rounder drainage basin. Various descriptive terms (dendritic, radial, rectangular, trellis) are used to describe these patterns. Drainage density (the sum of channel length divided by drainage area, in km km^{-2}) is a measure of how finely dissected the network is. Drainage density is highest in semiarid regions, which have the optimal combination of enough precipitation to create surface runoff and move sediment, but limited vegetation. Drainage density is lower in arid regions because of limited precipitation, and lower in humid regions because of greater vegetation cover and soil development that promote infiltration and limit surface runoff and erosion. Relief ratio is the elevation difference divided by river length along the main axis, and thus is related to gradient and the pathway that the river takes.

At the upstream origin of the drainage network, near the drainage divide, a very large number of very small channels carry water only during rainy episodes or during snowmelt, but at some point the upstream area of these small channels is sufficient to generate year-round flow. This is the point where perennial flow begins, and above lies an ephemeral stream flowing only during wet periods (and shown as dashed blue lines on topographic maps). The exact transition between an ephemeral and perennial channel is indistinct, and migrates up- and downslope depending on precipitation.

Stream order is a useful measure because it describes the position of a stream in the hierarchy of tributaries. As previously described in Chap. 1, a segment without tributaries is a first-order stream. The union of two first-order segments forms a second-order stream, defined as having only first-order channels as tributaries. A third-order stream is formed by the coalescence of two second-order streams, and so on (Fig. 3.1). This system, originated by Horton (1945) and later refined by Strahler (1957), is perhaps the most widely used classification system for streams and rivers, but it has limitations. The identification of first-order streams is challenging, and if done using maps, will vary with map scale (1:24,000 or 1:25,000 is recommended), thereby affecting all higher-order designations. Because of differences in drainage density across regions, stream size and stream order may not correlate well. In addition, this approach ignores the entry of streams of order n into order $n + 1$. Link classification (Shreve 1966) incorporates the addition of first-order streams into higher order branches of a drainage network, but shares the other failings.

Nonetheless, stream order has been a durable concept because of its simplicity and its usefulness. In addition, stream order classification is a correlate of other catchment variables. Mean length, total number, and to a lesser degree mean slope all form straight lines when their logarithmic values are plotted against stream order. There are usually some three to four times as many streams of order $n - 1$ as of order n (the bifurcation ratio), each of which is roughly less than half as long as the stream of next-higher order, and

---

[1]Drainage area is the term commonly used when discussing the river network, and refers to the total area drained by the multitude of tributaries that feed the main channel or set of channels. It is interchangeable with catchment and (in American usage) watershed area. River basin also can be substituted, although the convention is to restrict its use to very large rivers.

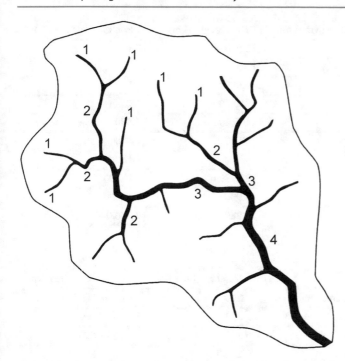

Fig. 3.1 A drainage network illustrating stream channel order within a fourth-order catchment. The terminus may be a lake or the junction with a larger river. Intermittent stream occur upstream of the first-order tributary, and often extend nearly to the catchment divide

drains somewhat more than one–fifth of the area. These ratios can be verified by inspection of Table 3.1, which summarizes the number, lengths, and drainage areas of US streams of order one through ten.

The great majority of the total length of river systems is comprised of lower-order or headwater systems, each of short length and small drainage area. Typically, first-through third-order streams comprise 70–80% of the total length of perennial channels in a catchment, emphasizing how land-water connections are especially tightly coupled in headwaters. Rivers that we might consider to be of medium size, fourth through sixth-order, are not uncommon – they include over 20,000 river channels in the contiguous US. About 250 US rivers are of order seven and higher. Only the Mississippi ranks among the fifteen largest rivers of the world based on the annual volume of its discharge. World-wide, rivers are classified by their hierarchic position within river basins and within regions. The US Geological Survey Hydrologic Unit Cataloging system (HUC, (Seaber et al. 1987)) catalogs watersheds at different geographical scales based on 1:100,000 mapping (Table 3.2). It first divides the US into 21 major regions that contain either the drainage area of a single river or the combined drainage area of a series of rivers. Smaller units are nested within regions. An 8-digit HUC is standard, and typically corresponds to a

drainage area in the order of 10^3–10^4 km^2. Further sub-divisions result in 11 or 14-digit watersheds.

3.1.2 The Stream Channel

The shape of the cross section of a stream channel is a function of the interaction between discharge and sediment, the erodibility of its bed and banks, the stabilizing influence of vegetation, and any structure (boulders, large wood) that can influence local channel conditions. A cross-sectional survey maps the shape of the channel and measures depth at multiple points, effectively creating a series of cells of known width and depth, whose product is summed to determine area, as described in Chap. 2 (Fig. 2.6). Mean depth can then be estimated as area divided by width. The location of maximum depth within the channel is known as the thalweg.

Channel shape and cross-sectional area will differ from transect to transect even within a reach, as some locations are wide and shallow, others narrow and fast. Water discharge must be the same at each transect, barring tributary inputs and groundwater exchange, but area and shape need not. This is known as the continuity of flow relationship. Channel cross sections are more regular, often trapezoidal, in straight stretches but are asymmetric at curves or bends, where the greatest depth and velocity usually occur at the outer bank (Fig. 3.2). Sediment deposition forms point bars along the inner bank due to reduced velocity and the helicoidal flow within the bend, in which near bed current flows from the outside toward the inside of the bend. Anglers make use of these shallow and gently sloping regions of streambed to cast towards the deeper water on the far bank. In steep, narrow valleys, channels are confined by topography, whereas flat, wide valleys allow more lateral movement and meandering (Fig. 3.3).

The bankfull stage, or depth of water at which overbank flooding occurs, can be determined by direct observation if a well-developed floodplain is present. Often this boundary is not apparent, and so it is estimated in various ways. A change in the topography, the elevation of point bars, the level of woody vegetation on the bank, and obvious signs of scouring are useful clues in establishing the bankfull channel. The dimensions of the wetted channel are of obvious importance to the aquatic biota and change frequently with fluctuations in discharge, whereas the bankfull dimensions are of particular importance in interpreting fluvial processes. The magnitude of the bankfull flood can also be estimated from discharge data (see Fig. 2.10), on the assumption that the flood with a recurrence interval of 1.5 years just overtops

Table 3.1 Number and lengths of river channels of various sizes in the United States (excluding tributaries of smaller order). Of the approximately 5,200,000 total river km in the contiguous US, nearly half are first-order, and the total for 1st- through 3rd-order combined is just over 85%. Examples of large rivers include the Allegheny (7th-order), the Gila (8th-order), the Columbia (9th-order) and the Mississippi (10th-order). (From Leopold et al. 1964)

Order	Number	Average length (km)	Total length (km)	Mean drainage area (km^2)
1	1,570,000	1.6	2,510,000	2.6
2	350,000	3.7	1,300,000	12.2
3	80,000	8.8	670,000	67
4	18,000	19	350,000	282
5	4,200	45	190,000	1,340
6	950	102	98,000	6,370
7	200	235	48,000	30,300
8	41	540	22,999	144,000
9	8	1,240	9.900	684,000
10	1	2,880	2,880	3,240,000

Table 3.2 Hydrologic Unit Cataloging illustrated for a specific watershed, Mill Creek (04173500), which enters the Huron River upstream of Ann Arbor, Michigan. The accounting code is sometimes referred to as basin, and the cataloging unit as sub-basin or watershed

Code segment	Name	Number in the US	Average area (km^2)
04	Water resource region	21	460,000
0417	Sub-region	222	43,500
041735	Accounting code	352	27,500
04173500	Cataloging unit	2,150	1,820

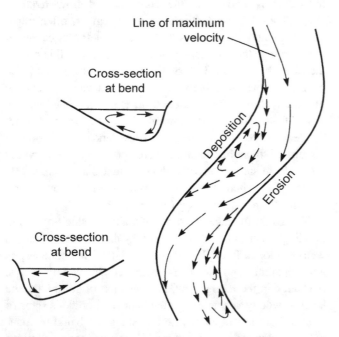

Fig. 3.2 A meandering reach, showing the line of maximum velocity and the separation of flow that produces areas of deposition and erosion. Cross sections show the lateral movement of water at bends (Reproduced from Morisawa 1968)

the banks. This assumption is most applicable in humid-temperate regions with limited hydrologic variability, and may be inappropriate for arid and tropical rivers.

3.1.3 Hydraulic Geometry

As discharge increases, either at a site due to hydrologic fluctuations or along a river's length as the river increases in size, because discharge $Q = w*d*v$ (see Eq. 2.3), any increase in discharge must result in an increase in width, depth, velocity, or some combination of these. Hydraulic geometry refers to the statistical relationships among these variables across a very wide range of discharges at a site and among sites along a river's length. These are referred to as at-a-station and downstream hydraulic geometries, respectively. Measured values are assumed to be representative of individual stream reaches (typically defined as a length ~ ten times the width), and equations for downstream hydraulic geometry use dimensions at bankfull discharge. First identified by Leopold and Maddock (1953), these relationships are linear on a log-log scale, and power equations provide good fits of width, depth, and velocity as a function of discharge.

$$w = aQ^b \qquad (3.1)$$

$$d = cQ^f \qquad (3.2)$$

$$v = kQ^m \qquad (3.3)$$

Because $Q = w \times d \times v$, it follows that $a \times c \times k = 1$ and $b + f + m = 1$.

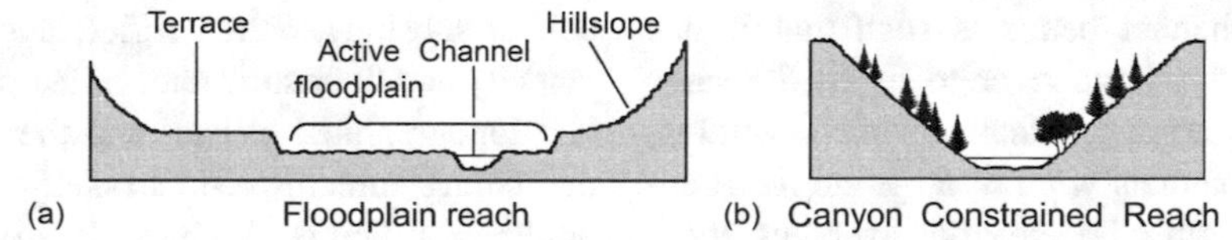

(a) Floodplain reach   (b) Canyon Constrained Reach

Fig. 3.3 (**a**) Diagrammatic cross-section of a valley showing present channel, the floodplain occupied in modern time, and a terrace representing a previous floodplain. (**b**) A constrained river channel with little opportunity to develop a floodplain (Reproduced from Ward et al. 2002)

Average values for at-a-station exponents in response to changes in discharge are b = 0.23, f = 0.42, and m = 0.35 (Park 1977), indicating that depth typically increases more rapidly than width or velocity, until the river overflows its banks (if there is a floodplain), and then width increases greatly. Width tends to increase faster than depth in wide, shallow channels, and the opposite tends to occur in deeper, narrower channels. Velocity increases least where bed roughness and flow resistance are greater.

As one proceeds downstream for a given flow stage, increases in Q are the result of tributary and groundwater inputs. Width, depth, and velocity all increase log-linearly with bankfull discharge (the preferred measure) or mean annual discharge. Exponent values typically are in the range of 0.4–0.5 for width, 0.3–0.4 for depth and 0.1–0.2 for velocity (Park 1977). The increase of width with discharge is greater than the increase of depth (note this is the opposite of the typical at-a-station response), while velocity increases least with discharge and can remain almost constant. The modest downstream increase in velocity may seem surprising, because we might expect velocity to decrease downstream due to a general decline in gradient. However, because channel depth generally is greater and substrates are finer as one proceeds downstream, resistance decreases longitudinally and this offsets the effects of reduction in slope. The River Tweed in Scotland illustrates this nicely (Ledger 1981). At most flows, the highest velocities are found at the lower and flatter end of the river system. Only in some situations involving floods does mean velocity not exhibit an increase in the downstream direction.

Coefficients and exponents of hydraulic geometry relationships share similar ranges across rivers and regions, as illustrated by a comparison of hydraulic geometries of four alluvial, single-thread gravel bed rivers from different regions (Parker et al. 2007). Bankfull depth, bankfull width, and down-channel slope showed consistency in their relationships with bankfull discharge and bed surface median sediment size, and these relations extended well to three independent data sets. However, it should be noted that hydraulic geometry equations are described as "quasi-universal relationships". Although remarkably constant across several orders of magnitude of discharge, there may be much variability at the local scale. Efforts to explain these relationships (and the range of coefficient estimates) invoke both external constraints, such as sinuosity and bank vegetation, and underlying physical relationships related to particle resistance and sediment transport.

While a full explanation of these empirically observed relationships based on underlying physical principles has not yet been satisfactorily attained, their empirical validity and utility are unquestioned (Gleason 2015). Hydraulic geometry relationships make it possible to estimate reach-averaged hydraulic variables across reaches, and within reaches across a range of discharge rates. This in turn forms the basis for engineering of river channel dimensions intended to be stable over the full range of expected flows, including floods, and so is widely used in channel design for river restoration. In addition, equations for depth and width can be used to generate equations for other hydraulic variables such as the dimensionless Froude number and Reynolds number (described in Chap. 5). This points to their relevance for understanding relationships with habitats and the distribution of organisms, and as a planning tool for river management and restoration (Lamouroux 2007).

3.1.4 Channel Pattern

Channel pattern is what one observes in a planform view at the scale of tens of bar or meander lengths. Early classifications distinguished single thread from braided rivers (Leopold et al. 1964), and straight from meandering (Schumm 1985). Most rivers are sinuous or meandering when viewed from above, but the degree of meandering varies considerably, from relatively straight channels with a sinuous thalweg, to channels with pronounced and regular curvature. The sinuosity of a reach is easily quantified as:

$$\text{Sinuosity} = \frac{\text{channel distance}}{\text{straightline downvalley distance}} \quad (3.4)$$

Many variables affect degree of sinuosity, and so values range from near one in simple, well-defined channels to 4 in highly meandering channels. Meandering usually is defined as an arbitrarily extreme level of sinuosity, typically >1.5 (Gordon et al. 2004).

The consistency of channel bends is such that if one scales a small stream and a large river to fit on the same page, their similarity is striking. Small channels wind in small curves and large channels wind in large curves (Leopold 1994). The wavelength of a meander averages about 10–14 channel widths, whether one measures a stream in a small experimental flume or the Gulf Stream meandering in the Atlantic Ocean. The radius of curvature of the channel bend averages 2–3 multiples of the channel width.

In addition to single thread channels, complex patterns of branching also occur where a section of stream, an anabranch, diverges from the main channel for some distance before rejoining the main channel. This is most often observed in lowlands and deltas. Historically, anabranching sections were common in many large rivers that over time have been engineered into single channels, as has been especially well documented for the Danube River near Vienna (Hohensinner et al. 2013). A braided river has its main channel divided into multiple channel threads. This often is seen where a mountain river carrying a large sediment load enters a flat valley, and the resultant decline in slope causes gravel to be deposited in bars within the channel. Stream power (defined as slope times discharge, Sect. 3.4), amount and size of bedload, width to depth ratio and channel instability provide a qualitative explanation of different channel patterns. With increasing stream power, sand bed streams exhibit straight, meandering and anabranching patterns. For gravel bed streams, straight, slightly sinuous with alternating bars, and braided rivers emerge (Kleinhans 2010).

The formation of a second channel may also be the precursor to channel migration. Avulsion is the rapid abandonment of a river channel and the formation of a new river channel that is less sinuous and of greater slope than the abandoned channel. It results in a shift of a river's course over the surrounding landscape and is most pronounced in low gradient rivers, where the old channel may be abandoned as the new channel captures more and more water and sediment, and the main branch migrates away. The new channel may be less sinuous, resulting in a slight gradient advantage, and be under-fed with sediment.

3.1.5 Pool-Riffle Features

Pool-riffle channels typically are found in moderate to low gradient, unconfined, gravel and sand-bed streams. At the scale of the stream reach, perhaps a few hundred meters in length, one can observe a more–or–less regular alternation between shallow areas of higher velocity and mixed gravel–

cobble substrate, called riffles, and deeper areas of slower velocity and finer substrate, called pools (Fig. 3.4). The riffle is a topographical hillock and the pool a depression in the undulating stream bed. In self-formed alluvial (but not boulder or bedrock) channels, riffles are formed by the deposition of gravel bars in a characteristic alternation from one side of the channel to the other, at a distance of approximately 5–7 channel widths (Leopold et al. 1964). Pool-riffle sequences are the result of particle sorting and require a range of sediment sizes to develop. At low flows, riffles have a high slope, tend to be shallow relative to pools, and have higher velocities. At high flows the water surface slope becomes more uniform between riffles and pools, although pools remain deeper, and velocities increase more in pools than in riffles. This results in changes in the distribution of forces on the stream bed. At flood stage, when flows are high enough to mobilize the bed, riffles are the locations of lowest transport capacity and thus the locations of gravel deposition.

Pools also form on the outside of bends and where large wood (LW) or other obstructions force pool development. In these circumstances the regular alternation of riffle-pool sequences may not be apparent. In high gradient, gravel bed streams of the Pacific Northwest, the presence of pools is strongly dependent on LW, and streams with a high loading of wood typically have closer pool spacing (Fig. 3.5). Large wood has its greatest influence on first through fourth-order streams, where it has been found to increase width, form waterfalls, and stabilize gravel bars as well as create pools (Bilby and Bisson 1998). Physical diversity in sand bed channels is often depressed in reaches where wood has been removed (Shields and Smith 1992).

3.1.6 The Floodplain

A floodplain is a level area near the stream channel that is inundated during moderate flow. It is constructed by the river under present climatic conditions by deposition of sediments during overbank flooding (Leopold 1994). Unconfined, flat valleys, which occur most commonly in lowland rivers, permit considerable meandering and lateral migration, and so tend to have well developed floodplains. In contrast, floodplain development in highly confined channels is correspondingly restricted (Fig. 3.2).

Channel movement and valley flooding are regular and natural behaviors of the river. The bankfull level of a river can be recognized by field observations as described previously or by directly observing the flood where the river just overflows its banks. In practice, however, this is not always

Fig. 3.4 A longitudinal profile (**a**) and a plan view (**b**) of a riffle-pool sequence. Water surface profiles in (**a**) depict high-, intermediate-, and low-flow conditions

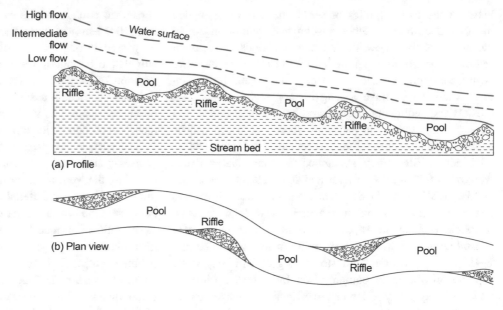

Fig. 3.5 Average pool spacing as a function of the frequency of pieces of wood in pool-riffle and plane-bed channels, from various locations in the Western US (Reproduced from Buffington et al. 2003)

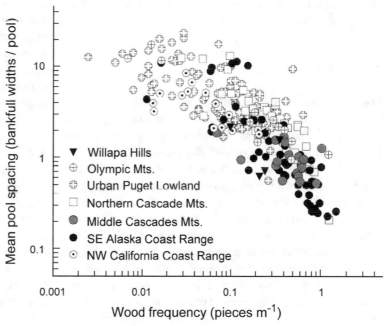

easily done. A widely used rule-of-thumb considers over-bank flows to occur about every one to two years (the 1.5 year recurrence event in Fig. 2.10). In actuality, the floodplain along a given stream reach may be inundated many times annually, or much less frequently. As a corollary to this statement, the river constructs a channel that is large enough to contain most discharges that it experiences; only less frequent, larger discharges spill out of the channel onto the floodplain.

Owing to changes in climate or basin conditions, a river can change its bed level upward (aggradation) or downward (degradation). Over long periods of degradation, the old floodplain, abandoned as the river cuts downward, remains as a terrace.

3.2 Sediments and Their Transport

The supply and transport of sediments are important because they strongly influence channel dynamics, affect habitat quality experienced by the biota, and can be costly to manage. The dynamic equilibrium that rivers seek is between the twin supplies of sediments and water, which together determine whether erosion and deposition are in balance and thus how the channel responds. Too much sediment, or too little, can be harmful to the biota and have costly consequences for human populations and infrastructure. Many rivers have a long history of human-induced erosion and sedimentation, causing habitat degradation and altering their ecology to the point where restoration will be extremely challenging (Gore and Shields 1995). In a 2008–2009 nationwide survey, streambed sediment characteristics were rated fair in 29% of river and stream lengths and poor in 15% (USEPA 2016). Sediments delivered from rivers also can cause serious damage to estuaries and reefs, whereas an insufficient sediment supply can cause river deltas and coastal shorelines to retreat, resulting in loss of habitat and inadequate protection from storm surges. Drinking water supplies also can be compromised by excess sediments, which is why many cities practice riparian management in the watersheds that supply its reservoirs, as a less costly solution than filtration.

3.2.1 Bed Material

The grain size of bed material in a stream reach is determined by the sizes introduced to the channel from upstream, from local tributaries and hillslopes, and by abrasion and sorting. Bed material is one of the imposed conditions to which channel form adjusts. A convenient size classification (Table 3.3), based on a progressive doubling of sizes, helps us be precise in our use of terms such as gravel, pebble, boulder, etc. Grain and particle describe a particle of any size, and clast usually refers to larger particles. Streambeds typically include a mix of particle sizes. The composition of the bed surface determines material available for transport, which is the main focus here. It also influences interstitial flow and interchange between hyporheic and surface waters, the near-bed hydraulic environment, and multiple aspects of habitat suitability for the biota including gravel-spawning fish, topics considered in future chapters.

Because the bed surface typically is composed of a mix of particle sizes, it is useful to quantify the average particle size as well as the size range. The pebble count is a simple and widely used method to quantify grain size of the surface layer and predict bed mobilization thresholds. The usual approach is to determine the intermediate axis of approximately 100 grains that are greater than 4 mm diameter by randomly selecting stones from the streambed or a gravel bar of a reasonably homogeneous stream reach, and measuring two axes with a ruler or by passing through a template corresponding to Wentworth units. However, when the streambed contains significant amounts of material smaller than 10–15 mm in diameter, samples of bed material must be passed through sieves of various sizes. A plot of the cumulative frequency distribution against a geometric progression of particle sizes (Fig. 3.6) allows one to quickly estimate the median particle size, or D50. The D16 and D84 often are reported as well, as they encompass one standard deviation on either side of the mean in a normal distribution.

The surface layer of a gravel bed stream usually has coarser grains than subsurface layers. Streambed armoring refers to the vertical layering of substrate, where coarser grains overlay finer material and may prevent the later from being entrained in the water column. This can be quantified as the ratio of surface D50: subsurface D50. Bed sediment measurement employs many methods in addition to the pebble count, include mapping, photographs (which can be repeated over time at fixed locations), and core sampling (to better quantify subsurface substrate) (Kondolf and Piegay 2016). If the interest is fish biology rather than sediment transport then the fraction of gravel within a certain size range preferred by fish for spawning may be the most useful measure, and if fine sediments are limiting the intergravel flow of oxygenated water necessary for incubating fish embryos or benthic invertebrates, then their quantification is important.

3.2.2 Bank and Bed Erosion

All sediments ultimately derive from erosion of basin slopes and water flowing across the land surface, but the immediate supply usually derives from the river bed and banks (Richards 2004). High flow episodes scour and transport sediments, eventually depositing them farther down the channel as flow subsides. For a stream section near equilibrium, these exported sediments should be replenished from upstream, and deposited in the scoured section as flows subside. Thus, bank erosion should be recognized as a natural process of a dynamic river, promoting riparian vegetation succession and creating habitats crucial for aquatic and riparian plants and animals (Florsheim et al. 2008). However, erosion of stream banks can become severe when peak flows are increased by human activities. Increases in impervious surface area and stormwater conveyance structures can greatly enhance runoff and thus cause erosion. The process begins when stream flows cause the bank to steepen by eroding material at its base (the "toe"). As tension cracks begin to form in the upper, horizontal bank surface and water infiltration raises pore-water pressure (and increases mass), shearing begins, leading to bank failure. Soil debris is deposited at the foot of the bank, and stream flow removes

Table 3.3 The Wentworth grain size scale defines size classes in intervals that increase by powers of 2

Size category		Particle diameter (range in mm)
Boulder		>256
Cobble		
	Large	128–256
	Small	64–128
Gravel		
	Very coarse	32–64
	Coarse	16–32
	Medium	8–16
	Fine	4–8
	Very fine	2–4
Sand		
	Very coarse	1–2
	Coarse	0.5–1
	Medium	0.25–0.5
	Fine	0.125–0.25
	Very fine	0.0625–0.125
Silt		<0.0625

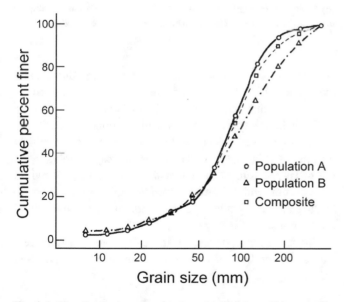

Fig. 3.6 The distribution of grain sizes obtained by pebble count for two locations on the stream bed identified as distinct local depositional environments from a site on Rush Creek, California. The median particle size is approximately 80 mm, in the size range of small cobbles, and slightly smaller at the location referred to as Population B (Reproduced from Kondolf et al. 2003)

corridor form and dynamics. The above-ground biomass of plants modifies flows and retains sediment, whereas plant below-ground biomass affects soil moisture and susceptibility to erosion. The influence of vegetation on channel shape is so pervasive that Gurnell (2014) refers to plants as "river system engineers". In multi-thread river channels, colonization of islands by woody vegetation can result in very long-term stabilization of channel planform. Single thread river channels are consistently found to be narrower and deeper in pasturelands, wider and shallower in forested reaches (Sweeney et al. 2004). This difference is the result of the deeper and denser root system of grasses relative to trees, and the bank stability that root systems confer. Rivers in catchments with a history of deforestation and subsequent forest recovery very likely have experienced these changes sequentially. This is seen in the channels of small streams in Vermont, US, which have widened in response to riparian reforestation based on measurements collected in 1966, 2004, and 2008 (McBride et al. 2010), and appear still to be actively widening, presumably recovering from historic deforestation (Fig. 3.7).

Bank stability can be increased by a number of management interventions. Re-vegetating of streambanks is a common practice in which the root system stabilizes the soil and evapotranspiration removes soil water. Drainage tiles can be installed to transport soil water directly to the stream channel, reducing the likelihood of mass wasting. Toe protection by various devices can prevent bank steepening. Bank stabilization is an important management activity, ranging from 'hard' solutions such as riprap (slabs of

the failed debris, increasing the sediment load of the stream. Bank steepening starts again, resulting in a vertical bank face and another cycle of bank erosion.

Most channel banks include finer material and this provides some degree of cohesion. In addition, vegetation within the river corridor can exert significant control on river

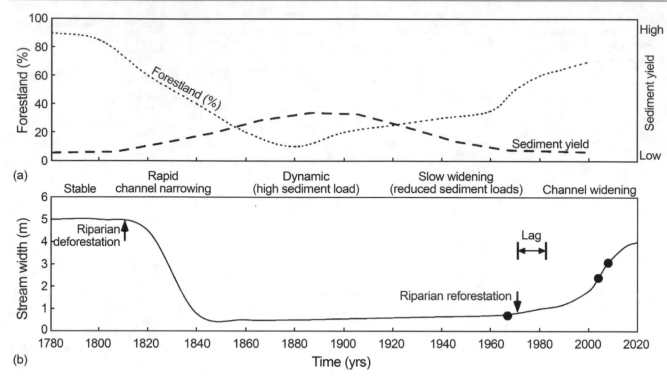

Fig. 3.7 Hypothesized historic channel response to riparian and catchment-scale deforestation and reforestation. (**a**) Presumed catchment-scale trends in forested area and sediment yield over time;

(**b**) Mean stream width versus time with known data points (·) along a plausible historic trend (Reproduced from McBride et al. 2010)

concrete) and gabions (wire baskets of stones) to more environment-friendly approaches using vegetation. However, efforts to stabilize banks will be ineffective if incision of the stream bed takes place.

Urban streams present a different scenario than agricultural settings. Peak flows typically increase with urban infrastructure, and banks often are hardened because channel shifts can undermine roads and houses. Bed erosion and downcutting are the expected consequences, with considerable sediment exported to lower reaches.

3.2.3 Particle Transport

Bed material is transported when discharge reaches a sufficient level to initiate motion and transport particles generally of larger size than fine sand. Not surprisingly, the size of particle that can be eroded and transported varies with current velocity (Fig. 3.8). The competence of a stream refers to the largest particle that can be moved along the streambed at some flow, and the critical erosion (competent) velocity is the lowest velocity at which a particle of a given size, resting on the stream bed, will move. Sand particles are the most easily eroded, having a critical erosion velocity of about 20 cm s^{-1}. Due to their greater mass, larger particles require higher current velocities to initiate movement, for example,

at least 1 m s^{-1} for coarse gravel. However, grains smaller than sands, including silts and clays, have greater critical erosion velocities because of their cohesiveness.

Once in transport, particles will continue in motion at somewhat slower velocities than was necessary to initiate movement (Fig. 3.8). As velocities decrease, grains settle out of suspension, beginning with the largest and heaviest. This occurs when discharge declines following a flood, in reaches of lower gradient, at the inside of bends, and behind obstructions.

The shear stress or tractive force (τ_o, force per unit area) exerted by the flow of water on the stream bed is estimated as:

$$\tau_o = \rho g R S \tag{3.5}$$

where ρ is fluid density, g is gravitational acceleration, the hydraulic radius R equals channel cross-sectional area divided by its wetted perimeter, and S is the water surface slope. For natural channels with a width much larger than mean flow depth, mean depth is a good approximation of the hydraulic radius.

This equation is important because it relates the resistance of the channel bed and banks to the downstream gravitational tractive force of the water: when the former is exceeded, sediment transport is initiated. Critical shear stress (τ_c) refers to the shear stress necessary to mobilize a given

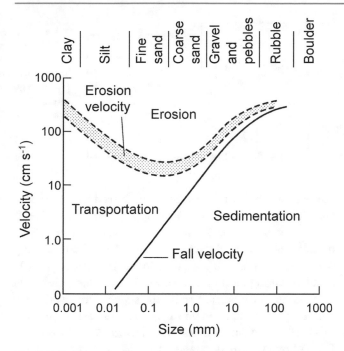

Fig. 3.8 Relation of mean current velocity in water at least 1 m deep to the size of mineral grains that can be eroded from a bed of material of similar size. Below the velocity sufficient for erosion of grains of a given size (shown as a band), grains can continue to be transported. Deposition occurs at lower velocities than required for erosion of a particle of a given size (Reproduced from Morisawa 1968)

grain size. For mobile, gravel-bed rivers with bed materials greater than 1 cm diameter, the particle size near the threshold of motion at bankfull flow is approximately equal to the median bed material size (cm). In other words, the D50 is a good indicator of the tractive force on the stream bed at bankfull flow.

3.2.4 Sediment Load

Sediment load is the amount of sediment passing a point over some time interval. It is estimated by multiplying sediment concentration by water discharge. Matter carried by fluvial systems can be separated into three components. These are the dissolved load, which consists of material transported in solution; the wash load, consisting of material between 0.5 μm (the upper limit for dissolved material) and 0.0625 mm (the boundary between silt and sand); and solid load, consisting of material greater than 0.0625 mm. Terms that describe the total sediment load refers either to the source of the material or the mode of transport (Hicks and Gomez 2016) (Fig. 3.9).

The dissolved load consists of solutes derived from chemical weathering of bedrock and soils, discussed more fully in Chap. 4. The relative amount of material transported as solute versus solid load depends upon basin

characteristics, lithology, and hydrologic pathways. In dry regions, sediments make up as much as 90% of the total load, whereas the contribution of solutes is substantially more in areas of very high runoff (Richards 2004).World–wide, it is estimated that rivers carry 15–20 billion metric tons of suspended materials annually to the oceans, which is roughly five times the dissolved load (Martin and Meybeck 1979).

By source, the total sediment load is split between wash load and bed material load (Hicks and Gomez 2016). The wash load (so named because this load is "washed" into the stream from banks and upland areas) consists of very fine particles including clay and silt up to very fine sand. It requires only low velocities and minor turbulence to remain in suspension, thus this material may never settle out. The amount of the wash load is determined by its supply from uplands and stream banks rather than by the stream's transport capacity, and is likely to be high where stream banks have a high clay and silt content. The bed material load is derived from the river bed, typically sand or gravel, and its concentration is directly related to the river's transport capacity.

By mode of transport, the sediment load is divided into suspended load and bedload. Flow competence refers to maximum size of sediment that can be moved by a given flow. The flow of water in rivers generally is turbulent, and exerts a shearing force that causes particles to move along the bed by pushing, rolling, and skipping, referred to as the bed load. This same shear causes turbulent eddies that entrain particles into suspension, called the suspended load. The distinction between bed load and suspended load is based on sampling method, and the same material that is transported as bed load at low discharge may become suspended load at higher discharge.

Suspended load is fairly easy to sample—a simple grab sample will suffice—but varies with depth and can change rapidly with discharge, and so sampling that integrates across depth and takes place frequently over the rise and fall of the hydrograph is preferred. Because fine sediments tend to be washed into the stream at the beginning of a rain event and entrained by rising water, their concentrations usually are greater during the rise of the hydrograph, and decline during the falling hydrograph due to exhaustion of the sediment supply. As a consequence, sediment concentrations can be different at identical discharges of the rising and falling hydrograph. This is referred to as hysteresis.

Suspended sediments cause turbidity by restricting the transmission of light through water due to scattering and absorption. By measuring light transmission through a water sample, turbidity meters provide a simple approximation of suspended sediment loads. These usually are reported as Nephalometric Turbidity Units (NTUs), which can be calibrated against measured sediment concentrations (mg L^{-1}).

Fig. 3.9 The components of stream sediment load shown in terms of sediment source and mode of transport (Reproduced from Hicks and Gomez 2016)

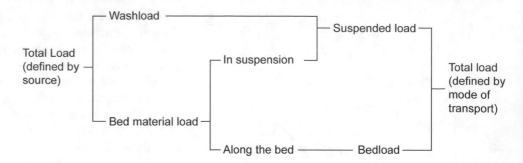

There are additional sources of turbidity, however, including algae and colloidal matter, and so turbidity is not solely a measure of suspended sediments.

The majority of sediment transport is due to the suspended load. Bedload transport rates are notoriously difficult to measure and often only suspended load data are available. Bedload traps installed below the bed surface and tracer particles such as painted stones are common methods. Literature estimates of bedload as a fraction of total load are often in the range of 10–20%, although some higher values have been given for mountain rivers. In large catchments the bedload fraction is insignificant, at about 1%. In sand-bed rivers, however, the bedload fraction may be substantial (30–50%) even for large catchments. In the absence of direct measurement, bedload transport if estimated at all depends largely on borrowed values from a location with as closely matching conditions as possible. However, results from an Austrian mountain stream, the Pitzbach, illustrate the risk of using any rule-of-thumb value. Estimates of the fraction of total load transported as bedload varied from zero to one, increased with rising discharge (Fig. 3.10), and varied from year to year and among floods (Turowski et al. 2010).

For a stream channel in equilibrium, the transport of bed material requires that it be replaced by material derived from the river's upstream banks and channel, in a cycle of scour and fill that accompanies the rise and fall of flood waters; if no supply of sediments from upstream is available, the bed will downcut. The Colorado River immediately downstream from Glen Canyon Dam provides an illustrative case study (Grams et al. 2007). Before the dam closure in 1963, the

downstream 25-km river section experienced a classic scour and fill cycle annually. Each spring, snowmelt floods scoured the bed some 2–7 m; then, sediment delivered from upstream accumulated through the summer and fall, roughly balancing the preceding scour. Modest incision apparently occurred during the first half of the 20th century, attributed to a decline in upstream sediment supply, but overall the year to year changes in mean annual bed elevation were much smaller than changes within each year. Closure of Glen Canyon Dam decreased the magnitude of the mean annual flood in Glen Canyon by 63% and the annual sediment load by 99%. The resulting sediment deficit caused channel incision and bed-sediment evacuation, lowering thalweg bed elevation by an average of 2.6 m, and deepening riffles and pools by up to 4 and 6 m, respectively (Fig. 3.11). Most of this scour occurred in 1965 during a series of pulsed high flows, and the bed elevation has persisted into the present.

3.2.5 Factors Influencing Sediment Concentrations and Loads

A stream's capacity is the total load of bed material it can carry. This increases with velocity and discharge unless the supply of sediment becomes depleted. The larger the flow, in general, the larger the quantity of sediment transported (Richards 2004). Throughout most of the year, discharge usually is too low to scour, shape channels, or move significant quantities of sediment; however, sand-bed streams can experience change much more frequently. Although one might suppose that extreme events also account for the greatest proportion of total sediment transport, flow events of intermediate frequency actually move more sediment over the years. The discharge at which sediment transport peaks is called the effective or dominant discharge, and it is estimated from the product of the discharge frequency curve and the curve describing sediment transport rate as a function of discharge (Fig. 3.12). Because the effective discharge accomplishes the most geomorphic work compared to other flows, it follows that fluvial landforms are shaped mainly by frequently occurring moderate floods, rather than by

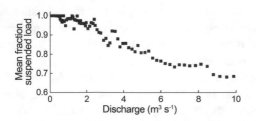

Fig. 3.10 Partitioning of the total sediment load for the Pitzbach, an alpine river in Austria. The mean fraction that is suspended load declines with increasing discharge, while bedload transport increases (Reproduced from Turowski et al. 2010)

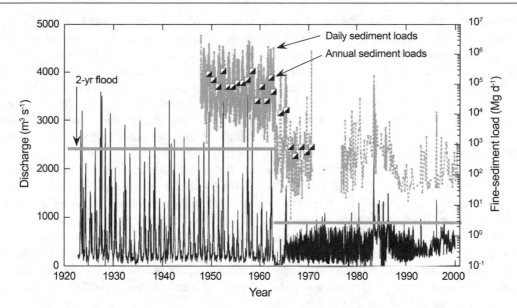

Fig. 3.11 The rate and pattern of bed incision and bank adjustment on the Colorado River in Glen Canyon downstream from Glen Canyon Dam. Shifts in instantaneous discharge of the Colorado River at Lees Ferry, Arizona, 1921–2000, and measured sediment load for the same location, 1947–2000, document influence of closing of Glen Canyon dam in 1963. The gray points connected by the dashed line are the computed loads for each day that sediment concentration was measured. The black and white boxes are the annual loads (expressed in Mg d^{-1}) for the years with sufficient data. The thick horizontal line indicates the magnitude of the pre-dam and post-dam 2-year recurrence flood (Reproduced from Grams et al. 2007)

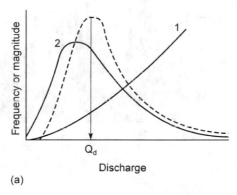

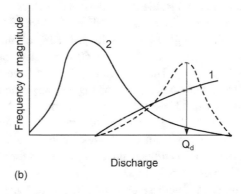

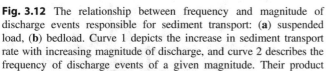

Fig. 3.12 The relationship between frequency and magnitude of discharge events responsible for sediment transport: (**a**) suspended load, (**b**) bedload. Curve 1 depicts the increase in sediment transport rate with increasing magnitude of discharge, and curve 2 describes the frequency of discharge events of a given magnitude. Their product (dashed line) is the discharge that transports the most sediment, referred to as the dominant or effective discharge. The effective discharge is approximately Q_{bkf} for suspended sediments, and is in the range $Q_{1.5}$ – Q_{10} for bedload (Reproduced from Richards 2004)

infrequent large floods. There are exceptions, of course, including channel shaping by very large paleo-floods that occurred in past millennia.

The concept of a channel-forming discharge is widely used in river restoration designs because it suggests ways to estimate, fairly easily, the equilibrium channel dimensions. Now a cornerstone of river channel restoration design (Doyle et al. 2007), in practice it is quantified in different ways. Effective discharge is computed by finding the maximum of the curve resulting from the product of the flow frequency curve and a sediment discharge rating curve (Fig. 3.12).

Bankfull discharge is estimated from various field indicators, or as the discharge magnitude that occurs, on average, once in 1–2 years. Use of effective discharge has the advantage that the sediment budget is quantified, but requires more data and analysis, and so surrogates such as the 1.5 year recurrence estimate often are the basis for restoration prescriptions. This is not unreasonable, at least as a first approximation, as analysis of suspended-sediment transport data from more than 2,900 sites across the United States, sorted into ecoregions, supported the use of the $Q_{1.5}$ as a measure of the effective discharge (Simon et al. 2004). Median values of the recurrence interval of the effective discharge for 17 ecoregions ranged from 1.1 to 1.7 years, and the detection of differences among regions argues for the use of regionalized curves. It is important to bear in mind, however, that natural stream morphology is the product of the entire range of discharges in the hydrograph, and effective discharge may not be adequately approximated by a return interval or bankfull estimate.

Sediment loads are calculated as the product of sediment concentration (mg L^{-1}) and water discharge (m^3 time^{-1}), and so are in the units of mass time^{-1} (note that 1,000 L = 1 m^3). Concentrations of suspended sediments vary greatly depending on the factors described above that influence sediment supply, and with discharge and velocity, which determine how much sediment is in transport at any time. Thus, accurate load estimation depends on adequate sampling of sediment concentrations over seasons and rising and falling flows. The load represents the total amount of sediment discharged annually or over some shorter time interval. Sediment yields from individual rivers are calculated as loads divided by catchment area, providing a useful comparison of variation in sediment export among rivers and over time. Sediment concentrations and yields vary greatly with geology and climate, the extent of human activities leading to erosion, and whether the channel is in a stable or unstable state.

Water discharge alone is a poor predictor of sediment loads except within a region. Rivers in just 10% of the world's drainage basins account for over 60% of sediment discharge (Milliman 1990). The Hwang Ho (Yellow River) of northern China is believed to carry the highest suspended load of any river, as much as 40% sand, silt, and clay by weight during high discharge. The great rivers of South America make a significant but nonetheless much smaller contribution to the world sediment flux, and large northern rivers account for considerably less.

Human activities can increase or reduce sediment loads, mainly by increasing erosion through deforestation and intensification of agriculture, and by trapping of sediments in reservoirs. Long term records of annual sediment load and runoff for world rivers provide many examples of non-stationary behavior (Walling and Fang 2003), including cases where loads are declining as a result of reservoir construction and sediment control programs (Fig. 3.13), and others where loads are increasing as a result of land clearance and land use intensification (Fig. 3.14). Trend analysis for 145 major rivers indicated that about half of the sediment load records showed evidence of statistically significant upward or downward trends, with the majority exhibiting declining loads. However, in many other cases, sediment load records were relatively stable, indicating an absence of environmental change or a buffering of different effects. Overall, however, reservoir construction is probably the most important influence on land-ocean sediment fluxes.

Renwick et al. (2005) estimate that, at least for the later part of the 20th century, much of the sedimentation in the US actually is occurring in impoundments. Both the Nile and the Colorado have experienced a complete cessation of sediment export, and the Rhône is estimated to export approximately 5% of its load of a century ago. Thus, in a number of large rivers we have the apparent paradox of increased erosion within the drainage basin coupled with reduced export to the oceans.

The global consequences of these trends can be seen in Table 3.4, which summarizes discharge and sediment fluxes for pre-human and modern times by continent. The global transport of sediments from land to sea is estimated to be in the range of 15–20 billion metric tons per year, but is not known with certainty due to incomplete data and changing environmental conditions (Walling and Fang 2003). Combining data and models, Syvitski et al. (2005) estimated the

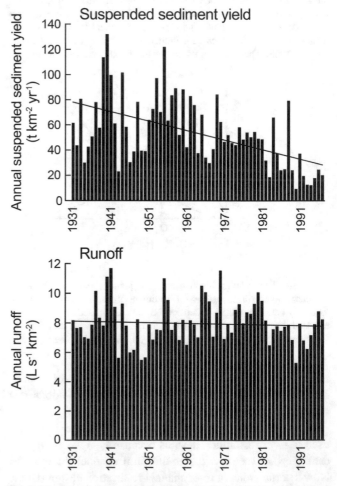

Fig. 3.13 Recent changes in the sediment load and annual runoff of the River Danube at Ceatal Izmail, Romania, close to its delta and its discharge into the Black Sea. The main reduction in sediment flux started during the mid-1960s, and resulted in a ca. 70% reduction in sediment flux by the 1990s. Reservoir and barrage construction are presumed causes (Reproduced from Walling and Fang 2003)

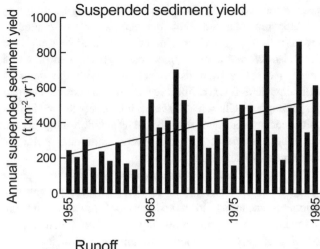

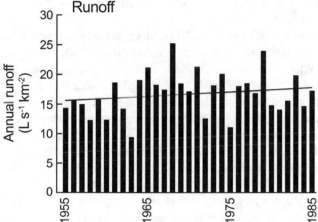

Fig. 3.14 Recent changes in the sediment load and annual runoff of the Hongshuihe River at Qianjiang, a tributary of the Pearl River, in Guangxi Province, China. Sediment loads increased by ca. 75% over the period of record. Population growth and the expansion and intensification of agriculture are presumed causes (Reproduced from Walling and Fang 2003)

global total prior to human influence to be 14 billion metric tons annually (15.5 BT/year when bedload is included). Asia produces the greatest quantity of fluvial sediment, whereas Oceania and Indonesia have the highest sediment yields as well as the highest runoff per area. By latitude, warm areas produce the highest sediment yields, accounting for nearly two-thirds of global delivery. The modern sediment flux exported by rivers is estimated to be about 10% less than the pre-human value, but in the absence of dams, sediment inputs would be at least 10% greater. Thus, sediment flux into global rivers due to erosion has increased, while sediment yields to the world's coasts have declined.

3.3 Fluvial Processes Along the River Continuum

We are now familiar with many of the principal features of rivers, including the shape of their channels, the lateral connection to the floodplain, the presence of riffles, pools, and meanders, and the sediments of the stream bed and its banks. These features vary along the river's course from headwaters to lowlands, and with regional differences in climate and terrain. As stated earlier, streams seek a state of dynamic equilibrium between the imposed conditions of valley slope, discharge, and sediment supply, and channel adjustments that can include width, depth, velocity, reach slope, roughness, and sediment size. Over historical time, the channel adjusts to changes in discharge and sediment supply due to human activities, climate change, and extreme events. Increasingly, human activities are responsible for destabilizing the equilibrium of rivers, often triggering a series of changes that pose problems for our built environment.

Table 3.4 Landmass area, discharge, predicted sediment flux to the world's coastal zones from world rivers under pre-human and modern conditions, and percentage of sediment load retained in reservoirs. Uncertainty estimates for sediment fluxes and sediment retained in reservoirs have been omitted for simplicity, but range from 15 to 30% of stated values. See Syvitski et al. (2005) for details

Landmass	Area (M km^2)	Discharge (km^3 yr^{-1})	Prehuman suspended load Qs (MT yr^{-1})	Modern suspended load Qs (MT yr^{-1})	Load retained in reservoirs (%)
Africa	20	3,800	1,310	800	25
Asia	31	9,810	5,450	4,740	31
Australasia	4	610	420	390	8
Europe	10	2,680	920	680	12
Indonesia	3	4,260	900	1,630	1
North America	21	5,820	2,350	1,910	13
Ocean Islands	0.01	20	4	8	0
South America	17	11,540	2,680	2,450	13
Total	106	38,540	14,030	12,610	20

The concept of stream power, and the relationships between the supply of water and sediments, nicely express the key processes at work. Stream power describes the ability of the stream to mobilize and transport material. It is the product of discharge and slope, and additional formulae estimate unit stream power per unit area or length of reach. Thus steeper slopes and higher discharges both increase stream power, but because slope tends to decrease downstream while discharge increases, a mountain stream in flood may generate much more power than a large lowland river, and a river that overflows its banks into the floodplain will have lower power than one that stays within its banks (Gordon et al. 2004). Stream power is related approximately to sediment load by:

$$Q_s D50 \sim Q_w S \qquad (3.6)$$

where Q_s is the sediment discharge (load), D50 is the median particle size, Q_w is water discharge and S is slope (Lane 1955). Referred to as Lane's law or Lane's balance, this relationship is qualitative, yet very useful for visualizing how stream channels are expected to respond to changes in water or sediment supply. Dust and Wohl (2012) have since expanded this relationship to incorporate channel adjustments to channel geometry.

Lane's balance implies that the channel will remain in equilibrium as long as no major change occurs in any of the variables, or if changes in one are balanced by changes in another (Fig. 3.15). However, if sediments are trapped behind a dam, the outflow usually has a low sediment load, resulting in bed coarsening and channel incision or widening as the sediment-hungry river entrains bed material from the bed or banks. If sediments are introduced by poor land-use practices in a tributary catchment, the stream lacks the power to transport that additional material, resulting in deposition and aggradation. If a stream is straightened, its slope is increased (because the same elevation drop now occurs over a shorter distance), and erosion (due to greater transport capacity) is expected. When as little as 10-20% of a catchment is covered with roofs, pavement, or other impervious surfaces, peak discharge typically increases and channels typically widen or deepen in response, a familiar sight in urban areas (Bledsoe and Watson 2001).

3.3.1 Fluvial Processes and Channel Morphologies

We now turn to a process-based analysis of river channels, linking the governing conditions of water and sediment supply to the channel features we have encountered and their changes along the river continuum. This will focus primarily on alluvial channels, defined as those with bed sediments that are transported by the stream. Rearranging relation 3.6, it is apparent that sediment transport is directly related to stream power ($Q_w * S$) and inversely related to grain size D50 (also referred to as sediment caliber). The interplay of a stream's capacity to transport sediment with the input of sediments and their caliber results in distinctive channel morphologies (Church 2002). Bank strength, which is influenced by sediment texture and vegetation, large wood, and other channel constraints, exerts additional influence over channel shape (Fig. 3.16).

As one proceeds downslope from uplands to upland valley to large river, the river changes from primarily a sediment-evacuating to a sediment-accumulating system, and from being coupled to hillslopes for its sediment supply to being largely uncoupled. The upland region is closely associated with hillslopes, from which upland streams receive sediments (Fig. 3.17). Much of these sediments, but not the largest clasts, are transported downstream. In the middle section (the upland valley), where gradients are lower, sediments mobilized in upland channels may be deposited, forming an alluvial channel and a floodplain. Material from this section is transported onward, and replenished from upstream, in episodes of erosion and deposition. In large lowland rivers near the distal end of the drainage system, sediment deposition is dominant, resulting in large floodplains, alluvial fans, and deltas.

A series of inter-related changes in stream flow and sediment character occur systematically along the river's length. As previously mentioned, channel gradient declines and stream discharge increases. The discharge-slope product, stream power, is greatest in the mid-range of the river system. Sediment grain size is largest in the headwaters, where large clasts introduced from hillslopes often have diameters equal to or greater than bankfull depth and are immobile even at highest flows. Smaller material is transported downstream in cycles of erosion and deposition that sort particles and carry farthest those of smallest size. Hence characteristic sediment size becomes finer as one proceeds downstream. Sorting of stones also can result in predicable arrangements that enhance their stability. Gradient and relative roughness (grain diameter divided by depth) can be combined to predict specific details of bed morphology (Montgomery and Buffington 1997; Church 2002). For example, the pool-riffle sequence of gravel bed streams, with point bars at bends, typically occurs at a slope near 0.01–0.03 and a relative roughness near 0.3. Near the downstream terminus, sand-bed channels are found; because of their small grain size, bed material is mobile over a wide range of flows. Visually this can produce a pleasing picture of ripples and dunes, but for most organisms, bed instability makes a hostile environment. As Benke et al. (1985) have shown, most biological production in sand-bed rivers is associated with submerged wood, the only stable habitat for invertebrates.

Fig. 3.15 The flow-sediment balance relationship depicts how channels adjusts in response to changes in water or sediment yield. A conceptualization by Lane (1955), it illustrates how channel equilibrium reflects a balance among stream discharge (Q_w), channel slope (S), sediment load (Q_s) and sediment particle size (D50) (Reproduced from Dust and Wohl 2012)

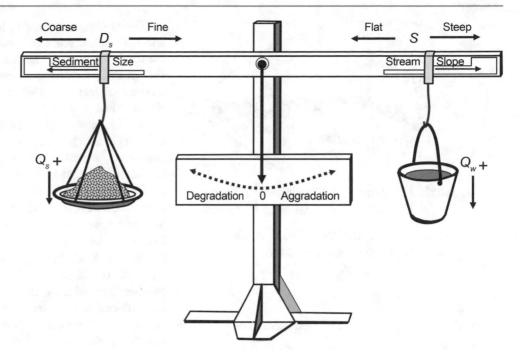

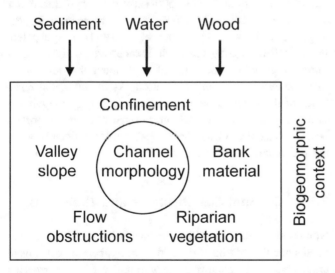

Past, present and future inputs of

Sediment   Water   Wood

Fig. 3.16 The principal controls on channel morphology include sediment supply (amount and size), transport capacity (discharge magnitude and frequency), and riparian vegetation. Channel morphology is further influenced by flow obstructions (bedrock outcrops, large wood), geomorphic context (confinement and valley slope), and disturbance history. Sediment supply and transport relationships govern channel type (Reproduced from Montgomery and MacDonald 2002)

These ideas underlie a process-based classification of channel morphologies for mountain drainages (Montgomery and Buffington 1997). In an idealized long profile from upper hillslopes to lowland river, these channel types occur in longitudinal sequence (Fig. 3.18). Cascade channels occur on steep slopes, are confined by valley walls, and have a substrate that typically consists of cobbles and boulders. Pools commonly are small and spaced less than one channel width apart. Cascade channels retain larger clasts but rapidly transport smaller sediments to lower-gradient channels. In step-pool channels, longitudinal steps are formed by larger clasts, resulting in discrete pools with a spacing of one to four channel widths. Gradients are steep, width-to-depth ratios are low, and valley wall confinement is evident. At moderate to high gradients and in relatively straight channels, the channel type known as plane bed develops. It usually includes a somewhat featureless combination of riffles, runs, and rapids. The substrate includes gravel, cobbles and small boulders, and may be armored or not. An armored substrate indicates a transport capacity greater than sediment supply, whereas a substrate that is not armored indicates a balance between transport capacity and sediment supply (Dietrich et al. 1989). Thus, the plane bed stream is transitional between supply-limited upper reaches and transport-limited lower reaches. Pool-riffle channels with pools spaced at five to seven channel widths occur at moderate gradients. They represent a shift to a more transport-limited system, although this varies with degree of armoring. Dune-ripple channels typically are found in low-gradient, sand-bed systems, experiencing significant sediment mobility under most flow conditions.

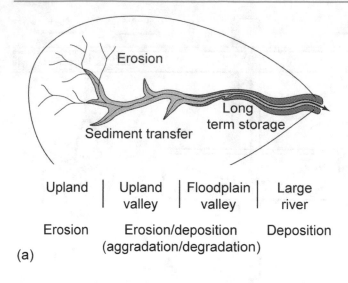

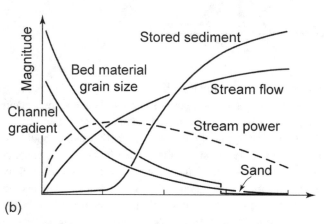

Fig. 3.17 A drainage basin, illustrating some of the principal longitudinal trends in stream channels. (**a**) Three principal longitudinal zones: an upland zone where the drainage forms and from which sediments are exported; a middle, transitional zone where erosion and deposition of sediments may be approximately in balance; and a lower floodplain where sediments may accumulate. (**b**) General patterns of sediment occurrence and transport. Note the longitudinal decrease in bed material grain size. Stream power, the product of slope and discharge, peaks in the middle zone, and the stream's competence (ability to move sediment of a given size) declines downstream (Reproduced from Church 2002)

3.3.2 Channel Dynamics Over Long Timeframes

The prior discussion describes how changes in geomorphic processes along the river's length help us understand the development of different channel configurations and features. It emphasizes how adjustments are continually occurring to maintain an approximate equilibrium, and it emphasizes events of intermediate magnitude—the effective or dominant discharge—as primarily responsible. Taking a historical view, we should also emphasize the importance of episodes of climate change over the past 15,000 years, and of even older tectonic events and glaciations. For example,

the Grand Canyon of the Colorado River is thought to be about 5 million years old and its establishment to depend ultimately on tectonic movements and the opening of the Gulf of California. Both ancient and more recent floods have on occasion had lasting effect on fluvial landscapes, carving channels and placing large clasts that subsequent floods are unable to substantially modify (Knighton 2014). River channels in a large area of the western US were shaped by a paleoflood estimated to have discharges as high as 10 million m^3 s^{-1} (Benito 1997), which occurred when an ice dam on glacial Lake Missoula failed some 15,000 years ago. Researchers in paleoflood hydrology have documented more than two dozen floods greater than 100,000 m^3 s^{-1} during Quaternary time, mainly from breaches of glacial-age dams that blocked large drainage basins, or from breaches of other types of natural dams (O'Connor and Costa 2004). Less spectacularly, climatic fluctuations since 10,000 years ago have affected water balances, vegetation patterns, flows, and the supply of materials, resulting in fluctuations in fluvial activity. Stratigraphic analysis of flood deposits left by the Colorado River in the Grand Canyon, Arizona, provided evidence of at least 15 large floods over the last 4500 years, including one flood some 1600-1200 years ago that had a discharge exceeding 14,000 m^3 s^{-1}, more than twice the largest gaged flood (O'Connor et al. 1994). Large paleofloods can have long-lasting influence over a river's geomorphology with timescales of adjustment that are longer than modern records of observation. As the erosional resistance of the channel boundaries increases, high magnitude, infrequent floods become progressively more important because only these floods are capable of significantly altering the channel configuration.

3.3.3 Channel Classifications and Their Uses

Numerous stream classification systems have been developed since the middle of the 20th century, both as a means to understand and categorize the variability of river systems seen in nature, and for practical uses including context for evaluating human impacts and developing management prescriptions (Melles et al. 2012; Kondolf and Piegay 2016). Early approaches from a biological perspective favored longitudinal classifications because distinctive faunal assemblages, especially fishes, could be associated with trends in stream size, slope and other physical variables along a river's length (Illies and Botosaneanu 1963). Some useful physical classifications have already been discussed, including stream order, the three longitudinal zones of erosion, transfer, and deposition, and whether a channel is straight, meandering or braided. Channels can be divided into bedrock, colluvial, and alluvial beds, and the latter further subdivided (at least for mountain drainages) into

Fig. 3.18 River channel types occur in succession along the river's profile due to complex interactions governed by slope (S), sediment supply, trapping of sediments by large wood in the channel, and other factors. Although thresholds may be difficult to detect, certain channel features prevail over a substantial distance, referred to as a process domain. This model was originated for small mountain streams in the Western US (Reproduced from Montgomery and Buffington 1997)

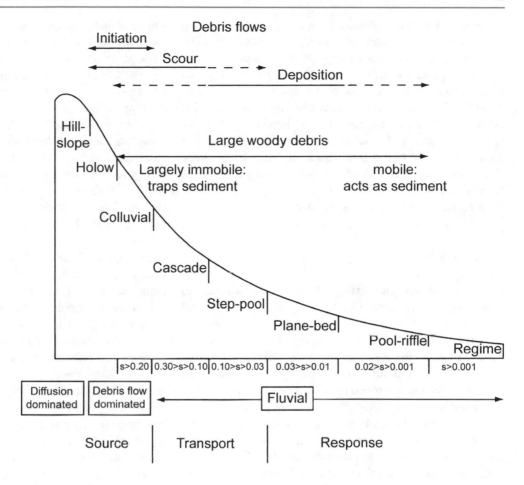

types determined by transitions in process domains and forcing by large wood. Floodplain rivers have been characterized based on river planforms (Kellerhals and Church 1989). A classification scheme devised by Rosgen (1994, 1996) and widely used in river restoration recognizes seven major and 42 minor channel types based on channel pattern, width-to-depth ratio, sinuosity, and bed material size. Most recently, a number of investigators have drawn from the fields of geographic information science and landscape ecology to create multi-scale, spatially homogenous aquatic ecoregions (Olden et al. 2012; Melles et al. 2014). These result in landscape-scale hierarchical classifications wherein, ideally at least, upper levels can be subdivided into smaller and smaller regions at finer spatial scales (e.g., Brierley and Fryirs 2000; Snelder and Biggs 2002).

Geomorphic classifications generally are hierarchical. From largest to smallest scale, the sequence is from ecoregion or landscape; then to valley segments, typically defined as the river segments between tributary confluences, and including both corridor and floodplain; then to channel reaches, frequently including pools, riffles, etc.; and at the finest scale, the microhabitats recognizable partly by physical conditions of substrate and flow, and partly by the organisms therein. The state of lower hierarchical units is influenced by controls at upper levels, but not the converse. Many useful classifications exist that cluster streams and rivers into types based on similarities in environmental setting, size, position in the river network, habitat types, species assemblages, and more, serving as useful descriptive typologies and also suggestive of some similarities in controlling variables. Such classifications benefit managers by providing geomorphologically based guidance for channel maintenance and by facilitating inventories of stream condition, for example to prioritize conservation and restoration actions. However, as will be discussed in more detail in a subsequent section, over-reliance on classification as a guide to restoration has received substantial criticism (Kondolf et al. 2016).

3.4 Applications of Fluvial Geomorphology

Applications of fluvial geomorphology to river restoration are many and diverse (Kondolf and Piegay 2016). Wherever human actions have altered the balance between river flow and sediment supply, an ecologically degraded river is the frequent outcome. The following sections review two of the most influential means by which this occurs: sediment trapping and flow modification by dams, and gravel mining in the channel and floodplain. Both alter the balance between discharge and sediment supply, typically causing a river to be starved of sediment. As a consequence, the fluvial system is likely to experience erosion of the channel bed and banks as well as channel incision and coarsening of bed material, potentially causing degradation of habitat quality for spawning fishes and benthic fauna (Kondolf 2004). But these are just two of the many human activities that degrade rivers, thus we conclude with a broader consideration of river restoration, evaluation of its effectiveness, and an exploration of how it may be improved.

3.4.1 Dams

Regardless of size and setting, all dams trap sediment to some degree, and most alter the magnitude and seasonal distribution of flows, thereby altering sediment transport dynamics (Kondolf 1997). Trapping of sediments and modification of peak flows are two main pathways by which dams and reservoirs alter geomorphic processes in rivers. Dam removal and controlled floods are two main approaches for eliminating or mitigating these impacts. Sediment deposition occurs upstream of a dam as all bedload sediment and much of the suspended load are retained within the reservoir and upstream in backwater reaches. Downstream, water released from the dam carries little or no sediment but has the energy to do so, hence is referred to as "sediment-hungry". Although effects vary widely among river systems, in general dams and their reservoirs bring about reduced sediment flux downstream of the dam, which may cause destabilization of banks, channel incision, and coarsening of the bed material until equilibrium is reached and the remaining material cannot be moved by the flows. The magnitude of incision can vary greatly depending upon reservoir operations, channel characteristics, and bed material size. Incision may be minimal if flood peaks are dramatically reduced, and as much as several meters or more in sand bed rivers. When significant down-cutting does occur, gravels and finer materials often are removed, leaving behind coarser deposits of large gravel, cobbles, or boulders in a compacted or armored layer that may be less suitable for nest construction by spawning fishes and as biological habitat.

Artificial ("induced") floods are a recent innovation for highly regulated rivers, intended to restore physical habitat and improve biological condition and water quality (Olden et al. 2014). These can be expensive experiments. Four large controlled floods from Glen Canyon Dam costing $12 million in foregone power revenue were undertaken primarily to rebuild sandbars along the river margins, thereby adding backwater habitat for juvenile fishes. Without an adequate

supply of fine sediments and annual floods sufficient to scour fine sediment from the bed, lateral sandbars gradually erode. Induced floods to redistribute sediment that seasonally accumulates on the river bed and build lateral sandbars have proven effective, but changes can be short-lived as interflood dam operations erode sandbars within months to years (Mueller et al. 2014). Thus, maintaining sandbars in the post-dam, sediment-depleted river will require frequent controlled floods, and flood frequency must be based on the supply rate of fine sediment from tributaries and upstream sources. Sediments to the Colorado River below Glen Canyon dam are mainly from two tributaries, the Paria and Little Colorado Rivers, which enter the River some 25 and 100 km downstream of the dam, and cumulatively provide about 6–16% of the pre-dam sand supply. Sediment inputs from these tributaries usually are driven by summer or fall thunderstorms rather than spring snowmelt, and so controlled floods should be timed accordingly for maximum effect in building sand bars.

Complete removal of a dam is another means to restore a more natural flow and sediment regime. Decommissioning of dams is increasing, primarily for safety and economic reasons. Of the more than 79,000 dams in the United States, 3,500 have been rated as unsafe, and over 600 dams have been removed (Doyle et al. 2008). The largest dam removal to date, the phased removal of the Elwha and Glines Canyon Dams (32 and 64 m in height, respectively, Fig. 3.19) on the Elwha River, Washington, is a remarkably well documented case study with extensive pre- and post-dam removal field studies, laboratory flume experiments, predictions of system response, a sediment budget and more (Duda et al. 2008; Warrick et al. 2015). During the first two years of the project (September 2011–September 2013), a sand and gravel sediment wave dispersed down the river channel, filling channel pools and floodplain channels, aggrading the river channel by about1 m, greatly reducing sediment grain sizes in the channel, and depositing over 2 million m^3 of sand and gravel on the seafloor offshore of the river mouth. The rate of sediment erosion as a percent of storage was greater in the Elwha River during the first two years of the project than in other dam removals, indicating that steep, high-energy rivers have enough stream power to export volumes of sediment deposited over several decades in only months to a few years. There is still much to learn from this real-world experiment, as roughly two-thirds of the combined reservoir sediments remained within the former reservoirs, and so sediments will continue to be released into the foreseeable future.

The effects of dam removal vary widely for a number of reasons, including whether dam removal is rapid or phased over months to years, the amount and nature of stored sediments, such as gravel versus sand and fines. and channel gradient. Rapid drawdown (complete evacuation in 90 min)

Elwha Dam
est. 1913, 32 m tall
~5 million m³ sediment

(a)

Glines Canyon Dam
est. 1927, 64 m tall
~16 million m³ sediment

(b)

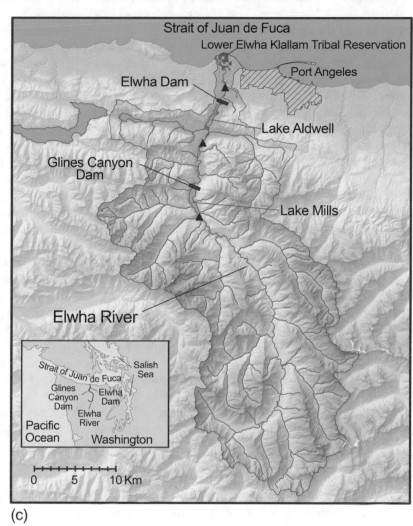

(c)

Fig. 3.19 (**a–b**) Photographs of the two former dams on the Elwha River, Washington USA, prior to decommissioning. (**c**) Map of the Elwha River watershed, showing the watershed, the locations of the dams, USGS gaging stations (triangles), the original setting (insert), and other boundaries (Reproduced from Warrick et al. 2015)

of the 38-m tall Condit dam on the White Salmon River, a tributary of the Columbia River in the Pacific Northwest, US, caused over a third of stored sediments to be exported in 6 days, and ~60% of the predominantly fine-grained material was exported by 15 weeks (Wilcox et al. 2014). In contrast, sediment export may be minimal in low gradient systems due to insufficient stream power, and go on for decades if exposed stream banks are highly erodible.

Although not in wide use, methods are available to pass sediment through or around reservoirs without dam removal. Depending on specifics of valley and dam shape, it may be possible to bypass sediment around the reservoir, employ rapid drawdown to transport sediment through the reservoir, and vent turbid waters by directing currents through the dam. Ideally, rates and timing of sediment passage would be similar to pre-dam conditions (Kondolf et al. 2014). To retrofit dams with sediment passage facilities after they are built is expensive or infeasible, so as dams continue to be constructed it will be important to make sediment management part of the design.

Addition of gravel to river channels, referred to as gravel augmentation or replenishment, is another strategy to compensate for the trapping of gravel in upstream reservoirs. The intent is to reduce the size of bed material to improve spawning habitat, rebuild bar-pool topography to increase habitat diversity, and increase the flow of oxygenated water into the sediments. Gravel augmentation is frequently done to improve spawning conditions for salmonids, usually in highly regulated rivers. In the Mokelumne River, California, Chinook salmon (*Oncorhynchus tshawytscha*) began spawning at a previously unused site within two months after gravel placement and continued to use the site during the three spawning seasons encompassed by the study (Merz and Setka 2004). Macroinvertebrates quickly colonized new gravel as well (Merz and Chan 2005).

Gravel augmentation presents a number of technical challenges, including the amount of gravel to be added, grain size, frequency and timing of augmentation, and whether the material is placed within the channel bed or along channel margins to allow floods to redistribute the material. The river's topography is expected to adjust over time, particularly to episodic high flows, and whether the rehabilitated spawning habitat will persist through a flood event is uncertain. Surveys of bed topography in the Mokelumne River following very high flows indicated a well-designed project, as roughly 75% of the spawning habitats were of equal or higher quality when compared with pre-flood conditions (Wheaton et al. 2010).

3.4.2 Gravel Mining

The extraction of sand and gravel from alluvial rivers is common practice in many parts of the world, owing to ease of access and the fact that material generally is well sorted, reducing costs to quarry operators. Gravel mining disrupts the balance between sediment supply and fluvial transport capacity, typically inducing incision upstream and downstream of the extraction site and potentially undermining bridges and other built infrastructure. It is prohibited or being reduced in a number of European countries, permitted under various federal and state regulatory procedures in the US, and either unregulated or carried out illegally in a number of countries (Kondolf 1997; Meador and Layher 1998).

An assessment of channel adjustments in five major rivers in southern Italy identified three distinct phases over the past 150 years, primarily in response to human disturbances (Scorpio et al. 2015). Land use change, dams, and other hydraulic interventions, as well as in-channel sediment mining, all contributed. Slight channel widening dominated from the last decades of the nineteenth century to the 1950s, likely because of deforestation and increased sediment delivery. From the 1950s to the end of the 1990s, decreased sediment availability brought about by in-channel mining and river control works resulted in pronounced channel narrowing (up to 96%) (Fig. 3.20), accompanied by moderate to very intense incision (up to 6–7 m), and changes in channel configuration from multi-threaded to single-threaded patterns. Most recently, the period from around 2000–2015 has been characterized by channel stabilization and local widening. The cessation of in-channel mining, implying higher in-channel sediment supply, coupled with increased sediment mobility due to episodic larger floods, likely is responsible for the trend towards channel stabilization and recovery.

Instream gravel mining is an important cause of channel degradation in some areas of South America, especially in rivers located near rapidly growing cities and where no river management strategies exist. The Maipo River flowing through the metropolitan region of Santiago, Chile, is an example where currently unregulated gravel mining is continuing to expand, causing incision and channel narrowing, and threatening not just aquatic habitat but valuable infrastructure in a city of 6.5 million (Arróspide et al. 2018). A sustainable level of gravel mining has sometimes been estimated to be all or some fraction of baseload transport, although this can be problematic in part because baseload transport is difficult to estimate (Sect. 3.2.4), and in part because capturing 100% of baseload transport may be excessive. An extraction rate many times higher than estimated bedload, as was documented for the Avon River in southeastern Australia, caused prolonged channel instability as the river replenished its gravel supply through bank erosion (Davis et al. 2000).

3.4.3 River Restoration

River restoration is a widely used umbrella term for a variety of management practices aimed at improving the natural state and functioning of river systems in support of ecological and social goals. Either implicitly or explicitly, the focus of river restoration often has been to return to a historic and presumed pristine state. However, a broader view might consider a range of possible outcomes based on human uses and benefits, cultural preferences, and a sustainable endpoint in light of both human and natural constraints (Dufour and Piegay 2009). This perspective, perhaps inevitable for landscapes with a long history of human intervention such as Western Europe, may be in tension with the meaning of restoration, and helps explain why some prefer the terms rehabilitation or simply river management to describe improvement over a degraded condition. But because even within this more limited (and realistic) vision it remains important to restore natural processes to the extent compatible with human uses, many prefer the continued use of the term restoration.

In recent years, the focus of river restoration has shifted from an emphasis on channel shape and form to one emphasizing restoration, to the extent feasible, of river function (Wohl et al. 2005). This latter perspective is clear in synthesis papers describing the natural flow regime (Poff et al. 1997), the natural sediment regime (Wohl et al. 2015), and the natural wood regime (Wohl 2019), and in the geomorphological toolkit of Kondolf and Piegay (2016). Most

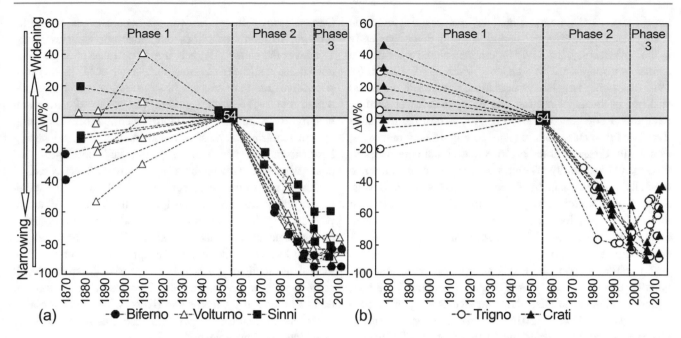

Fig. 3.20 Adjustments in channel width over 150 years in five Italian rivers, expressed as change from 1954 used as a reference point. Width estimates were derived from various sources including aerial photograph. Channel narrowing after the 1950s also transformed most channels from braided to sinuous single-thread channels (Reproduced from Scorpio et al. 2015)

restoration projects involve managing flows, channel dynamics, and habitat elements within the stream channel, often referred to as hydrogeomorphic river management. Flow management approaches, often referred to as environmental flows or e-flows, were discussed in Chap. 2. In this chapter, we provide some examples of management based on geomorphology and habitat enhancement. The number of projects described as river restoration has grown rapidly over the past several decades, and there has been considerable controversy regarding its successes and failures. We return to this subject in Chap. 15, in discussing river management broadly, and needs for the future.

Many restoration projects are intended to improve conditions for the biota, and seek to improve habitat by adding wood, boulders, or gravel to create flow heterogeneity, add shelter, stabilize the stream bed against scour, and so on.

Surveys of habitat quality, and the abundance and diversity of fish, macroinvertebrates, or aquatic macrophytes before and after project completion are common measures to evaluate success. At least in the short-term, results often are positive. An analysis of 24 individual studies of macroinvertebrate response to increasing habitat heterogeneity found significant, positive effects on number of species, although density increases were negligible (Miller et al. 2010). Additions of large wood produced the largest and most consistent responses, whereas responses to boulder additions and channel reconfigurations were positive but highly variable. Similarly, a review of wood addition to agricultural streams in Australia reported enhanced diversity of fish and macroinvertebrates, increased storage of organic material and sediment, and improved bed and bank stability (Lester and Boulton 2008). Most recently, Kail et al. (2015) summarized 91 river restoration projects, generally finding positive effects of river restoration on fish, macroinvertebrates, and aquatic macrophytes in response to instream habitat improvements and river widening, respectively. Most studies are of short duration, however, and longer-term studies have reported a tendency for benefits to fade over time (Feld et al. 2010; Kail et al. 2015). Variability in response may be associated with differences in land use, river width, and age of projects, as well as failing to take into account how catchment-scale processes may influence local outcomes. In a review focused on stream rehabilitation techniques (note how the terminology varies among authors), Roni et al. (2008) reported that reconnection of isolated habitats, floodplain rehabilitation, and instream habitat improvement have frequently proven effective for improving habitat and increasing local fish abundance. However, the short duration and limited scope of most published evaluations makes it difficult to reach firm conclusions, especially for other common practices including riparian rehabilitation, road improvements to reduce wash-in of sediments, dam removal, and restoration of natural flood regimes.

Because many lowland rivers have been channelized to recover land for agriculture and route floodwaters downstream, re-meandering of rivers is a frequent type of river restoration. The re-meandering of the River Skjern, the largest river in Denmark, illustrates a successful outcome (Feld et al. 2010). Prior to regulation works that began early in the 20th century, the River Skjern meandered dynamically across its floodplain with a channel width of 65–100 m. By 1968, following large-scale regulation, river width was fixed at 30 m in an upper section and 45 m in a lower section. Restoration completed in 2003 sought to turn the degraded river back to a more natural state, adding 46 new meanders; as a result, cross-sectional area decreased and current speed increased substantially. Within two years macroinvertebrate diversity and abundance had reached pre-restoration levels. Success of this project may be attributed to several factors: the original river form was meandering, the project was at large scale, and no other factors seriously impaired the river's response.

To restore a river one needs a target endpoint, and frequently this relies on identifying a reference condition, although what constitutes the reference is subject to controversy (Kondolf et al. 2016). A pristine reference site or historical condition serves the ideal of preserving natural conditions, and either implicitly or explicitly, the focus of river restoration often has been to return to a historic and presumed-pristine state. However, that target may be unattainable due to lack of historical information or a suitable reference site, as well as conflicting river and landscape uses. A broader view might consider a range of possible outcomes based on human uses and benefits, cultural preferences, and a sustainable endpoint in light of both human and natural constraints (Dufour and Piegay 2009).

One vision of these trade-offs is illustrated in Fig. 3.21 (Kondolf 2011), where the dynamic interplay of sediment supply and river discharge are the central challenges. The range of river management options varies along two axes, one that captures total erosion risk (sediment supply and stream power), and a second that captures human dominance of the corridor and floodplain. Where both the river's capacity to erode and human presence are greatest, society may elect hard engineering to protect infrastructure, and construct whitewater parks for recreation. Tamer rivers in urban settings may be restored for maximum human benefit with trails, canoe liveries, and so on. In more natural settings, high-energy rivers may be given space for natural processes of flooding, erosion, and channel migration to take place. Where flow regime and sediment supply are insufficient for self-repair, designed channels and habitat features may be the best approach. This illustrates that river management must address the context of the river or river segment and understand the geomorphologic processes at work to ensure a sustainable outcome (Piégay et al. 2016).

Despite shortcomings described above, river restoration is a rapidly advancing field that is informed by, and informs, fundamental river science in diverse fields. Lessons are learned from failures as well as successes as a rapidly growing literature describes outcomes and debates goals. Whatever the stated goal it is important to be clear on what constitutes a successful outcome. Palmer et al. (2005) proposed five criteria of success from an ecological perspective: a plan that begins with a guiding image of a dynamic and healthy river, measurable improvement, a self-sustaining system requiring at most minimal maintenance, no lasting harm from the construction phase, and both pre- and post-assessment with public sharing of data. Success can be evaluated from other perspectives as well, including stakeholder satisfaction and lessons learned that can inform future projects (Fig. 3.22). Jansson et al. (2005) added a criterion emphasizing science and hypothesis testing: the need for specific hypotheses and/or a conceptual model of the ecological mechanisms by which the proposed activities will achieve their target. Given the investment of billions of dollars in restoring streams and rivers, widespread adoption of targets, evaluation metrics, and standards for success are of undeniable importance.

3.5 Summary

Fluvial geomorphology emphasizes the dynamic interplay between rivers and landscapes in the shaping of river channels and drainage networks. This includes study of the linkages among channel, floodplain, network, and catchment using a diversity of approaches. These include stratigraphic analyses, experimental studies of sediment transport in flumes, modeling of physical processes, comparisons of landforms, and sophisticated statistical approaches to gain greater understanding of the physical dynamics of river systems. These analyses help make sense of the enormous variety exhibited among fluvial systems, and thus the habitat and environmental conditions experienced by the biota. Quantification of the relationships among river features and analysis of the underlying processes contribute to a deeper understanding of how rivers respond to human-induced changes in water and sediment supply that can cause rivers to change their shapes.

A central theme in fluvial geomorphology is that the development of stream channels and entire drainage networks, and the existence of various regular patterns in the shape of channels, indicate that rivers are in dynamic equilibrium between erosion and deposition, and governed by common hydraulic processes. A river system's discharge, sediment load, and elevational extent from origin to base are controlling or independent variables that the river cannot control, and therefore must adjust to. Channel width and depth, velocity, grain size of sediment load, bed roughness, and the degree of sinuosity and braiding are other variables

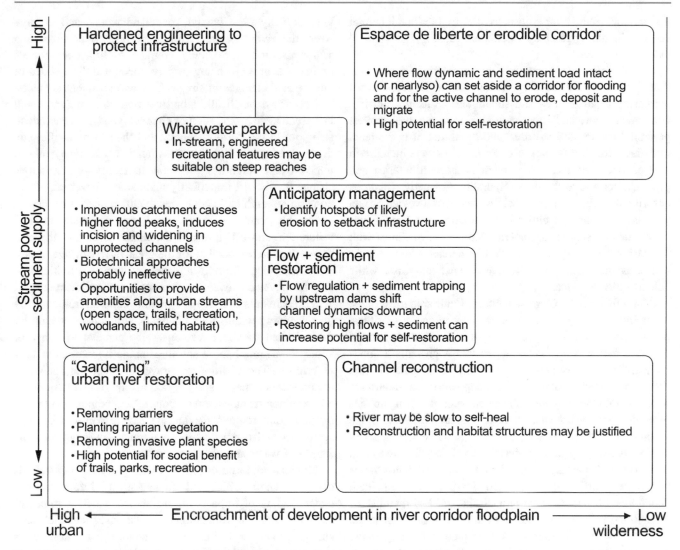

Fig. 3.21 Setting goals in river restoration, based on the degree to which the river still retains its dynamic flow regime and sediment supply, and the degree to which it is constrained or not by land use and infrastructure (Reproduced from Kondolf 2011)

Fig. 3.22 The three primary axes of effective river restoration. In addition to the five attributes of ecological success, it is important to consider stakeholder success reflecting human satisfaction with restoration outcome, and learning success reflecting advances in scientific knowledge and management practices that will benefit future restoration action (Reproduced from Palmer et al. 2005)

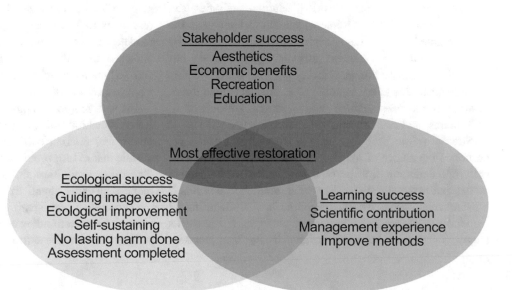

that interact as the river adjusts to the variables that it cannot control, including discharge, sediment load, and elevational extent.

The drainage basin encompasses a network of channels that join with others downstream in a progression of increasing drainage area and stream size. Stream order is a convenient shorthand for stream size, in which the smallest perennial stream is first-order, and the union of two stream of order n forms a stream of order $n + 1$. Rivers increase in size as one proceeds downstream because tributaries and groundwater add to the flow. Hydraulic geometry equations describe the relationships of width, depth, and velocity to an increase in discharge either downstream or as flow varies over time at a station. In general, depth increases most with increasing discharge at a station, whereas width usually increases more than depth in the downstream case, while velocity increases least.

Many features of river channels are familiar to most of us, including sinuosity or meandering, an alternation of riffles, pools, and runs, and the presence of a floodplain provided the river valley is not so V-shaped that it prevents a floodplain from being formed. These features are formed by the river through cycles of erosion and deposition that in turn are strongly influenced by the balance between the twin supplies of water and sediments. A river's sediment load is the amount of sediment passing a point over some time interval, and includes very fine material that likely is always in transport, and coarser material from the bed and banks that is transported either as suspended load or as bed load, depending on particle size and discharge. The quantity of transported sediment increases with velocity and discharge, but flow events of intermediate frequency actually move more sediment over the years, because extreme events are infrequent. The dominant or effective discharge is that at which sediment transport is greatest, and it often is approximately the bankfull flood. Human activities can increase or reduce sediment yield, which is the area-weighted load. Due to erosion brought about by changing land use, sediment flux into global rivers has increased, while sediment delivery to the world's coasts have declined due to the trapping of sediments behind dams. Some consequences include coastal retreat, subsidence of river deltas, and loss of coastal wetland habitat.

Stream power, the product of discharge and slope, describes the ability of the stream to mobilize and transport material. Sediment transport is directly related to stream power and inversely related to median grain size, and this is a useful relationship for understanding how a stream might respond to changes in sediment and water supply along its length, or due to human interference. As one proceeds longitudinally from upland, to upland valley, and to large river, the river changes from exporting to accumulating sediments, and from being coupled to hillslopes for its sediment supply, to being largely uncoupled. The series of inter-related changes in stream flow and sediment character that occur systematically along the river's length can result in a predictable progression of channel types from cascade to step-pool to planform channels, and then to pool-riffle and dune-ripple types. But this classification, like many others, imposes discontinuities on what in reality is continuous change that is still imperfectly understood. River classification as a research topic continues to attract interest because it undeniably can be very useful for management and restoration, and it is a test of our understanding of the processes that are responsible for the great variety of river types.

Rivers are profoundly affected by human activities that alter the balance between water supply and sediment supply. When sediments are trapped behind dams or removed by gravel mining within river channels, subsequent erosion by the sediment-starved river causes the channel to incise or widen, damaging habitat and threatening infrastructure such as bridges. The addition of gravel to meet the spawning requirements of fishes and provide habitat of invertebrates is one common management response. Less common but more dramatic, dam removal can allow the river to evacuate stored sediments and restore a more natural balance between the supply of water and sediments.

Fluvial geomorphology has many practical applications to river engineering and management, informing our understanding of impacts due to altered water or sediment supply, human modifications to the river channel, and changing land use within the catchment. River restoration is a widely used umbrella term for a variety of management practices aimed at improving the natural state and functioning of river systems in support of ecological and social goals. Whether the desired outcome is a river system that resembles its historical or natural state, or one that accommodates human presence in a developed landscape, an understanding of channel responses to the fluxes of water and sediments provides the underlying physical basis for management decisions.

Finally, geomorphology provides insight into the channel features, habitat units, surface and subsurface zones, floodplains, and riparian corridors that form a complex, shifting mosaic that in turn provides the setting in which biologically diverse communities flourish. When this complexity is reduced by dams, channelization, and regulation of river flow, the subsequent homogenization of habitat causes declines in taxon richness, reminding us of the importance of management actions that view the entire river basin as an integrated unit.

References

Arróspide F, Mao L, Escauriaza C (2018) Morphological evolution of the Maipo River in central Chile: influence of instream gravel mining. Geomorphology 306:182–197. https://doi.org/10.1016/j.geomorph.2018.01.019

Benito G (1997) Energy expenditure and geomorphic work of the cataclysmic Missoula flooding in the Columbia River Gorge, USA. Earth Surf Process Landforms 22:457–472

Benke AC, Henry RL, Gillespie DM, Hunter RJ (1985) Importance of snag habitat for animal production in southeastern streams. Fisheries 10:8–13. https://doi.org/10.1577/1548-8446(1985)010%3c0008:ioshfa%3e2.0.co;2

Bilby R, Bisson PA (1998) Function and distribution of large woody debris. In: Naiman RJ, Bilby R (eds) River ecology and management. Springer, New York, pp 324–346

Bledsoe BP, Watson CC (2001) Effects of urbanization on channel instability. J Am Water Resour Assoc 37:255–270

Brierley GJ, Fryirs K (2000) River styles, a geomorphic approach to catchment characterization: implications for river rehabilitation in Bega catchment, New South Wales, Australia. Environ Manage 25:661–679. https://doi.org/10.1007/s002670010052

Buffington JM, Woodsmith RD, Booth DB, Montgomery DR (2003) Fluvial processes in Puget Sound rivers and the Pacific Northwest. In: Montgomery DR, Bolton S, Booth DB, Wall L (eds) Restoration of Puget Sound rivers. University of Washington Press, Seattle, pp 46–78

Charlton R (2007) Fundamentals of fluvial geomorphology. Routledge, New York

Church M (2002) Geomorphic thresholds in riverine landscapes. Freshw Biol 47:541–557

Davis J, Bird J, Finlayson B, Scott R (2000) The management of gravel extraction in alluvial rivers: a case study from the avon river Southeastern Australia. Phys Geogr 21:133–154. https://doi.org/10.1080/02723646.2000.10642703

Dietrich W, Kirchner J, Ikeda H, Iseya F (1989) Sediment supply and the development of the coarse surface layer in gravel-bedded rivers. Nature 340:215–217

Doyle MW, Shields D, Boyd KF et al (2007) Channel-forming discharge selection in river restoration design. J Hydraul Eng 133:831–837. https://doi.org/10.1061/(ASCE)0733-9429(2007) 133:7(831)

Doyle MW, Stanley EH, Havlick DG, et al (2008) Environmental science: aging infrastructure and ecosystem restoration. Science (80-) 319:286–287. https://doi.org/10.1126/science.1149852

Duda J, Freilich J, Schreiner E (2008) Baseline studies in the Elwha River ecosystem prior to dam removal: Introduction to the special issue. Northwest Sci 82:1–12

Dufour S, Piegay H (2009) From the myth of a lost paradise to targeted river restoration: Forget natural references and focus on human benefits. River Res Appl 25:568–581

Dust D, Wohl E (2012) Conceptual model for complex river responses using an expanded Lane's relation. Geomorphology 139–140:109–121. https://doi.org/10.1016/j.geomorph.2011.10.008

Feld C, Birk S, Bradley DC, Other (2010) From natural to degraded rivers and back again: a test of restoration ecology theory and practice. Adv Ecol Res 44:119–210

Florsheim J, Mount JF, Chin A (2008) Bank erosion as a desirable attribute of rivers. Bioscience 59:519–529

Gleason CJ (2015) Hydraulic geometry of natural rivers: a review and future directions. Prog Phys Geogr 39:337–360. https://doi.org/10.1177/0309133314567584

Gordon ND, McMahon TA, Finlayson BL et al (2004) Stream hydrology: an introduction for ecologists. Wiley, Chichester, West Sussex

Gore JA, Shields FD Jr (1995) Can large rivers be restored? Bioscience 45:142–152

Grams PE, Schmidt JC, Topping DJ (2007) The rate and pattern of bed incision and bank adjustment on the Colorado River in Glen Canyon downstream from Glen Canyon Dam, 1956–2000. Bull Geol Soc Am 119:556–575. https://doi.org/10.1130/B25969.1

Gurnell A (2014) Plants as river system engineers. Earth Surf Process Landforms 39:4–25. https://doi.org/10.1002/esp.3397

Hicks DM, Gomez B (2016) Sediment transport. In: Kondolf GM, Piégay H (eds) Tools in fluvial geomorphology. Wiley, West Sussex, pp 324–356

Hohensinner S, Lager B, Sonnlechner C (2013) Changes in water and land: the reconstructed Viennese riverscape from 1500 to the present. Water Hist 5:145–172

Horton RE (1945) Erosional development of streams and their drainage basins: hydrophysical approach to quantitative morphology. Bull Geol Soc Am 56:275–370

Illies J, Botosaneanu L (1963) Problèmes et méthodes de la classification et de la zonation écologique des eaux courantes, considerées surtout du point de vue faunistique. Mitteilungen Int Vereinigung für Theor und Angew Limnol 12:1–57

Jansson R, Backx H, Boulton A, et al (2005) Stating mechanisms and refining criteria for ecologically successful river restoration: a comment on Palmer et al. (2005). J Appl Ecol 42:218–222

Kail J, Brabec K, Poppe M, Januschke K (2015) The effect of river restoration on fish, macroinvertebrates and aquatic macrophytes: A meta-analysis. Ecol Indic 58:311–321. https://doi.org/10.1016/j.ecolind.2015.06.011

Kellerhals R, Church M (1989) The morphology of large rivers: characterization and management. In: Dodge DP (ed) Proceedings of the International Large River Symposium. Canadian Special Publication in Fisheries and Aquatic Sciences 106, pp 31–48

Kleinhans MG (2010) Sorting out river channel patterns. Prog Phys Geogr 34:287–326. https://doi.org/10.1177/0309133310365300

Knighton D (2014) Fluvial forms and processes: a new perspective, 2nd edn. Routledge, London

Kondolf GM (1997) Hungry water: effects of dams and gravel mining on river channels. Environ Manage 21:533–551. https://doi.org/10.1007/s002679900048

Kondolf GM (2004) Assessing salmonid spawning gravel quality. Trans Am Fish Soc 129:262–281. https://doi.org/10.1577/1548-8659(2000)129%3c0262:assgq%3e2.0.co;2

Kondolf GM (2011) Setting goals in river restoration: when and where can the river "heal itself"? Geophys Monogr Ser 194:29–43. https://doi.org/10.1029/2010GM001020

Kondolf GM, Lisle TE, Wolman GM (2003) Bed sediment measurement. In: Kondolf GM, Piégay H (eds) Tools in Fluvial Geomorphology. Wiley, West Sussex pp 347–395

Kondolf GM, Gao Y, Annandale GW et al (2014) Sustainable sediment management in reservoirs and regulated rivers: experiences from five continents. Earth's Futur 2:256–280. https://doi.org/10.1002/2013ef000184

Kondolf GM, Piégay H (2016) Tools in Fluvial Geomorphology, 2nd edn. Wiley, West Sussex

Kondolf GM, Piégay H, Schmitt L, Montgomery DR (2016) Geomorphic classification of rivers and streams. In: Kondolf GM, Piégay H (eds) Tools in fluvial geomorphology. Wiley, West Sussex, pp 133–158

Lamouroux N (2007) Hydraulic geometry of stream reaches and ecological implications. Dev Earth Surf Process 11:661–675. https://doi.org/10.1016/S0928-2025(07)11153-6

Lane EW (1955) Design of stable alluvial channels. Trans Am Soc Civ Eng 120:1234–1260

Ledger DC (1981) The velocity of the River Tweed and its tributaries. Freshw Biol 11:1–10

Leopold LB (1994) A view of the River. Harvard University Press, Cambridge, MA

Leopold LB, Maddock T jr (1953) The hydraulic geometry of stream channels and some physiographic implications. USGS Prof Pap 242:57

Leopold LB, Wolman MG, Miller JP (1964) Fluvial processes in geomorphology. W H Freeman and Company, San Francisco

Lester RE, Boulton AJ (2008) Rehabilitating agricultural streams in Australia with wood: a review. Environ Manage 42:310–326. https://doi.org/10.1007/s00267-008-9151-1

Martin JM, Meybeck M (1979) Elemental mass balance of material carried by major world rivers. Mar Chem 7:173–206

McBride M, Hession WC, Rizzo DM (2010) Riparian reforestation and channel change: how long does it take? Geomorphology 116:330–340. https://doi.org/10.1016/j.geomorph.2009.11.014

Meador M, Layher A (1998) Instream sand and gravel mining: Environmental issues and regulatory process in the United States. Fisheries 23:6–13

Melles SJ, Jones NE, Schmidt B (2012) Review of theoretical developments in stream ecology and their influence on stream classification and conservation planning. Freshw Biol 57:415–434. https://doi.org/10.1111/j.1365-2427.2011.02716.x

Melles SJ, Jones NE, Schmidt BJ (2014) Evaluation of current approaches to stream classification and a heuristic guide to developing classifications of integrated aquatic networks. Environ Manage 53:549–566. https://doi.org/10.1007/s00267-014-0231-0

Merz JE, Chan LKO (2005) Effects of gravel augmentation on macroinvertebrate assemblages in a regulated California river. River Res Appl 21:61–74. https://doi.org/10.1002/rra.819

Merz JE, Setka JD (2004) Evaluation of a spawning habitat enhancement site for Chinook Salmon in a regulated California river. North Am J Fish Manag 24:397–407. https://doi.org/10.1577/m03-038.1

Miller SW, Budy P, Schmidt JC (2010) Quantifying macroinvertebrate responses to in-stream habitat restoration: Applications of meta-analysis to river restoration. Restor Ecol 18:8–19. https://doi.org/10.1111/j.1526-100X.2009.00605.x

Milliman JD (1990) Fluvial sediment in coastal seas: flux and fate. Nat Resour 26:12–22

Montgomery DR, Buffington JM (1997) Channel-reach morphology in mountain drainage basins. Bull Geol Soc Am 109:596–611. https://doi.org/10.1130/0016-7606(1997)109%3c0596:CRMIMD%3e2.3.CO;2

Morisawa M (1968) Streams: their dynamics and morphology. McGraw-Hill, New York

Mueller ER, Grams PE, Schmidt JC et al (2014) The influence of controlled floods on fine sediment storage in debris fan-affected canyons of the Colorado River basin. Geomorphology 226:65–75. https://doi.org/10.1016/j.geomorph.2014.07.029

O'Connor J, Costa JE (2004) The world's largest floods, past and present: their causes and magnitudes. USGS Circu:

O'Connor J, Ely L, Wohl E, Others (1994) A 4500-year record of large floods on the Colorado River in the Grand Canyon, Arizona. J Geol 102:1–9

Olden JD, Kennard MJ, Pusey BJ (2012) A framework for hydrologic classification with a review of methodologies and applications in ecohydrology. Ecohydrology 5:503–518. https://doi.org/10.1002/eco.251

Olden JD, Konrad CP, Melis TS et al (2014) Are large-scale flow experiments informing the science and management of freshwater ecosystems? Front Ecol Environ 12:176–185. https://doi.org/10.1890/130076

Palmer MA, Bernhardt ES, Allan JD, et al (2005) Standards for ecologically successful river restoration. J Appl Ecol 42. 10.1111/j.1365-2664.2005.01004.x

Park CC (1977) World-wide variations in hydraulic geometry exponents of stream channels: An analysis and some observations. J Hydrol 33:133–146. https://doi.org/10.1016/0022-1694(77)90103-2

Parker G, Wilcock PR, Paola C et al (2007) Physical basis for quasi-universal relations describing bankfull hydraulic geometry of single-thread gravel bed rivers. J Geophys Res Earth Surf 112:1–21. https://doi.org/10.1029/2006JF000549

Piégay H, Kondolf GM, Sear DA (2016) Integrating geomorphological tools to address practical problems in river management and restoration. In: Kondolf GM, Piegay H (eds) Tools in fluvial geomorphology, 2nd edn. Wiley, West Sussex, pp 509–532

Poff NL, Allan JD, Bain MB, et al (1997) The natural flow regime: a paradigm for river conservation and restoration. Bioscience 47

Renwick W, Smith SV, Bartley JD, Buddemeier RW (2005) The role of impoundments in the sediment budget of the conterminous United States. Geomorphology 71:99–111

Richards K (2004) Rivers: form and process in alluvial channels. Methuen, London

Roni P, Hanson K, Beechie T (2008) Global review of the physical and biological effectiveness of stream habitat rehabilitation techniques. North Am J Fish Manag 28:856–890. https://doi.org/10.1577/m06-169.1

Rosgen D (1994) A classification of natural rivers. CATENA 22:169–199

Rosgen D (1996) Applied river morphology. Wildland Hydrology, Pagosa Springs, CO

Schumm SA (1985) Patterns of alluvial rivers. Annu Rev Earth Planet Sci 13:5–27

Scorpio V, Aucelli PPC, Giano SI et al (2015) River channel adjustments in Southern Italy over the past 150 years and implications for channel recovery. Geomorphology 251:77–90. https://doi.org/10.1016/j.geomorph.2015.07.008

Seaber PR, Kapinos FP, Knapp GL (1987) Hydrologic Unit Maps: US Geological Survey Water-Supply Paper 2294

Shields FD jr, Smith RH (1992) Effects of large woody debris removal on physical characteristics of a sand-bed river. Aquat Conserv Mar Freshw Ecosyst 2:145–163

Shreve RL (1966) Statistical law of stream numbers. J Geol 74:17–37

Simon A, Dickerson W, Heins A (2004) Suspended-sediment transport rates at the 1.5-year recurrence interval for ecoregions of the United States: transport conditions at the bankfull and effective discharge? Geomorphology 58:243–262

Snelder TH, Biggs BJF (2002) Multiscale river environment classification for water resources management. J Am Water Resour Assoc 38:1225–1239. https://doi.org/10.1111/j.1752-1688.2002.tb04344.x

Strahler AN (1957) Quantitative analysis of watershed geomorphology. Trans Am Geophys Uniion 38:913–920

Sweeney BW, Bott TL, Jackson JK et al (2004) Riparian deforestation, stream narrowing, and loss of stream ecosystem services. Proc Natl Acad Sci U S A 101:14132–14137. https://doi.org/10.1073/pnas.0405895101

Syvitski JPM, Vörösmarty C, Kettner A, Green P (2005) Impact of humans on the flux of terrestrial sediment to the global coastal ocean. Science (80-) 308:376–380

Turowski JM, Rickenmann D, Dadson SJ (2010) The partitioning of the total sediment load of a river into suspended load and bedload: a review of empirical data. Sedimentology 57:1126–1146. https://doi.org/10.1111/j.1365-3091.2009.01140.x

USEPA (2016) National National Rivers and Streams Assessment 2008–2009: A Collaborative Survey. Washington, DC

Ward JV, Tockner K, Arscott DB, Claret C (2002) Riverine landscape diversity. Freshwat Biol 47:517–539

Walling DE, Fang D (2003) Recent trends in the suspended sediment loads of the world's rivers. Glob Planet Change 39:111–126. https://doi.org/10.1016/S0921-8181(03)00020-1

Warrick JA, Bountry JA, East AE et al (2015) Large-scale dam removal on the Elwha River, Washington, USA: source-to-sink sediment budget and synthesis. Geomorphology 246:729–750. https://doi.org/10.1016/j.geomorph.2015.01.010

Wheaton JM, Brasington J, Darby SE, Other (2010) Linking geomorphic changes to salmonid habitat at a scale relevant to fish. River Reseach Appl Res Appl 26:469–486

Wilcox AC, O'Connor JE, Major JJ (2014) Rapid reservoir erosion, hyperconcentrated flow, and downstream deposition triggered by breaching of 38 m tall Condit Dam, White Salmon River, Washington. J Geophys Res Earth Surf 119:1376–1394

Wohl E (2014a) Time and the rivers flowing: Fluvial geomorphology since 1960. Geomorphology 216:263–282. https://doi.org/10.1016/j.geomorph.2014.04.012

Wohl E, Angermeier PL, Bledsoe B, et al (2005) River restoration. Water Resour Res 41. 10.1029/2005WR003985

Wohl EE (2014b) Rivers in the Landscape: Science and Management. John Wiley and Sons, Chichester, West Sussex

Wohl E, Bledsoe BP, Jacobson RB et al (2015) The natural sediment regime in rivers: broadening the foundation for ecosystem management. Bioscience 65:358–371. https://doi.org/10.1093/biosci/biv002

Streamwater Chemistry

We all have an intuitive appreciation that river water contains a variety of dissolved and suspended constituents. Mountain streams appear pure, farm creeks often are muddy with sediments, and drainages in limestone-rich regions are fertile while those containing only granitic rocks usually are less so. Rainwater and rivers throughout the world are polluted by organic and inorganic materials generated through human activities that impact the structure and function of freshwater systems.

Many factors influence the chemical composition of river water, producing variation from place to place. Rain is one source of chemical inputs to rivers, and a stream flowing through a region of relatively insoluble rocks can be chemically similar to rainwater in its composition. However, most streams and rivers contain much more suspended and dissolved material than is typically found in rain. Ultimately, all of the constituents of river water originate from dissolution of the earth's rocks. Variation in water chemistry is driven by the world's heterogeneous geology and by the magnitude of chemical inputs from other pathways including, but not limited to, precipitation, volcanic activity, and anthropogenic activities. Chemical composition of stream water also is altered by physical processes, such as evaporation and by chemical and biological interactions within the stream.

The materials transported in river water can be organized into dissolved versus suspended and organic vs. inorganic constituents. They include water, suspended inorganic matter, dissolved major ions (e.g., Ca^{2+}, Na^+, Mg^{2+}, K^+, HCO_3^-, SO_4^{2-}, Cl^-), dissolved nutrients (e.g., N, P, Si), suspended and dissolved organic matter, gases (e.g., N_2, CO_2, O_2), and trace metals, both dissolved and suspended (e.g., Pb, Hg, Zn, Cu) (Berner and Berner 2012).

In this chapter, we focus on the major dissolved constituents and the gases in freshwaters. We will also discuss some of the interactions between water chemistry and the aquatic biota, and briefly review some of the chemical contaminants of freshwater systems. Readers wishing more detailed discussions of aquatic chemistry and geochemistry should consult Christensen and Li (2014), Weiner (2012), or other specialized volumes.

4.1 Dissolved Gases

Oxygen, carbon dioxide, and nitrogen occur in significant amounts as dissolved gases in river water. Although nitrogen gas (N_2) can be incorporated into nitrogen cycling within stream ecosystems through specialized bacteria, the concentration of dissolved N_2 in water is typically of little biological importance.

Both oxygen and carbon dioxide gas occur in the atmosphere and dissolve into water according to partial pressure and temperature (Table 4.1). The solubility of oxygen in freshwater is reduced at high elevations due to lower atmospheric partial pressure. Oxygen solubility also declines with increasing salinity. Air is nearly 21% O_2 and just 0.03% CO_2 by volume, but the latter is much more soluble in water. Hence, although saturated freshwater has higher concentrations of O_2 than CO_2, the difference is not as great in water as it is found in air. Groundwater frequently is very low in dissolved O_2 and enriched in CO_2 due to the microbial processing of organic matter as water passes through soil. Thus, streams receiving substantial groundwater inputs may have localized regions of high CO_2 concentrations in the water column, but equilibration with the atmosphere usually occurs once hyporheic water enters the stream.

Lotic systems can experience spatial and temporal shifts in dissolved gas concentrations due to weather and natural changes in photosynthesis and respiration. In highly productive waters, whole-system photosynthesis results in elevated concentrations of oxygen during the day, while whole-system respiration causes oxygen to decline at night. These diel (24 h) changes in oxygen concentration provide a means of estimating photosynthesis and respiration

© Springer Nature Switzerland AG 2021
J. D. Allan et al., *Stream Ecology*,
https://doi.org/10.1007/978-3-030-61286-3_4

Table 4.1 Concentration of dissolved oxygen and carbon dioxide in saturated pure water for atmospheric partial pressure at sea level

Temperature (°C)	O_2 (mg L^{-1})	CO_2 (mg L^{-1})
0	14.2	1.1
15	9.8	0.6
30	7.5	0.4

of the total ecosystem, further discussed in Chap. 14. First applied to productive, slow-moving rivers and lentic waters where diffusion is relatively low and more easily estimated, recent improvements in measuring diffusion rates and detecting small changes in oxygen concentrations are extending this approach even to small woodland streams (Young and Huryn 1998). However, especially in small, turbulent streams receiving limited amounts of pollution, diffusion maintains O_2 and CO_2 near saturation. Should biological or chemical processes create a demand for or generate an excess of either gas within the water column, exchange with the atmosphere usually maintains concentrations very near to equilibrium. Gas concentrations often fluctuate seasonally, and longitudinally as water moves downstream. For example, water leaving Lake Constance—the headwaters of the Rhine River—is depleted in CO_2 in summer months relative to atmospheric partial pressure due to the productivity of lake phytoplankton. However, inputs of organic pollution increase downstream, stimulating algal and microbial growth. High summer temperatures yield high respiration rates, producing downstream average pCO_2 values about twenty times greater than atmospheric pCO_2 (Kempe et al. 1991).

Supersaturation of CO_2 in rivers is not uncommon. For example, of the ~6700 streams evaluated by Raymond et al. (2013), the average pCO_2 was found to be almost eight times greater than atmospheric concentrations. Some of this carbon is lost from rivers and streams via evasion, or degassing. Recent modelling efforts suggest that the amount of carbon lost in CO_2 evasion from streams and rivers is comparable to the amount of carbon exported from rivers into the ocean (Lauerwald et al. 2015; Magin et al. 2017). However, the source of the CO_2 degassed from flowing waters seems to be related to the size of the system in question. Hotchkiss et al. (2015) suggest that CO_2 emissions from smaller streams are dominated by terrestrially-derived CO_2 entering streams from the surrounding landscape, whereas much of CO_2 evasion from larger systems is primarily derived from internally generated CO_2 produced through aquatic respiration. These, and other studies, indicate flowing waters may play an important but often underestimated role in global carbon cycling.

Though pCO_2 in freshwaters can be exceptionally high and variable, changes in natural patterns of pCO_2 in inland waters are expected to coincide with increasing concentrations of atmospheric CO_2. Nonetheless, it is still largely unclear how increases in anthropogenically-derived CO_2 emissions will influence the aquatic chemistry and ecology of freshwater systems (Hasler et al. 2015). Research suggests that the quality and quantity of both allochthonous (e.g., leaf litter) and autochthonous (e.g., algae) food resources will be impacted by increasing atmospheric CO_2. For example, the coupled effects of increasing temperature and CO_2 concentrations are expected to preferentially facilitate the growth of cyanobacteria over eukaryotic algae (Visser et al. 2016). Changes in the behavior and development of aquatic organisms are also expected if they experience stress derived from exposure to increasing pCO_2 in inland waters and/or increases in acidity associated with increasing pCO_2 (Hasler et al. 2015).

Animals and plants can be responsible for ecologically important shifts in dissolved gas concentrations in rivers. A fascinating example from the Kenyan portion of the Mara River demonstrates at least one way in which animals can influence oxygen dynamics in rivers (Dutton et al. 2018). Hippopotami are nocturnally-feeding animals that congregate in the Mara during daylight hours. During the dry season, excreting and egesting hippopotami become more densely packed in smaller and smaller volumes of water, producing a dense layer of organic-rich feces in the benthic regions of pools. In some cases, hippopotamus activity in pools can disturb this layer, aerating and mixing the water column. However, without enough aeration, anoxia can develop through the decomposition of organic matter (Fig. 4.1). When the rainy season begins, river discharge increases and flushes hippo pools; fish kills are not uncommon as volumes of anoxic water move downstream (Dutton et al. 2018).

Eutrophication occurs when a body of water is overly enriched with nutrients, such as nitrogen or phosphorus, which stimulate algal and plant growth and influence primary productivity and respiration. Though it is not as commonly studied as it is in lakes and reservoirs, eutrophication of rivers and streams is of global conservation concern (Dodds and Smith 2016; Smith 2016). When instream conditions facilitate the accumulation of algal biomass, harmful algal blooms (HABs) can occur (Downing 2014). Populations of eukaryotic algae or cyanobacteria can create HABs and may produce large shifts in dissolved gasses, via primary production and community respiration, which can

Fig. 4.1 Observations of hypoxic events in hippo pools in the Mara River in Kenya. (**a**) Robotic boat surveying a hippo pool (image credit: Amanda Subalusky). (**b**) Dissolved oxygen (grey) and (**c**) discharge (grey) for a 3-month subset of data between November 2014 through January 2015. Black indicates discharge that was >2 times the calculated baseflow. (**d**) 3-D interpolation of the conductivity in pools containing or lacking high densities of hippopotami (Reproduced from Dutton et al. 2018)

(a)

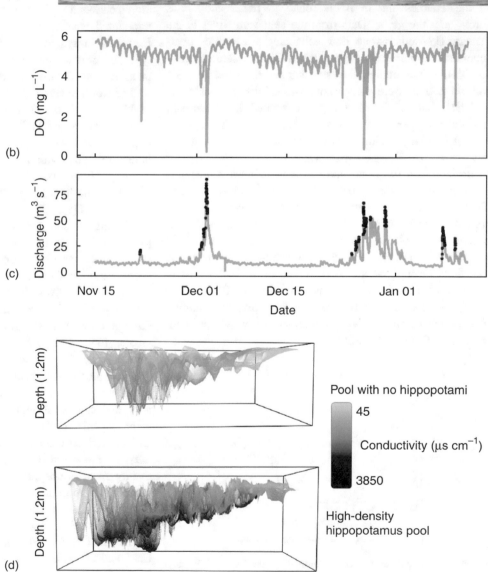

(b)

(c)

(d)

influence other aquatic organisms and biogeochemical processes. Human-induced environmental change, including increasing atmospheric temperatures and CO_2 concentrations, are expected to support increasing frequency of HABs (Reid et al. 2018).

Organic pollutants have a long and notorious history of altering oxygen and carbon dioxide concentrations in rivers and streams. As early as the 1600s, the River Thames rose to malodorous infamy, as human and animal waste discharged into the river were linked to "foulness" (Gameson and Wheeler 1977). In fact, 1858 was known as the "Year of the Great Stink" due to the ill-effects of organic pollution in the Thames. Wastewater treatment and more stringent environmental regulations have reduced sources of organic pollution in the Thames and in rivers found in richer economies throughout the globe. Data from the Delaware River in the eastern US demonstrate how enhancing wastewater treatment can support the recovery of heavily impacted systems. In 1950, ~ 280 municipal wastewater systems serving ~ 3.4 million people were discharging waste into the Delaware, and only half of the discharge was treated. Low instream dissolved oxygen (DO) concentrations reflected the negative impact that large volumes of organic waste were having in the system. By 1960, all wastewater was being treated prior to discharge in the river, and subsequent improvements in infrastructure in the 1970s and 80s were correlated with significant improvements in instream DO concentrations (Fig. 4.2, Marino et al. 1991). It is important to mention that organic pollution from human waste and other sources still remains an immense conservation challenge for resource managers throughout the world (Wen et al. 2017); this is especially true in poorer nations (Capps et al. 2016). As of 2017, the UN estimated more than 80% of global human wastewater is discharged into surface water without any treatment (Connor et al. 2017).

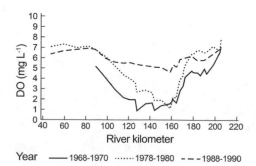

Fig. 4.2 Historic dissolved oxygen profiles for the Delaware Estuary during summer months (Reproduced from Marino et al. 1991)

4.2   Major Dissolved Constituents of River Water

Salinity is a term that refers to the sum of the concentrations of all of the dissolved ions in water. However, four major cations, sodium (Na^+), potassium (K^+), calcium (Ca^{2+}), magnesium (Mg^{2+}), and four major anions, bicarbonate (HCO_3^-), carbonate (CO_3^{2-}), sulfate (SO_4^{2-}), and chloride (Cl^-), are the dominant ions in fresh waters. Other ions, including those of nitrogen (N), phosphorus (P), and iron (Fe), are biologically important but contribute relatively little to total ionic concentrations. A similar metric used to describe the concentration of dissolved material is total dissolved solids (TDS), or sum of the concentrations of the aforementioned major ions plus organic matter. In relatively unpolluted waters, TDS is often very similar to salinity. However, in systems influenced by wastewater and other types of organic pollution, TDS can far exceed salinity measurements (Thompson et al. 2006).

The world average of TDS in rivers is about 100 mg L^{-1} (Table 4.2). Variation in the concentrations of individual constituents and of TDS among river systems results from regional and local variation in natural and anthropogenic inputs of ions within a watershed. However, the vast majority of the world's rivers are dominated by HCO_3^- (>50% of the TDS), Cl^-, and SO_4^{2-} ($Cl^- + SO_4^{2-} \sim 10$–30% of the TDS), reflecting the important contribution of the weathering of sedimentary rocks to riverine chemistry. This pattern is unsurprising, as three-quarters of the earth's land surface is covered by sedimentary rocks, which are rich in carbonate minerals and contribute the majority of total dissolved solids to rivers (Berner and Berner 2012).

Precipitation is the ultimate source of water flowing through rivers and streams; yet, river water typically has greater ionic concentrations than rainwater (Berner and Berner 2012). The ionic concentration of rainwater (Table 4.3) is typically low, with average concentrations of a few milligrams per liter. However, these values are highly variable, and can be strongly influenced by both natural and anthropogenic sources. The ions Na^+, K^+, Ca^{2+}, Mg^{2+}, and Cl^- are derived primarily from particles in the air, whereas SO_4^{2-}, ammonium (NH_4^+), and nitrate (NO_3^-) are derived mainly from atmospheric gases. Marine salts (NaCl) are especially important near the oceans, and a transition to calcium sulfate ($CaSO_4$)- or calcium bicarbonate (Ca(HCO_3)$_2$)-dominated rain occurs as one proceeds inland (Berner and Berner 2012). Roughly 10–15% of the Na^+, Ca^{2+}, and Cl^- in US river water comes from rain, compared to one-fourth of the K^+, and almost half of the SO_4^{2-}. In

Table 4.2 Chemical composition of river waters of the world. Values of individual cations and anions, and the sum of all of the cations and anions, or the total dissolved solids (TDS) are in mg L^{-1}. From Wetzel (2001) and sources therein

	Ca^{2+}	Mg^{2+}	Na^+	K^+	CO_3^{2-} HCO_3^-	SO_4^{2-}	Cl^-	NO_3^-	Fe (as Fe_2O_3)	SiO_2	TDS
North America	21.0	5.0	9.0	1.4	68.0	20.0	8.0	1.0	0.16	9.0	142
South America	7.2	1.5	4.0	2.0	31.0	4.8	4.9	0.7	1.4	11.9	65
Europe	31.1	5.6	5.4	1.7	95.0	24.0	6.9	3.7	0.8	7.5	182
Asia	18.4	5.6	5.5	3.8	79.0	8.4	8.7	0.7	0.01	11.7	142
Africa	12.5	3.8	11.0	–	43.0	13.5	12.1	0.8	1.3	23.2	121
Australia	3.9	2.7	2.9	1.4	31.6	2.6	10.0	0.05	0.3	3.9	59
World	15.0	4.1	6.3	2.3	58.4	11.2	7.8	1.0	0.67	13.1	120

Table 4.3 Concentrations of major ions in continental and marine coastal precipitation (mg L^{-1}). Modified from Table 3.3 in Berner and Berner (2012)

Ion	Continental rain	Marine coastal rain
Ca^{2+}	0.1–3.0[a]	0.2–1.5
Mg^{2+}	0.05–0.5	0.4–1.5
Na^+	0.01–4.0	1.0–24
K^+	0.1–0.3[a]	0.2–0.6
NH_4^+	0.1–0.5[b]	0.01–0.05
SO_4^{2-}	1–3[ab]	1–3
Cl^-	0.2–2	1–10
NO_3^-	0.4–1.3[b]	0.1–0.5
pH	4–6	5–6

[a]Concentrations from remote inland sites: K^+ = 0.02–0.07; Ca^{2+} = 0.02–0.20; SO_4^{2-} = 0.2–0.8
[b]Concentrations from polluted sites: NH_4^+ = 1–2; NO_3^- = 1–3; SO_4^{2-} = 3–8

contrast, almost none of the SiO_2 or HCO_3^- comes from rain. The relative importance of chemical inputs to rivers from precipitation can vary seasonally and over short distances, as Sutcliffe and Carrick (1983) document for streams of the English Lake District. These data emphasize the need to examine the origins of each of these major cations and anions in order to understand what influences their concentrations in rivers and streams.

Calcium is the most abundant cation in the world's rivers. It originates almost entirely from the weathering of sedimentary carbonate rocks, although pollution and atmospheric inputs constitute small sources of the element. Along with magnesium, the concentration of calcium is used to characterize "soft" versus "hard" waters, which are discussed more fully below. Magnesium also originates almost entirely from the weathering of rocks, particularly from magnesium-silicate minerals and dolomite. Atmospheric inputs are minimal, and pollution contributes only slightly to the concentration of magnesium in fresh water.

Sodium is generally found in association with chloride, reflecting their common origin. Weathering of NaCl-containing rocks accounts for most of the sodium and chloride found in river water. However, rainwater inputs from sea salts can contribute significantly to total sodium and chloride concentrations, especially near coasts. Berner and Berner (2012) estimate that worldwide, approximately 28% of the sodium and 30% of the chloride in rivers is derived from pollution. Though chloride is chemically and biologically unreactive, and often is used as a conservative tracer in experiments examining nutrient dynamics in streams, the impact of anthropogenically-derived sodium on aquatic communities can be profound. We will discuss some of the sources of anthropogenically-derived sodium and its impact on aquatic communities later in the chapter.

Potassium (K^+) is the least abundant of the major cations in river water, and its concentration is the least variable among systems. Roughly 90% of riverine potassium originates from the weathering of silicate materials, especially potassium feldspar and mica. Concentrations of potassium in river water vary with underlying geology, and tend to increase substantially when moving from polar latitudes toward the tropics. This pattern is apparently due to more complete chemical weathering at higher temperatures. Biological activity can also influence patterns in silica concentrations. Silica (also called silicon dioxide; SiO_2) is derived from the weathering of silica-rich minerals that are dominant parts of the earth's crust. Silica is used by diatoms in the formation of their external cell wall and can on occasion limit algal productivity.

Riverine bicarbonate (HCO_3^-) is derived almost entirely from the weathering of carbonate minerals. However, the immediate source of the majority of bicarbonate in river water is CO_2 produced by bacterial decomposition of organic matter that is dissolved in soil and groundwater. Bicarbonate is a biologically important anion. High concentrations of bicarbonate are reflected in measures of alkalinity and are indicative of fertile waters. The carbonate buffer system, alkalinity, and hardness are interrelated and will be discussed more fully below. Anthropogenic increases in acidity, caused by acid precipitation or mining, reduce bicarbonate levels through the formation of carbonic acid (H_2CO_3).

Sulphate has many sources, including the weathering of sedimentary rocks and pollution from fertilizers, organic waste, mining activities, and the burning of fossil fuels. Biogenically-derived sulphate in rain and volcanic activity are additional inputs. In areas of sulphuric acid rain, sulphate concentrations are high relative to overall ionic concentrations. In most river water, sulphate and bicarbonate concentrations are inversely correlated, especially in low alkalinity areas.

The concentration of hydrogen ions (H^+) is very important both chemically and biologically, because it determines the acidity of water. This is expressed as pH, on a logarithmic scale in which a tenfold change in H^+ activity corresponds to a change of 1 pH unit. A pH of 7 is neutral, higher values are alkaline, and lower values are acidic. CO_2 concentration and pH are interdependent. CO_2 readily dissolves in water to form carbonic acid (H_2CO_3). Carbonic acid dissociates first to a bicarbonate (HCO_3^-) ion and H^+, and then to a carbonate (CO_3^{2-}) ion and H^+; hence, changes in CO_2 concentrations can produce significant shifts in the acidity of stream water.

We will address two key nutrients, nitrogen and phosphorus, more fully in Chap. 13, including their forms (e.g., nitrate, ammonium, phosphate, etc.) and their influence on ecosystem processes. Dissolved inorganic phosphorus and nitrogen are primary nutrients that limit plant and microbial production, and cycle rapidly between their inorganic forms and their incorporation into the food web.

The concentration of major ions can be measured in many ways. For example, TDS is measured by evaporating an aliquot of filtered stream water and weighing the remaining residue. Additionally, colorimetric methods (specific ions react with specific chemicals to form colored compounds), ion chromatography, and ion-specific probes can be used to measure ionic concentrations. Dissolved constituents are reported as units of mass, mg L^{-1} (equal to parts per million, ppm), or as chemical equivalents. In the latter case, milliequivalents per liter are calculated from mg L^{-1}, by dividing the concentration by the equivalent weight of the ion (its ionic weight divided by its ionic charge).

Conductivity is a measure of electrical conductance of water, and is an approximate measure of total dissolved ions. Distilled water has a very high resistance to electron flow, and the presence of ions in water reduces that resistance. Differences in conductivity result mainly from the concentration of the charged ions in solution, and to a lesser degree from ionic composition and temperature. Conductivity measurements by probes can compensate for temperature (20 or 25 °C) and values are reported as $\mu S \cdot cm^{-1}$ (microSiemens per centimeter) or in the older literature as $\mu mho \cdot cm^{-1}$ (the reciprocal of ohms). Total dissolved solids can also be estimated from specific conductance (SC). The relationship between TDS and SC is typically linear (TDS = k * SC) with a value of k between 0.55 and 0.75. However, the value of the constant varies with location and must be established empirically (Walling 1984). In fresh water, salinity is usually estimated from conductivity and temperature by using salinity sensors and values are reported as ppt (parts per thousand).

4.2.1 Variability in Ionic Concentrations

The chemistry of fresh waters is quite variable, rivers usually more so than lakes. Natural spatial variation is determined mainly by the type of rocks available for weathering, the climate, and by the chemical composition of rain, which in turn is influenced by proximity to the sea and by anthropogenic activities. The ionic concentration of rivers draining igneous and metamorphic terrains is roughly half that of rivers draining sedimentary terrain, because of the differential resistance of rocks to weathering. All of these factors provide the opportunity for substantial local variation in river chemistry. As a consequence, the concentration of total dissolved ions can vary considerably amongst the headwater branches of a large drainage. However, these heterogeneities tend to average out and ionic concentration tends to increase as one proceeds downstream (Livingstone 1963).

Although small streams in the same region often are chemically similar, they can also differ markedly. In one of the first such studies, Walling and Webb (1975) reported a concentration range of total ions from 25 to 650 mg L^{-1} among a series of small streams in southwest England, resulting from small-scale shifts between igneous and sedimentary geology and variation in land use. Research from around the globe has continued to document similar chemical variation at relatively small spatial scales. Small streams draining a volcanic landscape in central Costa Rica exhibited pronounced differences in solute concentrations depending on geology, soil types, and elevation. Concentrations of phosphorus, several major cations and anions, and trace elements were higher in headwater streams draining younger volcanic landscapes and were much lower in streams draining older lava flows (Pringle and Triska 1991). Another intriguing example of small-scale spatial variation in stream chemistry comes from the Taylor Valley in Antarctica, where streams fed by glacial meltwater that are strongly influenced by chemical weathering exhibit a range of chemical compositions and TDS that are similar to ranges found in larger temperate and tropical systems (Welch et al. 2010).

Drastic differences in ion concentrations can also be seen at the confluence of rivers with disparate water chemistry. The blackwater Río Negro and whitewater Solimões that converge to form the Amazon River dramatically illustrate the chemical differences between tributaries draining distinctive landscapes. The Río Negro drains well-weathered, crystalline rock and is much lower in ions and much higher in organic acids, whereas the Amazon mainstem drains the comparatively young Andes and has a much higher dissolved load. At their confluence near Manaus, Brazil, known as "the meeting of waters", the tea-colored waters of the Rio Negro run parallel to the sediment-laden café-au-lait waters of the Solimões without mixing for about 6 km, because of differences in temperature and water density. The unique chemical signatures of both mighty rivers can be detected as much as 100 km below their confluence.

Within-site, temporal variation in water chemistry is also common in stream and rivers due to the influence of fluctuating discharge, precipitation, and biological activity. Discharge has especially strong effects on ionic concentrations in river water. Rivers are fed by a combination of groundwater and surface water, resulting in variation due to local geology and precipitation. Because of its longer association with rocks, the chemistry of groundwater is often more concentrated but less variable than surface waters. Increases in flow due to rain events typically dilute stream water, although it is not a simple relationship to predict (Livingstone 1963). Golterman (1975) argued that there are two broadly-defined patterns relating discharge with water chemistry. The TDS may be inversely related to discharge, which is the expected dilution effect when the input of materials is relatively constant and chemicals are homogenously mixed in the river. Alternatively, ion concentrations may remain relatively constant in response to fluctuations in discharge. This pattern can be explained by two mechanisms. First, the water may be reaching equilibrium with soils; therefore, ion concentration would remain constant regardless of flow volume. Second, chemical concentrations may remain relatively constant if they are approaching saturation values for the system. The response to rising discharge in the Orinoco River in northern South America differs depending on the source of ions. This very large river system is characterized by a very high runoff rate and very low concentrations of geologically-derived nutrients because a large fraction of its catchment is underlain by a shield rock that is resistant to weathering and covered with undisturbed

forest. Seasonal increases in discharge result in a dilution response of geologically-derived major ionic solids, including soluble silica and phosphorus. In contrast, biologically-derived substances, including organic carbon and nitrogen, exhibit a purging response in which their ion concentrations increase with increasing discharge (Lewis and Saunders 1989).

Studies of stream water draining the Hubbard Brook Experimental Forest in the northeastern US illustrate how ionic concentrations can change in response to seasonal variation in precipitation inputs, discharge, and the cycle of growth of the terrestrial vegetation (Likens and Bormann 1995). The relative constancy in stream chemistry evident in these data is notable and is probably typical of intact, undisturbed ecosystems (Fig. 4.3). Most dissolved substances vary within a narrow range (less than twofold), whereas stream flow can vary as much as four orders of magnitude over an annual cycle. In the Hubbard Brook Experimental Forest, virtually all drainage water must pass through its mature and highly permeable podzolic soils. This affords considerable buffering capacity, and accounts for the relatively constant chemical composition of stream water at these sites (Likens et al. 1970).

The relationship between discharge and ionic concentration may also vary among ions within a given site. For example, associations between cation concentrations and discharge were ion-specific over a two-year period in a single watershed in Hubbard Brook (Fig. 4.3). When all data were pooled in the analyses, there were no significant relationships between potassium, calcium, or magnesium and discharge. However, sodium concentrations were negatively related to flow. This pattern is most likely due to a dilution effect resulting from limited sodium availability in the system (Likens et al. 1967). In contrast, similar work conducted in mountain streams in California documented negative relationships between calcium and magnesium concentrations and discharge (Johnson and Needham 1966), highlighting how regional differences in the physicochemical environment can influence stream water chemistry.

Although solute concentrations may exhibit only modest variation in response to discharge fluctuations in an intact ecosystem, human activities in a watershed, such as timber harvest and road construction, are significant disturbances that influence solute export. Following deforestation and suppression of regrowth by herbicides in a catchment of Hubbard Brook, the concentrations of most major ions increased in stream water, and total output of ions increased six-fold. Only ammonium and carbonate remained low and constant, and sulphate declined because of reductions in sulphate generation by sources internal to the ecosystem. The average concentrations of calcium and magnesium increased by over 400%, sodium by 177%, and potassium concentrations increased over 18-fold. Altered ion

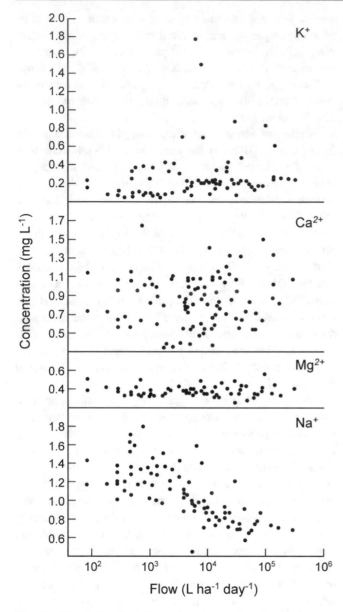

Fig. 4.3 The relationship between the concentration of major ions and stream discharge in a small forested catchment in the Hubbard Brook Experimental Forest, New Hampshire, between 1963 and 1965 (Reproduced from Likens et al. 1967)

concentrations were attributed to increased discharge, changes in the nitrogen cycle within the ecosystem, and higher temperatures (Likens and Bormann 1995). Use of best management practices (BMPs) can moderate the response of stream water chemistry to the disturbances associated with timber harvest. A long-term study of a well-managed clear-cut of a forested catchment found that almost all solutes and nutrients showed elevated concentrations and exports after harvest. However, overall losses were

relatively small and judged not detrimental to forest productivity because the forest was harvested using several BMPs including cable logging, which minimized the need for roads and dragging of felled trees (Swank et al. 2001).

Climate also exerts considerable influence over regional variation in the chemical composition of rivers. Across a gradient from arid to humid conditions, a general inverse relationship is seen between annual precipitation and total solute concentration. High concentrations of total dissolved ions are found in rivers draining arid areas due to the small volumes of precipitation and runoff, salt accumulation in the soil, and evaporation (Walling 1984). Changes in the timing and intensity of precipitation associated with anthropogenic climate change has the potential to influence the chemical composition of rivers and streams. For example, work in boreal streams illustrates the impact of increasing discharge and winter climate conditions on inter-annual variation in dissolved organic carbon concentrations in streams during snowmelt (Ågren et al. 2010).

Research from long-term (~20 year) studies demonstrates that concentrations of solutes in stream water seem to have seasonal patterns that are driven by geology, terrestrial vegetation, and temporal patterns in climate variables, including insolation (a measure of incident solar radiation per unit area and time), temperature, and precipitation (Lucas et al. 2013; Lutz et al. 2012; Navrátil et al. 2010). Long-term studies also provide insight into how environmental change may alter relationships between ion concentrations, especially Ca^{2+}, Mg^{2+}, and SO_4^{2-}, and other watershed attributes. In a 20-year study of the Walker Branch Watershed in Tennessee in the southeastern US, researchers documented a ~1.0 °C increase in mean annual temperature, a ~30% increase in the rates of forest evapotranspiration, and a ~20% decline in the amount of precipitation entering the watershed, which were linked to a ~34% decline in river runoff (Lutz et al. 2012). Concentrations of solutes changed significantly with discharge, showing dilution for Ca^{2+} and Mg^{2+}, and increasing concentrations for SO_4^{2-} (Fig. 4.4). This difference is because calcium and magnesium are controlled by bedrock weathering, whereas the main source for sulphate in this system is wet and dry deposition from nearby coal-fired power plants. In addition, over the 20-year record, concentrations of Ca^{2+} and Mg^{2+} increased, and this is attributable to reduced precipitation. As a result of climate-related declines in runoff, groundwater, rich in Ca^{2+} and Mg^{2+}, has become an increasing proportion of stream flow.

In response to the implementation of environmental regulations, such the 1970 Clean Air Act in the US, many, but not all watersheds, have experienced declines in SO_4^{2-} deposition (Lutz et al. 2012; Navrátil et al. 2010). Coal-fired

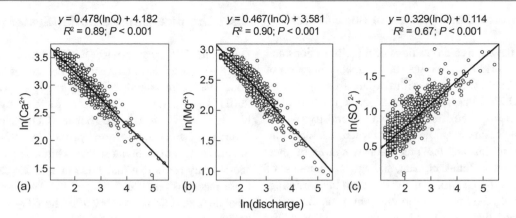

Fig. 4.4 Relationships between the concentration of geochemical solutes and discharge. Original concentration units are mg L^{-1} and discharge units are in L s^{-1}. Values are ln-transformed. Regression summaries are listed above each panel and all relationships are significant (Reproduced from Lutz et al. 2012)

power plants are still active in many regions throughout the globe, and nearby watersheds may still receive relatively large inputs of SO$_4^{2-}$ through deposition (e.g., Fig. 4.4) that can influence concentrations of other ions. Once deposited in the watershed, the impact of SO$_4^{2-}$ on stream water chemistry is variable and depends largely upon the volume of surface water runoff, soil chemistry, and underlying geology. Analyzing a 20-year dataset from 60 Swedish streams, Lucas and others (2013) documented declines in Ca^{2+}, Mg^{2+}, K$^+$, and Na$^+$ since the early 1990s that were related to declines in SO$_4^{2-}$ in streams in southern Sweden. In contrast, cation concentrations in streams in northern Sweden—a region that was not affected as strongly by anthropogenically derived SO$_4^{2-}$ deposition—were characterized by seasonal variability linked to climate variables, such as temperature and precipitation (Lucas et al. 2013).

Rising ion concentrations due to other anthropogenic activities, such as agricultural development, wastewater discharge, and natural resource extraction, also threaten the chemical integrity of freshwaters throughout the globe (Griffith 2014). Increases in some ions, particularly bicarbonate salts, chlorides, and sulphates, can alter stream community structure and function by decreasing the number of species able to live in a system and increasing the mortality of stream organisms (Johnson et al. 2015; Tyree et al. 2016).

4.2.2 The Dissolved Load

The dissolved load of a river is the product of concentration and discharge, and usually is expressed as kg day^{-1} or metric tons yr^{-1}. In comparing catchments or river basins it is helpful to present this as a yield per unit area, by dividing by the drainage area. Streamflow is more variable than ionic concentration, and so between-year variation in the export of ions can depend strongly on inter-annual variation in

discharge (Fig. 4.5). Because discharge and ionic concentration often are inversely related, the range of the dissolved load of ions transported by the world's major rivers varies over only two orders of magnitude, from around 3 to as high as 500 t · km^{-2} yr^{-1}, with highest values observed in small alpine rivers (Meybeck 1977). The dissolved load typically experiences less temporal variation than does the amount of suspended solids carried by rivers. Greater discharge of rivers in wetter regions of the globe more than compensates for lower ionic concentrations; therefore, the dissolved load is usually smaller in arid regions when compared to areas with more surface runoff.

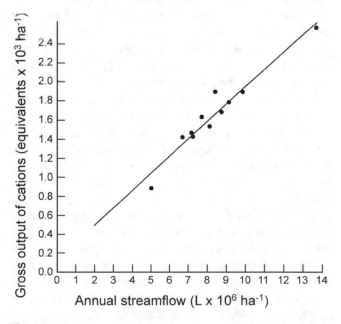

Fig. 4.5 Between-year variation in the gross output of calcium, sodium, magnesium, and potassium (kg ha^{-1}) primarily depends on between-year variation in discharge for undisturbed catchments of the Hubbard Brook Experimental Forest. Data were collected between 1963–1974 (Reproduced from Likens and Bormann 1995)

4.2.3 Chemical Classification of River Water

If one relates the relative proportions of principle anions and cations in the world's surface waters to the concentration of total dissolved solids, a curve with two arms emerges (Fig. 4.6). Gibbs (1970) interpreted this as evidence that three major mechanisms control surface water chemistry. At the left side of the "boomerang" lie systems where the rocks and soils of river basins are the predominant source of their dissolved materials. Relief, climate, and age and hardness of rocks determine the positions of rivers within this grouping. Proceeding along the lower arm to the right of the figure we encounter waters that are lower in ions and whose chemical composition most closely resembles the rain. These are mainly the tropical rivers of Africa and South America, with their sources in highly leached areas of low relief. In tropical humid regions, precipitation with a composition similar to seawater is an important source of ions. The average ionic concentrations of river water in these regions are typically less than 30 mg L^{-1}. Proceeding along the upper arm to the right, we encounter systems with high concentrations of dissolved ions and again a relative predominance of Na^+ and Cl^-. These are rivers of hot, arid regions, where the combined influence of evaporation and precipitation of $CaCO_3$ from solution accounts for their higher ionic concentrations. Total dissolved salts exceed 1000 mg L^{-1} in streams in these regions, and can be as high as several g L^{-1}, as seen in some rivers of central Asia (Crosa et al. 2006). Thus, three major mechanisms—atmospheric precipitation, dissolution of rocks, and the evaporation-crystallization process—are considered to account for the principal trends of dissolved ions in the world's surface waters. Other factors, including topography, vegetation, and the composition of rocks and soils then can be invoked to explain differences in stream chemistry within these major groupings.

Justifiably, critics of this classification scheme question the interpretation of control at the ends of the boomerang. For example, the Río Negro's chemistry is equally a consequence of its long history of intense weathering, and its basin, which is predominantly comprised of silicious rocks (Stallard and Edmond 1983). Similarly, saline rivers can be strongly influenced by near-surface halite deposits (Kilham 1990). These examples downplay the roles of precipitation and evaporation, and suggest that local geology can be of primary importance in determining river chemistry over all extremes. Regardless, most of the world's rivers are closer to the middle than the ends of this diagram, are low in Na^+/$(Na^+ + Ca^{2+})$, and are dominated by Ca^{2+} and HCO_3^- from carbonate dissolution. This is in accord with the long-held view that the weathering of sedimentary rocks provides most of the dissolved ions in most of the world's major rivers (Berner and Berner 2012).

4.3 The Bicarbonate Buffer System

Dissolved CO_2 reacts with H_2O to form carbonic acid (H_2CO_3), a weak inorganic acid that occurs at low concentrations relative to unhydrated CO_2 at pH values <8 (Wetzel 2001). H_2CO_3 further dissociates to form hydrogen (H^+), bicarbonate (HCO_3^-), and carbonate (CO_3^{2-}) ions. The latter in turn react with water to form hydroxyl ions (OH^-). When the natural content of carbonate rocks is high, as is common in the sedimentary rocks that make up much of the earth's surface, these reactions result in sufficient hydroxyl ions to produce alkaline waters. This is known as the CO_2–HCO_3^-–CO_3^{2-} buffering system, because it resists change in pH. Although addition of hydrogen ions will neutralize hydroxyl ions, more hydroxyl ions are formed immediately through the reaction of water with carbonate, and so pH resists change. However, once the supply of carbonate is exhausted, which occurs most rapidly in igneous drainages, pH can drop quickly.

The relative proportions of CO_2, HCO_3^-, and CO_3^{2-} are pH-dependent (Fig. 4.7). At a pH below 4.5 only CO_2 and H_2CO_3 are present, and almost no bicarbonate or carbonate can be found. Indeed, bicarbonate concentration commonly is measured by titration with a strong acid until reaching a pH of about 4.3. At higher pH values dissociation of carbonic acid occurs, bicarbonate and carbonate are present, and CO_2 and H_2CO_3 are no longer detectable. At intermediate pH values, HCO_3^- predominates. These dissociation dynamics are influenced by both temperature and ionic concentrations, and the relationships shown in Fig. 4.7 may not be valid for water of very high ionic concentration. The inorganic carbon required for photosynthesis is most easily obtained from CO_2, although many photosynthetic organisms can utilize bicarbonate. By influencing the forms of inorganic carbon available, shifts in pH can influence the efficiency of different primary producers in the ecosystem.

Fresh waters can vary widely in acidity and alkalinity due to both natural and anthropogenic factors. Very acid and very alkaline waters are harmful to most organisms, and so the buffering capacity of water is critical to the maintenance of life. The pH of water is a measure of the concentration of hydrogen ions, hence the strength and amount of acid present. Because the scale of pH is logarithmic to the base 10, a decline of one pH unit represents a tenfold increase in hydrogen ion concentration. In natural waters, pH is largely governed by H^+ ions from the dissociation of H_2CO_3 and OH^- ions from the hydrolysis of bicarbonate. Rainwater is usually acidic (typically near a pH of 5.7) because of its carbon dioxide content, and also due to naturally occurring sulphate. Normally these acids are neutralized as rainwater passes through the soil. However, in catchments primarily composed of hard rocks, little buffering capacity, and large surface water (as opposed to groundwater) inputs, stream

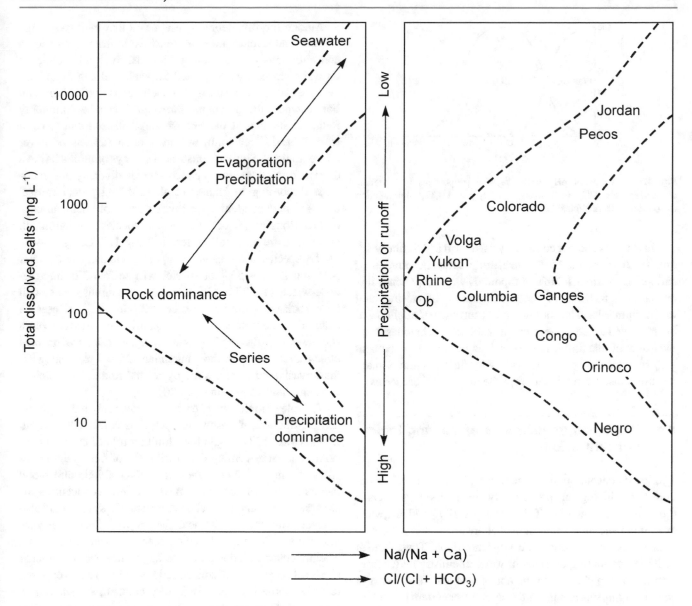

Fig. 4.6 A classification of surface waters of the world based on ratios of sodium to calcium and chloride to bicarbonate, in relation to total dissolved salts. As one proceeds from left to right along the lower arm, inputs shift from a dominance of rock dissolution to a dominance of precipitation. The majority of large tropical rivers are found to the lower right. As one proceeds from left to right along the upper arm, sodium and chloride increase. These high salinity rivers lie in arid regions where evaporation is great. Note the vertical axis also reflects a gradient from high precipitation and runoff at the base to arid regions at the top (Modified from Gibbs 1970)

water will be acidic even if pollution is absent. Organic acids also contribute to low pH values. Where decaying plant matter is abundant, especially in swamps, bogs, and peaty areas, humic acids result in "brown" or "black" waters, and a pH in the range of 4–5. In addition, volcanic fumes and local seepage from sulphurous or soda springs can produce natural extremes of pH.

Alkalinity refers to the quantity and kinds of compounds that collectively shift the pH into the alkaline range (>7.0 pH). It is a measure of the capacity of the solution to neutralize acid, and is determined by titration with a strong acid and expressed as milliequivalents per L or mg L^{-1}. The bicarbonate buffering system is mainly responsible for alkalinity. Hence, alkalinity is often measured as mg L^{-1} of $CaCO_3$, which measures the acid-neutralizing capacity due to carbonate and bicarbonate. Because alkalinity is estimated from a filtered sample, it measures the capacity of solutes to neutralize acid. Acid neutralizing capacity (ANC) is similar, but is determined for an unfiltered sample and thus measures the capacity of solutes plus particulates to neutralize acid.

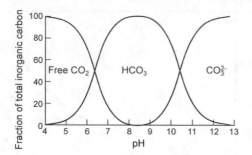

Fig. 4.7 Influence of pH on the relative proportions of inorganic carbon species, CO_2 (+ H_2CO_3), HCO_3^-, and CO_3^{2-} in solution (Reproduced from Wetzel 2001)

Hardness is another commonly used term to describe water quality. It is calculated by measuring the concentration of cations that form insoluble compounds with soap; hence it is primarily a measure of the concentration of calcium and magnesium salts. The common co-occurrence of calcium and bicarbonate has led some to equate estimates of hardness with measures of alkalinity. However, it is possible to document very high alkalinity in water with very little calcium or magnesium; thus, it can be inappropriate to equate these terms.

4.4 Biological Implications of Varying Ionic Concentrations

The ionic concentration of water can affect species distributions and biological productivity in rivers and streams throughout the world. This is especially evident when examining stream communities that are adapted to extremely high or extremely low ionic concentrations. Thus, human activities promoting changes in water chemistry may consequently lead to changes in the abundance and distribution of aquatic organisms and produce shifts in ecosystem processes.

4.4.1 Variation in Ionic Concentration

If one wished to develop an entirely artificial medium for the culture of freshwater invertebrates, as has been done for algae and zooplankton, a long list of chemicals would be needed in the mixture to successfully support population growth (Kilham et al. 1998). Unquestionably, stream-dwelling organisms require water of some minimal ionic concentration to survive. The majority of the evidence linking the ionic content of water to the stream biota comes from surveys, and streams with very low ionic concentrations often support smaller numbers of species and fewer organisms. Hynes (1970) described a number of natural examples where the species of algae, mosses, and higher plants differ between soft versus hard waters.

Among the invertebrates, it appears that mollusks, crustaceans, and leeches are more sensitive to the range of ionic concentrations than are aquatic insects. Such a dependence on water chemistry is expected for shell-building organisms because $CaCO_3$ is necessary for shell deposition and growth. For instance, the amphipod *Gammarus* can be commonly found in streams of the English Lake District that have at least 3 mg L^{-1} calcium, but are rare in streams of lower concentrations. Similarly, historical surveys of mollusks have described relationships between species diversity and hardness. According to Hunter et al. (1967), regional mussel diversity is dependent upon a large range of chemical conditions. Roughly 5% of the molluscan species of a region will occur in extremely soft waters (<3 mg L^{-1} Ca^{2+}), ~40% of the species are dependent upon moderately soft water (<10 mg L^{-1}), ~55% of species will be found in intermediate waters (10–25 mg L^{-1}), and the remaining species will be associated with hard water (>25 mg L^{-1}). Research examining the potential range expansion of invasive zebra (*Dreissena polymorpha*) and quagga (*D. rostriformis bugensis*) mussels in the Laurentian Great Lakes suggests that invasion success is strongly related to ambient calcium concentrations (Baldwin et al. 2012).

The role of ionic concentrations in governing the condition and distribution of freshwater species is context dependent. Lodge et al. (1987) argue for a limiting role of calcium concentrations in governing the distribution of freshwater snails when <5 mg L^{-1}, but suggest that other abiotic and biotic factors become more important when calcium concentrations are greater. The chemical characteristics of a system may also interact with local biological factors to influence species interactions. For example, snails reared in water of differing calcium concentrations produced larger, heavier, and thicker shells at higher calcium concentrations and also when exposed to water-borne cues of a predatory crayfish, *Procambarus*, indicative of the importance of calcium availability to predator avoidance (Bukowski and Auld 2014).

Alkalinity is widely used as a surrogate for stream fertility, linking alkalinity to productivity of fish populations via their food supply. Compiling data on salmonid production and stream alkalinity from numerous studies in the US, Kwak and Waters (1997) demonstrated a statistically significant but nonetheless imperfect relationship between the two variables (Fig. 4.8). However, their own data from 13 southeastern Minnesota streams did not show any dependence of salmonid production on stream alkalinity, which the authors suggested may be due to the modest range of alkalinity values among sites as well as the influence of additional variables including temperature and biological interactions. Recent work examining links between alkalinity and freshwater organisms has often focused on the relationships between alkalinity and toxicity of heavy metals in rivers and streams.

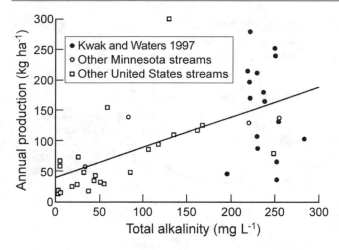

Fig. 4.8 Relationship between the annual production of salmonids and alkalinity for streams in southeastern Minnesota (black circles) and other locations throughout the US. (Reproduced from Kwak and Waters 1997)

4.4.2 Salinization and Alkalinization of Freshwater Systems

The natural range of salinity and alkalinity in inland waters is considerable. The impacts of salinization, or the process of increasing concentrations of dissolved salts in fresh waters, has received much attention by researchers. As described in Sect. 4.2, salinity differs from TDS in one important way: its measurement does not include the concentrations of organic compounds in water. Therefore, in relatively unpolluted water, TDS is often very similar to salinity and the terms are often used interchangeably. Alkalinization refers to the process of increasing concentrations of dissolved alkaline substances (e.g., bicarbonate, carbonate, and hydroxide) in freshwater systems. In comparison with acidification, alkalinization is not as well studied. Yet, increases in stream alkalinity are also associated with the application of road salts, accelerated weathering due to the deposition of acidic compounds, and altered microbial activities. In some cases, the processes are linked. For instance, the application or dissolution of salts containing strong bases and carbonates can increase both salinity and alkalinity in receiving waters (Kaushal et al. 2018).

Globally, average salinity in rivers is approximately 120 mg L^{-1}, a value thought to be elevated relative to natural conditions due to anthropogenically derived pollution (Berner and Berner 2012). Recent work has suggested that anthropogenic activities are largely responsible for increasing salinity and alkalinity in inland waters. This pattern, termed the freshwater salinization syndrome (Kaushal et al. 2018), can be seen over large spatial scales and across heterogeneous landscapes (Fig. 4.9). Specific drivers of the syndrome include salt pollution (irrigation runoff, road deicers, sewage), increased weathering of geological materials by strong acids (acid rain, acid mine drainage, fertilizers), and the abundance of easily weathered materials in agricultural (e.g., lime) and urban (e.g., concrete) environments (Kaushal et al. 2018).

In colder regions of the globe, salts are used on roads to melt ice and reduce vehicle accidents (Fay and Shi 2012). Runoff carrying salts and other de-icing compounds can significantly elevate the salinity of receiving waters, and cause large fluctuations in salinity over relatively short time scales. The volume of deicing salts used on roadways has increased dramatically through time (Schuler and Relyea 2018a). Between 1950 and 2014, the volume of road salt applied to roads in the US increased from less than 1 million tons to approximately 20 million tons. Similar patterns of road salt application have been documented in other regions of the world (Schuler and Relyea 2018b).

Runoff from road salts can negatively affect aquatic populations, communities, and ecosystems (Schuler and Relyea 2018a). Laboratory experiments have repeatedly documented how road salts are toxic to individual species and have negative effects on animal populations. For example, work with rainbow trout (*Oncorhynchus mykiss*) demonstrated the distinct effects of three, commonly-used types of road salt on trout growth and development. When scaled to the population level, researchers suggested that the reduced growth caused by salts at critical early-life stages has the potential to alter population structure by reducing trout recruitment (Hintz and Relyea 2017). Field-based observational and experimental research indicates exposure to increasing salinity from road salt can shift community composition, initiate trophic cascades, and alter important ecosystem processes, such as denitrification. Notably, many negative effects of road salt have been observed at concentrations below the chronic and acute thresholds (230 mg Cl$^-$ L^{-1} and 860 mg Cl$^-$ L^{-1}, respectively) that have been established by the US Environmental Protection Agency, suggesting that environmental regulations may not adequately protect the integrity of aquatic communities (Schuler and Relyea 2018b).

Results from a long-term ecological research project of urban areas as ecological systems, focused on the city of Baltimore in the eastern US, document the long-term trends and implications of increasing salinity in urban streams. Kaushal et al. (2005) reported chloride concentrations as high as 25% of seawater during winter, and this trend continues to increase (Fig. 4.10). Similarly, Bird et al. (2018) used 16 years of water chemistry data to investigate relationships between major ions and land cover in the Baltimore region through time. Though deicing salt and concrete were the principal nonpoint source contributions to ionic concentrations, higher concentrations of all major ions were recorded in regions with more urban land cover.

Fig. 4.9 A conceptual model of the freshwater salinization syndrome. At least three sets of processes contribute to freshwater salinization. They are listed in order from upstream to downstream including: (i) accelerated weathering throughout the network; (ii) human salt inputs; and (iii) enhanced biological alkalinization in larger systems in response to increased light and nutrient availability. Underlying geology results in different starting points in headwaters based on the potential for chemical weathering. Weathering processes introduce alkalinity, bicarbonate, and base cations along drainage networks. The relative influence of changes in weathering may respond to increasing disturbance along a river network, as salt and nutrient pollution typically change in response to hydrological modifications and stream size. The spatial and temporal extent of the freshwater salinization syndrome varies with land use, climate, and underlying geology (Reproduced from Kaushal et al. 2018)

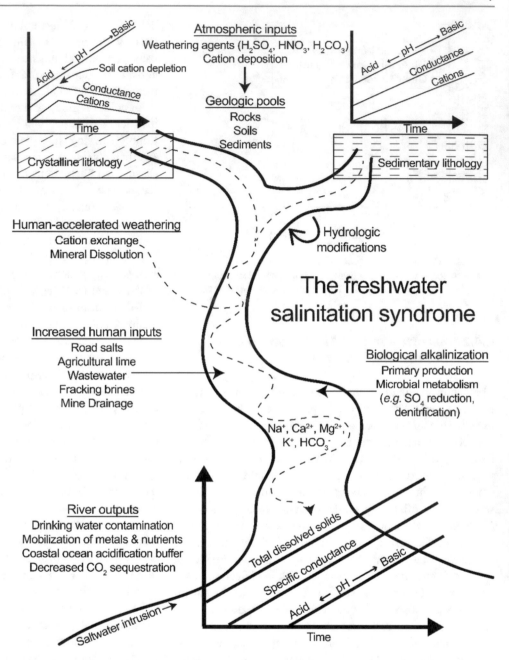

Unexpectedly, the concentrations of most major ions were also increasing in urban streams through time, even with no major changes in land cover during the study. This pattern was not evident in forested and rural sites (Fig. 4.11). Collectively, these results provide additional evidence that road salt application has large impacts on freshwater communities, and that human activity can have long-term effects on salinity and alkalinity in streams.

Agricultural development and resource extraction in a watershed also can influence freshwater salinity (Tyree et al. 2016). In the US, agricultural runoff is a major source of excess Na^+ and Cl^- to rivers and streams. Increasing K^+ concentrations are associated with rivers in the Midwestern US and most likely are due to the application of agricultural fertilizer (potash) rich in K^+. Agricultural liming to support greater crop production also may enhance stream alkalinity and the concentration of dissolved salts in rivers and streams (Kaushal et al. 2018). Secondary salinization, or the salinization of soil, surface water, or groundwater due to human activities, is a particular problem in arid and semi-arid areas due to high demands for both surface and groundwater for irrigation to support dryland agriculture. Irrigation in dryland regions concentrates salts in streams in two ways: by reducing total water volume entering streams

Fig. 4.10 The mean annual concentration of chloride increases with impervious surface area (an indicator of urbanization) for streams along a rural to urban gradient near Baltimore, Maryland. Dashed lines represent thresholds for damage to some land plants and chronic toxicity to sensitive aquatic organisms (Reproduced from Kaushal et al. 2005)

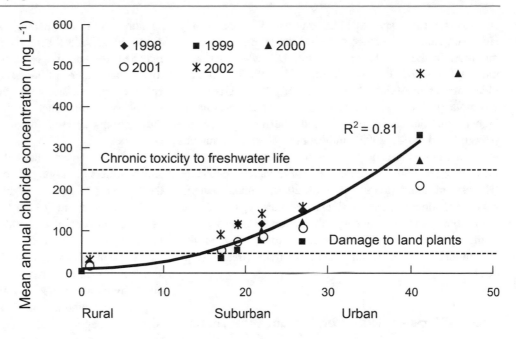

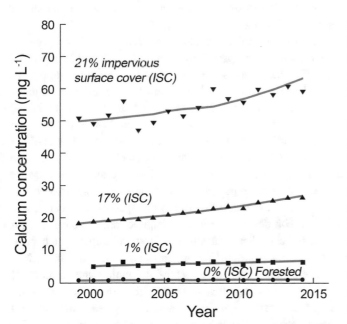

Fig. 4.11 Temporal changes in calcium concentrations in streams along a gradient of urbanization. Major ion concentrations were greater in areas with more urban land cover. Concentrations increased through time in areas with more urban development even through there were no corresponding major changes in land cover (Reproduced from Bird et al. 2018)

(Williams and Williams 1991). Salinization of the Pecos River typifies challenges faced in arid watersheds in the US and Mexico. Reduced flows, inputs of high salinity groundwater, and increased evapotranspiration have collectively increased its salinity (Hoagstrom 2009).

Mining activities in a watershed can also increase salinity and alkalinity in streams and rivers, as chemicals added to mine effluent to neutralize pH, precipitate metals, and oxidize sulfur compounds prior to discharge can introduce large volumes of water containing salts, lime, and other constituents (Kimmel and Argent 2010). For example, stream reaches downstream of discharges from a Pennsylvanian mine complex in the eastern US had TDS and specific conductance values more than an order of magnitude greater than reference sites. Fish community metrics, including species richness (the total number of species), species density, and species diversity (species richness and species evenness), were much lower in sites with the highest TDS and specific conductance values relative to the reference sites (Kimmel and Argent 2010).

The ecological response to anthropogenically-driven changes in salinity and alkalinity is varied. Many aquatic species, especially those adapted to life in arid regions, can tolerate large fluctuations in and high concentrations of ions. For example, in two river systems of Australia, researchers found no relationship between macroinvertebrate community assemblages and salinity levels that exceeded 2 g L^{-1}. Similarly, several fish species in the Murray River, also in Australia, survived laboratory exposures to salinities up to 30 g L^{-1}, possibly reflecting a relatively recent marine

due to increased rates of evapotranspiration, and by increasing the salts entering streams from soil leachate due to increased infiltration rates. In Australia, where salinization is widespread in semiarid agricultural areas, salinity in the lower South Australian Murray River averages ~0.5 g L^{-1}

ancestry for these species (Williams and Williams 1991). However, salinity tolerance can also structure aquatic communities. In the Red River in Texas, US, where salinities range from ~ 200 to $\sim 35,000$ mg L^{-1} TDS, fish communities are grouped into low-, medium-, and high-salinity assemblages. Conspicuously, the species most sensitive to higher salinities have experienced the greatest decline (Higgins and Wilde 2005), indicating that even in streams where organisms are adapted to higher ionic concentrations, increasing salinity presents a conservation challenge (Hoagstrom 2009; Hoagstrom et al. 2010). Similar patterns have been documented in rivers in other regions of the world. For instance, secondary salinization is responsible for changing diversity and abundance of the freshwater fish community rivers in southwestern Australia (Beatty et al. 2011).

4.4.3 Effects of Acidity on Stream Ecosystems

Fresh waters may be naturally acidic due to the decay of organic matter, and anthropogenically acidified by atmospheric deposition of strong inorganic acids formed from sulphate and nitrous oxides released in the burning of fossil fuels, or from acids leached from mining deposits. Naturally acidic waters, tea-colored from the breakdown of organic matter occur in diverse settings including northern peatlands, tropical regions such as the aptly named Río Negro, and blackwater rivers draining swamp forests such as the Ogeechee River in the southeastern US.

Acid precipitation is a phenomenon due to industrialization, and has its greatest influence in streams and rivers in regions of poor buffering capacity, especially those in granitic catchments. The strong inorganic acids H_2SO_4 and HNO_3, formed in the atmosphere from oxides of sulphur and nitrogen released in the burning of fossil fuels, have lowered surface water pH in many regions of Europe and North America. The deleterious effects of acidic stream water on freshwater organisms are well established, primarily in terms of reduced numbers of species and individuals, but there also is evidence of altered ecosystem processes in response to declining pH. The relative impact of increased acidity on aquatic biodiversity is influenced by the amount of deposition, the buffering capacity of the system, and the life history and physiology of aquatic organisms.

Direct physiological effects of acidity are implicated by both field and laboratory studies of aquatic organisms. Reductions of pH have been linked to increased mortality rates and issues with the development of animals (Willoughby and Mappin 1988). Most likely, the negative outcomes are due to an increasing inability to regulate ions as acidity increases, including declining sodium concentrations in body tissues, and increasing inability to obtain sufficient

calcium from surrounding waters (Økland and Økland 1986). Field collections indicate that the effects of lower pH may be taxon-specific, as plecopterans and tricopterans seem to tolerate waters of lower pH more effectively than do ephemeropterans and some dipterans. Hall and Ide (1987) speculate that differences in life cycle and respiratory style of these groups account for their differential susceptibility to acid stress.

Evidence suggests that the timing and intensity of exposure to waters of varying pH can produce varied outcomes in the behavior of aquatic organisms and aquatic community structure. For example, the availability and spatial location of refuge streams of moderate to high alkalinity influenced the habitat use of brook trout in streams of the central Appalachian Mountains of West Virginia, US. Spawning and recruitment occurred primarily in small tributaries with alkalinity above 10 mg L^{-1}, whereas large adults apparently dispersed throughout the catchment (Petty et al. 2005). Similarly, Lepori et al. (2003) were able to differentiate between macroinvertebrate assemblages in streams that became acidic during snowmelt (pH reduced to 5) and those in well-buffered streams (pH remained above 6.6) in alpine regions of Switzerland.

Ecosystem processes also are affected by anthropogenic changes in stream pH (Ferreira and Guérold 2017). Inputs of autumn-shed leaves are an important energy supply to woodland streams, and breakdown rates respond to a number of environmental variables. Breakdown rates of beech leaves *Fagus sylvatica* varied more than 20-fold between the most acidified and circumneutral sites in 25 woodland headwater streams along an acidification gradient in the Vosges Mountains, France (Dangles et al. 2004a). More acidic streams experienced lower rates of microbial respiration and a reduction of microbial species that were associated with decaying leaves. It is important to note that field data indicate that naturally acidic streams may not behave like streams experiencing human-driven increases in acidity. For example, even at a pH of 4, neither the number of microbial taxa nor leaf decomposition rates were strongly depressed relative to more alkaline systems in naturally acidic streams of northern Sweden (Dangles et al. 2004b). The streams in question contain a unique fauna, suggesting that communities found in naturally acidic systems were adapted to the physiological challenges of this environment.

Acidic precipitation promotes the leaching of metals from soils and increased concentrations of free metal ions in affected rivers and streams, and this can affect aquatic organisms and ecosystem processes. Increasing free metal concentrations can be very deleterious to aquatic organisms, as they can accumulate on the surface of gills and impair osmoregulatory processes. For example, aluminum commonly occurs at elevated concentrations in acidic waters (Fig. 4.12). Separate and combined additions of aluminum compounds and inorganic acids to stream channels have been used to distinguish the direct influence of hydrogen ion concentration from the effects of elevated aluminum. In a short-term (24 h) manipulation of a soft-water stream in upland Wales, two salmonid species exhibited far greater susceptibility to the combined effects of acid and aluminum versus sulphuric acid alone, apparently because of respiratory inhibition (Ormerod et al. 1987). Episodes of acidification and elevated aluminum concentrations restricted stream fishes from sites in the northeastern US that had suitable chemical conditions (pH > 6 and inorganic Al < 60 μg L^{-1}) at low flow (Baker et al. 1996). Abundances of the brook trout (*Salvelinus fontinalis*) were reduced and the blacknose dace (*Rhinichthys atratulus*) and sculpin (*Cottus bairdi* and *C. cognatus*) were eliminated from streams that episodically experienced pH < 5 and inorganic Al > 100–200 μg L^{-1}. Behavioral avoidance is one cause of decline, and offers the possibility of subsequent recolonization if alkaline refuge areas are available. Baker et al. suggested that lower mobility in sculpins relative to brook trout may explain why the former were eliminated from acid-pulsed streams. Ecosystem processes, such as leaf-litter decomposition, can also be affected by exposure to Al. In a decomposition study of leaf litter in eight, first- and second-order streams in northeastern France, Ferreira and Guérold (2017) documented reductions in leaf decomposition rates of three tree species with increasing acidity (i.e., decreasing pH) and with increasing Al concentrations (Fig. 4.13).

In response to increasingly stringent air pollution regulations in Europe and North America, there have been sharp reductions in the emission of chemicals contributing to acid rain in these regions. Recovery from acid rain is evident in the chemical characteristics of many systems. Though the biological response to the reductions has been slower than changes in water chemistry, evidence suggests that populations of freshwater organisms are beginning to recover in response to increasing pH in the water column (Warren et al. 2017).

Though acid rain is on the decline in many regions of the world, acidification of surface waters can be driven by other activities including mining. In addition to decreasing pH associated with acid mine drainage, mining activity in a watershed is often associated with increased concentrations of mixtures of heavy metals (e.g., lead, zinc, copper, and aluminum). Ambient metal concentration has been linked to declines in macroinvertebrate species richness and abundance. Notably, the effects of contaminants may influence organisms differently depending on their ontogeny, or life stage. For example, Schmidt et al. (2013) documented the impacts of metal bioavailability on larval and adult insect populations along a gradient of metal concentrations in streams in the Colorado Mineral Belt, a region that has been mined for the last two centuries. There was a significant negative relationship between metal concentrations and larval densities along the gradient, but densities fell precipitously when metal concentrations exceeded aquatic life criteria (cumulative criterion accumulation ratio (CCAR) ≥ 1). A contrasting pattern was seen in the relationship between metal concentration and the emergence of adult insects. Sharp declines in emergence were seen at lower concentrations, while smaller changes were seen in emergence rates in reaches with concentrations above the CCAR (Fig. 4.14). This work indicates that the emergence of adults was a more sensitive indicator of metal contamination, pointing to the need to consider environmental impacts across all stages of an organism's development in order to effectively protect freshwater communities.

Synergistic interactions between heavy metal pollution and land use have the potential to mediate the impact of metal contamination on aquatic ecosystems. Metal toxicity likely is less important in naturally brown- and blackwater streams (Collier et al. 1990; Winterbourn and Collier 1987) or in watersheds with wetlands, owing to the chelating abilities of humic acids, which bind metal ions mobilized at low pH. However, although some evidence indicates algal communities may be resistant to heavy metal contamination, research does suggest that microbial communities can be strongly impacted by metals (Schuler and Relyea 2018a).

Addition of lime to neutralize acid conditions is widely practiced. The River Auda in Norway had lost its anadromous salmon and sensitive mayflies due to anthropogenic

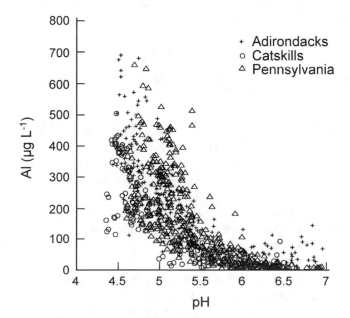

Fig. 4.12 Relationship between aluminum concentrations (measured as inorganic monomeric Aluminum, Alim) and pH for streams of northeastern US. Note the wide range in Alim at low pH values (Reproduced from Wigington et al. 1996)

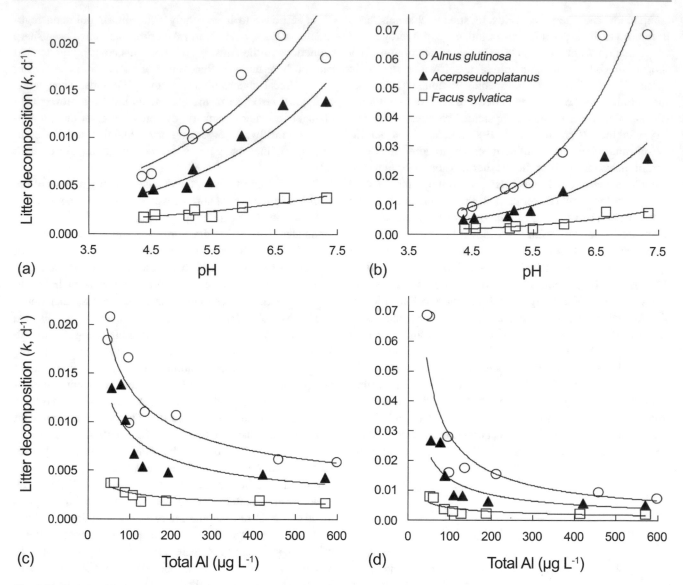

Fig. 4.13 Relationship between the decomposition rates of leaves of three tree species that were enclosed in fine (**a**, **c**) and coarse (**b**, **d**) mesh bags, and pH (**a**, **b**) and total Al concentrations (**c**, **d**) (Reproduced from Ferreira and Guérold 2017)

acidification when liming commenced in 1985. Within two years, sensitive mayflies had returned, and additional macroinvertebrates appeared over the following five-plus years. However, in some cases liming has not been sufficient to offset the effects of episodic acidification. In three acidified Welsh streams that were evaluated for 10 years following the liming of their catchments, pH increased to above 6 and the number of macroinvertebrate species increased, but relatively few acid-sensitive species recovered (Ormerod and Edwards 1987). The occasional appearance but limited persistence of acid-sensitive taxa in limed streams led the authors to suggest that episodes of low pH continued to affect acid-sensitive taxa even after liming. Whether it makes sense to add lime to naturally acidic streams presents a further complication. In Sweden, approximately US $25

million has been spent since 1991 to lime some 8000 lakes and 12,000 km of streams to restore their condition, and as Dangles et al. (2004b) point out, expending funds to lime naturally acidic systems may not be wise management.

4.5 Legacy and Emerging Chemical Contaminants

Globally, rivers and streams face pollution problems associated with both legacy and emerging contaminants. Legacy contaminants are chemicals often produced through industrial and agricultural activities that remain in streams and rivers long after they are first released into the environment (Macneale et al. 2010). Emerging freshwater contaminants

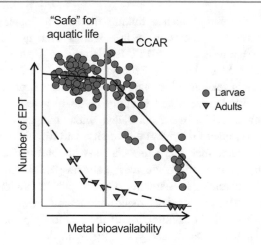

Fig. 4.14 Effects of metal bioavailability on adult and larval aquatic insects. The Cumulative Criterion Accumulation Ratio (CCAR) is derived from the US EPA Aquatic Life Criteria for metals, which is based on the concentration of metals that can be present in surface water before it is likely to harm plant and animal life (Reproduced from Schmidt et al. 2013)

can be defined as naturally occurring or manmade substances that present realized or potential risks to the structure and function of rivers and streams (Nilsen et al. 2019). For example, substances that have never previously been detected or substances that were previous found in much lower concentrations in flowing waters may be considered emerging contaminants because their impacts on human health and the environment are not well understood.

4.5.1 Legacy Contaminants in Rivers and Streams

Examples of legacy contaminants that threaten the structural and functional integrity of freshwaters include metals (e.g., mercury and lead) and pesticides (e.g., dichloro-diphenyl-trichloroethane (DDT)), and polychlorinated biphenyls (PCBs). Often, legacy contaminants collect in sediments that can be repeatedly resuspended in the water column to interact with other chemical constituents and aquatic biota. Some legacy contaminants such as DDT, PCBs, and mercury accumulate in animal tissues because they are differentially retained rather than excreted. Therefore, the longer animals are exposed to the contaminant, the higher the concentration of the contaminant in their tissues will be. Some legacy contaminants also biomagnify, meaning contaminant concentrations increase in tissues of animals feeding at higher trophic levels (i.e., a tertiary consumer would be expected to have higher tissue concentrations than a primary consumer in the same environment). Rachel Carson famously wrote about the environmental implications of the broad consequences of legacy contaminants in her book, *Silent Spring* (Carson 1962).

A notable example of a biomagnifying legacy contaminant is DDT, a synthetic insecticide that was widely and effectively used in the 1940s and 50s to control mosquito populations and agricultural pests. Three million tons of DDT had been produced and used by 1970 (Zimdahl 2015). After a time, mosquito populations evolved resistance to the pesticide, and coupled with increasing amounts of data highlighting the disastrous environmental effects of DDT application, the US Environmental Protection Agency delivered a cancellation order for DDT in the US in 1972. Like many pesticides, DDT can be toxic to non-target organisms (Macneale et al. 2010). It has been linked to egg-shell thinning and subsequent population declines in many North America bird species. This was especially true for birds of prey, including the osprey (*Pandion haliaetus*) and bald eagle (*Haliaeetus leucocephalus*), as they accumulated high concentrations of DDT through biomagnification.

Though use of DDT in the US declined precipitously after the cancellation order, application of the pesticide was still permitted with federal exemption. Additionally, until 1985 the US continued to produce and export the pesticide to countries where there were fewer restrictions on pesticide use (Turusov et al. 2002). As of 2003, only three countries continue to produce DDT—China, the Democratic People's Republic of Korea, and India. However, the emergence and spread of mosquito-borne illnesses, such as Zika and Chikungunya, and the continued persistence of malaria and dengue, support the continued use of the pesticide. For example, DDT continues to be used in many countries for the control of malaria and leishmaniosis; yet, data suggest that disease vectors in affected regions are developing resistance to the pesticide, and that DDT application may eventually decline further (Van Den Berg et al. 2017).

4.5.2 Emerging Contaminants in Running Waters

In a review assessing the potential risk of contaminants of emerging concern (CEC) on aquatic food web ecology, Nilsen et al. (2019) asserted there are currently five primary challenges in evaluating these risks. First, we lack detailed information about complexity of mixtures of CECs that exist in the environment. Second, there is insufficient understanding of the sublethal impacts of CECs on aquatic organisms. Third, research is needed to elucidate some of the biological consequences of the frequency and duration of exposure to CECs and whether these effects are evident across generations (e.g., maternal effects). Fourth, we need a better understanding of how exposure to other environmental stressors (e.g., changes in temperature or flow) may exacerbate or reduce the impact of exposure to CECs. Last, research is needed to understand how CEC exposure can influence trophic relationships. The term "emerging" implies a potential for change—a substance defined as a CEC may later be defined as a legacy contaminant or may be recognized as a new, but relatively benign addition to freshwater systems. In light of this uncertainty, our discussion focuses on CECs that have been observationally or experimentally linked to changes in the structure and function of aquatic systems, while acknowledging this is not an exhaustive list. For those interested in learning about the newest CECs and the methods to detect them in aquatic systems, Dr. Susan Richardson leads a biennial review of developments in water analysis for CECs in *Analytical Chemistry* (e.g., Richardson and Kimura 2016; Richardson and Temes 2018).

In response to the well-known, negative environmental effects of many pesticides and herbicides, there has been a large investment in developing new chemicals and methods to combat agricultural pests and arthropod vectors of human disease. Currently, glyphosate-based herbicides, marketed under the trade name Roundup®, are among the most widely used agricultural chemicals globally (Annett et al. 2014). There is a wide diversity of formulas and applications of glyphosate; hence, aquatic communities may be simultaneously exposed to a variety of formulas and concentrations of the herbicide, with poorly understood consequences. It is likely that glyphosate toxicity is species-specific, and may occur through multiple mechanisms, influenced by the timing, formulation, and quantity of the herbicide applied (Annett et al. 2014). Amphibians seem to be particularly sensitive to exposure to the herbicide due to aspects of their physiology and life history. In a mesocosm study investigating the impact of glyphosate on aquatic macroinvertebrates, initial evidence suggests that under experimental conditions, sedimentation, rather than pesticide addition, had a stronger impact on drift and adult emergence (Magbanua et al. 2016).

Fracking, the hydraulic fracturing of rock by injection of a pressurized liquid, has markedly increased the amount of natural gas used to produce energy in the US and elsewhere. Through this process, a mixture of water, sand, and other chemicals are injected at high pressure into a wellbore, or deep hole, to create cracks in deep rock formations to release natural gas and other materials. If not treated properly, the wastewater produced from fracking can contaminate surface and groundwater supplies. Reliance on coal may be reduced, improving air quality, but other environmental harm can occur due to high water consumption, contaminants in return water, and release of methane, a powerful greenhouse gas. In watersheds with newly drilled natural gas wells, streams and rivers are threatened with increases in sediment runoff, reduced streamflow, and contamination from introduced chemicals and the resulting wastewater (Entrekin et al.

2011). There is evidence linking fracking to increases in concentrations of ions in surface waters used as drinking water (Harkness et al. 2015). In a study of streams in the Marcellus Shale basin in Pennsylvania in the eastern US, researchers examined stream physiochemical properties, trophic relationships, and mercury levels to assess whether fracking activities were correlated with changes in water chemistry, macroinvertebrate diversity, and trophic ecology. Fracking activities were associated with elevated methyl mercury concentrations, decreased stream pH, elevated dissolved mercury in stream water, and reductions in environmentally-sensitive macroinvertebrate taxa (Grant et al. 2016). In an examination of the relationships between instream microbial diversity and well-pad density, Trexler et al. (2014) documented lower diversity in bacterial communities collected in sites with natural gas development relative to reference sites. Moreover, community structure was more similar among bacterial communities collected from sites with gas development than communities at reference sites. This may be because fracking results in a decline in pH, which in turn is responsible for influencing microbial diversity at the population and community level.

Pharmaceuticals and personal care products (PPCPs) occurring in the water column, sediments, and organisms found in freshwater systems are an important class of chemical of emerging concern (Ebele et al. 2017). The diversity and use of PPCPs is increasing in both human and veterinary medicine throughout the globe. Caffeine and other stimulants, antihistamines, amphetamines, antibiotics, and antidepressants are all PPCPs commonly found in streams and rivers (Richmond et al. 2017). Many PPCPs are persistent in surface water, and are not removed by conventional wastewater treatment; hence they can move from wastewater effluent into freshwater systems. In addition, PPCPs often enter the natural environment via veterinary pathways without treatment (Ebele et al. 2017). Therefore, even with enhanced treatment technologies in existing systems or the construction of new facilities, increasing concentrations of PPCPs in freshwater systems will most likely remain a challenge for decades to come. Though research on PPCPs is relatively limited, researchers have begun to assess the risk they pose to the environment by examining their persistence in the environment, whether or not they bioaccumulate in aquatic food webs, and if and in what concentrations they are toxic to aquatic organisms (Ebele et al. 2017; Nilsen et al. 2019).

Existing literature on the potential ecological impacts of PPCPs gives reason for concern. Evidence indicates that these compounds are being consumed by aquatic organisms, may be transferred across terrestrial-aquatic boundaries through trophic interactions, can alter the growth and behavior of aquatic organisms, and can influence ecosystem processes (Richmond et al. 2017). A study examining relationships between wastewater discharge and PPCP concentrations in consumers documented increased concentrations of PPCPs in both aquatic and riparian consumers at sites with increasing influence from wastewater inputs (Fig. 4.15a; Richmond et al. 2018). Comparable within-site concentrations of PPCPs in aquatic invertebrate larvae and riparian predators indicate there may be a direct trophic transfer from emerging insects to riparian predators. Additionally, vertebrates feeding on aquatic macroinvertebrates could be consuming a wide variety of PPCPs, and in the case

of some drugs (i.e. antidepressants), they may be ingesting as much as one-half of a recommended therapeutic dose for humans (Fig. 4.15b).

Exposure to PPCPs may influence ecosystem function, as Rosi et al. (2018) showed by exposing biofilm communities to drugs commonly found in wastewater. Biofilm respiration rates varied in response to exposure to caffeine, cimetidine (acid-reducer, commonly used to treat ulcers), ciprofloxacin (antibiotic), and diphenhydramine (antihistamine) among four streams along an urban gradient. Respiration rates were

Fig. 4.15 (**a**) Pharmaceutical concentrations in aquatic invertebrates and riparian spiders. Total pharmaceutical concentration (ng g^{-1} dry weight) in riparian web-building spiders (dark grey) and aquatic invertebrates (black), arranged in decreasing order of wastewater influence. Violin plots illustrate kernel probability density and horizontal lines within each plot reflect median concentrations. (**b**) Estimated dietary intake of pharmaceuticals by two invertebrate predators compared to recommended pharmaceutical doses for humans. Dietary intake rates are displayed as the percentage of recommended human doses by therapeutic class for platypus (black) and brown trout (grey) in Brushy Creek in southeastern Australia. (CCBs calcium channel blockers, GORD gastroesophageal reflux disease, NSAID non-steroidal anti-inflammatory drugs, RAS renin angiotensin system) (Modified from Richmond et al. 2018)

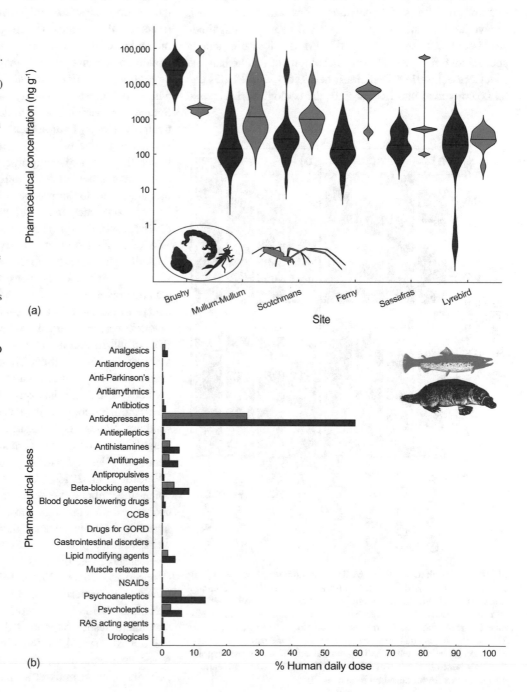

low in the least urban stream when exposed to the PPCPs, but respiration in the most urban stream was resistant to drug exposure (Fig. 4.16). This suggests that biofilm communities in more urban systems may be exposed to a suite of PPCPs that select for populations of bacteria that can maintain functional processes even when exposed to contaminants.

Engineered nanomaterials (ENMs) are small (1–100 nm) particles used in various manufacturing applications. They have high surface to volume ratios and can exhibit unique physical and chemical properties, making it difficult to predict their impact on stream ecosystems. Often, ENMs are engineered with surface coatings that can increase their bioavailability and bioreactivity by orders of magnitude (Reid et al. 2018). Sensitivity to ENM toxicity varies among species and between stages of development (Callaghan and MacCormack 2017). Pesticides, herbicides, and PPCPs can all be converted into ENMs; thus, the ecological risks posed

by ENMs also include the aforementioned risks presented by emerging contaminants already discussed.

4.6 Plastic Pollution in Freshwater Systems

Microplastics fragments (<5 mm) resulting from the fragmentation of plastic debris or consumer and industrial products are a newly recognized contaminant of aquatic ecosystems worldwide (Hoellein et al. 2014, 2019). First recognized in marine systems, an estimated 80% of plastic in the sea originates from inland sources and is transported by rivers to the oceans (Mani et al. 2015). A survey of microplastics over 820 km of the Rhine River from Basel to Rotterdam, the first and last major industrial cities along this large European river, found microplastic particles everywhere, but most abundantly near large cities (Fig. 4.17). Sources of variability include proximity to wastewater treatment plants, hydrological effects of mixing and turbulence, and timing of sampling relative to timing of releases. Given the origin of many microplastics from the fragmentation of synthetic fabrics and abrasives in personal care products, it is not surprising that concentrations are highest below wastewater treatment plants, as McCormick et al. (2016) reported from surveys upstream and downstream of WWTP effluent sites at nine rivers in Illinois, US.

Although research into ecological effects are at an early stage, negative consequences for growth and survival of aquatic organisms are a concern. Chemical contaminants can adhere to particles that are small enough to be ingested by invertebrates, and passed up the food chain. Ingestion poses risks of gut irritation and blockage, and of nutritional value that may be low or absent (Scherer et al. 2018). Transfer of toxins adhered to plastic into organisms may pose an even greater risk, but that has yet to be ascertained.

Macroinvertebrates collected from several rivers in southern Wales, UK, identified microplastic particles in roughly half of the individuals sampled (baetid and heptageniid mayflies and hydropsychid caddisflies) and at all sites (Windsor et al. 2019). Particles were significantly reduced when invertebrate gut contents were evacuated prior to analysis, indicating that most particles were acquired by ingestion. Microplastics are indeed entering freshwater food webs, with the potential for trophic transfers, and they have been observed in the guts of predatory fish in UK waters (Horton and Dixon 2018). Microplastic also serves as a substrate for bacterial assemblages, but whether this benefits particle consumers is uncertain. RNA sequencing revealed that microplastics supported distinct bacterial assemblages that differed from those found on natural particles in having fewer taxa, apparently those best able to degrade plastic polymers (McCormick et al. 2016).

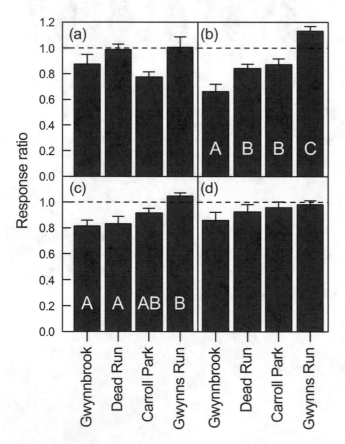

Fig. 4.16 Respiration rates of biofilm communities grown on organic substrates and exposed to pharmaceuticals in streams along an urbanization gradient (left to right) (**a**) caffeine, (**b**) ciprofloxacin, (**c**) cimetidine, and (**d**) diphenhydramine. Bars indicate response ratios of respiration rates for treatments with pharmaceuticals to no pharmaceutical controls. The dashed line is equal to a response ratio of 1, indicating no effect of exposure. Lowercase letters indicate significant differences between sites (Tukey's honestly significant difference; $P < 0.05$). Error bars represent standard error (Reproduced by Rosi et al. 2018)

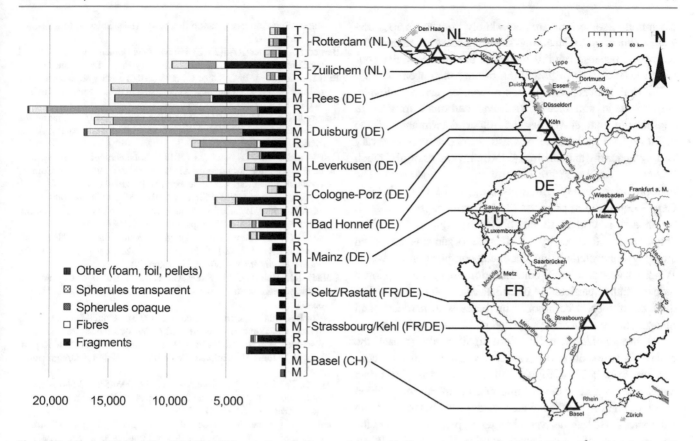

Fig. 4.17 Microplastics profile of the Rhine River in Europe. Horizontal columns indicate microplastic abundance 1000 m^{-3} and the fraction of the kinds of microplastic. Letters indicate the position in the cross section of the river (L: *left bank*, M: *mid-river*, R: *right bank*, T: *transect*) (Reproduced from Mani et al. 2015)

4.7 Summary

The constituents of river water include suspended inorganic matter, dissolved major ions, dissolved nutrients, suspended and dissolved organic matter, gases, and trace metals. The dissolved gases of importance are oxygen and carbon dioxide. Exchange with the atmosphere maintains the concentrations of both at close to the equilibrium determined by temperature and atmospheric partial pressure, especially in streams that are small and turbulent. Photosynthetic activity in highly productive settings can elevate oxygen to super-saturated levels during the day. Respiration has the opposite effect, reducing oxygen and elevating CO_2. High levels of organic waste can reduce oxygen concentrations below life-sustaining levels, and CO_2 can be elevated from groundwater inputs or biological activity.

Many factors influence the composition of river water and, as a consequence, it is highly variable in its chemical composition. The concentration of the dissolved major ions (Ca^{2+}, Na^+, Mg^{2+}, K^+, HCO_3^-, SO_4^{2-}, Cl^-) in river water is roughly 120 mg L^{-1} on a world average. However, the chemical makeup of river water is highly variable, ranging from a few mg L^{-1} where rainwater collects in catchments of very hard rocks to some thousands of mg L^{-1} in arid regions.

Chemical variation among streams is primarily governed by the type and composition of rocks and by the amount and chemical composition of precipitation in a given region. For instance, the concentration of total dissolved salts is roughly twice as great in rivers draining sedimentary terrain compared to rivers flowing through regions dominated by igneous and metamorphic rock, owing to differences in weathering and chemical composition of the parent material. Due to increased rates of dilution and decreased rates of evaporation, regions with large amounts of precipitation and surface water runoff typically have stream water with lower ionic concentrations relative to water flowing through arid climates. The influence of precipitation on water chemistry is typically minimal except in regions with high surface runoff flowing over recalcitrant rocks.

River chemistry changes temporally under the multiple influences of seasonal changes in discharge regime, precipitation inputs, and biological activity. Groundwater typically is both more concentrated and less variable in ionic concentration than surface waters, because of its longer association with the underlying geology. In undisturbed catchments, some ions are remarkably constant across discharge fluctuations spanning several orders of magnitude. However, because rainfall increases the surface water contribution, increases in flow often dilute ion concentration.

Natural waters contain a solution of carbon dioxide, carbonic acid and bicarbonate, and carbonate ions in an equilibrium that serves as the major determinant of the acidity-alkalinity balance of fresh waters. Rivers can vary widely in acidity and alkalinity. Extreme values of pH (much below 5 or above 9) are harmful to most organisms. The bicarbonate buffer system, consisting of the CO_2–HCO_3^-–CO_3^{2-} equilibrium, provides the buffering capacity of fresh water that is critical to the health of aquatic biota. The abundance and diversity of riverine organisms are often strongly connected to the chemical composition of water. Water of very low ionic concentration appears to support a reduced fauna, particularly of crustaceans and mollusks. The number of species commonly increases with hardness, and many taxa are distinctly "soft-water" or "hard-water" forms.

Anthropogenic activities can significantly impact the chemical composition of rivers and streams. Land use conversion from natural to agricultural, industrial, or urban landscapes can alter the volume and chemical composition of precipitation, surface water, and groundwater entering watersheds. Human-derived changes in pH, salinity, and the concentrations of heavy metals are linked to changes in the structure and function of aquatic communities. Research focused on emerging contaminants, such as PCPPs, plastics, and nanoparticles, has demonstrated the ubiquity of anthropogenically-derived chemical changes in the world's aquatic systems. However, much more research is needed to elucidate the response of aquatic organisms and ecosystem processes to acute and chronic exposure to known and emerging contaminants, and to the potentially interactive effects among anthropogenic stressors in freshwater systems.

References

Ågren A, Haei M, Kohler S et al (2010) Regulation of stream water dissolved organic carbon (DOC) concentrations during snowmelt: the role of discharge, winter climate and memory effects. Biogeosciences 7:2901–2913

Annett R, Habibi HR, Hontela A (2014) Impact of glyphosate and glyphosate-based herbicides on the freshwater environment. J Appl Toxicol 34:458–479

Baker J, Van Sickle J, Gagen C et al (1996) Episodic acidification of small streams in the northeastern United States: effects on fish populations. Ecol Appl 6:422–437

Baldwin BS, Carpenter M, Rury K et al (2012) Low dissolved ions may limit secondary invasion of inland waters by exotic round gobies and dreissenid mussels in North America. Biol Invasions 14:1157–1175

Beatty SJ, Morgan DL, Rashnavadi M et al (2011) Salinity tolerances of endemic freshwater fishes of south-western Australia: implications for conservation in a biodiversity hotspot. Mar Freshw Res 62:91–100

Berner EK, Berner RA (2012) Global environment: water, air, and geochemical cycles. Princeton University Press, Princeton, NJ

Bird DL, Groffman PM, Salice CJ et al (2018) Steady-state land cover but non-steady-state major ion chemistry in urban streams. Environ Sci Technol 52:13015–13026

Bukowski SJ, Auld JR (2014) The effects of calcium in mediating the inducible morphological defenses of a freshwater snail, Physa acuta. Aquat Ecol 48:85–90

Callaghan NI, MacCormack TJ (2017) Ecophysiological perspectives on engineered nanomaterial toxicity in fish and crustaceans. Compar Biochem Physi Part C: Toxic Pharma 193:30–41

Capps KA, Bentsen CN, Ramirez A (2016) Poverty, urbanization, and environmental degradation: urban streams in the developing world. Freshw Sci 35:429–435. https://doi.org/10.1086/684945

Carson R (1962) Silent spring. Houghton Mifflin Harcourt, New York

Christensen ER, Li A (2014) Physical and chemical processes in the aquatic environment. Wiley, Hoboken, NJ

Collier KJ, Ball OJ, Graesser AK et al (1990) Do organic and anthropogenic acidity have similar effects on aquatic fauna? Oikos 33–38

Connor R, Renata A, Ortigara C et al (2017) The United Nations World Water Development Report 2017. Wastewater: The untapped resource. The United Nations World Water Development Report. United Nations Educational, Scientific and Cultural Organization, Paris

Crosa G, Froebrich J, Nikolayenko V et al (2006) Spatial and seasonal variations in the water quality of the Amu Darya River (Central Asia). Water Res 40:2237–2245

Dangles O, Gessner MO, Guerold F et al (2004a) Impacts of stream acidification on litter breakdown: implications for assessing ecosystem functioning. J Appl Ecol 41:365–378

Dangles O, Malmqvist B, Laudon H (2004b) Naturally acid freshwater ecosystems are diverse and functional: evidence from boreal streams. Oikos 104:149–155

Dodds WK, Smith VH (2016) Nitrogen, phosphorus, and eutrophication in streams. Inland Waters 6:155–164. https://doi.org/10.5268/IW-6.2.909

Downing JA (2014) Limnology and oceanography: two estranged twins reuniting by global change. Inland Waters 4:215–232

Dutton CL, Subalusky AL, Hamilton SK et al (2018) Organic matter loading by hippopotami causes subsidy overload resulting in downstream hypoxia and fish kills. Nat Commun 9:1951

Ebele AJ, Abdallah MA-E, Harrad S (2017) Pharmaceuticals and personal care products (PPCPs) in the freshwater aquatic environment. Emerging Contaminants 3:1–16

Entrekin S, Evans-White M, Johnson B et al (2011) Rapid expansion of natural gas development poses a threat to surface waters. Front Ecol Environ 9:503–511

Fay L, Shi X (2012) Environmental impacts of chemicals for snow and ice control: state of the knowledge. Water Air Soil Pollut 223:2751–2770

Ferreira V, Guérold F (2017) Leaf litter decomposition as a bioassessment tool of acidification effects in streams: evidence from a field study and meta-analysis. Ecol Indic 79:382–390

Gameson A, Wheeler A (1977) Restoration and recovery of the Thames estuary. In: Cairns J, Dickinson KL, Herricks EE (eds) Recovery and restoration of damaged ecosystems. University of Virginia Press, Charlottesville, VA, pp 72–101

Gibbs R (1970) Mechanisms controlling world water chemistry. Science 170:1088–1090

Golterman H (1975) Chemistry. In: Whitton B (ed) River ecology. Volume 2. Studies in ecology. University of California Press, Berkeley, pp 39–80

Grant CJ, Lutz AK, Kulig AD et al (2016) Fracked ecology: response of aquatic trophic structure and mercury biomagnification dynamics in the Marcellus Shale Formation. Ecotoxicology 25:1739–1750

Griffith MB (2014) Natural variation and current reference for specific conductivity and major ions in wadeable streams of the conterminous USA. Freshw Sci 33:1–17

Hall RJ, Ide FP (1987) Evidence of acidification effects on stream insect communities in central Ontario between 1937 and 1985. Can J Fish Aquat Sci 44:1652–1657

Harkness JS, Dwyer GS, Warner NR et al (2015) Iodide, bromide, and ammonium in hydraulic fracturing and oil and gas wastewaters: environmental implications. Environ Sci Technol 49:1955–1963. https://doi.org/10.1021/es504654n

Hasler C, Butman D, Jeffrey J et al (2015) Freshwater biota and rising pCO$_2$? 19:98–108. https://doi.org/10.1111/ele.12549

Higgins CL, Wilde GR (2005) The role of salinity in structuring fish assemblages in a prairie stream system. Hydrobiologia 549:197–203

Hintz WD, Relyea RA (2017) Impacts of road deicing salts on the early-life growth and development of a stream salmonid: salt type matters. Environ Pollut 223:409–415

Hoagstrom CW (2009) Causes and impacts of salinization in the lower Pecos River. Great Plains Res 2009:27–44

Hoagstrom CW, Zymonas ND, Davenport SR et al (2010) Rapid species replacements between fishes of the North American plains: a case history from the Pecos River. Aquat Invasions 5:141–153

Hoellein T, Rojas M, Pink A et al (2014) Anthropogenic litter in urban freshwater ecosystems: distribution and microbial interactions. PLoS One 9:e98485

Hoellein TJ, Shogren AJ, Tank JL et al (2019) Microplastic deposition velocity in streams follows patterns for naturally occurring allochthonous particles. Sci Rep 9:1–11

Horton AA, Dixon SJ (2018) Microplastics: an introduction to environmental transport processes. Wiley Interdiscip Rev-Water 5:10. https://doi.org/10.1002/wat2.1268

Hotchkiss E, Hall R Jr, Sponseller R et al (2015) Sources of and processes controlling CO$_2$ emissions change with the size of streams and rivers. Nat Geosci 8:696

Hunter WR, Apley ML, Burky AJ et al (1967) Interpopulation variations in calcium metabolism in the stream limpet, *Ferrissia rivularis* (Say). Science 155:338–340

Hynes HBN (1970) The ecology of stream insects. Ann Rev Ent 15:25–42

Johnson C, Needham P (1966) Ionic composition of Sagehen Creek, California, following an adjacent fire. Ecology 47:636–639

Johnson E, Austin BJ, Inlander E et al (2015) Stream macroinvertebrate communities across a gradient of natural gas development in the Fayetteville Shale. Sci Total Environ 530:323–332

Kaushal SS, Groffman PM, Likens GE et al (2005) Increased salinization of fresh water in the northeastern United States. Proc Natl Acad Sci 102:13517–13520

Kaushal SS, Likens GE, Pace ML et al (2018) Freshwater salinization syndrome on a continental scale. Proc Nat Acad Sci 115:E574–E583

Kempe S, Pettine M, Cauwet G (1991) Biogeochemistry of European rivers. In: Degens ET, Kempe S, Richey JE (eds) Biogeochemistry of major world rivers. SCOPE 42. Scientific Committee on Problems of the Environment (SCOPE). Wilely, New York, pp 169–212

Kilham P (1990) Mechanisms controlling the chemical composition of lakes and rivers: data from Africa. Limnol Oceanogr 35:80–83

Kilham SS, Kreeger DA, Lynn SG et al (1998) COMBO: a defined freshwater culture medium for algae and zooplankton. Hydrobiologia 377:147–159

Kimmel WG, Argent DG (2010) Stream fish community responses to a gradient of specific conductance. Water Air Soil Pollut 206:49–56

Kwak TJ, Waters TF (1997) Trout production dynamics and water quality in Minnesota streams. Trans Amer Fishs Soci 126:35–48

Lauerwald R, Laruelle GG, Hartmann J et al (2015) Spatial patterns in CO$_2$ evasion from the global river network. Glob Biogeochem Cycle 29:534–554

Lepori F, Barbieri A, Ormerod SJ (2003) Effects of episodic acidification on macroinvertebrate assemblages in Swiss Alpine streams. Freshw Biol 48:1873–1885

Lewis WM, Saunders JF (1989) Concentration and transport of dissolved and suspended substances in the Orinoco River. Biogeochemistry 7:203–240

Likens GE, Bormann FH (1995) Biogeochemistry of a forested ecosystem. Springer, Berlin

Likens G, Bormann F, Johnson N et al (1967) The calcium, magnesium, potassium, and sodium budgets for a small forested ecosystem. Ecology 48:772–785

Likens GE, Bormann FH, Johnson NM et al (1970) Effects of forest cutting and herbicide treatmetn on nutrient budgets in a Hubbard Brook watershed-ecosystem. Ecol Monogr 40:23–47. https://doi.org/10.2307/1942440

Livingstone DA (1963) Chemical composition of rivers and lakes (vol. 440). US Government Printing Office, Washington, DC

Lodge DM, Brown KM, Klosiewski SP et al (1987) Distribution of freshwater snails: spatial scale and the relative importance of physicochemical and biotic factors. Am Malacol Bull 5:73–84

Lucas RW, Sponseller RA, Laudon H (2013) Controls over base cation concentrations in stream and river waters: a long-term analysis on the role of deposition and climate. Ecosystems 16:707–721

Lutz BD, Mulholland PJ, Bernhardt ES (2012) Long-term data reveal patterns and controls on stream water chemistry in a forested stream: Walker Branch, Tennessee. Ecol Monogr 82:367–387

Macneale KH, Kiffney PM, Scholz NL (2010) Pesticides, aquatic food webs, and the conservation of Pacific salmon. Front Ecol Environ 8:475–482

Magbanua FS, Townsend CR, Hageman KJ et al (2016) Individual and combined effects of fine sediment and glyphosate herbicide on invertebrate drift and insect emergence: a stream mesocosm experiment. Freshw Sci 35:139–151

Magin K, Somlai-Haase C, Schäfer RB et al (2017) Regional-scale lateral carbon transport and CO$_2$ evasion in temperate stream catchments. Biogeosciences 14:5003–5014

Mani T, Hauk A, Walter U et al (2015) Microplastics profile along the Rhine River. Sci Rep 5:17988

Marino GR, DiLorenzo JL, Litwack HS, Najarian TO, Thatcher ML (1991) General water quality assessment and trends analysis of the delaware estuary, part one: status and trend analysis. Najarian Associates, Eatontown, NJ. http://www.najarian.com/

McCormick AR, Hoellein TJ, London MG et al (2016) Microplastic in surface waters of urban rivers: concentration, sources, and associated bacterial assemblages. Ecosphere 7:22. https://doi.org/10.1002/ecs2.1556

Meybeck M (1977) Dissolved and suspended matter carried by rivers: composition, time and space variations and world balance. Paper presented at the Interactions between sediments and freshwater, Amsterdam

Navrátil T, Norton SA, Fernandez IJ et al (2010) Twenty-year inter-annual trends and seasonal variations in precipitation and stream water chemistry at the Bear Brook Watershed in Maine, USA. Environ Monit Assess 171:23–45

Nilsen E, Smalling KL, Ahrens L et al (2019) Critical review: grand challenges in assessing the adverse effects of contaminants of emerging concern on aquatic food webs. Environ Toxicol Chem 38:46–60. https://doi.org/10.1002/etc.4290

Økland J, Økland K (1986) The effects of acid deposition on benthic animals in lakes and streams. Experientia 42:471–486

Ormerod S, Edwards R (1987) The ordination and classification of macroinvertebrate assemblages in the catchment of the River Wye in relation to environmental factors. Freshw Biol 17:533–546

Ormerod S, Boole P, McCahon C et al (1987) Short-term experimental acidification of a Welsh stream: comparing the biological effects of hydrogen ions and aluminium. Freshw Biol 17:341–356

Petty JT, Lamothe PJ, Mazik PM (2005) Spatial and seasonal dynamics of brook trout populations inhabiting a central Appalachian watershed. Trans Am Fish Soc 134:572–587

Pringle CM, Triska FJ (1991) Effects of geothermal groundwater on nutrient dynamics of a Lowland Costa Rican stream. Ecology 72:951–965. https://doi.org/10.2307/1940596

Raymond PA, Hartmann J, Lauerwald R et al (2013) Global carbon dioxide emissions from inland waters. Nature 503:355

Reid AJ, Carlson AK, Creed IF et al (2018) Emerging threats and persistent conservation challenges for freshwater biodiversity. Biol Rev 94:849–873

Richardson SD, Kimura SY (2016) Water analysis: emerging contaminants and current issues. Anal Chem 88:546–582. https://doi.org/10.1021/acs.analchem.5b04493

Richardson SD, Temes TA (2018) Water analysis: emerging contaminants and current issues. Anal Chem 90:398–428. https://doi.org/10.1021/acs.analchem.7b04577

Richmond EK, Grace MR, Kelly JJ et al (2017) Pharmaceuticals and personal care products (PPCPs) are ecological disrupting compounds (EcoDC). Elementa-Sci Anthrop 5:8. https://doi.org/10.1525/elementa.252

Richmond EK, Rosi EJ, Walters DM et al (2018) A diverse suite of pharmaceuticals contaminates stream and riparian food webs. Nat Commun 9:9. https://doi.org/10.1038/s41467-018-06822-w

Rosi EJ, Bechtold HA, Snow D et al (2018) Urban stream microbial communities show resistance to pharmaceutical exposure. Ecosphere 9:16. https://doi.org/10.1002/ecs2.2041

Scherer C, Weber A, Lambert S et al (2018) Interactions of microplastics with freshwater biota. Freshw Microplast. Springer, Cham, pp 153–180

Schmidt TS, Kraus JM, Walters DM et al (2013) Emergence flux declines disproportionately to larval density along a stream metals gradient. Environ Sci Technol 47:8784–8792. https://doi.org/10.1021/es3051857

Schuler MS, Relyea RA (2018a) A review of the combined threats of road salts and heavy metals to freshwater systems. Bioscience 68:327–335

Schuler MS, Relyea RA (2018b) Road salt and organic additives affect mosquito growth and survival: an emerging problem in wetlands. Oikos 127:866–874

Smith VH (2016) Effects of eutrophication on maximum algal biomass in lake and river ecosystems. Inland Waters 6:147–154

Stallard R, Edmond J (1983) Geochemistry of the Amazon: 2. The influence of geology and weathering environment on the dissolved load. J Geophys Res: Oceans 88:9671–9688

Sutcliffe D, Carrick T (1983) Relationships between chloride and major cations in precipitation and streamwaters in the Windermere catchment (English Lake District). Freshw Biol 13:415–441

Swank WT, Vose J, Elliott K (2001) Long-term hydrologic and water quality responses following commercial clearcutting of mixed hardwoods on a southern Appalachian catchment. For Ecol Manage 143:163–178

Thompson K, Christofferson W, Robinette D et al (2006) Characterizing and managing salinity loadings in reclaimed water systems. American Water Works Association, Alexandria, VA

Trexler R, Solomon C, Brislawn CJ et al (2014) Assessing impacts of unconventional natural gas extraction on microbial communities in headwater stream ecosystems in Northwestern Pennsylvania. Front Microbiol 5:522

Turusov V, Rakitsky V, Tomatis L (2002) Dichlorodiphenyltrichloroethane (DDT): ubiquity, persistence, and risks. Environ Health Perspect 110:125–128

Tyree M, Clay N, Polaskey S et al (2016) Salt in our streams: even small sodium additions can have negative effects on detritivores. Hydrobiologia 775:109–122

Van Den Berg H, Manuweera G, Konradsen F (2017) Global trends in the production and use of DDT for control of malaria and other vector-borne diseases. Malaria J 16:401

Visser PM, Verspagen JM, Sandrini G et al (2016) How rising CO_2 and global warming may stimulate harmful cyanobacterial blooms. Harmful Algae 54:145–159

Walling D (1984) Dissolved loads and their measurement. Erosion and sediment yield: some methods of measurement and modelling. Geo Books, Regency House, Norwich, pp 111–177

Walling D, Webb B (1975) Spatial variation of river water quality: a survey of the River Exe. Trans Instit Brit Geogr 1975:155–171. https://doi.org/10.2307/621615

Warren DR, Kraft CE, Josephson DC et al (2017) Acid rain recovery may help to mitigate the impacts of climate change on thermally sensitive fish in lakes across eastern North America. Glob Change Biol 23:2149–2153

Weiner ER (2012) Applications of environmental aquatic chemistry: a practical guide, 3rd edn. CRC Press

Welch KA, Lyons WB, Whisner C et al (2010) Spatial variations in the geochemistry of glacial meltwater streams in the Taylor Valley, Antarctica. Antarct Sci 22:662–672

Wen Y, Schoups G, Van De Giesen N (2017) Organic pollution of rivers: combined threats of urbanization, livestock farming and global climate change. Sci Rep 7:43289

Wetzel RG (2001) Limnology: lake and river ecosystems. Academic Press, San Diego, CA

Wigington PJ Jr, DeWalle DR, Murdoch PS, et al (1996) Episodic acidification of small streams in the northeastern United States: ionic controls of episodes. Ecol Appl 6:389–407

Williams M, Williams W (1991) Salinity tolerances of four species of fish from the Murray-Darling river system. Hydrobiologia 210:145–150

Willoughby L, Mappin R (1988) The distribution of Ephemerella ignita (Ephemeroptera) in streams: the role of pH and food resources. Freshw Biol 19:145–155

Windsor FM, Tilley RM, Tyler CR et al (2019) Microplastic ingestion by riverine macroinvertebrates. Sci Total Environ 646:68–74

Winterbourn M, Collier K (1987) Distribution of benthic invertebrates in acid, brown water streams in the South Island of New Zealand. Hydrobiologia 153:277–286

Young RG, Huryn AD (1998) Comment: improvements to the diurnal upstream-downstream dissolved oxygen change technique for determining whole-stream metabolism in small streams. Can J Fish Aquat Sci 55:1784–1785

Zimdahl RL (2015) Six chemicals that changed agriculture. Academic Press, London

The Abiotic Environment

At all spatial scales in fluvial ecosystems, studies of stream-dwelling organisms support the expectation that greater physical complexity of the environment promotes increased biological richness. That organisms are adapted to aspects of habitat, such that the traits of organisms reflect features of the environment, is a fundamental idea in ecology referred to as the habitat template concept (Southwood 1988). The key habitat needs of a species commonly are identified from the subset of environmental variables that best correlate with its distribution and abundance. Although additional factors also influence the composition and diversity of biological assemblages, including interactions among species and taxon richness at the regional scale, the abiotic environment provides an important starting point for investigations of species distributions and abundances. This view has two important corollaries. First, environments that are either structurally simple or extreme tend to support fewer species, whereas more moderate and heterogeneous habitats support more species. Second, a high frequency of disturbance tends to diminish biological richness, although a moderate level of disturbance potentially may enhance diversity by maintaining an ever-changing spatial mosaic of conditions. These principles forecast the consequences of human disturbance: anthropogenic degradation and homogenization of habitat will lead to biodiversity decline with unpredictable consequences for ecosystem function.

Habitat is often described as where a species lives, and thus is an important component of a species' niche. The latter term is broader, describing a species' place in a biological community and incorporating all of the physical and biological conditions needed for a species to maintain its population in an area (Begon et al. 2005). The niche concept incorporates species interactions, distinguishing between the space that a species could occupy in the absence of predators and competitors (the fundamental niche) and the more restricted space where a species is found (the realized niche). In this chapter we focus on key abiotic aspects of habitat that influence the distribution and abundance of the biota of fluvial ecosystems; later, in Chap. 10, the influence of species interactions will be explored.

Habitat features vary across small to large spatial scales, referred to as micro-, meso-, and macrohabitat (Vinson and Hawkins 2003); and from very short to long time scales. Individual taxa are adapted to a specific range of habitat conditions, and will be more or less abundant depending on the matching of their morphological, behavioral, and physiological traits to environmental conditions. Thus the abiotic environment, acting on species traits, serves as a filter that determines which taxa of the region are likely to be found at the local scale (Poff 1997).

In fluvial ecosystems, key abiotic features of the environment usually are those related to current, substrate, temperature, and sometimes water chemistry variables such as alkalinity and dissolved oxygen. Water chemistry and dissolved oxygen are important under natural conditions only in some unusual environments and under low flows, but both factors can be very influential when human activities result in polluted waters. Current is the defining feature of rivers and streams. It conveys benefits, such as transport of resources to the organism and removal of wastes, but also risks, of which being swept away is the most obvious. Physical habitat includes the substrate composition of the stream bed, boulders and large wood that influence local current and scour patterns, and also larger-scale channel features including riffles, pools, bars, and channel meanders. The substrate of running waters differs greatly from place to place, and is important to algae and many invertebrates as the surface on which they dwell. For many organisms, substrate is where their food is found. Larger physical elements, such as large wood, are important to many fishes as the structure near which they find shelter from current or enemies. Temperature affects all life processes, and because most stream-dwelling organisms are ectothermic, growth rates, life cycles, and the productivity of the entire system are strongly under its influence. Thus, current, physical habitat, and temperature are three key aspects of the abiotic

© Springer Nature Switzerland AG 2021
J. D. Allan et al., *Stream Ecology*,
https://doi.org/10.1007/978-3-030-61286-3_5

Fig. 5.1 The relative abundance
of some macroinvertebrate taxa in
four large New Zealand rivers
within substrate size (left), depth
(right), and velocity ranges.
Aoteapsyche (Trichoptera);
Colobursicus humeralis,
Deleatidium, and *Nesameletus*
(Ephemeroptera). Error bars are 1
standard deviation (Reproduced
from Jowett 2003)

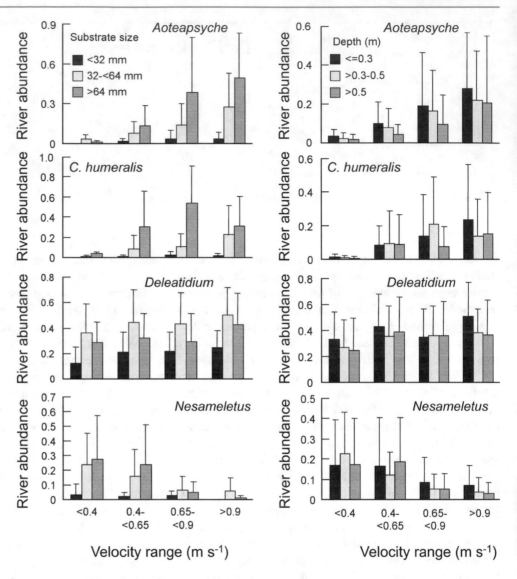

environment that we should understand in order to appreciate the functioning of a lotic ecosystem and the adaptations
of its denizens.

To decipher how organisms respond to individual habitat
variables is complicated because organisms are subject to the
simultaneous and interactive effectives of multiple abiotic
factors. The relationship of macroinvertebrate abundances to
velocity, substrate size, and depth (Fig. 5.1) nicely illustrates
differences in habitat preferences, but because these environmental factors are inter-related, it can be difficult to distinguish causal from correlated responses. Often, we focus
on the influence of average conditions on aquatic organisms,
but the variance in average conditions and the frequency and
magnitude of extremes may be equally important to organismal abundance and the structure of aquatic communities.
When environmental conditions episodically become unfavorable, such as an area of substrate that receives excessive
scour or a stream section that becomes too warm for days or

weeks, then patches of remaining suitable habitat provide
critical refuge until the disturbance passes and recolonization
can occur. Species traits, such as motility or tolerance to high
temperatures, determine how populations will respond to
unfavorable conditions and play an important role in shaping
aquatic communities.

5.1 The Flow Environment

In fluvial systems the flow of water is a dominant and
characterizing variable that influences diverse aspects of the
stream environment (Hart and Finelli 1999). It affects
channel shape and substrate composition, and episodically
disturbs both. Flow strongly influences the hydraulic forces
operating in the benthic and near-bed microhabitats occupied
by much of the biota, and is important to ecological interactions, rates of energy transfer, and material cycling

Fig. 5.2 Multiple causal
pathways by which flow can
affect organisms. Potential
interactions among pathways are
not shown (Reproduced from
Hart and Finelli 1999)

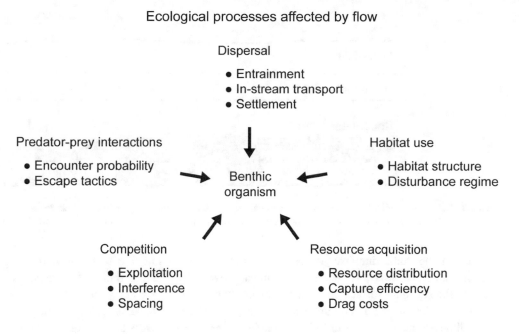

Ecological processes affected by flow

Dispersal
- Entrainment
- In-stream transport
- Settlement

Predator-prey interactions
- Encounter probability
- Escape tactics

Benthic
organism

Habitat use
- Habitat structure
- Disturbance regime

Competition
- Exploitation
- Interference
- Spacing

Resource acquisition
- Resource distribution
- Capture efficiency
- Drag costs

(Fig. 5.2). Current velocity is a physical force that organisms experience within the water column as well as at the substrate surface. Organisms are directly affected by current velocity when eroded from a substrate and displaced downstream, or when their energy reserves are depleted as they attempt to maintain position against current. They are indirectly affected when the delivery of food particles, nutrients, or dissolved gasses—factors that influence their metabolism and growth—change with changes in current velocity. Relevant definitions and methods of measurement were given in Chap. 2; recall that current is the speed of moving water (usually in cm s^{-1} or m s^{-1}), and flow or discharge is volume per unit time (usually m^3 s^{-1} or cfs). Turbulence refers to rapidly varying and unpredictable fluctuations in current velocity, and may be more influential than mean velocity. It can be quantified from the standard deviation of velocity divided by the mean, or from time series analyses.

Current velocity varies enormously, not only along a river's length and with the rise and fall of the hydrograph, but also from place to place within stream channels at meso- and micro-habitat scales owing to bed friction, topography, and bed roughness due to large substrate particles and wood. The vertical velocity profile (Fig. 2.7) is of fundamental importance to any consideration of the effects of current on organisms, as the flow conditions near the streambed may differ markedly from open-channel flow. When the depth of flow is substantially greater than the height of roughness elements, one expects an outer layer in which velocities vary little with depth (free stream velocity) and a logarithmic layer of declining velocity near the streambed (Fig. 5.3). In reality, the velocity profile is usually more complex.

Water in contact with the streambed has zero velocity, referred to as the no-slip condition, and velocity increases above the bed to its free stream value. The boundary layer is defined as a flow region where water velocity increases from zero in contact with the stream bed to the free stream velocity with increasing distance away from the streambed (Vogel 1996). An average velocity within this boundary layer can be estimated from fine-scale measurement of the velocity profile, provided that the profile is semi-logarithmic. This is based on a principle of fluid dynamics called the "law of the wall", which states that the average velocity of a turbulent flow at a certain point is proportional to the logarithm of the distance from the "wall". In laboratory settings the wall is the surface of a pipe or smooth surface of a flume, whereas in a natural stream that surface is usually rough (Fig. 5.3), creating turbulence and an irregular velocity profile. In theory this near-bed profile provides an estimate of shear velocity, which expresses the force, or shear stress, of near-bed flows in units of velocity. Shear stress is useful in predicting sediment mobilization and transport, and thus is also a plausible measure of the hydraulic forces experienced by an organism on the streambed.

Three fundamental types of flow characterize moving fluids: laminar, turbulent, and transitional. In laminar flow, fluid particle movement is regular and smooth, and particles can be thought of as "sliding" in parallel layers with little mixing. Turbulent flow is characterized by irregular movement with considerable mixing. Intermediate conditions are described as transitional. In order to understand the flow environment experienced by benthic organisms, much interest has focused on the nature of flows within the boundary layer down to within a few millimeters of a stone

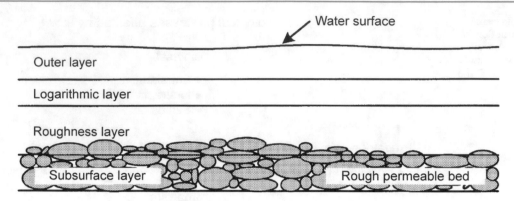

Fig. 5.3 Subdivision of hydraulically rough open-channel flow into horizontal layers. Flow velocities within the roughness layer are unpredictable based solely on knowledge of flow in the logarithmic layer. This figure is not drawn to scale (Reproduced from Hart and Finelli 1999)

surface. Flume studies indicate the possibility of much lower current velocities, less turbulence, and laminar flows very close to the substrate surface. In most natural circumstances, however, friction with an uneven streambed induces three-dimensional flows and turbulence within the near-bed environment where most stream organisms dwell (Hart and Finelli 1999).

5.1.1 Characterizing the Flow Environment

Table 5.1 summarizes some hydraulic variables in common use. Mean velocity (U) and depth (D) are the same variables described in Chap. 2, although here we use the symbols of hydraulic engineers rather than of hydrologists. Surface roughness (k_s) can be estimated in a number of ways, including from sediment dimensions such the D84 (Sect. 3.2.1) multiplied by an empirical roughness constant, and with bed profiler devices that use acoustic signals, lasers, or stereophotogrammetry to quantify topography (Bertin et al. 2014). Mean velocity, depth, and surface roughness are simple hydraulic variables that provide useful information about the flow environment. When channel depth is shallow relative to substrate roughness, such as in riffles and broken water, flow will be very complex.

The size and longitudinal spacing of roughness elements along the streambed influences the complexity of flow in the near-bed environment (Davis and Barmuta 1989). Large roughness elements such as pebble clusters and boulders are common features of poorly sorted gravel-bed rivers, where their presence generates intense turbulence downstream (Fig. 5.4). Velocity and turbulence estimates can differ appreciably on a centimeter scale downstream of these elements (Lacey and Roy 2008). Depending on their density and positioning, the wake behind each element may dissipate before the next element is encountered, which Davis and Barmuta (1989) called isolated roughness flow. When spacing between roughness elements is less, their wakes

interfere with one another, producing high local velocities and turbulence, termed wake interference flow. Lastly, skimming flow describes the circumstance when roughness elements are very closely spaced, which allows flow to skim across the tops of elements and produces a relatively smooth flow environment and slow eddies in the intervening spaces.

Bed surface roughness is due not only to stones of various sizes, but also wood and vegetation. Flow measured in and around a common lotic macrophyte *Ranunculus penicillatus* showed that velocities dropped to a low and constant value within 5 cm into the plant bed, forcing most of the flow over and around it. A dead-water zone formed immediately downstream, and then a region of high turbulence (Green 2005). As Nepf (2012) has shown, the presence of vegetation in stream channels alters the velocity field across several scales, ranging from individual branches and blades on a single plant, to vegetation patches, to the channel reach.

Using open-channel measurements and certain constants, one can estimate two widely used hydraulic parameters, channel Reynolds number (R_e) and Froude's number (Fr). The Reynolds number quantifies the ratio of inertial forces of the moving fluid to the viscous properties of a fluid that resist mixing (Newbury and Bates 2017). It is a dimensionless number that can be used to distinguish types of flow and the forces experienced by an organism. Depth is used to estimate R_e for the channel, and the length of a fish or insect can be used to estimate the forces that act directly on an organism. At low R_e flow is laminar and viscous forces predominate, whereas at high R_e turbulence occurs and inertial forces predominate. Laminar flow usually requires current velocities well below 10 cm s^{-1}, especially if depth exceeds 0.1 m; in short, quite shallow and slow-moving water. Hence turbulent flow is the norm in the channels of rivers and streams.

Fr is a dimensionless velocity-to-depth ratio, and it differentiates tranquil flow from broken and turbulent flow (Davis and Barmuta 1989). Low values of Fr are characteristic of pool habitats and higher values of riffle habitats. In

Table 5.1 Some terms and equations useful in describing streamflow. It is the convention of this literature to represent velocity with U, depth with D, and the constant for kinesmatic viscosity of water as v

Terms	Description	Units	Measurement		
$\bar{U}$	Mean velocity	cm s^{-1}	Measured at 0.4 depth from bottom or from open-channel velocity profile		
U_*	Shear velocity	cm s^{-1}	Estimated from fine-scale velocity plotted against log depth near the streambed		
D	Water depth	cm	Total depth, surface to bed		
k_s	Substrate roughness	cm	Height of surface roughness elements measured individually or with bed profiler		
k_s/D	Relative roughness	Dimensionless	Height of roughness elements in relation to stream depth		
g	Acceleration due to gravity		9.8 m s^{-2}		
v	Kinematic viscosity		1.004×10^{-6} m^2 s^{-1} at 20 °C		
Equations					
R_e	Bulk flow Reynolds number	Dimensionless	Re = $U\,D/v$	$R_e < 500 \rightarrow$ laminar flow	
				$500 < R_e < 10^3$ to $10^4 \rightarrow$ transitional flow	
				$R_e > 10^3$ to $10^4 \rightarrow$ turbulent flow	
Fr	Froude number	Dimensionless	$Fr = U(gD)^{-0.5}$	$Fr < 1 \rightarrow$ subcritical flow	
				$Fr = 1 \rightarrow$ critical flow	
				$Fr > 1 \rightarrow$ super-critical flow	
R_{e*}	Roughness Reynolds number	Dimensionless	$Re_* = U_*\,k/v$	$R_{e*} < 5 \rightarrow$ hydraulically smooth flow	
				$>5\ R_{e*} < 70 \rightarrow$ transitional flow	
				$R_{e*} > 70 \rightarrow$ hydraulically rough flow	

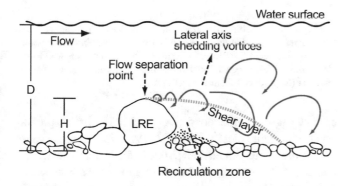

Fig. 5.4 Sideview schematic of a typical large roughness element (LRE) and associated wake characteristics, showing flow separation at a downstream point on the stone surface, flow reversal, and shear. H is large roughness element protrusion height, D is water depth (Reproduced from Lacey and Roy 2008)

some New Zealand streams, Fr generally was less than 0.18 and rarely as high as 0.4 in pools, greater than 0.41 and as high as 1 in riffles, and intermediate in runs (Jowett 1993).

Using an estimate of shear velocity (U_*), which can be derived from the velocity profile near the streambed, and substituting the height of roughness elements for water depth, one can estimate boundary (roughness) Reynolds number (R_{e*}) (Table 5.1). This variable and the dimensionless shear stress τ describe the conditions under which

particle movement is likely to be initiatcd (Scct. 3.2.3). Both near-bed velocity and bed shear stress increase with increasing relative roughness (k_s/D) and mean velocity (U).

5.1.2 Flows at the Scale of Organisms

Spatial variation in hydraulic parameters has been shown to correlate with the local distribution of stream macroinvertebrates. Parameters including those based on the main channel such as mean velocity, Froude number, and Reynolds number (Statzner and Muller 1989; Mérigoux and Dolédec 2004), and those based on near-bed measurements of velocity, turbulence, and shear stress, have been found to be important (Hart et al. 1996; Bouckaert and Davis 1998; Enders et al. 2003). Such relationships are unsurprising as micro-scale studies in laboratory flumes reveal how crawling movements and posture shifts by benthic invertebrates respond to very fine-scale velocity and turbulence patterns (Weissenberger et al. 1991; Rice et al. 2008).

Efforts to understand the actual forces experienced by organisms whose profile extends less than 10 mm into the water column have been informed by fluid dynamics theory, studies of sediment dynamics, and laboratory flume studies conducted with both inanimate and living invertebrates.

From the former, we know that flow separation occurs when frictional shearing between the layers of water closest to the substrate surface becomes too great for the layer closest to the surface to remain attached. Simply put, the fluid's flow is not able to follow the shape of the stone surface or body of an organism, but instead becomes detached and reverses flow, creating eddies and vortices (Fig. 5.4). The force of fluid moving in opposite direction to the main flow is referred to as resistance or drag, and increases with the square of velocity when flow is turbulent. Even low-profile larval mayflies on a flat plate experience lift and drag as flows impinge on and separate from their bodies (Weissenberger et al. 1991). Collectively these findings indicate that most of our knowledge about the association of organisms with current are useful correlations, but a mechanistic understanding of how they cope with the forces of current requires improved methods of measurement.

Investigators have had some success in explaining macroinvertebrate distributions using average shear velocity and boundary Reynolds number estimated from the near-bed velocity profile, provided the stream bed is not too irregular and the vertical velocity profile is semi-logarithmic. Using an electromagnetic current meter to within approximately 2 cm of the stream bed in two New Zealand rivers, Quinn and Hickey (1994) found good log-normal velocity profiles at most sites sampled, and strong relationships between benthic invertebrate distribution patterns and bed hydraulic variables (shear velocity and boundary Reynolds number) under baseflow conditions. Similarly, in riffle microhabitats of the Kangaroo River of southeastern Australia, the majority of the macroinvertebrate community was associated with riffle areas of lowest near-bed turbulence (Brooks et al. 2005). Macroinvertebrate abundance and number of taxa were negatively related to velocity, roughness Reynolds number, shear velocity, and Froude number. In particular, some mayflies of the families Leptophlebiidae and Baetidae, and the water penny Psephenidae, were associated with low Reynolds numbers (Fig. 5.5).

Commonly used current meters such as propeller and electromagnetic flow meters have a spatial resolution of one to many centimeters, and bed roughness elements in natural streams will generally cause near-bed currents to be highly variable and to deviate from log-normal (Hart et al. 1996; Hoover and Ackerman 2004). Together these considerations point to the need for fine-scale measurements that can describe the spatial and temporal variability of near-bed flows. Several new approaches have been explored in recent years in hope of more accurately characterizing flows at the scale of an invertebrate on the streambed, including the acoustic Doppler velocimeter (ADV), hot-film anemometry, syringe-like pressure devices, and a series of hemispheres of standard size but different mass placed on a flat plate on the streambed. The ADV emits sound energy that is reflected back by particles in the water, whose movement with the

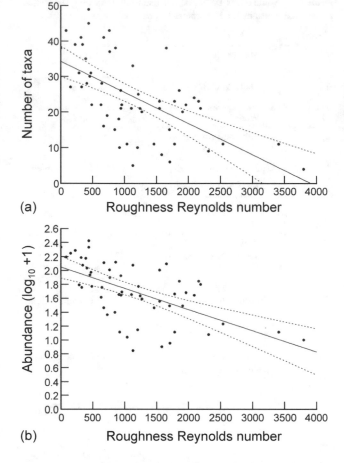

Fig. 5.5 Relationship between roughness Reynolds number and (**a**) number of invertebrate taxa and (**b**) macroinvertebrate abundance in sampled areas of 0.07 m^2 within three riffles in the Kangaroo River, New South Wales, Australia. Dotted lines indicate 95% confidence intervals (Reproduced from Brooks et al. 2005)

current causes a frequency (Doppler) shift that is proportional to current velocity. Use of ADVs is increasing because they allow three-dimensional, fine-scale velocity measurements in field settings and thus estimation of shear stress and turbulence. Hot-film anemometry is based on heat transfer and voltage drop recorded on a very fine sensor covered with a heat conducting film, calibrated against current speed for a given temperature. Using this instrument, Hart et al. (1996) were able to measure velocities only millimeters above stone surfaces, documenting extensive spatial and temporal variation at very fine scales and short intervals. Finally, methods have existed for some years to estimate current velocity using a small capillary tube and pressure differences near stone surfaces (Vogel 1996). Ackerman and Hoover (2001) have elaborated on this approach, referred to as the Preston-static tube, employing a small diameter syringe connected to a differential pressure transducer, oriented with a micro-positioning device, and positioned using dye release and an underwater periscope viewer.

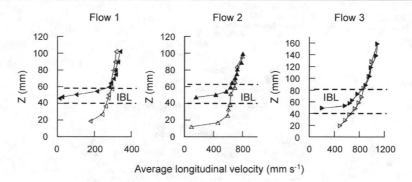

Fig. 5.6 Average longitudinal velocity over cobbles with (closed triangles) and without (open triangles) the moss *Frissidens rigidulus* in a laboratory flume at three flow levels (1 is lowest, 3 is highest). The vertical axis is distance above the stream bed. The existence of the internal boundary layer (IBL) and influence of the moss are clearly evident (Reproduced from Nikora et al. 1998)

While the above methods and devices have shown promise for measuring velocity and turbulence at centimeter and finer scales, their use is still relatively limited. ADVs and hot film anemometry are expensive instruments and the latter is vulnerable to damage and fouling under field conditions. The Preston tube does not have these limitations, but detailed measurements are laborious, and spatial and temporal variation seems almost without limit. Despite these challenges, careful studies of near-bed flows have led to very useful insights.

Nikora et al. (1998) used an ADV to examine how the presence of the aquatic moss *Fissidens rigidulus* influenced near-bed flow environment in a stream flume by measuring flow around cobbles with moss, and then repeating the measurements after removing the moss without disturbing the cobbles' position. Velocities in the upper layer followed the standard logarithmic profile, but within the lower sublayer the interaction of flow with roughness due to moss had a marked effect, reducing velocity, stress measures, and turbulence (Fig. 5.6). Using an ADV to measure flows in a flume with a constructed gravel bed, Rice et al. (2008) found that the crawling behavior of the cased caddisfly *Potamophylax latipennis* was associated with low elevations, low flow velocities, and low turbulent kinetic energies. Discrimination among locations was greater at higher discharges, indicating that caddisfly movement was contingent upon flow conditions.

Using hot-film anemometry to study the spatial positioning of blackfly larvae (*Simulium vittatum*) in a Pennsylvania stream, Hart et al. (1996) showed that larval abundance was positively related to current speed at 2 mm above the stone surface (the approximate height of the feeding appendages), but not to velocity at a height of 10 mm. In addition, velocities at 2 and 10 mm were not significantly correlated. Semi-logarithmic velocity profiles were observed in only a few instances, making estimating of shear stress from the law of the wall impractical. Near-continuous measurement of current velocity within

4-second time series revealed large fluctuations, often by at least 30 cm s^{-1} and occasionally by as much as 100 cm s^{-1}. The conclusions from this study were that velocities of 30–50 cm s^{-1} occur at 2 mm height above a stone surface, and flow is turbulent, as shown by rapid and chaotic changes in flow velocity. Hart et al. (1996) further concluded that complex bed topography was responsible for highly turbulent flow, and that much of the turbulence they observed very close to the surface of individual stones was produced not by local shear but was inherited from upstream roughness elements.

Following a month of colonization of experimentally deployed stones in a Canadian Rocky Mountain stream, Ackerman and Hoover (2001) found more algae on rougher, higher areas of the substrate and very high densities of mayfly nymphs (*Epeorus longimanus*) on high-shear regions of the upper, exposed stone surfaces. Nymphs avoided regions of flow separation at the downstream end of stones and appeared less tolerant of spatially variable or oscillatory flows. When stone orientation was reversed by the investigators, nymphs repositioned, indicating a proximate response to near-bed flows. Shear stress was measured with a Preston-static tube at grid points across the surface of each stone, and 15 measurements over a 30-s interval were time-averaged. Similar to results with hot-film anemometry, data from the Preston-static tube showed pronounced differences among points 2.5 cm apart, and vertical velocity profiles that often were not log-normal (Hoover and Ackerman 2004).

A surrogate method to estimate shear stress on the stream bed uses a set of 24 FST (FliesswasserStammTisch) hemispheres of identical size (diameter 7.8 cm) and surface texture, but different densities, calibrated against shear stress in flumes (Statzner and Muller 1989). Hemispheres are exposed sequentially on a small, weighted plexiglass plate on the stream bed near the location of invertebrate sampling, and the heaviest hemisphere moved by the flow provides an estimate of hydraulic forces experienced by the organisms.

Fig. 5.7 Ordination of the fauna collected from the Ardèche River, France during spring sampling versus a hydraulic axis constructed from hydraulic parameters including shear stress estimated using the FST hemisphere method, Froude number, and depth and substrate measures. The bottom axis denotes the hydraulic axis. Taxa are positioned according to their locations along the axis, and the area of each circle is proportional to taxon abundance. Horizontal lines represent the standard deviation of the hydraulic score (Reproduced from Mérigoux and Dolédec 2004)

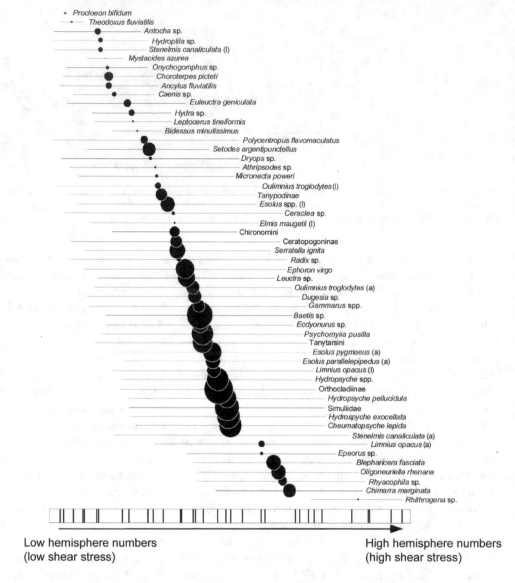

Low hemisphere numbers
(low shear stress)

High hemisphere numbers
(high shear stress)

Their relative ease of use is an advantage, although given their size and placement on a weighted plate on the streambed, the hemispheres obviously are not a direct measure of the forces that organisms experience. Nonetheless, studies have shown that the distribution or abundances of the majority of taxa show a significant relationship to FST hemisphere numbers (Mérigoux and Dolédec 2004, Fig. 5.7). It should be noted that the shear stress axis in this figure is a linear combination of coefficients including those for FST number, depth, substrate particle size, Froude number, and bed roughness. FST number and Froude number were the most influential parameters on the sheer stress axis in this study. In another FST application, Sagnes et al. (2008) showed that some species of macroinvertebrates were found at similar shear velocities, regardless of developmental stage and body size, while others showed differing velocity preferences with size changes. Statistical

associations can be quite strong for some taxa, but for many others FST number explains only about one-fourth of the total variation in densities (e.g., Dolédec et al. 2007). The FST method has been criticized for its modest level of predictive power (Frutiger and Schib 1993), which presumably indicates the importance of other environmental variables, some imprecision in the hydraulic estimates, or that some combination of factors influence habitat selection by macroinvertebrates.

The above studies demonstrate that benthic invertebrates can experience high current velocities, turbulence, and shear stress within 5–10 mm above the surface of a stone. They also draw our attention to the three-dimensional, rapid, and often extreme velocity fluctuations that occur around the time-averaged velocity across multiple scales, that is, to turbulence rather than mean velocity. Turbulence can be characterized statistically using the ratio of the standard deviation of velocity to its mean, although this also is a kind of averaging. A time series of near-bed velocities measured with an ADV more fully describes turbulence.

Turbulence is strongly influenced by bed roughness, as illustrated in Fig. 5.4. Estimating turbulent flow from current velocity measured with an ADV at 10 cm above the bed of a gravel-bed river, Roy et al. (2010) concluded that standard habitat variables had a relatively low capacity to explain turbulent properties. Mean flow velocity explained the largest proportion of the turbulent flow variation, which is unsurprising as the Reynolds number increases linearly with U, while depth and substrate showed less influence. In this study, turbulence was confined to a relatively small, localized zone in the lee of large roughness elements.

Fisheries scientists have long known that fish take advantage of roughness elements and uneven flows, sheltering behind large clasts on the stream bed. When water velocity microhabitats were quantified at the scale of millimeters, rainbow darters (*Etheostoma caeruleum*) in the Mad River, Ohio, were consistently found in microhabitat shelters where velocities were significantly lower than at adjacent (<5 cm distance) sites (Harding et al. 1998). Similarly, by holding positions in low-velocity water behind current obstructions, stream-dwelling salmonids optimize the tradeoff between the energy supply from drifting invertebrates and the energy cost of swimming (Fausch 1984). Position choice in drift-feeding Arctic grayling (*Thymallus arcticus*) was explained by a model in which net energy intake depended on capture rate, which was a function of visual reaction distance, depth, and velocity; and on the velocity-dependent cost of swimming (Hughes and Dill 1990). Because grayling must intercept prey entering their field of view before the prey is swept downstream, velocity increases the encounter rate but decreases the proportion of macroinvertebrates captured by fish.

Precisely how fish utilize or avoid turbulent conditions, which can vary at the scale of cm downstream of a roughness element, is not fully resolved. Laboratory studies show that turbulence can induce higher swimming costs (Enders et al. 2003), but also reveal the adoption of energy-saving swimming synchronized to shedding vortices (Liao et al. 2003). Brown trout in a Michigan stream take advantage of spatial variation in flows, based on turbulence measured during the day when trout presumably are in resting locations. Individuals selected positions near cover with low average velocities and low turbulence values, and avoided low velocity, high-turbulence locations (Cotel et al. 2006). In a flume experiment with hemispheric boulders that generated a range of flow velocities and turbulence, measured using an ADV, guppies exhibited different microhabitat preferences depending on size, sex, and parasite load, presumably reflective of swimming costs (Hockley et al. 2014). Larger guppies spent more time in areas of high velocity and low turbulence beside boulders, whereas smaller guppies frequented areas behind boulders with lower velocities and higher turbulence. Males, with their larger fins, spent more time in low velocities, and individuals infected with a parasite frequented low turbulence locations. In addition to differences among species and size classes in swimming ability, other factors including food availability and safety from predators is likely to influence flow microhabitat selection.

A detailed understanding of the physical forces experienced by organisms in flowing water remains a challenge for freshwater scientists. However, this field is advancing due to the arrival of more sophisticated measurement and analytic tools, as well as greater cross-fertilization among disciplines. Recent years have seen an integration of hydraulic and biological studies creating the new, interdisciplinary field of ecohydraulics (Maddock et al. 2013), which seeks to predict ecological responses to hydraulic conditions and inform management regarding habitat restoration, hydropower operations, and other flow-related impacts on the ecosystem. Although we lack a full understanding of the relationships between hydraulic characteristics and organismal behavior, there is no question that the distribution and abundance of many and perhaps most organisms exhibit a statistical association with a subset of measurable hydrodynamic parameters. These findings can be shown to have practical use in predicting how the fauna will respond to human-induced changes in flow conditions.

5.1.3 Influence of Extreme Flows

In addition to the challenge of determining the flow-associated habitat preferences of benthic invertebrates and fish under 'normal' flows, we would also like to know how organisms cope with extremes. This includes floods of sufficient magnitude to scour the bed, as well as sudden and rapid increases in flow (often referred to as spates or freshets) that follow a heavy rain or are due to rapid snowmelt. Recall from Sect. 3.2.5 that floods of modest magnitude are frequent, may result in episodes of streambed scour and fill, and are likely to be affect some areas of streambed more than others. At water velocity and shear stress below the levels that cause sediment entrainment, benthic organisms are affected only by the shear force exerted by flowing water. As water velocity and shear stress increase, movement of fine sediments and then coarse sediments takes place, affecting the biota through abrasion, bed scour, and habitat disruption. Floods sufficient to induce bedload transport can dramatically alter the composition, density, and biomass of benthic invertebrate communities (Holomuzki and Biggs 2000; Death 2008), and periphyton (Biggs et al. 1998).

During high-flow events, some locations may serve as refuges for aquatic organisms. These include floodplains, stream margins, depositional areas, and debris dams at the meso-scale (Lancaster and Hildrew 1993; Palmer et al. 1995; Francoeur et al. 1998), and crevices and surface roughness at the micro-scale (Dudley and D'Antonio 1991; Bergey 2005). Studies in New Zealand streams have examined the stability of local habitats at the stone or patch scale. For instance, Matthaei et al. (2000) compared invertebrate densities on stones that were well-embedded and stable, versus less stable stones lying loosely on top of the stream bed, in response to flooding. There were no differences in invertebrate densities before a moderate flood, but there were significantly higher densities of macroinvertebrates on stable stones after the event. Notably, changes were ephemeral as most differences disappeared within less than 3 weeks, largely due to animals leaving stable substrates to colonize other parts of the streambed. Effenberger et al. (2006) also documented similar, short-lived "refuge effects" in macroinvertebrate communities subjected to souring events. Freshwater mussel distribution is thought to be strongly linked to hydraulic variables because they are long-lived, filter-feeding organisms that colonize the surface substrate of the streambed. Supporting this expectation, Strayer (1999) found that mussel beds occurred in flow-protected locations of two rivers in New York, US.

Occupation of flow refuges may be the result of strong habitat preferences, simple chance in which some fraction of the population happens to be in a more protected location, or active response to changing flows. In flume studies, snails and the mayfly *Deleatidium* moved into low-velocity crevices on all substrates as current velocities increased, and caddisfly larvae *Pycnocentrodes* unreeled their silken drag-lines to reach more sheltered locations (Holomuzki and Biggs 2000). In one intriguing example, the giant water bug *Abedus herberti* apparently uses rainfall as a cue to avoid flash floods in its desert stream habitat (Lytle 1999). Fish, being larger and more mobile, are likely to shift habitat in response to rising flows. Using prepositioned electrofishing devices, Schwartz and Herricks (2005) showed that the fish assemblage of small, low-gradient Illinois streams occupied different habitats depending on flood stage. At near-bankfull flows, fish were associated with vegetated point bars and concave-bank benches, at half-bankfull conditions fish abundance and biomass were greatest in low-velocity eddies, and at base flow the main channel habitat of pools, riffles, and glides contained higher numbers and greater biomass than did lateral habitat units. Juvenile fish may be especially vulnerable to displacement and thus especially dependent on flow refuges. When juvenile rainbow trout *Oncorhynchus mykiss* were acclimated to low current speeds of 0.2 m s^{-1} in a flume with a cobble bed and then subjected to much higher currents, individuals sought cover in cobble interstices as current increased, generally to the deepest extent possible. The number of deep interstices (>20 cm depth, measured with a fine probe) determined the number of individuals that were able to find refuge, suggesting that the availability of suitable interstitial habitat may limit abundances of juvenile fishes in natural streams (Ligon et al. 2016).

Studies of streams experiencing large floods have found major changes in aquatic community distribution and structure in some instances, while others have found that stream communities were highly resistant and resilient to major floods. Comparing invertebrate and fish abundance before and after a major flood event in the northeastern US among sites that differed with respect to flood intensity, Nislow et al. (2002) observed the smallest change in fish and invertebrate abundance at a site experiencing the lowest-magnitude flood (approximately bankfull). Two other sites experienced more intense, overbank flooding and greater subsequent changes in species abundance, particularly where geomorphic change was greatest. In response to greater flood intensity, the abundance of macroinvertebrates and yearling fish declined, while over-yearling salmonids exhibited normal or greater than normal abundance.

Mortality events can also occur when streams experience intense flooding. A radiotelemetry study of tagged brown trout in a New Zealand stream, designed to study movement patterns, was interrupted by a severe, 50-year flood that caused substantial scouring, bed load movement, and removal of riverbed and bank vegetation (Young et al. 2009). Subsequent location of radio signals originating from beneath gravel banks, within debris piles, and out in the flood plain indicated that the flood killed 60–70% of the tagged fish. Though hardly fortuitous from the perspective of the investigators or the fish, this rare example confirms that flood-induced mortality can affect a substantial proportion of adult fishes in a population. This is noteworthy, as most studies of the effects of floods on fishes have shown that relative to juveniles, adults typically are impacted less.

Resistance and recovery of benthic invertebrates and fishes to flow-induced disturbance clearly depends on a number of variables. For example, the physical effects of a pronounced increase in flow will depend on the rate and magnitude of increase, channel morphology, the availability of refuges, and overall geomorphic disturbance. Impacts on the biota likely are affected by the availability of stable patches of substrate, the organisms' ability to find refuge and/or withstand increased current forces, and the timing of the event with respect to growth and recruitment. This serves as a reminder that in order to persist in a given location, populations of organisms must be able to cope with extreme events in addition to average conditions, and that episodic extreme conditions can have a large effect on the composition of stream assemblages.

5.1.4 Flow Management Applications

Knowledge of hydraulic habitat preferences and species traits that are associated with flow (e.g., Reynold's number, swimming ability, etc.) inform our ability to anticipate changes in faunal assemblages in response to flow manipulation, and our design of infrastructure (e.g., fish passages at dams) and management actions (e.g., mimicking natural flow regimes) to mitigate the effects of flow modification on aquatic communities. Fishways to allow passage by migrating fish around weirs, dams, and natural barriers have been in use in Europe since as early as the mid-eighteenth century (Katopodis and Williams 2012). However, many were ineffective, and efforts to improve fishway design after the mid-20th century benefited from testing of hydraulic relationships among various fishway types and biological assessment of their effectiveness. Studies of fishway effectiveness often have focused on salmonids due to their high economic and recreational value, and passage through dams by fish of different swimming behaviors and low commercial value has been largely neglected. Surveying 37 fishways in Portugal, Santos et al. (2012) found that more than half were unsuitable for fish species for which passage was required under the European Water Framework Directive.

In response to the widespread reduction in current velocities and total discharge due to impoundments and water withdrawals, a number of methods have been developed to estimate the necessary amount of water, termed instream flows, to meet the basic habitat needs of the biota. These methods vary in the complexity of the models used to predict changes in hydrology and in the number of habitat variables, species, and life-stages that are included. The widely used Physical Habitat Simulation System (PHABSIM) was developed as a flow assessment tool to ensure sufficient flows for aquatic life (Bovee 1982). Field sampling identifies those habitat conditions where higher densities of fish are found, and statistical models then relate preferred fish habitat to flow. PHABSIM provided managers with a basis for identifying the flows that provided sufficient amounts of the preferred habitat to sustain desired population densities of focal species. PHABSIM is relatively simple to apply, as it is based on univariate curves relating the abundance of individual species to the amount of preferred habitat (expressed as current velocity, depth, and substrate), which can be combined to develop a habitat index that varies with discharge. Applying this approach to predict spawning habitat for chinook salmon (*Oncorhynchus tschawytscha*) and steelhead trout (*O. mykiss*) in several California rivers, Gard (2009) found that occupied nest locations accorded well with model predictions. As discussed below, however, this approach has critics as well as supporters.

Throughout the 20th century, numerous channel segments in the mainstem of the French Rhône River experienced flow reductions due to diversion dams that directed river discharge through artificial channels to hydroelectric plants. After 2000, restoration efforts were undertaken to increase flows in by-passed channels. From published information about flow preferences of many species (e.g. Fig. 5.8), it was possible to predict how the species assemblage would respond to flow restoration. In reaches where the increase in minimum flow was greatest, the abundance of fish species preferring fast-flowing and deep microhabitats roughly doubled, whereas the abundance of other species declined (Lamouroux and Olivier 2015). Macroinvertebrate taxa found near the banks prior to restoration tended to decline; however, approximately equal numbers of species found in the main channel experienced either increases or decreases in abundance (Mérigoux et al. 2015). The coupling of habitat preference with hydraulic data in this study effectively predicted the ecological response of individual species, measured as the natural log of change in density. Because the habitat preference models were developed at other locations, these results also provide some confidence that statistical habitat models are transferable across sites. Lastly, when the responses were re-examined using traits

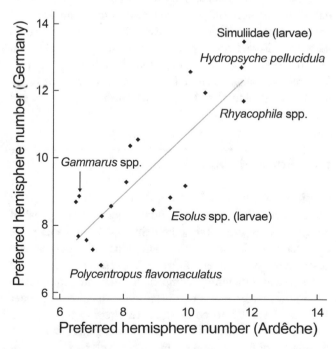

Fig. 5.8 The average 'preferred' bottom shear stress in the Ardèche River in France (data from Mèrigoux and Dolèdec 2004) predicts the average 'preferred' bottom shear stress observed in German streams (data from Dolèdec et al. 2007), for 20 taxa involved in both studies ($R^2 = 0.68$, $p < 0.001$). Some taxa, defined at different biological levels, are indicated by labels (Reproduced from Lamouroux et al. 2010)

rather than taxa, traits related to locomotion and attachment, as well as general biology and physiology, proved useful in understanding the effects of hydraulic restoration on aquatic communities (Dolédec et al. 2015). After restoration, clingers and passive filter feeders dominated invertebrate communities in the main channels. Within fish communities, species exhibiting the life-history strategy characteristic of downstream river segments (long life span, large body, late sexual maturity) increased with restoration.

Managing rivers to ensure that flow conditions support diverse assemblages of native species is an on-going, global challenge. Early models such as PHABSIM remain in use, but its shortcomings are widely acknowledged, including that it is too simple both in its description of habitat and in its modeling approach (Railsback 2016). More detailed and dynamic models that can simulate more complex flow conditions and more fully represent population processes are becoming more widely used, and may well become the standard approach in the future. This topic shares many considerations with discussions about natural flow regimes and environmental flows (Sect. 2.5) although they emphasize different outcomes. Environmental flow modelling tends to emphasize restoring the overall hydrograph in the direction of its unaltered or reference state to benefit ecosystem processes. In comparison, the motivation behind integrating habitat preferences and flow modelling is to ensure adequate environmental conditions for particular species in specific locations. Sound management strategies obviously will require both approaches applied together to support flow-related management decisions that restore or maintain the desired structural and functional attributes of rivers and streams.

5.2 Physical Habitat

Physical habitat structure is widely held to be a major determinant of assemblage composition in all types of ecosystems, as structural complexity and heterogeneity are considered to influence both individual abundances and taxon richness (number of species). Heterogeneous habitat mosaics and structurally complex habitats are expected to provide a greater range of niches, thus enabling a greater number of species to co-exist; more effective refuge from predators and physical disturbance; and a greater range and abundance of food resources. In streams and rivers, physical habitat structure includes a wide variety of inorganic and organic substrates of varying size, large objects such as boulders and submerged wood, and channel units such as pools, riffles, and bends. Substrate is especially important to macroinvertebrates, small in size and dwelling on or within the stream bed. Larger elements, including boulders and wood, are of particular importance to fish, especially larger

species found in the water column. Large substrate elements in a segment of stream influence flow and scour that affect the characteristics and diversity of substrate, so they are strongly interrelated. This makes it challenging to quantify and compare the influence of physical habitat structure on organisms (Kovalenko et al. 2012). First, there are many types of substrate and habitat elements, both inorganic (bedrock, silt, sand, gravel, pebbles, cobbles, up to large boulders) and organic (leaf litter, tree roots, wood of various sizes and textures). Second, terms such as heterogeneity and complexity are often used interchangeably and are loosely defined. Third, relationships between taxon richness and habitat structure often are further confounded by variations in surface area. Complex habitats typically have a larger surface area that supports more individuals, and because the number of species generally increases logarithmically with number of individuals, more complex habitats may support more species as a consequence of greater numbers, referred to as a passive sampling effect.

Because much of the focus on habitat complexity concerns its relationship with assemblage diversity, it is useful to define a few terms. Habitat heterogeneity can be defined as both the composition (number and relative abundance) and configuration (spatial arrangement) of habitat patches or types. Habitat complexity can be defined as the total abundance of structural features, such as crevices on a stone surface or number of accumulations of large wood, in relation to surface area. A variety of statistical approaches have been employed to describe habitat heterogeneity and complexity, including fractal dimensions (a statistical index of complexity that describes how some pattern changes with the scale of measurement), and various concepts from landscape ecology that examine the surrounding mosaic in terms of the diversity, distribution, and patch size attributes of surrounding habitat types (Palmer et al. 2000; Kovalenko et al. 2012). Unfortunately, because the terms "heterogeneity" and "complexity" have been used inconsistently in published studies, it is not feasible to distinguish between them in reviewing the literature; hence, in our discussion we generally follow the terminology used by the authors of the original studies.

Species diversity generally refers to the number of species at a location, typically referred to as species or taxon richness. The latter is more general, as it can be used for datasets using genus- or family-level classification. There are also formulae that take into account the relative abundance of species, based on the premise that an assemblage in which one species makes up the majority of total individuals is less diverse than one in which dominance is reduced and a greater number of species are well represented. Total diversity can be sub-divided into the diversity within each habitat type (α-diversity; e.g., the diversity within just riffle or just pool habitats in a given reach), and the additional diversity resulting from differences in assemblage composition between habitat types or locations (β-diversity; e.g., the number of species found only in riffles, plus the number of species found only in pools). β-diversity is sometimes referred to as dissimilarity or species turnover. In practice, field studies often collect macroinvertebrates by disturbing the substrate at a number of locations and/or habitat types, and collect fish by electroshocking an entire stream reach, combining data across habitats and effectively making it impossible to distinguish α- from β-diversity. Although less commonly measured, β-diversity is a useful metric for evaluating the extent to which habitat diversity is linked to species diversity in a given system, as it identifies the total number of species that are uniquely associated with each habitat.

5.2.1 Inorganic Substrates

Many of the important features of inorganic substrate were described previously from the perspective of fluvial geomorphology (Sect. 3.2.1), where the emphasis was on the interaction of sediment supply and flow on particle transport and channel form. Relevant measurements and concepts include the size categories of inorganic particles (Table 3.3) and their quantification using pebble counts to determine the median size (D50) and range (D16 and D84); the relationship between particle size eroded and current velocity (Fig. 3.8); and the development of channel features including riffles, pools, point bars, and undercut banks, which are meso-scale physical features often referred to as habitat units. Surface substrate usually is coarser than sub-surface substrate and at least partially protects this finer material from transport, resulting in vertical heterogeneity within the stream bed. The stability of the substrate depends on the magnitude and frequency of hydrological events and particle size. Permeability of the sub-surface region (the interstitial zone or hyporheos) adds a vertical dimension to available habitat by allowing water to circulate and transport gases, nutrients, and fine organic material. Texture and the availability of crevices also can influence a particle's suitability as habitat. Low levels of siltation may be beneficial, particularly for species adapted to consuming silt for its organic content, but high silt levels usually have a negative influence on habitat for surface-dwelling organisms by reducing habitat heterogeneity, filling interstitial spaces, and coating consumers and their food resources.

Stream beds of gravel, cobble, and boulders occur in a great many regions around the world, harboring a diverse fauna of lithophilous taxa that Hynes (1970) remarks is broadly similar almost everywhere. Sand generally is considered to be a poor substrate, especially for macroinvertebrates, due to its instability and because tight packing of sand grains reduces the trapping of detritus and can limit the availability of oxygen. Nevertheless, a variety of taxa, termed psammophilous, are specialists of this habitat. The meiofauna, defined as invertebrates passing a 0.5 mm sieve but retained on a smaller sieve of 0.05 mm, can be very abundant, dwelling interstitially to considerable depth (Palmer 1990). Burrowing taxa can be quite specific in the particle size of substrate they inhabit. The mayflies *Ephemera danica* and *E. simulans* burrow effectively in gravel. *Hexagenia limbata* cannot, but does well in fine sediments. Substrates composed of finer sediments generally are low in oxygen, and *H. limbata* meets this challenge by beating its gills to create a current through its U-shaped burrows (Eriksen 1964). Invertebrates living in poorly oxygenated environments have a variety of means to create current and move oxygenated water over their bodies, including gill beating, body undulations, and other movements (Resh et al. 2008).

A number of species of fishes and other vertebrates of rivers also occur on or near particular substrates, and some fishes are quite specialized in their affinities. For example, the mud darter *Etheostoma asprigene* is restricted primarily to the backwaters of larger tributaries of the Mississippi River, the southern sand darter *Ammocrypta meridiana* to clean, sandy substrates of the Mobile River basin, and the Blenny darter *Etheostoma blennius* to the gravel and rubble bottom of fast riffles in Tennessee River tributaries in the southeastern US (Lee et al. 1980). For gravel-spawning fish, gravels of an appropriate size that are neither compacted nor embedded with fine particles are essential for reproduction so that water flows into interstices and oxygen is transported to buried eggs. The requirements of salmonids have been especially well-studied. Substrate material must be moveable to allow the female to excavate a nest, termed a redd, and the size of stones that can be moved varies with size of fish (Kondolf 2004). Successful incubation requires circulation of water to supply oxygen, and so an excess of fines within the interstitial matrix can be harmful to egg and larval survival.

Aquatic insects also select particular substrates for oviposition. Egg masses of the mayfly *Baetis* were highly aggregated under protruding stones with specific characteristics associated with lower probabilities of desiccation in a Rocky Mountain stream (Encalada and Peckarsky 2006). Similarly, hydrobiosid caddis flies were observed lay their eggs in single masses beneath emergent rocks in an upland Australian stream, and "landing pad size" was thought to influence oviposition choice (Reich and Downes 2003).

5.2.2   Organic Substrates

Organic substrates including algae, moss, macrophytes, dead leaves, and wood vary greatly in size, the conditions where they occur (depth, current, stream size), and in their temporal persistence. Small organic particles less than 1 mm usually serve as food rather than as substrate, except perhaps for the smallest invertebrates and microorganisms. Macroinvertebrates generally are more abundant where greater amounts of fine organic matter occur on the surfaces of mineral substrates, within their interstices, and in depositional zones behind obstructions. Autumn-shed leaves and the fungi and bacteria they support are a major energy source for invertebrate consumers, especially in woodland streams, and often are most abundant in depositional zones where fine particles often are trapped. Thus, aggregations of leaves on the streambed provide food as well as habitat and typically support a high abundance and diversity of invertebrates (Mackay and Kalff 1969). On the other hand, higher plants and submerged wood are consumed by only a few specialists, and support high animal abundances mainly because these large organic substrates serve as perches from which to capture food items transported in the water column, as sites where fine detrital material accumulates, and as surfaces for algal and biofilm growth. The presence of wood in streams also adds substantially to meso-scale habitat complexity, acting both as a geomorphic agent influencing channel shape and pool formation (Sect. 3.1.5), and as important habitat.

The invertebrate taxa that live in association with aquatic plants are referred to as phytophilous. A number of species are found primarily on moss, including the free-living caddis larva *Rhyacophila verrula* and a number of mayflies with backward-directed dorsal spines that are thought to minimize entanglement (Hynes 1970). Most commonly, mosses and filamentous algae provide habitat rather than food, serving as a refuge and a trap for silt and organic matter (Steinman and Boston 1993). Macrophytes add to the physical complexity of aquatic environments, creating habitat that algae, microbes, and invertebrates may colonize (Tokeshi and Pinder 1985), and providing refuge for fishes from high flows and predators (Grenouillet et al. 2000). Some fish species preferentially spawn on submerged vegetation, attaching adhesive eggs to live or dead plants and submerged roots (Balon 1981). For examples, certain darters in the very diverse genus *Etheostoma* specialize in spawning on rotting vegetation (*E. exile*), the macroalga *Cladophora* (*E. blennioides*), and other rooted plants (*E. lepidum*, *E. punctulata*).

Xylophilous, or wood-dwelling, taxa attest that wood constitutes yet another substrate category of lotic environments. In the headwater streams of forested areas, as much as one-quarter to one-half of the streambed can be wood and wood-created habitat (Anderson and Sedell 1979). Similar to aquatic mosses, wood appears to be substrate more often than it is food. Some taxa, such as the beetle *Lara avara*, feed mainly on wood. However, this beetle has an exceptionally slow growth rate and long life cycle among stream insects (Huryn and Wallace 2000). Many wood-associated taxa actually obtain their nourishment from biofilms occurring on wood surfaces, rather than the wood itself (Hax and Golliday 1993). In lowland rivers where the substrate is largely sand, fallen trees are especially important as a substrate coated with periphyton and biofilm, and as a perch from which to collect particles in suspension (Benke et al. 1985).

5.2.3   The Influence of Physical Habitat on Stream Assemblages

The density and richness of invertebrates have been shown to correlate with amount of detritus, algal biomass, substrate stability and complexity, depth, and velocity (Rabeni and Minshall 1977; Barmuta 1989; Quinn and Hickey 1994); and the strength of correlation has been found to depend upon the spatial scale at which substrate-related variables are measured (Beisel et al. 2000). In gravel-bed rivers, a diverse macroinvertebrate fauna exhibits a patchy spatial distribution that surely is determined at least in part by the heterogeneity of the substrate. In fact, abundance and taxon richness typically are low in fine substrates and increase with substrate size at least up to gravel and cobble (Minshall 1984; Mackay 1992). Substrate size tends to decline downstream, for reasons discussed in Chap. 3, but tributaries can interrupt the longitudinal fining of substrate with inputs of coarse material. In gravel-bed rivers of western Canada, Rice et al. (2011) found an increased abundance of taxa that prefer coarse substrate at these points of sediment recruitment, as well as an overall increase in diversity.

Studies of substrate-induced habitat complexity conducted in Steavenson River, a stony upland stream in southeastern Australia, found that stone surface area accounted for some 70–80% of variation in species richness, small stones had fewer species because they had less surface roughness, and the filamentous red alga *Audouinella hermannii* enhanced roughness and the presence of macroinvertebrates (Downes et al. 1998). Using clay bricks as experimental substrate, Downes et al. manipulated three aspects of habitat structure: large surface pits and cracks, surface texture (small pits), and abundance of macroalgae. Sampled after 14 and 28 days of macroinvertebrate colonization, the majority of common species reached higher abundances on rough substrates, there was a disproportionate accumulation of small individuals, and each of the three manipulated elements of habitat structure had separate, additive effects on both the total abundance of individuals and the total number of species. Because the number of taxa increased disproportionately relative to increases in the numbers of individuals, it appeared that species richness was augmented by habitat complexity. Although this study provided strong evidence for the importance of substrate roughness, the mechanisms by which crevices and surface roughness affected the biota were unclear. Surface roughness appears to be influential for algal colonization in the Downes et al. study, but because *A. hermannii* responded strongly to surface texture, it was not possible to separate the effects on the fauna of increased algal cover alone from that of increased algae in combination with a rough surface. Similarly, following 45 days of colonization in a 4th-order stream in southern Brazil, total species richness of algae was higher on rough than on smooth substrates, and species composition differed between substrates, likely due to differences in species' capabilities to colonize substrates with or without crevices (Schneck et al. 2011).

The frequently observed positive relationship between taxon richness and measures of habitat complexity could be mediated through a positive effect of complexity on habitat area and total abundances, i.e., a passive sampling effect. In a study of macroinvertebrate richness and abundance in small patches (<0.1 m^2) in small streams in Wales, UK, surface area had the greatest influence on both abundance and richness, and richness was largely the consequence of abundance (Barnes et al. 2013). The authors detected a modest effect of complexity (estimated by a fractal dimension of surface area) but not of heterogeneity (estimated from the mix of surrounding habitats). Interestingly, habitat type (bedrock, silt, sand, gravel, pebbles and cobbles) was the single best explanatory variable (Fig. 5.9) and rendered complexity and surface area redundant in predictive models. Habitat type also was the most important variable determining assemblage composition of macroinvertebrates in three streams of Brazil. Estimating β-diversity to evaluate the relative importance of between-stream versus between-habitat effects, Costa and Melo (2008) found that microhabitat was most important in determining community composition. In essence, macroinvertebrate assemblages from adjacent but different microhabitats (stones in riffles, submerged roots of terrestrial plants, mosses at the air-water interface zone, and litter deposited in pools) in a single stream site were more dissimilar than those found in a single microhabitat at different stream sites.

5.2.3.1 Fine Particles

Fine sediments <1–2 mm in diameter are a natural component of aquatic ecosystems, but they have become a world-wide environmental concern due to their accumulation within stream beds and clogging of substrate interstices. An increase in fine sediments within gravel-bed streams

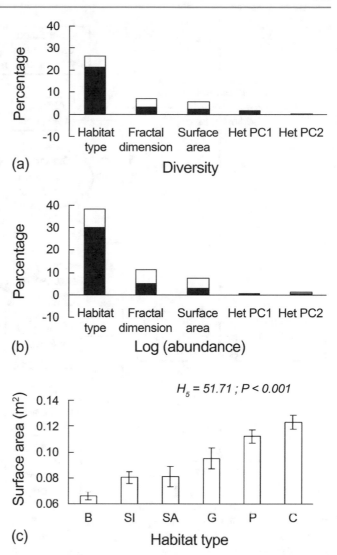

Fig. 5.9 Percentage of the variance in macroinvertebrate diversity and abundance found on different habitat types (bedrock, silt, sand, gravel, pebbles, cobbles) in a stream in Wales, UK. (**a**) diversity, (**b**) logarithmically transformed abundance. Fractal dimension and surface area were derived from profile measurements of bed surface. A Principal Components Analysis of patch size attributes and surrounding habitat types resulted in two measures of heterogeneity (Het PC1, Het PC2). Bars show amount of variance explained independently (□) and jointly (■) by habitat type. (**c**) Surface area is highly dependent on habitat types (mean ± 95% confidence interval) B, bedrock; SI, silt; SA, sand; G, gravel; P, pebbles; C, cobbles (Reproduced from Barnes et al. 2013)

typically results in declines in total abundance and taxonomic richness for both the benthic and hyporheic faunas, with pronounced declines of aquatic insects in the Ephemeroptera, Plecoptera, and Trichoptera (EPT) groups, and increasing dominance by Oligochaeta and Chironomidae (Lenat et al. 1981; Waters 1995). By clogging interstitial pore-spaces, fine sediments restrict the circulation of water, resulting in less oxygen exchange and a shift from mostly aerobic to anaerobic biogeochemical processes.

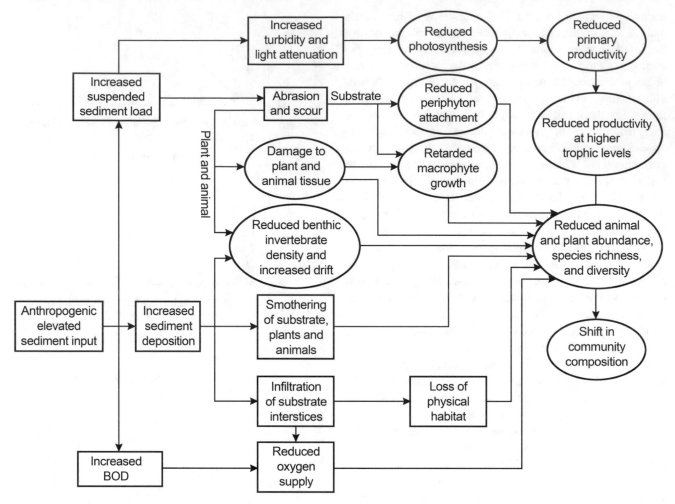

Fig. 5.10 Negative impacts of anthropogenically enhanced sediment input on lower trophic levels in stream ecosystems. Rectangles and ovals respectively denote physicochemical effects and direct and long-term biological and ecological responses. BOD is a measure of oxygen demand by aerobic organisms (Reproduced from Kemp et al. 2011)

Substrates with a high proportion of fine sediment affect lower trophic levels through a number of mechanisms (Fig. 5.10). The associated fauna frequently is dominated by taxa with low dissolved oxygen requirements (e.g., Oligochaeta and Chironomidae). Burrowing taxa of small size and deposit feeders tend to be common, while taxa sensitive to fine sediments, because of their filter-feeding or respiratory adaptations, may be absent. Surveys of macroinvertebrates and fine sediments in tributaries of the River Usk, a temperate, montane catchment in rural Wales, UK, showed pronounced changes in invertebrate composition at the patch scale ($\sim$3 m^2). Numbers of EPT taxa decreased by 20 and 25% at the most sediment-impacted sites relative to sediment-free sites (Larsen et al. 2009). Comparing lightly, moderately, and heavily clogged locations in three rivers of the Rhône basin of eastern France, Descloux et al. (2013) found that numbers of individuals and taxa of benthic invertebrates were significantly lower in the heavily clogged reaches than in the lightly clogged ones. Total density,

number of taxa, and number of EPT taxa all exhibited a negative relationship with increasing fines following colonization of perforated cylinders that were filled with a coarse gravel matrix and 10–60% fines (<2 mm) by volume, and inserted into the stream bed for 40 days (Fig. 5.11). While benthic (the first 5 cm of the streambed) and hyporheic (deeper sediment layers) densities and taxonomic richness both were affected, the decline in taxonomic richness with an increasing percentage of fine sediment was especially pronounced for the hyporheic fauna. In a similar experiment using faunal colonization columns inserted vertically into the bed of two small lowland rivers in Rutland, UK, macroinvertebrate communities after 14 days strongly differed between those filled with washed gravel >8 mm diameter vs those with gravel plus fine sand <2 mm sufficient to fill all interstices (Mathers et al. 2017). Results varied with the date when the columns were deployed, suggesting that life cycle timing affected whether the presence of fines strongly influenced colonization.

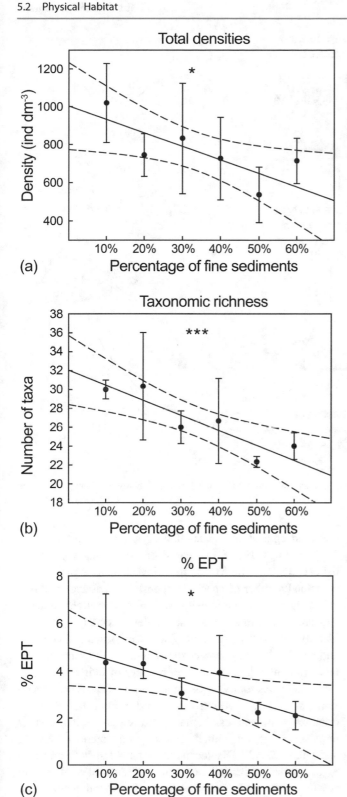

(a)

(b)

(c)

Fig. 5.11 Invertebrates colonizing vertical columns of artificial substrate containing coarse gravel and varying amounts of fine (<2 mm) sediments exhibited declines in (**a**) total densities, (**b**) taxonomic richness, (**c**) % Ephemeroptera, Plecoptera and Trichoptera (% EPT) with increases in percentage of fines. (dots: mean density for each percentage of fine sediment, n = 3 bars are standard deviation, dash line: 95% confident band). (***P < 0.001, **P < 0.01, *P < 0.05, ns non significant) (Reproduced from Descloux et al. 2013)

Studies describing indirect and direct impacts of fine sediment on freshwater fish have focused on relatively few families, notably the Salmonidae, but negative effects have been reported across a broad range of fish families (Kemp et al. 2011). Increased loading of fine sediments can have negative effects on fish via their prey populations, indirectly through changes to water quality, directly by coating gills and abrading tissue, and by degrading interstitial habitat for gravel-spawned embryos, developing eggs, and embryonic stages of fish species that excavate gravel nests. Degradation of spawning habitat due to sedimentation of fine particles is generally considered a key impact for fishes. In addition, turbidity due to fine particles in suspension has been shown to have negative effects on feeding behavior, including feeding rate and prey detection, although the strength of effect varied between species known to be turbidity-tolerant versus intolerant (Chapman et al. 2014). Fish managers frequently use various sediment control devices to manage sediments, but studies of their effectiveness are few.

Compiling lists of sediment-sensitive macroinvertebrate taxa is a useful tool in characterizing biological impairment due to excess sediments. For example, Turley et al. (2016) developed a biomonitoring tool to identify the impacts of fine sediments using data from 835 field sampling sites throughout the UK. Sensitive taxa at the family level were those whose 75th percentile of abundance corresponded with a fine sediment value of <33%. Because species assemblages differ geographically, sensitive taxa lists will require some level of regional specificity, as Relyea et al. (2012) did for separate ecoregions of the western US. One difference between that study and Turley's is that the latter had good success with family-level classification, whereas Relyea et al. (2012) found that sensitivities varied within invertebrate families and concluded that family level was insufficient for effective classification. Recognizing that the identification of sediment-sensitive taxa is for practical management applications, ease of use and accuracy of classification may be the most important factors to consider when developing classification schemes.

5.2.3.2 Macrophytes

Aquatic macrophytes acting as physical structure increase habitat complexity and heterogeneity, thereby affecting numerous species of invertebrates, fishes, and water birds (Thomaz and Cunha 2010). They do so through a chain of mechanisms that involve the availability of shelter and feeding sites (Fig. 5.12). Macrophytes of greater structural complexity typically support more abundant and richer communities of invertebrates (Taniguchi et al. 2003). As Fig. 5.13 illustrates, habitat complexity provided by macrophytes operates over a range of scales, and so can be important to epiphytic algae, to the smallest invertebrates, and to larger organisms including fishes. Greater taxon

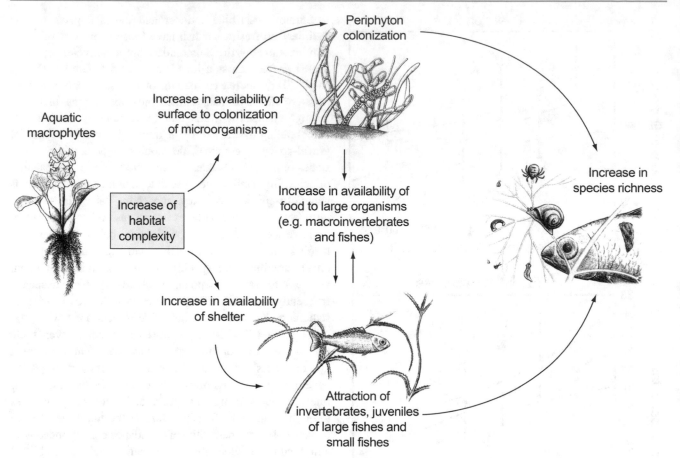

Fig. 5.12 A conceptual model depicting how the structural complexity provided by macrophytes may increase the diversity of other species in the aquatic assemblage (Reproduced from Thomaz and Cunha 2010)

richness can be explained by a greater variety of niches, whereas greater abundances may be due to more space for attachment, avoidance of predators and negative effects of current, and greater food supply for consumers of epiphytes and smaller invertebrates.

Because macrophyte complexity is influenced by details of plant architecture and is manifested at different scales, investigators have found it challenging to develop standardized measurements for cross-study comparisons. Area-adjusted number of leaves and stems, size and frequency of interstitial spaces, and fractal dimensions have been explored in various studies (Kovalenko et al. 2012). St. Pierre and Kovalenko (2014) quantified macrophyte complexity by measuring vertical and horizontal interstitial distances, finding that space-size heterogeneity was a more important contributor to macroinvertebrate taxon richness than overall complexity and the other complexity attributes examined. In a study of six representative plant species with varying physical features from lagoons and backwaters of the Upper Paraná River, Thomaz et al. (2008) examined invertebrate response as a function of plant area, a fractal dimension derived from a regression of number of objects

observed against the scale of analysis, and plant identity. Their findings (Fig. 5.14), based on comparing several statistical models, indicated that invertebrate density best explained number of species, suggesting a passive sampling effect. However, complexity was also important since it appeared in the second-best model selected, and plant identity was least important. Plant area, fractal dimension, and species identity shared approximately the same importance in explaining variation in number of individuals.

Aquatic macrophytes become invasive species in many settings, as they have been dispersed around the world for ornamental objectives, human food supply, mitigation of impacted areas, and other anthropogenic interests (Schultz and Dibble 2012). Displacement of native species and biological homogenization are common outcomes of a successful invasion, but studies have also found neutral and positive effects of non-native macrophytes on macrophyte-associated species, depending upon environmental setting (Strayer et al. 2003). In some circumstances, macrophyte introduction can play a constructive role in restoring degraded aquatic environments by adding habitat complexity, which in turn supports biodiversity recovery. In

Fig. 5.13 Different hierarchal scales within the aquatic macrophyte *Myriophyllum spicatum* showing different structural complexity at different scales (Reproduced from Thomaz and Cunha 2010)

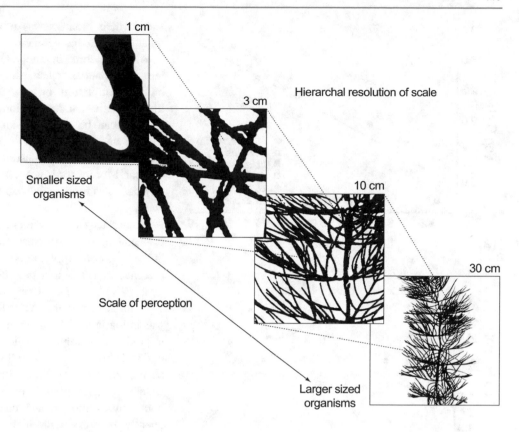

5.2.3.3 Wood

Downed wood in aquatic systems is a major component of habitat complexity, provides an important substrate for biological activity, and increases the diversity of organisms (Benke and Wallace 2003). Dudley and Anderson (1982) considered 52 taxa in the northwestern US to be closely associated with wood, and another 129 to be facultatively associated. A survey of invertebrates associated with wood in Central Europe concluded that 15 taxa inhabiting freshwater ecosystems were obligate xylophagous, 22 taxa were facultatively xylophagous, and additional taxa potentially fed on wood but had not been confirmed (Hoffman and Hering 2000). Even in agricultural streams in the Midwestern US where wood was not abundant, the majority of the recorded taxa (~90%) used wood as habitat, and the presence of wood substantially increased the number of taxa at a site (Johnson et al. 2003). Wood plays key roles in streams by influencing velocity and sedimentation profiles, forming pools, and strengthening banks. For the biological community, wood provides habitat for fauna, substrate for biofilms, refuge from predators and flow extremes, and enhances in-stream diversity of fish and macroinvertebrates. It can be

especially important in the simplified channels typical of many agricultural streams (Lester and Boulton 2008).

A number of studies have documented high levels of invertebrate biomass and diversity on wood (Benke et al. 1985; Scholz and Boon 1993). Wood affects a variety of invertebrate habitat conditions in streams and rivers (Benke and Wallace 2003). Dams formed by wood in small streams create pools and eddies, influencing fine-scale patterns of substrate and trapping organic matter. Trees from undercut banks that fall into the main channel may be the only stable habitat for invertebrates, especially in low-gradient, sand-bed rivers, and benefit fishes as flow refuge and source of invertebrate prey. In lowland rivers where the substrate is largely sand, fallen trees, also called snags, are especially important as a substrate for aquatic organisms. In the Satilla River, Georgia, Benke et al. (1985) estimated that snag, mud, and sand substrates occurred in the ratio 1:1.4:14 at an upriver site, and 1:3.6:18 at a downriver site. Though snag substrates were relatively limited, compared to mud and sand, snags supported more taxa and a far higher biomass of invertebrates than both mud and sand—more than half of the estimated invertebrate biomass in the river channel (Table 5.2). Interestingly, total numbers of organisms per unit area did not differ markedly between snags and sand. However, the invertebrates in the sand substrate were mostly oligochaetes and psammophilous midges of very small size,

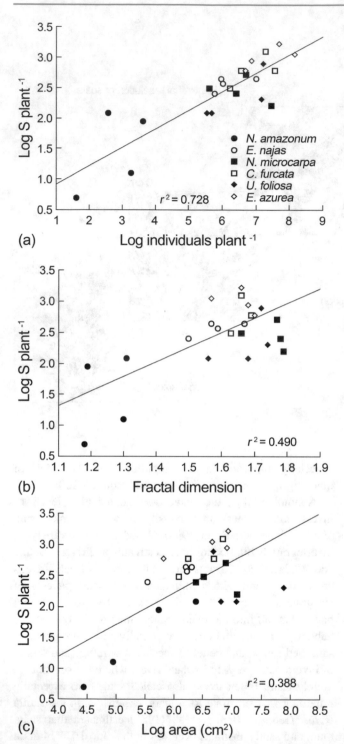

Fig. 5.14 Relationships between number of invertebrate taxa (S) and invertebrate abundances (cm⁻²) from six plant species from lagoons and backwaters of the Upper Paraná River exhibiting different architectures: *Nymphaea amazonum*, *Egeria najas*, *Cabomba furcata*, *Najas microcarpa*, *Utricularia foliosa*, and *Eichhornia azurea* (Reproduced from Thomaz et al. 2008)

and so their biomass was modest relative to many of the organisms colonizing snags.

Although some invertebrates use wood as food or tunnel into it as habitat, most reside on its surface and feed on biofilm, periphyton, or accumulated organic particles. Biofilm (bacteria and fungi) biomass and periphyton on wood surfaces can be an important food source for detritivores (Eggert and Wallace 2007). Due to its slow breakdown, wood serves as a long-lasting substrate for biofilm development. More work is needed to understand the contribution of wood, as a food source and as a substrate, to aquatic food webs.

Numerous factors influence the recruitment and storage of wood in river systems, including forest dynamics that influence wood supply, flood and channel dynamics, and the presence of snags that provide an anchoring point for log jams (Wohl 2017). More than half of large wood in a 177-km survey of the lower Roanoke River, Virginia, US, was in log jams (Moulin et al. 2011). Individual large wood is produced mainly by bank erosion and is isolated on the mid and upper banks at low flow, where it does not appear to be important as aquatic habitat. Log jams occur near or at water level, creating bank complexity in an otherwise homogenous fine-grained channel. They occur most frequently in areas with high snag concentrations, low to intermediate bank heights, high sinuosity, high local wood recruitment rates, and narrow channel widths.

5.2.3.4 Physical Habitat and Ecosystem Processes

In addition to its effects on species abundances and diversity, physical habitat heterogeneity also may influence ecosystem processes. Following experimental manipulation of substrate heterogeneity in a low gradient stream in northern Virginia, US, Palmer et al. (2002) observed significant increases in the primary productivity of stream algae and the respiration of the benthic biofilm in high-heterogeneity riffles. Increased near-bed flow velocity and turbulence intensity, and their influence on the supply of nutrients, gasses, and organic matter, were suggested mechanisms. Habitat complexity also increases the opportunity for water exchange between the channel and interstitial spaces within the substrate. Using the transit time for dye in streams of simple versus complex habitat, Kaufmann and Faustini (2012) found that transient storage, a measure that reflects the retention of water mass due to interstitial mixing, eddies and so on, was greater in channels with wood habitat and in more complex relative to less complex channels.

Increased hydraulic complexity is expected to provide more favorable conditions for ecosystem processes. Spatial

Table 5.2 The number of taxa and standing crop biomass of invertebrates found in snag, sand and mud habitats in the Satilla River, Georgia. Wood was a small percentage of habitat but contributed over half of the total biomass to the river reach (From Benke et al. 1985)

	Wood substrates			Sand			Mud		
	No. of genera	Biomass (mg m^{-2})		No. of genera	Biomass (mg m^{-2})		No. of genera	Biomass (mg m^{-2})	
		Lower site	Upper site		Lower site	Upper site		Lower site	Upper site
Diptera	17	243	696	15	64	124	11	148	309
Trichoptera	9	4222	1581	0	–	–	3	24	30
Ephemeroptera	5	97	56	0	–	–	0	–	–
Plecoptera	2	137	109	0	–	–	0	–	–
Coleoptera	3	218	117	1	8	11	0	–	–
Megaloptera	1	379	259	0	–	–	0	–	–
Odonata	3	529	578	1	–	–	0	–	–
Oligochaeta	0	–	–	3	22	22	0	420	290
Totals	40	5825	3396	20	94	157	17	592	629

heterogeneity of flow conditions influenced bacterial diversity and ecosystem processes within 40-m-long streamside flumes constructed with different bedform heights so as to create landscapes of differing velocity variance (Singer et al. 2010). Flow heterogeneity, quantified by Acoustic Doppler Velocimetry, was positively associated with bacterial biodiversity within biofilms on the substrate as quantified by RNA fingerprinting, and also with the diversity of dissolved organic carbon (DOC) compounds removed from the water (Fig. 5.15). While the mechanisms are not yet fully understood, bacterial uptake of glucose, a highly bioavailable compound, was thought to be related primarily to turbulence-enhanced diffusion into the boundary layer overlaying the biofilm. In contrast, because DOC leached from terrestrial leaves is thought to be less easily assimilated, the influence of flow heterogeneity on its uptake may have been mainly the result of enhanced bacterial biodiversity.

5.2.4 Physical Habitat Restoration

Rapidly expanding interest in stream restoration has led to many attempts to improve stream condition and benefit biological assemblages via improvements to physical habitat. The expectation is that, by providing more habitat elements and presumably more habitat complexity, these structural improvements will benefit individual species and potentially support higher taxon richness. Weirs, flow deflectors, cover structures, boulder placements, and large woody debris are common practices for restoring habitat in rivers, as well as gravel additions for gravel-spawning fish (Roni et al. 2008). By altering flow and scour patterns, these physical structures are expected to result in more diversified physical habitat, and thereby bring about increases in fish abundance and biomass. Compiling data from 211 stream restoration projects intended to benefit habitat and salmonid populations, Whiteway et al. (2010) quantified the effect size for a number of physical and biological variables in a statistical analysis of this large data set, called a meta-analysis. Following installation of in-stream structures, pool area, average depth, large woody debris, and percent cover all increased significantly (Fig. 5.16), while riffle area decreased.

Today, many streams and rivers have greatly reduced amounts of wood, as human activities have removed wood to benefit navigation, reduce damage to bridges and other infrastructure, and constrained channels to make more riparian land available for development and agriculture. Empirical evidence attests to the positive impact wood has on habitat complexity that supports biodiversity at all trophic levels, justifying management programs to re-introduce wood to rivers with simplified channels (Roni et al. 2008). Following the placement of engineered log jams in two large Pacific Northwest, US, river systems, periphyton biomass and invertebrate densities were significantly higher on log jams than on cobbles within the same reach (Coe et al. 2009). Because these rivers experience high flows capable of moving relatively large substrate, wood serves as a more complex and stable colonization surface compared to cobble. However, invertebrate communities on wood were dominated by meiofauna, whereas larger chironomids dominated on cobbles, evidently reflecting different preferences for substrate type among these taxa.

Placement of wood in streams typically leads to improvements in physical habitat characteristics including increases in pools, cover, and habitat complexity known to

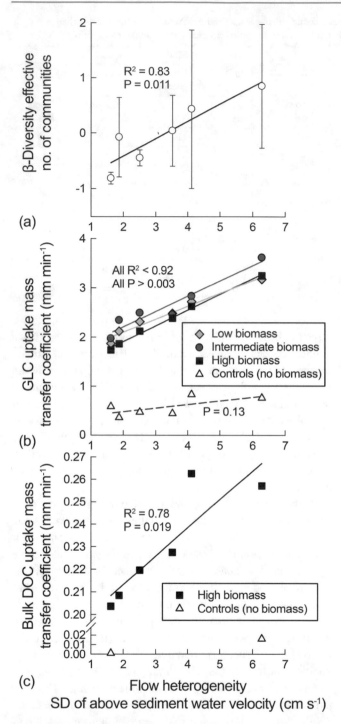

Fig. 5.15 Bacterial diversity, glucose uptake and leaf leachate DOC uptake all increased as a function of flow heterogeneity in experimental flumes. Flow heterogeneity was quantified as the standard deviation of 3-dimensional velocity measured with an Acoustic Doppler Velocimeter. (**a**) β-diversity is a measure of diversity of bacterial types among microhabitats. Diversity was classified using RNA fingerprinting. (**b**) Glucose is a highly bioavailable compound to bacteria. Its uptake is estimated by a mass transfer coefficient derived from declines in concentration over time. (**c**) Uptake of dissolved organic carbon (DOC) derived from terrestrial leaf leachate. DOC is a mix of compounds considered much less biologically available than glucose. Experimental flumes employed a range of bedform topographies to create flow heterogeneity. A flume without bedforms served as a control. Triangles indicate controls with no biofilms in (**b**) and (**c**) (Reproduced from Singer et al. 2010)

be important to fish (Roni et al. 2014). Although a few studies have reported high structural failure rates, most additions of wood remain stable in stream channels for more than a decade. A long-term study of five small streams in Colorado, US, illustrates the benefits (White et al. 2011). All streams supported wild populations of primarily brook trout (*Salvelinus fontinalis*), brown trout (*Salmo trutta*), or a mixture of brook, brown, and rainbow trout (*Oncorhynchus mykiss*). Log weirs installed as trout habitat survived well over more than two decades and brought about rapid and long-lasting increases in trout abundance. Pool volume was more than three times higher in treatment sections of a 500-m study reach, and mean depth was also greater. Adult trout benefited but no response was detected for juveniles, probably because their recruitment is strongly influenced by the variable effects of snowmelt runoff. In the more than 200 projects reviewed by Whiteway et al. (2010), salmonid local density and biomass both increased significantly, with 73% of projects resulting in increased densities and 87% in increased biomass. Results differed among size classes and species, with greater response by larger individuals and by rainbow trout. Studies of non-salmonid fishes, while fewer and limited in scope and duration, suggest a positive response of species richness in more diverse systems (Roni et al. 2008). Wood addition is not always a panacea for enhancing native biodiversity, however. An experiment involving the placement of 20 engineered log jams over an 1100 m reach of a river in south-eastern Australia substantially improved sediment storage and somewhat improved pool and bar areas, but after four years there was no significant increase in richness or abundance of fishes in the test reach compared to the control (Brooks et al. 2006).

Often it is unclear whether to attribute increased numbers and biomass to population gains resulting from higher recruitment, survival, and growth, or to population redistribution due to movements of individuals into the restored reach (Gowan and Fausch 1996; Roni et al. 2008). The population in question may not be limited primarily by habitat, or different life stages may be limited by different factors, creating a population bottleneck that better habitat alone cannot fix. Immigration almost certainly contributes to rapid increases in fish abundance following habitat enhancement, especially by large, dominant individuals seeking more complex habitat or more profitable feeding positions.

Studies of the response of macroinvertebrates, while fewer in number, nonetheless are adequate to indicate that the meso-scale physical habitat elements provided to benefit fish or create a more natural appearance, at best have mixed success for benthic invertebrates. The review by Roni et al. (2008) found that effects on macroinvertebrates were highly variable based on a number of response metrics, including abundance, diversity, and traits such as functional feeding groups. Although a few studies reported positive responses,

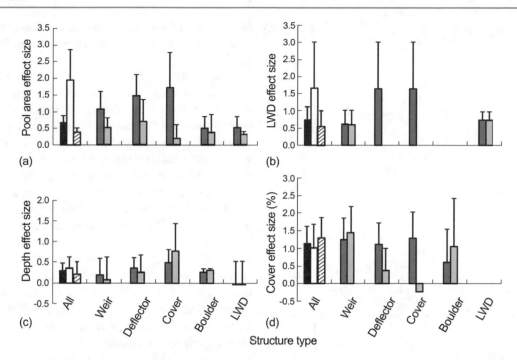

Fig. 5.16 Effect size (mean +95% confidence interval) estimated as the natural log of (treatment mean/control mean) of (**a**) pool area, (**b**) pieces of large woody debris (LWD), (**c**) stream depth, and (**d**) cover. Within the "all" bars, the solid bar represents the average effect for all structure types, the open bar represents projects that utilized only one type of structure, and the hatched bar represents projects that used two or more structure types. Within each structure type, the darker shaded bar represents the mean for all projects that used that structure (whether or not another type of structure was used) and the lighter shaded bar represents the mean for projects that only used that type of structure (Reproduced from Whiteway et al. 2010)

many did not, leading Roni et al. to suggest that macroinvertebrates may be neither sensitive to nor appropriate as success indicators for fish habitat enhancement projects. A review of 78 individual projects found that most were successful in enhancing physical habitat, but only two showed statistically significant increases in macroinvertebrate biodiversity that made them more similar to reference sites (Palmer et al. 2010). However, a meta-analysis that included only a small number of well replicated studies did find significant, positive effects of habitat restoration on macroinvertebrate richness relative to unrestored control reaches or pre-restoration conditions (Miller et al. 2010). Density also increased in restored reaches, but not significantly so due to variability in the direction and magnitude of density responses. Addition of large wood resulted in significantly greater increases in macroinvertebrate richness than did boulder additions, by adding pool–riffle morphologies and increasing the proportion of low velocity depositional habitats characterized by finer particle sizes, organic matter retention, and favorable conditions for shredders, collector-gatherers, and predatory macroinvertebrates that otherwise are rare or absent within channelized reaches. Using land use as a proxy for watershed-scale conditions, restoration projects implemented in forested upland environments exhibited more consistent, positive density and richness responses than projects located in

agricultural or urban areas (Fig. 5.17). Similarly, channel re-configuration in four degraded urban streams in the Piedmont region of North Carolina resulted in no improvement in habitat or biological condition (Violin et al. 2011). Restored urban streams were indistinguishable from those not restored, and did not transition in the direction of forested streams chosen as the reference or desired condition. The authors inferred that watershed-level hydrologic processes prevented change.

Habitat improvements to benefit macrophytes have received less study, but the evidence that exists is encouraging. Surveys of macrophyte communities of 40 restored river reaches in the lowland and lower mountainous areas of Germany when compared with upstream, unrestored reaches documented significant responses to a mix of hydrologic and morphological channel improvements (Lorenz et al. 2012). In comparison with the deep, uniformly flowing character of the unrestored reaches, restored reaches had wider and shallower stream channels that were less shaded, thus enhancing macrophyte growth, and more diverse flow and depth patterns, providing conditions for a more diverse assemblage. Macrophyte cover, abundance, and diversity all were greater in restored reaches (Fig. 5.18), despite the fact that restored and unrestored reaches were on average 0.5 km apart, and no restoration at the watershed scale took place. More natural and diverse substrates and an increased

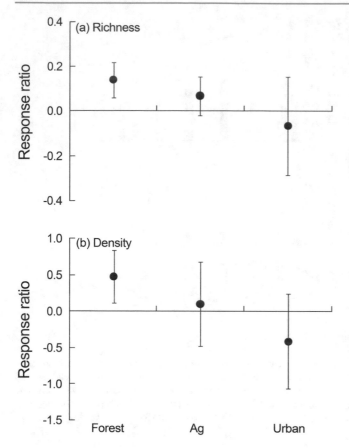

Fig. 5.17 Average response ratios to restoration with 90% confidence intervals for (**a**) macroinvertebrate richness and (**b**) macroinvertebrate density compared among forested, agricultural, and urban reaches (Reproduced from Miller et al. 2010)

effectiveness of instream devices is because instream devices tend to be less effective in larger streams (Stewart et al. 2009).

Despite the mixed verdict on the success of stream restoration to date, and in particular the uneven benefits to macroinvertebrates, improvements to physical habitat frequently benefit fish—usually the intended target—and may benefit other taxa as well. Loss of habitat complexity in streams and rivers is widespread, the result of wood removal, channel alteration, the floating of logs by river to market, altered hydrology, and more. Efforts to increase habitat complexity in streams are based on scientific understanding of organism-habitat relationships, and result in significant improvement when well designed and well matched to the geomorphic processes at work (Roni et al. 2015). Consideration of other factors such as water quality, hydrology, and habitat connectivity acting at larger scale than the typical single project can do much to ensure a successful biological outcome. As with all aspects of river restoration, learning from experience is critical to improving practices for the deployment of physical habitat structures and to ensure that the return on investment justifies the effort.

5.3 Temperature

floodplain area in the restored reaches were primarily responsible, as well as a greater variability of current and depth patterns.

The mixed success of restoration projects has multiple plausible explanations, including how well executed and long-lasting is the project, conditions in the upper watershed, and whether the outcome was evaluated based on population response or taxon richness, among others (Roni et al. 2008; Stewart et al. 2009; Kail et al. 2015). Because restoration measures often are reach-scale, covering short stretches of rivers, it is not surprising that outcomes can be offset by problems at the watershed scale that remain unresolved. The most important predictors of restoration success in a meta-analysis by Kail et al. (2015) were project age and agricultural land use in the upstream catchment. The negative influence of these two variables indicate that restoration actions often deteriorate over time, and that other environmental variables for which agricultural land use is a proxy, including water quality, sediment loading, and hydrologic variability, may offset the anticipated benefits of instream habitat improvements. Some of the variability in the

Figure 5.19 illustrates the main factors that influence stream water temperatures. The heat flux at the air-water interface results from energy exchange mainly from solar and long-wave radiation, evaporation, and convective heat transfer resulting from temperature differences between the river and the atmosphere (Caissie 2006; Olden and Naiman 2010). Detailed heat budgets find that radiative fluxes account for most (>70%) of heat inputs, but friction of the water with the bed and the banks and heat transfer from the atmosphere were also significant sources of heat energy (Webb et al. 2008). Evaporative heat transfers can account for significant cooling. Topography or geographical setting is important because it influences atmospheric conditions, and stream discharge because it influences the volume of water to be heated or cooled. Heat flux at the streambed, though considerably smaller than at the air-water interface, may be important in some settings, the result of geothermal heating and of heat transfer through groundwater inputs and hyporheic exchange.

The processes that influence stream temperature vary in their relative importance along a river's length from headwaters to mouth (Poole and Berman 2001). Mean daily water temperature generally is close to the groundwater temperature in headwater streams, and increases in the downstream direction. In temperate climates, groundwater inputs usually are cooler than channel water in spring and summer, and

Fig. 5.18 Macrophyte quantity, number of taxa, and number of growth forms were significantly higher in restored reaches than in unrestored reaches. Box–Whisker plots show differences in restored (re) and unrestored (un) mountain and lowland reaches in macrophyte quantity (**a**, **d**), richness (**b**, **e**), and number of growth forms (**c**, **f**). Left panel (**a–c**) submerged and emergent macrophytes, right panel (**e–f**) only submerged macrophytes. sm, submerged macrophytes; em, emergent macrophytes, *P < 0·05, **P < 0·01, ***P < 0·001 (Reproduced from Lorenz et al. 2012)

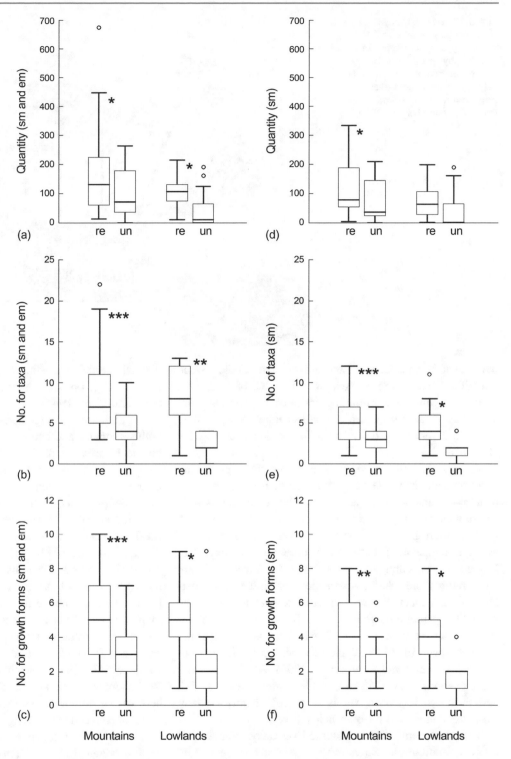

warmer than channel water in fall and winter. As a consequence, headwater streams that have strong groundwater influence commonly are cooler than would otherwise be expected in spring and summer, and warmer in fall and winter. Additionally, in alluvial streams with highly permeable gravel and cobble streambeds, there can be considerable bidirectional exchange of water between the channel and alluvial aquifer, referred to as hyporheic exchange

(Arrigoni et al. 2008). When surface water flows into the hyporheic zone it is referred to as downwelling or hyporheic recharge; the reverse is upwelling or hyporheic discharge. Because the annual range in hyporheic water temperature is typically less than that of channel water, hyporheic discharge commonly reduces the diel range in channel water temperature. In streams and rivers with substantial but spatially variable hyporheic exchange, considerable thermal

Fig. 5.19 Heat exchange processes responsible for variability in water temperatures and the physical drivers that control the rate of heat and water delivery to stream and river ecosystems (italics) (Reproduced from Olden and Naiman 2010)

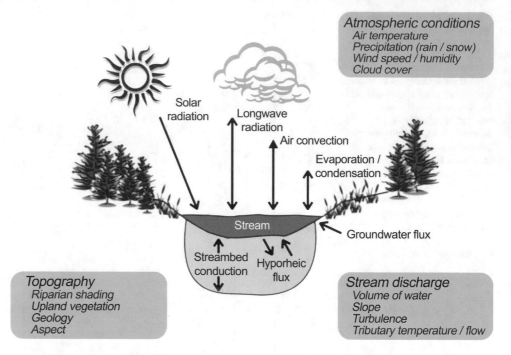

heterogeneity can occur. In the gravel- and cobble-bedded Umatilla River, Oregon, Arrigoni et al. (2008) recorded diverse daytime and nighttime mosaics of surface water temperatures across main and small channels, despite only minor differences in daily mean temperatures among the channels.

As a small stream becomes a larger river, increasing width lessens the moderating influence of riparian shade on heat inputs, and the influence of groundwater lessens as well. Tributaries entering the main channel may be warmer or colder, interrupting the gradual downstream increase in average temperature that would otherwise be expected. Floodplain pools and lakes when isolated can reach different temperatures and when re-connected can influence temperature in the main channel. In most temperate rivers, the annual temperature range is between 0 and 25 °C, but desert streams can reach nearly 40 °C, which is near the thermal tolerance even of fishes adapted to these extreme environments (Matthews and Zimmerman 1990). At high latitudes and elevations, maximum temperatures rarely exceed 15 °C, and they can be cooler yet in very cold climates where ice-cover can extend for over half the year.

Temporal variability is observed on daily, annual, and longer timeframes. Streamwater temperature generally reaches a daily minimum in the early morning and a maximum in late afternoon to early evening, and this diel variation is most pronounced in wide, shallow rivers of medium size. Groundwater dominance in small streams and thermal inertia in large rivers are responsible for dampening diel fluctuations. Tropical rivers can have very constant river temperatures owing to the constancy of solar radiation throughout the year and, in the case of large rivers, their

thermal inertia. The Amazon River at Manaus, Brazil, at 29 ± 1 °C, is one of most thermally stable water masses in the world (Sioli 1984).

Air and stream temperatures usually are well correlated, both seasonally and across locations, allowing air temperature to be used in predictive modeling. Simple regression models using weekly or monthly data have proven effective in predicting water temperature from air temperature, and more complex models may include heat flux estimates and some measure of long-term temperature fluctuations (Caissie 2006). Because the water–air temperature relationship departs from linearity both at low and high air temperatures, such models may be linear using a limited range (roughly 0–25 °C), or sigmoid using a wider range of air temperatures. For temperatures above freezing, Crisp and Howson (1982) found that mean weekly water temperatures (and the growth rate of brown trout) could be predicted from air temperatures using a 5–7 day lag. Some 60% of their estimates were within ±1 °C, and 80% within ±1.5 °C, of the measured stream temperatures. Despite the frequent use of correlations between air and water temperatures, however, solar radiation, not convective warming of water by the air, is the main heat input to streams, and so air temperature is better viewed as a surrogate rather than as a causal variable (Johnson 2003). In fact, statistical models to predict stream temperatures based on more readily available data on air temperatures, although widely used, have limited ability to project stream temperatures over time. This is because the underlying processes governing heat budgets of air and water are distinctive in each medium (Arismendi et al. 2014).

Comparison of modeled to observed streamwater temperatures also provides insight into departures from expectations that implicate anthropogenic influences. Using weather data to estimate the water temperature at which the sum of all heat fluxes through the water surface is zero, termed the equilibrium temperature, Bogan et al. (2003) compared stream and equilibrium temperatures for 596 stream gaging stations in the eastern and central US as a way to identify moderating influences. Weekly equilibrium temperature was a good estimator of weekly stream temperature for approximately one-fourth of stream gaging stations, those with no or minimal wind sheltering or sun shading. For the remaining three-fourth of streams, weekly equilibrium temperature was still a good predictor of weekly stream temperature, but with a significantly reduced slope, indicating the importance of other factors in addition to surface heat exchange.

In geologically complex regions, the source water of the tributaries of a single river system can be very different, resulting in a wide range of thermal habitats among headwaters that in turn may favor greater biodiversity as well as provide unique thermal niches for endemic taxa. In alpine streams of the French Pyrenees, where source waters include glaciers, snowpack, karst groundwater, and hillslope aquifers, karstic groundwater streams are coolest and most stable, hillslope groundwater streams are warmest and most variable, and glacial streams warm and become more variable downstream (Brown et al. 2007). Temperature variation due to source water and flow paths can be observed at much finer scales as well. In floodplain sections of the Tagliamento, a large, braided river of the Swiss Alps, cool-water habitats governed by groundwater inputs differ by as much as 15 °C from warm-water habitats of semi-isolated backwaters (Arscott et al. 2011). Microhabitat-scale temperature variation also is observed when stream water is forced into or drawn out from the streambed due to topographic undulations, meanders, bars, or other channel obstructions (White et al. 1987).

In addition to atmospheric, groundwater, topographic, and other physical controls on stream temperatures, various human influences significantly affect water temperatures. Most often the result is to warm rather than to cool, with the notable exception of bottom release reservoirs. Human influences likely to result in stream temperature warming include loss or reduction of riparian and upland forest cover and stream widening, which increase heat flux into the stream; reduced stream flows, which lower the volume of water to absorb heat; reduced groundwater exchange, usually a source of cooler water; discharge of warm-water effluents from municipal and industrial sources and from runoff over paved surfaces; and the effects of global warming, including increased air temperatures. Lakes and small impoundments are likely to have a warming effect. Large reservoirs that release cold bottom water typically lower stream temperatures in summer and warm them in winter.

Hester and Doyle (2011) concluded that most human influences typically altered stream or river temperatures by 5 °C or less, but warming due to loss of riparian shading and cooling brought about by deep-release dams could exceed 10 °C. Most effects are relatively localized, but the effects of deep-release dams can extend for many kms. In contrast, thermal heterogeneity resulting from hyporheic flows may occur at the scale of only a few m^2.

5.3.1 Shade

The presence of a forest canopy is known to modify the amount of solar radiation and other meteorological factors influencing stream temperatures (Fig. 5.19). In addition to blocking solar radiation from reaching the channel, riparian vegetation reduces near-stream windspeed and traps air against the water surface, thereby reducing heat exchange with the atmosphere (Poole and Berman 2001). Numerous studies have documented the impact of streamside forest removal on river water temperature (Beschta et al. 1987). Following timber harvest of a headwater stream in British Columbia, Canada, maximum daily temperatures increased by as much as 5 °C, and were positively associated with maximum daily air temperature and negatively with discharge (Moore et al. 2005a). Even greater increases in summer maximum temperature of about 8 °C were observed for two streams in the western Cascades in Oregon, US. Stream temperatures often recover to pre-harvest levels within 10 years but may take longer. When riparian buffers are left in place following timber harvest, stream warming is not observed or greatly reduced, although what buffer width is sufficient depends on such conditions as stream width and forest type (Moore et al. 2005b). Based on a review of the literature that evaluated the effectiveness of streamside forest in protecting water quality, habitat, and biota for small streams, Sweeney and Newbold (2014) concluded that riparian buffers should be at least 30 m in width. Benefits are most pronounced in small streams and lessen with increasing stream width because more stream surface is exposed to direct sun.

Stream reaches of alternating open and closed canopy have been observed to warm and cool over distances of less than one km, but not in all cases, indicating that not only changing solar radiation but also cooler groundwater inputs may be involved. A small stream in the central interior of British Columbia, Canada, cooled by approximately 3 °C as it passed from an open into a shaded reach, and groundwater inflow was responsible for about 40% of this cooling (Story et al. 2003). Thermal patches at least 3 °C colder than ambient stream water were identified at multiple sites of the Grand Ronde basin in northern Oregon, US, associated with side channels, alcoves, seeps, and floodplain spring brooks

(Ebersole et al. 2003). Experimental shading cooled maximum daily temperatures within cold patches by 2–4 °C, demonstrating the influence of riparian shade on the expression of cold-water micro-habitats.

5.3.2 Hydrologic Influences

Any reduction in river discharge due to water withdrawals or water diversion projects for municipal, hydroelectric power, or agricultural use will decrease the volume of water receiving heat inputs. This can result in substantial warming during summer when heat flux is greatest and flows may be naturally low. Water withdrawals can also deplete groundwater resources, thereby reducing cool inflows during warm seasons. The presence of lakes and impoundments along a river's course has a further influence on seasonal and diel water temperatures (Jones 2010). Due to the thermal inertia of large standing bodies of water, lakes tend to reduce diel variation downstream. At its outlet a lake drains from surface waters that often are warmer than river temperatures in summer and into the autumn, but cooler in the spring because of thermal inertia and the time required to warm lake surface waters.

Following a century of dam construction, many of the world's rivers are regulated rivers (Stanford et al. 1996). Water released from large reservoirs as well as from small impoundments usually modify downstream water temperatures by releasing water that is colder or warmer than would otherwise be the case (Olden and Naiman 2010). In temperate climates, large, deep reservoirs exhibit thermal stratification similar to large, deep lakes. The density of water is greatest at 4 °C, and so bottom water in deep reservoirs during winter is warmer than the near 0 °C temperature of the reservoir's upper layers, and of river water. During summer, reservoir surface waters may warm into the 20 °C range, but bottom waters typically are in the 4–8 °C range, hence bottom-release dams can cool river water substantially and for many km downstream. Temperatures in the lower Colorado River below Glen Canyon Dam are 9–12 ° C year-round, compared to the historic temperature range of 2–26 °C. The Green River, Utah, below the Flaming Gorge Dam (Fig. 5.20) illustrates the extent to which temperatures downstream from bottom-release dams are much cooler during summer and exhibit greatly dampened seasonal amplitude. A number of management options exist for mitigating the thermal impacts of dams, most commonly by means of multi-level water withdrawals that exploit the reservoir's temperature stratification by selectively withdrawing water of the desired temperature. This was implemented for the Flaming Gorge Dam, resulting in significant increases in spring–summer temperatures toward

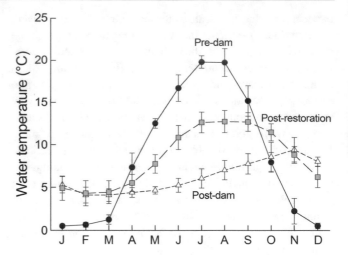

Fig. 5.20 Differing annual temperature regimes of the Green River, Utah, US, below Flaming Gorge Dam. Comparisons of monthly water temperature during pre-dam (1958–62, circles) and post-dam (1963–77, triangles) years shows the thermal dampening effect of the release of deep, cold water. Use of a multi-level water release structure during thermal restoration years (1978–2007, squares) resulted in significant increases in spring–summer temperatures toward unregulated conditions, but not during winter months (Reproduced from Olden and Naiman 2010)

unregulated conditions, but without reducing thermal alteration during the winter months.

The majority of impoundments are the result of small, surface release dams, and they are most likely to raise downstream temperatures during summer. Small impoundments act like lakes and beaver ponds that tend to increase stream temperatures because they increase the residence time of water and the surface area exposed to solar radiation. By sampling upstream and downstream of small impoundments on ten rivers throughout Michigan, US, Lessard and Hayes (2003) observed temperature increases in nine of the ten locations, by more than 5 °C at some sites and an overall average near 3 °C. As expected, the densities of several cold-water fish species including brown trout (*Salmo trutta*), brook trout (*Salvelinus fontinalis*), and slimy sculpin (*Cottus cognatus*) declined.

5.3.3 Urbanization

Anthropogenic factors associated with urban and industrial development can influence stream temperatures, usually causing warming. Thermal pollution refers to water released as industrial effluent, including from thermal-electric power generating plants, wastewater treatment facilities, construction holding ponds, and urban stormwater runoff (Webb et al. 2008). The release of heated, once-through condenser cooling water from power plants to the mainstem significantly increased summer temperatures in the Missouri River (Wright et al. 1999). Stream temperatures in the Ara River system, an urban river flowing through central Tokyo, increased in winter and early spring by 0.11–0.21 °C per year from 1978 through 1998 in segments that experienced a substantial increase in heat inputs from urban wastewater treatment plants over the same time period (Kinouchi et al. 2007). Changes in effluent temperature rather than air temperature best explained the result. Urban stream temperatures were observed to warm by 1–10 °C as a result of water retained in construction site sedimentation basins in Pennsylvania (Ehrhart et al. 2013). Localized rainstorms at urbanized sites in small watersheds of the Piedmont region of Maryland, US, resulted in temperature surges that averaged about 3.5 °C and dissipated over about three hours, presumably due to stormwater runoff over warm pavement (Nelson and Palmer 2007). At the most urbanized sites, these surges could occur on up to 10% of summer days and could briefly increase maximum temperature by 7 °C. A comparison of two watersheds in Puerto Rico shows that warming due to urbanization also affects tropical streams. Water temperatures in the Rio Piedras and the Espiritu Santo were similar during the early 1980s, but average annual temperatures are now higher in the urbanized Rio Piedras watershed, which experiences temperatures above 28 °C with greater frequency than its more forested counterpart (Ramírez et al. 2009).

5.3.4 Climate Change

Anthropogenic climate warming is likely to warm stream and rivers. However, although temperature records may span as long as a century, and some rivers indeed show a warming trend, it is difficult to attribute any observed changes to human influences on climate, in part because of large-scale climatic oscillations, and especially because urbanization and reservoir construction throughout the 20th century are additional drivers of warming water temperatures. No trend was observed for 90 years of water temperature data from north-central Austria (Webb and Nobilis 1997) but a significant increase of 0.8 C was seen over a similar time period in the River Danube, with greatest increases in autumn and early winter months (Webb and Nobilis 1994). Because no statistically significant trends were evident for air temperature or river discharge, the increase in river temperature does not appear to be purely a function of changing climatic conditions since 1900. Instead, rising water temperatures likely reflect increasing human modification of this heavily altered river system, including increases in the volume of heated effluent discharges and construction schemes that canalize and regulate the Danube. Compiling historical time series of water temperatures from 40 different stream and river sites located throughout the US, Kaushal et al. (2010) documented statistically significant, long-term warming of mean annual water temperatures of 0.009–0.077 °C per year. The most rapid rates of increase were observed for streams and rivers near urban areas of the mid-Atlantic US. These trends likely have multiple causes, including air temperature warming, urbanization, and potentially other factors including loss of riparian shade and reduced groundwater inputs. Thus, while it is highly likely that observed increases are due mostly to human activities, the amount that can be attributed to climate change is uncertain.

Clearly, rivers can be warmed by human influences other than climate change, but future climate warming is likely to contribute further to temperatures increases. Over the next several decades, air temperatures are projected to warm by about 0.2 °C per decade for a range of projected emissions scenarios (IPPC 2007). Mohseni et al. (1996) projected weekly stream temperatures under a climate scenario of a doubling of atmospheric CO_2 based on nonlinear models relating air to stream temperatures developed for 803 stream gaging stations in the contiguous US. Their projections showed that for all but 5% of the sites, mean annual stream temperatures would increase by 2–5 °C, and maximum and minimum weekly stream temperatures were projected to experience a 1–3 °C warming on average. Largest weekly changes were forecast for spring, and less change was projected for winter and summer.

5.3.5 Temperature and Ecological Processes

Temperature is a critical environmental variable determining the metabolic rates of organisms, their distribution along a river's length and over geographic regions, and their success in interacting with other species. Because species composition and biological rates are temperature dependent, ecosystem processes including leaf breakdown, nutrient uptake, and biological production are affected as well. Stream temperature changes in response to a variety of human actions, and so management intervention may be required to maintain a natural range of stream temperatures.

Every species is restricted to some temperature range that also limits its geographic distribution to a certain range of latitude and elevation. Species that occupy a narrow temperature range are referred to as stenothermal, while those that thrive over a wide range are called eurythermal. In addition, a species may be considered adapted to cold, cool-, or warm-water thermal environments. Few taxa are able to cope with very high temperatures, however. Coldwater fishes cannot survive water temperatures above 25 °C for very long, and most warmwater fishes including the pike family (Esocidae) and many minnows (Cyprinidae) have upper limits near 30 °C. Some fishes of desert streams can tolerate nearly 40 °C, a few invertebrates live at up to 50 °C, and specialized Cyanobacteria of hot springs survive 75 °C (Hynes 1970). In the heterogeneous glaciated landscape of Michigan, streams exhibit substantial regional variation in weekly mean temperature and in temperature fluctuation during warm seasons, allowing Wehrly et al. (2004) to determine the realized thermal niche of stream fishes based on three temperature categories (cold, less than 19 °C; cool, 19–22 °C; and warm, greater than or equal to 22 °C) and three temperature fluctuation categories (stable, less than 5 °C; moderate, 5–10 °C; and extreme, greater than 10 °C). The brook trout *Salvelinus fontinalis* and smallmouth bass *Micropterus dololieu* are good examples of cold-water and warm-water species, respectively. Overall fish diversity increased with mean water temperature across some 300 Michigan sites, documenting the well-established higher diversity of warmwater over coldwater streams. This pattern has often been reported as a longitudinal gradient in which downstream temperature increase is accompanied by an increase in river size and change in many other variables. However, because the Michigan study provided a wide range of temperature regimes within a relatively modest range of stream sizes, these results strongly implicate temperature as the causal variable.

There are a number of reasons for the specific temperature requirements of a particular species. The taxonomic lineage to which it belongs may have originated and diversified in cool waters at high latitudes (e.g., Plecoptera, Hynes 1988), or in warmer water at low latitudes (e.g., Odonata, Corbet 1980). The timing of an insect's life cycle, which often is both cued and regulated by temperature, determines the seasons when it most actively grows, consumes resources, and is exposed to predators. Thus, resource availability and predation risk, two topics that will re-appear in later chapters, can be important evolutionary pressures that are responsible for a particular life cycle and suite of temperature adaptations. A number of studies document how closely related species hatch, grow, and emerge in such a neatly staggered sequence that their life cycle separation appears to ameliorate competition (Sweeney and Vannote 1981; Elliott 1987) (Fig. 5.21). By determining when an insect hatches and grows, temperature synchronizes the life cycle to changing seasonal conditions, coordinating growth with resource availability and ensuring the availability of mates. When the growing season is very short, a common life cycle is to alternate short periods of rapid growth separated by long periods of dormancy or diapause (Danks 1992). Such life cycles often are seen at high latitudes and may require exposure to near-freezing temperatures followed by a rapid temperature rise in order to break egg diapause. When a deep release dam was built on the Saskatchewan River, Canada, water temperatures were maintained near 4 °C throughout the year, and so the cue to ending egg diapause was

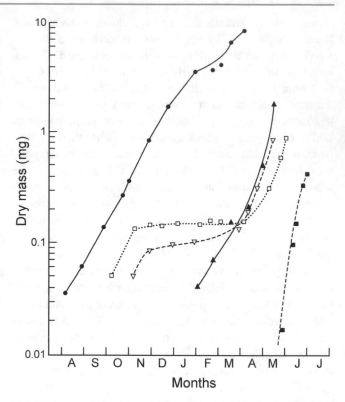

Fig. 5.21 Larval growth period for five species of riffle-inhabiting ephemerellid mayflies in White Clay Creek, Pennsylvania. (●) *Ephemerella subvaria*; (▲) *E. dorothea*; (□) *Seratella deficiens*; (■) *S. serrata*; (inverted open triangle) *Euryophella verisimilis* (Reproduced from Sweeney and Vannote 1981)

eliminated. Virtually all insect taxa disappeared from the stretch of river with modified temperatures. A fauna that previously included 12 orders, 30 families, and 75 species was reduced to only the midge family Chironomidae (Lehmkuhl 1974).

The influence of temperature on stream biota has been demonstrated in the laboratory, from life cycle studies, and from their distributions. For 12 invertebrate taxa from New Zealand, the lethal temperature that killed half of the individuals in 96 h (the LT_{50}) ranged from 22.6 to 32.6 °C (Quinn et al. 1994). Laboratory results on thermal tolerances also were consistent with field observations that Plecoptera and Ephemeroptera were much less abundant in rivers where typical summer temperatures exceeded 19 and 21 °C, respectively (Quinn and Hickey 1990). Sampling of 20 sites in streams located in the lower mountainous area of western Germany revealed that summer temperature variation explained more of the variability in the macroinvertebrate assemblage among sites than did other environmental factors including conductivity, substrate type, and the percent coverage of local riparian forest (Haidekker and Hering 2008). Temperature was more important for the macroinvertebrate composition of smaller than of larger streams, indicating that the latter possessed more tolerant, eurythermic species.

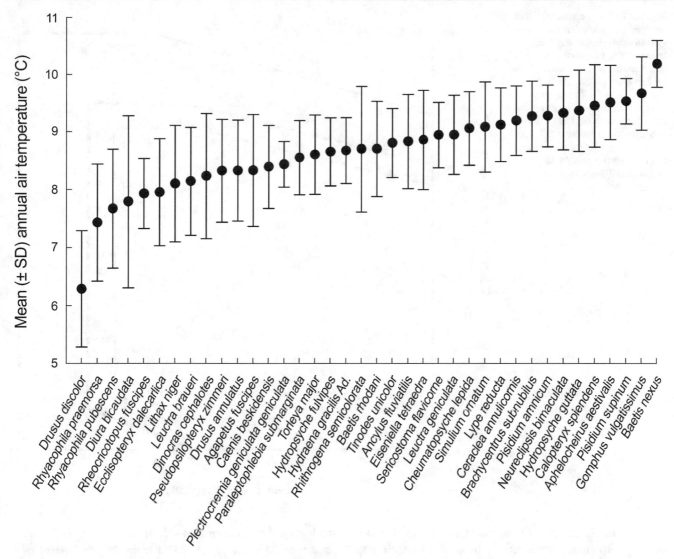

Fig. 5.22 Mean (± SD) annual air temperatures corresponding to the occurrence of 38 species of stream macroinvertebrates, based on presence-absence data for species occurring in submontane streams in Germany. Species were selected to represent those occurring only in upper reaches, hence with a preference for cooler temperatures (12 species); those occurring only in the lower reaches, hence with a preference for warmer temperatures (12); and species occurring over a wide range of zones and thereby exhibiting a broad temperature range preference (14). Air temperature was used as a surrogate for water temperature because data were more readily available, and the air-water temperature correlation is well established (Reproduced from Domisch et al. 2011)

Using climate warming scenarios and the known temperature range (Fig. 5.22) associated with the distribution of 38 species of benthic stream macroinvertebrates for the same region, species ranges were predicted to shift upward in elevation by approximately 80–120 m, contract for species occupying cool environments, and expand for those occupying warmer environments. Assemblage composition was expected to change, and headwater species to decline (Domisch et al. 2011).

Much research has been done on the thermal requirements and life cycles of salmonid species, making them excellent candidates to evaluate how future warming may affect populations. Temperature influences all aspects of the life cycle, including growth, time of spawning, egg hatching, and larval emergence, with best performance at optimal temperatures. Within a species' thermal tolerance zone, rates increase up to an optimal temperature, above which rising metabolic costs take a toll (Fig. 5.23). Under global warming, salmonid populations at the southern end of their range in the northern hemisphere are likely to be extirpated, while more northerly populations likely will benefit and expand northward (Jonsson and Jonsson 2009). For cool-water, northern hemisphere species at their southern-most range extent and thus near their upper thermal limits in warm water streams, canopy shade may be particularly important to their continued survival. In lowland rivers in the New Forest of

Fig. 5.23 Temperature tolerance polygon for *Salmo trutta* showing growth zone (inside the thin broken line), tolerance zone inside incipient lethal level (thick broken line) within which *S. trutta* feed, and ultimate lethal level where death is almost instantaneous (solid line). Growth at temperatures below 4 °C can occur during winter but usually not from spring to autumn (Reproduced from Jonsson and Jonsson 2009)

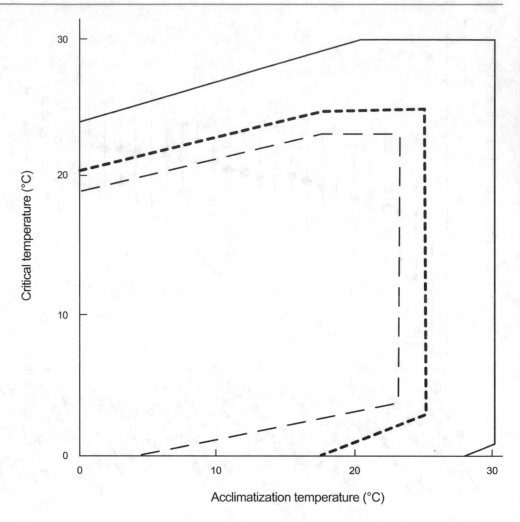

southern England, brown trout occur in streams that occasionally experience temperatures in excess of their upper limit for growth of 19.1 °C and their incipient lethal limit of 24.7 °C, depending upon extent of shade (Broadmeadow et al. 2011). A relatively low level of shade (20–40%) sufficed to keep summer temperatures below the incipient lethal limit for brown trout, but more extensive shade of at least 80% was necessary to ensure that water temperatures did not exceed the range for optimum growth. Expansion of riparian woodland thus is a management opportunity to protect temperature-sensitive species from anticipated climate warming.

By shifting both southern and northern distributional limit, climate warming is expected to shift species ranges to higher latitudes or elevations, causing species replacements. Locally, there will be both winners and losers, assuming that habitat is suitable as species disperse to follow their migrating thermal niche. Unfortunately, some species will have nowhere to go, including some adapted to the cool waters found at high elevations. Limited headwater habitat area may result in substantial reduction in population size

and even local extirpation of some species. Using data spanning 25 years (1981–2005) from the headwaters of a stream in the uplands of Wales, UK, Durance and Ormerod (2007) observed a warming trend that was detectable after accounting for effects of the North Atlantic Oscillation, implicating directional climate change. As stream temperatures increased, macroinvertebrate abundances fell and assemblage composition changed significantly, although this result was not detected in species-poor streams of low pH. Extrapolating their finding to future increases of 1, 2, and 3 ° C, reasonable given current model scenarios, total macroinvertebrate abundance could decline by up to 60%.

A warming environment has implications beyond individual species, of course, as multiple species disperse into new thermal conditions and encounter a different mix of competitors and predators. At the southern limit of brown trout in Europe, warm-adapted species such as pike, cyprinids, and percids may move into habitat from which salmonids have been displaced. At the northern limit of trout in Europe, arctic rivers, which are currently unsuitable or marginally suitable for salmonids, may become habitable

(Jonsson and Jonsson 2009). At present, these locations are more suitable for the arctic char *Salvelinus alpinus* because of its high growth efficiency at low water temperatures, but warming will set the stage for competition between brown trout and char. Temperature changes also can have indirect effects on populations by providing more favorable conditions for their enemies. The northern pikeminnow *Ptychocheilus oregonensis* is an important predator of juvenile Pacific salmon migrating to sea through the Columbia River. Using data from previous warm and cold periods associated with climate oscillations in the North Pacific Ocean and a bioenergetics model, Petersen and Kitchell (2001) predicted that predation on salmonids would have been 26–31% higher during warm than cold periods, and 68–96% higher when comparing the warmest with the coldest year. Climate regime shifts have the potential to significantly alter predation rates, and presumably yearly recruitment success of young salmon in this system.

Small-scale thermal heterogeneity can also be important to organisms, especially to fish that are able to seek out and reside in patches of water with more favorable temperatures. When surface waters are too warm, fish are able to avoid stressful high temperatures and behaviorally thermoregulate by moving into area where cool groundwater enters the surface channel, thereby maintaining optimal body temperatures. In winter, when groundwater is often warmer than surface waters, areas of groundwater upwelling are likely to allow faster growth rates and stable, warmer conditions for developing embryos of winter-spawning species. Thermal refuges can be very important to species near the limits of their thermal niche, such as the brown trout in southern England described above. Juvenile rainbow trout are able to persist in northeast Oregon streams where maximum temperatures exceeded 26 °C for several hours a day, although abundances were negatively correlated with maximum stream temperatures (Ebersole et al. 2001). Reaches of the Klamath River below the Iron Gate Dam in northern California, US, often have daily maximum temperatures during summer that exceed 25 °C as a result of reduced flows and other climatic and within-basin anthropogenic changes. This is well above the optimal temperatures for juvenile salmonids, including coho salmon (*Oncorhynchus kisutch*), Chinook salmon (*O. tshawytscha*) and migratory rainbow trout that occur in this system. By monitoring temperatures and counting fish using daytime snorkeling, Sutton et al. (2007) showed that most juvenile salmonids moved into thermal refuges associated with a tributary junction when main-stem temperatures exceeded 22–23 °C.

Finally, temperature controls the metabolism of all producers and ectothermic consumers in fluvial ecosystems, thus climate warming is likely to cause a number of changes in stream ecosystem structure and function, including greater productivity by benthic algae and microbes, which in turn may influence nutrient demand and efficiency of use. Laboratory incubation of stones taken from the Thur River in Switzerland showed that respiration of the benthic biofilm increased with incubation temperature similarly across locations with mean temperatures ranging from 8 to 19 °C (Acuña et al. 2008). Extrapolating from these results indicates that an increase of 2.5 °C will increase river respiration by an average of 20%. Ecosystem production and respiration increased with water temperature along a gradient of mean water temperatures (7.5–23.6 °C) in small streams of a geothermally active area of southwestern Iceland (Williamson et al. 2016), and in response to a ~ 3.3 °C experimental warming of a 35 m reach of one stream using a geothermal heat exchanger (Hood et al. 2018). Because these streams were low in nutrients, especially nitrogen, nutrient supply might have been expected to limit system productivity. This was not the case, apparently due to greater efficiency of nutrient uptake and especially to N-fixation by cyanobacteria. The broader implication is that future climate change may alter the relationship between photosynthetic carbon fixation and nutrient dynamics in unanticipated ways.

5.4 Summary

Abiotic factors include all physical and chemical variables that influence the distribution and abundance of organisms. Current, physical habitat, and temperature often are the most important variables in fluvial environments, and all organisms show adaptations that limit them to a subset of conditions. Species differ in the specific environmental conditions under which they thrive, and whether those conditions are narrow or comparatively broad. Habitat preferences can be inferred from the subset of environmental variables that best correlate with a species' distribution and abundance, with the important caveat that interactions with other species may further restrict the habitat occupied. Environments that are either structurally simple or extreme tend to support fewer species, and habitats that are more complex tend to support more species. How patterns in the abiotic environment are manifested across spatial scale, and the persistence of suitable conditions over temporal scale, add challenge to the task of deciphering organism-habitat relationships.

Current and related hydraulic forces affect diverse aspects of the stream environment including channel shape and substrate composition, the physical structure and hydraulic forces operating in the benthic and near-bed microhabitats, and the balance between physical drag processes and the benefits due to the delivery of food, nutrients, and gasses and the removal of wastes. Most organisms live attached to, within, or associated with the streambed, where flows are turbulent and difficult to measure, especially at relevant scales. Considerable effort has gone into theoretical approaches and methods of measurement that can better characterize the flow environment experienced by organisms. Both simple and complex hydraulic variables can be effective predictors of the distribution of individual species and the overall abundance and richness of the invertebrate fauna.

Even under 'normal' flows, many organisms will experience the varied effects of current, including shear stress from current velocity at the scale and location of the organism, and the rapid changes in velocity that create turbulence. Episodic increases in flow, termed spates and floods, can cause abrasion by suspended sediments and erosion of the substrate on which organisms reside, with more serious and even catastrophic effects on organisms. Thus, the association of organisms with particular currents and substrates can reflect the ability of particular habitats to serve as refuges. Some invertebrates and fishes have been shown to move quickly into more sheltered habitats as flows increase, but in other cases it may be happenstance whether an organism is protected from flow during a spate.

Physical habitat includes a wide variety of inorganic and organic substrates of varying size, large objects such as boulders and submerged wood, and channel units such as pools, riffles, and bends. Thus, physical habitat varies from the micro to the meso and macro scales. It interacts strongly with current, which both influences substrate stability and is in turn influenced by bed friction; and with roughness, which creates complex, near-bed flow environments. The quantification of physical habitat requires multiple approaches, including size analyses for inorganic substrates, counts of wood and estimates of biomass or volume, and descriptive categories such as leaf accumulations and debris dams. At the mesohabitat scale, channel units such as riffles and pools, and channel depth, while not strictly substrate categories, are important habitat categories for larger organisms, especially fishes. Among inorganic substrates, gravel of intermediate size supports a diverse, lithophilous flora and fauna that have received a great deal of study by lotic ecologists. Terrestrial leaves that accumulate in streams and wood that modifies habitat and channel shape are important and well-studied organic substrates.

Studies of the importance of physical habitat frequently are simultaneously studies of current, as these two abiotic factors are linked. Substrate size and heterogeneity appear to promote species richness, at least to a degree, and surface texture and roughness additionally promote abundance and diversity. The stability of the substrate and the protection it affords from the forces of current clearly are critical aspects as well.

Temperature is a key environmental variable determining the metabolic rates of organisms, their distributions, and quite possibly their success in interacting with other species. Stream temperature usually varies on seasonal and daily time scales and among locations due to climate, extent of streamside vegetation, and the relative importance of groundwater inputs. For these reasons thermal regimes are highly diverse, and can vary on all spatial scales from micro-scale patches to the longitudinal gradient from headwaters to river mouth. Human activities can alter the natural temperature regime in many ways, including removal of shade-enhancing vegetation, changes to flow paths such as increased impervious surface, construction of impoundments, and of course by influencing the climate.

Freshwater organisms exhibit a wide range of thermal tolerances that correspond to the thermal environment they inhabit, which may be cool or warm, constant or fluctuating. With few exceptions the organisms of running waters are ectotherms, and so external temperature governs their metabolism and growth. Although warmer temperatures generally allow greater activity, they also impose greater metabolic costs. Whether the evidence is based on lethal temperatures in the laboratory, detailed analysis of energy budgets, or field surveys, it is evident that temperature strongly affects distributions and abundances. The thermal tolerances of fishes have received a great deal of study, and it is common in the temperate zone to speak of coldwater and warmwater fishes, which typically are arrayed along latitudinal and elevational gradients. The extreme sensitivity of coldwater fishes to micro-patch differences in temperature is evident in their ability to at least temporarily reside in warmwater systems by using locations where groundwater inputs and shade provide suitable thermal conditions.

References

Ackerman JD, Hoover TM (2001) Measurement of local bed shear stress in streams using a Preston-static tube. Limnol Oceanogr 2080–2087

Acuña V, Wolf A, Uehlinger U, Tockner K (2008) Temperature dependence of stream benthic respiration in an Alpine river network under global warming. Freshw Biol 53:2076–2088. https://doi.org/10.1111/j.1365-2427.2008.02028.x

Anderson NH, Sedell JR (1979) Detritus processing by macroinvertebrates in stream ecosystems. Annu Rev Entomol 24:351–377

Arismendi I, Safeeq M, Dunham JB, Johnson SL (2014) Can air temperature be used to project influences of climate change on stream temperature? Environ Res Lett 9:084015

Arrigoni AS, Poole GC, Mertes LAK et al (2008) Buffered, lagged, or cooled? Disentangling hyporheic influences on temperature cycles in stream channels. Water Resour Res 44:1–13. https://doi.org/10.1029/2007wr006480

Arscott DB, Tockner K, Ward JV (2011) Thermal heterogeneity along a braided floodplain river (Tagliamento River, northeastern Italy).

Can J Fish Aquat Sci 58:2359–2373. https://doi.org/10.1139/f01-183

Balon E (1981) Additions and amendments to the classification of reproductive styles in fishes. Environ Biol Fishes 6:377–389

Barmuta LA (1989) Habitat patchiness and macrobenthic community structure in an upland stream in temperate Victoria, Australia. Freshw Biol 21:223–236. https://doi.org/10.1111/j.1365-2427.1989.tb01361.x

Barnes JB, Vaughan IP, Ormerod SJ (2013) Reappraising the effects of habitat structure on river macroinvertebrates. Freshw Biol 58:2154–2167. https://doi.org/10.1111/fwb.12198

Begon M, Townsend CR, Harper J (2005) Ecology: from individuals to ecosystems. Blackwell, Oxford

Beisel J, Usseglio-polatera P, Moreteau J (2000) The spatial heterogeneity of a river bottom: a key factor determining macroinvertebrate communities. Hydrobiologia 422(423):163–171

Benke AC, Henry RL, Gillespie DM, Hunter RJ (1985) Importance of snag habitat for animal production in southeastern streams. Fisheries 10:8–13. https://doi.org/10.1577/1548-8446(1985)010%3c0008:ioshfa%3e2.0.co;2

Benke AC, Wallace JB (2003) Influence of wood on invertebrate communities in streams and rivers. In: Gregory S, Boyer K, Gurnell A (eds) The ecology and management of wood in world rivers. Symposium 37. American Fisheries Society, Bethesda MA, pp 149–177

Bergey EA (2005) How protective are refuges? Quantifying algal protection in rock crevices. Freshw Biol 50:1163–1177. https://doi.org/10.1111/j.1365-2427.2005.01393.x

Bertin S, Friedrich H, Delmas P et al (2014) DEM quality assessment with a 3D printed gravel bed applied to stereo photogrammetry. Photogramm Rec 29:241–264. https://doi.org/10.1111/phor.12061

Beschta R, Bilby R, Brown G et al (1987) Stream temperature and aquatic habitat: fisheries and forestry interactions. In: Salo E, Cundy T (eds) Streamside management: forestry and fishery interactions. Institute of Forest Resources, Contribution No. 57. University of Washington, Seattle, Washington, pp 191–232

Biggs BJF, Stevenson RJ, Lowe RL (1998) A habitat matrix conceptual model for stream periphyton. Arch Fur Hydrobiol 143:21–56. https://doi.org/10.1127/archiv-hydrobiol/143/1998/21

Bogan T, Mohseni O, Stefan HG (2003) Stream temperature-equilibrium temperature relationship. Water Resour Res 39https://doi.org/10.1029/2003WR002034

Bouckaert WF, Davis J (1998) Microflow regimes and the distribution of macroinvertebrates around stream boulders. Freshw Biol 40:77–86

Bovee KD (1982) A guide to stream habitat analysis using the instream flow incremental methodology. U. S. Fish and Wildlife Service, Office of Biological Services, Instream Flow Information Paper 12, FWS/OBS-82/26

Broadmeadow SB, Jones JG, Langford TEL et al (2011) The influence of riparian shade on lowland stream water temperatures in southern England and their viability for brown trout. River Res Appl 27:226–237. https://doi.org/10.1002/rra.1354

Brooks AJ, Haeusler T, Reinfelds I, Williams S (2005) Hydraulic microhabitats and the distribution of macroinvertebrate assemblages in riffles. Freshw Biol 50:331–344. https://doi.org/10.1111/j.1365-2427.2004.01322.x

Brooks AP, Howell T, Abbe TB, Arthington AH (2006) Confronting hysteresis: wood based river rehabilitation in highly altered riverine landscapes of south-eastern Australia. Geomorphology 79:395–422. https://doi.org/10.1016/j.geomorph.2006.06.035

Brown LE, Hannah DM, Milner AM (2007) Vulnerability of alpine stream biodiversity to shrinking glaciers and snowpacks. Glob Chang Biol 13:958–966. https://doi.org/10.1111/j.1365-2486.2007.01341.x

Caissie D (2006) The thermal regime of rivers: a review. Freshw Biol 51:1389–1406. https://doi.org/10.1111/j.1365-2427.2006.01597.x

Chapman JM, André M-È, Bliss S et al (2014) Clear as mud: a meta-analysis on the effects of sedimentation on freshwater fish and the effectiveness of sediment-control measures. Water Res 56:190–202. https://doi.org/10.1016/j.watres.2014.02.047

Coe HJ, Kiffney PJ, Pess GR, Kloehn KK, McHenry ML (2009) Periphyton and invertebrate response to wood placement in large Pacific coastal rivers. River Res Appl 25:1025–1035. https://doi.org/10.1002/rra.1201

Corbet PS (1980) Biology of Odonata. Annu Rev Entomol 25:189–217

Costa SS, Melo AS (2008) Beta diversity in stream macroinvertebrate assemblages: among-site and among-microhabitat components. Hydrobiologia 598:131–138. https://doi.org/10.1007/s10750-007-9145-7

Cotel AJ, Webb PW, Tritico H (2006) Do brown trout choose locations with reduced turbulence? Trans Am Fish Soc 135:610–619. https://doi.org/10.1577/t04-196.1

Crisp DT, Howson G (1982) Effect of air temperature upon mean water temperature in streams in the north Pennines and English Lake District. Freshw Biol 12:359–367. https://doi.org/10.1111/j.1365-2427.1982.tb00629.x

Danks HV (1992) Long life cycles in insects. Can Entomol 124:167–187

Davis JA, Barmuta LA (1989) An ecologically useful classification of mean and near-bed flows in streams and rivers. Freshw Biol 21:271–282. https://doi.org/10.1111/j.1365-2427.1989.tb01365.x

Death R (2008) Effects of floods on aquatic invertebrate communities. In: Lancaster J, Briers R (eds) Aquatic insects: challenges to populations. CAB International, Wallingford, Oxfordshire, pp 103–121

Descloux S, Datry T, Marmonier P (2013) Benthic and hyporheic invertebrate assemblages along a gradient of increasing streambed colmation by fine sediment. Aquat Sci 75:493–507. https://doi.org/10.1007/s00027-013-0295-6

Dolédec S, Castella E, Forcellini M et al (2015) The generality of changes in the trait composition of fish and invertebrate communities after flow restoration in a large river (French Rhône). Freshw Biol 60:1147–1161. https://doi.org/10.1111/fwb.12557

Dolédec S, Lamouroux N, Fuchs U, Mérigoux S (2007) Modelling the hydraulic preferences of benthic macroinvertebrates in small European streams. Freshw Biol 52:145–164. https://doi.org/10.1111/j.1365-2427.2006.01663.x

Domisch S, Jähnig SC, Haase P (2011) Climate-change winners and losers: stream macroinvertebrates of a submontane region in Central Europe. Freshw Biol 56:2009–2020. https://doi.org/10.1111/j.1365-2427.2011.02631.x

Downes BJ, Lake PS, Schreiber ESG et al (1998) Habitat structure and regulation of local species diversity in a stony, upland stream. Ecol Monogr 68:237–257

Dudley T, Anderson N (1982) A survey of invertebrates associated with wood debris in aquatic habitats. Melandaria 39:1–21

Dudley TL, D'Antonio CM (1991) The effects of substrate texture, grazing, and disturbance on macroalgal establishment in streams. Ecology 72:297–309

Durance I, Ormerod SJ (2007) Climate change effects on upland stream macroinvertebrates over a 25-year period. Glob Chang Biol 13:942–957. https://doi.org/10.1111/j.1365-2486.2007.01340.x

Ebersole JL, Liss WJ, Frissell CA (2003) Cold water patches in warm streams: physicochemical characteristics and the influence of shading. J Am Water Resour Assoc Resour 39:355–368

Effenberger M, Sailer G, Townsend CR, Matthaei CD (2006) Local disturbance history and habitat parameters influence the microdistribution of stream invertebrates. Freshw Biol 51:312–332. https://doi.org/10.1111/j.1365-2427.2005.01502.x

Eggert SL, Wallace JB (2007) Wood biofilm as a food resource for stream detritivores. Limnol Oceanogr 52:1239–1245. https://doi.org/10.4319/lo.2007.52.3.1239

Ehrhart BJ, Shannon RD, Jarrett AR (2013) Effects of construction site sedimentation basins on receiving stream ecosystems. Trans ASAE 45https://doi.org/10.13031/2013.8833

Elliott JM (1987) Egg hatching and resource partitioning in stoneflies: the six British Leuctra Spp. (Plecoptera: Leuctridae). J Anim Ecol 56:415–426

Encalada AC, Peckarsky BL (2006) Selective oviposition of the mayfly *Baetis bicaudatus*. Oecologia 148:526–537. https://doi.org/10.1007/s00442-006-0376-5

Enders EC, Boisclair D, Roy AG (2003) The effect of turbulence on the cost of swimming for juvenile Atlantic salmon (*Salmo salar*). Can J Fish Aquat Sci 60:1149–1160. https://doi.org/10.1139/f03-101

Eriksen C (1964) The influence of respiration and substrate upon the distribution of burrowing mayfly naiads. Verhandlungen der Int Vereinigung fur Theor und Angew Limnol 15:903–911

Fausch KD (1984) Profitable stream positions for salmonids: relating specific growth rate to net energy gain. Can J Zool 62:441–451. https://doi.org/10.1139/z84-067

Francoeur SN, Biggs BJF, Lowe RL (1998) Microform bed clusters as refugia for periphyton in a flood-prone headwater stream. New Zeal J Mar Freshw Res 32:363–374. https://doi.org/10.1080/00288330.1998.9516831

Frutiger A, Schib J-L (1993) Limitations of FST hemispheres in lotic benthos research. Freshw Biol 30:463–474. https://doi.org/10.1111/j.1365-2427.1993.tb00829.x

Gard M (2009) Comparison of spawning habitat predictions of PHABSIM and River2D models. Int J River Basin Manag 7:55–71. https://doi.org/10.1080/15715124.2009.9635370

Gowan C, Fausch KD (1996) Long-term demographic responses of trout populations to habitat manipulation in six Colorado streams. Ecol Appl 6:931–946

Green JC (2005) Velocity and turbulence distribution around lotic macrophytes. Aquat Ecol 39:1–10. https://doi.org/10.1007/s10452-004-1913-0

Grenouillet G, Pont D, Olivier JM (2000) Habitat occupancy patterns of juvenile fishes in a large lowland river: interactions with macrophytes. Arch Fur Hydrobiol 149:307–326. https://doi.org/10.1127/archiv-hydrobiol/149/2000/307

Haidekker A, Hering D (2008) Relationship between benthic insects (Ephemeroptera, Plecoptera, Coleoptera, Trichoptera) and temperature in small and medium-sized streams in Germany: a multivariate study. Aquat Ecol 42:463–481. https://doi.org/10.1007/s10452-007-9097-z

Harding JM, Burky AJ, Way CM (1998) Habitat preferences of the Rainbow Darter, *Etheostoma caeruleum*, with regard to microhabitat velocity shelters. Copeia 1998:988–997

Hart DD, Clark BD, Jasentuliyana A (1996) Fine-scale field measurement of benthic flow environments inhabited by stream invertebrates. Limnol Oceanogr 41:297–308. https://doi.org/10.4319/lo.1996.41.2.0297

Hart DD, Finelli CM (1999) Physical-biological coupling in streams: the pervasive effects of flow on benthic organisms. Annu Rev Ecol Syst 30:363–395

Hax CL, Golliday SW (1993) Macroinvertebrate colonization and biofilm development on leaves and wood in a boreal river. Freshw Biol 29:79–87. https://doi.org/10.1111/j.1365-2427.1993.tb00746.x

Hester ET, Doyle MW (2011) Human impacts to river temperature and their effects on biological processes: a quantitative synthesis. J Am Water Resour Assoc 47:571–587. https://doi.org/10.1111/j.1752-1688.2011.00525.x

Hockley FA, Wilson CAME, Brew A, Cable J (2014) Fish responses to flow velocity and turbulence in relation to size, sex and parasite load. J R Soc Interface 11:20130814 https://doi.org/10.1098/rsif.2013.0814

Hoffman A, Hering D (2000) Wood-associated macroinvertebrate fauna in Central European streams. Int Rev Hydrobiol 85:25–49

Holomuzki JR, Biggs BJF (2000) Eutrophication of streams and rivers: dissolved nutrient-chlorophyll relationships for benthic algae. J North Am Benthol Soc 19:670–679

Hood JM, Benstead JP, Cross WF et al (2018) Increased resource use efficiency amplifies positive response of aquatic primary production to experimental warming. Glob Chang Biol 24:1069–1084. https://doi.org/10.1111/gcb.13912

Hoover TM, Ackerman JD (2004) Near-bed hydrodynamic measurements above boulders in shallow torrential streams: implications for stream biota. J Environ Eng Sci 3:365–378. https://doi.org/10.1139/s04-012

Hughes NF, Dill LM (1990) Position choice by drift-feeding salmonids: model and test for Arctic Grayling (*Thymallus arcticus*) in subarctic mountain streams, interior Alaska. Can J Fish Aquat Sci 47:2039–2048. https://doi.org/10.1139/f90-228

Huryn AD, Wallace JB (2000) Life history and production of stream insects. Annu Rev Entomol 45:83–110. https://doi.org/10.1146/annurev.ento.45.1.83

Hynes H (1970) The ecology of running waters. University of Toronto Press, Toronto

Hynes HBN (1988) Biogeography and origins of the North American stoneflies (Plecoptera). Mem Entomol Soc Canada 144:31–37

IPPC (2007) Climate change 2007: the physical science basis. Contribution of working group I to the fourth assessment report of the Intergovernmental Panel on Climate Change. Cambridge University Press, Cambridge, UK

Ebersole JL, Liss WJ, Frissell CA (2001) Relationship between stream temperature, thermal rerugia and rainbow trout Oncorhynchus mykiss abundance in arid-land streams in northwestern United States. Ecol Freshw Fish 10:1–10

Johnson LB, Breneman DH, Richards C (2003) Macroinvertebrate community structure and function associated with large wood in low gradient streams. River Res Appl 19:199–218. https://doi.org/10.1002/rra.712

Johnson SL (2003) Stream temperature: scaling of observations and issues for modelling. Hydrol Process 17:497–499. https://doi.org/10.1002/hyp.5091

Jones NE (2010) Erratum: incorporating lakes within the river discontinuum: longitudinal changes in ecological characteristics in stream–lake networks. Can J Fish Aquat Sci 67:2058–2058. https://doi.org/10.1139/f10-142

Jonsson B, Jonsson N (2009) A review of the likely effects of climate change on anadromous Atlantic salmon *Salmo salar* and brown trout *Salmo trutta*, with particular reference to water temperature and flow. J Fish Biol 2381–2447. https://doi.org/10.1007/978-94-007-1189-1

Jowett IG (1993) A method for objectively identifying of turbulent eddies. Pool, run, and riffle habitats from physical measurements. N Z J Mar Freshw Res 27:241-248

Jowett IG (2003) Hydraulic constraints on habitat suitability for benthic invertebrates ingravel-bed rivers. River Res Appl 19:495–507

Kail J, Brabec K, Poppe M, Januschke K (2015) The effect of river restoration on fish, macroinvertebrates and aquatic macrophytes: a meta-analysis. Ecol Indic 58:311–321. https://doi.org/10.1016/j.ecolind.2015.06.011

Katopodis C, Williams JG (2012) The development of fish passage research in a historical context. Ecol Eng 48:8–18. https://doi.org/10.1016/j.ecoleng.2011.07.004

Kaufmann PR, Faustini JM (2012) Simple measures of channel habitat complexity predict transient hydraulic storage in streams. Hydrobiologia 685:69–95. https://doi.org/10.1007/s10750-011-0841-y

Kaushal SS, Likens GE, Jaworski NA et al (2010) Rising stream and river temperatures in the United States. Front Ecol Environ 8:461–466. https://doi.org/10.1890/090037

Kemp P, Sear D, Collins A et al (2011) The impacts of fine sediment on riverine fish. Hydrol Process 25:1800–1821. https://doi.org/10.1002/hyp.7940

Kinouchi T, Yagi H, Miyamoto M (2007) Increase in stream temperature related to anthropogenic heat input from urban wastewater. J Hydrol 335:78–88. https://doi.org/10.1016/j.jhydrol.2006.11.002

Kondolf GM (2004) Assessing salmonid spawning gravel quality. Trans Am Fish Soc 129:262–281. https://doi.org/10.1577/1548-8659(2000)129%3c0262:assgq%3e2.0.co;2

Kovalenko KE, Thomaz SM, Warfe DM (2012) Habitat complexity: approaches and future directions. Hydrobiologia 685:1–17. https://doi.org/10.1007/s10750-011-0974-z

Lacey RWJ, Roy AG (2008) The spatial characterization of turbulence around large roughness elements in a gravel-bed river. Geomorphology 102:542–553. https://doi.org/10.1016/j.geomorph.2008.05.045

Lamouroux N, Mérigoux S, BH C et al (2010) The generality of abundance-environment relationships in microhabitats: a comment on Lancaster and Downes (2009). River Res Appl 26:915–920

Lamouroux N, Olivier JM (2015) Testing predictions of changes in fish abundance and community structure after flow restoration in four reaches of a large river (French Rhône). Freshw Biol 60:1118–1130. https://doi.org/10.1111/fwb.12324

Lancaster J, Hildrew AG (1993) Characterizing in-stream flow refugia. Can J Fish Aquat Sci 50:1663–1675

Larsen S, Vaughan IP, Ormerod SJ (2009) Scale-dependent effects of fine sediments on temperate headwater invertebrates. Freshw Biol 54:203–219. https://doi.org/10.1111/j.1365-2427.2008.02093.x

Lee D, Gilbert C, Hocutt C et al (1980) Atlas of north american freshwater fish. North Carolina State Museum of Natural History, Raleigh, NC

Lehmkuhl D (1974) Thermal regime alteration and vital environmental physiological signals in aquatic organisms. In: Gibbons JW, Scharits RR (eds) Thermal ecology, AEC symposium series, CONF 730505, pp 216–222

Lenat DR, Penrose DL, Eagleson KW (1981) Variable effects of sediment addition on stream benthos. Hydrobiologia 79:187–194. https://doi.org/10.1007/BF00006126

Lessard JL, Hayes DB (2003) Effects of elevated water temperature on fish and macroinvertebrate communities below small dams. River Res Appl 19:721–732. https://doi.org/10.1002/rra.713

Lester RE, Boulton AJ (2008) Rehabilitating agricultural streams in Australia with wood: a review. Environ Manage 42:310–326. https://doi.org/10.1007/s00267-008-9151-1

Liao J, Beal D, Lauder G, Triantafyllou M (2003) Fish exploiting vortices decrease muscle activity. Science (80) 302:1566–1569

Ligon FK, Nakamoto RJ, Harvey BC, Baker PF (2016) Use of streambed substrate as refuge by steelhead or rainbow trout Oncorhynchus mykiss during simulated freshets. J Fish Biol 88:1475–1485. https://doi.org/10.1111/jfb.12925

Lorenz AW, Korte T, Sundermann A et al (2012) Macrophytes respond to reach-scale river restorations. J Appl Ecol 49:202–212. https://doi.org/10.1111/j.1365-2664.2011.02082.x

Lytle DA (1999) Use of rainfall cues by Abedus herberti (Hemiptera: Belostomatidae): a mechanism for avoiding flash floods. J Insect Behav 12:1–12. https://doi.org/10.1023/A:1020940012775

Mackay RJ (1992) Colonization by lotic macroinvertebrates: a review of processes and patterns. Can J Fish Aquat Sci 49:617–628. https://doi.org/10.1139/f92-071

Mackay RJ, Kalff J (1969) Seasonal variation in standing crop and species diversity of insect communities in a small Quebec stream. Ecology 50:101–109

Maddock I, Harby A, Kemp P, Wood P (2013) Ecohydraulics: an integrated approach. Wiley-Blackwell, Chichester, West Sussex, U. K.

Mathers KL, Rice SP, Wood PJ (2017) Temporal effects of enhanced fine sediment loading on macroinvertebrate community structure and functional traits. Sci Total Environ 599–600:513–522. https://doi.org/10.1016/j.scitotenv.2017.04.096

Matthaei CD, J AC, Townsend C R, (2000) Stable surface stones as refugia for invertebrates during disturbance in a New Zealand stream. J North Am Benthol Soc 19:82–93

Matthews WJ, Zimmerman EG (1990) Potential effects of global warming on native fishes of the Southern great plains and the Southwest. Fisheries 15:26–32. https://doi.org/10.1577/1548-8446(1990)015%3c0026:peogwo%3e2.0.co;2

Mérigoux S, Dolédec S (2004) Hydraulic requirements of stream communities a case study on invertebrates. Freshw Biol 49:600–613

Mérigoux S, Forcellini M, Dessaix J et al (2015) Testing predictions of changes in benthic invertebrate abundance and community structure after flow restoration in a large river (French Rhône). Freshw Biol 60:1104–1117. https://doi.org/10.1111/fwb.12422

Miller SW, Budy P, Schmidt JC (2010) Quantifying macroinvertebrate responses to in-stream habitat restoration: applications of meta-analysis to river restoration. Restor Ecol 18:8–19. https://doi.org/10.1111/j.1526-100X.2009.00605.x

Minshall GW (1984) Aquatic insect–substratum relationships. In: Resh V, Rosenberg D (eds) The ecology of aquatic insects. Praeger, New York, pp 358–400

Mohseni O, Erickson TR, Stefan HG (1996) Sensitivity of stream temperatures in the United States to air temperatures projected under a global warming scenario. Water Resour Res 35:3723–3733

Moore RD, Spittlehouse DL, Story A (2005) riparian microclimate and stream temperarure response to forest harvesting: a review. J Am Water Resour Assoc 41:813–834

Moore RD, Sutherland P, Gomi T, Dhakal A (2005) Thermal regime of a headwater stream within a clear-cut, coastal British Columbia, Canada. Hydrol Process 19:2591–2608. https://doi.org/10.1002/hyp.5733

Moulin B, Schenk ER, Hupp CR (2011) Distribution and characterization of in-channel large wood in relation to geomorphic patterns on a low-gradient river. Earth Surf Process Landforms 36:1137–1151. https://doi.org/10.1002/esp.2135

Nelson KC, Palmer MA (2007) Stream temperature surges under urbanization. J Am Water Resour Assoc 43:440–452. https://doi.org/10.1111/j.1752-1688.2007.00034.x

Nepf HM (2012) Hydrodynamics of vegetated channels. J Hydraul Res 50:262–279. https://doi.org/10.1080/00221686.2012.696559

Newbury RW, Bates DJ (2017) Chapter 4–dynamics of flowing water. Methods stream ecology, vol 1, pp 71–87

Nikora VI, Suren AM, Brown SLR, Biggs BJF (1998) The effects of the moss Fissidens rigidulus (Fissidentacea: Musci) on near-bed flow structure in an experimental cobble bed flume. Limnol Oceanogr 43:1321–1331

Nislow KH, Magilligan FJ, Folt CL, Kennedy BP (2002) Within-basin variation in the short-term effects of a major flood on stream fishes and invertebrates. J Freshw Ecol 17:305–318. https://doi.org/10.1080/02705060.2002.9663899

Olden JD, Naiman RJ (2010) Incorporating thermal regimes into environmental flows assessments: modifying dam operations to restore freshwater ecosystem integrity. Freshw Biol 55:86–107. https://doi.org/10.1111/j.1365-2427.2009.02179.x

Palmer MA (1990) Temporal and spatial dynamics of meiofauna within the hyporheic zone of Goose Creek, Virginia. J North Am Benthol Soc 9:17–25

Palmer MA, Arensburger PA, Silver Botts P et al (1995) Disturbance and the community structure of stream invertebrates: patch-specific effects and the role of refugia. Freshw Biol 34:343–356. https://doi.org/10.1111/j.1365-2427.1995.tb00893.x

Palmer MA, Menninger HL, Bernhardt E (2010) River restoration, habitat heterogeneity and biodiversity: a failure of theory or practice? Freshw Biol 55:205–222. https://doi.org/10.1111/j.1365-2427.2009.02372.x

Palmer MA, Swan CM, Brooks S et al (2002) The influence of substrate heterogeneity on biofilm metabolism in a stream ecosystem. Ecology 83:412. https://doi.org/10.2307/2680024

Palmer MA, Swan CM, Nelson K et al (2000) Streambed landscapes: evidence that stream invertebrates respond to the type and spatial arrangement of patches. Landsc Ecol 15:563–576. https://doi.org/10.1023/A:1008194130695

Petersen JH, Kitchell JF (2001) Climate regimes and water temperature changes in the Columbia River: bioenergetic implications for predatorsof jevenile salmon. Can J Fish Aquat Sci 58:1831–1841. https://doi.org/10.1139/cjfas-58-8-1831

Poff NL (1997) Landscape filters and species traits: towards mechanistic understanding and prediction in stream ecology. J North Am Benthol Soc 16:391–409

Poole GC, Berman CH (2001) An ecological perspective on in-stream temperature: natural heat dynamics and mechanisms of human-caused thermal degradation. Environ Manage 27:787–802. https://doi.org/10.1007/s002670010188

Quinn JM, Hickey CW (1994) Hydraulic parameters and benthic invertebrate distributions in two gravel-bed New Zealand rivers. Freshw Biol 32:489–500. https://doi.org/10.1111/j.1365-2427.1994.tb01142.x

Quinn JM, Hickey CW (1990) Characterisation and classification of benthic invertebrate communities in 88 New Zealand rivers in relation to environmental factors. N Z J Mar Freshw Res 24:387–409. https://doi.org/10.1080/00288330.1990.9516432

Quinn JM, Steele GL, Hickey CW, Vickers ML (1994) Upper thermal tolerances of twelve New Zealand stream invertebrate species. N Z J Mar Freshw Res 28:391–397. https://doi.org/10.1080/00288330.1994.9516629

Rabeni CF, Minshall GW (1977) Factors affecting microdistribution of stream benthic insects. Oikos 29:33–43

Railsback SF (2016) Why it is time to put PHABSIM out to pasture. Fisheries 41:720–725. https://doi.org/10.1080/03632415.2016.1245991

Ramírez A, De Jesús-Crespo R, Martinó-Cardona DM et al (2009) Urban streams in Puerto Rico: what can we learn from the tropics? J North Am Benthol Soc 28:1070–1079. https://doi.org/10.1899/08-165.1

Reich P, Downes BJ (2003) The distribution of aquatic invertebrate egg masses in relation to physical characteristics of oviposition sites at two Victorian upland streams. Freshw Biol 48:1497–1513. https://doi.org/10.1046/j.1365-2427.2003.01101.x

Relyea CD, Minshall GW, Danehy RJ (2012) Development and validation of an aquatic fine sediment biotic index. Environ Manage 49:242–252. https://doi.org/10.1007/s00267-011-9784-3

Resh V, Buchwalter D, Lamberti G, Eriksen C (2008) Aquatic insect respiration. In: Merritt R, Cummins K, Berg M (eds) An introduction to aquatic insects of North America, 4th edn. Kendall/Hunt, Dubuque Iowa, pp 39–54

Rice S, Buffin-Bélanger T, Lancaster J, Reid I (2008) Movements of a macroinvertebrate (Potamophylax latipennis) across a gravel-bed substrate: effects of local hydraulics and microtopography under increasing discharge. In: Habersack H, Piegay H, Rinaldi M (eds) Gravel-bed rivers VI: from process understanding to river restoration. Elsevier, Amsterdam, pp 637–660

Rice SP, Greenwood MT, Joyce CB (2011) Tributaries, sediment sources, and the longitudinal organisation of macroinvertebrate fauna along river systems. Can J Fish Aquat Sci 58:824–840. https://doi.org/10.1139/f01-022

Roni P, Beechie T, Pess G, Hanson K (2014) Wood placement in river restoration: fact, fiction, and future direction. Can J Fish Aquat Sci 72:466–478. https://doi.org/10.1139/cjfas-2014-0344

Roni R, Beechie T, Pess G, Hanson K. (2015) Wood placement in river restoration: fact, fiction, and future. Can J Fish Aquat Sci 72:466–478. https://doi.org/10.1139/cjfas-2014-0344

Roni P, Hanson K, Beechie T (2008) Global review of the physical and biological effectiveness of stream habitat rehabilitation techniques. North Am J Fish Manag 28:856–890. https://doi.org/10.1577/m06-169.1

Roy ML, Roy AG, Legendre P (2010) The relations between "standard" fluvial habitat variables and turbulent flow at multiple scales in morphological units of a gravel-bed river. River Res Appl 26:439–435.https://doi.org/10.1002/rra.1281

Sagnes P, Mérigoux S, Péru N (2008) Hydraulic habitat use with respect to body size of aquatic insect larvae: case of six species from a French Mediterranean type stream. Limnologica 38:23–33. https://doi.org/10.1016/j.limno.2007.09.002

Santos JM, Silva A, Katopodis C et al (2012) Ecohydraulics of pool-type fishways: getting past the barriers. Ecol Eng 48:38–50. https://doi.org/10.1016/j.ecoleng.2011.03.006

Schneck F, Schwarzbold A, Melo AS (2011) Substrate roughness affects stream benthic algal diversity, assemblage composition, and nestedness. J North Am Benthol Soc 30:1049–1056. https://doi.org/10.1899/11-044.1

Scholz O, Boon PI (1993) Biofilms on submerged River Red Gum (Eucalyptus camaldulensis Dehnh. Myrtaceae) wood in billabongs: an analysis of bacterial assemblages using phospholipid profiles. Hydrobiologia 259:169–178. https://doi.org/10.1007/BF00006596

Schultz R, Dibble E (2012) Effects of invasive macrophytes on freshwater fish and macroinvertebrate communities: the role of invasive plant traits. Hydrobiologia 684:1–14. https://doi.org/10.1007/s10750-011-0978-8

Schwartz JS, Herricks EE (2005) Fish use of stage-specific fluvial habitats as refuge patches during a flood in a low-gradient Illinois stream. Can J Fish Aquat Sci 62:1540–1552. https://doi.org/10.1139/f05-060

Singer G, Besemer K, Schmitt-Kopplin P et al (2010) Physical heterogeneity increases biofilm resource use and its molecular diversity in stream mesocosms. PLoS One 5https://doi.org/10.1371/journal.pone.0009988

Sioli H (1984) The Amazon: limnology and landscape ecology of a mighty tropical river and its Basin. Dr. W. Junk, Dordrecht

Southwood TRE (1988) Tactics, strategies and templets. Oikos 52:3. https://doi.org/10.2307/3565974

St. Pierre JI, Kovalenko KE (2014) Effect of habitat complexity attributes on species richness. Ecosphere 5:1–10

Stanford JA, Ward JV, Liss WJ et al (1996) A general protocol for restoration of regulated rivers. Regul Rivers Res Manag 12:391–413

Statzner B, Muller R (1989) Standard hemispheres as indicators of flow characteristics in lotic benthos research. Freshw Biol 21:445–459

Steinman AD, Boston HL (1993) The ecological role of aquatic Bryophytes in a woodland stream. J North Am Benthol Soc 12:17–26

Stewart GB, Bayliss HR, Showler DA et al (2009) Effectiveness of engineered in-stream structure mitigation measures to increase salmonid abundance: a systematic review. Ecol Appl 19:931–941

Story A, Moore RD, Macdonald JS (2003) Stream temperatures in two shaded reaches below cutblocks and logging roads: downstream cooling linked to subsurface hydrology. Can J for Res 33:1383–1396. https://doi.org/10.1139/x03-087

Strayer DL (1999) Use of flow refuges by Unionid mussels in rivers. J North Am Benthol Soc 18:468–476

Strayer DL, Lutz C, Malcom HM et al (2003) Invertebrate communities associated with a native (*Vallisneria americana*) and an alien (*Trapa natans*) macrophyte in a large river. Freshw Biol 48:1938–1949. https://doi.org/10.1046/j.1365-2427.2003.01142.x

Sutton RJ, Deas ML, Tanaka SK et al (2007) Salmonid observations at a Klamath River thermal refuge under various hydrological and meteorological conditions. River Res Appl 23:775–785. https://doi.org/10.1002/rra

Sweeney BW, Newbold JD (2014) Streamside forest buffer width needed to protect stream water quality, habitat, and organisms: a literature review. J Am Water Resour Assoc 50:560–584. https://doi.org/10.1111/jawr.12203

Sweeney BW, Vannote RL (1981) *Ephemerella* mayflies of White Clay Creek: bioenegetic and ecological relationships among six coexisting species. Ecology 62:1353–1369

Taniguchi H, Nakano S, Tokeshi M (2003) Influences of habitat complexity on the diversity and abundance of epiphytic invertebrates on plants. Freshw Biol 48:718–728. https://doi.org/10.1046/j.1365-2427.2003.01047.x

Thomaz SM, Cunha ER (2010) The role of macrophytes in habitat structuring in aquatic ecosystems: methods of measurement, causes and consequences on animal assemblages' composition and biodiversity. Acta Limnol Bras 22:218–236. https://doi.org/10.4322/actalb.02202011

Thomaz SM, Dibble ED, Evangelista LR et al (2008) Influence of aquatic macrophyte habitat complexity on invertebrate abundance and richness in tropical lagoons. Freshw Biol 53:358–367. https://doi.org/10.1111/j.1365-2427.2007.01898.x

Tokeshi M, Pinder LCV (1985) Microhabitats of stream invertebrates on two submersed macrophytes with contrasting leaf morphology. Oikos 8:313–319

Turley MD, Bilotta GS, Chadd RP et al (2016) A sediment-specific family-level biomonitoring tool to identify the impacts of fine sediment in temperate rivers and streams. Ecol Indic 70:151–165. https://doi.org/10.1016/j.ecolind.2016.05.040

Vinson MR, Hawkins CP (2003) Broad-scale geographical patterns in local stream insect genera richness. Ecography (Cop) 26:751–767https://doi.org/10.1111/j.0906-7590.2003.03397.x

Violin CR, Cada P, Sudduth EB et al (2011) Effects of urbanization and urban stream restoration on the physical and biological structure of stream ecosystems. Ecol Appl 21:1932–1949

Vogel S (1996) Life in moving fluids, 2nd edn. Princeton University Press, Princeton, NJ

Waters T (1995) Sediment in streams: sources, biological effects and control. American Fisheries Society, Bethesda, MD

Webb BW, Hannah DM, Moore RD et al (2008) Recent advances in stream and river temperature research. Hydrol Process 22:902–918

Webb BW, Nobilis F (1994) Water temperature behaviour in the River Danube during the twentieth century. Hydrobiologia 291:105–113. https://doi.org/10.1007/BF00044439

Webb BW, Nobilis F (1997) Long-term perspective on the nature of the water-air temperature relationship-a case study. Hydrol Process 11:137–147

Wehrly KE, Wiley MJ, Seelbach PW (2004) Classifying regional variation in thermal regime based on stream fish community patterns. Trans Am Fish Soc 132:18–38. https://doi.org/10.1577/1548-8659(2003)132%3c0018:crvitr%3e2.0.co;2

Weissenberger J, Spatz H, Emanns A, Schwoerbel J (1991) Measurement of lift and drag forces in the m N range experienced by benthic arthropods at flow velocities below 1.2 m s^{-1}. Freshw Biol 25:21–31. https://doi.org/10.1111/j.1365-2427.1991.tb00469.x

White DS, Elzinga CH, Hendricks SP (1987) Temperature patterns within the hyporheic zone of a northern Michigan river. J North Am Benthol Soc 6:85–91

White SL, Gowan C, Fausch KD et al (2011) Response of trout populations in five Colorado streams two decades after habitat manipulation. Can J Fish Aquat Sci 68:2057–2063. https://doi.org/10.1139/f2011-125

Whiteway SL, Biron PM, Zimmermann A et al (2010) Do in-stream restoration structures enhance salmonid abundance? A meta-analysis. Can J Fish Aquat Sci 67:831–841. https://doi.org/10.1139/f10-021

Williamson TJ, Cross WF, Benstead JP et al (2016) Warming alters coupled carbon and nutrient cycles in experimental streams. Global 22:2152–2164. https://doi.org/10.1111/gcb.13205

Wohl E (2017) Bridging the gaps: an overview of wood across time and space in diverse rivers. Geomorphology 279:3–26. https://doi.org/10.1016/j.geomorph.2016.04.014

Wright SA, Holly Jr FM, Allen Bradley A, Krajewski W (1999) Long-term simulation of thermal regime of Missouri River. J Hydraul Eng 125:242–252

Young RG, Hayes JW, Wilkinson J, Hay J (2009) Movement and mortality of adult brown trout in the Motupiko River, New Zealand: effects of water temperature, flow, and flooding. Trans Am Fish Soc 139:137–146. https://doi.org/10.1577/t08-148.1

Autotrophs are organisms that acquire energy from nonliving sources. Most autotrophs in running waters use sunlight and inorganic components to build biomass. Conversely, heterotrophs obtain energy and materials by consuming living or dead organic matter. All animals of course are heterotrophic, but so also are fungi and many protists and bacteria that gain nourishment through the processing of organic matter. By colonizing and processing organic matter, heterotrophic microorganisms (e.g., bacteria and fungi) often make organic matter more nutrient rich and more accessible to other consumers. Together, autotrophs and microbial heterotrophs (the latter will be discussed in more detail in Chap. 8) constitute the basal energy resources that support higher trophic levels in lotic food webs.

The major autotrophs of streams and rivers include algae, a diverse group of organisms from multiple taxonomic groups (Sheath and Wehr 2015); and higher plants, collectively referred to as macrophytes due to their larger size. The algae include prokaryotic (lacking a cell nucleus) and eukaryotic (with a cell nucleus) organisms that use chlorophyll a as their main photosynthetic pigment, have unicellular reproductive cells, and lack roots, stems, and leaves (Stevenson 1996a; Lee 2008).

Freshwater algae come in many growth forms and are found in many habitats. They can be unicellular, but they can also grow in colonies, or in multicellular filamentous and tissue forms (Sheath and Wehr 2015). In streams and rivers, algae are typically microscopic, but some filamentous and tissue growth forms can be much larger. Because currents can displace organisms suspended in the water column, the majority of stream autotrophs are attached to the stream bed or to other structures within the channel. Interchangeably referred to as benthic algae, periphyton, or *aufwuchs*, attached algae occur on virtually all surfaces within rivers, typically in intimate association with heterotrophic microbes and an extracellular matrix. Algae suspended in the water column are referred to as phytoplankton. Phytoplankton are unable to maintain populations in fast–flowing streams, but they can become abundant in slowly moving rivers and backwaters where their doubling rates exceed the rate of cells lost downstream.

Algae are classified into groups or divisions that are based on photosynthetic pigments and membranes, flagella, cell wall components, and/or storage product chemistry. Cyanobacteria (Cyanophyta) are prokaryotic autotrophs that belong to the Eubacteria and have existed since Precambrian times. They were the first photosynthetic organisms on earth, colonizing many aquatic and terrestrial habitats. Cyanobacteria continue to be a very important group of primary producers in aquatic environments (Komárek and Johansen 2015), and they can achieve high biomass in flowing waters. Many are "nitrogen-fixers", are able to convert atmospheric nitrogen (N_2) into more-reactive nitrogen compounds such as ammonia, nitrate, or nitrite that are biologically available to other organisms. Cyanobacteria can also produce toxins that can render water poisonous for other life forms, including humans. We discuss cyanobacterial blooms and relationships with human activities later in the chapter.

All other algae are eukaryotes. The most common groups found in running waters are the green algae (Chlorophyta and Streptophyta), diatoms (Bacillariophyceae), red algae (Rodophyta), yellow-green algae (Xanthophyceae), and chrysomonads (Chrysophyceae). Other groups can also be found, including brown algae (Phaeophyceae), euglenoids (Euglenophyta), Synurophyceae, and dinoflagellates (Pyrrophyta), but these are much less common in lotic environments.

Green algae comprise one of the most diverse algal groups. Originating between 1.5 and 1 billion years ago when a heterotrophic eukaryote incorporated a cyanobacterium capable of photosynthesis into its cell, green algae formed the first photosynthetic eukaryote cell (Leliaert et al. 2012). Green algae include the Chlorophyta, which have both microscopic and filamentous forms, and the Streptophyta, including the larger, branching Charales (stoneworts) (John and Rindi 2015).

Diatoms are another dominant and diverse group of algae in rivers and streams. Rich in silicon dioxide (SiO_2), diatoms have myriad growth forms ranging from sessile species growing close to the substrate to motile, stalked, and planktonic forms. Compared to the cyanobacteria and green algae, diatoms evolved later, becoming a dominant component of oceanic algae approximately 145 million years ago. Diatom cells (called frustules) are composed of two SiO_2 valves, which are highly ornamented and provide the basis for diatom taxonomy. Some diatoms display radial symmetry (centric diatoms), while others are bilaterally symmetrical (pennate diatoms). Some diatoms are motile and have a slit in their valves (called raphe) that allows them to move over surfaces. Diatoms are also sensitive to changes in the aquatic environment; therefore, diatom abundance and diversity are common metrics used to assess the biological condition of flowing waters (Kociolek et al. 2015a, b; Benoiston et al. 2017).

Although not as common, other kinds of algae are found in rivers and streams. Oftentimes, populations of other taxonomic groups are regulated by physicochemical conditions, such as ambient water chemistry and temperature or habitat availability. For example, red algae are often found in unpolluted streams, and some species also are used as indicators of water quality (Sheath and Vis 2015). Though rare, photosynthetic euglenoids typically inhabit small, eutrophic water bodies (Triemer and Zakryś 2015). Temperature seems to be a dominant factor regulating populations of yellow-green algae; they are unicellular or colonial coccoids (round shape) and prefer cold waters. Other taxa, such as brown algae, are associated with specific habitat features like cobbled stream beds.

There are algal groups that are found primarily in the water column, and this is especially true in larger systems. For instance, Synurophytes are motile unicellular or colonial algae, with siliceous scales. They are mostly planktonic and can occur in large rivers and pools of streams (Sheath and Wehr 2015). The Pyrrhophyta or dinoflagellates are unicellular motile organisms that can dominate among suspended algae at warmer temperatures (Carty and Parrow 2015). Similarly, many chrysomonad species are motile, and occur in the water column rather than on substrates. Although present in streams and rivers, chrysomonads have received limited study (Nicholls and Wujek 2015).

Macrophytes make up another important group of primary producers in rivers. This includes angiosperms, ferns, bryophytes (mosses and liverworts), and some members of the benthic algae, such as *Cladophora*. Angiosperms and ferns generally require moderate depths and slow currents; thus, they are common in springs, rivers of intermediate size, and along the margins and in backwaters of larger rivers. Bryophytes, such as mosses, are restricted in distribution, but they can be abundant in cool climates and in shaded headwater streams, although also occurring in floodplain lakes. Macrophytes are frequently found in large rivers and floodplain lakes, and they exhibit different growth forms—emergent, free-floating, floating-leaved, or submerged. Aquatic organisms, ranging from periphyton to fishes, are associated with macrophytes, which they use for substrate, refuge, and food resources. Macrophytes can also influence nutrient cycling, current velocity, and sediment deposition in streams and rivers.

Community composition of autotrophs can change with river size. According to a somewhat idealized view of the longitudinal profile of a river system (Vannote et al. 1980), benthic algae and the occasional bryophyte dominate in headwater and upper stream sections, while benthic algae become more abundant farther downstream where the river widens and is less shaded by riparian vegetation. Macrophytes occur mainly in mid–sized rivers and along the margins of larger rivers in regions of slower currents, and substantial phytoplankton populations develop only in large, lowland rivers. Benthic algae are found in nearly all running waters, and they are often important components of fluvial food webs. Therefore, benthic algae have been a primary focus of study in stream ecology.

6.1   Benthic Algae

Virtually all substrates receiving light, whether in small streams or large rivers, sustain a benthic algal community. Benthic algae support fluvial food webs, engage in nutrient cycling, and can attenuate the current and stabilize sediments, thereby modifying the aquatic habitat (Stevenson 1996b; Lowe and LaLiberte 2017). Benthic algae can be further categorized by their size. Macroalgae are benthic forms having a mature thallus visible to the naked eye, in comparison with smaller microalgae that cannot be distinguished without a microscope. Algal communities in the benthos are also categorized by the substrate where they grow. Epilithon is found on stones, epipelon is found on sediments, and epipsammon is found growing on sand. Benthic algae growing on wood are epidendric and algae growing on other plants are called epiphyton. Epipelic and epipsammic taxa form films or mats on silt, mud, or sand bottoms, and typically are motile and easily swept away by increases in current. Epiphytic taxa occur on macrophytes, particularly angiosperms, where epiphytic coating can be detrimental to the host plant. Unlike epipelic species, epiphytic and epilithic taxa usually are firmly attached by mucilaginous secretions or via a basal cell and stalk, and thus are less likely to be dislodged by the current.

Some algal species are in contact with the substrate along the entire cell wall, colony, or filamentous system. This growth form is termed adpressed, and contrasts with erect forms in which only a basal cell or basal mucilage contacts the substrate. The wide variety of growth forms of benthic algal species produces important structural diversity in the benthic habitats of rivers and streams (Fig. 6.1). Growth forms are the result of habitat conditions such as nutrient availability, water turbulence, and grazing pressure that can influence periphyton composition and biomass. Vadeboncoeur and Power (2017) described three types of algal assemblages. First, where the scouring effect of storms and grazing pressures are strong, a thin film (<0.1 mm thick) formed by small, adnate or motile diatoms, commonly is found. Second, in high light and low nutrient environments, layers (0.1–2 mm thick) of nitrogen-fixing cyanobacteria usually develop, with diatom overgrowth, providing forage for algivorous fishes. Third, mats of filamentous Chlorophytes, usually with an epiphytic layer of diatoms, generally occur in areas with low grazing pressure or few flood events.

Periphyton assemblages typically are comprised primarily of diatoms, green algae, and cyanobacteria (Sheath and Wehr 2015). Diatoms often dominate periphyton in temperate, sub-tropical, and tropical streams because their high species richness and variety of growth forms allows colonization of a range of aquatic habitats (Yang et al. 2009; Wehr and Sheath 2015). Although some diatoms can form mats, gelatinous colonies, or filaments, they are primarily microscopic. In contrast, green algae, cyanobacteria, and red algae generally dominate stream macroalgae. In an extensive survey of the macroalgae of 1,000 stream reaches in North America, including major biomes from tropical to arctic latitudes, Sheath and Cole (1992) recorded 259 taxa, of which 35% were green algae, 24% were cyanobacteria, 21%

were diatoms and chrysophytes, and 20% were red algae. Similar patterns have been documented in a study from central Mexico (Bojorge-Garcia et al. 2014). In tropical streams of Hawaii, green algae (Chlorophyta and Streptophyta) represented 50% of the macroalgae taxa collected, cyanobacteria 31%, red algae 8%, and diatoms, yellow-green algae, and euglenoids 10% (Sherwood et al. 2014).

The composition, abundance and biomass of benthic algae can be measured in several ways (Lowe and LaLiberte 2017; Steinman et al. 2017). Typically, for algae attached to hard substrates (e.g., epilithon), a known area of substrate is sampled and the resulting slurry can be used to identify and count the algae and estimate biomass through biovolume, or it can be analyzed for chlorophyll, which is extracted from the slurry and assayed using spectrophotometry, fluorometry, or high-performance liquid chromatography (HPLC). Artificial substrates, including clay tiles and glass slides, can also be deployed in rivers and streams to collect algae that colonize the substrate surface. To sample epipelon and episammon, algal communities found in stream sediments, the top half of a petri dish can be inserted into the sediment, and the resulting volume of sediment is collected by sliding a glass plate underneath the dish. The sediment sample can be placed in alcohol, resuspended through agitation, and the resulting mixture can be analyzed for chlorophyll concentrations. Epipelon also can be sampled by directly collecting surface sediments with a pipette (Moulton et al. 2002; Lowe and LaLiberte 2017).

Because chlorophyll *a* is the most abundant pigment in plants, its concentration is most frequently used as a proxy

Fig. 6.1 Hypothetical representations of major growth forms of periphyton assemblages. Different modes of herbivory are expected to be most effective with particular growth forms (Reproduced from Steinman 1996)

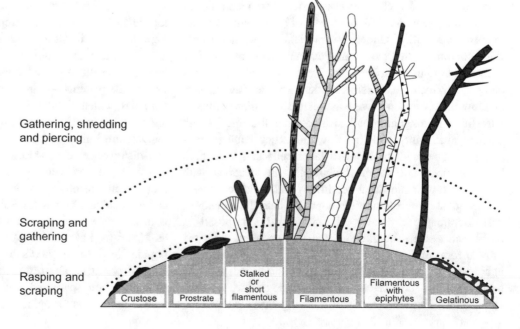

for algal biomass. Cell counts usually are reported as numbers per cm^2, or if cell volume also is estimated, as biovolume per cm^2. Ash-free dry mass (AFDM) can also be used as a proxy to estimate the biomass of algae. An algal sample is first dried (~ 60 °C) and then ashed (at ~ 500 °C). The difference in mass between the dried and ashed sample is the mass lost from combustion of organic matter. Biomass measured as AFDM includes the biomass of algae and heterotrophs such as bacteria and fungi, so usually is accompanied by chlorophyll estimates to estimate algal contribution to the organic matter. New portable fluorometric instruments allow in situ chlorophyll measurement without pigment extraction because the instrument can be placed directly on substrates such as stones and sediments to estimate chlorophyll levels (Steinman et al. 2017). The measurement of primary production, or the rate of formation of new autotroph biomass, is also an important metric for stream ecologist to quantify. Common methods to estimate primary productivity are detailed in Chap. 14.

Because algae are so diverse, when describing algal communities it can be helpful to use classification systems that recognize groups of organisms with similar resource requirements and/or similar ability to respond to disturbance. This type of classification system is referred to as algal functional group classification, or occasionally as algal guilds. Functional group classifications are based on traits— the morphological, physiological, behavioral, and life history features that characterize individual species. Cell size, growth form, the capability to fix nitrogen, and cell motility are examples of algal traits that form the basis for this type of classification. Although this approach has been applied since the late 1970s to phytoplankton, functional group classification is a relatively new way to distinguish among benthic algae (Tapolczai et al. 2016). For example, Biggs et al. (1998b) proposed a classification system based on traits of dominant algae under different levels of flood frequency and nutrient availability. Under low nutrient supply and stable flood conditions, this model predicts periphyton mats of low to medium biomass, mostly with filamentous and erect forms and small to medium cell size. Moderate nutrient availability and low flood frequency were associated with filamentous or unicellular forms, and small to medium cell sizes. In high nutrient availability and low flow frequency habitats, filamentous algae with medium to large cells that form high biomass mats will be most abundant. In locations with moderate to unstable flood conditions and variable nutrient supply, taxa that colonize denuded habitats, like diatoms, will dominate. Similarly, Passy (2007) proposed a trait-based system for diatoms that was centered upon the growth form of diatoms. Diatoms found in resource-limited habitats with minimal disturbance are classified as "low profile" and of short stature. They include adnate, prostate, erect, solitary centric, and slow-moving taxa. In contrast,

high profile species are of tall stature (e.g., erect, filamentous, branched, chain-forming, tube forming, stalked, and colonial centrics). High profile species are not typically limited by resource availability, but are usually regulated by disturbance events. Motile species are capable of fast movement and tolerate eutrophic and polluted conditions (Passy 2007). Lange et al. (2016) also developed a trait-based classification system that relied on seven algal traits—cell size, life form, attachment morphology, nitrogen fixation ability, motility, primary reproductive method, and spore formation—to classify periphyton communities in response to farming intensity and hydrological alteration. Functional trait classification can be especially useful when studying the responses of periphyton to anthropogenic stressors; however, classifications to date have focused on responses to nutrients and flow alterations. Other factors such as water chemistry, seasonality, and habitat diversity need to be considered in future studies to use these classifications in stream and river assessments (Tapolczai et al. 2016).

Functional group classification may not provide enough resolution for researchers to answer some ecological questions. In these cases, species-level identification may be necessary. Algae taxonomic identification is usually performed by microscopy, although new molecular methods using short gene sequences to identify species are becoming more common (Groendahl et al. 2017).

6.1.1 Factors Influencing Benthic Algae

The distribution and abundance of benthic algae is influenced by many proximate physicochemical variables, including light availability, water temperature, current velocity, substrate type, and ambient water chemistry. Biological factors, such as grazing pressure, may also play an important role in structuring algal communities. Regional and catchment characteristics such as topography, geology, land use, vegetation, and climate are the ultimate factors driving the patterns and interactions among these proximate variables, which directly regulate biomass accrual of benthic algae (Biggs 1996). Interactions among light, nutrients, and temperature influence biomass accrual rates, while grazing and disturbance (e.g., abrasion by sediment, substrate erosion and transport, current velocity, and turbulence) remove biomass and dislodge benthic algae (Fig. 6.2). Algal growth can be regulated by whatever resource (e.g., light, nutrients) is in the shortest supply relative to other resources. Known as "Liebig's Law of the Minimum", this idea was made popular in the mid-1800 s by Justus von Liebig, an agricultural scientist. The resource limiting algal growth often changes in both space and time, and the interactions between factors limiting autotrophic growth can be complex.

Fig. 6.2 Factors controlling the biomass and physical structure of periphyton in streams (Reproduced from Biggs 1996)

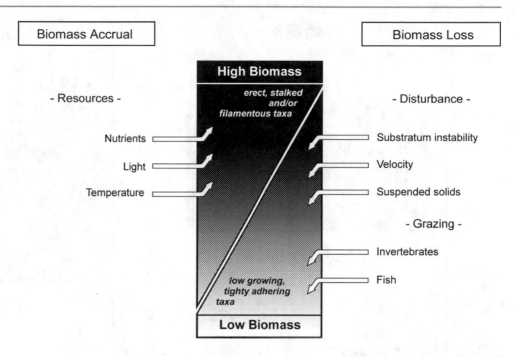

6.1.1.1 Light

Available light can influence the community composition, biomass, and productivity of benthic algae. The wavelength of solar radiation ranges from less than 300 nm to more than 5000 nm; however, photoautotrophs conduct photosynthesis in the 400–700 nm range, referred to as Photosynthetic Active Radiation (PAR). In shallow streams, measurements of PAR above the water can be determined using a quantum sensor, with the assumption that all radiation reaches the stream bed. For deeper streams, an underwater sensor is more appropriate. Typically, sensors are connected to data loggers to provide continuous measurements (Hill 2017). Alternative methods involve the use of dyes (fluorescein and rhodamine) that decompose when exposed to light. Dyes are incubated in clear glass vials at a single or several locations on the stream bed and concentrations are measured at the beginning and at the end of the experiment, which can last for hours for fluorescein or days for rhodamine. This method can provide integrated spatial and temporal light exposure measurements at a lower cost than using in-stream sensors (Bechtold et al. 2012; Warren et al. 2013).

Evidence suggests that the major groups of algae respond differently to irradiance. Green algae usually are associated with high light levels, while diatoms and cyanobacteria appear to require lower light intensities than do green algae (Hill 1996). Motile algae can avoid extremes by movement along the light gradient. Non-motile, prostrate taxa that grow near the substrate may decrease in abundance when light levels decline as a result of shading by overstory forms within the algal community. Becoming dormant or heterotrophic may allow some non-motile species to persist during conditions of very low light (Larned 2010). In tropical streams exposed to a range of irradiance levels, green algae showed higher maximum photosynthetic rates than did cyanobacteria, Rhodophyta, and Xanthophyta, indicating they can be more productive in streams with higher irradiance (Branco et al. 2017). Compensation irradiance (Ic) is the irradiance at which photosynthesis is equal to respiration, and is lower in taxa adapted to shaded environments. In this study, the Ic values for Rhodophyta and Xanthophyta were lower than for green algae. Cyanobacteria showed a wide range of responses with species adapted to both high and low irradiance, which probably accounts for their contribution to benthic algal production over a broad range of conditions.

Riparian vegetation plays an important role in regulating the amount of light that reaches the stream channel, and can influence the composition and growth of benthic algae (Hill 2017). In smaller systems, the tree canopy can block more than 95% of PAR, and benthic algae commonly experience irradiances below the level at which photosynthesis is maximal (Hill 1996). Although some physiological acclimation by algae to low light is possible, it is typically insufficient to compensate for light scarcity (Steinman 1992; Hill et al. 1995). Repeated studies measuring periphyton response to increased light after forest harvest have demonstrated the important role that land use, especially in the riparian zone, plays in controlling instream algal communities (Keithan and Lowe 1985; Lowe et al. 1986). For instance, in a comparison of light reaching headwater streams in a temperate deciduous forest and a site where riparian canopy had been removed, light levels quickly fell in the forested site in response to leaf emergence (Fig. 6.3) and there was a subsequent rapid decline in primary

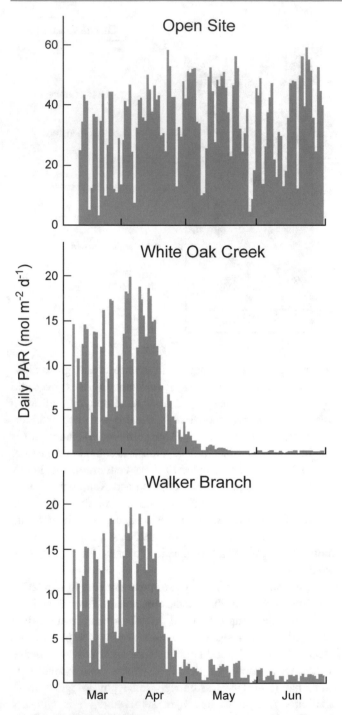

Fig. 6.3 Daily photosynthetically active radiation (PAR) at three sites in eastern Tennessee. The open site is for reference. White Oak Creek and Walker Branch are small headwater streams in deciduous forest (Reproduced from Hill et al. 2001)

production by periphyton (Fig. 6.4) (Hill et al. 2001). Other researchers have experimentally provided artificial light to forested streams and shaded open canopies to estimate the strength of light limitation. Gjerlov and Richardson (2010) found that gross primary production (GPP) and periphyton biomass increased relative to ambient light levels under

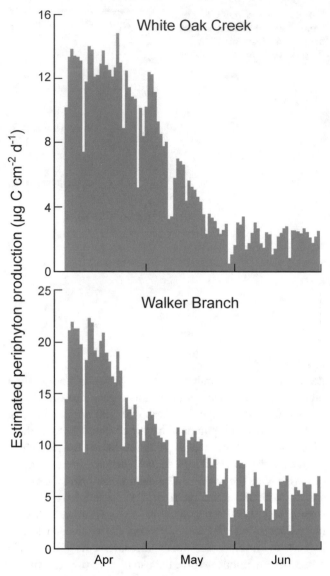

Fig. 6.4 Daily primary production by periphyton for two small forested streams over the period of spring leafout. Primary production was estimated from equations relating production to incident light based on experiments in which stones with periphyton were incubated in chambers placed in the stream and differentially shaded (Reproduced from Hill et al. 2001)

artificial light, while GPP declined in open reaches in response to artificial shading. Forest regeneration in logged regions of the Pacific Northwest, US, is linked to reductions in PAR at the stream surface (Kaylor et al. 2017), with cascading effects on a number of ecosystem attributes (Warren et al. 2016).

Periphyton growth may be limited by interactions between light and other factors, such as ambient nutrient availability. A number of studies have found that under low-light conditions, light availability is more important than nutrient supply in regulating algal biomass and GPP. For example, Mosisch et al. (2001) compared the relative

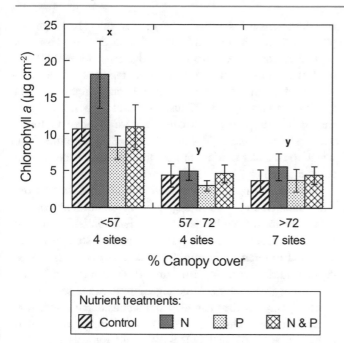

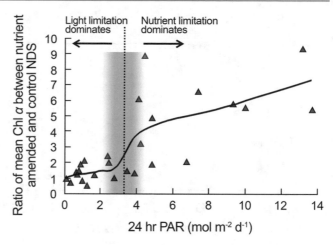

Fig. 6.6 Ratios of chlorophyll *a* in nutrient-amended and control (no nutrients added) diffusing substrates (NDS) over a range of light availability in stream channels covered by either old-growth or secondary forest. The fitted curved was determined using a non-parametric (loess) function. The inflection point (threshold) of the curve is indicated by dotted line and represents the transition from light to nutrient limitation. Gray area represents the error values around the threshold (Reproduced from Warren et al. 2017)

Fig. 6.5 Effects of riparian canopy cover and nutrient enrichment on periphyton chlorophyll content. In the legend, nutrient treatments are: control (no nutrients added), N (nitrogen added), P (phosphorus added), and N & P (nitrogen and phosphorus added). Mean and ±1 standard deviation are shown. Letters indicate significant differences among canopy cover categories (Tukey's test, *P* < 0.05) (Reproduced from Mosisch et al. 2001)

influence of light and nutrient additions in sites that differed in their riparian canopy cover. They observed greater periphyton biomass at sites with lower canopy cover, irrespective of nutrient addition (Fig. 6.5). Additionally, they documented taxon-specific responses—filamentous green algae generally dominated in open canopies and diatoms were typically dominant in shaded sites. By deploying nutrient-diffusing substrata at sites in a forested stream that differed in forest type and extent of riparian cover, Warren et al. (2017) found that algae were light-limited at locations with low light, but nutrient-limited at sites with high light (Fig. 6.6). Notably, the influence of increasing light levels on algal growth may attenuate at higher light levels. In response to increasing irradiance in laboratory streams, accrual of algal biovolume initially increased with increasing light, but leveled off when exposed to greater light intensity (Fig. 6.7a), indicating light saturation (Hill et al. 2009). The same study also manipulated phosphorus concentrations, but light had a stronger influence, and phosphorus was responsible mainly for a lower plateau at very low phosphorus concentrations (Fig. 6.7b); however, phosphorus influenced the response of algae to light because algal biovolume varied with phosphorus concentrations (Fig. 6.7c). Similar results were documented in a field study manipulating light and nutrient availability (Rier et al. 2014). Periphyton ash-free

biomass, chlorophyll, and bacterial biomass were limited by high-intensity light conditions, while the addition of phosphorus had no impact on periphyton abundance.

The aforementioned examples clearly demonstrate the influence light availability can have on the structure and function of algal communities. However, its effects can be over-ridden when other resources are in short supply or when algal biomass is controlled by other factors (Fig. 6.2), such as grazing. In a meta-analysis of the effects of light and grazers on benthic algae, Hillebrand (2005) found that either decreases in grazing pressure or an increase in light had positive effects on algal biomass; however, the stimulating effect of light on biomass was typically observed in the absence of grazers.

6.1.1.2 Nutrients

Dissolved inorganic phosphorus (P), nitrogen (N) as nitrate and ammonia, and silica generally are considered the most critical nutrients for algal production, although other chemical constituents can also limit growth. We begin this section by discussing the large body of research that has been devoted to understanding how the availability of nitrogen and phosphorus limit the growth, influence the productivity, and alter the community structure of stream algal communities.

In the mid-1900s, Dr. Alfred C. Redfield, a physiologist from Harvard University, reported an astounding relationship between the stoichiometry, or elemental ratios, of carbon, N, and P of seawater and of marine phytoplankton. After analyzing thousands of samples of seawater and plankton collected from all over the world, he reported that

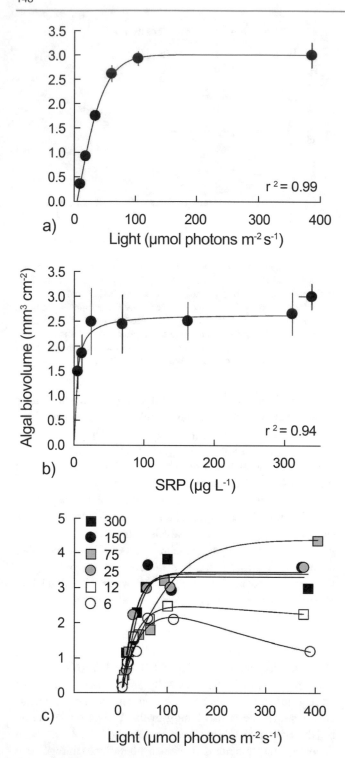

Fig. 6.7 Light and phosphorus effects on algal biovolume. (**a**) Biovolume versus light intensity; (**b**) Biovolume versus phosphorus concentration. (**c**) Biovolume versus light intensity by phosphorus concentration. In (**a**) and (**b**) mean ±1 standard error of total biovolumes are shown; in (**c**) values represent means (Reproduced from Hill et al. 2009)

concentrations of seawater are greater than 16:1, phytoplankton should be limited by P, and when they are less than 16:1, they should be limited by N. Referred to as the Redfield ratio, this is an average value reported for a wide variety of marine phytoplankton; N:P ratios for individual species can vary widely (Klausmeier et al. 2004). Though it should be interpreted with caution when applied to freshwater systems, Redfield's work created a baseline from which to consider nutrient limitation of primary producers in aquatic systems.

When assessing nutrient limitation in flowing waters from N:P ratios, taxon- and site-specific algal nutrient content should be considered because the nutrient content of the cell may differ from ambient levels. Phosphorus storage within cells and differences in nutrient uptake rates can influence cellular nutrient content (Borchardt 1996; Dodds and Welch 2000). For freshwater benthic algae, these ratios seem to be much higher, but they can also show considerable variation (Steinman and Duhamel 2017). According to Kahlert (1998), C:P > 369 and N:P > 32 could indicate strong P limitation while C:N > 11 could indicate N limitation in freshwater benthic algae. Most streams of the eastern United States have quite high (ca. 70:1) ratios of N:P, suggesting that P limitation is common. Nitrate levels and N:P ratios are low in some mountainous regions, especially in the Pacific Northwest and in desert streams of the southwestern US, where N limitation has been reported (Fisher et al. 1982; Tank and Dodds 2003).

In some cases, however, N:P ratios have been found to be poor predictors of nutrient limitation. Despite very low concentrations of P, and N:P ratios between 20 and 36, P limitation was not observed for a subset of North American streams ranging from Puerto Rico (18° N) to Alaska (68° N) (Tank and Dodds 2003). Additional evidence suggests the ratios of elements (i.e., N:P) or the stoichiometry of stream water may not always be a good indicator of nutrient limitation in streams and rivers. When Stelzer and Lamberti (2002) manipulated N:P ratios (65:1, 17:1, 4:1) and total nutrient concentration (low and high) in a factorial experiment using streamside flumes, chlorophyll concentration and algal biovolume responded to nitrogen addition, but not to changes in N:P. The N:P ratio of the source water was high; hence, this outcome was unexpected. Similarly, the TN:TP ratio only predicted algal responses to nutrient addition in three out of 11 streams in a study conducted in Southern California (Klose et al. 2012). From a synthesis of 382 nutrient enrichment studies conducted in streams, Keck and Lepori (2012) found that absolute concentrations of N or P best predicted the magnitude of response to N or P enrichment, and the N:P ratio was an effective predictor only at extremely low and high N:P values.

Differing nutrient requirements among algal species, nutrient storage in cells, and variation in experimental

seawater had an atomic C:N:P of 106:16:1, and the N:P ratio of sea water mirrored the average atomic N:P of marine phytoplankton. Redfield's findings suggest that when N:P conditions (e.g., light intensity, grazing, and physical disturbance) are possible explanations for the discrepancy between ambient nutrient concentrations and ratios and algal responses (Francoeur et al. 1999; Keck and Lepori 2012). Therefore, it is difficult to predict spatial or temporal patterns in nutrient limitation, even if ambient nutrient concentrations are known.

The study of nutrient limitation usually involves the addition of one or more nutrients, alone and in combination, in some experimental design. Whole-stream additions are feasible in small streams, but flow-through chambers or channels and nutrient-releasing substrates also are widely used. Algal responses can be measured from changes in biomass, stoichiometry, enzyme activity, and assemblage composition. Enzymatic assays include the measurement of the enzyme nitrogenase, which is involved in N-fixation conducted by cyanobacteria, and its activity is related to N deficiency (Scott et al. 2009; Lang et al. 2012). The activity of alkaline phosphatase, an extracellular enzyme that cleaves phosphate from large molecules, is often measured to assess P availability. Extracellular enzymes are released by microbes (including bacteria, protists, and fungi) into the water to cleave nutrients from large polymeric compounds (Chróst 1990; Sinsabaugh and Follstad Shah 2012). Enzyme activity shows an inverse relationship to nutrient availability and thus, enzyme reactions can act as indicator of nutrient deficiency (Lang et al. 2012; Hiatt et al. 2019).

Nutrient diffusing substrates (NDS) have become increasingly popular because they afford considerable experimental flexibility. In this approach, clay pots or plastic containers are filled with an agar solution containing the desired mix of nutrients, which leach through the clay walls or the material used to cap the plastic container. Materials used to cap the NDS are usually fritted-glass or filter paper (Tank et al. 2017), although organic material such as cellulose sponge cloth has also been used (Johnson et al. 2009). Although a common technique, there are many different ways to construct and deploy different types of NDS. Notably, construction methods can produce different results in the same system (Capps et al. 2011), thus researchers interested in estimating limitation using this approach alone should proceed with caution. In a metanalysis of NDS studies, Beck et al. (2017) echoed this concern and emphasized the need to standardize NDS methods because results vary with the type of substrate (clay, plastic), nutrient concentrations, composition of PO_4 compounds used in the experiments, and duration of the incubation.

General statements about nutrient limitation patterns have proven difficult; yet, a multitude of studies have reported experimental and observational results of adding nutrients to running waters. For instance, P additions produced a strong algal response in artificial channels receiving water from a river with very low (<4 μg L^{-1}) concentrations of soluble

reactive phosphorus (Davies and Bothwell 2012). Whether phosphate was delivered in brief pulses or continuously, algal biomass increased in proportion to the amount added over time (Fig. 6.8a), indicating that P controlled algal growth. As the supply of P increased, activity of the enzyme alkaline phosphatase declined (Fig. 6.8b), indicating phosphorus limitation was being alleviated. Nitrogen has also been shown to be the primary factor limiting benthic algae in a number of streams and rivers. For example, the addition of nitrate to a tropical stream in the foothills of the Venezuelan Andes resulted in major increases in algal biomass whereas P addition had no effect (Fig. 6.9) (Flecker et al. 2002). In the same system, nutrient uptake rates were significantly higher for nitrate-N and ammonium-N than for phosphate-P. Additionally, in a subtropical river in Queensland, Australia, chlorophyll increased on nitrogen-enriched substrates in comparison to controls and P-enriched treatments (Fig. 6.10) (Mosisch et al. 1999).

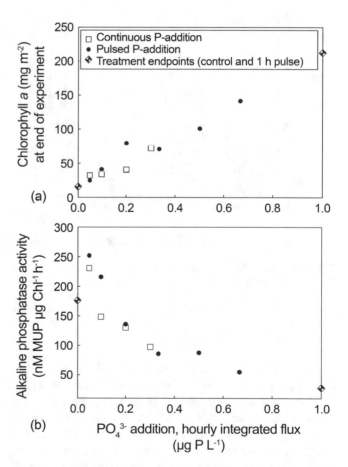

Fig. 6.8 Responses of chlorophyll *a* (**a**) and alkaline phosphatase activity (**b**) to phosphorus additions. Phosphorus was added either continuously or in pulsed doses. In continuous additions, phosphorus was added each minute for an hour in concentrations ranging from 0.05 to 1 μg L^{-1}. In pulsed-P additions, the channels received 1.0 μg L^{-1} phosphorus from 3 to 60 min per hour (Reproduced from Davies and Bothwell 2012)

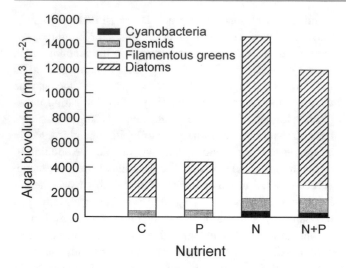

Fig. 6.9 Algal community response after eight days of nutrient additions into flow-through channels placed in Río Las Marias, Venezuela (Reproduced from Flecker et al. 2002)

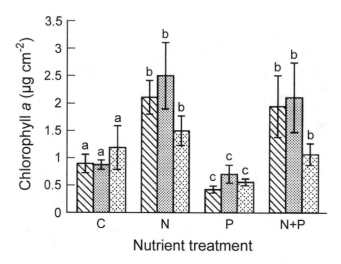

Fig. 6.10 Chlorophyll a values of periphyton in experiments testing the influence of light and nutrients on periphyton growth. The data presented are the average value of five replicate samples and ± 1 standard deviation. Letters over bars indicate significant differences among means (Treatments: no shade, 50% shade, 90% shade) (Reproduced from Mosisch et al. 1999)

Nutrient limitation studies have often documented colimitation by N and P, as algal community assemblages may be comprised of a diversity of species that have different nutrient requirements (Tank and Dodds 2003). Colimitation is demonstrated when the combined addition of N and P produces an algal response, when there is no response from the addition of a single element, or when the combined influence of elements is synergistic relative to the responses of additions of each element. In 38 NDS experiments conducted in 13 streams of the Pacific Northwest, which were generally low in nutrients, NP colimitation occurred in 39% of the experiments, N limitation in 18%, and primary N and

secondary P limitation in 11% (Sanderson et al. 2009). Using NDS in four streams in Texas, Lang et al. (2012) observed the strongest response of algae to N and P together, some response to N alone, and none to P alone. Alkaline phosphatase and nitrogenase activity were higher at low nutrient sites, indicating deficiencies in ambient N and P, and decreased as N and P were added in the NDS experiments. Chlorophyll responses indicated N limitation or NP colimitation, while enzyme activity indicated that some components of periphyton were also experiencing P deficiency.

Nutrient addition also can bring about changes in the composition of benthic algae. For example, sixteen years of whole-stream additions of summer phosphorus (as phosphoric acid) to the Kuparuk River on the North Slope of Alaska resulted in a change in the community composition of primary producers (Slavik et al. 2004). With phosphorus addition, bryophytes replaced epilithic diatoms as the dominant primary producers. Positive responses to phosphorus fertilization were observed at all trophic levels in the river, including increases in the stocks of epilithic algae, the densities of some insect species, and the growth rates of some fish. Similarly, the addition of phosphorus (as solid $CaHPO_4$) to karstic streams in China resulted in shifts in the dominant species present (Pan et al. 2017). Relative to untreated reaches, sections that received P additions saw the community composition of primary producers shift to species that were tolerant of high nutrient levels, such as filamentous green algal taxa. Using NDS to investigate nitrogen limitation in Sycamore Creek, a desert stream in Arizona, Peterson and Grimm (1992) observed dominance by an N-fixing diatom on unenriched substrates, whereas enriched substrates had a higher diversity of non N-fixing diatoms and a delayed transition to an N-fixing cyanobacterium that typically became abundant later in algal succession.

In sum, the above studies make clear than both N and P can be limiting, individually or in combination. They also caution against making strong inferences regarding nutrient limitation from ambient nutrient levels, and highlight the fact that regional conditions influence nutrient limitation patterns in ways that are not yet fully understood. This has motivated several efforts to directly search for geographic patterns in limitation of periphyton by N and P alone and in combination. A field study performed with nutrient diffusing substrates in ten rivers across the US found evidence for nitrogen limitation alone, nitrogen with phosphorus as a secondary limiting element, and nitrogen and phosphorus colimitation depending on location (Fig. 6.11) (Tank and Dodds 2003). Positive responses were mainly to nitrogen enrichment, and in fact no responses to P alone were observed. Although the streams of Fig. 6.11 all had relatively low nutrient levels, only half seemed to be primarily nutrient limited; the others probably were limited by other factors such as light or grazing. N limitation or N and P

Fig. 6.11 Chlorophyll *a* values of periphyton growing on nutrient diffusing substrata from four nutrient treatments: C, control (no nutrients added); N, nitrogen (nitrate added); P, phosphorus (phosphate added); N + P, nitrogen and phosphorus (nitrate and phosphate added). The mean response from 5 replicates plus one standard error are shown. Asterisks indicate significant effects of N or P alone and I indicates a significant interaction between the two nutrients (Reproduced from Tank and Dodds 2003)

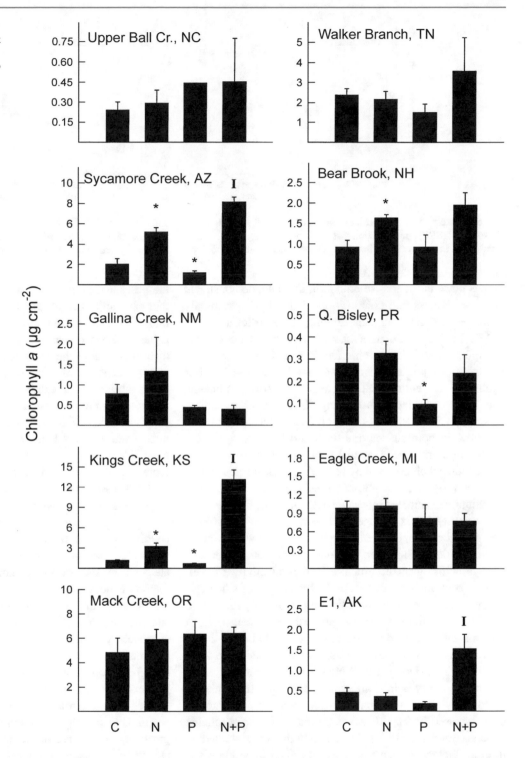

colimitation was more common in mountain rivers of western US than in rivers of the midwest or arid west, where nutrient concentrations are higher (Reisinger et al. 2016). In a review of 158 studies to assess the extent of nutrient limitation of periphyton, Dodds and Welch (2000) found that 13% showed stimulation by N, 18% by P, 44% by both, and 26% did not respond to nutrient additions. A similar but independent analysis by Francoeur (2001) of 237 experiments conducted in the US, Australia, New Zealand, and India found N-limitation in 17%, P-limitation in 19%, and co-limitation in 23% of the cases. A few studies reported inhibition by N or P, and 43% of the studies found neither stimulation nor inhibition of periphyton by either nutrient.

Other studies have documented interactions between nutrient availability and light controlling algal growth. For example, experimental manipulation of P and light levels in

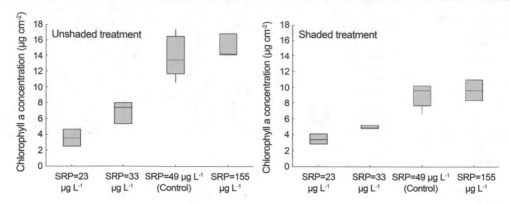

Fig. 6.12 Responses of algal biomass (as chlorophyll *a*) to light and phosphorus (SRP) manipulations at the end of the 10-day flume experiment. Light treatments included unshaded and shaded conditions, and phosphorus (SRP) levels included control (ambient SRP concentration), addition (SRP added), and reduction (SRP concentrations reduced by precipitation) (Reproduced from McCall et al. 2017)

flumes place in the River Lambourn, England, demonstrated that the biomass and composition of the algal community depended on the dominant limiting factor (McCall et al. 2017). In this study, researchers added and removed (via precipitation with ferrous chloride) P from the water column. Phosphorus addition did not increase periphyton biomass relative to ambient nutrient treatments (Fig. 6.12), indicating that some other factor is limiting algae above ambient P; however, a 50% decrease in SRP reduced chlorophyll by 75 and 60% in both shaded and unshaded treatments, respectively. In flumes with ambient and augmented P levels, shade reduced chlorophyll by 35%. Augmented P resulted in a lower proportion of nutrient sensitive diatoms and in a community dominated by cyanobacterial algae, likely related to its ability to fix nitrogen, while nano- and picochlorophytes and cryptophytes increased when SRP was reduced.

In addition to nitrogen and phosphorus, other elements may limit the growth, structure, and productivity of periphyton communities. For example, the supply of silica also might be expected to become limiting, especially in systems dominated by diatoms because the frustules, or cell walls, of diatoms are composed of siliceous material. In lakes, diatom population dynamics have been shown to depend upon silica concentrations (Dodds and Whiles 2010). In contrast, silica in streams rarely is in short supply; consequently, it should seldom limit benthic diatom growth. However, few studies have assessed its importance as a limiting factor in running waters (Borchardt 1996). Iron, although rarely studied as limiting factor, can have an effect on algal biovolume and species richness once nitrogen and phosphorus are supplied (Larson et al. 2015).

6.1.1.3   Current

Stream current can have opposing effects on the accrual of benthic algae, best described as a 'subsidy-stress' response owing to both beneficial and detrimental effects of flow (Biggs et al. 1998b). The flow of water continually renews the gases and nutrients delivered to periphyton, and so current can support algal growth by enhancing nutrient uptake rates. However, current also exerts a shear stress on benthic algae, which can cause cell sloughing. High flows can also disturb and scour the substrate, influencing the growth form and architecture of benthic algal assemblages. Adherent forms are less vulnerable to sloughing than filamentous forms, and may be especially dependent upon diffusion to supply needed resources. Thus, adherent forms would benefit most from increased current velocity. Filamentous forms most likely benefit from higher rates of diffusion, but are more vulnerable to shear due to increasing flow.

Traits associated with either the "subsidy" or "stress" response associated with flow can be taxon-specific, and many species have been found to occupy specific flow conditions. Under high current velocities, unicellular algae usually are represented by small, adherent diatoms that are attached to the substrate along their length (Stevenson 1996b). Keithan and Lowe (1985) found different algal taxa associated with particular regions in small Tennessee streams, and their description of individual growth forms is consistent with a direct effect of current speed. In slower currents they found diatoms to be more densely packed, with a higher proportion growing in an erect position and a greater abundance of large colonial forms. Many of the same species also were found at faster currents, but in the prostrate position. At the highest velocities most of the diatoms were prostrate, many were in crevices, and tightly adherent species were prevalent. Similarly, in an experiment conducted in an indoor flume with artificial cobbles, differences in algal physiognomy and composition were found among different velocities (Graba et al. 2013). A thick biofilm composed of long (up to 10 cm) and very thick filaments developed under low current conditions, while thinner and shorter filaments (around 3 cm) were observed under the intermediate and high velocity treatments. *Melosira moniliformis* was the dominant species in all samples; however, low velocity assemblages were characterized by multicellular forms, while unicellular prostate or adnate forms dominated under intermediate and high velocity conditions. In a study conducted in streams of the Bode catchment, Germany, to investigate the influence of near bed turbulence on the composition and architecture of epilithic biofilms, results indicated that biofilms under low turbulence were dominated by adnate diatoms, while at high turbulence sites the small diatom *Cocconeis* dominated, with cells laying side by side over the stone surface (Risse-Buhl et al. 2017).

Current velocity can influence the arrival rate of algal cells and therefore the process of colonization. Algal community development proceeds more slowly at high velocities, evidently because algal cells are less able to become established, but higher biomass accumulation occurs under intermediate velocities rather than in low velocities due to the positive effect of current on nutrient uptake (DeNicola and McIntire 1990; Poff et al. 1990). Diatom immigration rates (Stevenson 1983) and emigration rates (Stevenson and Peterson 1991) have been found to vary with current velocity. Small species appear to be most effective at colonizing fast current sites, whereas both large and small species were present at sites with slower current velocities (Stevenson 1996b). In an study conducted in laboratory flumes and using artificial cobbles, colonization of biofilm started in two symmetrical areas located at the front of the cobbles, which showed low to moderate shear stress, and then progressed to the rest of the substrate surface (Coundoul et al. 2015).

The continual renewal of nutrients provided by flowing water, along with turbulence and mixing that favor the diffusion of nutrients, can stimulate algal growth, respiration, and reproduction (Wehr and Sheath 2015). Whitford and Schumacher (1964) conducted one of the first studies of such physiological enrichment, showing that rates of ^{32}P uptake and CO_2 liberation in *Spirogyra* and *Oedogonium* increased with current up to 40 cm s^{-1}, the highest velocity tested. Growth of filamentous green algae *Stigeoclonium* and *Spirogira* in a laboratory flume over a range of current and shear velocities (a measurement of turbulence near the bed) showed greater biomass accumulation at higher velocities up to a maximum, then growth declined as velocity increased further (Fig. 6.13). Moderate flows likely facilitated growth through the effect of turbulence on nutrient availability, while at higher velocities growth was limited by algal detachment (Hondzo and Wang 2002). Hiatt et al. (2019) observed a decrease in the activity of alkaline phosphatase in streams in central Texas at higher current velocities, indicating greater P availability, and the effect was most pronounced at low ambient P concentrations, lending credence to the view that increasing current conveys a nutrient subsidy to periphyton.

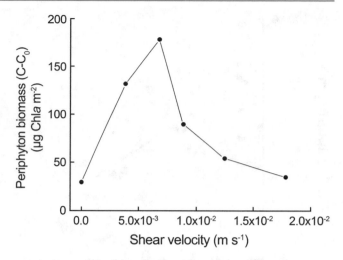

Fig. 6.13 Accumulated biomass of periphyton on clay tiles after 14 days of colonization under a range of shear velocity at the sediment-water interface. Biomass was estimated from the difference in chlorophyll concentrations between initial and final measurements (C-C$_o$) (Reproduced from Hondzo and Wang 2002)

Extreme discharge can have a strongly negative impact on lotic algal populations. Scouring of cells from surfaces can result simply from increases in current velocity, from overturning substrates (Robinson and Rushforth 1987), from abrasion due to tumbling (Power and Stewart 1987), and/or by abrasion from suspended sediments. The effects of high flows on benthic algae depend on growth form, the senescence of the mat, and substrate (Biggs et al. 1998a). Thus "dense and coherent" communities such as mucilagous diatom-cyanobacterial mats tend to increase their biomass as current increases within the interflood period, while stalked-short filamentous diatom communities show a unimodal relationship with current, and communities dominated by long filamentous green algae exhibit a negative response to increases in current (Fig. 6.14).

Thick periphyton mats are particularly vulnerable to dislodgement due to senescence of the bottom-most layers, which weakens their attachment to the substrate and renders the entire mat vulnerable to sloughing. Shading, the buildup of metabolites, and reduced rates of exchange of gases and nutrients all can contribute to the lower–most layer being unable to support the weight of the overlying mat. In experimental channels with constant discharge, periphyton biomass commonly goes through a rapid increase followed by a precipitous decline over a time span of 3–6 weeks (Stockner and Shortreed 1978; Triska et al. 1983). In Sycamore Creek, Arizona, a substantial periphyton mat was observed to increase steadily in biomass and chlorophyll *a* for approximately 60 days following a flood event, until a second flash flood virtually eliminated periphyton and the process re-commenced (Fisher et al. 1982). However, some algae can modify hydrodynamic conditions to enhance their

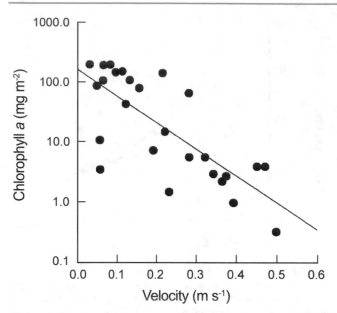

Fig. 6.14 Response of long, filamentous green algal communities measured as chlorophyll *a* to variation in water column velocities in the Waiau River, New Zealand (Reproduced from Biggs et al. 1998a)

survival. For example, the structure of the mat-forming benthic diatom *Didymosphenia geminate* can reduce near-bed turbulence, thereby decreasing the risk of dislodgement. Compared to bare cobbles, the zone of maximum turbulent kinetic energy shifted its position from behind the cobbles to several centimeters above the mat. These hydrodynamic changes are probably common to other algal mats, and although they may influence *Didymosphenia* proliferation, other factors related to its structure and growth are also involved (Larned et al. 2011).

In addition to dislodging and burying benthic algae, storm events can shift the substrate to which they are attached and cause abrasion by sediment mobilization. Katz et al. (2018) reported an inverse relationship between algal biomass (as chlorophyll *a*) and shear stress (a measure of bed disturbance) in an Oregon stream, where shear stress explained 49% of the spatial variation in algal abundance on the stream bed. The causal role of sediment abrasion and saltation was indicated by a negative relationship of chlorophyll with the quantity of bed particles mobilized. Luce et al. (2013) determined that sand and small gravel transported by saltation (bouncing along the stream bed) can produce greater periphyton losses than abrasion by suspended material. The effects of abrasion on algal removal is influenced by community physiognomy and composition, duration of the flood event, composition of suspended sediments, and position on the substrate (Francoeur and Biggs 2006).

Benthic algal species richness declined following flooding in New Zealand streams that received high sediment inputs and had mobile beds, but floods had no effect on species richness in streams with an armored (compacted) substrate (Biggs and Smith 2002). Algal taxon richness did not correlate with time post-disturbance in streams that experienced frequent, bed-moving flood events, whereas a positive relationship was observed in streams where such floods were less frequent. Biggs and Smith (2002) proposed that initial recovery of benthic algae proceeds rapidly (within a week) and involves taxa with high tolerance to disturbance as well as taxa with high reproduction and immigration rates. Subsequently, taxa appear that are less resistant to flood disturbance and have lower rates of immigration and growth. The duration of this second phase (one to several months) is more variable because it is more susceptible to resource supply, which likely varies among streams.

Seasonal changes in discharge may also impact patterns in algal accrual. For example, in tropical streams, flood disturbance during the wet season can reduce algal biomass compared to the dry season. Following a storm-flow event in savanna streams of the Daly River in tropical Australia, over 90% of epilithic algal biomass was dislodged (Townsend and Douglas 2014). Similarly, in the arid Gila River in the southwestern US, summer monsoonal floods strongly affected diatom abundance and composition, particularly in small streams (Tornes et al. 2014). Highest resistance to repeated scour was seen in tightly adhered, adnate, and prostrate species of *Achnanthidium* and *Cocconeis*. In contrast, loosely attached diatoms such as *Nitzschia* and *Navicula* were most susceptible to flood loss.

6.1.1.4 Substrate

Substrate, whether living or nonliving, provides a surface for benthic algal growth that creates physical and chemical conditions that are different from the surrounding water. The physical structure and the stability of the substrate can influence algal colonization. Algae that grow into large mats are usually found on larger stones, while small, motile algae colonize sediments and small particles (Burkholder 1996). In streams of Eastern Ontario and Western Quebec, Canada, finer substrates were dominated by motile diatoms and cyanobacteria, whereas prostrate and filamentous algae occurred on larger substrata (Cattaneo et al. 1997). Substrate roughness is another important factor in algal colonization rates. In a study conducted in a stream in southern Brazil, greater chlorophyll and ash free dry mass were found on rougher substrates, likely due to greater refuge from grazing and current on rough surfaces (Schneck et al. 2013). Crevices on rough substrate can also influence colonization, and more complex surfaces tend to facilitate establishment of species with varying habitat requirements. This can result in higher overall algal diversity than is found on more uniform substrates (Petsch et al. 2017).

6.1.1.5 Temperature

Temperature directly affects the biomass and production of benthic algae, and because algal species may differ in their thermal tolerance, temperature also can influence assemblage composition. Though it is not a universal pattern, major divisions of algae tend to dominate at different temperature ranges: diatoms between 5 and 20 °C, green and yellow-brown algae between 15 and 30 °C, and cyanobacteria above 30 °C (DeNicola 2007). Notably, high algal diversity and patches of increased algal biomass can develop at high altitudes and at low temperatures (Rott et al. 2006). In a study comparing algal populations in geothermally-heated streams along an in-stream temperature gradient in Iceland, diatom and green alga taxa composition varied among streams with different temperatures; however, variation in epilithic chlorophyll was not related to temperature (Gudmundsdottir et al. 2011).

Seasonal changes in the taxonomic composition of benthic algal assemblages can be observed in temperate rivers. Often, there is a greater representation of green algae and cyanobacteria during summer months. By assembling comparative data on primary productivity for stream periphyton, and lake and ocean phytoplankton, Morin et al. (1999) developed empirical models to predict primary production from chlorophyll *a* and water temperature. Although production was lower for stream periphyton than for lake or marine phytoplankton, presumably because of reduced nutrient diffusion into algal mats, production was more strongly related to water temperature in stream periphyton than for phytoplankton of either lakes or oceans. Temperature can also influence the nutritional quality of periphyton. In headwater streams in Pennsylvania, US, seasonal changes in algae composition were accompanied by changes in periphyton fatty acid content, which can be a measure of basal food quality. This variation was related to seasonal variation in temperature, probably because some fatty acids are involved in algal adaptation to temperature changes (Honeyfield and Maloney 2015).

6.1.1.6 Grazers

Numerous studies provide strong evidence of the important influence of grazers on benthic algae in fluvial ecosystems (Feminella and Hawkins 1995). Grazers can reduce algal biomass and influence community composition by selectively eliminating certain species and growth forms (Steinman 1996; Rosemond et al. 2000). For example, in a coastal stream in Southeastern Brazil, periphyton biomass was reduced by half through the grazing of baetid (Ephemeroptera) nymphs, as determined by in-stream exclusion experiments (Moulton et al. 2015).

More complex relationships among stream taxa demonstrate the strong impact grazing species can have on periphyton biomass. By restocking fishes in a fish-free section of the River Elbe in Germany, Winkelmann et al. (2014) showed the effects of a trophic cascade on stream periphyton. Invertebrate grazers (mostly ephemeropterans) decreased by 87% and benthic algae biomass increased by 25%. Similar patterns have been documented in tropical streams. For example, foraging by omnivorous consumers (fishes and crabs) in Trinidadian streams decreased chlorophyll in controls, compared to treatments where fishes were excluded (Marshall et al. 2012). Similarly, five months after the decline of a tadpole population in a tropical stream due to a fungal infection, periphyton biomass increased 2.8-fold in pools and 6.3-fold in riffles (Connelly et al. 2014). Invasive grazing species can also have large impacts on algal communities. Using mesocosm and in situ experiments, Capps et al. (2015) determined that populations of the invasive, grazing armored catfish *Pterygoplichthys* (Loricariidae) reduced algal biomass in streams in southern Mexico. However, in mesocosms when algae were protected from grazing, but exposed to nutrient re-supply through fish excretion, algal biomass increased by 42% compared to the control. This work and others demonstrate that grazers can exert a negative effect on algal biomass through algal consumption, and a positive effect on algal biomass via fertilization through nutrient mineralization via excretion and egestion.

Interactions between the effects of grazers and nutrients have been reported in many studies. From a literature survey of 85 experiments that examined herbivore presence and nutrient supply, Hillebrand (2002) found grazing and nutrients had a strong influence on algal biomass, but grazer effects were stronger than nutrient effects. Manipulation of nutrients and grazing snails (*Elimia clavaeformis*) in Walker Branch, Tennessee, showed that algal species most reduced by herbivores were those that increased most in response to nutrient addition, and vice versa, suggesting a trade-off between resistance to herbivory and nutrient-saturated growth rates (Rosemond et al. 1993). Grazed communities were dominated by chlorophytes and cyanobacteria, which were overgrown by diatoms when herbivores were removed. Flecker et al. (2002) observed strong effects of both nitrogen addition and herbivorous fishes on algal standing crop in a tropical river, but top-down grazing pressure from consumers on algal biomass and composition was found to be considerably stronger than bottom-up resource limitation. Hydrologic variability can also be an important mediator of grazer-algal-nutrient interactions. Algal biomass can be more responsive to nutrients in frequently scoured streams, whereas in more stable streams grazers can suppress algae irrespective of nutrient concentration (Riseng et al. 2004). Interactions between herbivores and algal communities in lotic ecosystems are described in greater detail in Chaps. 9 and 10.

6.1.2   Temporal and Spatial Variation in Benthic Algae

The interactions among multiple environmental factors just described is responsible for the diverse temporal and spatial patterns observed in algal composition and biomass of rivers and streams. Temporal patterns in algal biomass have been describe in three categories (Biggs 1996): relatively low but constant biomass when disturbance is frequent, cycles of accumulation and loss when disturbances are less frequent, and seasonal cycles owing to seasonal change in environmental factors. Spatial variation is likewise associated with changing physicochemical variables that are associated with different habitats and/or systems of varying size (e.g., between riffles and pools and/or among streams that differ in size or landscape position).

Seasonal abundance and diversity data for algae from the tropics are relatively limited, but numerous studies of stony streams from the temperate zone reveal seasonal changes in algal abundance and diversity. For instance, diatoms dominate during winter and continue to be a major component of the flora in spring and early summer, although the species composition changes. Other groups can become abundant during summer, particularly green algae and cyanobacteria, likely in response to temperature as previously mentioned. In temperate rivers, total abundance and biomass generally is greatest in the spring, and a secondary peak can occur in autumn. This is thought to be related to increasing light and/or nutrient availability. For example, in streams of Tennessee, daily chlorophyll-specific primary production peaked in the spring, and then chlorophyll concentrations decreased after leaf emergence, reached a minimum in summer due to shading, and then increased again in the autumn. In these rivers, seasonal changes were primarily related to changes in light availability (Hill and Dimick 2002). In streams of Hubbard Brook Experimental Forest, New Hampshire, both light and nitrogen concentrations are higher in the spring before leaf-out, which likely explains higher biomass accrual rates in spring compared to summer and autumn (Bernhardt and Likens 2004). A factorial manipulation of grazers, nutrients, and irradiance demonstrated how factors limiting periphyton can change seasonally (Rosemond et al. 2000). Periphyton biomass was limited by light in fall and summer but not in spring, while nutrients limited periphyton when light availability was higher. In alpine streams, higher biomass in autumn and winter likely are related to lower turbidity and more stable flows during these seasons than in summer, when biomass is low (Uehlinger et al. 2010).

In tropical rivers, seasonality in hydrology can be important to algal biomass and composition. In the Daly River, northern Australia, annual variation in benthic algae biomass is primarily controlled by seasonal changes in flows (Townsend and Padovan 2005). During the dry season, low velocities favored the colonization and growth of *Spirogyra*, which appeared in the river in mid-May, increased in biomass through July and August, and disappeared from the river at the beginning of the wet season in late October-early November. The frequent high flow disturbances that occurred during the wet season probably limited the colonization and growth of *Spirogyra*, primarily due to abrasion and dislodgment. In subtropical mountain streams in Taiwan, very low algae biomass coincided with the typhoon season (July–October), while algal biomass responded positively to moderate increases in discharge during spring and early summer (Tsai et al. 2014). In contrast, streams in Hong Kong showed only modest dry-season increases in algal biomass, relative to the wet season, apparently because grazers limited algae growth during the dry season (Yang et al. 2009).

The spatial distribution of benthic algae at local, reach, and river basin scales reflects spatial patterns in environmental variables. As previously discussed, spatial differences in algal composition and biomass at the reach scale can be related to variation in light, substrate, and current velocity (Warren et al. 2013; Katz et al. 2018). In a New Zealand stream, the distribution of two types of macroscopic algal groupings was related to current velocity. Locations dominated by filamentous green algae (*Mougeotia*, *Cladophora*) had low near-bed velocities while locations dominated by prostate taxa of cyanobacteria (*Phormidium*, *Oscillatoria*) were areas of higher velocity (Hart et al. 2013).

Nutrient availability can also influence spatial distribution of algae at the reach scale. In a desert stream, higher abundance of N-fixing cyanobacteria was observed at the upstream edge of sandbars, where nitrate concentration was lower, compared with the downstream edge, where non-N-fixing taxa dominated the community (Henry and Fisher 2003). In fast-flowing, V-shape alpine streams, the effects of water level fluctuation and spray can result in transverse periphyton zonation: the green alga *Trentepohlia*, tolerant of desiccation, is common in the spray zone in the outer part of the channel (Fig. 6.15), an amphibian zone is dominated by species that tolerate flow and desiccation conditions, while diatoms and macroalgae that tolerate high flows dominate the wetted zone (Rott et al. 2006). Algal species richness in these streams was lower in the more glaciated parts of the catchment and increased with distance from glaciers (Gesierich and Rott 2012).

Fig. 6.15 Microhabitat in a channel cross-section of a cascading alpine or subalpine stream illustrating, (**a**) spray zone with the green alga *Trentepohlia*, (**b**) amphibian grazing zone with *Gloeocapsa*, (**c**) wetted zone with epilithic taxa *Achnanthes*, *Cymbella*, *Cocconeis*, *Homoeothrix* and (**d**) macroalgae *Hydrurus* and *Lemanea* with epiphytic algae (Reproduced from Rott et al. 2006)

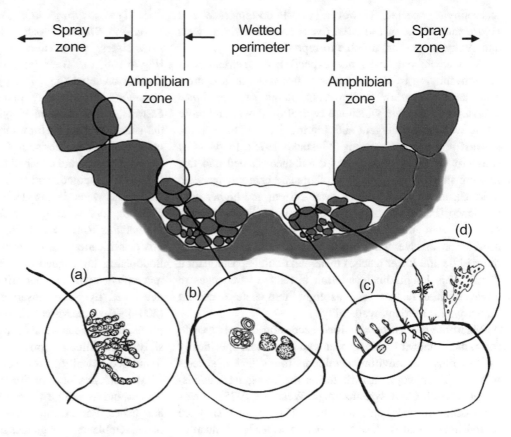

At larger scales, patterns in diatoms assemblages have been related to position in the river network, catchment land cover, and water chemistry. In a study analyzing 897 sites in the US, diatom composition showed a gradient from mountain areas to lowland sites. Species present in mountain regions were associated with rock surfaces, low temperatures, and fast currents, while species in the lowlands were associated with silt sediments and higher alkalinity (Potapova and Charles 2002). An increase in diatom species richness with stream order has been observed in both temperate and tropical streams (Stenger-Kovács et al. 2014; Jyrkänkallio-Mikkola et al. 2018).

6.2 Macrophytes

Macrophytes are aquatic plants, and include the bryophytes (mosses and liverworts), ferns, and flowering plants. Because of their size, some would also include the Charales and other large algal species among the macrophytes of flowing waters. Bryophytes are small plants lacking lignin and transport structures (xylem and phloem) as are found in vascular plants, although they do have tissues with similar functions. Bryophytes also do not have true leaves or roots. Remaining macrophytes excluding macro-algae are considered vascular plants (Bowden et al. 2017; Glime 2017b).

Bryophytes occur world–wide in relatively cool streams and characteristically are found in headwater regions associated with high currents, high substrate stability, and low light, although the free-floating liverwort *Ricciocarpus natans* is common in floodplain lakes of large rivers (de Tezanos Pinto and O'Farrell 2014). In the Tiber basin, Italy, bryophytes showed preferences for habitats with coarse substrate, fast currents, and clear, oxygenated, low temperature and oligotrophic (low nutrient) waters, although some species were associated with turbid, eutrophic, and slow current conditions (Ceschin et al. 2012). Evidence does suggest that stream bryophyte populations respond negatively to human disturbance. For instance, in streams of the Madeira Island, Portugal, flow disturbance and degraded bank conditions were linked with a downstream decline in bryophyte species richness (Luis et al. 2015).

Relatively few species of macrophytes can withstand fast currents. Two flowering plant families, the Podostemaceae and Hydrostachyaceae, and a number of bryophytes are associated with faster flows. However, in general, macrophytes exhibit few adaptations to life in flowing water, and most are successful in slow currents and backwaters. Certain characteristics permit establishment and maintenance of populations in appreciable current. Tough, flexible stems and leaves, firm attachment by adventitious roots, rhizomes, or stolons, and vegetative reproduction typify most macrophytic species. However, the Podostemaceae and Hydrostachyaceae of torrential rivers possess aerial flowers and sticky seeds, and are able to reproduce sexually.

At a catchment scale, one expects a downstream succession, moving from bryophytes to fast-water angiosperms such as *Ranunculus*, to flowering plants such as *Potamogeton* and *Elodea*, which are typical of slower and more fertile waters, to emergent and floating–leaved plants in the slowest and deepest sections (Westlake 1975). In rivers of southern Germany, bryophytes dominated shaded and fast flowing streams while vascular plants and charophytes were more common in sites with slower current and higher light incidence (Passauer et al. 2002). Similarly, liverworts and mosses were more abundant in the headwaters of a Mediterranean river while emergent macrophytes dominated the middle and lower reaches (Manolaki and Papastergiadou 2013). This longitudinal variation is related to changes in environmental factors such as light, temperature, current velocity, and nutrient availability.

Macrophytes can be classified according to their growth form, their manner of attachment, and even more specifically by the range of environmental conditions that a species inhabits. Four major growth forms are recognized in historical classification systems (e.g., Westlake 1975). Emergents occur on river banks and shoals (Fig. 6.16a). They are rooted in soil that is close to or below water level much of the year, and their leaves and reproductive organs are aerial

(e.g., *Typha, Phragmites, Cyperus, Lycopus, Polygonum, Saggitaria, Thalia*). Floating–leaved taxa occupy margins of slow rivers, where they are rooted in submerged soils (Fig. 6.16b, c). Their leaves and reproductive organs are floating or aerial (e.g., *Ludwigia, Nymphaea, Nymphoides, Nelumbo*). Free-floating plants (e.g., *Azolla, Eichhornia, Salvinia, Pistia, Lemna, Wolffia*) usually are not attached to the substrate and can form large mats, often entangled with other species and debris, in slow tropical rivers (Fig. 6.16d). Submerged taxa are attached to the substrate, their leaves are entirely submerged, and they typically occur in midstream unless the water is too deep (e.g., *Cabomba, Vallisneria, Potamogeton, Cycnogeton, Elodea, Stuckenia, Veronica, Ranunculus, Callitriche*). Bowden et al. (2017) supported this classification for macrophytes in large rivers and their floodplains, but argued for a habitat-based classification for macrophytes of smaller rivers and streams, originally developed by Sand-Jensen et al. (1992) and Riis et al. (2001). That classification recognizes obligate submerged plants, amphibious plants found in water and on land (possibly with different growth forms), and terrestrial plants found occasionally in the wetted channel.

Macrophytes play an important role in lotic systems as habitat and as refuge from predators for both invertebrates and fishes. For example, the composition and diversity of fish assemblages was greater in macrophyte-rich areas of the Paraná floodplain, Brazil, compared with areas lacking

Fig. 6.16 (**a**) *Sphenoclea zeylanica* (Sphenocleaceae), (**b**) *Nelumbo lutea* (Nelumbonaceae), (**c**) *Nymphaea* sp. (Nymphaeaceae), (**d**) *Eichhornia* sp. (Pontederiaceae) Photographs by M M Castillo

(a)   (b)

(c)   (d)

vegetation (Gomes et al. 2012). Similarly, in agricultural streams of Western Australia, reaches with macrophytes showed greater richness and abundance of macroinvertebrates, particularly grazers, than reaches without macrophytes (Paice et al. 2017). Macrophytes may also play an important role in aquatic food webs as they can be consumed by aquatic and terrestrial animals (Thomaz et al. 2008; Wood and Freeman 2017).

Macrophytes can also increase habitat heterogeneity for aquatic organisms by modifying water velocity across the stream channel and increasing the range of water velocities found in a river or stream (Champion and Tanner 2000). By slowing the current, macrophytes can trap sediments and particulate organic matter, affecting sediment deposition, resuspension, and stream bed erosion, and also influence nutrient release into the water column (Madsen et al. 2001). The effects of macrophytes on the habitat of benthic organisms depends on stream flow, varying from stabilizing the stream bed under flood conditions to reducing free-flowing areas under low flows (Holomuzki et al. 2010; Suren and Riis 2010). Effects of macrophytes on hydrodynamics likely vary among macrophyte species, as Sand-Jensen (1998) reported for *Callitriche cophocarpa* and *Sparganium emersum* in Danish streams. *Callitriche cophocarpa* forms a dense canopy whereas *Sparganium emersum* has long and flexible leaves and forms more open patches. Current velocity decreased markedly near the bed of *C. cophocarpa*, which enhanced the accumulation of fine sediments and elevated the substrate surface (Fig. 6.17). In contrast, the open canopy of *S. emersum* had less effect on current velocity, sediment composition, and topography. Organic matter and nutrient enrichment in the sediments was also higher in *C. cophocarpa* than in *S. emersum* patches, emphasizing that macrophytes should not be viewed as a homogenous ecological unit.

6.2.1 Limiting Factors for Macrophytes

In contrast to terrestrial habitats, where coverage of the soil surface by vegetation often is near 100%, only a small percentage of the stream bed supports growth of plants. As Hynes (1970) points out, aquatic botanists have an understandable bias toward studying areas where plants are fairly abundant. Even so, studies of macrophyte production in two mid–size Appalachian rivers estimated percent cover between 27 and 42% (Hill and Webster 1983; Rodgers et al. 1983), and in Bavarian streams 37% of the area studied was found to have <10% cover of macrophytes(Gessner 1955). In a small, Danish lowland stream, the combined area of *Callitriche cophocarpa* patches covered up to 70% of the streambed. The remaining area apparently was unsuitable habitat due to a strong shear effect and coarse substrate that limited the growth

of the plant (Sand-Jensen et al. 1999). In a nutrient-rich chalk river in the UK, *Ranunculus* covered more than 70% of the streambed during the summer (Cotton et al. 2006). In streams of the Western Alps and Central Apennines in Italy, bryophytes were very common, covering over 10% of the sampled area in 50% of the sites (Ceschin et al. 2015). While macrophytes can be abundant in larger rivers and where currents are slow, their extent of coverage is variable and usually modest even in locations where they have attracted study.

Where macrophytes do occur, the growing season can be quite long if water temperatures stay above freezing throughout winter. In British rivers, many species simply grow slowly or cease growth during winter, although others, emergent plants in particular, shed leaves and die back to rhizomes and stolons (Hynes 1970). In subarctic streams of Iceland, bryophyte biomass, dominated by the moss *Fontinalis antipyretica,* increased with water temperature while the liverwort *Jungermannia exsertifolia* was found in some cold streams but was absent from the warmest sites (Gudmundsdottir et al. 2011). Macrophytes usually die or enter in dormancy below 3 °C (Lacoul and Freedman 2006). In areas of the Upper Mississippi River, including main and side channels and backwaters, biomass of free-floating macrophytes was very low in late-spring and early-summer, reached its maximum in August, and disappeared in autumn due to flooding (Giblin et al. 2014). In tropical waters there likely is little seasonality to growth, unless seasonal changes in flow regulate macrophyte growth. In two lakes of the Paraná River floodplain, greater leaf biomass in *Eichhornia crassipes* was found during the period when lakes and river were connected rather than during the isolation phase, presumably because of greater nutrient availability induced by flooding (Neiff et al. 2008).

Water chemistry may also play a role in defining when and where macrophyte growth is common in rivers and streams. Hardness of water, or its correlates including calcium, alkalinity, and pH, influences the distribution of particular macrophyte species and also limits the occurrence of bryophytes, probably by affecting the availability of free CO_2. In streams of Denmark, variation in the composition of macrophyte assemblages was associated with differences in alkalinity among the study regions (Baattrup-Pedersen et al. 2008). In streams, high turbulence can lead to CO_2 evasion, decreasing its concentration in the water and potentially affecting photosynthesis. To avoid C limitation, many macrophytes can use bicarbonate as a source of inorganic carbon for photosynthesis (Maberly et al. 2015). However, aquatic mosses lack this ability, thus they usually are found at sites with low pH, where the main form of inorganic carbon is CO_2, and in fast currents where CO_2 renewal is favored (Bowden et al. 2017; Glime 2017a).

High flow events strongly influence macrophyte distribution in flowing waters. For example, Riis and Biggs

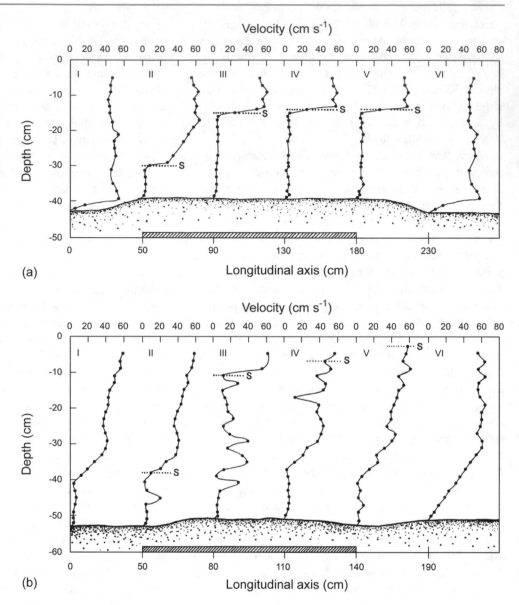

Fig. 6.17 Vertical velocity profiles along a transect passing through the center of patches of (**a**) *Callitriche cophocarpa* and (**b**) *Sparganium emersum*. Velocity was measured in six locations: upstream from the patch (I), within the patch (II–V), and downstream from the patch (VI). The hatched line on the x-axis denotes the patch length, and the location of patch surface (S) is also shown (Reproduced from Sand-Jensen 1998)

(2001, 2003) reported a negative correlation between macrophyte abundance and frequency of flood disturbance in streams of New Zealand. The abundance and diversity of macrophytes decreased as the frequency of floods increased. In streams with very high flood frequency (more than 13 per year), macrophytes were not found. Similarly, the biomass of *Najas*, a cosmopolitan angiosperm, declined after floods in an intermittent stream in Brazil, and larger floods resulted in longer periods before *Najas* populations recovered (Maltchik and Pedro 2001).

Connectivity between floodplain areas and river mainstems can influence macrophyte assemblages in large river floodplains. In the Middle Paraná, greater river-floodplain connectivity was negatively correlated with the biomass of emergent and submerged macrophytes, apparently because greater connectivity resulted in higher water levels and

fluctuations in both light and turbulence (Schneider et al. 2018a). Connectivity with the main channel, water permanence in the floodplain, and water drainage were important factors influencing macrophyte diversity, composition, and zonation in the Lower Paraná floodplain (Morandeira and Kandus 2015). Similarly, in the Upper Mississippi floodplain, free-floating macrophyte biomass was lower in backwaters with greater connection with the river, relative to more isolated backwaters, due to differences in water velocity and nutrient availability (Giblin et al. 2014).

As reported for benthic algae, higher current velocities can stimulate macrophyte photosynthetic rates above rates observed at low current velocities, presumably by increasing the supply of nutrients and CO_2. However, this is another example of the subsidy-stress relationship, where greater water velocities can cause mechanical stress or alter the substrate, and negatively affect macrophyte growth (Madsen et al. 2001; Franklin et al. 2008). In streams of New Zealand, macrophyte cover increased with water velocity up to about 0.4 m s^{-1}, and then declined at higher current speeds (Riis and Biggs 2003). Spatial variation in aquatic macrophytes assemblages in the Mary River, Australia, was explained by discharge intensity and variability, among other variables that include nutrient levels, substrate composition, and riparian cover (Mackay et al. 2003). In headwater streams in Portugal, bryophyte richness and diversity were strongly influenced by discharge and degree of turbulence at the segment scale, possibly because of the role of turbulence in CO_2 acquisition, substrate size, and stability (Monteiro and Vieira 2017). At the microhabitat scale, however, high current velocity had a negative influence likely related to plant detachment.

Along with current, light is one of the most important factors limiting macrophytes growth and distribution in streams. Light attenuates with depth, which inhibits the establishment of submerged macrophytes in deeper rivers, although the precise depth at which photosynthesis can no longer balance respiration varies with turbidity and species–specific light requirements. Reduction in light intensity above 55% by river bank vegetation decreased macrophyte biomass by 20% in the Lower River Spree, Germany, suggesting light limitation at the shaded sites (Köhler et al. 2010). In floodplain lakes of the Upper Paraná River, canopy openness was an important factor influencing macrophyte species richness. However, species composition was influenced by depth, which affects abiotic factors such as sediment grain size and turbidity and thus plant zonation (Bando et al. 2015). Riparian shading affected macrophyte assemblage composition in agricultural streams in Australia, where assemblages dominated by *Potamogeton* spp. and *Ottelia ovalifolia* were associated with reaches with greater light incidence, while assemblages dominated by *Cycnogeton* spp. occurred primarily under shaded conditions (Paice et al. 2017).

As seen for algae, nutrients can also be limiting to macrophytes. The relationship between nutrients and macrophytes is influenced by macrophyte growth form because the latter determines the primary nutrient sources for the plants. In the case of rooted macrophytes, sediment N and P may be a better predictor of growth than water nutrient concentrations because these plants acquire nutrients primarily from the sediments (Carr and Chambers 1998). In some rivers and streams, macrophyte biomass is positively related to nutrient availability (O'Hare et al. 2010; Mebane et al. 2014). In the Upper Mississippi River, changes in tissue N:P and C:N ratios indicate that P is limiting to free-floating macrophytes during June, and N is limiting in late summer and early fall (Giblin et al. 2014). A mesocosm study of the submerged macrophyte *Myriophyllum spicatum* observed higher plant biomass and longer shoots and shoot internodes in plants growing in the high nutrient treatment. In contrast, plants provided a lower nutrient supply showed greater root biomass and greater storage of carbohydrates in shoots and roots (Xie et al. 2013).

Herbivory on freshwater macrophytes generally has been viewed as unimportant in limiting their growth and abundance, but recent studies have shown strong impact of grazers on the abundance, biomass, and composition of aquatic plants (Bakker et al. 2016; Wood et al. 2017). Terrestrial herbivores consume emergent vegetation, some river-dwelling vertebrates including waterfowl, manatees, and grass carp (see Chap. 11) consume submerged aquatic macrophytes, and others such as muskrats harvest plant material for construction of lodges (Wood et al. 2018). In the River Frome, England, swans reduced the biomass of *Ranunculus pseudofluitans* in grazed areas, affecting habitat for fishes and macroinvertebrates (O'Hare et al. 2007). Although it is generally assumed that few representatives of the major groups of aquatic invertebrates are able to graze on macrophytes until after death and decomposition of the plant, recent evidence indicate that insects and crustaceans feed on macrophytes. In an Australian stream, macrophytes represented more than 60% of the diet of the Trichoptera and Ephemeroptera examined (Watson and Barmuta 2011). In a stream in northern Belgium, macrophytes represented between 31 and 49% of the diet of immature insects (Lepidoptera, Ephemeroptera, Diptera) and the crayfish *Orconectus limosus,* suggesting that these organisms can impact macrophyte growth (Wolters et al. 2018).

6.3 Phytoplankton

The phytoplankton are algae suspended in the water column and transported by currents. Whether a river phytoplankton could be self-sustaining was in doubt for some time, because downstream flow would seem to prevent the persistence of their populations. It was suggested that any river plankton was the result of displacement of cells from the benthos or standing waters along the river's course, and reflected wash–out and export rather than a true "potamoplankton". However, there are species that can grow and maintain populations in the flow environment (Reynolds and Descy 1996). Thus, river phytoplankton likely is composed of species originated from different sources, including those that grow in the river channel, the true potamoplankton.

In upstream reaches, sloughing of benthic algae likely is the primary source of suspended algae, and any cells in the water column are simply eroded material in transit. This accidental occurrence of benthic algae in the plankton is referred to as tychoplanktonic, and differs from meroplankty, which refers to planktonic algae that spend part of their life cycle on river sediments, as do some diatoms (Reynolds and

Descy 1996; Istvánovics et al. 2014). In sluggish, lowland streams, in side channels, and in rivers of considerable length, under favoring conditions of flow, channel retentiveness, light, temperature, and nutrient availability, phytoplankton can develop substantial populations. Some species found in rivers are drawn from the same pool of species found in standing water. Thus, the presence of lakes, ponds, and backwaters, and the creation of impoundments, can be of great importance in seeding the river with some truly planktonic algal groups, including chlorophytes. River sediments can also be an important source of certain algae including diatoms, which likely maintain their populations in the river channel through meroplankty (Istvánovics and Honti 2011).

Diatoms, particularly centric diatoms, have been found to dominate the composition of phytoplankton in many rivers, likely related to their capacity to live in turbid and turbulent waters. In the Tigris River, Turkey, diatoms dominated in terms of species richness and abundance, followed by green algae and cyanobacteria (Varol and Şen 2018). In the Po River, Italy, diatoms represented between 55 and 95% of the biomass, followed by cryptophytes and chlorophytes (Tavernini et al. 2011). Despite the abundance of diatoms, green algae can dominate the lower reaches of large temperate rivers during the summer (Abonyi et al. 2014). In tropical rivers, desmids (a green alga) contribute the largest number of species while diatoms dominate the biomass, although to a lesser degree than in temperate rivers (Rojo et al. 1994). In the Baia River, a tributary of the Paraná, cyanobacteria and diatoms dominated phytoplankton biomass whereas chlorophytes had the largest number of species (Train and Rodrigues 1998). Cyanobacteria have been reported to form dense blooms in the Nile and during summer in temperate rivers (Talling and Rzóska 1967; Bennett et al. 1986), apparently in response to nitrate depletion and because of their nitrogen-fixing capability. Bacillariophyceae, Chlorophyceae, and Euglenophyceae were predominant amongst the phytoplankton of the Upper Paraguay of the Pantanal region (de Domitrovic et al. 2014). In the Congo River in Africa, chlorophytes dominated during the period of high water while diatoms dominated in falling water, when cryptophytes and cyanobacteria were also abundant (Descy et al. 2017). In the tropical Daly River, Australia, diatoms showed the greater number of species while the biovolume was dominated by Cryptophyta and Pyrrophyta (Townsend and Douglas 2017). These studies establish that phytoplankton occur in virtually all major rivers examined, although there appears to be considerable variation in which groups predominate.

In addition to taxonomic groupings, a functional classification of freshwater phytoplankton has been proposed based on 31 categories, with a description of the habitat, main representative genera, tolerance, and sensitivities for each group (Reynolds et al. 2002; Padisák et al. 2009). In the Lower Salado River, Argentina, groups C (centric diatoms), D (small centric diatoms), X1(small Chloroccocales) and X2 (mainly volvocaleans) dominated the phytoplankton, indicating the prevalence of small cells with high metabolism and tolerance to mixed environments. These groups are common in enriched lowland rivers, and they can tolerate hydraulic stress and low light conditions (Devercelli and O'Farrell 2013). In the Mudanjiang River, China, functional groups varied seasonally. Group Y (Cryptophyceae) dominated in winter and spring; Group MP (meroplanktonic diatoms) showed greater biomass in the summer and Group P (diatoms) in the autumn (Yu et al. 2012). In the Mura, Drava, Danube, and Sava rivers in Croatia, diatom Groups C, D and TB (benthic diatoms) dominated the phytoplankton; however, when water residence time increased, Group T (chlorococcalean green algae) increased (Stanković et al. 2012).

Variation in temperature, light availability, nutrients, and discharge influence seasonal variation in phytoplankton abundance. In temperate and subtropical rivers, phytoplankton are most abundant in spring and summer. In the Columbia River of western North America, higher nutrient levels, lower flows, and higher light availability favor diatom abundance during the spring (Sullivan et al. 2001). In the St. Johns River, Florida, US, seasonal variation is related to light availability, which is primarily influenced by the color of the water. Less color and lower light attenuation are observed during spring and summer, and these conditions correspond with greatest abundance of phytoplankton (Phlips et al. 2000). In the lower Mississippi, chlorophyll increased from spring to summer, in response to reduced discharge, depth, and turbidity (Ochs et al. 2013). In the Thames River, spring and summer blooms were related to higher temperature and light levels, and lower flows (Waylett et al. 2013).

In tropical rivers, phytoplankton abundance responds to hydrologic seasonality, and is greatest during periods of low water. In the Orinoco River and some of its tributaries, highest production and biomass of phytoplankton were observed during the periods of falling and low water, likely due to greater transparency, shallower depths, and lower flows (Lewis 1988). In the Congo River, greater biomass and diversity were observed during the period of falling water compared to high water (Descy et al. 2017). In the Baia River, phytoplankton biomass also peaked during low water and was dominated by cyanobacteria. This was followed by an increase in diatoms, probably favored by an increase in rainfall and turbulence (Train and Rodrigues 1998). In the Upper Paraguay River, greater phytoplankton biomass occurred during the summer and both biomass and cell density showed a significant correlation with water temperature (de Domitrovic et al. 2014).

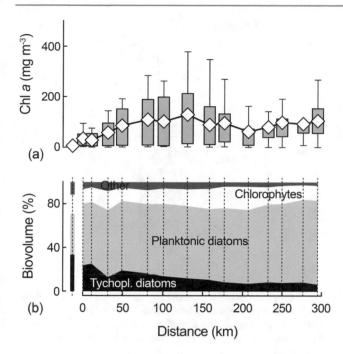

Fig. 6.18 Phytoplankton from the Somes River, Romania-Hungary, in the summer of 2012. (**a**) Chlorophyll *a* concentration (diamonds are average values; boxes are quartile plots of minima and maxima); (**b**) Composition of phytoplankton community by biovolume along the Somes River. The Somes River is formed by the confluence of the Someşul Mare and the Someşul Mic Rivers. The leftmost symbol in (**a**) and the left column in (**b**) represent data for the Someşul Mic River. Algal profiles start at the 0 km sampling site in the Someşul Mare River (Reproduced from Istvánovics et al. 2014)

Phytoplankton biomass can also show longitudinal variation along the river channel. In the Somes River, Romania, chlorophyll increased in the middle sections, related to proliferation of centric diatoms. In these reaches, lower channel slope enhanced benthic retention of meroplanktonic diatoms, reducing their flushing during periods of high discharge. This retention favors algal recruitment from the sediments, explaining the observed increase in biomass. Tychoplanktonic diatom species decreased in the downstream direction, showing greater contribution to biomass in a tributary of the Somes River, Romania-Hungary (Istvánovics et al. 2014) (Fig. 6.18). In the agricultural Le Sueur River watershed, Minnesota, US, greater chlorophyll concentrations were measured at sites located in mid-order reaches (4th to 6th order) compared to smaller streams; mid-size rivers also have the largest proportion of lakes and wetlands, potentially important sources of algae to lotic environments (Dolph et al. 2017).

6.3.1 Limiting Factors for Phytoplankton

Factors affecting the growth of phytoplankton in running waters include all the same variables that limit algal growth in lakes, such as light, temperature, and nutrients. However, discharge regime has a profound influence over river phytoplankton, and the influence of light and nutrients differ in some ways from what is seen in standing waters (Reynolds 2000). In addition, adjacent stagnant waters are critical to the establishment of river phytoplankton, through their influence upon the size of the inoculum. This can be of considerable importance, especially when residence time of the water mass is short enough to limit the build-up of populations.

An inverse relationship between river discharge and phytoplankton abundance is perhaps the most common finding of detailed investigations of river phytoplankton (Sullivan et al. 2001; Tavernini et al. 2011). Discharge can affect water residence time, phytoplankton abundance, and water column conditions such as light availability. As a mass of water moves downstream and the entrained plankton multiply, one expects maximal abundances to be associated with a water mass that is traveling slowly and is uninterrupted over a long distance. Talling and Rzóska (1967) estimated that a water mass traversed the 357 km section of the Blue Nile between Sennar Reservoir and Khartoum in 40 days at low flows, but required only 2 days at high flood. Since phytoplankton populations are capable of a maximum of about 1–2 doublings per day, the consequences for eventual population size are considerable. In the River Thames, a strong positive correlation between river length and chlorophyll is the result of the influence of water residence time, measured as distance to river source, on algal biomass (Bowes et al. 2012a). In the Somes River, water residence time also influenced phytoplankton blooms in the middle sections during periods of low flow and in the summer (Istvánovics et al. 2014). In tidal rivers, longer water residence times during period of low flow can increase phytoplankton biomass (Ensign et al. 2012).

Flow variability also is associated with changes in turbidity and light availability. In the Adige, a river draining the Alps in northern Italy, chlorophyll and phytoplankton biovolume were negatively correlated with discharge, showing lower biomass during spring and summer. Increased flows not only diluted phytoplankton abundance, which was dominated by small diatoms, but also increased turbidity and thus reduced light penetration. Greater biomass was observed during periods of low flow in fall and winter, despite low temperatures (Salmaso and Zignin 2010). In the Po River, also in Italy, discharge was negatively correlated to species richness of the phytoplankton. When discharge was high, large diatoms dominated the phytoplankton because of their capacity to live in turbid waters (Tavernini et al. 2011).

In rivers of the Tisza catchment, a tributary of the Danube, an inverse relationship between phytoplankton chlorophyll, and daily mean discharge was observed; however, when responses to flow were examined at a hourly

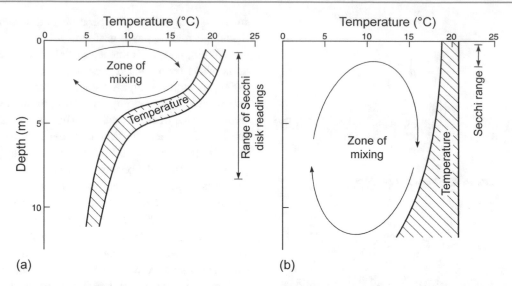

Fig. 6.19 Schematic diagram comparing effect of depth of mixing on primary production in phytoplankton of a lake versus a river. In a lake (**a**), establishment of a temperature barrier between surface and deep waters restricts mixing to the upper few meters. In a river (**b**), temperature stratification is impeded by turbulence of flow, and the water column typically mixes from top to bottom. Depths of 5–20 m are common in large rivers. Rivers often carry substantial sediment loads, usually restricting light penetration to the upper 1–2 m

scale, chlorophyll and diatoms increased with the flow pulse (Istvánovics and Honti 2011). In addition, diatoms, cyanobacteria, and chlorophytes responded differently to flow pulses. The dominant diatoms showed an increase in biomass, while chlorophytes always showed dilution, suggesting that diatoms are meroplanktonic and chlorophytes are planktonic. Large flood pulses decreased abundances of both chlorophytes and diatoms, but increased cyanobacteria, which only appeared in the water column during flow pulses, probably because of their association with sediment particles.

In a large river with ample nutrients and a transit time long enough to permit multiplication of phytoplankton, it is likely that phytoplankton abundance is limited by light via interactions among turbidity, depth, and turbulence. If the water column mixes to a depth greater than the photic zone, then an individual cell will spend part of the day at light levels too low to support photosynthesis (Fig. 6.19). Cole et al. (1991) estimated that the average phytoplankton cell in the Hudson River would spend from 18 to 22 h below the 1% light level. Rather than growing, the cell would be expected to lose biomass. This is a real puzzle, because phytoplankton biomass does increase during the spring and summer. One possible explanation is that phytoplankton blooms originate only in river sections <4 m in depth (Cole et al. 1992). In the Mississippi River, phytoplankton were light limited because irradiance in the water column was not sufficient to saturate the photosynthetic rate. In addition, the maximum exposure of phytoplankton to sunlight was only 3 h per day. This resulted in negative net primary production over the study period, despite the observed increase in chlorophyll during the summer. This again suggests that the river main channel is receiving phytoplankton biomass subsidies from lateral and backwater areas (Ochs et al. 2013).

Although the usual effects of high flows are dilution and downstream transport, under certain circumstances floods might augment river plankton by washing in populations from stagnant areas. Lewis (1988) found that primary production per unit volume was greatest during the period of low water, but nonetheless was quite low, due to a combination of light limitation and the short residence time of river water. Total phytoplankton transport exhibited a minimum just as discharge began its seasonal increase and was maximal at high water. Lewis concluded that flushing of backwaters in or adjacent to the river channel accounted for increased transport during the rise of flood waters, and the floodplain contributed phytoplankton during the flushing and draining periods associated with peak and declining flood stages. In the Daly River, Australia, chlorophyll concentrations were lower during the wet than the dry season; however, chlorophyll increased during rising water and phytoplankton biovolume was greater during the wet season when main channel and floodplain lakes were connected, suggesting that the most probable sources of algae are located in floodplain water bodies and enter the river main channel during periods of high flood (Townsend and Douglas 2017).

Although downstream transport and light typically are over-riding variables, nutrient concentration can limit phytoplankton in some rivers. Comparing the more free-flowing but more turbid Ohio River to the more regulated and less turbid Cumberland and Tennessee rivers, Koch et al. (2004) found frequent light limitation of the phytoplankton of Ohio

River. At higher irradiance, however, P limitation was observed for the Cumberland phytoplankton, while N and P were co-limiting in the Tennessee, and silicate was limiting in the Ohio River. In the River Rhine, a spring bloom of diatom-dominated phytoplankton reached very high abundances until dissolved silicate became depleted, which led to a population collapse (De Ruyter Van Steveninck et al. 1992). In the Thames basin, England, water residence time is a key factor influencing phytoplankton biomass (Bowes et al. 2012a); however, when spring blooms are well established, phosphorus and silica become depleted, suggesting that these elements could limit or co-limit algae growth. In this basin, phosphorus concentrations have declined since the late 1990s due to improved wastewater treatment, and further reduction may lead to spring blooms of lesser magnitude. Similarly, reduction in the concentrations of phosphorus in the River Rhine resulted in lower mid-summer chlorophyll concentrations (Van Nieuwenhuyse 2007). In the river Loire, a 24-year data record provided evidence of a decrease in phytoplankton biomass and biovolume and an increase in taxon diversity that correlated with a reduction in phosphorus concentration (Larroudé et al. 2013).

Grazing by zooplankton does not appear to constitute a major loss term for phytoplankton under most conditions (Reynolds and Descy 1996). In the Danube (Bothar 1987), the Apure (a whitewater tributary of the Orinoco, Saunders and Lewis 1988), and the Hudson (Pace et al. 1992), zooplankton grazing was believed to have little impact on phytoplankton. However, the presence of invasive clams can have major impacts on river phytoplankton. The zebra mussel *Dreissena polymorpha*, which first appeared in the Hudson River in 1991, quickly became so abundant that their populations were capable of filtering the entire volume of the Hudson estuary every 1–4 days during summer. Phytoplankton biomass fell by 85% due to direct consumption (Caraco et al. 1997), and the entire ecosystem was affected through associated changes in nutrients, water clarity, other grazers, and due to increases in submersed macrophytes (Strayer et al. 1999). In addition, zebra mussel grazing rate explained 90% of the interannual variation in phytoplankton biomass in the Hudson river (Caraco et al. 2006). In the River Meuse, *Corbicula* spp., which first appeared in 1990s, has led to declines in phytoplankton biomass since 2004, and estimated annual biomass losses of 70% (Pigneur et al. 2014).

6.4 Human Influence

Human activities can alter the factors that influence autotrophs, including light levels, temperature, nutrients, water quality, flow, and grazers, thereby impacting biomass and assemblage composition. Frequently, several environmental variables are affected simultaneously, making it difficult to attribute causality. Studies of the impacts of anthropogenic activities often compare impacted to reference sites as well as conduct experiments to test the effects of a single factor or combinations of more than one stressor. Algal bioassessments use various attributes including biomass, taxa composition, diversity, and tolerance to disturbance. Diatoms are widely as indicators, although Chlorophyta and Cyanobacteria have recently been considered in environmental assessments (Stevenson and Smol 2015; Stevenson and Rollins 2017).

Acidification of stream water has altered the composition and biomass of periphyton over broad regions of North America, Europe, and Asia (Schneider et al. 2018b). Sources include atmospheric emissions of SO_2, NO_x and NH_3, which form acids in combination with water vapor and enter surface waters as both wet (rain and snow) and dry deposition. Acid precipitation from atmospheric pollution lowers the pH of surface waters and also mobilizes aluminum ions from soils. Aluminum is toxic to algae because it binds to phosphorus, reducing its availability (Burns et al. 2008; Schneider et al. 2018b). Although water quality has improved in the past decade in many areas due to regulations that lower the emission of acid-forming gases, the recovery of the biota has been slower. This may be the result of episodic acidification from precipitation events, and the sensitivity of algal assemblages to acidification (Kowalik et al. 2007; MacDougall et al. 2008). Diatoms show species-specific responses to lowered pH alteration, resulting in changes in dominant taxa and decreased diversity, and so are widely use in studies of acidification and recovery (Hirst et al. 2004; Passy 2006). However, other algal taxa have also been used to develop indices to assess acidification (Schneider and Lindstrøm 2009).

Mine drainage is also an important cause of stream acidification in many areas, due to the exposure of sulfides in exposed rocks to air and water. In addition to lowered pH, acid mine drainage typically increases heavy metal concentrations in stream water. Periphyton species diversity declines, and a few acid-tolerant diatoms or filamentous green algae become dominant (Sabater et al. 2003; Hogsden and Harding 2012). Acid mine drainage can also affect periphyton growth via nutrient availability. Precipitation of metals such as iron and aluminum can adsorb phosphorus, reducing its concentration in solution up to 90% compared to sites located upstream from mine discharges (Simmons 2010). This decrease in P availability affected the magnitude of periphyton P limitation in impacted streams of the Ohio River watershed, where responses to P addition increased along a gradient of increasing influence of acid mining drainage (DeNicola and Lellock 2015).

Increased water temperature also affects stream periphyton assemblages. Thermoelectric power plants can warm rivers by taking in river water for cooling and then releasing warmer water back into the river (Madden et al. 2013). Human-induced climate warming will have a much broader and longer-term impact (Van Vliet et al. 2013). Modest warming may have positive effects, increasing photosynthetic and growth rates. A 1.4 °C rise in temperature increased algal cell density and biomass, and resulted in higher densities of the cyanobacterium *Cylindrospermum* in experimental channels (Piggott et al. 2012). When stream water temperature was increased by 3 °C in one half of a stream channel, periphyton biomass, chlorophyll concentration, and diatom density increased relative to the control half, particularly during the colder months, suggesting that warming can stimulate algal growth (Delgado et al. 2017). The effect of increased temperature can be modified by other stressors such as sediments and nutrients, resulting in complex responses (Piggott et al. 2015). For example, higher temperatures under increased sediment supply can enhance the negative effect of sediments on algae, although positive effects have been detected in some algae groups.

Nutrient inputs can increase greatly due to human activities, especially from fertilizer runoff from agricultural lands, areas of high livestock densities, and from wastewater treatment plants (Carpenter et al. 2011; Withers et al. 2014). High nutrient loading to streams and rivers can result in proliferation of algae and macrophytes, referred to as eutrophication, and a major cause of aquatic ecosystem impairment. Although lake eutrophication has received extensive study, the process is less well understood in streams and rivers (Hilton et al. 2006; Dodds and Smith 2016). Certainly, all categories of autotrophs can be affected, including periphyton, macrophytes, and phytoplankton, commonly with increases in overall biomass but declines in tax on richness and diversity. Greater responses to nutrient enrichment usually occur under stable flow, longer water retention times, and increased light availability, the latter of which can be affected by land use change when riparian vegetation is altered (Biggs 2000; Hilton et al. 2006; Johnson et al. 2009). Eutrophication symptoms can affect a number of human benefits including recreational use by altering the aesthetic appeal of a river and producing foul odors, interfering with water purification and supply, and increasing the risk of floods by excessive growth of macrophytes in the river channel (Hilton et al. 2006; O'Hare et al. 2010; Le Moal et al. 2019). For example, under high nutrient concentrations, cyanobacterial blooms can occur in drinking water sources, producing toxins that cannot be removed completely by conventional water treatment processes. This has affected drinking water safety from sources as diverse as the Nile River, Egypt (Mohamed et al. 2015) and western Lake Erie (Wynne and Stumpf 2015).

Eutrophication mitigation usually involves setting nutrient criteria and reducing nutrient loading through voluntary and regulatory means. As described above, reduction in phosphorus concentrations has resulted in decreased phytoplankton biomass in several European rivers. However, the reduction of nutrient inputs to prevent blooms can be expensive or less effective than other management practices such as riparian shading. In the Ouse catchment, England, modelling studies indicated that phosphorus reduction would lower peak phytoplankton chlorophyll concentration by 10%, while implementing riparian reforestation would reduce chlorophyll levels by more than 40% (Hutchins et al. 2010). Experimental studies in the Upper River Thames, also suggest that increasing riparian shading of the channels will decrease periphyton growth substantially (Bowes et al. 2012b).

Algal assays can be used to identify threshold nutrient values. Stevenson et al. (2008) used several diatom attributes or metrics (number of observed taxa, evenness, proportion of native taxa, among others) to develop total phosphorus criteria recommendations for the Mid-Atlantic Region, US. Black et al. (2011) used the relationship between 11 algal attributes, including diatom and non-diatom species, and nutrient concentrations to identify nutrient thresholds for streams in the western US. Thresholds obtained from each algal attribute or metric were consistent and did not vary between fine and coarse substrates. Schneider and Lindstrøm (2011) developed a trophic index for Norwegian rivers (periphyton index of trophic status, PIT) using algae other than diatoms, which were less common in their system, by using the average of total phosphorus concentration where each algal species occurred. The relationship between total phosphorus and the trophic index indicated that concentrations below 10 µg L^{-1} did not result in major changes in periphyton (Fig. 6.20).

Agricultural and domestic wastewaters contribute additional pollutants to streams including pesticides, pharmaceuticals, and a wide range of contaminants. Herbicides have been found to reduce diatom growth, although some species tolerant to eutrophication like *Melosira varians*, *Nitzschia dissipata*, and *Cocconeis placentula* are not affected (Debenest et al. 2009). Another study found pesticide effects to be more apparent using diatom guilds and life forms rather than individual taxa (Rimet and Bouchez 2011). The abundance of high-profile diatoms decreased while motile, low-profile, and mucous tubule diatoms increased in experimental channels exposed to pesticides, likely due to the presence of structures that protect cells from contaminants. Antibiotic exposure affected taxon richness and assemblage composition by decreasing the relative abundances of high-profile and motile guilds. Interesting, periphyton biomass was not reduced and the assemblage recovered after 16 days, likely due to the short duration of the antibiotic pulse (Winkworth et al. 2015). Titanium dioxide nanoparticle (Nano-TiO$_2$) is an engineered

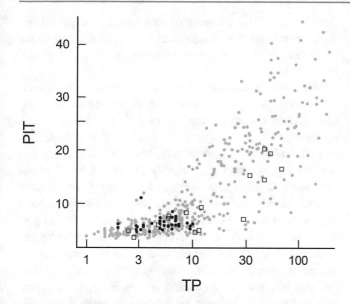

Fig. 6.20 Relationship between periphyton index of trophic status (PIT) and total phosphorus (TP, in μgL^{-1}). Grey dots represent the data used to develop the index, the black dots represent reference sites, and squares represent the independent data used to validate the index (Reproducede from Schneider and Lindstrøm 2011)

nanomaterial commonly used in personal care products and food, and it can be toxic to the aquatic biota. Inputs of this compound in streams through domestic wastewater can decrease biofilm algal density and alter assemblage composition (Binh et al. 2016; Wright et al. 2018).

Metal pollution is another cause of anthropogenic harm to aquatic autotrophs. Diatoms are very sensitive to metal pollution (Roig et al. 2016). Macrophytes can accumulate large amounts of metals, making them good indicators of metal exposure. In the Imera Meridionale River, Italy, metal concentrations in river water and sediments were significant and positively correlated to concentrations in roots, leaves, and stems of *Phragmites australis* (Bonanno and Lo Giudice 2010). In the Ctalamochita River, Argentina, heavy metal concentrations in submerged macrophytes *Myriophyllum aquaticum* and *Potamogeton pusillus* were correlated with metal levels in water and sediments, reflecting the variation in metal levels along the river and also indicating the plants' potential for biomonitoring (Harguinteguy et al. 2016).

Invasive species are yet another potential pressure on the algae and macrophytes of running waters. The diatom *Didymosphenia geminata* is associated with massive blooms in temperate, oligotrophic running waters. Evidence indicates that "didymo", as it is widely known, is native to mountain streams of the Northern Hemisphere but has recently expanded to the Southern Hemisphere, where it is considered an invasive species (Blanco and Ector 2009). However, there is some controversy concerning whether this diatom is an ubiquitous species that has benefited from changes in environmental conditions favoring its growth, or whether it is a novel species to Chile, Argentina, and New Zealand, where it is also a nuisance (Reid and Torres 2014; Taylor and Bothwell 2014; Kilroy and Novis 2018). Didymo has large cells and forms branched stalks attached to substrates by an extracellular polymeric substance, resulting in thick mats that can cover large areas of stream channels (Kilroy and Bothwell 2011; Bray et al. 2016). Its proliferation increases algal biomass and usually changes the composition of the benthic algal assemblage. In Patagonian streams, didymo can impact the native benthic algae by replacing low biomass communities dominated by nitrogen-fixing cyanobacteria and macrophytes with didymo's high biomass, stalk-forming morphology with a high capacity to trap sediments (Reid and Torres 2014).

However, some studies have reported increased algae species richness in didymo mats (Sanmiguel et al. 2016), or higher diatom diversity, suggesting that the polymeric matrix may enhance algae colonization (Gillis and Lavoie 2014). The growth of small pioneer diatoms like *Achnanthidium minutissimum* can be enhanced by the habitat afforded by *Dydymosphenia* filaments, while larger attached diatom species probably are displaced by the rapid growth and substrate coverage of *D. geminata,* producing a decline in diatom diversity (Ladrera et al. 2018). Explanations for the proliferation of this diatom under oligotrophic conditions include its uptake capacity for phosphorus and its greater photosynthetic production of polymeric substrate under P limitation (Kilroy and Bothwell 2011; Sundareshwar et al. 2011). It is known that P limitation, high light levels, and flow stability are related to didymo blooms (Miller et al. 2009; Bray et al. 2016) but it is not clear why *D. geminata* is spreading so quickly and why its nuisance proliferation is becoming more common (Taylor and Bothwell 2014; Kilroy and Novis 2018).

Some macrophyte species can become very abundant in new environments, often but not invariably with detrimental effects. The invasion of the Asian macrophyte *Hydrilla verticillata* (Hydrocharitaceae) to the Paraná River Basin in Brazil has affected native macrophytes, particularly the South American Hydrocharitaceae (*Egeria najas* and *E. densa*) since they are ecologically similar. Because *Hydrilla* can develop under a wide range of habitat conditions and exhibit high growth rates, it can exert strong pressure on native macrophytes (Sousa 2011). In the Upper Paraná, *Hydrilla* was absent from lakes but developed high biomass in rivers, reaching more than four times the biomass of *Egeria najas* (Sousa et al. 2010). Experimentally, Silveira et al. (2018) found that high densities of *Hydrilla* reduced the biomass of *E. najas,* decreasing its capacity to colonize sites with *Hydrilla*.

6.5 Summary

Autotrophs acquire their energy from sunlight and their materials from nonliving sources. The major autotrophs of streams and rivers include algae, a diverse group of organisms from multiple taxonomic groups, and higher plants, collectively referred to as macrophytes due to their larger size. Algae occur in many growth forms and habitats. When attached to surfaces in intimate association with heterotrophic microbes, they are referred to as periphyton, biofilm, and sometimes as *aufwuchs*. Phytoplankton are algae that develop in the water column. Most common groups of algae in streams and rivers are cyanobacteria (Cyanophyta), green algae (Chlorophyta and Streptophyta), diatoms (Bacillariophyceae), red algae (Rodophyta), yellow-green algae (Xanthophyceae) and chrysomonads (Chrysophyceae). Although less common, brown algae (Phaeophyceae), euglenoids (Euglenophyta), Synurophyceae, and dinoflagellates (Pyrrophyta) can also be found.

Periphyton are important in fluvial food webs, especially in headwater and midsized streams, and they also influence benthic habitat and nutrient cycling. Proximate factors that may limit benthic algal communities include light and nutrients, which, along with temperature, influence biomass accrual; and disturbance and grazing, which are the factors that lead to algal dislodgement and biomass loss. The importance of any one factor to algal growth depends upon whether some other factor is in even shorter supply, and these environmental conditions vary with location and by season. Light often is the limiting factor in streams shaded by forest cover, as evidenced by lower benthic algal biomass in shaded versus unshaded streams, and by seasonal peaks before leaf-out and after leaves are shed. Phosphorus and nitrogen have each been shown to limit algal growth, sometimes in combination, but nutrient limitation can be over-ridden by low light or intense herbivory. Stream current has opposing effects on benthic algae accrual depending on growth form and architecture and is best described as a 'subsidy-stress' response. The flow of water brings continual renewal of gases and nutrients, and so current benefits algal growth by enhancing nutrient uptake. However, current also exerts a shear stress on benthic algae, which can cause cell sloughing, and high flows disturb and scour the substrate. Different temperatures apparently favor particular algal taxa, with filamentous green algae and cyanobacteria more abundant in warmer rivers; and growth rate of course increases with increasing temperatures. Substrate is a factor through its interaction with current, in that different areas of a stone may be more or less exposed to flow and smaller particles are more likely to be dislodged. Grazers can significantly limit benthic algae and influence community composition by selectively eliminating certain species and growth forms. The composition and biomass of the benthic algal assemblage varies temporally and spatially due to the interaction of these multiple environmental factors at different scales.

Macrophytes include the bryophytes (mosses and liverworts), vascular plants (flowering plants, ferns) and the Charales. Macrophytes can be placed in four categories according to their growth form. These are emergent, submersed, and floating–leaved taxa, which are all rooted, and free-floating plants, which usually are not attached to the substrate and often form large mats. Angiosperms require moderate depths and slow currents, and so are most common in springs, rivers of intermediate size, and along the margins and in backwaters of larger rivers. Bryophytes are restricted in distribution but can be abundant in cool climates and in shaded headwater streams. Macrophytes are important to fish and invertebrates as habitat and as refuge from predators, and they can increase habitat heterogeneity for aquatic organisms by modifying water velocity and trapping sediments and organic matter. Current and flow, light, and nutrients are probably the most important factors influencing macrophyte growth.

The phytoplankton consist of cells and colonies of algae suspended in the water column and transported by currents. Light, nutrients, flow, and temperature all influence seasonal variation in phytoplankton abundance. Unable to maintain populations in fast–flowing streams, phytoplankton can become abundant in slow-flowing rivers and backwaters where their doubling rates exceed downstream losses due to current. Discharge has a strong influence on phytoplankton by affecting residence time and light conditions. In large rivers, vertical mixing within a deep and turbid water column further limits the opportunity for photosynthesis. Although invasive mollusks have been observed to dramatically reduce phytoplankton populations in some large rivers, grazing on phytoplankton by zooplankton is low.

Anthropogenic stressors such as acidification, increased water temperature, eutrophication, pollution, and invasive species can have major impacts on autotroph biomass and diversity in running waters. Because algae and macrophytes are sensitive to changes in environmental conditions, they can be a useful tool in stream bioassessments to determine human impacts on a river ecosystem.

References

Abonyi A, Leitão M, Stanković I et al (2014) A large river (River Loire, France) survey to compare phytoplankton functional approaches: do they display river zones in similar ways? Ecol Indic 46:11–22. https://doi.org/10.1016/J.ECOLIND.2014.05.038

Baattrup-Pedersen A, Springe G, Riis T et al (2008) The search for reference conditions for stream vegetation in northern Europe.

Freshw Biol 53:1890–1901. https://doi.org/10.1111/j.1365-2427.2008.02003.x

Bakker ES, Wood KA, Pagès JF et al (2016) Herbivory on freshwater and marine macrophytes: a review and perspective. Aquat Bot 135:18–36. https://doi.org/10.1016/J.AQUABOT.2016.04.008

Bando FM, Michelan TS, Cunha ER et al (2015) Macrophyte species richness and composition are correlated with canopy openness and water depth in tropical floodplain lakes. Brazilian J Bot 38:289–294. https://doi.org/10.1007/s40415-015-0137-y

Bechtold HA, Rosi-Marshall EJ, Warren DR, Cole JJ (2012) A practical method for measuring integrated solar radiation reaching streambeds using photodegrading dyes. Freshw Sci 31:1070–1077. https://doi.org/10.1899/12-003.1

Beck WS, Rugenski AT, Poff NLR (2017) Influence of experimental, environmental, and geographic factors on nutrient-diffusing substrate experiments in running waters. Freshw Biol 62:1667–1680. https://doi.org/10.1111/fwb.12989

Bennett J, Woodward J, Shultz D (1986) Effect of discharge on the chlorophyll a distribution in the tidally-influenced Potomac River. Estuaries Coasts. 9:250–260. https://doi.org/10.2307/1352097

Benoiston A-S, Ibarbalz FM, Bittner L et al (2017) The evolution of diatoms and their biogeochemical functions. Philos Trans R Soc London Ser B 372:20160397. https://doi.org/10.1098/rstb.2016.0397

Bernhardt ES, Likens GE (2004) Controls on periphyton biomass in heterotrophic streams. Freshw Biol 49:14–27. https://doi.org/10.1046/j.1365-2426.2003.01161.x

Biggs BJF (1996) Patterns in benthic algae of streams. In: Stevenson RJ, Bothwell ML, Lowe R (eds) Algal ecology: freshwater benthic ecosystems. Academic Press, San Diego, pp 31–56

Biggs BJF (2000) Eutrophication of streams and rivers: dissolved nutrient-chlorophyll relationships for benthic algae. J North Am Benthol Soc 19:17–31. https://doi.org/10.2307/1468279

Biggs BJF, Smith RA (2002) Taxonomic richness of stream benthic algae: effects of flood disturbance and nutrients. Limnol Oceanogr 47:1175–1186. https://doi.org/10.4319/lo.2002.47.4.1175

Biggs BJF, Goring DG, Nikora VI (1998a) Subsidy and stress responses of stream periphyton to gradients in water velocity as a function of community growth form. J Phycol 34:598–607. https://doi.org/10.1046/j.1529-8817.1998.340598.x

Biggs BJF, Stevenson RJ, Lowe RL (1998b) A habitat matrix conceptual model for stream periphyton. Arch Fur Hydrobiol 143:21–56. https://doi.org/10.1017/CBO9781107415324.004

Binh CTT, Adams E, Vigen E et al (2016) Chronic addition of a common engineered nanomaterial alters biomass, activity and composition of stream biofilm communities. Environ Sci 3:619–630. https://doi.org/10.1039/c5en00274e

Black RW, Moran PW, Frankforter JD (2011) Response of algal metrics to nutrients and physical factors and identification of nutrient thresholds in agricultural streams. Environ Monit Assess 175:397–417. https://doi.org/10.1007/s10661-010-1539-8

Blanco S, Ector L (2009) Distribution, ecology and nuisance effects of the freshwater invasive diatom Didymosphenia geminata (lyngbye) M. Schmidt: a literature review. Nov Hedwigia 88:347–422. https://doi.org/10.1127/0029-5035/2009/0088-0347

Bojorge-Garcia M, Carmona J, Ramirez R (2014) Species richness and diversity of benthic diatom communities in tropical mountain streams of Mexico. Inl Waters 4:279–292. https://doi.org/10.5268/IW-4.3.568

Bonanno G, Lo Giudice R (2010) Heavy metal bioaccumulation by the organs of Phragmites australis (common reed) and their potential use as contamination indicators. Ecol Indic 10:639–645. https://doi.org/10.1016/j.ecolind.2009.11.002

Borchardt MA (1996) Nutrients. In: Stevenson RJ, Bothwell ML, Lowe RL (eds) Algal ecology: freshwater benthic ecosystems. Academic Press, San Diego, pp 183–227

Bothar A (1987) The estimation of production and mortality of Bosmina longirostris (OF Müller) in the River Danube (Daubialia Hungarica, CIX). Hydrobiologia 145:285–291

Bowden WB, Glime JM, Riis T (2017) Macrophytes and bryophytes. In: Hauer FR, Lamberti GA (eds) Methods in stream ecology, vol 1: ecosystem structure, 3rd edn. Academic Press, Boston, pp 243–271

Bowes MJ, Gozzard E, Johnson AC et al (2012a) Spatial and temporal changes in chlorophyll-a concentrations in the River Thames basin, UK: are phosphorus concentrations beginning to limit phytoplankton biomass? Sci Total Environ 426:45–55. https://doi.org/10.1016/j.scitotenv.2012.02.056

Bowes MJ, Ings NL, McCall SJ et al (2012b) Nutrient and light limitation of periphyton in the River Thames: implications for catchment management. Sci Total Environ 434:201–212. https://doi.org/10.1016/j.scitotenv.2011.09.082

Branco CCZ, Riolfi TA, Crulhas BP et al (2017) Tropical lotic primary producers: Who has the most efficient photosynthesis in low-order stream ecosystems? Freshw Biol 62:1623–1636. https://doi.org/10.1111/fwb.12974

Bray J, Harding JS, Kilroy C et al (2016) Physicochemical predictors of the invasive diatom Didymosphenia geminata at multiple spatial scales in New Zealand rivers. Aquat Ecol 50:1–14. https://doi.org/10.1007/s10452-015-9543-2

Burkholder JM (1996) Interactions of benthic algae with their substrata. In: Stevenson RJ, Bothwell ML, Lowe RL (eds) Algal ecology: freshwater benthic ecosystems. Academic Press, San Diego, pp 253–297

Burns DA, Riva-Murray K, Bode RW, Passy S (2008) Changes in stream chemistry and biology in response to reduced levels of acid deposition during 1987–2003 in the Neversink River Basin, Catskill Mountains. Ecol Indic 8:191–203. https://doi.org/10.1016/j.ecolind.2007.01.003

Capps KA, Ulseth A, Flecker AS (2015) Quantifying the top-down and bottom-up effects of a non-native grazer in freshwaters. Biol Invasions 17:1253–1266. https://doi.org/10.1007/s10530-014-0793-z

Capps KA, Booth MT, Collins SM et al (2011) Nutrient diffusing substrata: a field comparison of commonly used methods to assess nutrient limitation. J North Am Benthol Soc 30:522–532. https://doi.org/10.1899/10-146.1

Caraco NF, Cole JJ, Strayer DL (2006) Top down control from the bottom: regulation of eutrophication in a large river by benthic grazing. Limnol Oceanogr 51:664–670. https://doi.org/10.4319/lo.2006.51.1_part_2.0664

Caraco NF, Cole JJ, Raymond PA et al (1997) Zebra mussel invasion in a large, turbid river: Phytoplankton response to increased grazing. Ecology 78:588–602. https://doi.org/10.1890/0012-9658(1997)078%5b0588:zmiial%5d2.0.co;2

Carpenter SR, Stanley EH, Vander Zanden MJ (2011) State of the world's freshwater ecosystems: physical, chemical, and biological changes. Ann Rev Environ Resour 36:75–99. https://doi.org/10.1146/annurev-environ-021810-094524

Carr GM, Chambers PA (1998) Macrophyte growth and sediment phosphorus and nitrogen in a Canadian prairie river. Freshw Biol 39:525–536. https://doi.org/10.1046/j.1365-2427.1998.00300.x

Carty S, Parrow MW (2015) Dinoflagellates. In: Wehr JD, Sheath RG, Kociolek JP (eds) Freshwater algae of North America, 2nd edn. Academic Press, Boston, pp 773–807

Cattaneo A, Kerimian T, Roberge M, Marty J (1997) Periphyton distribution and abundance on substrata of different size along a gradient of stream trophy. Hydrobiologia 354:101–110. https://doi.org/10.1023/A:1003027927600

Ceschin S, Aleffi M, Bisceglie S et al (2012) Aquatic bryophytes as ecological indicators of the water quality status in the Tiber River basin (Italy). Ecol Indic 14:74–81. https://doi.org/10.1016/j.ecolind.2011.08.020

Ceschin S, Minciardi MR, Spada CD, Abati S (2015) Bryophytes of Alpine and Apennine mountain streams: floristic features and ecological notes. Cryptogam Bryol 36:267–283. https://doi.org/10.7872/cryb/v36.iss3.2015.267

Champion PD, Tanner CC (2000) Seasonality of macrophytes and interaction with flow in a New Zealand lowland stream. Hydrobiologia 441:1–12. https://doi.org/10.1023/A:1017517303221

Chróst RJ (1990) Microbial ectoenzymes in aquatic environments. In: Overbeck J, Chróst RJ (eds) Aquatic microbial ecology: biochemical and molecular approaches. Springer, New York, pp 47–78

Cole JJ, Caraco NM, Peierls BL (1991) Phytoplankton primary production in the tidal, freshwater Hudson River, New York (USA). Internationale Vereinigung für Theoretische und Angewandte Limnologie: Verhandlungen 24:1715–1719

Cole JJ, Caraco NF, Peierls BL (1992) Can phytoplankton maintaina positive carbon balance in a turbid, freshwater, tidal estuary? Limnol Oceanogr 37:1608–1617. https://doi.org/10.4319/lo.1992.37.8.1608

Connelly S, Pringle CM, Barnum T et al (2014) Initial versus longer-term effects of tadpole declines on algae in a Neotropical stream. Freshw Biol 59:1113–1122. https://doi.org/10.1111/fwb.12326

Cotton JA, Wharton G, Bass JAB et al (2006) The effects of seasonal changes to in-stream vegetation cover on patterns of flow and accumulation of sediment. Geomorphology 77:320–334. https://doi.org/10.1016/j.geomorph.2006.01.010

Coundoul F, Bonometti T, Graba M et al (2015) Role of local flow conditions in river biofilm colonization and early growth. River Res Appl 31:350–367. https://doi.org/10.1002/rra.2746

Davies J-M, Bothwell ML (2012) Responses of lotic periphyton to pulses of phosphorus: P-flux controlled growth rate. Freshw Biol 57:2602–2612. https://doi.org/10.1111/fwb.12032

de Domitrovic YZ, Devercelli M, Forastier ME (2014) Phytoplankton of the Paraguay and Bermejo rivers. Adv Limnol 65:67–80. https://doi.org/10.1127/1612-166X/2014/0065-0034

De Ruyter Van Steveninck ED, Admiraal W, Breebaart L et al (1992) Plankton in the River Rhine: structural and functional changes observed during downstream transport. J Plankton Res 14:1351–1368. https://doi.org/10.1093/plankt/14.10.1351

de Tezanos PP, O'Farrell I (2014) Regime shifts between free-floating plants and phytoplankton: a review. Hydrobiologia 740:13–24. https://doi.org/10.1007/s10750-014-1943-0

Debenest T, Pinelli E, Coste M et al (2009) Sensitivity of freshwater periphytic diatoms to agricultural herbicides. Aquat Toxicol 93:11–17. https://doi.org/10.1016/J.AQUATOX.2009.02.014

Delgado C, Almeida SFP, Elias CL et al (2017) Response of biofilm growth to experimental warming in a temperate stream. Ecohydrology 10. https://doi.org/10.1002/eco.1868

DeNicola DM (2007) Periphyton responses to temperature at different ecological levels. In: Stevenson RJ, Bothwell ML, Lowe RL (eds) Algal ecology: freshwater benthic ecosystems. Academic Press, San Diego, pp 149–181

DeNicola DM, Lellock AJ (2015) Nutrient limitation of algal periphyton in streams along an acid mine drainage gradient. J Phycol 51:739–749. https://doi.org/10.1111/jpy.12315

DeNicola DM, McIntire CD (1990) Effects of substrate relief on the distribution of periphyton in laboratory streams. I. Hydrology. J Phycol 26:624–633. https://doi.org/10.1111/j.0022-3646.1990.00624.x

Descy J-P, Darchambeau F, Lambert T et al (2017) Phytoplankton dynamics in the Congo River. Freshw Biol 62:87–101. https://doi.org/10.1111/fwb.12851

Devercelli M, O'Farrell I (2013) Factors affecting the structure and maintenance of phytoplankton functional groups in a nutrient rich lowland river. Limnologica 43:67–78. https://doi.org/10.1016/J.LIMNO.2012.05.001

Dodds WK, Smith VH (2016) Nitrogen, phosphorus, and eutrophication in streams. Inl Waters 6:155–164. https://doi.org/10.5268/IW-6.2.909

Dodds WK, Welch EB (2000) Establishing nutrient criteria in streams. J North Am Benthol Soc 19:186–196. https://doi.org/10.2307/1468291

Dodds WK, Whiles MR (2010) Freshwater ecology: concepts and environmental applications of limnology, 2nd edn. Academic Press, San Diego

Dolph CL, Hansen AT, Finlay JC (2017) Flow-related dynamics in suspended algal biomass and its contribution to suspended particulate matter in an agricultural river network of the Minnesota River Basin, USA. Hydrobiologia 785:127–147. https://doi.org/10.1007/s10750-016-2911-7

Ensign SH, Doyle MW, Piehler MF (2012) Tidal geomorphology affects phytoplankton at the transition from forested streams to tidal rivers. Freshw Biol 57:2141–2155. https://doi.org/10.1111/j.1365-2427.2012.02856.x

Feminella JW, Hawkins CP (1995) Interactions between stream herbivores and periphyton: a quantitative analysis of past experiments. J North Am Benthol Soc 14:465–509. https://doi.org/10.2307/1467536

Fisher SG, Gray LJ, Grimm NB, Busch DE (1982) Temporal succession in a desert stream ecosystem following flash flooding. Ecol Monogr 52:93–110. https://doi.org/10.2307/2937346

Flecker AS, Taylor BW, Bernhardt ES et al (2002) Interactions between herbivorous fishes and limiting nitrients in a tropical stream ecosystem. Ecology 83:1831–1844

Francoeur SN (2001) Meta-analysis of lotic nutrient amendment experiments: detecting and quantifying subtle responses. J North Am Benthol Soc 20:358–368. https://doi.org/10.2307/1468034

Francoeur SN, Biggs BJF (2006) Short-term effects of elevated velocity and sediment abrasion on benthic algal communities. Hydrobiologia 561:59–69. https://doi.org/10.1007/s10750-005-1604-4

Francoeur SN, Biggs BJE, Smith RA, Lowe RL (1999) Nutrient limitation of algal biomass accrual in streams: seasonal patterns and a comparison of methods. J North Am Benthol Soc 18:242–260

Franklin P, Dunbar M, Whitehead P (2008) Flow controls on lowland river macrophytes: a review. Sci Total Environ 400:369–378. https://doi.org/10.1016/j.scitotenv.2008.06.018

Gesierich D, Rott E (2012) Is diatom richness responding to catchment glaciation? A case study from Canadian headwater streams. J Limnol 71:7. https://doi.org/10.4081/jlimnol.2012.e7

Gessner F (1955) Hydrobotanik I: energiehaushalt. Veb Deutsch Ver Wissensch, Berlin

Giblin SM, Houser JN, Sullivan JF et al (2014) Thresholds in the response of free-floating plant abundance to variation in hydraulic connectivity, nutrients, and macrophyte abundance in a large floodplain river. Wetlands 34:413–425. https://doi.org/10.1007/s13157-013-0508-8

Gillis CA, Lavoie I (2014) A preliminary assessment of the effects of Didymosphenia geminata nuisance growths on the structure and diversity of diatom assemblages of the Restigouche River basin, Quebec, Canada. Diatom Res 29:281–292. https://doi.org/10.1080/0269249X.2014.924437

Gjerlov C, Richardson JS (2010) Experimental increases and reductions of light to streams: effects on periphyton and macroinvertebrate assemblages in a coniferous forest landscape. Hydrobiologia 652:195–206. https://doi.org/10.1007/s10750-010-0331-7

Glime JM (2017a) Nutrient relations: CO_2. In: Glime J (ed) Bryophyte ecology. Volume 1. Physiological ecology. Ebook 2-1-1 sponsored by Michigan Technological University and the International Association of Bryologists

Glime JM (2017b) Meet the bryophytes. In: Glime JM (ed) Bryophyte ecology. Volume 1. Physiological ecology. Ebook 2-1-1 sponsored by Michigan Technological University and the International Association of Bryologists

Gomes LC, Bulla CK, Agostinho AA et al (2012) Fish assemblage dynamics in a Neotropical floodplain relative to aquatic macrophytes and the homogenizing effect of a flood pulse. Hydrobiologia 685:97–107. https://doi.org/10.1007/s10750-011-0870-6

Graba M, Sauvage S, Moulin FY et al (2013) Interaction between local hydrodynamics and algal community in epilithic biofilm. Water Res 47:2153–2163. https://doi.org/10.1016/j.watres.2013.01.011

Groendahl S, Kahlert M, Fink P (2017) The best of both worlds: a combined approach for analyzing microalgal diversity via metabarcoding and morphology-based methods. PLoS One 12:1–15. https://doi.org/10.1371/journal.pone.0172808

Gudmundsdottir R, Olafsson JS, Palsson S et al (2011) How will increased temperature and nutrient enrichment affect primary producers in sub-Arctic streams? Freshw Biol 56:2045–2058. https://doi.org/10.1111/j.1365-2427.2011.02636.x

Harguinteguy CA, Cofré MN, Fernández-Cirelli A, Pignata ML (2016) The macrophytes *Potamogeton pusillus* L. and *Myriophyllum aquaticum* (Vell.) Verdc. as potential bioindicators of a river contaminated by heavy metals. Microchem J 124:228–234. https://doi.org/10.1016/j.microc.2015.08.014

Hart DD, Biggs BJF, Nikora VI, Flinders CA (2013) Flow effects on periphyton patches and their ecological consequences in a New Zealand river. Freshw Biol 58:1588–1602. https://doi.org/10.1111/fwb.12147

Henry JC, Fisher SG (2003) Spatial segregation of periphyton communities in a desert stream: causes and consequences for N cycling. J North Am Benthol Soc 22:511–527. https://doi.org/10.2307/1468349

Hiatt DL, Back JA, King RS (2019) Effects of stream velocity and phosphorus concentrations on alkaline phosphatase activity and carbon:phosphorus ratios in periphyton. Hydrobiologia 826:173–182. https://doi.org/10.1007/s10750-018-3727-4

Hill WR (1996) Effects of light. In: Stevenson RJ, Bothwell ML, Lowe R (eds) Algal ecology: freshwater benthic ecosystems. Academic Press, San Diego, pp 121–148

Hill WR (2017) Light. In: Hauer FR, Lamberti GA (eds) Methods in stream ecology, vol 1: ecosystem structure, 3rd edn. Academic Press, Boston, pp 121–127

Hill WR, Dimick SM (2002) Effects of riparian leaf dynamics on periphyton photosynthesis and light utilisation efficiency. Freshw Biol 47:1245–1256. https://doi.org/10.1046/j.1365-2427.2002.00837.x

Hill BH, Webster JR (1983) Aquatic macrophyte contribution to the New River organic matter budget. In: Fontaine T, Bartell S (eds) Dynamics of lotic ecosystems. Ann Arbor Science, Ann Arbor, p 273–282

Hill WR, Ryon MG, Schilling EM (1995) Light limitation in a stream ecosystem: responses by primary producers and consumers. Ecology 76:1297–1309. https://doi.org/10.2307/1940936

Hill WR, Mulholland PJ, Marzolf ER (2001) Stream ecosystem responses to forest leaf emergence in spring. Ecology 82:2306–2319. https://doi.org/10.1890/0012-9658(2001)082%5b2306:sertfl%5d2.0.co;2

Hill WR, Fanta SE, Roberts BJ (2009) Quantifying phosphorus and light effects in stream algae. Limnol Oceanogr 54:368–380. https://doi.org/10.4319/lo.2009.54.1.0368

Hillebrand H (2002) Top-down versus bottom-up control of autotrophic biomass—a meta-analysis on experiments with periphyton. J North Am Benthol Soc 21:349–369. https://doi.org/10.2307/1468475

Hillebrand H (2005) Light regime and consumer control of autotrophic biomass. J Ecol 93:758–769. https://doi.org/10.1111/j.1365-2745.2005.00978.x

Hilton J, O'Hare M, Bowes MJ, Jones JI (2006) How green is my river? A new paradigm of eutrophication in rivers. Sci Total Environ 365:66–83. https://doi.org/10.1016/j.scitotenv.2006.02.055

Hirst H, Chaud F, Delabie C et al (2004) Assessing the short-term response of stream diatoms to acidity using inter-basin transplantations and chemical diffusing substrates. Freshw Biol 49:1072–1088. https://doi.org/10.1111/j.1365-2427.2004.01242.x

Hogsden KL, Harding JS (2012) Consequences of acid mine drainage for the structure and function of benthic stream communities: a review. Freshw Sci 31:108–120. https://doi.org/10.1899/11-091.1

Holomuzki JR, Feminella JW, Power ME (2010) Biotic interactions in freshwater benthic habitats. J North Am Benthol Soc 29:220–244. https://doi.org/10.1899/08-044.1

Hondzo M, Wang H (2002) Effects of turbulence on growth and metabolism of periphyton in a laboratory flume. Water Resour Res 38:1277. https://doi.org/10.1029/2002wr001409

Honeyfield DC, Maloney KO (2015) Seasonal patterns in stream periphyton fatty acids and community benthic algal composition in six high-quality headwater streams. Hydrobiologia 744:35–47. https://doi.org/10.1007/s10750-014-2054-7

Hutchins MG, Johnson AC, Deflandre-Vlandas A et al (2010) Which offers more scope to suppress river phytoplankton blooms: reducing nutrient pollution or riparian shading? Sci Total Environ 408:5065–5077. https://doi.org/10.1016/j.scitotenv.2010.07.033

Hynes HBN (1970) The ecology of running waters. University of Toronto Press, Toronto

Istvánovics V, Honti M (2011) Phytoplankton growth in three rivers: the role of meroplankton and the benthic retention hypothesis. Limnol Oceanogr 56:1439–1452. https://doi.org/10.4319/lo.2011.56.4.1439

Istvánovics V, Honti M, Kovács Á et al (2014) Phytoplankton growth in relation to network topology: time-averaged catchment-scale modelling in a large lowland river. Freshw Biol 59:1856–1871. https://doi.org/10.1111/fwb.12388

John DM, Rindi F (2015) Filamentous (nonconjugating) and plantlike green algae. In: Wehr JD, Sheath RG, Kociolek JP (eds) Freshwater algae of North America, 2nd edn. Academic Press, Boston, pp 375–427

Johnson LT, Tank JL, Dodds WK (2009) The influence of land use on stream biofilm nutrient limitation across eight North American ecoregions. Can J Fish Aquat Sci 66:1081–1094. https://doi.org/10.1139/F09-065

Jyrkänkallio-Mikkola J, Siljander M, Heikinheimo V et al (2018) Tropical stream diatom communities—the importance of headwater streams for regional diversity. Ecol Indic 95:183–193. https://doi.org/10.1016/j.ecolind.2018.07.030

Kahlert M (1998) C:N: P ratios of freshwater benthic algae. Arch Hydrobiol Spec Issues Adv Limnol 51:105–114

Katz SB, Segura C, Warren DR (2018) The influence of channel bed disturbance on benthic chlorophyll a: a high resolution perspective. Geomorphology 305:141–153. https://doi.org/10.1016/j.geomorph.2017.11.010

Kaylor MJ, Warren DR, Kiffney PM (2017) Long-term effects of riparian forest harvest on light in Pacific Northwest (USA) streams. Freshw Sci 36:1–13. https://doi.org/10.1086/690624

Keck F, Lepori F (2012) Can we predict nutrient limitation in streams and rivers? Freshw Biol 57:1410–1421. https://doi.org/10.1111/j.1365-2427.2012.02802.x

Keithan ED, Lowe RL (1985) Primary productivity and spatial structure of phytolithic growth in streams in the Great Smoky Mountains National Park, Tennessee. Hydrobiologia 123:59–67. https://doi.org/10.1007/BF00006615

Kilroy C, Bothwell M (2011) Environmental control of stalk length in the bloom-forming, freshwater benthic diatom *Didymosphenia geminata* (Bacillariophyceae). J Phycol 47:981–989. https://doi.org/10.1111/j.1529-8817.2011.01029.x

Kilroy C, Novis P (2018) Is *Didymosphenia geminata* an introduced species in New Zealand? Evidence from trends in water chemistry, and chloroplast DNA. Ecol Evol 8:904–919. https://doi.org/10.1002/ece3.3572

Klausmeier CA, Litchman E, Daufresne T, Levin SA (2004) Optimal nitrogen-to-phosphorus stoichiometry of phytoplankton. Nature 429:171–174. https://doi.org/10.1038/nature02454

Klose K, Cooper SD, Leydecker AD, Kreitler J (2012) Relationships among catchment land use and concentrations of nutrients, algae, and dissolved oxygen in a southern California river. Freshw Sci 31:908–927. https://doi.org/10.1899/11-155.1

Koch RW, Guelda DL, Bukaveckas PA (2004) Phytoplankton growth in the Ohio, Cumberland and Tennessee Rivers, USA: inter-site differences in light and nutrient limitation. Aquat Ecol 38:17–26. https://doi.org/10.1023/B:AECO.0000021082.42784.03

Kociolek JP, Spaulding SA, Lowe RL (2015a) Bacillariophyceae: the raphid diatoms. In: Wehr JD, Sheath RG, Kociolek JP (eds) Freshwater algae of North America, 2nd edn. Academic Press, Boston, pp 709–772

Kociolek JP, Theriot EC, Williams DM et al (2015b) Centric and araphid diatoms. In: Wehr JD, Sheath RG, Kociolek JP (eds) Freshwater algae of North America, 2nd edn. Academic Press, Boston, pp 653–708

Köhler J, Hachoł J, Hilt S (2010) Regulation of submersed macrophyte biomass in a temperate lowland river: interactions between shading by bank vegetation, epiphyton and water turbidity. Aquat Bot 92:129–136. https://doi.org/10.1016/j.aquabot.2009.10.018

Komárek J, Johansen JR (2015) Coccoid Cyanobacteria. In: Wehr JD, Sheath RG, Kociolek JP (eds) Freshwater algae of North America, 2nd edn. Academic Press, Boston, pp 75–133

Kowalik RA, Cooper DM, Evans CD, Ormerod SJ (2007) Acidic episodes retard the biological recovery of upland British streams from chronic acidification. Glob Chang Biol 13:2439–2452. https://doi.org/10.1111/j.1365-2486.2007.01437.x

Lacoul P, Freedman B (2006) Environmental influences on aquatic plants in freshwater ecosystems. Environ Rev 14:89–136. https://doi.org/10.1139/a06-001

Ladrera R, Gomà J, Prat N (2018) Effects of *Didymosphenia geminata* massive growth on stream communities: smaller organisms and simplified food web structure. PLoS ONE 13:e0193545. https://doi.org/10.1371/journal.pone.0193545

Lang DA, King RS, Scott JT (2012) Divergent responses of biomass and enzyme activities suggest differential nutrient limitation in stream periphyton. Freshw Sci 31:1096–1104. https://doi.org/10.1899/12-031.1

Lange K, Townsend CR, Matthaei CD (2016) A trait-based framework for stream algal communities. Ecol Evol 6:23–36. https://doi.org/10.1002/ece3.1822

Larned ST (2010) A prospectus for periphyton: recent and future ecological research. J North Am Benthol Soc 29:182–206. https://doi.org/10.1899/08-063.1

Larned ST, Packman AI, Plew DR, Vopel K (2011) Interactions between the mat-forming alga *Didymosphenia geminata* and its hydrodynamic environment. Limnol Oceanogr Fluids Environ 1:4–22. https://doi.org/10.1215/21573698-1152081

Larroudé S, Massei N, Reyes-Marchant P et al (2013) Dramatic changes in a phytoplankton community in response to local and global pressures: a 24-year survey of the river Loire (France). Glob Chang Biol 19:1620–1631. https://doi.org/10.1111/gcb.12139

Larson CA, Liu H, Passy SI (2015) Iron supply constrains producer communities in stream ecosystems. FEMS Microbiol Ecol 91:fiv041. https://doi.org/10.1093/femsec/fiv041

Le Moal M, Gascuel-Odoux C, Ménesguen A et al (2019) Eutrophication: a new wine in an old bottle? Sci Total Environ 651:1–11. https://doi.org/10.1016/j.scitotenv.2018.09.139

Lee RE (2008) Phycology, 4th edn. Cambridge University Press, Cambridge

Leliaert F, Smith DR, Moreau H et al (2012) Phylogeny and molecular evolution of the green algae. CRC Crit Rev Plant Sci 31:1–46. https://doi.org/10.1080/07352689.2011.615705

Lewis WM (1988) Primary production in the Orinoco River. Ecology 69:679–692. https://doi.org/10.2307/1941016

Lowe RL, LaLiberte GD (2017) Benthic stream algae: distribution and structure. In: Hauer, FR and Lamberti G (ed) Methods in stream ecology, vol 1: ecosystem structure, 3rd edn. Academic Press, Boston, pp 193–221

Lowe RL, Golladay SW, Webster JR (1986) Periphyton response to nutrient manipulation in streams draining clearcut and forested watersheds. J North Am Benthol Soc 5:221–229. https://doi.org/10.2307/1467709

Luce JJ, Lapointe MF, Roy AG, Ketterling DB (2013) The effects of sand abrasion of a predominantly stable stream bed on periphyton biomass losses. Ecohydrology 6:689–699. https://doi.org/10.1002/eco.1332

Luis L, Bergamini A, Sim-Sim M (2015) Which environmental factors best explain variation of species richness and composition of stream bryophytes? A case study from mountainous streams in Madeira Island. Aquat Bot 123:37–46. https://doi.org/10.1016/j.aquabot.2015.01.010

Maberly SC, Berthelot SA, Stott AW, Gontero B (2015) Adaptation by macrophytes to inorganic carbon down a river with naturally variable concentrations of CO_2. J Plant Physiol 172:120–127. https://doi.org/10.1016/J.JPLPH.2014.07.025

MacDougall SE, Carrick HJ, DeWalle DR (2008) Benthic algae in episodically acidified Pennsylvania streams. Northeast Nat 15:189–208. https://doi.org/10.1656/1092-6194(2008)15%5b189:baieap%5d2.0.co;2

Mackay SJ, Arthington AH, Kennard MJ, Pusey BJ (2003) Spatial variation in the distribution and abundance of submersed macrophytes in an Australian subtropical river. Aquat Bot 77:169–186. https://doi.org/10.1016/S0304-3770(03)00103-7

Madden N, Lewis A, Davis M (2013) Thermal effluent from the power sector: an analysis of once-through cooling system impacts on surface water temperature. Environ Res Lett 8. https://doi.org/10.1088/1748-9326/8/3/035006

Madsen JD, Chambers PA, James WF et al (2001) The interaction between water movement, sediment dynamics and submersed macrophytes. Hydrobiologia 444:71–84. https://doi.org/10.1023/A:1017520800568

Maltchik L, Pedro F (2001) Responses of aquatic macrophytes to disturbance by flash floods in a Brazilian semiarid intermittent stream. Biotropica 33:566. https://doi.org/10.1646/0006-3606(2001)033%5b0566:roamtd%5d2.0.co;2

Manolaki P, Papastergiadou E (2013) The impact of environmental factors on the distribution pattern of aquatic macrophytes in a middle-sized Mediterranean stream. Aquat Bot 104:34–46. https://doi.org/10.1016/j.aquabot.2012.09.009

Marshall MC, Binderup AJ, Zandonà E et al (2012) Effects of consumer interactions on benthic resources and ecosystem processes in a Neotropical stream. PLoS One 7:e45230. https://doi.org/10.1371/journal.pone.0045230

McCall SJ, Hale MS, Smith JT et al (2017) Impacts of phosphorus concentration and light intensity on river periphyton biomass and community structure. Hydrobiologia 792:315–330. https://doi.org/10.1007/s10750-016-3067-1

Mebane CA, Simon NS, Maret TR (2014) Linking nutrient enrichment and streamflow to macrophytes in agricultural streams. Hydrobiologia 722:143–158. https://doi.org/10.1007/s10750-013-1693-4

Miller MP, McKnight DM, Cullis JD et al (2009) Factors controlling streambed coverage of *Didymosphenia geminata* in two regulated streams in the Colorado Front Range. Hydrobiologia 630:207–218. https://doi.org/10.1007/s10750-009-9793-x

Mohamed ZA, Deyab MA, Abou-Dobara MI et al (2015) Occurrence of cyanobacteria and microcystin toxins in raw and treated waters of the Nile River, Egypt: implication for water treatment and human health. Environ Sci Pollut Res 22:11716–11727. https://doi.org/10.1007/s11356-015-4420-z

Monteiro J, Vieira C (2017) Determinants of stream bryophyte community structure: bringing ecology into conservation. Freshw Biol 62:695–710. https://doi.org/10.1111/fwb.12895

Morandeira NS, Kandus P (2015) Multi-scale analysis of environmental constraints on macrophyte distribution, floristic groups and plant diversity in the Lower Paraná River floodplain. Aquat Bot 123:1–25. https://doi.org/10.1016/j.aquabot.2015.01.006

Morin A, Lamoureux W, Busnarda J (1999) Empirical models predicting primary productivity from chlorophyll a and water temperature for stream periphyton and lake and ocean phytoplankton. J North Am Benthol Soc 18:299–307. https://doi.org/10.2307/1468446

Mosisch TD, Bunn SE, Davies PM, Marshall CJ (1999) Effects of shade and nutrient manipulation on periphyton growth in a subtropical stream. Aquat Bot 64:167–177. https://doi.org/10.1016/S0304-3770(99)00014-5

Mosisch TD, Bunn SE, Davies PM (2001) The relative importance of shading and nutrients on algal production in subtropical streams. Freshw Biol 46:1269–1278. https://doi.org/10.1046/j.1365-2427.2001.00747.x

Moulton S, Kennen J, Goldstein R, Hambrook J (2002) Revised protocols for sampling algal, invertebrate, and fish communities as part of the National Water-Quality Assessment Program. USGS Open-File Rep 02:36–54

Moulton TP, Lourenço-Amorim C, Sasada-Sato CY et al (2015) Dynamics of algal production and ephemeropteran grazing of periphyton in a tropical stream. Int Rev Hydrobiol 100:61–68. https://doi.org/10.1002/iroh.201401769

Neiff JJ, Casco SL, de Neiff AP (2008) Response of *Eichhornia crassipes* (Pontederiaceae) to water level fluctuations in two lakes with different connectivity in the Parana River floodplain. Rev Biol Trop 56:613–623. https://doi.org/10.15517/rbt.v56i2.5612

Nicholls KH, Wujek DE (2015) Chrysophyceae and Phaeothamniophyceae. In: Wehr JD, Sheath RG, Kociolek JP (eds) Freshwater algae of North America, 2nd edn. Academic Press, Boston, pp 537–586

O'Hare MT, Stillman RA, Mcdonnell J, Wood LR (2007) Effects of mute swan grazing on a keystone macrophyte. Freshw Biol 52:2463–2475. https://doi.org/10.1111/j.1365-2427.2007.01841.x

O'Hare MT, Clarke RT, Bowes MJ et al (2010) Eutrophication impacts on a river macrophyte. Aquat Bot 92:173–178. https://doi.org/10.1016/j.aquabot.2009.11.001

Ochs CA, Pongruktham O, Zimba PV (2013) Darkness at the break of noon: Phytoplankton production in the Lower Mississippi River. Limnol Oceanogr 58:555–568. https://doi.org/10.4319/lo.2013.58.2.0555

Pace ML, Findlay SEG, Lints D (1992) Zooplankton in advective environments: the Hudson River community and a comparative analysis. Can J Fish Aquat Sci 49:1060–1069. https://doi.org/10.1139/f92-117

Padisák J, Crossetti LO, Naselli-Flores L (2009) Use and misuse in the application of the phytoplankton functional classification: a critical review with updates. Hydrobiologia 621:1–19. https://doi.org/10.1007/s10750-008-9645-0

Paice RL, Chambers JM, Robson BJ (2017) Native submerged macrophyte distribution in seasonally-flowing, south-western Australian streams in relation to stream condition. Aquat Sci 79:171–185. https://doi.org/10.1007/s00027-016-0488-x

Pan Y, Deng G, Wang L et al (2017) Effects of in situ phosphorus enrichment on the benthos in a subalpine karst stream and implications for bioassessment in nature reserves. Ecol Indic 73:274–283. https://doi.org/10.1016/j.ecolind.2016.09.055

Passauer B, Meilinger P, Melzer A, Schneider S (2002) Does the structural quality of running waters affects the occurrence of macrophytes? Acta Hydrochim Hydrobiol 30:197–206. https://doi.org/10.1002/aheh.200390003

Passy SI (2006) Diatom community dynamics in streams of chronic and episodic acidification: The roles of environment and time. J Phycol 42:312–323. https://doi.org/10.1111/j.1529-8817.2006.00202.x

Passy SI (2007) Diatom ecological guilds display distinct and predictable behavior along nutrient and disturbance gradients in running waters. Aquat Bot 86:171–178. https://doi.org/10.1016/j.aquabot.2006.09.018

Peterson CG, Grimm NB (1992) Temporal variation in enrichment effects during periphyton succession in a nitrogen-limited desert stream ecosystem. J North Am Benthol Soc 11:20–36. https://doi.org/10.2307/1467879

Petsch DK, Schneck F, Melo AS (2017) Substratum simplification reduces beta diversity of stream algal communities. Freshw Biol 62:205–213. https://doi.org/10.1111/fwb.12863

Phlips EJ, Cichra M, Aldridge FJ et al (2000) Light availability and variations in phytoplankton standing crops in a nutrient-rich blackwater river. Limnol Oceanogr 45:916–929. https://doi.org/10.4319/lo.2000.45.4.0916

Piggott JJ, Lange K, Townsend CR, Matthaei CD (2012) Multiple stressors in agricultural streams: a mesocosm study of interactions among raised water temperature, sediment addition and nutrient enrichment. PLoS One 7:e49873. https://doi.org/10.1371/journal.pone.0049873

Piggott JJ, Salis RK, Lear G et al (2015) Climate warming and agricultural stressors interact to determine stream periphyton community composition. Glob Chang Biol 21:206–222. https://doi.org/10.1111/gcb.12661

Pigneur L-M, Falisse E, Roland K et al (2014) Impact of invasive Asian clams, *Corbicula* spp., on a large river ecosystem. Freshw Biol 59:573–583. https://doi.org/10.1111/fwb.12286

Poff NL, Voelz NJ, Ward JV, Lee RE (1990) Algal colonization under four experimentally-controlled current regimes in high mountain stream. J North Am Benthol Soc 9:303–318. https://doi.org/10.2307/1467898

Potapova MG, Charles DF (2002) Benthic diatoms in USA rivers: distributions along spatial and environmental gradients. J Biogeogr 29:167–187

Power ME, Stewart AJ (1987) Disturbance and recovery of an algal assemblage following flooding in an Oklahoma stream. Am Midl Nat 117:333–345. https://doi.org/10.2307/2425975

Reid B, Torres R (2014) *Didymosphenia geminata* invasion in South America: ecosystem impacts and potential biogeochemical state change in Patagonian rivers. Acta Oecol 54:101–109. https://doi.org/10.1016/j.actao.2013.05.003

Reisinger AJ, Tank JL, Dee MM (2016) Regional and seasonal variation in nutrient limitation of river biofilms. Freshw Sci 35:474–489. https://doi.org/10.1086/685829

Reynolds CS (2000) Hydroecology of river plankton: the role of variability in channel flow. Hydrol Process 14:3119–3132. https://doi.org/10.1002/1099-1085(200011/12)14:16/17%3c3119:AID-HYP137%3e3.0.CO;2-6

Reynolds CS, Descy J-P (1996) The production, biomass and structure of phytoplankton in large rivers. Arch Hydrobiol Suppl 113:161–187. https://doi.org/10.1127/lr/10/1996/161

Reynolds CS, Huszar V, Kruk C et al (2002) Towards a functional classification of the freshwater phytoplankton. J Plankton Res 24:417–428. https://doi.org/10.1093/plankt/24.5.417

Rier ST, Shirvinski JM, Kinek KC (2014) In situ light and phosphorus manipulations reveal potential role of biofilm algae in enhancing enzyme-mediated decomposition of organic matter in streams. Freshw Biol 59:1039–1051. https://doi.org/10.1111/fwb.12327

Riis T, Biggs BJF (2001) Distribution of macrophytes in New Zealand streams and lakes in relation to disturbance frequency and resource supply—a synthesis and conceptual model. New Zeal J Mar Freshw Res 35:255–267. https://doi.org/10.1080/00288330.2001.9516996

Riis T, Biggs BJF (2003) Hydrologic and hydraulic control of macrophyte establishment and performance in streams. Limnol Oceanogr 48:1488–1497. https://doi.org/10.4319/lo.2003.48.4.1488

Riis T, Sand-Jensen K, Larsen SE (2001) Plant distribution and abundance in relation to physical conditions and location within Danish stream systems. Hydrobiologia 448:217–228. https://doi.org/10.1023/A:1017580424029

Rimet F, Bouchez A (2011) Use of diatom life-forms and ecological guilds to assess pesticide contamination in rivers: Lotic mesocosm approaches. Ecol Indic 11:489–499. https://doi.org/10.1016/j.ecolind.2010.07.004

Riseng CM, Wiley MJ, Stevenson RJ (2004) Hydrologic disturbance and nutrient effects on benthic community structure in midwestern US streams: a covariance structure analysis. J North Am Benthol Soc 23:309–326. https://doi.org/10.1899/0887-3593(2004)023%3c0309:HDANEO%3e2.0.CO;2

Risse-Buhl U, Anlanger C, Kalla K et al (2017) The role of hydrodynamics in shaping the composition and architecture of epilithic biofilms in fluvial ecosystems. Water Res 127:211–222. https://doi.org/10.1016/j.watres.2017.09.054

Robinson CT, Rushforth SR (1987) Effects of physical disturbance and canopy cover on attached diatom community structure in an Idaho stream. Hydrobiologia 154:49–59. https://doi.org/10.1007/BF00026830

Rodgers JJ, McKevitt M, Hammerlund D et al (1983) Primary production and decomposition of submergent and emergent aquatic plants of two Appalachian rivers. In: Fontaine T, Bartell S (eds) Dynamics of lotic ecosystems. Ann Arbor, pp 283–301

Roig N, Sierra J, Moreno-Garrido I et al (2016) Metal bioavailability in freshwater sediment samples and their influence on ecological status of river basins. Sci Total Environ 540:287–296. https://doi.org/10.1016/j.scitotenv.2015.06.107

Rojo C, Cobelas MA, Arauzo M (1994) An elementary, structural analysis of river phytoplankton. Hydrobiologia 289:43–55. https://doi.org/10.1007/BF00007407

Rosemond AD, Mulholland PJ, Elwood JW (1993) Top-down and bottom-up control of stream periphyton: effects of nutrients and herbivores. Ecology 74:1264–1280. https://doi.org/10.2307/1940495

Rosemond AD, Mulholland PJ, Brawley SH (2000) Seasonally shifting limitation of stream periphyton: response of algal populations and assemblage biomass and productivity to variation in light, nutrients,

and herbivores. Can J Fish Aquat Sci 57:66–75. https://doi.org/10.1139/f99-181

Rott E, Cantonati M, Füreder L, Pfister P (2006) Benthic algae in high altitude streams of the Alps—a neglected component of the aquatic biota. Hydrobiologia 562:195–216. https://doi.org/10.1007/s10750-005-1811-z

Sabater S, Buchaca T, Cambra J et al (2003) Structure and function of benthic algal communities in an extremely acid river. J Phycol 39:481–489. https://doi.org/10.1046/j.1529-8817.2003.02104.x

Salmaso N, Zignin A (2010) At the extreme of physical gradients: phytoplankton in highly flushed, large rivers. Hydrobiologia 639:21–36. https://doi.org/10.1007/s10750-009-0018-0

Sanderson BL, Coe HJ, Tran CD et al (2009) Nutrient limitation of periphyton in Idaho streams: results from nutrient diffusing substrate experiments. J North Am Benthol Soc 28:832–845. https://doi.org/10.1899/09-072.1

Sand-Jensen K (1998) Influence of submerged macrophytes on sediment composition and near-bed flow in lowland streams. Freshw Biol 39:663–679. https://doi.org/10.1046/j.1365-2427.1998.00316.x

Sand-Jensen K, Pedersen MF, Nielsen SL (1992) Photosynthetic use of inorganic carbon among primary and secondary water plants in streams. Freshw Biol 27:283–293. https://doi.org/10.1111/j.1365-2427.1992.tb00540.x

Sand-Jensen K, Andersen K, Andersen T (1999) Dynamic properties of recruitment, expansion and mortality of macrophyte patches in streams. Int Rev Hydrobiol 84:497–508. https://doi.org/10.1002/iroh.199900044

Sanmiguel A, Blanco S, Álvarez-Blanco I et al (2016) Recovery of the algae and macroinvertebrate benthic community after *Didymosphenia geminata* mass growths in Spanish rivers. Biol Invasions 18:1467–1484. https://doi.org/10.1007/s10530-016-1095-4

Saunders JF, Lewis WM (1988) Zooplankton abundance and transport in a tropical white-water river. Hydrobiologia 162:147–155. https://doi.org/10.1007/BF00014537

Schneck F, Schwarzbold A, Melo AS (2013) Substrate roughness, fish grazers, and mesohabitat type interact to determine algal biomass and sediment accrual in a high-altitude subtropical stream. Hydrobiologia 711:165–173. https://doi.org/10.1007/s10750-013-1477-x

Schneider S, Lindstrøm EA (2009) Bioindication in Norwegian rivers using non-diatomaceous benthic algae: the acidification index periphyton (AIP). Ecol Indic 9:1206–1211. https://doi.org/10.1016/j.ecolind.2009.02.008

Schneider SC, Lindstrøm EA (2011) The periphyton index of trophic status PIT: a new eutrophication metric based on non-diatomaceous benthic algae in nordic rivers. Hydrobiologia 665:143–155. https://doi.org/10.1007/s10750-011-0614-7

Schneider B, Cunha ER, Marchese M, Thomaz SM (2018a) Associations between macrophyte life forms and environmental and morphometric factors in a large sub-tropical floodplain. Front Plant Sci 9:195. https://doi.org/10.3389/fpls.2018.00195

Schneider SC, Oulehle F, Krám P, Hruška J (2018b) Recovery of benthic algal assemblages from acidification: how long does it take, and is there a link to eutrophication? Hydrobiologia 805:33–47. https://doi.org/10.1007/s10750-017-3254-8

Scott JT, Lang DA, King RS, Doyle RD (2009) Nitrogen fixation and phosphatase activity in periphyton growing on nutrient diffusing substrata: evidence for differential nutrient limitation in stream periphyton. J North Am Benthol Soc 28:57–68. https://doi.org/10.1899/07-107.1

Sheath RG, Cole KM (1992) Biogeography of stream macroalgae in North America. J Phycol 28:448–460. https://doi.org/10.1111/j.0022-3646.1992.00448.x

Sheath RG, Vis ML (2015) Red algae. In: Wehr JD, Sheath RG, Kociolek JP (eds) Freshwater algae of North America, 2nd edn. Academic Press, Boston, pp 237–264

Sheath RG, Wehr JD (2015) Introduction to the freshwater algae. In: Wehr JD, Sheath RG, Kociolek JP (eds) Freshwater algae of North America, 2nd edn. Academic Press, Boston, pp 1–11

Sherwood AR, Carlile AL, Neumann JM et al (2014) The Hawaiian freshwater algae biodiversity survey (2009–2014): systematic and biogeographic trends with an emphasis on the macroalgae. BMC Ecol 14:1–23. https://doi.org/10.1186/s12898-014-0028-2

Silveira MJ, Alves DC, Thomaz SM (2018) Effects of the density of the invasive macrophyte *Hydrilla verticillata* and root competition on growth of one native macrophyte in different sediment fertilities. Ecol Res 33:927–934. https://doi.org/10.1007/s11284-018-1602-4

Simmons JA (2010) Phosphorus removal by sediment in streams contaminated with acid mine drainage. Water Air Soil Pollut 209:123–132. https://doi.org/10.1007/s11270-009-0185-7

Sinsabaugh RL, Follstad Shah JJ (2012) Ecoenzymatic stoichiometry and ecological theory. Ann Rev Ecol Evol Syst 43:313–343. https://doi.org/10.1146/annurev-ecolsys-071112-124414

Slavik K, Peterson BJ, Deegan LA et al (2004) Long-term responses of the Kuparuk river ecosystem to phosphorus fertilization. Ecology 85:939–954. https://doi.org/10.1890/02-4039

Sousa WTZ (2011) *Hydrilla verticillata* (Hydrocharitaceae), a recent invader threatening Brazil's freshwater environments: a review of the extent of the problem. Hydrobiologia 669:1–20

Sousa WTZ, Thomaz SM, Murphy KJ (2010) Response of native *Egeria najas Planch.* and invasive *Hydrilla verticillata* (L.f.) Royle to altered hydroecological regime in a subtropical river. Aquat Bot 92:40–48. https://doi.org/10.1016/j.aquabot.2009.10.002

Stanković I, Vlahović T, Gligora Udovič M et al (2012) Phytoplankton functional and morpho-functional approach in large floodplain rivers. Hydrobiologia 698:217–231. https://doi.org/10.1007/s10750-012-1148-3

Steinman AD (1992) Does an increase in irradiance influence periphyton in a heavily-grazed woodland stream? Oecologia 91:163–170. https://doi.org/10.1007/BF00317779

Steinman AD (1996) Effects of grazers on freshwater benthic algae. In: Stevenson RJ, Bothwell ML, Lowe RL (eds) Algal ecology: freshwater benthic ecosystems. Academic Press, San Diego, pp 341–373

Steinman AD, Duhamel S (2017) Phosphorus limitation, uptake, and turnover in benthic stream algae. In: Lamberti GA, Hauer FR (eds) Methods in stream ecology, vol 2: ecosystem function, 3rd edn. Academic Press, Boston, pp 197–218

Steinman AD, Lamberti GA, Leavitt PR, Uzarski DG (2017) Biomass and pigments of benthic algae. In: Hauer FR, Lamberti G (ed) Methods in stream ecology, vol 1: ecosystem structure, 3rd edn. Academic Press, Boston, pp 223–241

Stelzer RS, Lamberti GA (2002) Ecological stoichiometry in running waters: Periphyton chemical composition and snail growth. Ecology 83:1039–1051. https://doi.org/10.1890/0012-9658(2002)083%5b1039:esirwp%5d2.0.co;2

Stenger-Kovács C, Tóth L, Tóth F et al (2014) Stream order-dependent diversity metrics of epilithic diatom assemblages. Hydrobiologia 721:67–75. https://doi.org/10.1007/s10750-013-1649-8

Stevenson RJ (1983) Effects of current and conditions simulating autogenically changing microhabitats on benthic diatom immigration. Ecology 64:1514–1524. https://doi.org/10.2307/1937506

Stevenson RJ (1996a) An introduction to algal ecology in freshwater benthic ecosystems. In: Stevenson J, Bothwell ML, Lowe RL (eds) Algal ecology: freshwater benthic ecosystems. Academic Press, San Diego, pp 3–30

Stevenson RJ (1996b) The stimulation and drag of current. In: Stevenson RJ, Bothwell ML, Lowe RL (eds) Algal ecology: freshwater benthic ecosystems. Academic Press, San Diego, pp 321–340

Stevenson RJ, Peterson CG (1991) Emigration and immigration can be important determinants of benthic diatom assemblages in streams. Freshw Biol 26:279–294. https://doi.org/10.1111/j.1365-2427.1991.tb01735.x

Stevenson RJ, Rollins SL (2017) Ecological assessment with benthic algae. In: Hauer FR, Lamberti GA (eds) Methods in stream ecology, vol 2: ecosystem function, 3rd edn. Academic Press, Boston, pp 277–292

Stevenson RJ, Smol JP (2015) Use of algae in ecological assessments. In: Wehr JD, Sheath RG, Kociolek JP (eds) Freshwater algae of North America, 2nd edn. Academic Press, Boston, pp 921–962

Stevenson RJ, Pan Y, Manoylov KM et al (2008) Development of diatom indicators of ecological conditions for streams of the western US. J North Am Benthol Soc 27:1000–1016. https://doi.org/10.1899/08-040.1

Stockner JG, Shortreed KRS (1978) Enhancement of autotrophic production by nutrient addition in a coastal rainforest stream on Vancouver Island. Can J Fish Aquat Sci 35:28–34. https://doi.org/10.1139/f78-004

Strayer DL, Caraco NF, Cole JJ et al (1999) Transformation of freshwater ecosystems by bivalves: a case study of zebra mussels in the Hudson River. Bioscience 49:19–27. https://doi.org/10.1525/bisi.1999.49.1.19

Sullivan BE, Prahl FG, Small LF, Covert PA (2001) Seasonality of phytoplankton production in the Columbia River: a natural or anthropogenic pattern? Geochim Cosmochim Acta 65:1125–1139. https://doi.org/10.1016/S0016-7037(00)00565-2

Sundareshwar PV, Upadhayay S, Abessa M et al (2011) *Didymosphenia geminata*: algal blooms in oligotrophic streams and rivers. Geophys Res Lett 38:10405. https://doi.org/10.1029/2010GL046599

Suren AM, Riis T (2010) The effects of plant growth on stream invertebrate communities during low flow: a conceptual model. J North Am Benthol Soc 29:711–724. https://doi.org/10.1899/08-127.1

Talling JF, Rzóska J (1967) The development of plankton in relation to hydrological regime in the Blue Nile. J Ecol 55:637–662

Tank JL, Dodds WK (2003) Nutrient limitation of epilithic and epixylic biofilms in ten North American streams. Freshw Biol 48:1031–1049. https://doi.org/10.1046/j.1365-2427.2003.01067.x

Tank JL, Reisinger AJ, Rosi EJ (2017) Nutrient limitation and uptake. In: Lamberti GA, Hauer FR (eds) Methods in stream ecology, vol 2: ecosystem function, 3rd edn. Academic Press, pp 147–171

Tapolczai K, Bouchez A, Stenger-Kovács C et al (2016) Trait-based ecological classifications for benthic algae: review and perspectives. Hydrobiologia 776:1–17. https://doi.org/10.1007/s10750-016-2736-4

Tavernini S, Pierobon E, Viaroli P (2011) Physical factors and dissolved reactive silica affect phytoplankton community structure and dynamics in a lowland eutrophic river (Po river, Italy). Hydrobiologia 669:213–225. https://doi.org/10.1007/s10750-011-0688-2

Taylor BW, Bothwell ML (2014) The origin of invasive microorganisms matters for science, policy, and management: the case of *Didymosphenia geminata*. Bioscience 64:531–538. https://doi.org/10.1093/biosci/biu060

Thomaz SM, Dibble ED, Evangelista LR et al (2008) Influence of aquatic macrophyte habitat complexity on invertebrate abundance and richness in tropical lagoons. Freshw Biol 53:358–367. https://doi.org/10.1111/j.1365-2427.2007.01898.x

Tornes E, Perez MC, Duran C, Sabater S (2014) Reservoirs override seasonal variability of phytoplankton communities in a regulated Mediterranean river. Sci Total Environ 475:225–233. https://doi.org/10.1016/j.scitotenv.2013.04.086

Townsend SA, Douglas MM (2014) Benthic algal resilience to frequent wet-season storm flows in low-order streams in the Australian tropical savanna. Freshw Sci 33:1030–1042. https://doi.org/10.1086/678516

Townsend SA, Douglas MM (2017) Discharge-driven flood and seasonal patterns of phytoplankton biomass and composition of an Australian tropical savannah river. Hydrobiologia 794:203–221. https://doi.org/10.1007/s10750-017-3094-6

Townsend SA, Padovan AV (2005) The seasonal accrual and loss of benthic algae (Spirogyra) in the Daly River, an oligotrophic river in tropical Australia. Mar Freshw Res 56:317–327. https://doi.org/10.1071/MF04079

Train S, Rodrigues LC (1998) Temporal fluctuations of the phytoplankton community of the Baia River, in the upper Parana River floodplain, Mato Grosso do Sul, Brazil. Hydrobiologia 361:125–134. https://doi.org/10.1023/A:1003118200157

Triemer RE, Zakryś B (2015) Photosynthetic euglenoids. In: Wehr JD, Sheath RG, Kociolek JP (eds) Freshwater algae of North America, 2nd edn. Academic Press, Boston, pp 459–483

Triska FJ, Kennedy VC, Avanzino RJ, Reilly BN (1983) Effect of simulated canopy cover on regulation of nitrate uptake and primary production by natural periphyton assemblages. In: Fontaine T, Bartell S (eds) Dynamics of lotic ecosystems. Ann Arbor Science, Ann Arbor, pp 129–159

Tsai JW, Chuang YL, Wu ZY et al (2014) The effects of storm-induced events on the seasonal dynamics of epilithic algal biomass in subtropical mountain streams. Mar Freshw Res 65:25–38. https://doi.org/10.1071/MF13058

Uehlinger U, Robinson CT, Hieber M, Zah R (2010) The physico-chemical habitat template for periphyton in alpine glacial streams under a changing climate. Hydrobiologia 657:107–121. https://doi.org/10.1007/s10750-009-9963-x

Vadeboncoeur Y, Power ME (2017) Attached algae: the cryptic base of inverted trophic pyramids in freshwaters. Ann Rev Ecol Syst 48:255–279. https://doi.org/10.1146/annurev-ecolsys-121415

Van Nieuwenhuyse EE (2007) Response of summer chlorophyll concentration to reduced total phosphorus concentration in the Rhine River (Netherlands) and the Sacramento—San Joaquin Delta (California, USA). Can J Fish Aquat Sci 64:1529–1542. https://doi.org/10.1139/f07-121

Van Vliet MTH, Franssen WHP, Yearsley JR et al (2013) Global river discharge and water temperature under climate change. Glob Environ Chang 23:450–464. https://doi.org/10.1016/j.gloenvcha.2012.11.002

Vannote RL, Minshall GW, Cummins KW et al (1980) The river continuum concept. Can J Fish Aquat Sci 37:130–137. https://doi.org/10.1139/f80-017

Varol M, Şen B (2018) Abiotic factors controlling the seasonal and spatial patterns of phytoplankton community in the Tigris River, Turkey. River Res Appl 34:13–23. https://doi.org/10.1002/rra.3223

Warren DR, Keeton WS, Bechtold HA, Rosi-Marshall EJ (2013) Comparing streambed light availability and canopy cover in streams with old-growth versus early-mature riparian forests in western Oregon. Aquat Sci 75:547–558. https://doi.org/10.1007/s00027-013-0299-2

Warren DR, Keeton WS, Kiffney PM et al (2016) Changing forests-changing streams: riparian forest stand development and ecosystem function in temperate headwaters. Ecosphere 7:e01435. https://doi.org/10.1002/ecs2.1435

Warren DR, Collins SM, Purvis EM et al (2017) Spatial variability in light yields colimitation of primary production by both light and nutrients in a forested stream ecosystem. Ecosystems 20:198–210. https://doi.org/10.1007/s10021-016-0024-9

Watson A, Barmuta LA (2011) Feeding-preference trials confirm unexpected stable isotope analysis results: freshwater macroinvertebrates do consume macrophytes. Mar Freshw Res 62:1248–1257. https://doi.org/10.1071/MF10298

Waylett AJ, Hutchins MG, Johnson AC et al (2013) Physico-chemical factors alone cannot simulate phytoplankton behaviour in a lowland river. J Hydrol 497:223–233. https://doi.org/10.1016/j.jhydrol.2013.05.027

Wehr JD, Sheath RG (2015) Habitats of freshwater algae. In: Wehr JD, Sheath RG, Kociolek JP (eds) Freshwater algae of North America, 2nd edn. Academic Press, Boston, pp 13–74

Westlake DF (1975) Macrophytes. In: Whitton BA (ed) River ecology. University of California Press, Berkeley, pp 106–128

Whitford LA, Schumacher GJ (1964) Effect of a current on respiration and mineral uptake in Spirogyra and Oedogonium. Ecology 45:168. https://doi.org/10.2307/1937120

Winkelmann C, Schneider J, Mewes D et al (2014) Top-down and bottom-up control of periphyton by benthivorous fish and light supply in two streams. Freshw Biol 59:803–818. https://doi.org/10.1111/fwb.12305

Winkworth CL, Salis RK, Matthaei CD (2015) Interactive multiple-stressor effects of the antibiotic monensin, cattle effluent and light on stream periphyton. Freshw Biol 60:2410–2423. https://doi.org/10.1111/fwb.12666

Withers PJA, Neal C, Jarvie HP, Doody DG (2014) Agriculture and eutrophication: where do we go from here? Sustainability 6:5853–5875. https://doi.org/10.3390/su6095853

Wolters J-W, Verdonschot RCM, Schoelynck J et al (2018) Stable isotope measurements confirm consumption of submerged macrophytes by macroinvertebrate and fish taxa. Aquat Ecol 52:269–280. https://doi.org/10.1007/s10452-018-9662-7

Wood J, Freeman M (2017) Ecology of the macrophyte Podostemum ceratophyllum Michx. (Hornleaf riverweed), a widespread foundation species of eastern North American rivers. Aquat Bot 139:65–74. https://doi.org/10.1016/j.aquabot.2017.02.009

Wood KA, O'Hare MT, McDonald C et al (2017) Herbivore regulation of plant abundance in aquatic ecosystems. Biol Rev 92:1128–1141. https://doi.org/10.1111/brv.12272

Wood KA, Stillman RA, Clarke RT et al (2018) Water velocity limits the temporal extent of herbivore effects on aquatic plants in a lowland river. Hydrobiologia 812:45–55. https://doi.org/10.1007/s10750-016-2744-4

Wright MV, Matson CW, Baker LF et al (2018) Titanium dioxide nanoparticle exposure reduces algal biomass and alters algal assemblage composition in wastewater effluent-dominated stream mesocosms. Sci Total Environ 626:357–365. https://doi.org/10.1016/j.scitotenv.2018.01.050

Wynne TT, Stumpf RP (2015) Spatial and temporal patterns in the seasonal distribution of toxic cyanobacteria in western Lake Erie from 2002–2014. Toxins (Basel) 7:1649–1663. https://doi.org/10.3390/toxins7051649

Xie D, Yu D, You W-H, Wang L-G (2013) Morphological and physiological responses to sediment nutrients in the submerged macrophyte Myriophyllum spicatum. Wetlands 33:1095–1102. https://doi.org/10.1007/s13157-013-0465-2

Yang GY, Tang T, Dudgeon D (2009) Spatial and seasonal variations in benthic algal assemblages in streams in monsoonal Hong Kong. Hydrobiologia 632:189–200. https://doi.org/10.1007/s10750-009-9838-1

Yu H-X, Wu J-H, Ma C-X, Qin X-B (2012) Seasonal dynamics of phytoplankton functional groups and its relationship with the environment in river: a case study in northeast China. J Freshw Ecol 27:429–441. https://doi.org/10.1080/02705060.2012.667371

Detrital Energy and the Decomposition of Organic Matter

The decomposition of organic carbon provides key energy inputs to most food webs, and this is especially true in fluvial ecosystems. These energy pathways are referred to as detrital or detritus-based, and the immediate consumers of this material are decomposers and detritivores. Detritus includes all forms of non-living, organic carbon including fallen leaves, the waste products and carcasses of animals, fragments of organic material of unknown origin, and organic compounds (Table 7.1). Some of this material originates within the stream (autochthonous), such as dying macrophytes, animal carcasses and feces, and extracellular release of dissolved compounds, and some is transported into the stream from outside the channel (allochthonous), such as leaf fall, soil particulates, and compounds dissolved in soil water. Carbon entering streams through detrital pathways can be stored, exported, or respired from a given reach. Estimates suggest that microbial decomposition of allochthonous detritus is responsible for a net efflux of more than 2 Pg C yr^{-1} globally from freshwater systems (Raymond et al. 2013; Battin et al. 2009). In many streams, energy derived from detrital resources can substantially exceed the energy fixed within streams by photosynthesis.

The division of non-living organic energy sources into size classes is widely employed in studying detrital dynamics in streams. The usual categories are coarse particulate organic matter (CPOM, > 1 mm), fine particulate organic matter (FPOM, < 1 mm and > 0.45 μm) and dissolved organic matter (DOM, < 0.45 μm). Each category can be divided further, but the dividing lines are arbitrary. Processes influencing the breakdown of forest leaves that enter steams, a dominant category of CPOM, have received extensive study and are comparatively well understood. FPOM can originate in a number of different ways but its origin and quality as a food resource can be difficult to establish. DOM, often referred to as dissolved organic carbon (DOC), likewise can be challenging to study; its entry into food webs is largely due to uptake by microbes in biofilms occurring on the stream bed and other surfaces,

discussed in detail in Chap. 8. Regardless, it is clear that the dynamics of organic matter in streams are complex, microorganisms are critical mediators of organic matter processing, and climate conditions and the surrounding landscape significantly influence what takes place within the stream.

Detrital inputs moving across ecosystem boundaries from terrestrial, marine, and other freshwater habitats generate substantial sources of energy and nutrients in streams. Often referred to as "spatial subsidies" (Polis et al. 1997), the physical and chemical characteristics, quantity, timing, and duration of resources entering streams from other systems generate spatial and temporal variation in the availability of nutrients and energy, the distribution and abundance of aquatic organisms, and the rates of biogeochemical processes in rivers and streams (Marcarelli et al. 2011; Subalusky and Post 2019). Subalusky and Post (2019) developed an effective conceptual framework with which to examine interactions between spatial subsidies and the characteristics of donor and recipient systems that can be modified for our detrital discussion (Fig. 7.1). This figure highlights how the net impact of allochthonous subsidies on stream ecology is influenced by the size, species diversity, productivity, resource availability, and seasonality of both the donor and recipient systems, and the characteristics of the resources crossing ecosystem boundaries. Examining the large body of research about detrital resources through this conceptual lens is an effective way to begin to understand how the structural and functional integrity of streams depends on upstream, downstream, and lateral connectivity with other systems.

In this chapter we review the principal factors influencing the physical and chemical characteristics, quantity, timing, and duration of detrital inputs into streams, and the physical, chemical, and biological characteristics of streams that mediate the decomposition of organic matter (Fig. 7.2). Some of the concepts in this chapter are revisited or described in further detail in other sections of the book,

Table 7.1 Sources of organic matter (OM) to fluvial ecosystems. Much OM originates outside the stream reach where it is measured. Some (sources marked with an asterisk) is produced by photosynthesis within the stream and subsequently enters the pools of dissolved or particulate OM

Sources of Input	Comments
Coarse particulate organic matter (CPOM)	
Leaves and needles	Major input in woodland streams, typically pulsed seasonally
Macrophytes during dieback*	Locally important
Woody debris	May be major biomass component, very slowly utilized
Other plant parts (flowers, fruit, pollen)	Relatively little information available
Other animal parts (feces and carcasses	Relatively little information available
Fine particulate organic matter (FPOM)	
Breakdown of CPOM	Major input where leaf fall or macrophytes provide CPOM
Feces of small consumers	Important transformation of CPOM
From DOM by microbial uptake	Organic microlayers on stones and other surfaces
From DOM by physical-chemical processes	Flocculation and adsorption, probably less important than microbial uptake route
Sloughing of algae*	Of local importance, may show temporal pulses
Sloughing of organic layers	Relatively little information available
Forest floor litter and soil	Influenced by storms causing increased channel width and inundation of floodplain, affected by overland versus sub-surface flow
Stream bank and channel	Little known, likely related to storms
Dissolved organic matter (DOM)	
Groundwater	Major input, relatively constant over time, often highly recalcitrant
Sub-surface or interflow	More important during storms
Surface flow	Possibly important during storms causing overland flow
Leachate from detritus of terrestrial origin	Major input, pulsed depending upon leaf fall
Throughfall	Smaller input, dependent on contact of precipitation and clouds with canopy
Extracellular release and leachate from algae*	Of local importance, may show seasonal and diel pulses
Extracellular release and leachate from macrophytes*	Of local importance, may show seasonal and diel pulses

especially the segments devoted to carbon dynamics, microbial ecology, and trophic interactions.

7.1　Inputs, Storage and Transport of CPOM

7.1.1　Sources of CPOM

Leaves and wood are the principal inputs of allochthonous CPOM to many streams, although other plant products, aquatic macrophytes, terrestrial invertebrates, and decomposing animal tissue also contribute to the total volume (Wenger et al. 2019; Webster et al. 1995; Wallace et al. 1995; Wipfli and Baxter 2010; Dalu et al. 2016). In some systems, macrophytes and algae from upstream may also represent important sources of energy and nutrients under certain conditions. The lability (referring to ease and speed of chemical transformation) and origin of particulate organic matter (POM), and the spatial and temporal distribution of particulate inputs, are highly variable among systems (Tank et al. 2010). Stream ecologists use the terms "standing stocks" or "standing crops" to describe the volume of POM in stream at a given time. The climate and land use history of a watershed strongly influence the lability, quantity, timing, and duration of allochthonous inputs entering a system (Thomas et al. 2016). Streams moving through forested landscapes in relatively undisturbed systems typically have greater inputs of POM (Golladay 1997; Webster et al. 1990). Heterotrophic microorganisms and other consumers utilize these carbon sources, and in some settings, allochthonous detritus is what fuels stream metabolism. Instream primary production

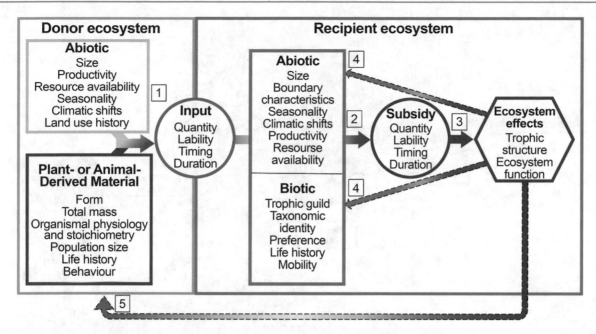

Fig. 7.1 Conceptual framework to consider the effects of detrital subsidies on ecosystem processes. (1) The abiotic features of the donor ecosystem will interact with characteristics of the plant- or animal-derived material and affect the quantity, lability, timing and duration (QLTD) of a given input to the system to the recipient ecosystem. (2) Resource inputs interact with the biotic and abiotic characteristics of the recipient systems to produce the QLTD of the resource subsidy. (3) The QLTD of the subsidy governs the ecosystem effects in the recipient ecosystem. (4) Changes in ecosystem processes in the recipient system alter the conditions that influence the QLTD of future subsidies, as there may be important feedbacks in certain abiotic and biotic conditions in the recipient system in response to the initial changes in ecosystem processes. (5) Changes to ecosystem processes in the recipient ecosystem may also generate feedback into the donor system that influence the QLTD of future inputs to the recipient system. (Modified from Subalusky and Post 2019)

may contribute to DOM pools by extracellular release after sloughing and dieback, and be incorporated into POM by microbial uptake, so the separation of allochthonous versus autochthonous sources can be imperfect.

Research examining CPOM dynamics has focused primarily on estimating changes in direct (vertical and upstream) inputs to stream surfaces. In contrast, much less work has been devoted to understanding within and among system variability in inputs from beyond the stream's banks (lateral); yet, in some systems they can represent a large proportion of the total. The timing, duration, and volume of lateral inputs depend on many factors including, but not limited to, wind patterns, precipitation regimes, bank slope, underlying geology, groundcover, litter accumulation, litter humidity, and the distance between the stream and the forest (Kochi et al. 2010; Tonin et al. 2017). Physical, chemical (e.g., leaching), and biological (e.g., microbial conditioning) processes in the terrestrial environment influence the characteristics of lateral inputs entering streams and may have subsequent impacts on ecosystem-level processes such as litter decomposition, ecosystem metabolism, and biomass production of higher trophic levels, termed secondary production (Tonin et al. 2017). In a study comparing leaf-litter decomposition rates between lateral and vertical inputs of

leaf litter in third order stream in central Portugal, Abelho and Descals (2019) documented negative relationships between the length of litter exposure in terrestrial habitats and in-stream decomposition rates. Invertebrate colonization also tended to decline with increasing terrestrial exposure; however, the richness and biomass of invertebrates were more influenced by litter type than by whether litter originated from lateral or vertical inputs. These authors also measured the response of aquatic fungi, finding that the aquatic hyphomycete species richness of the community colonizing leaf litter and fungal sporulation rates were negatively related to the length of time spent by litter in the terrestrial environment.

In a review of litter inputs to streams of the eastern US by Webster et al. (2006), lateral inputs were estimated to contribute roughly one-fourth of the total CPOM. This is consistent with an estimate of about 30% of total inputs from lateral sources in a review of 18 streams from different biomes and continents by Benfield (1997). Lateral inputs can be especially dominant in some systems. For example, in the Ogeechee River, a blackwater river in the southeastern US, lateral inputs were fourfold greater than direct inputs, probably because the width of the river minimized direct litterfall, and floodplain inundation maximized inputs from

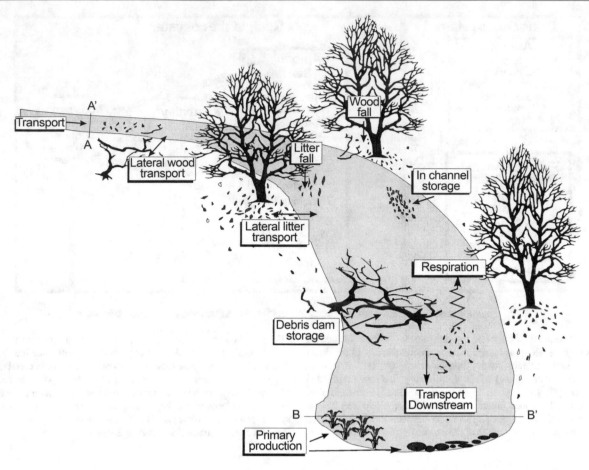

Fig. 7.2 Inputs, outputs and standing stocks of coarse particulates for a forest stream segment defined by the transects A–A' and B–B' (Reproduced from Minshall 1996)

outside of the channel. Lateral inputs, which were associated with increased precipitation and steep bank slopes, dominated CPOM inputs in a third-order stream on Santa Catarina Island in southern Brazil (Lisboa et al. 2015). The contribution of lateral inputs to CPOM can vary widely, even at smaller spatial scales. Work by Tonin et al. (2017) in streams from different tropical biomes emphasized the strong influence that local factors, including the density and diversity of riparian vegetation and site-specific topography, can have on the quantity, timing, and duration of lateral inputs.

The relative contribution of lateral and direct CPOM inputs often are temporally variable among systems. In temperate systems, seasonal variation in leaf inputs are common, with the greatest inputs occurring in the autumn, although lateral inputs of CPOM typically occur throughout the year. In an Austrian stream, direct CPOM inputs during the autumn contributed 61–65% of total inputs (Artmann et al. 2003). Similar results were obtained for a forest stream in Hokkaido, Japan, where leaf litter during October represented 58% of the annual inputs (Kochi et al. 2004). Additional seasonal variation in leaf fall can be attributed to diversity of terrestrial vegetation. In several small New

Zealand streams, inputs of leaf litter peaked in the summer in streams running through native forest, but maxima were observed in autumn in pasture areas. In pine forests, litter inputs were lowest in the winter and peaked during the spring when large inputs of pollen heads were observed (Scarsbrook et al. 2001).

Litter inputs expressed per unit area of streambed are expected to decline as stream width increases, and be greater in warmer and wetter climates because of higher forest productivity. Inputs of CPOM are expected to decrease with increasing stream order owing to increasing stream width and lower retention. This is supported by data from first-through fourth-order streams throughout the US including Georgia-North Carolina (Wallace et al. 1982a), New Hampshire (Meyer and Likens 1979; Bilby and Likens 1980), and from sites in Michigan and Pennsylvania (Minshall et al. 1983). Litterfall inputs were less at fifth-order sites of a Pennsylvania stream compared with lower-order sites (Bott et al. 1985). However, the expected downstream decline in CPOM inputs has not been apparent in many comparisons, evidently because differences among sites in climate, floodplain connectivity, and anthropogenic influence exert a larger influence on CPOM dynamics. Blackwater streams of the southeastern US received very high litter inputs as a consequence of floodplain interactions. In contrast, litter inputs to desert and boreal streams can be relatively low.

Allochthonous sources of energy and nutrients are fundamental components of food webs and ecosystem processes in both temperate and tropical systems (Lamberti et al. 2017; Neres-Lima et al. 2017). In a comparison of litterfall estimates from 33 sites ranging from 78°S to 75°N and in six different biomes, Benfield (1997) concluded that litter inputs were primarily related to the presence of forested versus non–forested vegetation. However, litterfall was also positively related to annual precipitation and decreased with increasing latitude, reflecting the influence of climate on overall terrestrial productivity (Fig. 7.3). In a study of streams across three tropical biomes (Atlantic forest, Amazon, Cerrado) in Brazil, Tonin et al. (2017) found that larger litter inputs were associated with wetter climates among seasons, and litterfall in streams located in the two biomes with dry seasons (Amazon and Cerrado) was negatively related to increases in precipitation, indicating there was more litterfall in the dry season. Notably, this work highlighted the variability in the magnitude and timing of direct and lateral inputs of litter and the storage rates of OM among tropical biomes (Tonin et al. 2017).

Though dominant components, leaf litter and woody debris are not the only forms of OM inputs to rivers and streams. In some systems, macrophytes are important sources of OM in stream food webs. For example, OM surveys

conducted seasonally in the Kowie River, a small temperate river, indicated that macrophytes significantly contributed to the pool of suspended particulates in the upper reaches of the stream (Fig. 7.4; Dalu et al. 2016). Animal movement may also introduce large pulses of OM into lotic systems from other habitats. A large body of work has examined the effects of marine- and lake-derived organic matter entering streams through the excretion, egestion, and the decomposition of gametes and carcasses of migrating fishes (e.g., Tiegs et al. 2011). Similarly, mass drownings of wildebeest (*Connochaetes taurinus*) commonly occur in the Mara River in Kenya during their annual migration, introducing more than 1,000 tons of biomass each year (Subalusky et al. 2017). In the same river system, hippos (*Hippopotamus amphibius*) contribute approximately 8,500 kg of terrestrially-derived OM to the river each day through their feces (Dutton et al. 2018). Though not as commonly studied, direct inputs of OM occur when animals in the riparian zone directly fall into streams, or via egestion when frass produced by insects in overhanging vegetation falls into the water below (Kochi et al. 2004). For instance, researchers have found significant changes in CPOM inputs and subsequent changes in nutrient dynamics and stream metabolism in response to periodic emergence of cicadas (Cicadidae: *Magicicada*) every 13–17 years (Menninger et al. 2008; Pray et al. 2009).

7.1.2 Storage of CPOM

In relatively undisturbed systems, CPOM tends to be retained near the point of entry in the stream (Webster et al. 1994). This is particularly true in small streams (Golladay 1997). The retention of OM is affected by a number of stream features that vary with location and thus potentially determine the efficiency of a stream reach in processing inputs. Retention is likely to be greatest when current velocity is low, when boulders or other channel features create depositional locations and cause accumulation of organic material, when macrophyte beds reduce water velocity, and when floodplain connectivity allows flooding rivers to overflow their banks, slowing the passage of water and material downstream. By increasing the retentiveness of stream reaches, such features should increase the amount of organic matter respired by the consumer community and decrease the amount exported downstream. As retention varies among locations, seasons, or stream types, so should the relationship between processing and export.

The benthic storage of OM on or within the stream bed is influenced by the magnitude of terrestrial organic matter inputs and by the retention capacity of the river channel (Jones 1997). Wood generally comprises the largest proportion of stored OM (Pfeiffer and Wohl 2018), followed by

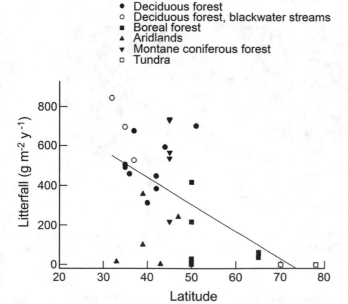

Fig. 7.3 A linear regression of litterfall versus latitude for 33 stream locations across six biomes (Reproduced from Benfield 1997)

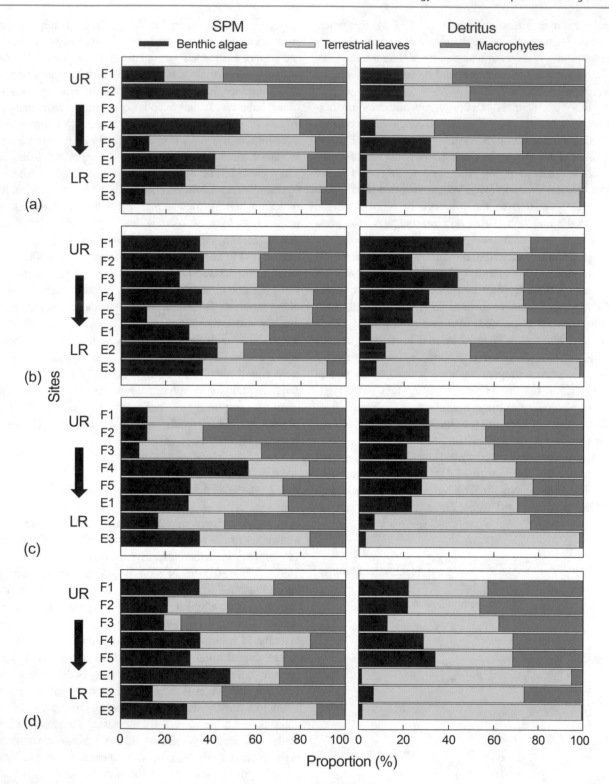

Fig. 7.4 Percentage contribution (relative mean) of benthic algae, macrophytes, and terrestrial plants to suspended particulate matter (SPM) and detritus in upper (UR) and lower reaches (LR) of the Kowie River, a small, temperate river in southern Africa in (**a**) early spring, (**b**) late spring, (**c**) summer, and (**d**) winter (Reproduced from Dalu et al. 2016)

coarse and fine materials; however, in some systems fines may be the largest component (Newbold et al. 1997; Martinez et al. 1998; Galas and Dumnicka 2003). In first- and second-order streams of the eastern US, FPOM can be more than 50% of the total amount of organic matter in the stream (Webster et al. 1995). Over 80% of stored OM was estimated to be buried in a Virginia coastal streams (Smock 1990), evidence of how challenging it is to make robust estimates of the standing stock of organic matter. High temporal variability was reported for the Njoro River in Kenya, where streambed OM accumulation ranged from approximately 93 to 6,700 g ash free dry weight m^{-2}, probably due to variability in litter inputs and discharge (Magana and Bretschko 2003).

Channel morphology can influence the retention of CPOM within a given reach. Retention rates of CPOM were higher in natural meandering sections than in straightened sections of a third-order stream in New Zealand (James and Henderson 2005). Similarly, Hoover et al. (2006) documented a number of effects of streambed geometry on leaf retention depending on flow and channel features. Protruding boulders were critical in retaining leaves in riffles, but not in pools, where leaves simply settled to the bottom. Leaf retention was greater in locations of greater depth and lower water velocity relative to reference streambed measurements across the river channel. The relative importance of pool versus riffle retention changed seasonally in the Njoro River, Kenya, where more CPOM was retained in pools than riffles during high flows (Magana and Bretschko 2003). Nakajima et al. (2006) also noted that CPOM accumulated in pools during periods of high flow, probably due to lower velocities near the streambed in pools.

7.1.3   Anthropogenic Activities and CPOM Inputs

Organic matter inputs are strongly influenced by anthropogenic activities. Logging and other activities that disrupt riparian vegetation can change the magnitude of leaf litter and wood inputs. Following logging in the catchment of a small forested stream in the southeastern US, leaf inputs fell to less than 2% of previous levels (Webster and Waide 1982). Allochthonous inputs recovered to near reference values after 5 to 10 years of regrowth and forest succession, but inputs were still detectably below reference levels after 20 years (Webster et al. 2006). Although litter inputs from early successional trees were less than observed prior to logging, these leaves were more rapidly broken down. Agricultural development disrupts links between aquatic and terrestrial environments by altering the timing and supply of terrestrial POM. Daily POM inputs and the standing stock of benthic OM were greater in reference headwater streams than agricultural streams in central Germany (Wild et al. 2019). Organic matter dynamics varied predictably with season in reference streams, but similar changes were more variable in agricultural systems (Fig. 7.5). Urbanization can also transform CPOM dynamics in many ways including, but not limited to, stream burial and channelization, changes in the abundance and diversity of riparian vegetation,

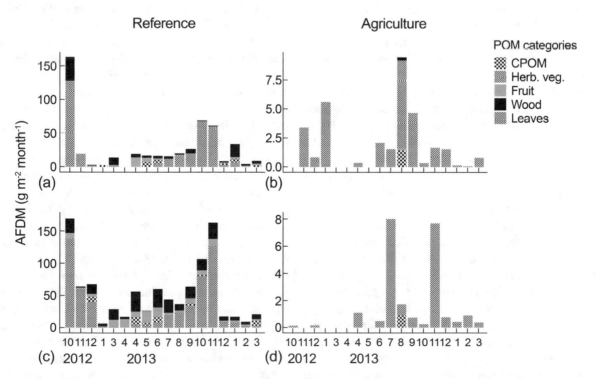

Fig. 7.5 Temporal variation in the inputs of particulate organic matter (POM) between October 2012 and March 2014 in reference (**a, c**) and agricultural (**b, d**) streams in north-central Germany (g AFDM m^{-2} d^{-1}). Values on the x-axes are months of the year beginning with October = 10. Please note that values on the y-axes are on different scales. Samples were taken twice in April (month 4) (Reproduced from Wild et al. 2019)

wastewater discharge, and the alteration of the volume, timing, and duration of surface runoff entering streams (Singer and Battin 2007; Kaushal et al. 2014; Smith and Kaushal 2015; Park et al. 2018).

Non-native riparian species can affect CPOM dynamics by altering the quantity and timing of leaf litter entering a stream reach. In Northern Spain, Pozo et al. (1997) noted that leaf litter input peaked during the summer in a stream dominated by non-native *Eucalyptus*, but similar OM peaks occurred during the autumn in a reference stream flowing through a native deciduous forest. Russian olive (*Elaeagnus angustifolia*) invasion has significantly changed the riparian habitats of many streams in the western US. Mineau et al. (2012) used a before-after-invasion comparison to document a ~25-fold increase in allochthonous inputs in sites where Russian olive was established. Introduced forest disease or parasites may also generate short- and long- term shifts in OM entering rivers and streams. Decades after chestnut blight (*Cryphonectria parasitica*) decimated the American chestnut populations from forests in the eastern US, dead chestnuts still contributed a large proportion of coarse woody debris in streams (Wallace et al. 2001). Similarly, eastern hemlock (*Tsuga canadensis*) populations have drastically declined in response to the invasion of the hemlock woolly adelgid (*Adelgis tsugae*), a parasitic insect, which was introduced from Japan to eastern North America in the 1950s. The adelgid has reduced populations of this once-dominant riparian evergreen tree and left large swaths of dead hemlocks in its wake, which are expected to influence the dynamics of wood entering streams. Riparian tree composition is expected to change after invasion, subsequently transforming CPOM dynamics (Pitt and Batzer 2015; Webster et al. 2012).

Climate change is expected to influence CPOM inputs through changes in temperature and precipitation, which are dominant controls on litterfall (Webster and Meyer 1997). In a nine-year study of a Mediterranean stream in the Montnegre-Corredor Natural Park in northeastern Spain, Sanpera-Calbet et al. (2016) demonstrated that changes in precipitation associated with the El Niño southern oscillation affected temporal dynamics in allochthonous inputs. Detrital inputs were influenced by the previous number of cumulative periods of no flow, as successive periods with no water flow progressively decreased the amount of riparian inputs entering streams. Instream flow events were also related to the timing of riparian inputs throughout the year. In years with permanent flow, streams were characterized by a single pulse of inputs in the autumn (~50% of total annual inputs) that was similar to values reported for streams in temperate climates. However, if a period of no flow occurred during the summer, the inputs were bimodal, occurring during the

no flow event and in the autumn (Sanpera-Calbet et al. 2016).

Changes in the timing, duration, and intensity of forest fires and storms associated with climate change are also expected to have large impacts on allochthonous resources entering streams and rivers. Feedback between forest condition and fire management strategies and the severity, size, and configuration of forest fires is expected to alter allochthonous subsides through changes in riparian vegetation, hydrology, and post-fire debris flow (Bixby et al. 2015; Harris et al. 2015). Intense storms can also affect CPOM inputs to rivers and streams (Ramos Scharrón et al. 2012). Wohl and Ogden (2013) reported that bank erosion and landslides contributed to changes in wood export from the Upper Río Chagres watershed in central Panama after intense storms in 2010. They estimated that the carbon exported in wood during the storms (9.6–16 Mg C km^{-2}) was an order of magnitude larger than estimated background rates. Strikingly, their estimates were two orders of magnitude less than similar values of wood export from watersheds in Taiwan (Tsengwen Reservoir: 7800 ± 2200; Kaoping River: 4500–9200 Mg C km^{-2}) after Typhoon Morakot (West et al. 2011).

7.1.4 Transport of CPOM

Transport rates are expected to vary with particle size, flows, and the retentiveness of the stream channel. High discharges will entrain and transport even large particles, as can be seen from the export versus discharge relationship for a headwater stream in North Carolina (Fig. 7.6). Based on a small number of direct measurements, however, the distances traveled by CPOM are surprisingly short. Leaves typically are trapped by obstructions within a few meters of their point of entry to the channel (Webster et al. 1994) and often are broken down in place without further transport, although they can move tens of meters in storms. Using rectangles of waterproof paper, Webster et al. estimated an average distance of about 1.5 m from first entry to the stream, depending mainly on depth and likelihood of encountering an obstacle. Wallace et al. (1995) recorded movements of spray-painted red maple leaves and small pieces of colored plastic transparency sheets for up to four years. Although few natural leaves were recovered after about five months, no differences were observed between natural and artificial leaves, which, over the four-year study, moved on average 10–20 m yr^{-1}. The comparison of leaf breakdown to transport rates makes a strong case that, at least for the small streams where most research has been conducted, CPOM is

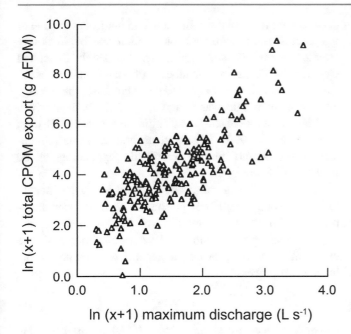

Fig. 7.6 CPOM export and maximum discharge in a headwater stream of the Cowceta Hydrologic Laboratory, North Carolina (Reproduced from Wallace et al. 1995)

transformed into other organic matter size classes or mineralized rather than exported.

7.2 The Decomposition of Organic Matter

Terrestrial leaf litter that enters streams from the surrounding landscape undergoes physical breakdown and decomposition, and in the process generates basal food resources for microbial and animal consumers. Leaf decomposition is a complex process that includes leaching of dissolved organic carbon (DOC) into the water column and leaf fragmentation into fine particulate organic matter (FPOM). Microbial colonization is rapid, and their secretion of extracellular enzymes transforms the organic matter into assimilable forms while their growth adds microbial biomass. Detritivorous animals of various feeding modes ingest both coarse and fine particles, obtaining nourishment from the leaf substrate and associated microbes. Organic inputs as leaf litter, often an abundant energy input to streams, are processed along multiple pathways (Marks 2019). Carbon compounds may be respired by microbes, ingested by detritivores, and transferred to higher trophic levels, or if especially recalcitrant, exported downstream. With the exception of OM buried in sediments or exported to the oceans, all constituent elements are mineralized and carbon is respired into CO_2.

It has long been recognized that leaf litter originating from a mixture of plant species includes material that undergoes rapid loss of mass, versus species that decompose much more slowly (Cummins et al. 1973; Kaushik and Hynes 1971). Typically quantified using an exponential decay coefficient obtained from measuring the loss of mass over time, and referred to as decomposition rate or breakdown rate (Benfield et al. 2017), species that decompose quickly are referred to as "labile", and species that decompose more slowly are often labeled "refractory" or "recalcitrant". The ease of measurement of leaf breakdown rate has resulted in a wealth of information on the differences among plant species and environmental conditions, and allowed global comparisons reviewed later in this chapter. Decomposition rates vary among plant species due to their structural and chemical traits. Leaves with higher concentrations of lignins and tannins and a tougher cuticle are more recalcitrant, and leaves with higher concentrations of sugars and nutrients are more labile (Gessner et al. 2010; Enríquez et al. 1993).

Rapidly decomposing leaves have traditionally been viewed as providing the greatest nutritional benefit to consumers, implying that a ranking of plant species by breakdown rate was equivalent to a ranking of food quality. Despite the appeal of such a convenient and easily obtained metric, Marks (2019) argues that equating breakdown rate with resource quality oversimplifies, and fails to distinguish the multiple pathways and fates of elements in leaf litter and their effects on organisms. A single measure of decomposition rate may fail to differentiate between resource availability to fungi versus to invertebrates, or recognize that different organisms may be limited by different elemental resources. During the early phases of decomposition, slowly decomposing litter retains most elements bound in more refractory litter, which subsequently becomes available to consumers (Fuller et al. 2015; Siders et al. 2018). By labeling trees with stable isotopes of carbon and nitrogen, and then incubating labeled leaves in a stream, Siders et al. (2018) showed that more carbon was assimilated by a shredding caddisfly from slowly decomposing oak litter than from rapidly decomposing cottonwood litter. However, carbon loss to leachate and to microbial biomass showed the opposite result, being higher from cottonwood leaf packs than from the oak litter.

In sum, the decomposition of detrital resources is a complex set of processes that support stream food webs and mediate biogeochemical cycling (Marks 2019). Elements bound in organic matter may cycle at different rates and elements limiting the growth and reproduction of organisms may shift with ontogeny or under different environmental conditions (Halvorson et al. 2018; Halvorson et al. 2015b). While rapidly decomposing leaves typically accrue microbial biomass more rapidly and are often preferred by macroinvertebrate consumers in feeding trials, equating leaf breakdown rate with food quality can mask the variety of the fates of the elements moving from detritus into stream food webs. Losses of the constituent elements of a leaf or other sources of OM follow multiple pathways and are influenced by structural and chemical characteristics of leaves that vary with tree species of origin and environmental conditions. Indeed, the presence of a mix of tree species with range of breakdown rates benefits consumers and the ecosystem, because it results in a more continuous resource supply for macroinvertebrates throughout their life cycles (Marks 2019). Though it does have its limitations, the relative ease and low cost of deploying leaf decomposition assays has allowed researchers to replicate experiments in streams and laboratories throughout the world. By employing loss of leaf mass as a proxy for OM decomposition, stream ecologists have made great advances in our understanding of the conditions governing the pathways and rates by which detrital resources are integrated into stream ecosystems.

7.2.1 Breakdown and Decomposition of Coarse Particulates

The rate of OM decomposition has been well studied because it can be measured easily and many of the physical, chemical, and biological factors that regulate decomposition can be detected within and among systems (Boyero et al. 2016; Tiegs et al. 2019). Decomposition is commonly estimated by measuring the change in mass of wood veneers or leaves or leaf discs (Benfield et al. 2017; Graça et al. 2005; Gregory et al. 2017). Briefly, the change in mass is calculated by drying and weighing OM, deploying the pre-weighed OM in streams or mesocosms for a known period of time, and again drying and re-weighing the OM to determine loss of mass. One can also measure changes in more homogenous organic substrates, such as strips of cotton, to reduce the influence of individual leaf and species-level trait diversity on decomposition rates (Tiegs et al. 2013; Tiegs et al. 2019). Once CPOM enters streams it undergoes a breakdown process or is exported (Webster et al. 1999). The sequence in the breakdown of CPOM, well documented for autumn-shed leaves in temperate streams (see reviews by Bärlocher 1985; Webster and Benfield 1986; Marks 2019), is illustrated in Fig. 7.7. Leaves fall directly or are windblown into streams, become wetted, and commence to leach soluble organic and inorganic constituents. Most of the leaching occurs within a few days and is followed by a period of microbial colonization and growth, causing numerous changes in leaf condition. The next stage, fragmentation by mechanical means and invertebrate activity, usually follows some period of softening of tissue by microbial enzymes, and is complete when no large particles remain. Although this model suggests sequential stages, leaf decomposition is a complex process and some of the events can occur simultaneously. For example, fragmentation can occur during microbial colonization and not just at the end of the process, and invertebrate colonization may begin shortly after leaves enter the stream (Gessner et al. 1999; Hieber and Gessner 2002). From a synthesis of many studies of 11 streams at the Coweeta Hydrologic Laboratory, the average leaf breakdown rate was found to be 0.0098/day, which implies the loss of 50% of initial mass after 71 days (Webster et al. 1999). Dissolved material and fine particulates can undergo further microbial degradation or be transported downstream.

The loss of leaf mass over time is approximately log-linear (Fig. 7.8), although some data have been interpreted as linear or as consisting of two or more distinct stages. Webster and Benfield (1986) argue that a simple exponential model provides a general description of the breakdown process

$$Wt = Wie^{-kt} \tag{7.1}$$

where W_t = dry mass at time t, W_i = initial dry mass, and t is time, measured in days. The statistic k (in units days^{-1}), which is the slope of the plot of the natural logarithm of leaf mass versus time, provides a single measure of breakdown rate.

7.2.2 Physical and Chemical Conditions Influencing OM Decomposition

A number of physicochemical factors, including temperature, hydrology, and water chemistry, influence breakdown rates. Graca et al. (2015) outlined a useful framework to describe how physical, chemical, and biological factors interact across spatial scales in low-order streams (Fig. 7.9), and we discuss these themes below. The over-riding influence, however, is leaf type (Fig. 7.10), which results in a fast-to-slow continuum of leaf breakdown rates. Although leaf breakdown can occur at near-zero temperatures, breakdown rates generally are predicted to be faster at warmer temperatures (Follstad Shah et al. 2017). In an effort to estimate global patterns in OM processing, Tiegs et al. (2019) deployed cotton strips in streams and adjacent riparian areas at more than 1,000 sites throughout the globe, finding that the mean rate and variability of decomposition declined with latitude. Their findings suggested that lower temperatures may constrain rates of decomposition toward the poles, but that other factors, such as nutrient availability, may regulate rates of decomposition in tropical regions where warmer climates support greater rates microbial productivity.

Changing hydrology may also influence litter breakdown rates. Hydrologic fluctuations can cause abrasion and fragmentation, which may expose more surface area to microbial action (Benfield et al. 2001), and burial, which can reduce microbial activity by reducing the availability of oxygen

Fig. 7.7 Leaf litter bags deployed in a stream in the southeastern US. Each mesh bag contains the same quantity of previously dried and weighed leaves. Leaved are retrieved, dried, and re-weighed at subsequent time intervals to estimate the rate of leaf decomposition. Inset (**a**) shows bag dimensions. Photo by Carolyn Cummins

(Sponseller and Benfield 2001). Decomposition in the hyporheic zone can also be a substantial contributor to carbon dynamics in streams, and this may be especially true in intermittent systems. In streams in eastern Australia, Burrows et al. (2017) found that leaf litter processing rates were almost 50% greater and cotton strip processing rates were approximately 125% greater in the hyporheic zone when compared to surface environments under multiple saturation conditions. Furthermore, they documented similar rates of microbial respiration on substrates incubated in both habitats.

Ambient chemistry conditions can also influence breakdown rates. Using data from almost three decades of nutrient addition experiments in headwater streams in the southeastern US, Rosemond et al. (2015) found that that average residence time of terrestrial organic carbon declined by approximate 50% in streams subjected to added nutrients. They argued that nutrient pollution may alleviate nitrogen

and phosphorus limitation of detrital food webs, and that increased carbon processing rates in polluted streams may negatively affect detrital food webs and alter rates of nutrient spiraling in streams. Variation in leaf breakdown rates along a stream in Costa Rica positively correlated with a natural gradient in phosphorus concentrations (Rosemond et al. 2002). Experimental addition of nitrogen and phosphorus to a stream in the southern Appalachians in the US enhanced the loss of wood mass and increased microbial respiration and fungal biomass (Gulis et al. 2004). However, nutrient additions to a stream in the Caribou National Forest in south-east Idaho in the western US did not affect leaf breakdown; the authors suggested this may have been because microorganisms were not nutrient limited due to relatively high ambient nutrient concentrations (Royer and Minshall 2001). Low pH and increased salinization have been shown to retard decomposition by inhibiting the activity of microorganisms and invertebrates (Dangles et al.

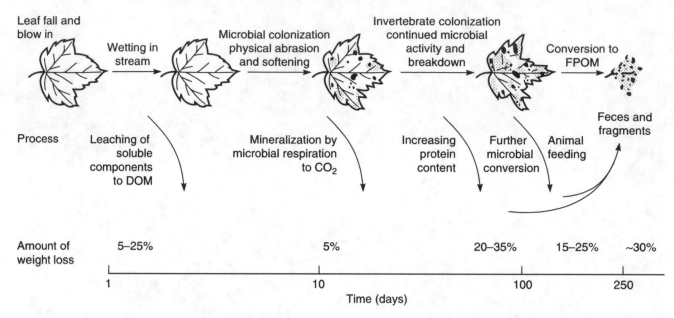

Fig. 7.8 The processing or "conditioning" sequence for a medium-fast deciduous tree leaf in a temperate stream. Leached DOM is thought to be rapidly transferred into biofilms by microbial uptake. Studies of organic matter breakdown start with the source material, often using leaves picked from riparian trees near the time of abscission, and follow its disappearance over time. The original leaf is transformed into several products including microbial and shredder biomass, FPOM, DOM, nutrients, and CO_2 (Gessner et al. 1999). Physical abrasion, softening of leaf tissue by wetting and the initial leaching of DOM, colonization by fungi and bacteria, and feeding by consumers all contribute to the conversion of CPOM into FPOM and DOM. Leaf breakdown can occur at very low temperatures, but rates typically increase with temperature, and are influenced by the presence of shredders, availability of nutrients, and whether leaves are buried or exposed

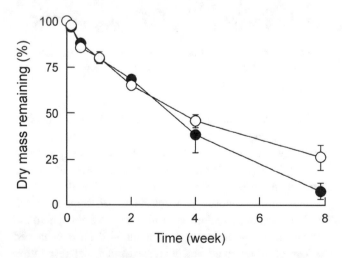

Fig. 7.9 Leaf dry mass remaining (as %) from alder (●) and willow (○) leaf packs in an experiment conducted in a Black Forest stream, Germany. Error bars represent 95% confidence intervals (Reproduced from Hieber and Gessner 2002)

2004; Canhoto et al. 2017). Metal pollution can decrease decomposition rates by negatively affecting shredders and microorganisms (Niyogi et al. 2001; Duarte et al. 2004; Carlisle and Clements 2005).

Addition of labile carbon, referred to as priming, may also influence decomposition rates of more recalcitrant organic matter. In streams, algae produce labile exudates, which may produce a priming response in OM decomposition by stimulating microbial activity. However, the relatively few results from experimental manipulations are equivocal (Bengtsson et al. 2018). Danger et al. (2013) suggested that the presence of diatoms had a positive priming effect on leaf decomposition rates in experimental microcosms. In contrast, Halvorson et al. (2019) tested algal-induced priming of decomposition of leaves from two species of trees, tulip poplar (*Liriodendron tulipifera*) and water oak (*Quercus nigra*), under light or dark conditions. Light enhanced algal biomass and production, thereby increasing bacterial abundance by 141%–733% and fungal production rates by 20%–157%. However, algal-stimulated fungal production rates were not related to long-term increases in litter decomposition rates, which were 164%–455% greater in the dark. Similar relationships were found with both leaf types, suggesting that the lability of leaves did not influence priming. Researchers suggested that algae may have supplied fungi with labile carbon and decoupled fungal activity from decomposition rates, and the fungi invested this carbon into growth and reproduction, instead of using the energy to

Fig. 7.10 Conceptual model of the drivers of litter breakdown. Arrows reflect the influence of one factor over the next one. The biological pathways are ultimately controlled by underlying geology and climate. (Modified from Graca et al. 2015)

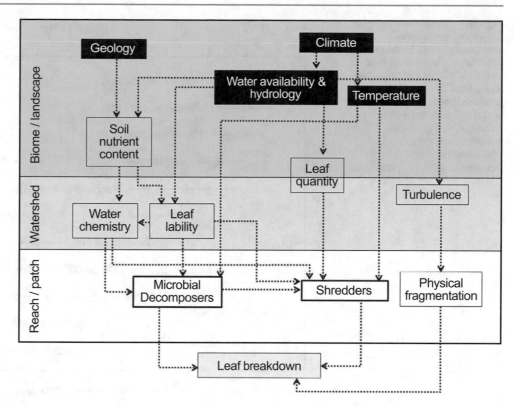

obtain nutrients and energy from leaves (Halvorson et al. 2019). Understanding functional feedback between autotrophic and heterotrophic communities is essential as human activities continue to modify the relative abundance of algal, fungal, and bacterial populations in streams.

7.2.3 The Lability of Organic Matter

Species-specific differences in physical, chemical, and ecological traits influence the initial lability of litter produced by plants, and affect the rate at which constituents are lost from the litter and move through aquatic food webs (Fig. 7.11; Irons et al. 1994; Boyero et al. 2017; Siders et al. 2018; Ferreira et al. 2016; Leroy and Marks 2006). Early research by Petersen and Cummins (1974) suggested a continuum of decomposition rates from slow to fast, based on the breakdown of leaves from six deciduous tree species in a small Michigan stream. They also recognized that this variation in leaf decomposition rates, which they termed a "processing continuum", had consequences for invertebrate consumers by extending the time interval over which microbially-colonized leaf litter was available. The wide variation in the breakdown rate of the leaves of different plant species has now been amply documented (Fig. 7.10). Non-woody plant leaves decompose much more quickly, on average, than do leaves of woody plants (mean half-lives in Fig. 7.10 are approximately 65 days and 100–150 days

respectively). Submerged and floating macrophytes are among the fastest to decay, presumably because they contain the least amount of support tissue and often the greatest concentration of potentially limiting elements such as nitrogen and phosphorus.

The physical structure of leaf litter affects the rate at which elements are lost through leaching. As much as 25% of the initial dry mass of freshly-abscised leaves is lost due to leaching in the first 24 h. Constituents lost during leaching are primarily soluble carbohydrates and polyphenols (Suberkropp et al. 1976). Leaves of different plants show species-specific leaching rates: alder (*Alnus rugosa*) lost only about 4% of dry mass over several days whereas elm (*Ulnus americana*) lost 16% in an early study by Kaushik and Hynes (1971). Release of DOC by leaves of several plants in a stream in British Columbia, Canada, also revealed substantial differences in leaching rates (McArthur and Richardson 2002). During the first day, Western hemlock needles (*Tsuga heterophylla*) lost 14% of the total DOC released over a 7-day period, compared with 30% for western red cedar (*Thuja plicata*) and 74% for red alder (*Alnus rubra*). By the end of the experiment, hemlock and cedar had released 40% and 20% respectively of the DOC released by alder (Fig. 7.12). More recent work has also provided evidence that rapidly decomposing litter tends to lose more carbon and nitrogen through leaching relative to more recalcitrant species (Siders et al. 2018; Wymore et al. 2015).

Fig. 7.11 The breakdown rates for various woody and non-woody plants, based on 596 estimates compiled from field studies in all types of freshwater ecosystems. Means ±1 standard error are shown, and the variation is due to (at least) effects of site, technique, and numerous environmental variables. The number of individual rate estimates is shown in parentheses (Reproduced from Webster and Benfield 1986)

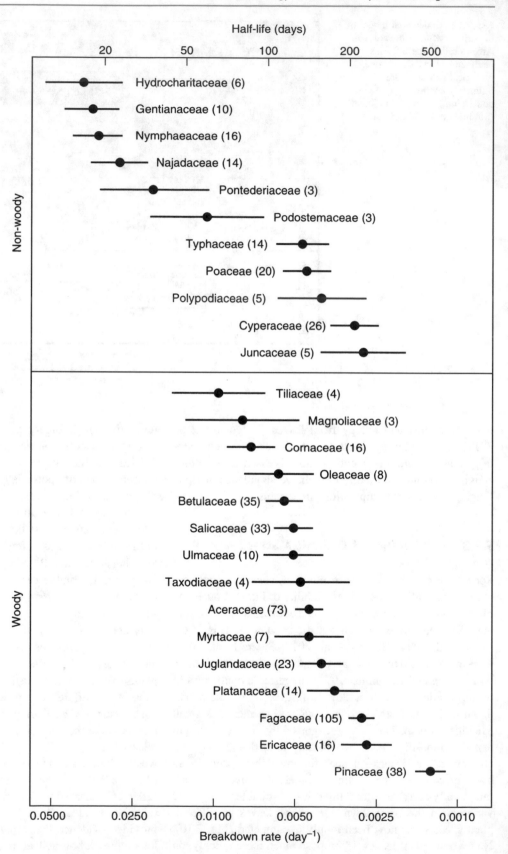

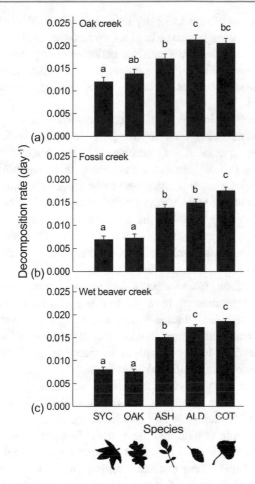

Fig. 7.12 Rates of leaf-litter decomposition for five species in three streams in the verde River catchment of Arizona, US (SYC *Platanus wrightii*, OAK *Quercus gambelii*, ASH *Fraxinus velutina*, ALD *Alnus oblongifolia*, COT *Populus fremontii*). Plotted values are the regression slopes ±1 SE for the ln-transformed regression model of the rate of decomposition. Significant differences among leaf types are represented by lower-case letters (Reproduced from Leroy and Marks 2006)

Differences in leaf chemistry and structure result in wide variation in decomposition rates (Webster and Benfield 1986; García-Palacios et al. 2016). Complex compounds including lignin, tannin, and phenols tend to reduce decomposition, whereas OM rich in labile carbohydrates, such as sugars, decompose more rapidly (Marks 2019). Similarly, leaves with higher initial nutrient concentrations tend to decompose more rapidly than leaves of lower nutrient content. Leaves that contain lower lignin and condensed tannin concentrations tend to be initially preferred by leaf-shredding invertebrates. However, as leaf conditioning proceeds in less labile, more slowly decomposing leaves, loss by leaching of some constituents that initially rendered the leaf less labile, as well as microbial growth, may reduce or even reverse initial palatability differences. Compson et al. (2018) demonstrated that, over time, a shredding caddisfly (*Hesperophylax magnus*, Limnephilidae) shifted

its preference from a fast to a slow-decomposing leaf type, assimilating more carbon and nitrogen from the latter. Leaf chemistry also explained differences in decomposition of *Croton gossypifolius* (Euphorbiaceae) and a species of *Clidemia* (Melastomataceae) in an Andean stream (Mathuriau and Chauvet 2002). Higher decomposition rates in *Croton* than in *Clidemia* appeared to be related to lower tannin and higher N content in *Croton*, which resulted in earlier peaks in ergosterol (a compound found primarily in fungi), sporulation activity, and macroinvertebrate colonization.

Intraspecific variation in leaf structure and chemical composition may also affect patterns in decomposition. In a study of the decomposition of five genotypes of each of four different cottonwood cross types (*Populus* sp.) that were grown under common conditions, LeRoy et al. (2007) found that polymorphism markers in the genotype and differences in genetically-controlled chemical characteristics explained a large proportion of the variation in leaf-litter decomposition rates. Similarly, Compson et al. (2016) studied how plant genetics influenced the composition and timing of emerging insects using a subset of the same cross types. They found that litter from more closely related genotypes of *P. angustifolia* were more similar in decomposition rates, leaf thickness, and litter nitrogen concentrations. Additionally, they had more similar communities of emerging insects. Conversely, the researchers found only marginally significant differences in emerging insect composition between *P. fremontii* and *P. angustifolia*. They also documented reductions in the influence of genetic effects in mixed litter packs relative to leaves from single genotypes. Their results provide additional evidence that intraspecific variation in riparian forests can affect leaf decomposition rates, and suggest that the genetic structure of tree communities may influence characteristics of aquatic insect emergence.

Evolutionary relationships between tree species and detritivores may help explain patterns in decomposition, as detritivore communities may be adapted to break down litter with which they have co-evolved (Jackrel and Wootton 2014; Jackrel et al. 2016). Similar to cottonwood studies, Jackrel and Wootton (2014) documented significant intraspecific variation in decomposition rates for red alder (*Alnus rubra*) in reciprocal transplant experiments in two smaller and two medium-sized rivers in the western US. Detritivores processed local litter more rapidly that litter from other riparian habitats, suggesting that stream consumers are adapted to local intraspecific variation in litter. Studies in the same rivers suggested that intraspecific variation in leaf chemistry may drive local detritivore adaptation to food resources (Jackrel et al. 2016). Concentrations of plant defensive compound such as ellagitannins, and flavonoids influenced decomposition rates, a pattern that may have been driven by local adaptation of detritivores to local chemical conditions in leaf litter. Collectively, these results suggest that land cover change may influence the chemistry of OM entering streams, subsequently altering the rates at which OM is converted into animal tissue.

In a global survey of the physical and chemical characteristics of leaves and leaf litter decomposition, Boyero et al. (2017) found that mean annual temperature and temperature seasonality, and thus latitude, were related to the stoichiometry, or ratio, of nitrogen to phosphorus in litter. In warmer and less variable climates, litter typically had greater amounts of nitrogen relative to phosphorus. Nutrient concentrations of leaves are thought to be governed by soil nutrient concentrations, and tropical soils are thought to be depleted in phosphorus relative to their temperate counterparts because they are often older and have been subjected to more weathering and leaching (Reich and Oleksyn 2004; Chadwick et al. 1999). However, nitrogen content is thought to be relatively similar between temperate and tropical regions as it is regulated, in large part, by biological nitrogen fixation (Houlton et al. 2008; Vitousek et al. 2010). Notably, Boyero et al. (2017) found that tropical litter was not particularly nutrient poor, but their data suggested that relative to nitrogen, phosphorus was limiting in tropical litter, which may cause stoichiometric imbalances that influence the diversity and abundance of aquatic consumers (Boyero et al. 2017; Frost et al. 2006).

Plants invest in chemical inhibitors that may also impede leaf decay. Tough outer coatings such as the cuticle of conifer needles slow fungal invasion (Bärlocher and Oertli 1978), and complexing of protein to tannins is a principal cause of slow breakdown in many broad-leafed woody plants. Toxic chemical constituents also may influence breakdown rates (Webster and Benfield 1986), just as secondary plant compounds defend against terrestrial herbivores, although evidence is scant. Somewhat surprisingly, chemical measures of tannins (total phenolics and condensed tannins) of 48 deciduous trees were unrelated to published breakdown rates (Ostrofsky 1993). However, Canhoto et al. (2002) found that oils of *Eucalyptus globulus* inhibited the growth of hyphomycetes and the activity of their enzymes, which could explain the delayed decomposition of eucalyptus in rivers. In the aforementioned global survey, Boyero et al. (2017) found that investment in plant defenses typically declined with increasing latitude, and tougher and more chemically defended litter is generally found near the equator.

Microbial colonization plays a fundamental role in altering the palatability of leaves for detritivores (Arsuffi and Suberkropp 1985) and in the fragmentation of leaf material. This colonization is primarily by fungi and bacteria, although protists also can be substantial contributors to leaf fragmentation (Ribblett et al. 2005). Microbial activity also softens plant tissue, favoring the release of compounds that can be incorporated into microbial biomass (Gessner et al. 1999; Graca 2001). In general, nitrogen typically increases

as a percent of remaining dry mass during leaf conditioning, and sometimes increases in absolute terms as well. Because protein complexed to lignin and cellulose is very resistant to breakdown, nitrogen compounds remain while other leaf constituents are lost, resulting in a relative increase of nitrogen.

Microbial immobilization of nutrients from external sources can also lead to increases in either relative or absolute quantities of nitrogen or phosphorus. When increases in total nitrogen are recorded, this immobilization of nitrogen usually is attributed to an increase in microbial biomass and incorporation of nitrogen from the surrounding water into new protein. In a headwater stream reach in Coweeta Hydrologic Laboratory, North Carolina, that received experimental additions of ammonium, nitrate, and phosphate, the nitrogen content of maple (*Acer rubrum*) and rhododendron (*Rhododendron maxima*) leaves increased significantly relative to a control (Fig. 7.13), suggesting higher microbial biomass under enrichment conditions (Gulis and Suberkropp 2003). The slower breakdown of rhododendron leaves compared to maple is attributable to their lower initial nitrogen content and lower surface area relative to leaf volume. In a study of oak and hickory leaves incubated in a Michigan stream over the winter (Suberkropp et al. 1976), cellulose and hemicellulose declined at about the same rate as total leaf mass, while lignin was processed more slowly and increased as a percentage of remaining weight. Lipids were lost more rapidly than total mass, and thus were a declining fraction of remaining dry mass of leaf material.

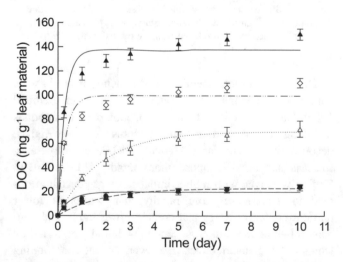

Fig. 7.13 Cumulative release of dissolved organic carbon (DOC) leachate over 10-day period for four tree species and grass cuttings (a common yard waste) in urban streams of southern Australia. Grass cuttings (●), English elm (*Ulmus procera*) (△), London plane (*Platanus acerifolia*) (■), river red gum (*Eucalyptus camaldulensis*) (◇) and white poplar (*Populus alba*) (▲) (Reproduced from Wallace et al. 2008)

Tracking the flow of energy and elements from individual riparian species into consumers may provide insights about links between biodiversity and ecosystem function in riparian and stream habitats. All elements stored in OM will eventually be mineralized and will cycle through various components of stream food webs. However, the rate at which elements cycle is governed by site-specific conditions, such as ambient nutrient concentrations, and the route by which elements move through organisms and abiotic pools. Transformations from one pool to another can occur rapidly —elements leached from leaf litter can be integrated into microbial biomass and respired within hours of entering a stream (Marks 2019).

Stable isotope labeling has generated a greater understanding about the rates and pathways of elements moving from CPOM into stream food webs (Cheever et al. 2013). Labeling allows scientists to quantitatively track carbon and nitrogen as they flow from leaf litter into microorganisms and macroinvertebrates. Unlike many commonly employed approaches, it can be used to compare the rates of elemental flux from the litter of different plant species and estimate the contribution of a given plant species to the productivity of higher trophic levels such as microbes and macroinvertebrates (Compson et al. 2015). Tracer studies have demonstrated that the carbon and nitrogen from slowly decomposing litter (frequently considered to be of lower quality) often supports biomass production in higher trophic levels comparable to faster decomposing species (Compson et al. 2015; Siders et al. 2018).

In more recalcitrant species, leaf chemistry influences assimilation rates of elements into higher-order consumers. Data from tracer experiments suggest that litter from rapidly decomposing species tends to support more microbial biomass and higher bacterial productivity, while resources from slowly decomposing litter tends to support the growth of fungi (Pastor et al. 2014; Wymore et al. 2013). Compson et al. (2018) found that lignin concentrations were positively correlated with nitrogen assimilation rates in macroinvertebrates, but that soluble condensed tannins were negatively correlated with tannin concentrations. They demonstrated that a 1% increase in lignin concentrations was expected to increase assimilation rates by 23% of its daily mean, and a similar increase in tannin concentrations was expected to reduce daily mean assimilation of nitrogen by 57% (Compson et al. 2018). Collectively, this body of work suggests that increased concentrations of plant structural compounds may initially impede nutrient cycling from litter, but the slower release rates of energy and nutrients from more recalcitrant species may increase the efficiency and longer-term transfer rates of elements to higher trophic levels (Marks 2019; Compson et al. 2018).

7.2.4 Microbial Succession and Decomposition Rates

Microbes typically have lower ratios of carbon to limiting nutrients when compared to the substrate they are colonizing, a pattern that supports the "peanut butter and cracker" analogy in which microbes function as the nutritious peanut butter covering a less nutritious cracker (Cummins 1974). However, this analogy fails to acknowledge some of the aforementioned interactions between microbes and macroconsumers that occur during leaf decomposition (Marks 2019). Though invertebrates often prefer to consume and grow faster when feeding on leaves that have been colonized by microbes, microbes are also consuming leaves and competing with macroinvertebrates for limiting resources. Microbes can produce chemical deterrents that retard macroinvertebrate consumption rates (Danger et al. 2016). Additionally, rapid microbial processing can enhance elemental loss from OM to the water column or atmosphere, preventing the resources from supporting productivity of higher trophic levels (Marks 2019).

Studies are in agreement that leaves submerged in a stream for some days to allow microbial colonization support higher invertebrate growth rates. Referred to as "conditioned leaves", microorganisms, especially aquatic hyphomycetes, increase in mass over a period of days to weeks, contributing to leaf breakdown and influencing the palatability of leaves to shredders (Cummins and Klug 1979; Kaushik and Hynes 1971). Fungal mycelia are much more readily assimilated than leaf tissue, which explains why conditioned leaves invariably are preferred over unconditioned or sterile leaves. Colonization of aspen leaves by aquatic hyphomycetes resulted in weight loss, leaf tissue softening, and increases in the ATP (adenosine triphosphate, an indicator of microbial biomass) and nitrogen content of leaves (Arsuffi and Suberkropp 1984). However, the timing of these changes varied among species, as some initiated leaf degradation in 5–10 days, and other fungal species required as much as 20 days to reach comparable stages of degradation. Feeding preferences of larval caddisflies varied with fungal species composition and the duration of fungal colonization. For any given fungus-leaf combination, optimum palatability generally accords with the period of greatest microbial growth, as measured by nitrogen and ATP content and activity of degradative enzymes (Bärlocher 1985). The amphipod *Gammarus pseudolimnaeus* assimilated the mycelia of 10 fungal species with average efficiencies of 64%, compared with 11% average efficiency with unconditioned leaves of elm (*Ulmus americana*) and maple (*Acer saccharum*) (Barlocher and Kendrick 1974). Two crustacean shredders, *Gammarus pulex* and *Asellus aquaticus* preferentially fed on conditioned rather than on unconditioned elm leaves, and conditioned leaves were highly assimilated by both shredders (Graça et al. 1993). Fungal carbon accounted for 100% of the growth of third instar larvae of the limnephilid caddis *Pycnopsyche gentilis*, and 50% of the growth of fifth instar larvae, when fed a diet of tulip poplar (*Liriodendron tulipifera*) leaf discs colonized by a radiolabeled aquatic hyphomycete, *Anguillospora filiformis*. When fed a diet of sterile leaves, however, both instars lost weight (Chung and Suberkropp 2009).

Fungal colonization of leaves takes place primarily in the water, because freshly abscised leaves exhibit low fungal biomass (measured as ergosterol content) before entering the stream (Gessner and Chauvet 1997; Hieber and Gessner 2002). Rates of fungal growth and production on leaf litter peaks soon after submergence, and growth rates are typically greatest when fungal biomass is relatively low (Kuehn 2016). Fungi can degrade the polysaccharides present in the cell walls of the leaves by the production of extracellular enzymes (Jenkins and Suberkropp 1995), and their hyphae also penetrate leaf tissue, facilitating the softening process (Wright and Covich 2005b). Propagules of soil fungi, although commonly carried into the stream on shed leaves, appeared to contribute little to decomposition (Suberkropp and Klug 1976). Bärlocher (1982) reported that typically 4–8 species of aquatic fungi dominate throughout the decomposition of leaves, while a similar or larger number of rare species appear erratically. Apparently, no particular succession occurs on a single leaf, and whichever fungal species arrives first as a waterborne spore establishes numerical dominance. Hieber and Gessner (2002) identified 30 species of hyphomycetes on decomposing leaves of alder and willow, but two species, *Flagellospora curvula* and *Tetrachaetum elegans*, were dominant. Bacteria from biofilms growing on decomposing leaves in a stream in Ohio were mostly of one type known as α-*Proteobacteria*, although representatives of β-*Proteobacteria* were occasionally abundant (McNamara and Leff 2004).

The relative influence of fungi and bacteria upon the decomposition process likely varies with substrate, habitat, and time. Suberkropp and Klug (1976) followed in detail the succession of dominant microorganisms on oak and hickory in a Michigan stream from November until June. Fungi, primarily aquatic hyphomycetes, dominated during the first half (12–18 weeks) of the processing period. Bacteria, whose numbers gradually increased throughout, dominated the terminal processing stage and perhaps were benefited by fungal-induced changes in leaf surface area or by the release of labile compounds. Although fungal biomass was found to be several times greater than bacterial biomass on alder and willow leaves in a stream in Germany's Black Forest region (Fig. 7.14), the fungal contribution to leaf mass loss was only about twice that of bacteria, whose shorter turnover times partly compensated for their lower biomass (Hieber

and Gessner 2002). Findlay et al. (2002) found that fungi dominated microbial biomass in large organic matter such as wood and leaves, whereas bacterial biomass was dominant in fine benthic organic matter (Fig. 7.15). Because fine benthic organic matter was more abundant than coarse detritus at the study sites, when bacterial biomass was weighted for the abundance of detritus in the reach, it was similar to or higher than that of fungi. In addition, bacterial biomass was less variable per unit of organic matter mass than fungal biomass, suggesting that bacteria could be a more reliable food resource than fungi. In a study of seasonal variation in fungal and bacterial colonization of leaf litter, Mora-Gómez et al. (2016) documented that the relative influence of microbes in decomposition shifted seasonally; fungi were more dominant decomposers in spring, but bacteria seemed to become more important in summer months.

Bacteria and fungi may interact synergistically and antagonistically during POM decomposition. In their study of decomposing leaf litter under low and high nutrient levels, Gulis and Suberkropp (2003) reported that microbial

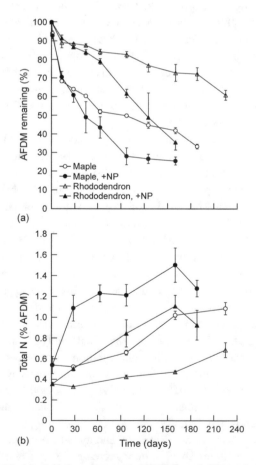

Fig. 7.14 (**a**) Ash free dry mass (AFDM) and (**b**) total nitrogen content in a decomposition experiment using maple and rhododendron leaves in control and nutrient-enriched reaches of a headwater stream at the Coweeta Hydrologic Laboratory, North Carolina, US (Reproduced from Gulis and Suberkropp 2003)

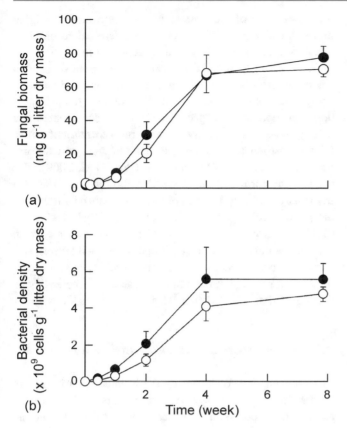

Fig. 7.15 (**a**) Fungal and (**b**) bacterial biomass in a leaf decomposition experiment conducted with alder (●) and willow (○) leaves in a Black Forest stream, Germany. Error bars represent 95% confidence intervals (Reproduced from Hieber and Gessner 2002)

biomass and production were always dominated by fungi. The bacterial contribution increased in treatments where fungi were excluded, suggesting competition between bacteria and fungi for resources. This interaction may be symbiotic as well, as researchers also found evidence that bacteria may benefit from fungal activity during leaf decay (Gulis and Suberkropp 2003). When leaf disks treated with antibacterial and antifungal solutions were incubated in two tropical streams in Venezuela, results indicated that fungi contributed more to total microbial biomass than did bacteria, while bacterial were responsible for a higher proportion of total microbial (Abelho et al. 2005). In a tropical headwater stream in Puerto Rico, leaf decomposition rates were faster when both bacteria and fungi were present, compared to treatments in which one was excluded. However, fungi reached higher biomass alone than in the presence of bacteria, suggesting an antagonistic effect (Wright and Covich 2005b). These results suggest that the interaction between bacteria and fungi can be synergistic in facilitating leaf decomposition but antagonistic in the mutual demand for a carbon source.

Even when encountered on a leaf, microbes may not be actively contributing to decomposition. New techniques are allowing stream ecologists to understand more about when microbes are actively dividing and consuming resources. For instance, quantitative isotope probing, a technique that measures the isotopic composition of specific nucleic acid sequences to differentiate between dormant and actively replicating cells, has begun to be used to study the bacterial and fungal contribution to leaf-litter decomposition in streams. In a pilot study, researchers found that of the 150 fungal species found on submersed leaves, only ∼35% were actively dividing (Marks 2019).

7.2.5 Decomposition of Macrophytes, Wood, and Other Sources of CPOM

Thought not as well-studied as leaf-litter decomposition, the decomposition of macrophytes and wood also generates sources of carbon and nutrients for streams. Macrophytes can be key sources of detritus where they are abundant, typically in larger rivers and in floodplains. Polunin (1984) reviewed studies of the decomposition and fate of this material, which is similar to that of terrestrial leaves. Breakdown rates are relatively fast, although less so for emergent macrophytes that contain more support tissue. Bacteria appear to play a greater role in macrophyte decomposition than is true for leaves of terrestrial origin (Webster and Benfield 1986).

Wood ranging from small branches to large tree trunks is abundant in small to mid-sized streams flowing through forested landscapes (Pfeiffer and Wohl 2018; Sutfin et al. 2016). Wood influences channel structure and stream habitat in a number of ways, and also contributes to the nutrition of some consumers (Anderson and Sedell 1979). Not surprisingly, wood decays very slowly. The high lignin and cellulose content of wood fiber, combined with low concentrations of N and P, relatively small surface area, and low penetrability, results in very slow breakdown with microbial activity confined to surface layers. Even small wood chips (0.75–1.5 cm size range) placed in coarse mesh bags in Quebec streams showed very slow loss rates (Melillo et al. 1983). Alder chips (*Alnus rugosa*) had a half-life of about 7 months, while spruce (*Picea mariana*) chips would require roughly 17 years to achieve a 50% reduction in weight. Anderson et al. (1978) retrieved sticks (2.5 × 2.5 92 cm) of alder, hemlock (*Tsuga*) and Douglas fir (*Pseudotsuga*) after 15 months in an Oregon stream, with similar results. Webster et al. (1999) estimated breakdown rates for sticks of different diameter (<20 mm, 20–25, and >25 mm) of yellow poplar (*Liriodendron tulipifera*), white pine (*Pinus strobus*), and red oak (*Quercus rubra*) in a 5-year experiment conducted in streams at Coweeta Hydrologic Laboratory. Breakdown rates were more rapid for sticks of smaller diameter, and for poplar relative to the pine and oak. Breakdown rates for sticks were one to two orders of magnitude slower than those reported for leaves of woody plants. Logs (20–32 cm diameter) that were placed in this stream lost their bark after 4–5 years and showed some signs of surface decomposition after 8 years, but still no differences in density were detected. Natural wood pieces, such as branches and twigs, have slower decomposition rates than commercial wood substrates such as chips, sticks, cubes, and disks, probably because the latter have a higher surface to volume ratio and because leaching losses are higher from the processed surfaces of wood products (Spanhoff and Meyer 2004).

Nutrient availability may have some influence on wood decomposition in streams. Comparison of alder, pine, oak, and eucalyptus branches at three sites in the Agüera River, Spain, that differed in nutrient levels suggested phosphorus limitation of wood breakdown (Diez et al. 2002). An initial weight loss (5–9%) due to leaching occurred during the first few weeks, and phosphorus was rapidly lost during the first six weeks in all but eucalyptus. Breakdown rates were most rapid for alder, followed by oak and eucalyptus, and then pine, which also showed the lowest values of ergosterol. In two streams in Germany, the decomposition rate for black alder (*Alnus glutinosa*) was higher and the half-life shorter at a site with higher nitrate and phosphate levels (Spanhoff and Meyer 2004). Experimental addition of nutrients enhanced the decay of wood (oak veneers and natural maple sticks), more so for veneers than sticks (Gulis et al. 2004). The faster

breakdown rate of the veneers was accompanied by higher nitrogen and ergosterol content and microbial respiration, which was thought to be primarily fungal (Fig. 7.16).

When present, animal tissue can also represent substantial sources of organic matter in streams. For example, in a study comparing detrital inputs of cicadas to leaf litter from a dominant riparian tree species, researchers documented that cicada tissue had relatively lower carbon to nutrient ratios, released carbon and nutrients at a faster rate, and decomposed more quickly than leaf litter (Pray et al. 2009). The content and form of nutrients and carbon stored in animal tissue may vary among types of tissue and this may influence biogeochemical cycling. For example, Subalusky et al. (2018) found varied rates of decomposition associated with different animal tissues (e.g., hippo feces, wildebeest soft tissue and bone). Feces and soft tissue had relatively quick turnover times and rapid nutrient leaching, whereas bone decomposed at much slower rates.

7.2.6   Macroconsumers and Detritivory

The fragmentation of leaves by detritivores is an important process in leaf breakdown. Aquatic insects and crustaceans are the most common consumers of CPOM. During the decomposition of autumn-shed leaves in temperate woodland streams, microbial populations play a central role not only in decomposing the leaf substrate, but also in altering the chemical nature of the leaf material, rendering it more palatable and nutritious to consumers. In turn, the feeding activities of detritivores significantly accelerate the decomposition process. Their contribution to the fragmentation of coarse particles through feeding activities and production of feces significantly accelerates breakdown rates and influences subsequent biological processing of the original CPOM.

Several lines of evidence indicate that insects that shred larger pieces of leaves, or shredders, accelerate the breakdown of leaves in streams (Webster and Benfield 1986). The finding that leaf packs in mesh bags decomposed more slowly than those tethered to bricks with fishing line indicated that the former method underestimated breakdown rate, presumably because of the exclusion of detritivores. In addition, leaf breakdown was more rapid when bags with larger mesh size were used, presumably because invertebrate access was greater. Comparison of decay rates in experiments with and without insect detritivores provides further evidence that as much as half of leaf degradation can be attributed to the presence of animals in some systems. Processing rates in two experimental streams, one lacking invertebrates and another stocked with detritivores (*Tipula*, *Pycnopsyche*, and *Pteronarcys*) at densities believed to represent natural maxima, indicated that 21–24% of the loss

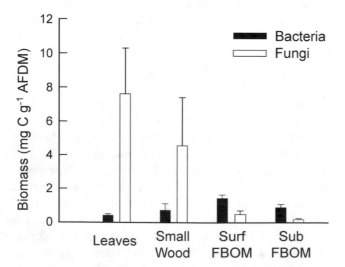

Fig. 7.16 Biomass of fungi and bacteria on leaves, wood, and surface and subsurface fine benthic organic matter (FBOM). Means and standard deviations of 7–9 streams are shown (Reproduced from Findlay et al. 2002)

of hickory leaves was due to the influence of detritivores (Petersen and Cummins 1974). The contribution of macroinvertebrates to the breakdown of *Phragmites*, a macrophyte, was comparable (Polunin 1982). In an Appalachian stream, exclusion of macroconsumers, primarily crayfish, using electric fences resulted in lower breakdown rates of rhododendron leaves (Schofield et al. 2001). When present, crayfish were responsible for 33% of leaf breakdown in summer and 54% in autumn. Sponseller and Benfield (2001) observed faster leaf breakdown with higher shredder density and biomass (Fig. 7.17). In addition to direct consumption, possible influences of detritivore feeding include release of nutrients and dissolved organic matter, comminution of litter, and modification of water circulation (Polunin 1984). The freshwater shrimp *Xiphocaris elongata*, a consumer of large leaves in tropical streams, increased the concentration of both total dissolved nitrogen and dissolved organic carbon, and also the concentration and transport of POM (Crowl et al. 2001).

The experimental removal of detritivorous insects from a small mountain stream in North Carolina provides a particularly convincing demonstration that animal consumers regulate rates of litter decomposition. Wallace et al. (1991,

1982b) added the insecticide methoxychlor to one small stream in February, with supplemental treatments in May, August, and November. Massive downstream drift of invertebrates occurred, and insect densities subsequently were reduced to less than 10% of numbers in an adjacent, untreated reference stream, while oligochaetes increased roughly threefold. Leaf breakdown rates were significantly slower in the treated stream, presumably due to the great reduction in insect density, and the magnitude of the effect was greatest for the most recalcitrant leaf species. Export of suspended fine particulates also was reduced in the treated stream, consistent with the finding of reduced leaf processing.

Studies conducted in Colombia, Costa Rica, Venezuela, Papua New Guinea, and Kenya found that insect shredders were scarce, despite the fact that densities of other invertebrates were similar to those reported in the temperate zone (Yule 1996; Dobson et al. 2002; Rincon et al. 2005; Wantzen and Wagner 2006). In these tropical locales, shredders represented less than 7% of total macroinvertebrate abundance, while in European rivers this value ranged from 10–43% (Dobson et al. 2002; Hieber and Gessner 2002). Nonetheless, leaf breakdown rates in streams in Costa

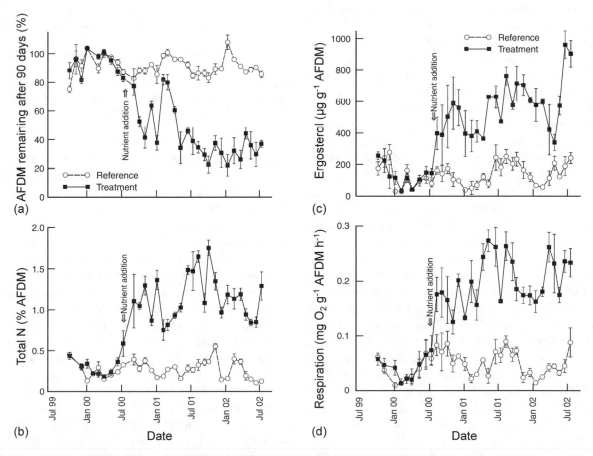

Fig. 7.17 Variation in (**a**) AFDM, (**b**) total N (as % of AFDM), (**c**) ergosterol and (**d**) respiration during a 90-day decomposition experiment using oak veneers incubated in a treatment stream (N and P additions) and a reference stream (no nutrient addition) (Reproduced from Gulis et al. 2004)

Rica (Rosemond et al. 1998) and in Colombia (Mathuriau and Chauvet 2002) were rapid, suggesting a greater role of microorganisms in tropical streams. Possible explanations for these findings include higher temperatures in the tropics, reduced lability of leaves due to the presence of defensive compounds, incorrect assignment of tropical invertebrates to trophic roles, or failure to detect the presence of large shredders (Irons et al. 1994; Schofield et al. 2001; Wantzen et al. 2002). In contrast to some reports of low shredder density in tropical streams, Cheshire et al. (2005) found that shredders represented 20% of total macroinvertebrate abundance in the Australian tropics, similar to values reported from the temperate zone. These authors suggested that shredder scarcity in tropical streams may reflect historical biogeography rather than a strict latitudinal effect, because in some temperate regions such as New Zealand, shredder insects also are scarce. Studies in streams in Colombia (Chará-Serna et al. 2012), Malaysia (Yule et al. 2009), and Panama (Camacho et al. 2009) also suggest that tropical shredders may be key in some locations, and that in tropical systems, site-specific shredder diversity may vary predictably with elevation, water temperature, and water chemistry.

Understanding how the lability, quantity, timing, and duration of allochthonous inputs to streams influence the structure aquatic food webs is a complex undertaking, but the importance of detrital resources in supporting stream food webs is well documented. Studies that have experimentally manipulated the quantity of subsidy inputs have enhanced our understanding of the importance of detritus in supporting ecosystem productivity (Venarsky et al. 2018). Drawing upon decades of observational and experimental data of research in forested, headwater streams in the southeastern US, Wallace et al. (2015) describe how the abundance and diversity of aquatic macroinvertebrates changes in response to whole-stream manipulations of benthic organic matter. Long-term data comparing reference streams to those subjected to whole-stream organic matter manipulations showed that in mixed substrate habitats, the amount of benthic OM was positively related to monthly invertebrate biomass and annual secondary production (Fig. 7.18), and this was especially pronounced in shredder populations. Their work emphasized that the organic matter remaining in the stream following litter exclusion and wood removal was more recalcitrant than that in the reference streams, whereas the litter inputs to reference streams was more labile and more easily incorporated into invertebrate biomass. Predator production was strongly correlated with total invertebrate production, providing evidence that terrestrially-derived resources are essential in supporting freshwater food web structure.

Investigations into CPOM dynamics have frequently been conducted in forested streams, where detrital inputs typically

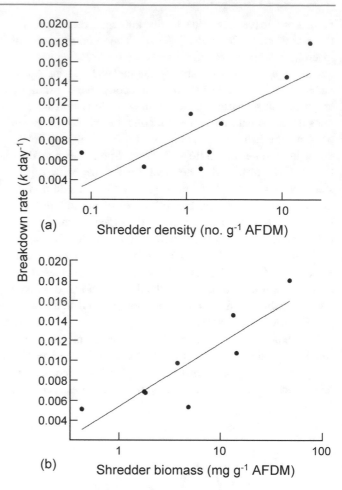

Fig. 7.18 Correlations between leaf breakdown rates and (**a**) density and (**b**) biomass of shredders expressed per g of leaf AFDM (Reproduced from Sponseller and Benfield 2001)

exceed consumer energy demands and where many populations of freshwater organisms have evolved to take advantage of specific temporal and spatial characteristics of subsidies. However, in streams with limited or no riparian vegetation, including cave, tundra, and desert systems, CPOM inputs may be reduced and seasonality not apparent (Tank et al. 2010). For instance, cave systems typically have relatively limited autochthonous production, and allochthonous inputs may be scarce as well; therefore, consumer demand for carbon is predicted to be high relative to carbon availability. Upon addition of maize to an experimental reach of the Bluff River Cave in the southeastern US, the abundance and biomass of organisms increased at all trophic levels, suggesting that carbon from the maize was rapidly assimilated by fungi and bacteria (Fig. 7.19), and that this energy was transferred to higher trophic levels (Venarsky et al. 2018). Intriguingly, only facultative cave taxa, which were also found in the connected surface streams, contributed to the community-level response and increased in abundance. In contrast, obligate cave species showed no

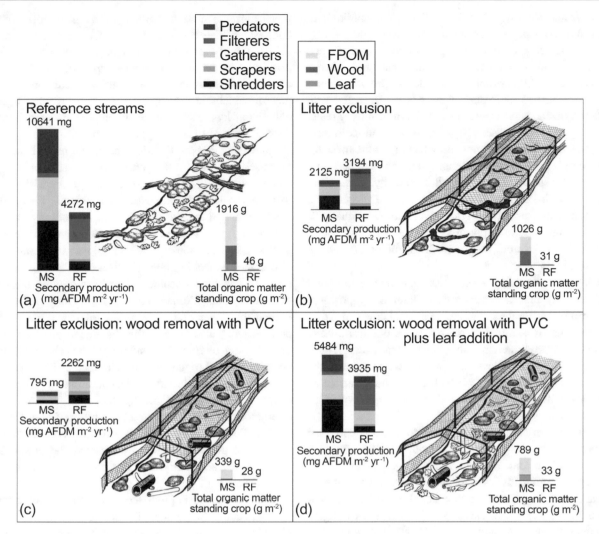

Fig. 7.19 Organic matter standing crops (g m^{-2}) and secondary production (mg AFDM m^{-2} yr^{-1}) in mixed substrate (MS) and rockface (RF) habitats, for (**a**) reference streams (n − 24 years); (**b**) end of third year of litter exclusion period; (**c**) end of small- and large-wood removal (n = 4 years) and the first year of PVC pipe addition period, and (**d**) end of fifth year of leaf addition. The litter exclusion canopy was maintained during the experimental period over the treatment stream. The values reflect total secondary production and organic matter standing crops for the reference streams, and the totals at the end of each experimental period for the litter exclusion stream. Rock face habitats characterized by low organic matter standing crops are drawn at a different scale from the mixed substrate organic matter standing crops. In contrast, secondary production values are drawn at the same scale for both habitats (Reproduced from Wallace et al. 2015)

response to the experimental treatment, indicating that evolutionary history may constrain the response of taxa to novel subsidies (Venarsky et al. 2018). In other words, this work suggests that species may have evolved under conditions of fluctuating nutrients and energy in order to be able to effectively respond to changes in resource availability.

Because of variation in the timing of leaf fall, species-specific rates of leaf conditioning by fungi, and the diversity of fungi present, leaves on the streambed are a mosaic of patches of microbial populations. The extent of this variation is shown by Bärlocher's (1983) study of a Swiss stream flowing through an alder-willow-maple forest. The standing crop of CPOM was maximal in October-November, and by April only veins and petioles remained for an 85% loss of leaf mass. Soluble protein also decreased after November, indicating that nutritional value declined from that time on. However, on any one date, the chemical characteristics of individual leaves were so variable that the amount of soluble protein in the richest 10% of leaves in mid-April exceeded the median value of leaves sampled in mid-November. Detritivores capable enough or fortunate enough in patch choice could enjoy food quality well above average.

Studies have found that shredders can discriminate among leaves colonized by different fungal species, showing preference for the fungi that induced the greatest growth. *Gammarus pseudolimnaeus* preferentially consumed leaf disks colonized by three of ten fungal species examined

(Bärlocher and Kendrick 1973). Similarly, limnephilid caddisflies showed a preference for aspen leaves colonized by fungal species linked to higher growth rates (Arsuffi and Suberkropp 1986). However, animals may not always effectively select higher quality food items. Growth of third and the fifth instars of *Pycnopsyche gentilis* larva fed with fungal-colonized leaves was always significantly greater than the growth of larvae fed with diets of un-colonized leaves; however, larvae did not show feeding preferences for leaves colonized with aquatic fungi (Chung and Suberkropp 2009). Final instars of the limnephilid caddisfly, *Clistoronia magnifica* require triglycerides for reproductive success, and preferentially consumed alder leaf-disks coated with lipids extracted from aquatic hyphomycetes, but third and fourth instars did not exhibit preference (Cargill et al. 1985a; Cargill et al. 1985b).

Where leaf shredding insects are rare or absent, other organisms may play a similar role in litter decomposition. Fish, shrimp, and crabs enhanced leaf breakdown in Costa Rican streams (Rosemond et al. 2002), as did shrimp and other macroinvertebrates in a stream in Puerto Rico (Wright and Covich 2005a). In New Zealand headwater streams, the crayfish *Paranephrops zealandicus* has a strong influence on the decomposition of leaf litter and the production of POM (Usio and Townsend 2001).

Though not as commonly studied, detritivores may also contribute to the processing of wood material via consumption (Benke and Wallace 2010; Eggert et al. 2007). Wood-eating organisms (xylophages), appear to be relatively rare, especially in the temperate zone. However, some caddisflies, beetles, and flies have been classified as obligate xylophages (Hoffmann and Hering 2000). In Neotropical rivers, several members of the suckermouth-armored catfishes (Loricariidae) feed on coarse woody debris. Interestingly, the ability to consume wood appears to have evolved multiple times, as species from two unrelated evolutionary lineages have evolved jaw morphologies that allow them to gouge and eat wood (Lujan et al. 2011).

7.2.7 Anthropogenic Change and Decomposition

Dominant land use may influence in-stream rates of decomposition, but findings are inconsistent among studies. Paul et al. (2006) found that leaves decomposed faster in agricultural (0.0465 day^{-1}) and urban (0.0474 day^{-1}) streams than in suburban (0.0173 day^{-1}) and forested (0.0100 day^{-1}) streams in the southeastern US. Faster breakdown in agricultural streams was attributed to enhanced biological activity supported by higher ambient nutrient concentrations, whereas increased rates of decomposition in urban settings likely were due to enhanced

physical fragmentation resulting from higher storm runoff. In contrast, in Puerto Rican streams, Classen-Rodríguez et al. (2019) found that urbanization negatively affected decomposition rates relative to forested streams, possibly because of enhanced sedimentation and burial of leaf bags and fewer macroinvertebrate decomposers in urban streams. Organisms colonizing leaf packs in the urban system were limited to species tolerant of low-oxygen conditions.

Increasing carbon dioxide concentrations and subsequent increases in temperature associated with anthropogenically-driven climate change are also expected to influence OM decomposition in streams. Elevated temperature may increase decomposition rates by increasing metabolic rates of microbial and macroinvertebrate decomposers (Flury and Gessner 2011). Higher temperatures may also enhance leaching rates (Batista et al. 2012), which can result in fast litter decomposition by removing the recalcitrant compounds (Amani et al. 2019). As long as they do not exceed the upper limits of species tolerance, elevated temperatures may also increase the abundance, growth rates, and reproduction of microbes, subsequently enhancing decomposition rates (Moghadam and Zimmer 2016). In contrast, elevated carbon dioxide concentrations decrease litter decomposition rates by shifting the carbon to nutrient ratios of organic matter and reducing the lability of litter (Tuchman et al. 2003). Higher carbon dioxide concentrations, induce changes in plant physiology, increasing the production of carbon-based secondary and structural compounds (Stiling and Cornelissen 2007).

Climate change is predicted to increase average temperatures and the frequency of extreme weather events, and affect the composition of freshwater communities through changes in temperature and hydrology (Wenisch et al. 2017). Therefore, it is essential to understand how environmental variables influence the distribution and density of organisms through periods of climatic variability (Gutierrez-Fonseca et al. 2018). In a 15-yr study examining changes in stream macroinvertebrate communities in lowland Costa Rica, Gutierrez-Fonseca et al. (2018) demonstrated that macroinvertebrate richness and abundance declined with increasing discharge and were positively related to the number of days since the last high flow event. Their work suggests that macroinvertebrate community structure is ultimately the result of large-scale climatic phenomena, such as El Nino/Southern Oscillation (ENSO), and changes in hydrology associated with climate change have the potential to restructure the invertebrate community and alter their functional contribution to ecosystem-level processes (Gutierrez-Fonseca et al. 2018). In two streams in the Palatinate forest, south-western Germany, Wenisch et al. (2017) compared bulk leaf decomposition rate and the leaf processing efficiency of shredders in enclosures containing three shredder diversity treatments, where species loss was simulated based on their sensitivity to climate change. They found that litter decomposition rates were strongly affected by changes in the macroinvertebrate community, with a 33% increase and 41% decrease in decomposition following species loss at the first and second site, respectively. Researchers attributed the conflicting results to the traits of sensitive taxa. In the first site, the least sensitive taxa had more biomass that may have compensated for the loss of sensitive taxa. However, in the second site, species that were sensitive to shifts in climate played a larger role in decomposition that was not compensated for by the remaining species. Their findings highlight how local diversity in species trait composition may buffer the potential effects of climate change on ecosystem processes (Wenisch et al. 2017).

Temperature-mediated shifts in microbial respiration may enhance the flux of carbon dioxide from streams, reducing the capacity of streams to store carbon. Tropical watersheds emit large amounts of carbon dioxide to the atmosphere (Raymond et al. 2013); therefore, shifts from macroinvertebrate shredding to microbially-mediated decomposition has the potential to alter carbon dynamics at larger spatial scales (Boyero et al. 2012). The interactive effects of increasing temperature and carbon dioxide on decomposition in streams are still emerging. From a meta-analysis of data from about 40 field and laboratory studies conducted between 1993 and 2017, Amani et al. (2019) found that elevated temperature significantly increased litter decomposition rates, but elevated carbon dioxide concentrations did not.

Anthropogenically-derived changes in nutrient concentrations also may have large effects on OM decomposition rates (Rosemond et al. 2015). At lower concentrations, increasing nutrient levels can enhance the nutrient content of leaf litter and accelerate litter breakdown by microbes and macroinvertebrates (Demi et al. 2018; Cross et al. 2006). This can be especially pronounced in more recalcitrant species, and reduce species-specific differences in decomposition (Manning et al. 2015). Though carbon storage in systems might initially buffer subsequent changes to stream food webs, persistent increases in microbial respiration and carbon export could have significant impacts on the duration of litter resources in streams. This may negatively affect the quantity of carbon stored in litter and the carbon transferred to macroinvertebrates within a given reach (Rosemond et al. 2015; Wallace et al. 2015; Kominoski et al. 2018), suggesting that nutrient pollution may limit macroinvertebrate productivity as carbon resources become limited (Halvorson et al. 2017).

Introduced plant species can influence the relative lability of leaf litter entering streams. In an investigation comparing CPOM in streams dominated by native riparian vegetation, pasturelands, or introduced rhododendron (*Rhododendron ponticum*), Hladyz et al. (2011) reported that CPOM in rhododendron-dominated sites had much greater carbon to nutrient ratios, suggesting they were more recalcitrant. Introduced species may also affect interactions between detritivores and organic matter. Kiffer et al. (2018) compared the effect of leaves from native species and non-native *Eucalyptus globulus* on the feeding activity and performance of larvae of a common caddisfly shredder, *Triplectides gracils*, in an Atlantic Forest stream in Brazil. The larvae preferred to feed on softer leaves, regardless of the nutrient content and concentrations of secondary compounds. When softer, native species were not present, caddisfly larvae preferred the exotic species to tougher native species. However, larva that were fed the non-native species lost biomass through time, and larval survival was lower on *Eucalyptus* compared to four of the five native species. These data provide further evidence that species-specific traits of litter, such as leaf toughness, influence the behavior of detritivores, and highlight that changes in the composition of riparian vegetation may negatively impact the growth and survival of native taxa (Kiffer et al. 2018).

Native biodiversity loss is a threat to aquatic communities and ecosystems throughout the world, and may influence OM processing rates in rivers and streams. Though the effects of plant species richness on decomposition are variable among studies, litter mixing studies suggest the greater species richness of litter tends to accelerate decomposition rates (Handa et al. 2014) and the rate of FPOM production (López-Rojo et al. 2018). Within streams, the effects of increased macroinvertebrate diversity tend to be greater in enhancing rates of litter decomposition when litter packs are more diverse. As consumers specialize on different types and sizes of organic matter, increasing trophic diversity in streams tends to enhance the volume and rate of energy moving to higher trophic levels (Tonin et al. 2018). Hence, anthropogenically-induced changes in the spread of disease, land use, and climate are expected to alter detritivory in streams.

The decline of stream-breeding amphibians in response to the spread of Chytridiomycosis, a disease caused by infections of the *Batrachochytrium dendrobatidis* fungus, has ecosystem-level consequences. Experimental manipulations of *Smilisca*, a widespread and abundant genus of frog in Central American streams, demonstrated that tadpoles affect leaf decomposition by influencing microbial community dynamics through their excretion of nutrients and feeding activities (Rugenski et al. 2012). This research, and the collective body of work examining amphibian declines in streams, suggests that the loss of a suite of species may have widespread consequences on fundamental ecosystem processes, such as OM decomposition (Whiles et al. 2013; Colon-Gaud et al. 2010).

7.3 Sources and Processing of Fine Particulate Organic Matter

Fine particulate organic matter (FPOM) is found suspended in the water column as seston, and in the benthos of streams as fine benthic organic matter (FBOM). It includes all of the fine material, including algae, bacteria, detritus, and sediment, that ranges from 0.45 μm to 1 mm in size (Hutchens et al. 2017). The origin of FPOM in streams can vary widely, and it includes both allochthonous and autochthonous sources. Fine particulates arise from the physical breakdown of instream CPOM, microbial processing and animal waste, flocculation or complexation of dissolved substances, and runoff from the surrounding terrestrial environment.

7.3.1 Temporal and Spatial Variation in FPOM

Much less is known about the energy pathways involving FPOM than CPOM. One source of FPOM obviously is the breakdown of leaf litter (Fig. 7.7) as fragments from mechanical breakdown and shredder activity enter the water column. When one includes the production of feces and the eventual contribution of leached DOC to formation of fine particles, it is apparent that a large fraction of leaf litter eventually becomes fine particulate matter. In addition, microbial activity on the leaf can result in the release of leaf and bacterial cells and the hyphae and conidia of fungi (Gessner et al. 1999). Fine particulates are also transported into streams from the terrestrial landscape by wind and runoff, and can be formed from dissolved organic matter primarily by the incorporation of DOC into microbial biomass. These sources are difficult to trace and potentially of greater magnitude than FPOM derived from leaf fragmentation.

The amounts and characteristics of FPOM fluctuate along and within a river network, and they are linked to season, climate, dominant land use, and underlying geology. Particulate concentrations often show a positive relationship with regional precipitation, probably due to the influence of precipitation on terrestrial production and the subsequent effect on the supply of CPOM. Floodplains can be a large source of organic matter, and FPOM concentrations often increase in lowland rivers during inundation (Golladay 1997). Terrestrially-derived FPOM inputs are typically low during base flow and increase during storms and seasonal high flows as rising flows entrain particles from stream banks and side channels (Mulholland 1997a). A comparison of 31 streams and medium-sized rivers in North America, Europe, the Arctic, and Antarctica found FPOM to range between 0.14 and 15.30 mg L^{-1} (Golladay 1997). In streams and rivers of eastern US, POM estimates ranged between 0.5 and 52 mg L^{-1} (Webster et al. 1995); in undisturbed forested catchments, mean annual concentrations often are less than 2 mg L^{-1} (Fisher and Likens 1973; Naiman and Sedell 1979).

The concentration of FPOM is also affected by changes in particle availability. Controls of particle availability include terrestrial production, instream seasonal change in biological processing, and discharge, which varies both seasonally and unpredictably. In streams of the eastern United States during normal flows, FPOM concentrations are higher during spring and summer than during autumn and winter (Fig. 7.20), probably because lower biological activity during the colder months results in less instream particle generation (Webster et al. 2006). If the particle supply is relatively constant, then an increase in discharge will cause a dilution effect, even if the total amount of FPOM in transport is greater.

An increase in streamflow in response to a rainstorm results in a corresponding increase in particle concentrations as POM generated during low flows and stored in depositional areas is entrained by rising stream levels (Fisher and Likens 1973; Meyer and Likens 1979). This indicates that the major pool of particulate organic matter lies in areas already wet or adjacent to the stream's wetted perimeter, where FPOM accumulates during low flows. Inputs of POM from outside of the stream also are likely to be greatest during the rising limb of the hydrograph, due to the erosive effects of rainfall on soil and streambank litter and generation of flow in previously dry channels. During the descending limb, water enters the stream principally by subsurface flow, and carries little or no POM. Thus, concentrations are highest on the rising limb of the hydrograph (Fig. 7.21) and then decline due to exhaustion of the particle supply and dilution of the entrained material.

Because particulate concentrations usually are higher when discharge is greater, most FPOM is transported during episodic and seasonal floods, and thus flow conditions that occur during only a small fraction of the annual discharge cycle can account for a very large fraction of annual transport. Some 75–80% of POM transport in small streams in the southeastern US occurred during storms (Webster et al. 1990), demonstrating the importance of accurately sampling these episodic events. In fact, unless sampling is continuous or captures high flows very thoroughly, total transport may be seriously under-estimated.

In addition to weather and climate, other factors, including the composition of riparian vegetation, land use, and the presence of impoundments are expected to alter the quantity and timing of FPOM in streams. Forested streams frequently exhibit higher FPOM concentrations that non-forested streams, as was observed in a comparison of native forest to pasture sites in a New Zealand stream (Young and Huryn 1999), and in the aforementioned 31

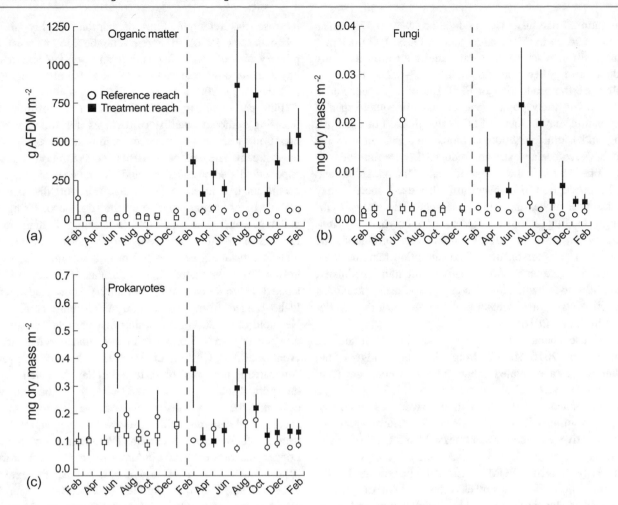

Fig. 7.20 Average values of (**a**) organic matter, (**b**) fungal biomass, and (**c**) prokaryotic biomass in the Bluff River Cave stream in Alabama in the southern US. Samples collected in the reference sites are represented with open circles and samples collected from the treatment reach are represented with open (pre-litter addition) and closed (post-litter addition) squares. The vertical, dashed line in each plot indicates when the litter addition began. Error bars are standard error (Reproduced from Venarsky et al. 2018)

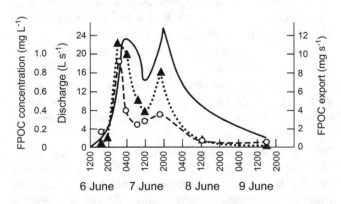

Fig. 7.21 Changes in discharge, FPOC concentrations, and FPOC transport during a summer storm in a small forested catchment in New Hampshire. Note that FPOC concentrations peak on the rising limb of the hydrograph, indicating rapid entrainment of small particulates. A second hydrograph peak resulted in a much smaller FPOC concentration peak, evidence that wash-out rapidly depletes the available FPOC supply. Solid line indicates discharge; O—O denotes FPOC concentration; and, ▲…▲ denotes FPOC export (Reproduced from Bilby and Likens 1979)

streams from different regions (Golladay 1997). However, high FPOM concentrations also are seen in low gradient streams flowing through agricultural or multiple-use catchments (Malmqvist et al. 1978), and especially in larger lowland rivers (e.g., Berrie 1972; Ward 1974).

Animals can also directly contribute to the FPOM pool through their feces (Wotton and Malmqvist 2001). Litter consumption rates and subsequent FPOM production can be influenced by the diversity of leaf litter (Fernandes et al. 2015), as macroinvertebrates that specialize in consuming leaf litter appear to be sensitive to changes in litter species and leaf conditioning by microbes (Santonja et al. 2018). Organisms specializing in FPOM consumption, such as black flies (Simuliidae, Diptera), can also contribute to the FPOM pool. Black fly larvae ingest FPOM and DOM and produce larger particulate material in the form of fecal pellets (Wotton et al. 1998). Hershey et al. (1996) observed a 28% increase in the mass of FPOM and an alteration of the particle size distribution downstream of a filtering black fly aggregation. Each larva can produce on average 575 pellets per day, and so in dense aggregations ($\sim$ 600,000 per m^{-2} within a 40 m reach of an Arctic tundra stream), the daily production of pellets was estimated at 1.3–9.2×10^9.

Because the production of FPOM is often dependent on in-stream consumer populations, changes in aquatic species composition may influence FPOM dynamics. For example, Greig and McIntosh (2006) demonstrated that an invasive trout in New Zealand streams reduced invertebrate density and, consequently, the rates of leaf decomposition and FPOM production. However, an introduced species may increase OM processing when the invader generates FPOM through feeding activities. Red-swamp crayfish *Procambarus clarkii* have introduced populations throughout the world. In a comparison of leaf consumption rates between the invasive crayfish and a native shredding caddisfly, researchers documented that decomposition rates and FPOM production were much greater in the presence of the invader (Carvalho et al. 2016).

Land use change also can influence FPOM inputs, as Epstein et al. (2016) showed from the carbon budget of the Jordan River, a regulated urban river that connects Utah Lake to the Great Salt Lake in the western US. Stable isotope analyses indicated that FPOM in the river was more isotopically similar to autochthonous source materials, including productive, eutrophic water released from Utah Lake and in-stream primary producers, than it was to terrestrially-derived CPOM. Temporal changes in FPOM concentration in the river reflected the flow release patterns from Utah Lake, indicating that eutrophication and water regulation associated with urban development can have strong effects on FPOM dynamics in streams.

7.3.2   Storage and Transport of FPOM

Depositional areas, especially pools and around wood accumulations, are key locations of FPOM retention in the benthos. Logs are a particularly effective retention device in low-order streams of forested catchments. Accumulations of organic matter formed when wood becomes lodged against obstructions trapping smaller material and leaves into a nearly watertight structure. Sediments and organic matter settle in the pools formed upstream of these dams, creating potential hotspots of detrital processing. Following experimental removal of all organic dams from a 175-m stretch of a small New Hampshire stream, organic matter export increased several-fold (Bilby 1981; Bilby and Likens 1980). When beaver *Castor canadensis* were unexploited they must have contributed greatly to organic matter storage over large areas of the north temperate zone. Where beaver occur at natural densities today, their activities influence 2–40% of the length of second- to fifth-order streams, and

increase the retention time of carbon roughly six-fold (Naiman et al. 1986). In large rivers, the floodplain can be a primary site of POM deposition and storage and macrophyte beds are important retention features for POM (Wanner et al. 2002).

Fine particulate transport distance can be estimated by releasing a known quantity of particles into the stream and measuring water column concentrations at various distances downstream. The decline in particle concentration is fit to an exponential decay equation, and the inverse of the decay coefficient is a measure of average transport distance of a particle before being retained on the streambed. Using corn pollen as a surrogate for FPOM (it is similar in diameter but less dense), Miller and Georgian (1992) estimated mean transport distances of 100–200 m in a second-order stream in New York. Estimated transport distance for natural FPOM labeled with radiocarbon in the Salmon River headwaters of Idaho ranged from 150 to 800 m (Cushing et al. 1993; Newbold et al. 2005). Assuming that particle resuspension occurs every 1.5 to 3 h and an average transport distance per event of 500 m, Cushing et al. (1993) calculated an average downstream transport of 4 to 8 km day^{-1}. Much greater transport distances, between 3,000 and 10,660 m, were estimated for a sixth-order lowland river, using spores of *Lycopodium clavatum* with a fluorescent label, and distances were greater under faster currents (Wanner and Pusch 2001). Longer transport distances in larger rivers may reflect fewer opportunities for particle entrapment, whereas the extent of water exchange between surface flows and the hyporheic zone has been shown to correlate with transport distances in smaller streams (Minshall et al. 2000). Owing to the relative slow rate of utilization of FPOM estimated from respiration measurements and the combination of relatively long transport distances with frequent re-suspension, export rather than mineralization appears to be the dominant fate of FPOM from studies of smaller streams and rivers.

Effectively measuring the inputs and standing stocks of FPOM can be challenging, as they require estimates of concentrations in suspension and standing stocks on the stream bed (Hutchens et al. 2017). Instantaneous seston concentrations (mg L^{-1}) can be measured by filtering a known volume of stream water through a glass fiber filter that has been weighed and ashed to remove organics. The filters are dried, weighed, ashed, and re-weighed to calculate ash-free dry mass, an estimate of organic matter content. In larger systems, seston concentration estimates may need to be depth-integrated and velocity-weighted to account for differential settling rates and current velocities along the width and depth of the river. Newer methods to estimate FPOM concentrations integrate data collected with sensors or through remote sensing tools that assess turbidity or total suspended solids (measurements that include organic and inorganic constituents) with the aforementioned methods to account for the organic content of the suspended material. Sampling for benthic FPOM should account for the diversity and distribution of habitat types within a reach. Specifically, characteristics of the stream, including presence or absence of retention features (e.g., large wood, floodplains, dams), current velocity, and substrate particle size need to be taken into account when estimating the standing stock of fine particles in the benthos (Pfeiffer and Wohl 2018). In smaller systems, substrate cores can be filtered through sieves to estimate benthic FPOM. However, larger sites may require the use of SCUBA equipment or dredges and may require trained operators to safely support sample collection (Hutchens et al. 2017). Fluorescently labeled latex particles (Harvey et al. 2012) and titanium dioxide particles (Karwan and Saiers 2009) have been used to investigate deposition and entrainment of ultrafine particles (Hutchens et al. 2017). A review of FPOM sampling and processing methods and detailed protocols can be found in Hutchens et al. (2017).

7.3.3 Lability of FPOM

As discussed previously, the lability of organic matter is only approximated by some aggregate measure of the rate at which it is processed, as in reality, a number of different measures are informative. Lability of FPOM can be assessed by such characteristics as stoichiometry, enzymatic activity, respiration, and extractable amounts of nitrogen and phosphorus, among others. The physical and chemical aspects of fine particulates originating from terrestrial ecosystems are affected by the composition of the terrestrial vegetation, position along the river network, and temporal variation in discharge. Several characteristics of benthic FPOM differed among Oregon stream sites in old growth forest dominated by Douglas fir and western hemlock, and young growth stands of Douglas fir and herbaceous vegetation with abundant deciduous trees in the riparian zone (Bonin et al. 2003). The FPOM of streams flowing through young growth stands had lower carbon to nitrogen ratios and higher denitrification potential, as well as greater extractable ammonium, phosphatase activity, and respiration rates, suggesting greater lability and microbial activity compared with streams in older stands. The lability and amount of inputs of terrestrial POM likely are higher at young growth sites, and an observed increase in microbial activity following a storm is evidence of system response to a pulsed input. Benthic respiration and thus microbial activity have also been observed to increase as one proceeds downstream (Webster et al. 1999), which may be the result of changes in organic matter lability as well as higher temperatures, greater nutrient availability and increases in substrate lability related to higher algal POM inputs (Fig. 7.22).

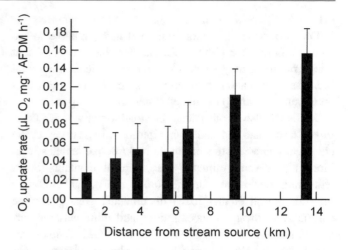

Fig. 7.22 Downstream variation in benthic FPOM respiration in Ball Creek-Coweeta Creek at the Coweeta Hydrologic Laboratory. Values shown are means and standard errors for all sampling dates for each site (Reproduced from Webster et al. 1999)

Even in headwater streams, FPOM originates from sources other than fragmentation of leaves, and this influences the palatability and physical and chemical recalcitrance of FPOM. Based on rough calculations of the magnitudes of inputs attributable to soil organic matter and the breakdown of wood, Ward and Aumen (1986) concluded that leaves and needles were minor sources of FPOM. Using flotation to separate organo-mineral particles from fragmented plant material collected from small forested streams in Oregon, Sollins et al. (1985) found that the majority of detrital carbon and nitrogen was present as organic material adsorbed on mineral surfaces, rather than as plant fragments. In an analysis of FPOM in the Amazon River, Hedges et al. (1986) compared the signatures of river particles to various potential organic sources, finding that the majority originated as soil humic material. At least for large rivers, this conclusion appears to be general (Bernardes et al. 2004; Onstad et al. 2000; Townsend-Small et al. 2005). Based on the ratio of carbon to nitrogen atoms in river seston world-wide, the majority of riverine FPOM most closely resembles soil organic matter. For instance, Spencer et al. (2016) demonstrated that organic particulates in the Congo River appeared to be sourced from soil-derived, mineral-associated organic matter. However, the relative proportion of fresh vascular plant material in the particulates increased with increasing discharge.

Lower C:N in FPOM relative to CPOM and DOM could be related to the presence of microbial biomass or clay minerals (Hedges et al. 1986; Devol and Hedges 2001). The ratio of C:N in particulates that have been measured in three of the world's largest rivers are similar (Congo River, 9.9–12.4 (Spencer et al. 2016); Amazon River, 6.8–13.2 (Hedges et al. 1986; Hedges et al. 1994); lower Mississippi River,

9.1–11.6 (Bianchi et al. 2007)). The values reported for FPOM are lower (i.e., more enriched in N) relative to the values reported for CPOM. Researchers have asserted that this enrichment may be due to nitrogen-enriched microbial biomass in soils and the preferential sorption of nitrogen-rich compounds to fine particulates (Spencer et al. 2016).

Evidence also suggests that the stoichiometry of FPOM is linked to climate and underlying geology. In a survey of the chemical characteristics of coarse and fine particulates from biomes of diverse climates (i.e., tropical montane forest, temperate deciduous forest, boreal forest, and tallgrass prairie), Farrell et al. (2018) documented greater average carbon and nitrogen concentrations and more variable carbon to nutrient ratios in CPOM that in FPOM. When compared to FPOM chemistry, phosphorus content was comparatively reduced and nitrogen to phosphorus ratios were greater in CPOM from tropical montane and temperate deciduous forests. However, CPOM was richer in phosphorus concentrations and had similar nitrogen to phosphorus ratios as observed for fine particulates in grassland and boreal systems (Fig. 7.23). The authors suggest that the microbial processing that transforms CPOM to FPOM results in more tightly constrained elemental composition, and subsequently less variability in the stoichiometry of FPOM. Stable isotope data collected in the same study suggested that fine particulates originated primarily from CPOM in tropical montane and temperate deciduous climates, but were derived from a mixture of detrital and non-detrital resources in the other biomes (Farrell et al. 2018). These observed patterns in CPOM and FPOM

dynamics may influence trophic dynamics, as shredding invertebrates that depend on relatively carbon-rich and nutrient-poor food resources may more commonly experience nutrient limitation. In contrast, organisms feeding on FPOM may be more prone to carbon limitation, as their food sources are relatively nutrient rich.

The ingestion of particles by collector-gatherers and filter-feeders acts both as retention and utilization, as some fraction of the ingested material is digested and metabolized by the animal consumer. There does not appear to be any estimate of the potential magnitude of this effect owing to benthic collector-gatherers. However, filter-feeders typically consume only a small fraction of transported particles, less than 1% of annual transport in one estimate (Webster 1983). Thus microorganisms, mainly bacteria, appear to be responsible for most of the breakdown and re-mineralization of the organic carbon of FPOM that occurs within stream ecosystems. However, this interpretation may reflect inadequate study of benthic detritivores. Rosi-Marshall and Wallace (2002) estimated considerable ingestion of amorphous detritus by macroinvertebrates in a mid-order river in North Carolina, but the influence on system-level degradation of benthic FPOM is largely unknown.

While the quantity of FPOM consumed is poorly documented, it can be approximated by the measurement of respiration rates. Using laboratory measurements of benthic FPOM respiration, Webster et al. (1999) estimated breakdown rates 0.00104/day, for a half-life of about 1.8 years. Respiration rates are expected to decline over time as FPOM mass is lost, leaving more recalcitrant material behind, but

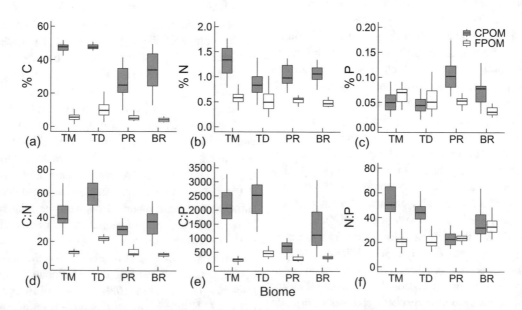

Fig. 7.23 Box and whisker plot of the elemental mass (%) of (**a**) carbon, (**b**) nitrogen, and (**c**) phosphorus, and the stoichiometric ratios (molar) for (**d**) carbon/nitrogen, (**e**) carbon/phosphorus, and (**f**) nitrogen/phosphorus. Coarse particulate matter (CPOM) is denoted by gray bars and fine particulate matter (FPOM) is denoted by white bars. All organic matter was collected from benthic habitats within stream networks in four biomes: Tropical Montane (TM), Temperate Deciduous (TD), Prairiegrass (PR), Boreal Forest (BF). Boxes reflect the 25th and 75th percentiles and whiskers reflect the 10th and 90th percentiles. (Modified from data presented in Farrell et al. 2018)

supporting evidence is weak (Sinsabaugh et al. 1992). Seasonal and latitudinal variation in temperature has a strong influence on respiration rates and thus on the utilization of FPOM (Webster et al. 1999). Microbial decomposition of FPOM correlates with rates of the original CPOM, but FPOM decomposes more slowly (Marks 2019). This may be because macroinvertebrate processing removes nutrients and labile carbon from detritus, producing FPOM with higher carbon to nutrient ratios and greater lignin concentrations (Santonja et al. 2018; Yoshimura et al. 2008).

Though aquatic shredders produce large quantities of FPOM as fragments and egested material, the comparative significance of shredder FPOM as a pathway of nutrient transformation remains understudied. In an experiment conducted with three invertebrate shredders, *Pycnopsyche lepida*, *Lepidostoma* sp., and *Tipula abdominalis*, Halvorson et al. (2015a) fed organisms conditioned oak or maple litter of varying dietary carbon to nutrient ratios. FPOM production differed significantly among species. Additionally, insects fed high-nutrient diets produced FPOM with greater microbial decomposition rates, which varied with types of leaf litter. This work suggests that changes in litter type and ambient nutrient concentrations may modify production rates, elemental composition and stoichiometry, and decomposition rates of FPOM produced by detritivores that may vary among shredder species.

7.3.4 The Influence of Anthropogenic Activities on Fine Particulates

Human-mediated environmental pressures may alter the physical and chemical characteristics, quantity, timing, and duration of fine particulates entering the environment, potentially with profound effects on stream ecology. For example, changes in water chemistry may influence the lability and toxicity of FPOM. Nutrient enrichment resulted in greater increases in both fungal and bacterial biomass for CPOM than FPOM (Tant et al. 2013). The potential for FPOM to interact with chemical stressors in streams is also exceedingly high due to sorption and complexation processes (Bundschuh and McKie 2016). When coupled with intense rain events in the Amazon, deforestation and petroleum development have resulted in major changes in the timing, quantity, and composition of fine particulates entering the watershed. Increased sedimentation buried CPOM and negatively affected all macroinvertebrates, even the filter-feeding invertebrates that may have benefitted from increased organic particulates (Couceiro et al. 2011). In a study examining downstream changes of POC composition and biodegradability in a mixed land-use watershed in South Korea, Jung et al. (2015) documented significant differences in the concentrations of particulates within the watershed;

particulates were greatest in the agricultural stream, but declined in downstream reaches.

The intensification of storms associated with anthropogenic climate change may also affect the lability, timing, and quantity of particulates entering streams. In a study of particulate exports in runoff from storm events in a small (~ 12 ha) watershed in Maryland in the eastern US, Dhillon and Inamdar (2013) documented that a single storm, Hurricane Irene, generated 56% of all of the organic particulates for 2011. This work suggests that the intensification of storm events may govern fine particulate dynamics in streams.

7.4 Sources and Processing of Dissolved Organic Matter

Dissolved organic matter (DOM) typically is the largest pool of organic carbon in running waters (Fisher and Likens 1973; Karlsson et al. 2005), and fluctuations in quantity and lability can strongly affect microbial metabolism and community structure (Bott et al. 1984; Judd et al. 2006). Dissolved organics comprise a variety of organic compounds that are potential carbon sources for microorganisms, and for this reason and because many studies now rely on automated carbon analysis by combustion of water samples, it often is reported as dissolved organic carbon (DOC). For all practical purposes these terms can be interconverted by assuming that DOM is 45–50% organic carbon by mass. Dissolved organics include labile organic compounds that are available to microorganisms for biological uptake as well as recalcitrant material that is less easily assimilated. Organic forms of nitrogen and phosphorous can be quantitatively significant fractions of DOM, and potentially are available for uptake by algae and heterotrophs. Groundwater, rainfall, throughfall, and leaching of leaves are primary sources of DOM in many systems, but this varies with climate, geology, and land use. Streams and rivers transport large quantities of DOM for great distances (Table 7.2), and some of the world's biggest rivers transport large amounts of DOC to the ocean. The 30 rivers with the greatest annual discharge export 90.2 Tg C year^{-1} as DOC, which represents 36% of global DOC flux by rivers to the ocean (250 Tg DOC-Cyear^{-1}). The Amazon River alone exports 11% of the total DOC flux to the oceans (Raymond and Spencer 2015).

The number of investigations focused on DOM in rivers has increased as new analytical approaches, continuous sensor technology, and advances in statistical analysis have allowed stream ecologist to address questions that were once untenable. This research has established that DOM is comprised of a mixture of compounds whose makeup significantly affects stream food webs, biogeochemical cycling, and the transport and uptake of contaminants (Creed et al. 2015). Methods used to evaluate changes in the lability, quantity, and composition

Table 7.2 Discharge, watershed area, and DOC fluxes and yields for the top 30 rivers ranked by discharge globally. (Reproduced from Raymond and Spencer 2015)

River Rank by Discharge	River Name	Discharge (km³ yr⁻¹)	Area (Mkm²)	DOC Flux (Tg C yr⁻¹)	Global DOC Flux (%)	DOC Yield (gC m⁻² yr⁻¹)
1	Amazon	6590	6.112	26.900	10.8	4.4
2	Congo	1325	3.698	12.400	5.0	3.4
3	Orinoco	1135	1.100	4.98	2.0	4.5
4	Changjiang (Yangtze)	928	1.808	1.58	0.6	0.9
5	Yenisey	673	2.540	4.65	1.9	1.8
6	Lena	588	2.460	5.68	2.3	2.3
7	Mississippi	580	2.980	2.10	0.8	0.7
8	Parana	568	2.783	5.92	2.4	2.1
9	Brahmaputra	510	0.580	1.90	0.8	3.3
10	Ganges	493	1.050	1.70	0.7	1.6
11	Irrawaddy (Ayeyarwady)	486	0.410	0.89	0.4	2.2
12	Mekong	467	0.795	1.11	0.4	1.4
13	Ob'	427	2.990	4.12	1.6	1.4
14	Tocantins	372	0.757	1.12	0.4	1.5
15	Amur	344	1.855	2.50	1.0	1.3
16	St. Lawrence	337	1.780	1.55	0.6	0.9
17	Mackenzie	316	1.780	1.38	0.6	0.8
18	Zhujiang (Pearl)	280	0.437	0.40	0.2	0.9
19	Magdalena	237	0.235	0.47	0.2	2.0
20	Columbia	236	0.669	0.40	0.2	0.6
21	Salween (Thanlwin)	211	0.325	0.23	0.1	0.7
22	Yukon	208	0.830	1.47	0.6	1.8
23	Danube	207	0.817	0.59	0.2	0.7
24	Essequibo	178	0.164	0.89	0.4	5.4
25	Niger	154	1.200	0.53	0.2	0.4
26	Ogooue	150	0.205	1.25	0.5	6.1
27	Uruguay	145	0.240	0.50	0.2	2.1
28	Fly	141	0.064	0.55	0.2	8.6
29	Lkolyma	136	0.650	0.82	0.3	1.3
30	Pechora	131	0.324	1.66	0.7	5.1
Total		**18,553**	**41.64**	**90.24**	**36.30**	**70.20**
Average		*618*	*1.39*	*3.01*	*1.21*	*2.34*

of DOM are varied, and are reviewed in Findlay and Parr (2017) and Ruhala and Zarnetske (2017). The installation of wells or lysimeters, the use of continuous sensors, and the collection of water samples are all methods used to measure DOM inputs to streams. Optical properties of DOM can be evaluated using fluorescence, which can be used to characterize the origin and lability of DOM. Advances in spectroscopic techniques, including absorbance and fluorescence, have also advanced our understanding of the role of DOM in supporting aquatic communities and ecosystems. Biochemical characteristics of DOM can be related to optical properties; hence, the characterization of DOM fluorescence can generate information about the source, redox state, and biological reactivity of DOM (Fellman et al. 2010).

7.4.1 Sources of DOM

Concentrations of DOM vary with terrestrial vegetation, soil flow pathways, and the presence of wetlands, and may be enriched by domestic sewage or agricultural runoff. Allochthonous DOM enters streams primarily through groundwater inputs at low flows, while more DOM enters via lateral flows during storms (Mulholland 1997b). Autochthonous sources of DOM are often seasonally dominant (Raymond and Spencer 2015) and are added to the organic pool via algae (Adams et al. 2018) and macrophytes (Kautza and Sullivan 2016). Wastewater treatment plants can also be a locally important source of DOC, where instream processing can generate abundant DOC from an allochthonous source, human waste (Yates et al. 2019). Much of the DOC exported from smaller systems, especially headwaters, is of terrestrial origin (Royer and David 2005), and this is especially true in forested regions of the temperate zone. In a review of rivers in the US, Spencer et al. (2012) suggest that the DOM in the great majority of rivers is predominantly from allochthonous sources. Exceptions to this pattern, the Colorado, Colombia, Rio Grande, and St. Lawrence, are highly modified with large impoundments that generate large inputs of DOM derived from photosynthetic plankton.

Dissolved organic matter concentrations ranged between 0.5 and 36.6 mg L^{-1} in a survey of 33 streams and rivers in the Caribbean, North America, Europe, and Antarctica (Mulholland 1997b). On a world-wide basis, DOC exceeds POC by approximately 2:1, but this depends on river type and discharge regime. Based on annual means, reported DOC:POC ratios for North American streams range from 0.09:1–70:1 (Moeller et al. 1979). The total export of organic matter from more than 80 Finnish catchments was dominated by DOC, which made up 94% of total organic carbon (Mattsson et al. 2005). In headwaters streams of British Columbia, DOM also represented a high percentage (84%) of the total organic matter export, and the remainder was mostly FPOM (Karlsson et al. 2005). In larger rivers, however, POC and DOC concentrations are similar, and at high discharge POC can exceed DOC (Thurman 2012).

The size division between FPOM and DOM is one of convenience, usually determined by what passes through a 0.45 μm filter. In reality, the dissolved fraction is likely to include some smaller bacteria, viruses, and some colloidal organic matter. Lock et al. (1977) used ultra-centrifugation to examine the colloidal fraction, which was defined by a sedimentation coefficient and estimated to correspond to a spherical diameter between 0.021 and 0.45 μm (perhaps 0.01- 0.5 μm should be considered the general size range for colloidal organic matter). In water from a variety of sources in Canada, the colloidal fraction constituted between 29 and 53% of total DOC.

Between 10–25% of DOM consists of identifiable molecules of known structure: carbohydrates and fatty, amino, and hydroxy acids. The remainder (50–75%, up to 90% in colored waters) can be placed in general categories such as humic and fulvic acids and hydrophilic acids. Humic acids separate from fulvic acids by precipitating at a pH < 2 while fulvic acids remain in solution. Fulvic acids also are smaller than humic acids, which often form colloidal aggregates of high molecular weight and may be associated with clays or oxides of iron and aluminum. Fulvic acids generally are the majority of humic substances (Thurman 2012). In the Amazon, for example, fulvic acids were approximately 50% and humic acids 10% of riverine DOC (Ertel et al. 1986).

Leachate from leaf litter and other POM is often a labile source of DOM. There are species-specific differences in DOC leached from leaves. As discussed previously, leaf leachate often is generated rapidly, incorporated into microbial biomass, and respired by microbes. Labile organic molecules are frequently lost from litter within the first few days after entering streams (Meyer et al. 1998). Typically, they are low molecular weight compounds that often account for a small proportion of pool of DOC but may play a disproportionally large role in supporting bacterial production in aquatic systems (Marks 2019). Some 42% of the autumnal DOC inputs to a small New England stream were attributed to this source (McDowell and Fisher 1976). Exclusion of leaf litter inputs to a stream at the Coweeta Hydrological Laboratory resulted in lower DOC concentrations than in a nearby, untreated reference stream (Meyer et al. 1998). Instream generation of DOC from leaf litter was estimated to contribute approximately 30% of daily DOC exports, and to be greatest in autumn and winter and during periods of increasing discharge rather than at baseflow. DOC concentrations were higher during the fall and early winter in a deciduous woodland stream in Tennessee (Mulholland 2003).

Inputs of DOC are also influenced by catchment characteristics such as geology, soils, and topography. The size of soil carbon pool is a strong predictor of stream DOC concentration (Aitkenhead et al. 1999). Water moving through shallow flowpaths has greater contact with the organic horizon of the soils, and so generally has higher DOC concentrations than are found in groundwater (Frost et al. 2006). Shallow flowpaths can be the result of steep slopes, shallow soils, and the presence of infiltration barriers or of saturated soils, such as wetlands and peatlands (Aitkenhead-Peterson et al. 2003). Deeper flowpaths increase the exposure of DOC to microbes and mineral soils, which can assimilate and adsorb DOC respectively. In the Amazon, differences in texture between oxisols and spodosols, the dominant soil types in clearwater and blackwater river catchments, result in different concentrations of DOC in groundwater and pronounced differences in DOC concentration and composition between these two types of rivers (McClain et al. 1997).

Soil organic matter originating in above-ground and below-ground terrestrial production is a quantitatively dominant source of DOC to fluvial ecosystems (Demars et al. 2018; Demars et al. 2019). Grasslands contain the highest soil organic matter, deserts the least, and forests are intermediate. The interstitial water of soils usually contains higher DOC concentrations due to solubilization of organic litter (Thurman 2012). Most soil DOC is produced in the organic horizon and from leaf litter and root exudate and decay. Enzymes released by soil microorganisms also contribute to the soil DOC pool (Aitkenhead-Peterson et al. 2003). Dissolved OM that reaches stream channels by surface and shallow sub-surface flowpaths is frequently more labile and in higher concentrations than DOM in groundwater, where low concentrations are a consequence of biological and chemical degradation of organic matter and physical adsorption. Mineral soils can also adsorb organic molecules, and an increase in the content of clay and aluminum and iron oxides is usually accompanied by higher adsorption of DOC (Aitkenhead-Peterson et al. 2003).

The contribution of groundwater to instream DOC concentrations is often underestimated (Webster and Meyer 1997; Tank et al. 2010). However, research has demonstrated that connections between groundwater supplies and streams can mediate patterns in stream carbon concentrations (Lupon et al. 2019; Dick et al. 2015). Median values for groundwater DOC are usually less than DOC in shallow soil water (Thurman 2012). In small streams in North Carolina, Meyer and Tate (1983) recorded DOC concentrations of 2–12 mg L^{-1} in soil water in contact with the active root zone, compared to 0.2–0.7 mg L^{-1} in subsurface seeps. Similarly, in the catchment of an Alberta stream, the median DOC concentration in soil interstitial waters was 7 (range 3–35) mg L^{-1}, whereas shallow groundwater in the saturated zone contained 3 mg L^{-1} DOC (Wallis et al. 1981). DOC of terrestrial origin is rich in aromatic components such as lignin and tannins because these compounds are abundant in terrestrial vegetation (Benner 2003), and as we shall see, these compounds are less accessible to microorganisms.

In catchments with a substantial area of wetlands, streams have elevated DOC concentrations due to the accumulation of organic acids that reduce pH, thus slowing bacterial decay (Thurman 2012; Lottig et al. 2013). The proportion of wetlands in their catchments explained up to 70% of the variation in DOC concentrations of Wisconsin rivers (Fig. 7.24), and was a better predictor than riparian wetland extent (Gergel et al. 1999). In catchments of the Adirondack Park of New York state, wetlands contribute 30% of DOC inputs but occupy just 12% of the surface area (Canham

et al. 2004). Rivers draining permafrost catchments like the Kuparuk in Alaska also exhibit high DOC concentrations due to surface flowpaths through organic-rich soil layers (Mulholland 1997b). Based on radiocarbon analysis, DOC exported from wetlands in Ontario, Canada, DOC was of recent origin despite the fact that peat at a depth of 50 cm had an age between 1000 and 2000 years, suggesting that carbon was exported from wetlands primarily through shallow flowpaths (Schiff et al. 1998).

Precipitation is a highly variable source of DOC, influenced by contact with dust and pollen (Aitkenhead-Peterson et al. 2003), and by the intensity of a storm event (Chen et al. 2019; Wise et al. 2019). When rainwater is intercepted by leaves of the forest canopy, leaching removes significant amounts of organic matter. Fisher and Likens (1973) estimated an average value of 17.8 mg L^{-1} for canopy drip in a hardwood forest in New England. Precipitation indirectly affects riverine DOM through its influence over soil moisture and hydrologic flowpaths. Water that moves near the soil surface has greater contact with the organic horizon of soils, resulting in higher DOC concentrations (Mulholland 2003).

A large fraction of seasonal or annual DOC export can occur over short periods of time during high flows (Raymond and Spencer 2015; Yoon and Raymond 2012). In the Yukon River, Canada, DOC concentrations were highest in May and 50% of annual DOC transport took place during spring under high flow (Guéguen et al. 2006). High DOC concentrations were also observed at the beginning of the snowmelt season in a Colorado headwater stream (Hood et al. 2005). Even when DOC concentrations do not increase with discharge, the total annual transport is strongly influenced by hydrological regime because seasonal variation in discharge usually exceeds seasonal variation in DOC concentrations. In pristine boreal forest streams in Quebec,

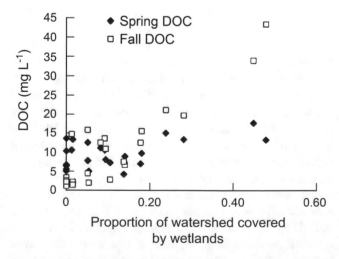

Fig. 7.24 The relationship between DOC concentrations and the proportion of wetlands in catchments for rivers in Wisconsin (Reproduced from Gergel et al. 1999)

Canada, DOC concentrations were relatively low and constant during the spring freshet, and higher during the productive summer and autumn (Naiman 1982). Nonetheless, the two-month stream freshet accounted for roughly half of annual discharge and 50% of DOC export.

Inputs of DOC increase during storms due to flushing of locations of DOC accumulation (particularly in organic-rich riparian zones and surface soils on hillslopes), canopy throughfall, and possibly due to leaching of newly entrained material (Meyer and Tate 1983; McDowell and Likens 1988). In well-drained soils, an increase in stream water DOC during a storm is expected due to shallow sub-surface flowpaths and flushing of soil DOC. In contrast, in streams draining wetlands, surface flow is dominant and increased rainfall may produce a decrease in DOC concentrations as a dilution effect (Mulholland 2003). In two small catchments in Ontario, Canada, DOC transport during storms represented 57–68% of the total during autumn and 29–40% of the total during spring (Hinton et al. 1997). During tropical storm Irene, a 500-year storm event in the northeastern US, Yoon and Raymond (2012) estimated that 40% of average annual DOC export occurred in the few days during and after the storm.

Primary producers can generate large amounts of DOM, releasing exudates from cell lysis that are typically of low molecular weight and labile (Bott et al. 1984; Bertilsson and Jones 2003). This organic carbon source may be most available during episodes of high primary production when exudates are produced, and within surface biofilms where exudates and products of cell lysis become concentrated. During springtime periphyton blooms, stream DOC concentrations have been noted to increase as much as 37% from a pre-dawn minimum to a late afternoon maximum, apparently due to extracellular release by algae (Kaplan and Bott 1989).

Animals also generate DOM in streams and rivers (Parr et al. 2019). A large body of work has examined the influence of animal consumers on the cycling of inorganic nitrogen and phosphorus in lotic systems (Atkinson et al. 2017; Vanni et al. 2017), but less is known about the influence of animals on DOM dynamics. Animal consumers may mediate the composition of dissolved organics directly via the excretion and egestion of more labile forms of organic matter, or indirectly by consuming photosynthetic organisms that produce labile exudates (Parr et al. 2020). In a headwater stream in the eastern US, Parr et al. (2019) demonstrated that the flux of DOM through animal communities via excretion can be substantial and that the energy derived from animal-mediated excretion may be important to microbial consumers, as it is more labile relative to other common sources of DOM in streams.

7.4.2 Spatial and Temporal Variability in DOM

The amount of DOM in river water varies on daily, seasonal, and yearly time scales; spatially, in accord with local geology, vegetation, rainfall and temperature; and with human activities in a watershed (Table 7.2). The river continuum concept (Vannote et al. 1980) assumed that DOM dynamics were in a steady state and functioned independently of time (Creed et al. 2015). However, subsequent research has demonstrated how the temporal dynamics of DOM are fundamental in supporting stream food webs and biogeochemical cycling (Raymond and Saiers 2010). Changes in flow affect DOM in distinct ways and at distinct time scales (hours to years) along a river continuum (Creed et al. 2015). Over longer time periods, DOM dynamics are affected by climatic cycles (e.g., El Nino–Southern Oscillation) and vary with long term changes in the physical and chemical environment (e.g., global warming, acid rain deposition). Over shorter times hydrological events, such as snowmelt and storms, influence the transport of DOM from terrestrial to aquatic environments.

Changes in DOM concentrations along the river continuum can be explained largely by changes in the nature of terrestrial-aquatic connectivity and the flow of terrestrially-derived DOM moving from smaller to larger systems. Lower-order systems are often characterized by high spatial and temporal variation in DOM concentrations. Site-specific variation can be explained by the diverse sources of DOM and their ability to respond rapidly to short-term events. For example, storm flows often shift flowpaths from deeper mineral soils to more shallow organic-rich soils, resulting in substantial variation in DOM between baseflow and stormflow events.

7.4.3 Lability and Uptake of DOM

The origin of DOM influences the chemical structure of the organic matter. In addition to influencing the concentration of DOM entering streams, factors including, but not limited to, weather and climate, land use, and changes in hydrology can alter the lability of DOM and cause its chemical characteristics to change seasonally The chemical composition of DOM derived from leachate varies among tree species. Additionally, microbial species can, in part, determine the chemical characteristics of DOC that is respired, integrated into higher trophic levels in stream food webs, or transported downstream (Wymore et al. 2018).

The composition of DOM governs the role of DOM in supporting biological communities and biogeochemical cycling (Raymond and Spencer 2015). The physical and chemical structure and volume of DOM influences the structure and function of the microbial community, with subsequent effects on upper trophic levels and ecosystem processes (Battin et al. 2016; Findlay 2010; Findlay and Parr 2017). Additionally, DOM composition can influence the formation of carcinogenic and mutagenic disinfection byproducts that threaten drinking water quality, and mediate the transport and reactivity of toxic substances, such as metals (Raymond and Spencer 2015).

Over the past few decades, researchers have employed many methods to characterize the composition of riverine DOM (Raymond and Spencer 2015). They include methods relying on carbon to nitrogen ratios, stable isotopes (e.g., $\delta^{13}C$, $\delta^{15}N$), and radiocarbon isotopes ($\Delta^{14}C$). Analytical techniques, such as advanced nuclear magnetic resonance spectroscopy and Fourier transform ion cyclotron mass spectrometry, are commonly employed to gain high-resolution data about the composition of DOM (Raymond and Spencer 2015). Optical properties of DOM, including florescence and UV absorbance at specific wavelengths, can help to distinguish specific natural and anthropogenic sources (Fellman et al. 2010). Using the US Geological Survey's Water Data for the Nation to analyze the composition of DOM in rivers in the US, Creed et al. (2015). observed a shift in UV absorbance suggesting that DOC rich in humic matter was characteristic of smaller rivers. However, there were exceptions to this pattern as humic-rich DOC also has been documented in larger black water rivers (Sun et al. 1997). Fluorescence components are typically classified as protein-like, humic-like, or fulvic-like (Fellman et al. 2010). Compounds including lignin, tannins, polyphenols, and melanins most likely generate most of the humic DOM fluorescence, which comprise a large portion of the fluorescence occurring in natural rivers. Research in aquatic systems has demonstrated that protein-like fluorescence can be correlated with bacterial production, bacterial respiration, and community respiration. Furthermore, protein-like fluorescence may also reflect DOM lability, as laboratory incubation experiments have documented relationships between biodegradable DOC and protein-like fluorescence (Fellman et al. 2010).

As previously mentioned, the contribution of aquatic animals to the pool of labile DOM pool via excretion and egestion may be functionally relevant in some systems. For instance, research in an Alaskan stream demonstrated that salmon-derived DOM was chemically distinct from the humic-rich, wetland-derived DOM that typically dominated the river, and was characterized by protein-like fluorescence (Fellman et al. 2010). In a forested, headwater stream in the US, Parr et al. (2019) found that the DOM excreted by stream invertebrates was two to five times more bioavailable to microbial heterotrophs than ambient stream water DOM, and could meet a significant proportion ($40 \pm 7\%$) of the microbial demand of labile carbon in the stream.

Dissolved organic matter is removed from stream water by both abiotic and biotic processes. Dahm (1981) estimated that adsoption onto clays and chemical complexing with oxides of aluminum and iron accounted for up to one-third of the initial removal of DOC from the water column. Over a period of several days, however, microbial uptake was responsible for the majority of DOC disappearance from the water column into the sediment layer. Photochemical degradation results in the transformation of DOC into other inorganic and organic compounds. Although it is not clear whether these organic products are more or less available to bacteria than the initial DOC, most studies conducted in freshwater systems using humic compounds or DOC from vascular plants have found that photochemical degradation enhances biological availability (Moran and Covert 2003). Increased biotic sorption and transformation into other compounds by photodegradation and relatively reduced concentrations of DOM tend to occur in watersheds dominated by soils that have high adsorption capacity (Tank et al. 2010).

The uptake of DOC by stream organisms is a key part of OM processing in streams, and it is a fundamental parameter to quantify in order to understand the contribution of rivers and streams to the global carbon budget (Mineau et al. 2016). Uptake of dissolved organic matter is primarily by microbes to sustain their respiration and growth (Bott et al. 1984), but research has suggested that some aquatic macroinvertebrates including black fly larvae and zebra mussels may also be able to directly utilize DOM (e.g., Ciborowski et al. 1997; Roditi et al. 2000). The bioavailability of ambient DOC varies among streams (Meyer 1994), flow conditions (Wilson et al. 2013), and seasons (Fellman et al. 2009). Hence, the parameters governing uptake are expected to vary accordingly (Mineau et al. 2016).

The incorporation of DOC into microbial biomass is of interest because of its potential as an energy input into stream food webs. It is a central tenet of this chapter that detrital energy pathways can be as or more important than primary production, and DOC can be a major carbon source for heterotrophic microorganisms. Bacteria likely play an even greater role in this regard than do fungi, but it also is apparent that different microorganisms are intimately intertwined with various OM sources, as well as with algae in complex energy-processing sites known as biofilms, the subject of Chap. 8. Experimental studies of the response of bacterial abundance and biomass to different carbon sources and nutrient levels are the basis for most current knowledge of DOC uptake by microbes. However, rapid advances in molecular microbial ecology including the ability to assay for key enzymes and to survey microorganisms for the functional genes that encode particular enzymes promise new insights into DOM dynamics.

Nutrients, temperature, oxygen, and many other environmental factors will influence the incorporation of these carbon sources into microbial biomass. High uptake rates for dissolved organic carbon are reported for labile molecules including leachate from leaves and highly productive algal mats, and the addition of nutrients often increases DOC uptake. However, because the majority of DOC enters streams from soil and groundwater and includes a heterogeneous mix of bioavailable, recalcitrant, and perhaps inhibitory compounds, total DOC is a poor predictor of microbial metabolism.

By measuring uptake coefficients of DOC using ^{13}C-labeled leaf leachate in mesocosms, Wiegner et al. (2006) estimated that the most readily assimilated DOC fraction would travel 175 m in White Clay Creek before being immobilized, and a second DOC pool they described as of intermediate lability would travel 3,692 m. These distances represented 7% and 150% of the third-order reach length, respectively, suggesting that readily available DOC was an energy input at the reach scale, whereas more recalcitrant material was exported and potentially served as a subsidy to downstream ecosystems. Because this experiment used fresh leachate rather than material aged by passage through the soil, uptake distances may be underestimated. Other studies have found that DOC in transport can support 11 to 55% of the benthic bacterial metabolism in streams and rivers (Bott et al. 1984; Findlay et al. 1993; Sobczak and Findlay 2002; Fischer et al. 2002). In the mesocosms studied by Wiegner et al. (2006), DOC met from one-third to one-half of the bacterial carbon demand, depending on the relative contribution of algal production.

Despite many measures of DOC concentrations and a reasonably good understanding of factors that influence spatial and temporal variation, neither input nor utilization rates are well quantified on an areal basis in streams. The cellular respiration resulting from the consumption of DOC contributes to the net flux of carbon dioxide from aquatic ecosystems to the atmosphere (Hotchkiss et al. 2015), and research suggests that a large proportion (27 to 45%) of DOC exported from terrestrial ecosystems to streams may be removed in watersheds before reaching coastal areas (Mineau et al. 2016). At least in small streams, downstream transport rather than utilization appears to be the fate of most DOC entering stream reaches.

7.4.4 Anthropogenic Influences on DOM

Anthropogenic activities may influence the lability, quantity, timing, and duration of DOM in streams and rivers. Increases in terrestrially-derived DOC concentrations have been measured in aquatic ecosystems over the last few decades (Marx et al. 2017). Anthropogenically-derived changes in precipitation, temperature, and land use have all been attributed to these changes.

Uptake rates of organic matter are expected to change with changing climate. By adding acetate, a labile form of DOC, to European streams from different ecoregions, Catalán et al. (2018). found that climate and the composition of DOC, rather than seasonal variation, had large effects on uptake. Specifically, mean annual precipitation explained half of the variability of acetate uptake velocity, a measurement that estimates demand for an element. Streams characterized by higher rates of precipitation, more recalcitrant DOM, and high respiration rates had the lowest uptake velocities. Semiarid streams that were characterized by more labile organic compounds and lower rates of respiration had higher uptake velocities, suggesting they were carbon limited. These findings emphasized the importance of interactions between climate and DOM composition in mediating DOC uptake in streams.

Changes in precipitation are linked to alterations in the chemical composition of DOC that reflect changes in the origin of the organic matter (Hood et al. 2006). In Mediterranean streams, drought can shift systems from lotic to lentic environments, fragmenting hydrologic connectivity and increasing autochthonous DOM relative to allochthonous DOM (Vazquez et al. 2011). Extreme flows associated with tropical storms and hurricanes are also expected to interact with land use to influence DOC dynamics. In the wake of hurricanes Harvey and Irma in 2017, high flows in streams in the southeastern US were linked to an increase in terrestrially-derived DOM relative to instream DOM. However, the total flux of more labile DOC was much higher at peak discharge, indicating materials transported by large storm flows could enhance microbial activity in streams. Watersheds with more urban development exported large loads of nutrients and labile DOC, and the watershed that had greatest percentage of wetland cover had a prolonged, but relatively subdued, export of DOC and nutrients (Chen et al. 2019).

Changes in temperature can alter the flow and lability of DOM entering lotic ecosystems, as temperature controls dissolution (Raymond and Spencer 2015). As temperatures warm in cooler climates, the retreat of glaciers may also influence DOM fluxes into riverine environments. Though relic glacial water has relatively low concentrations of DOM, the land that is exposed after glacier retreat exports much more DOM than previously ice-covered landscapes. Additionally, the DOM composition of glacier-derived water may be compositionally unique. Hence, glacier decline may affect the volume and lability of DOC entering certain aquatic environments.

Dominant land use often governs the source of DOM, with implications for the lability and processing rates of organic matter in streams (da Costa et al. 2017; Stanley et al. 2012).

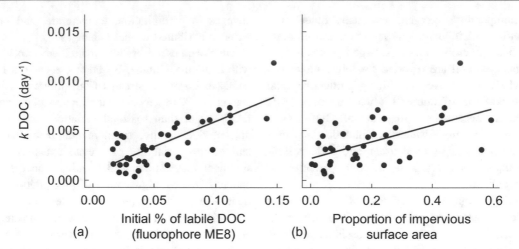

Fig. 7.25 (**a**) The relationship between the breakdown rates of dissolved organic carbon (k DOC) and the initial % dominance of a relatively labile form of DOC (fluorophore ME8), and (**b**) the relationship between the breakdown rate of DOC and impervious surface area (proportion ISA) in the study region (Reproduced from Parr et al. 2015)

Multiple studies have reported declines in the contribution of terrestrially-derived DOM relative to the contribution of DOM from autochthonous production and instream microbial sources in response to urban and agricultural development (Lu et al. 2014; Westerhoff and Anning 2000). As more land is converted to agricultural and urban landscapes, increases in the lability of DOM may fuel increases in microbial activity, subsequently altering rates of carbon cycling and the amount of carbon that is exported downstream (Williams et al. 2010; Parr et al. 2015). In headwater streams, urbanization did not have a strong effect on instream DOM concentrations, but did significantly increase the relative amount of bioavailable, hence labile, DOM derived from autochthonous microbial sources (Fig. 7.25). In a study examining the impacts of agricultural expansion in the Amazon on riverine carbon dynamics, Spencer et al. (2019) reported that agriculturally-impacted streams received lower total quantities of relatively labile DOC when compared to reference sites, resulting in lower instream DOC concentrations and greater rates of microbial degradation of DOM. Seasonal variation in the origin of DOM can also be affected by land use. In the Mara River in Kenya, Maese et al. (2017) reported that forest streams primarily transported DOM derived from terrestrial sources, whereas streams draining agricultural areas transported a mixture of DOM that was dominated by autochthonous sources during the dry season, but primarily DOM from terrestrial sources in the wet season (Masese et al. 2017). The increased soil erosion, organic matter oxidation rates, and relatively shallow soil-to-stream flow paths that often characterize agricultural landscapes can lead to higher DOC concentrations with large proportions of soil-derived, humic-like DOM, and greater DOC bioavailability (Shang et al. 2018).

Wastewater effluent can also introduce large volumes of chemically-distinct DOM (Singer and Battin 2007; Raymond and Spencer 2015). Yet, it is very challenging to estimate anthropogenic inputs of DOC from sewage because our knowledge of the state and structure of wastewater infrastructure is limited (WWP 2017; ASCE 2017, 2011). As Webster et al. (1995) emphasize, with over 24,000 sewage treatment plants in the eastern US, natural levels of DOC are typically unknown for larger rivers (>3rd order) in that region. Wastewater inputs to streams are commonplace and poorly quantified (Capps 2019); thus, the concern voiced by Webster et al. (1995) most likely applies to watersheds throughout the globe. Organic matter concentrations may also affect how chemical contaminants enter and subsequently influence aquatic ecosystems (Wohl 2015; Schmidt et al. 2012). Metals, such as zinc, copper, and mercury, bind to DOC and can enter aquatic food webs (Tomczyk et al. 2018a; Tomczyk et al. 2018b; Chaves-Ulloa et al. 2016).

7.5　Outgassing of Carbon from River Networks

The amount of CO_2 evasion currently estimated from inland waters and wetlands (2.9 Pg C yr$^{-1;}$ Sawakuchi et al. 2017) is the same order of magnitude as uptake of carbon by marine environments (2.4 ± 0.6 Pg C year^{-1}), uptake of carbon by terrestrial environments (2.7 ± 1.2 Pg C year^{-1}), deforestation and land use change (1.0 ± 0.7 Pg C year^{-1}), and anthropogenic CO_2 emissions from the burning of fossil fuels (7.9 ± 0.5 Pg C year^{-1}) (Ward et al. 2017). Because headwaters make up a large proportion of river networks (> 96% of the total number of streams globally), they potentially contribute large amounts of CO_2 to the atmosphere. Conservative estimates suggest 0.93 Pg C yr^{-1}, or 36% of all CO_2 outgassing from rivers and streams comes from headwater systems (Marx et al. 2017). Evasion estimates are poorly characterized for many of the world's rivers, and this is especially true for medium and large systems. However, existing estimates suggest that individual systems can emit large volumes of CO_2 into the atmosphere (Ward et al. 2017; Reiman and Xu 2019). For example, Sawakuchi et al. (2017) estimated that the Amazon River contributed 54% of global river and stream CO_2 outgassing. This is a remarkable estimation because the surface area of the Amazon only corresponds to 3% of the estimated surface area covered by all rivers globally.

In a study of the lower Mississippi River in the US, Reiman and Xu (2019) found that water in the river was constantly supersaturated with CO_2. In general, pCO_2 was much higher in the wet and the dry season and was positively associated with discharge and temperature (Fig. 7.26). Similar seasonal trends in pCO_2 have been documented in other rivers of the world, a pattern that could be explained by increasing heterotrophy associated with higher temperatures, or increased amounts of soil pore-water CO_2 entering the river due to increased mineralization and leaching of organic matter in the soil. This study documented an increase in riverine DOC concentrations (607 $\pm$ 158 μmol L^{-1}) from those that had been reported 5–15 years ago (296–489 μmol L^{-1}), reflective of the increase in air and water temperatures in the Mississippi River Basin. Coupled with greater discharge over the past few years, annual export of total dissolved carbon (16.2 Tg C) from the Mississippi River Basin to the Gulf of Mexico was higher than just seven years previously (15.5 Tg C). High partial pressures of CO_2 in the river water also resulted in a large quantity of carbon released into the atmosphere annually from the lower portion of the Mississippi. Both discharge and temperature have increased in the Mississippi River during the past few decades, and so both likely have contributed to changes in DOC levels. Specifically, discharge has increased more than 30% in the past 80 years and average annual air temperatur during the study were 0.2–0.4 °C warmer than their respective 5-year average ten years prior to the study, and 0.8–1.1 °C warmer than twenty years ago. Coupled with increasing DOC concentrations, the predicted changes in future temperature and discharge are expected to increase the rate of outgassing from the Mississippi (Reiman and Xu 2019).

7.6 Summary

Particulate and dissolved organic matter originating both within the stream and in the surrounding landscape are substantial basal resources to fluvial food webs. Detritus-based energy pathways can be particularly important relative to pathways originating from living primary producers in small streams shaded by a terrestrial canopy and in large, turbid rivers with extensive floodplains. Coarse, fine, and dissolved organic matter originate from a myriad of sources. Leaves, fruits, and other plant products that fall or are transported by wind and gravity into the stream are major CPOM inputs, and the carcasses and feces of insects and larger animals also contribute to this pool. Most FPOM originates from the fragmentation of larger particles, particularly terrestrial vegetation, and is likewise transported into the channel or is produced by the breakdown of larger particles within the stream. Water that has been filtered through soil is usually the major source of DOM, which is a heterogeneous pool of molecules of widely varying bioavailability. The lability of non-living organic matter as a basal resource typically depends on the presence of bacteria or fungi whose degradative activity can alter palatability or accessibility of OM to consumers. Microorganisms are critical mediators of organic matter pathways, aiding in the processing of POM and uptake of DOM, and markedly increasing the energy available to consumers both small and large.

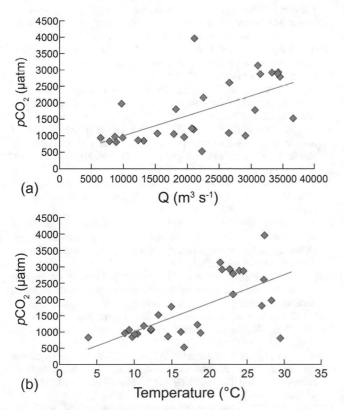

Fig. 7.26 Relationship between pCO_2 and (**a**) discharge (Q) and (**b**) water temperature in the Mississippi at Baton Rouge (Reproduced from Reiman and Xu 2019)

Autumn-shed leaves are a primary CPOM input to forested streams in temperate regions, and their decay serves as the primary model of CPOM processing. Leaves, which serve as the organic matter or carbon substrate, quickly release DOM when wetted, and soon are colonized by microorganisms and invertebrates, which enhance fragmentation and mineralization. The original leaf is transformed into several products including microbial and shredder biomass, FPOM, DOM, nutrients, and CO_2. Leaf breakdown rates vary considerably and, as a consequence, the supply of CPOM to the stream food web, although pulsed seasonally, is less so than would be the case if all leaves had similar breakdown rates. Temperature, oxygen availability, and nutrient supply are key environmental variables that influence the decomposition process. Colonization by microorganisms, particularly fungi, is critical to leaf decomposition as well as the leaf's nutritional value, although bacteria play a greater role as particle size diminishes, and the presence of leaf-shredding invertebrates also is functionally relevant. When detritivorous invertebrates are excluded, the breakdown process is significantly slowed.

Fine particulate matter is an amorphous collection of particles less than 1 mm, originating from instream CPOM breakdown, sloughed cells of algae, invertebrate fecal pellets, and fragments derived from the terrestrial environment. The uptake of DOM within biofilms provides another avenue for FPOM production. Black fly larvae, by consuming bacteria and very small FPOM and producing fecal pellets that are larger than some of the material that they ingest, illustrate yet another pathway for the generation of organic particles. The sources, processing, and eventual fate of FPOM is less well accounted for than is the case for CPOM. Although the breakdown of CPOM within the stream is the best-studied pathway, it seems likely that FPOM originates from many other sources. Because of the small size of FPOM, bacteria likely are more important than fungi in its microbial processing.

Dissolved organic matter typically is the largest pool of organic carbon in running waters and is incorporated into POM due primarily to uptake by microorganisms. Soil and groundwater are major pathways of DOM transport from terrestrial vegetation and wetlands into river water, whereas leaf leachate from leaves and from extracellular release by algae and higher plants can be temporally dominant instream sources. Because it comprises a heterogeneous mix of bioavailable, recalcitrant, and perhaps inhibitory compounds, total DOM is a poor predictor of microbial metabolism. DOM is removed from stream water by both abiotic and biotic processes.

Particulate and dissolved organic matter are key energy sources in almost all lotic ecosystems and frequently can be the dominant energy source. Detritus and the associated microbial biomass, along with algae and other primary producers, form the basal resources for the various trophic levels occupied by invertebrates, fishes and other consumers in stream food webs. We turn now to the diversity of consumers and the feeding adaptations that govern their effectiveness with the diverse producer and detrital resources described in this and the preceding chapter.

References

Abelho M, Cressa C, Graça MA (2005) Microbial Biomass, Respiration, and Decomposition of Hura crepitans L. (Euphorbiaceae) Leaves in a Tropical Stream. Biotropica 37:397–402

Abelho M, Descals EJ (2019) Litter movement pathways across terrestrial–aquatic ecosystem boundaries affect litter colonization and decomposition in streams. Ecology 33:1785–1797

Adams JL, Tipping E, Feuchtmayr H et al (2018) The contribution of algae to freshwater dissolved organic matter: implications for UV spectroscopic analysis. Inland Waters 8:10–21

Aitkenhead-Peterson JA, McDowell WH, Neff JC (2003) Sources, production, and regulation of allochthonous dissolved organic matter inputs to surface waters. Findlay SEG. Sinsabaugh RL. Aquatic Ecosystems. Elsevier, Amsterdam, pp 25–70

Aitkenhead J, Hope D, Billett M (1999) The relationship between dissolved organic carbon in stream water and soil organic carbon pools at different spatial scales. Hydrol Process 13:1289–1302

Amani M, Graça MA, Ferreira V (2019) Effects of elevated atmospheric CO2 concentration and temperature on litter decomposition in streams: a meta-analysis. Int Rev Hydrobiol 104:14–25

Anderson NH, Sedell JR (1979) Detritus processing by macroinvertebrates in stream ecosystems. An Rev Ento 24:351–377

Anderson NH, Sedell JR, Roberts LM et al (1978) The role of aquatic invertebrates in processing of wood debris in coniferous forest streams. Amer Mid Nat 1974:64–82

Arsuffi T, Suberkropp K (1986) Growth of two stream caddisflies (Trichoptera) on leaves colonized by different fungal species. J N Am Benthol Soc 5:297–305

Arsuffi TL, Suberkropp K (1984) Leaf processing capabilities of aquatic hyphomycetes-interspecific differences and influence on shredder feeding preferences. Oikos 42:144–154. https://doi.org/10.2307/3544786

Arsuffi TL, Suberkropp K (1985) Selective feeding by stream caddisfly (Trichoptera) detritivores on leaves with fungal-colonized patches. Oikos 50–58

Artmann U, Waringer JA, Schagerl M (2003) Seasonal dynamics of algal biomass and allochthonous input of coarse particulate organic matter in a low-order sandstone stream (Weidlingbach, Lower Austria). Limnologica 33:77–91

ASCE (2011) Failure to act: The economic impact of current investment trends in water and wastewater treatment infrastructure. American Societity of Civil Engineers Press, Washington, D.C

ASCE (2017) America's infrastructure report card. American Society of Civil Engineers. https://www.infrastructurereportcard.org/. Accessed 03 September 2019 2019

Atkinson CL, Capps KA, Rugenski AT et al (2017) Consumer-driven nutrient dynamics in freshwater ecosystems: from individuals to ecosystems. Biol Rev 92:2003–2023. https://doi.org/10.1111/brv.12318

Bärlocher F (1982) Conidium production from leaves and needles in four streams. Can J Bot 60:1487–1494

Bärlocher F (1983) Seasonal variation of standing crop and digestibility of CPOM in a Swiss Jura stream. Ecology 64:1266–1272

Bärlocher F (1985) The role of fungi in the nutrition of stream invertebrates. Bot J Linn Soc 91:83–94

Barlocher F, Kendrick B (1974) Dynamics of fungal population on leaves in a stream. J Ecol 62:761–791. https://doi.org/10.2307/2258954

Bärlocher F, Kendrick B (1973) Fungi and food preferences of *Gammarus pseudolimnaeus*. Arch Hydrobiol 72:501–516

Bärlocher F, Oertli J (1978) Inhibitors of aquatic hyphomycetes in dead conifer needles. Mycologia 70:964–974

Batista D, Pascoal C, Cassio E (2012) Impacts of warming on aquatic decomposers along a gradient of cadmium stress. Environ Pollut 169:35–41. https://doi.org/10.1016/j.envpol.2012.05.021

Battin TJ, Besemer K, Bengtsson MM et al (2016) The ecology and biogeochemistry of stream biofilms. Nat Rev Microb 14:251

Battin TJ, Luyssaert S, Kaplan LA et al (2009) The boundless carbon cycle. Nat Geosci 2:598

Benfield E (1997) Comparison of litterfall input to streams. J N Am Benthol Soc 16:104–108

Benfield E, Webster J, Tank J et al (2001) Long-term patterns in leaf breakdown in streams in response to watershed logging. Int Rev Hydrobio 86:467–474

Benfield EF, Fritz KM, Tiegs SD (2017) Chapter 27—Leaf-litter breakdown. In: Lamberti GA, Hauer, FR. Methods in Stream Ecology (Third Edition). Academic Press, Amsterdam pp 71–82. https://doi.org/10.1016/B978-0-12-813047-6.00005-X

Bengtsson MM, Attermeyer K, Catalan N (2018) Interactive effects on organic matter processing from soils to the ocean: are priming effects relevant in aquatic ecosystems? Hydrobiologia 822:1–17. https://doi.org/10.1007/s10750-018-3672-2

Benke A, Wallace JB (2010) Influence of wood on invertebrate communities in streams and rivers. In: Gregory SV, Boyer KL, Gurnell AM (eds), The ecology and management of wood in world rivers. American Fisheries Society, Symposium 37: Bethesda, Maryland, pp 149–177

Benner R (2003) Molecular indicators of the bioavailability of dissolved organic matter. Findlay SEG. Sinsabaugh RL. Aquatic Ecosystems. Elsevier, Amsterdam, pp 121–137

Berrie A (1972) The occurrence and composition of seston in the River Thames and the role of detritus as an energy source for secondary production in the river. In: Melchiorri-Santolini U, Hopton JW (eds) Detritus and its role in aquatic ecosystems. Springer, New York, pp 473–483

Bertilsson S, Jones JB (2003) Supply of dissolved organic matter to aquatic ecosystems: autochthonous sources. Findlay SEG. Sinsabaugh RL. Aquatic Ecosystems. Elsevier, Amsterdam, pp 3–24

Bianchi TS, Wysocki LA, Stewart M et al (2007) Temporal variability in terrestrially-derived sources of particulate organic carbon in the lower Mississippi River and its upper tributaries. Geochim Cosmochim Acta 71:4425–4437. https://doi.org/10.1016/j.gca.2007.07.011

Bilby RE (1981) Role of organic debris dams in regulating the export of dissolved and particulate matter from a forested watershed. Ecology 62:1234–1243

Bilby RE, Likens GE (1979) Effect of hydrologic fluctuations on the transport of fine particulate organic carbon in a small stream 1. Limnol Oceanog 24:69–75

Bilby RE, Likens GE (1980) Importance of organic debris dams in the structure and function of stream ecosystems. Ecology 61:1107–1113

Bixby RJ, Cooper SD, Gresswell RE et al (2015) Fire effects on aquatic ecosystems: an assessment of the current state of the science. Freshw Sci 34:1340–1350. https://doi.org/10.1086/684073

Bonin H, Griffiths R, Caldwell B (2003) Nutrient and microbiological characteristics of fine benthic organic matter in sediment settling ponds. Freshw Biol 48:1117–1126

Bott T, Brock J, Dunn C et al (1985) Benthic community metabolism in four temperate stream systems: an inter-biome comparison and evaluation of the river continuum concept. Hydrobiologia 123:3–45

Bott TL, Kaplan LA, Kuserk FT (1984) Benthic bacterial biomass supported by streamwater dissolved organic matter. Microb Ecol 10:335–344

Boyero L, Graça MA, Tonin AM et al (2017) Riparian plant litter quality increases with latitude. Sci Repor 7:1–10

Boyero L, Pearson RG, Dudgeon D et al (2012) Global patterns of stream detritivore distribution: implications for biodiversity loss in changing climates. Glob Ecol Biogeog 21:134–141. https://doi.org/10.1111/j.1466-8238.2011.00673.x

Boyero L, Pearson RG, Hui C et al (2016) Biotic and abiotic variables influencing plant litter breakdown in streams: a global study. Proc R Soci B 283:20152664

Bundschuh M, McKie BG (2016) An ecological and ecotoxicological perspective on fine particulate organic matter in streams. Freshw Biol 61:2063–2074. https://doi.org/10.1111/fwb.12608

Burrows RM, Rutlidge H, Bond NR et al (2017) High rates of organic carbon processing in the hyporheic zone of intermittent streams. Sci Repo 7:13198

Camacho R, Boyero L, Cornejo A et al (2009) Local variation in shredder distribution can explain their oversight in tropical streams. Biotropica 41:625–632. https://doi.org/10.1111/j.1744-7429.2009.00519.x

Canham CD, Pace ML, Papaik MJ et al (2004) A spatially explicit watershed-scale analysis of dissolved organic carbon in Adirondack lakes. Ecol Appl 14:839–854

Canhoto C, Simões S, Gonçalves AL et al (2017) Stream salinization and fungal-mediated leaf decomposition: a microcosm study. Sci Tot Envi 599:1638–1645

Capps KA (2019) Wastewater infrastructure and the ecology and management of freshwater systems. Acta Limnologica Brasiliensia 31

Cargill A, Cummins K, Hanson B et al (1985a) The role of lipids as feeding stimulants for shredding aquatic insects. Freshw Biol 15:455–464

Cargill AS, Cummins KW, Hanson BJ et al (1985b) The role of lipids, fungi, and temperature in the nutrition of a shredder caddisfly, *Clistoronia magnifica*. Freshw Invert Biol 4:64–78

Carlisle DM, Clements WH (2005) Leaf litter breakdown, microbial respiration and shredder production in metal-polluted streams. Freshw Bio 50:380–390

Carvalho F, Pascoal C, Cássio F et al (2016) Direct and indirect effects of an invasive omnivore crayfish on leaf litter decomposition. Sci Tot Environ 541:714–720. https://doi.org/10.1016/j.scitotenv.2015.09.125

Catalán N, Casas-Ruiz JP, Arce MI et al (2018) Behind the scenes: mechanisms regulating climatic patterns of dissolved organic carbon uptake in headwater streams. Glob Biogeochem Cyc 32:1528–1541

Chadwick OA, Derry LA, Vitousek PM et al (1999) Changing sources of nutrients during four million years of ecosystem development. Nature 397:491

Chará-Serna AM, Chara JD, del Carmen Zúñiga M et al. Diets of leaf litter-associated invertebrates in three tropical streams. In: Annales de Limnologie-International Journal of Limnology, 2012. vol 2. EDP Sciences, pp 139–144

Chaves-Ulloa R, Taylor BW, Broadley HJ et al (2016) Dissolved organic carbon modulates mercury concentrations in insect subsidies from streams to terrestrial consumers. Ecol Appl 26:1771–1784

Cheever B, Webster J, Bilger E et al (2013) The relative importance of exogenous and substrate-derived nitrogen for microbial growth during leaf decomposition. Ecology 94:1614–1625

Chen S, Lu YH, Dash P et al (2019) Hurricane pulses: Small watershed exports of dissolved nutrients and organic matter during large storms in the Southeastern USA. Sci Tot Environ 689:232–244. https://doi.org/10.1016/j.scitotenv.2019.06.351

Cheshire K, Boyero L, Pearson RG (2005) Food webs in tropical Australian streams: shredders are not scarce. Freshw Biol 50:748–769

Chung N, Suberkropp K (2009) Effects of aquatic fungi on feeding preferences and bioenergetics of *Pycnopsyche gentilis* (Trichoptera: Limnephilidae). Hydrobiologia 630:257–269

Ciborowski JJ, Craig DA, Fry KM (1997) Dissolved organic matter as food for black fly larvae (Diptera: Simuliidae). J N Am Benthol Soc 16:771–780

Classen-Rodríguez L, Gutiérrez-Fonseca PE, Ramírez A (2019) Leaf litter decomposition and macroinvertebrate assemblages along an urban stream gradient in Puerto Rico. Biotropica 51:641–651

Colon-Gaud C, Whiles MR, Brenes R et al (2010) Potential functional redundancy and resource facilitation between tadpoles and insect grazers in tropical headwater streams. Freshw Biol 55:2077–2088. https://doi.org/10.1111/j.1365-2427.2010.02464.x

Compson ZG, Hungate BA, Koch GW et al (2015) Closely related tree species differentially influence the transfer of carbon and nitrogen from leaf litter up the aquatic food web. Ecosystems 18:186–201

Compson ZG, Hungate BA, Whitham TG et al (2018) Linking tree genetics and stream consumers: isotopic tracers elucidate controls on carbon and nitrogen assimilation. Ecology 99:1759–1770

Compson ZG, Hungate BA, Whitham TG et al (2016) Plant genotype influences aquatic-terrestrial ecosystem linkages through timing and composition of insect emergence. Ecosphere 7:e01331

Couceiro SR, Hamada N, Forsberg BR et al (2011) Trophic structure of macroinvertebrates in Amazonian streams impacted by anthropogenic siltation. Austral Ecol 36:628–637

Creed IF, McKnight DM, Pellerin BA et al (2015) The river as a chemostat: fresh perspectives on dissolved organic matter flowing down the river continuum. Can J Fish Aquat Sci 72:1272–1285. https://doi.org/10.1139/cjfas-2014-0400

Cross WF, Wallace JB, Rosemond AD et al (2006) Whole-system nutrient enrichment increases secondary production in a detritus-based ecosystem. Ecology 87:1556–1565

Crowl TA, McDowell WH, Covich AP et al (2001) Freshwater shrimp effects on detrital processing and nutrients in a tropical headwater stream. Ecology 82:775–783

Cummins KW (1974) Structure and function of stream ecosystems. Bioscience 24:631–641

Cummins KW, Klug MJ (1979) Feeding ecology of stream invertebrates. Ann Rev Eco Syst 10:147–172

Cummins KW, Petersen RC, Howard FO et al (1973) Utilization of leaf litter by stream detritivores. Ecology 54:336–345

Cushing CE, Minshall GW, Newbold JD (1993) Transport dynamics of fine particulate organic matter in two Idaho streams. Limnol Oceanog 38:1101–1115

da Costa END, de Souza JC, Pereira MA et al (2017) Influence of hydrological pathways on dissolved organic carbon fluxes in tropical streams. Ecol Evol 7:228–239

Dahm CN (1981) Pathways and mechanisms for removal of dissolved organic carbon from leaf leachate in streams. Can J Fish Aquat Sci 38:68–76

Dalu T, Richoux NB, Froneman PW (2016) Nature and source of suspended particulate matter and detritus along an austral temperate river–estuary continuum, assessed using stable isotope analysis. Hydrobiologia 767:95–110

Danger M, Cornut J, Chauvet E et al (2013) Benthic algae stimulate leaf litter decomposition in detritus-based headwater streams: a case of aquatic priming effect? Ecology 94:1604–1613

Danger M, Gessner MO, Bärlocher F (2016) Ecological stoichiometry of aquatic fungi: current knowledge and perspectives. Fungal Ecology 19:100–111

Dangles O, Gessner MO, Guerold F et al (2004) Impacts of stream acidification on litter breakdown: implications for assessing ecosystem functioning. J Appl Ecol 41:365–378

Demars B, Friberg N, Kemp J et al. (2018) Reciprocal carbon subsidies between autotrophs and bacteria in stream food webs under stoichiometric constraints. bioRxiv:447987

Demars BO, Friberg N, Thornton B (2019) Pulse of dissolved organic matter alters reciprocal carbon subsidies between autotrophs and bacteria in stream food webs. Ecol Monogr 90:e01399

Demi LM, Benstead JP, Rosemond AD et al (2018) Litter P content drives consumer production in detritus-based streams spanning an experimental N: P gradient. Ecology 99:347–359

Devol A, Hedges J (2001) Organic matter and nutrients in the mainstem Amazon River. In: Victoria RL, Richey JE (eds) McClain ME. The Biogeochemistry of the Amazon Basin Oxford University Press, New York, pp 275–306

Dick J, Tetzlaff D, Birkel C et al (2015) Modelling landscape controls on dissolved organic carbon sources and fluxes to streams. Biogeochemistry 122:361–374

Diez J, Elosegi A, Chauvet E et al (2002) Breakdown of wood in the Aguera stream. Freshw Biol 47:2205–2215

Dobson M, Magana A, Mathooko JM et al (2002) Detritivores in Kenyan highland streams: more evidence for the paucity of shredders in the tropics? Freshw Biol 47:909–919

Duarte S, Pascoal C, Cássio F (2004) Effects of zinc on leaf decomposition by fungi in streams: studies in microcosms. Micro Eco 48:366–374

Dutton CL, Subalusky AL, Hamilton SK et al (2018) Organic matter loading by hippopotami causes subsidy overload resulting in downstream hypoxia and fish kills. Nat Comm 9:1951

Eggert S, Wallace JB (2007) Wood biofilm as a food resource for stream detritivores. Limnol Oceanog 52:1239–1245

Enríquez S, Duarte CM, Sand-Jensen KJ (1993) Patterns in decomposition rates among photosynthetic organisms: the importance of detritus C: N: P content. Oecologia 94:457–471

Epstein DM, Kelso JE, Baker MA (2016) Beyond the urban stream syndrome: organic matter budget for diagnostics and restoration of an impaired urban river. Urban Ecosyst 19:1041–1061. https://doi.org/10.1007/s11252-016-0557-x

Ertel JR, Hedges JI, Devol AH et al (1986) Dissolved humic substances of the Amazon River system 1. Limnol Oceanog 31:739–754

Farrell KJ, Rosemond AD, Kominoski JS et al (2018) Variation in detrital resource stoichiometry signals differential carbon to nutrient limitation for stream consumers across biomes. Ecosystems 21:1676–1691

Fellman JB, Hood E, D'amore DV et al. (2009) Seasonal changes in the chemical quality and biodegradability of dissolved organic matter exported from soils to streams in coastal temperate rainforest watersheds. Biogeochemistry 95:277–293

Fellman JB, Hood E, Spencer RGM (2010) Fluorescence spectroscopy opens new windows into dissolved organic matter dynamics in freshwater ecosystems: A review. Limnol Oceanog 55:2452–2462. https://doi.org/10.4319/lo.2010.55.6.2452

Fernandes I, Duarte S, Cássio F et al (2015) Plant litter diversity affects invertebrate shredder activity and the quality of fine particulate organic matter in streams. Mari Freshw Res 66:449–458

Ferreira V, Castela J, Rosa P et al (2016) Aquatic hyphomycetes, benthic macroinvertebrates and leaf litter decomposition in streams naturally differing in riparian vegetation. Aquat Ecol 50:711–725

Findlay S (2010) Stream microbial ecology. J N Am Benthol Soc 29:170–181. https://doi.org/10.1899/09-023.1

Findlay S, Strayer D, Goumbala C et al (1993) Metabolism of streamwater dissolved organic carbon in the shallow hyporheic zone. Limnol Oceanog 38:1493–1499

Findlay S, Tank J, Dye S et al (2002) A cross-system comparison of bacterial and fungal biomass in detritus pools of headwater streams. Microb Ecol 43:55–66

Findlay SEG, Parr TB (2017) Chapter 24—Dissolved Organic Matter. In: Lamberti GA, Hauer FR (eds) Methods in Stream Ecology (Third Edition). Academic Press, Amsterdam pp 21–36. https://doi.org/10.1016/B978-0-12-813047-6.00002-4

Fischer H, Wanner SC, Pusch M (2002) Bacterial abundance and production in river sediments as related to the biochemical composition of particulate organic matter (POM). Biogeochemistry 61:37–55

Fisher SG, Likens GE (1973) Energy flow in Bear Brook, New Hampshire-integrative approach to stream ecosystem metabolism. Ecol Mono 43:421–439. https://doi.org/10.2307/1942301

Flury S, Gessner MO (2011) Experimentally simulated global warming and nitrogen enrichment on microbial litter decomposers in a marsh. Appl Environ Microbiol 77:803–809. https://doi.org/10.1128/aem.01527-10

Follstad Shah JJ, Kominoski JS, Ardón M et al (2017) Global synthesis of the temperature sensitivity of leaf litter breakdown in streams and rivers. Glob Chan Bio 23:3064–3075

Frost PC, Benstead JP, Cross WF et al (2006) Threshold elemental ratios of carbon and phosphorus in aquatic consumers. Ecol Lett 9:774–779. https://doi.org/10.1111/j.1461-0248.2006.00919.x

Fuller CL, Evans-White MA, Entrekin SA (2015) Growth and stoichiometry of a common aquatic detritivore respond to changes in resource stoichiometry. Oecologia 177:837–848. https://doi.org/10.1007/s00442-014-3154-9

Galas J, Dumnicka E (2003) Organic matter dynamics and invertebrate functional groups in a mountain stream in the West Tatra mountains, Poland. Int Rev Hydrobio 88:362–371

García-Palacios P, McKie BG, Handa IT et al (2016) The importance of litter traits and decomposers for litter decomposition: a comparison of aquatic and terrestrial ecosystems within and across biomes. Funct Ecol 30:819–829

Gergel SE, Turner MG, Kratz TK (1999) Dissolved organic carbon as an indicator of the scale of watershed influence on lakes and rivers. Ecol Appl 9:1377–1390

Gessner M, Chauvet E (1997) Growth and production of aquatic hyphomycetes in decomposing leaf litter. Limnol Oceanog 42:496–505

Gessner MO, Chauvet E, Dobson M (1999) A perspective on leaf litter breakdown in streams. Oikos 85:377–384

Gessner MO, Swan CM, Dang CK et al (2010) Diversity meets decomposition. Trends Ecol Evol 25:372–380. https://doi.org/10.1016/j.tree.2010.01.010

Golladay SW (1997) Suspended particulate organic matter concentration and export in streams. J N Am Benthol Soc 16:122–131

Graça M, Maltby L, Calow P (1993) Importance of fungi in the diet of Gammarus pulex and Asellus aquaticus I: feeding strategies. Oecologia 93:139–144

Graça MA, Bärlocher F, Gessner MO (2005) Methods to study litter decomposition: a practical guide. Springer Science & Business Media, New York

Graca MAS (2001) The role of invertebrates on leaf litter decomposition in streams—A review. Int Rev Hydrobiol 86:383–393

Graca MAS, Ferreira V, Canhoto C et al (2015) A conceptual model of litter breakdown in low order streams. Int Rev Hydrobiol 100:1–12. https://doi.org/10.1002/iroh.201401757

Gregory SV, Gurnell A, Piégay H et al (2017) Chapter 29—Dynamics of Wood. In: Lamberti GA, Hauer FR (eds) Methods in Stream Ecology (Third Edition). Academic Press, Amsterdam, pp 113–126. https://doi.org/10.1016/B978-0-12-813047-6.00007-3

Greig H, McIntosh A (2006) Indirect effects of predatory trout on organic matter processing in detritus-based stream food webs. Oikos 112:31–40

Guéguen C, Guo L, Wang D et al (2006) Chemical characteristics and origin of dissolved organic matter in the Yukon River. Biogeochemistry 77:139–155

Gulis V, Rosemond AD, Suberkropp K et al (2004) Effects of nutrient enrichment on the decomposition of wood and associated microbial activity in streams. Freshw Biol 49:1437–1447

Gulis V, Suberkropp K (2003) Effect of inorganic nutrients on relative contributions of fungi and bacteria to carbon flow from submerged decomposing leaf litter. Microb Ecol 45:11–19

Gutierrez-Fonseca PE, Ramirez A, Pringle CM (2018) Large-scale climatic phenomena drive fluctuations in macroinvertebrate assemblages in lowland tropical streams, Costa Rica: The importance of ENSO events in determining long-term (15y) patterns. PLoS ONE 13:e0191781

Halvorson HM, Barry JR, Lodato MB et al (2019) Periphytic algae decouple spatial activity from leaf litter decomposition via negative priming. Funct Ecol 33:188–201

Halvorson HM, Fuller C, Entrekin SA et al (2015a) Dietary influences on production, stoichiometry and decomposition of particulate wastes from shredders. Freshw Biol 60:466–478. https://doi.org/10.1111/fwb.12462

Halvorson HM, Fuller CL, Entrekin SA et al (2018) Detrital nutrient content and leaf species differentially affect growth and nutritional regulation of detritivores. Oikos 127:1471–1481

Halvorson HM, Scott JT, Sanders AJ et al (2015b) A stream insect detritivore violates common assumptions of threshold elemental ratio bioenergetics models. Freshw Sci 34:508–518. https://doi.org/10.1086/680724

Halvorson HM, Sperfeld E, Evans-White MA (2017) Quantity and quality limit detritivore growth: mechanisms revealed by ecological stoichiometry and co-limitation theory. Ecology 98:2995–3002

Handa IT, Aerts R, Berendse F et al (2014) Consequences of biodiversity loss for litter decomposition across biomes. Nature 509:218. https://doi.org/10.1038/nature13247

Harris HE, Baxter CV, Davis JM (2015) Debris flows amplify effects of wildfire on magnitude and composition of tributary subsidies to mainstem habitats. Freshw Sci 34:1457–1467. https://doi.org/10.1086/684015

Harvey JW, Drummond JD, Martin RL et al (2012) Hydrogeomorphology of the hyporheic zone: Stream solute and fine particle interactions with a dynamic streambed. J Geophys Res: Biogeosciences 117

Hedges JI, Clark WA, Quay PD et al (1986) Compositions and fluxes of particulate organic material in the Amazon River1. Limol Oceanog 31:717–738

Hedges JI, Cowie GL, Richey JE et al (1994) Origins and processing of organic matter in the Amazon River as indicated by carbohydrates and amino acids. Limol Oceanog 39:743–761. https://doi.org/10.4319/lo.1994.39.4.0743

Hershey AE, Merritt RW, Miller MC et al. (1996) Organic matter processing by larval black flies in a temperate woodland stream. Oikos 524–532

Hieber M, Gessner MO (2002) Contribution of stream detrivores, fungi, and bacteria to leaf breakdown based on biomass estimates. Ecology 83:1026–1038

Hinton M, Schiff S, English M (1997) The significance of storms for the concentration and export of dissolved organic carbon from two Precambrian Shield catchments. Biogeochemistry 36:67–88

Hladyz S, Abjornsson K, Giller PS et al (2011) Impacts of an aggressive riparian invader on community structure and ecosystem functioning in stream food webs. J Appl Ecol 48:443–452. https://doi.org/10.1111/j.1365-2664.2010.01924.x

Hood E, Gooseff MN, Johnson SL (2006) Changes in the character of stream water dissolved organic carbon during flushing in three small watersheds. Oregon. J Geophys Res-Biogeosci 111:8. https://doi.org/10.1029/2005jg000082

Hood E, Williams MW, McKnight DM (2005) Sources of dissolved organic matter (DOM) in a Rocky Mountain stream using chemical fractionation and stable isotopes. Biogeochemistry 74:231–255

Hoover TM, Richardson JS, Yonemitsu N (2006) Flow-substrate interactions create and mediate leaf litter resource patches in streams. Freshw Biol 51:435–447

Hotchkiss E, Hall R Jr, Sponseller R et al (2015) Sources of and processes controlling CO 2 emissions change with the size of streams and rivers. Nat Geosci 8:696

Hutchens JJ, Wallace JB, Grubaugh JW (2017) Transport and storage of fine particulate organic matter. In: Lamberti GA, Hauer FR (eds) Methods in Stream Ecology (Third Edition). Academic Press, Amsterdam, pp 113–126. https://doi.org/10.1016/B978-0-12-813047-6.00007-3

Irons JG, Oswood MW, Stout RJ et al (1994) Latitudinal patterns in leaf-litter breakdown—is temperature really important. Freshw Biol 32:401–411

Jackrel SL, Morton TC, Wootton JT (2016) Intraspecific leaf chemistry drives locally accelerated ecosystem function in aquatic and terrestrial communities. Ecology 97:2125–2135

Jackrel SL, Wootton JT (2014) Local adaptation of stream communities to intraspecific variation in a terrestrial ecosystem subsidy. Ecology 95:37–43

James AB, Henderson IM (2005) Comparison of coarse particulate organic matter retention in meandering and straightened sections of a third-order New Zealand stream. Riv Res App 21:641–650

Jenkins CC, Suberkropp K (1995) The influence of water chemistry on the enzymatic degradation of leaves in streams. Freshwa Biol 33:245–253

Jones J (1997) Benthic organic matter storage in streams: influence of detrital import and export, retention mechanisms, and climate. J N Am Benthol Soc 16:109–119

Judd KE, Crump BC, Kling GW (2006) Variation in dissolved organic matter controls bacterial production and community composition. Ecology 87:2068–2079

Jung B-J, Jeanneau L, Alewell C et al (2015) Downstream alteration of the composition and biodegradability of particulate organic carbon in a mountainous, mixed land-use watershed. Biogeochemistry 122:79–99

Kaplan LA, Bott TL (1989) Diel fluctuations in bacterial activity on streambed substrata during vernal algal blooms: effects of temperature, water chemistry, and habitat. Limnol Oceanog 34:718–733

Karlsson OM, Richardson JS, Kiffney PM (2005) Modelling organic matter dynamics in headwater streams of south-western British Columbia, Canada. Ecol Model 183:463–476

Karwan DL, Saiers JE (2009) Influences of seasonal flow regime on the fate and transport of fine particles and a dissolved solute in a New England stream. Water Resour Res 45

Kaushal SS, Delaney-Newcomb K, Findlay SEG et al (2014) Longitudinal patterns in carbon and nitrogen fluxes and stream metabolism along an urban watershed continuum. Biogeochemistry 121:23–44. https://doi.org/10.1007/s10533-014-9979-9

Kaushik SJ, Hynes H (1971) The fate of autum-shed leaves that fall into streams. Archiv Für Hydrobiologie 68:465–515

Kautza A, Sullivan SMP (2016) The energetic contributions of aquatic primary producers to terrestrial food webs in a mid-size river system. Ecology 97:694–705. https://doi.org/10.1890/15-1095.1

Kiffer WP, Mendes F, Casotti CG et al (2018) Exotic Eucalyptus leaves are preferred over tougher native species but affect the growth and survival of shredders in an Atlantic Forest stream (Brazil). PLoS ONE 13:e0190743

Kochi K, Mishima Y, Nagasaka A (2010) Lateral input of particulate organic matter from bank slopes surpasses direct litter fall in the uppermost reaches of a headwater stream in Hokkaido, Japan. Limnology 11:77–84

Kochi K, Yanai S, Nagasaka A (2004) Energy input from a riparian forest into a headwater stream in Hokkaido, Japan. Archiv für Hydrobio 160:231–246

Kominoski JS, Rosemond AD, Benstead JP et al (2018) Experimental nitrogen and phosphorus additions increase rates of stream ecosystem respiration and carbon loss. Limnol Oceanog 63:22–36

Kuehn KA (2016) Lentic and lotic habitats as templets for fungal communities: traits, adaptations, and their significance to litter decomposition within freshwater ecosystems. Fungal Ecology 19:135–154

Lamberti GA, Entrekin SA, Griffiths NA et al. (2017) Coarse Particulate Organic Matter: Storage, Transport, and Retention. In: Methods in stream ecology. Elsevier, pp 55–69

Leroy CJ, Marks JC (2006) Litter quality, stream characteristics and litter diversity influence decomposition rates and macroinvertebrates. Freshw Biol 51:605–617

LeRoy CJ, Whitham TG, Wooley SC et al (2007) Within-species variation in foliar chemistry influences leaf-litter decomposition in a Utah river. J N Am Benthol Soc 26:426–438. https://doi.org/10.1899/06-113.1

Lisboa LK, da Silva ALL, Siegloch AE et al (2015) Temporal dynamics of allochthonous coarse particulate organic matter in a subtropical Atlantic rainforest Brazilian stream. Mar Freshw Res 66:674–680

Lock M, Wallis P, Hynes H (1977) Colloidal organic carbon in running waters. Oikos 1–4

López-Rojo N, Martínez A, Pérez J et al (2018) Leaf traits drive plant diversity effects on litter decomposition and FPOM production in streams. PLoS ONE 13:e0198243

Lottig NR, Buffam I, Stanley EH (2013) Comparisons of wetland and drainage lake influences on stream dissolved carbon concentrations and yields in a north temperate lake-rich region. Aquat Sci 75:619–630

Lu YH, Bauer JE, Canuel EA et al (2014) Effects of land use on sources and ages of inorganic and organic carbon in temperate headwater streams. Biogeochemistry 119:275–292. https://doi.org/10.1007/s10533-014-9965-2

Lujan NK, German DP, Winemiller K (2011) Do wood grazing fishes partition their niche?: morphological and isotopic evidence for trophic segregation in Neotropical Loricariidae. Funct Ecol 25:1327–1338

Lupon A, Denfeld BA, Laudon H et al (2019) Groundwater inflows control patterns and sources of greenhouse gas emissions from streams. Limnol Oceanog 64:1545–1557

Magana AM, Bretschko G (2003) Retention of coarse particulate organic matter on the sediments of Njoro River, Kenya. Int Rev Hydrobio 88:414–426

Malmqvist B, Nilsson LM, Svensson BS (1978) Dynamics of detritus in a small stream in southern Sweden and its influence on the distribution of the bottom animal communities. Oikos 1978:3–16

Manning DW, Rosemond AD, Kominoski JS et al (2015) Detrital stoichiometry as a critical nexus for the effects of streamwater nutrients on leaf litter breakdown rates. Ecology 96:2214–2224

Marcarelli AM, Baxter CV, Mineau MM et al (2011) Quantity and quality: unifying food web and ecosystem perspectives on the role of resource subsidies in freshwaters. Ecology 92:1215–1225

Marks JC (2019) Revisiting the fates of dead leaves that fall into streams. Ann Rev Ecol Evol System Review of Ecology, Evolution, and Systematics 50:547–568. https://doi.org/10.1146/annurev-ecolsys-110218-024755

Martinez B, Velasco J, Suarez ML et al (1998) Benthic organic matter dynamics in an intermittent stream in South-East Spain. Archiv für Hydrobiol 141:303–320

Marx A, Dusek J, Jankovec J et al (2017) A review of CO2 and associated carbon dynamics in headwater streams: A global perspective. Rev Geophys 55:560–585

Masese FO, Salcedo-Borda JS, Gettel GM et al (2017) Influence of catchment land use and seasonality on dissolved organic matter composition and ecosystem metabolism in headwater streams of a Kenyan river. Biogeochemistry 132:1–22

Mathuriau C, Chauvet E (2002) Breakdown of leaf litter in a neotropical stream. J N Am Benthol Soc 21:384–396

Mattsson T, Kortelainen P, Räike A (2005) Export of DOM from boreal catchments: impacts of land use cover and climate. Biogeochemistry 76:373–394

McArthur MD, Richardson JS (2002) Microbial utilization of dissolved organic carbon leached from riparian litterfall. Can J Fish Aquat Sci 59:1668–1676

McClain ME, Richey JE, Brandes JA et al (1997) Dissolved organic matter and terrestrial-lotic linkages in the central Amazon basin of Brazil. Glob Biogeochem Cycl 11:295–311

McDowell WH, Fisher SG (1976) Autumnal processing of dissolved organic matter in a small woodland stream ecosystem. Ecology 57:561–569

McDowell WH, Likens GE (1988) Origin, composition, and flux of dissolved organic carbon in the Hubbard Brook Valley. Ecol Mono 58:177–195

McNamara CJ, Leff LG (2004) Bacterial community composition in biofilms on leaves in a northeastern Ohio stream. J N Am Benthol Soc 23:677–685

Melillo JM, Naiman RJ, Aber JD et al (1983) The influence of substrate quality and stream size on wood decomposition dynamics. Oecologia 58:281–285

Menninger HL, Palmer MA, Craig LS et al (2008) Periodical cicada detritus impacts stream ecosystem metabolism. Ecosystems 11:1306–1317

Meyer J (1994) The microbial loop in flowing waters. Microb Ecol 28:195–199

Meyer JL, Likens GE (1979) Transport and transformation of phosphorus in a forest stream ecosystem. 60:1255–1269

Meyer JL, Tate CM (1983) The effects of watershed disturbance on dissolved organic carbon dynamics of a stream. Ecology 64:33–44

Meyer JL, Wallace JB, Eggert SL (1998) Leaf litter as a source of dissolved organic carbon in streams. Ecosystems 1:240–249. https://doi.org/10.1007/s100219900019

Miller J, Georgian T (1992) Estimation of fine particulate transport in streams using pollen as a seston analog. J N Am Benthol Soc 11:172–180

Mineau MM, Baxter CV, Marcarelli AM et al (2012) An invasive riparian tree reduces stream ecosystem efficiency via a recalcitrant organic matter subsidy. Ecology 93:1501–1508. https://doi.org/10.1890/11-1700.1

Mineau MM, Wollheim WM, Buffam I et al (2016) Dissolved organic carbon uptake in streams: A review and assessment of reach-scale measurements. J Geophys Res Biogeo 121:2019–2029

Minshall GW (1996) Organic matter budgets. In: Hauer ER, Lamberti G (eds) Methods in Stream Ecology. Academic Press, San Diego, pp 591–607

Minshall GW, Petersen RC, Cummins KW et al (1983) Interbiome comparison of stream ecosystem dynamics. Ecol Mono 53:1–25

Minshall GW, Thomas SA, Newbold JD et al (2000) Physical factors influencing fine organic particle transport and deposition in streams. J N Am Benthol Soc 19:1–16

Moeller JR, Minshall GW, Cummins KW et al (1979) Transport of dissolved organic carbon in streams of differing physiographic characteristics. Organ Geochem 1:139–150

Moghadam FS, Zimmer M (2016) Effects of warming, nutrient enrichment and detritivore presence on litter breakdown and associated microbial decomposers in a simulated temperate woodland creek. Hydrobiologia 770:243–256. https://doi.org/10.1007/s10750-015-2596-3

Mora-Gómez J, Elosegi A, Duarte S et al. (2016) Differences in the sensitivity of fungi and bacteria to season and invertebrates affect leaf litter decomposition in a Mediterranean stream. FEMS Micro Ecol 92:fiw121

Moran M, Covert J (2003) Photochemically mediated linkages between dissolved organic matter and bacterioplankton. In: Aquatic Ecosystems. Elsevier, pp 243–262

Mulholland P (1997a) Organic matter dynamics in the west fork of Walker Branch, Tennessee, USA. Journal of the North American Benthological Society 16:61–67

Mulholland P (2003) Large-scale patterns in dissolved organic carbon concentration, flux, and sources. Findlay SEG. Sinsabaugh RL. Aquatic Ecosystems. Elsevier, Amsterdam, pp 139–159

Mulholland PJ (1997b) Dissolved organic matter concentration and flux in streams. Journal of the North American Benthological Society 16:131–141

Naiman RJ (1982) Characteristics of sediment and organic carbon export from pristine boreal forest watersheds. Can J Fish Aquat Sci 39:1699–1718

Naiman RJ, Melillo JM, Hobbie JE (1986) Ecosystem alteation of boreal forest streams by beaver (Castor canadensis). Ecology 67:1254–1269

Naiman RJ, Sedell JR (1979) Characterization of particulate organic matter transported by some Cascade Mountain streams. J Fish Board of Canada 36:17–31

Nakajima T, Asaeda T, Fujino T et al (2006) Coarse particulate organic matter distribution in the pools and riffles of a second-order stream. Hydrobiologia 559:275–283

Niyogi DK, Lewis Jr WM, McKnight DM (2001) Litter breakdown in mountain streams affected by mine drainage: biotic mediation of abiotic controls. Ecol App 11:506–516

Neres-Lima V, Machado-Silva F, Baptista DF et al (2017) Allochthonous and autochthonous carbon flows in food webs of tropical forest streams. Freshw Biol 62:1012–1023

Newbold J, Bott T, Kaplan L et al (1997) Organic matter dynamics in White Clay Creek, Pennsylvania, USA. J N Am Benthol Soc 16:46–50

Ostrofsky ML (1993) Effect of tannins on leaf processing and conditioning rates in aquatic ecosystems: an empirical approach. Can J Fish Aquat Sci 50:1176–1180

Park JH, Nayna OK, Begum MS et al (2018) Reviews and syntheses: Anthropogenic perturbations to carbon fluxes in Asian river systems—concepts, emerging trends, and research challenges. Biogeosciences 15:3049–3069. https://doi.org/10.5194/bg-15-3049-2018

Parr TB, Capps KA, Inamdar SP et al (2019) Animal-mediated organic matter transformation: Aquatic insects as a source of microbially bioavailable organic nutrients and energy. Funct Ecol 33:524–535. https://doi.org/10.1111/1365-2435.13242

Parr TB, Cronan CS, Ohno T et al (2015) Urbanization changes the composition and bioavailability of dissolved organic matter in headwater streams. Limnol Oceanog 60:885–900

Parr TB, Vaughn CC, Gido KB (2020) Animal effects on dissolved organic carbon bioavailability in an algal controlled ecosystem. Freshw Biol 65:1298–1310

Pastor A, Compson ZG, Dijkstra P et al (2014) Stream carbon and nitrogen supplements during leaf litter decomposition: contrasting patterns for two foundation species. Oecologia 176:1111–1121

Paul MJ, Meyer JL, Couch CA (2006) Leaf breakdown in streams differing in catchment land use. Freshw Biol 51:1684–1695

Petersen R, Cummins K (1974) Leaf processing in a woodland stream. Freshw Biol 4:343–368

Pfeiffer A, Wohl E (2018) Where does wood most effectively enhance storage? Network-scale distribution of sediment and organic matter stored by instream wood. Geophys Res Lett 45:194–200. https://doi.org/10.1002/2017gl076057

Pitt DB, Batzer DP (2015) Potential impacts on stream macroinvertebrates of an influx of woody debris from eastern hemlock demise. Fores Sci 61:737–746. https://doi.org/10.5849/forsci.14-069

Polis GA, Anderson WB, Holt RD (1997) Toward an integration of landscape and food web ecology: The dynamics of spatially subsidized food webs. Annu Rev Ecol Syst 28:289–316. https://doi.org/10.1146/annurev.ecolsys.28.1.289

Polunin N (1982) Processes contributing to the decay of reed (*Phragmites australis*) litter in freshwater. Arch Hydrobiol 94:182–209

Polunin NV (1984) The decomposition of emergent macrophytes in fresh water. In: MacFayden A, Ford Ed (eds) Advances in Ecological Research, vol 14. Elsevier, Amsterdam, pp 115–166

Pozo J, González E, Díez J et al (1997) Leaf-litter budgets in two contrasting forested streams. Limnetica 13:77–84

Pray CL, Nowlin WH, Vanni MJ (2009) Deposition and decomposition of periodical cicadas (Homoptera: Cicadidae: Magicicada) in woodland aquatic ecosystems. J N Am Benthol Soc 28:181–195. https://doi.org/10.1899/08-038.1

Ramos Scharrón CE, Castellanos EJ, Restrepo C (2012) The transfer of modern organic carbon by landslide activity in tropical montane ecosystems. J Geophys Res: Biogeosciences 117

Raymond PA, Hartmann J, Lauerwald R et al (2013) Global carbon dioxide emissions from inland waters. Nature 503:355

Raymond PA, Saiers JE (2010) Event controlled DOC export from forested watersheds. Biogeochemistry 100:197–209

Raymond PA, Spencer RGM (2015) Riverine DOM. In: Hansell DA, Carlson CA. (eds) Biogeochemistry of Marine Dissolved Organic Matter, 2nd Edition. Academic Press Ltd-Elsevier Science Ltd, London. https://doi.org/10.1016/b978-0-12-405940-5.00011-x

Reich PB, Oleksyn J (2004) Global patterns of plant leaf N and P in relation to temperature and latitude. Proc Nat Acad Sci 101:11001–11006

Reiman J, Xu YJ (2019) Dissolved carbon export and CO2 outgassing from the lower Mississippi River-Implications of future river carbon fluxes. J Hydrol 578:124093

Ribblett SG, Palmer MA, Wayne Coats D (2005) The importance of bacterivorous protists in the decomposition of stream leaf litter. Freshw Biol 50:516–526

Rincon JE, Martinez I, Leon E et al (2005) Leaf litter processing of Anacardium excelsum in a tropical intermittent stream of northwestern Venezuela. Interciencia 30:228–234

Roditi HA, Fisher NS, Sanudo-Wilhelmy SA (2000) Uptake of dissolved organic carbon and trace elements by zebra mussels. Nature 407:78–80

Rosemond AD, Benstead JP, Bumpers PM et al (2015) Experimental nutrient additions accelerate terrestrial carbon loss from stream ecosystems. Science 347:1142–1145

Rosemond AD, Pringle CM, Ramirez A (1998) Macroconsumer effects on insect detritivores and detritus processing in a tropical stream. Freshw Biol 39:515–523

Rosemond AD, Pringle CM, Ramirez A et al (2002) Landscape variation in phosphorus concentration and effects on detritus-based tropical streams. Limnol Oceanog 47:278–289

Rosi-Marshall EJ, Wallace JB (2002) Invertebrate food webs along a stream resource gradient. Freshw Biol 47:129–141

Royer TV, David MB (2005) Export of dissolved organic carbon from agricultural streams in Illinois, USA. Aquat Sci 67:465–471. https://doi.org/10.1007/s00027-005-0781-6

Royer TV, Minshall GW (2001) Effects of nutrient enrichment and leaf quality on the breakdown of leaves in a hardwater stream. Freshw Biol 46:603–610

Rugenski AT, Murria C, Whiles MR (2012) Tadpoles enhance microbial activity and leaf decomposition in a neotropical headwater stream. Freshw Biol 57:1904–1913. https://doi.org/10.1111/j.1365-2427.2012.02853.x

Ruhala SS, Zarnetske JP (2017) Using in-situ optical sensors to study dissolved organic carbon dynamics of streams and watersheds: A review. Sci Tot Environ 575:713–723

Sanpera-Calbet I, Acuña V, Butturini A et al (2016) El Niño southern oscillation and seasonal drought drive riparian input dynamics in a Mediterranean stream. Limnol Oceanog 61:214–226

Santonja M, Pellan L, Piscart C (2018) Macroinvertebrate identity mediates the effects of litter quality and microbial conditioning on leaf litter recycling in temperate streams. Ecol Evol 8:2542–2553

Sawakuchi HO, Neu V, Ward ND et al (2017) Carbon dioxide emissions along the lower Amazon River. Front Mar Sci 4:76

Scarsbrook M, Quinn J, Halliday J et al (2001) Factors controlling litter input dynamics in streams draining pasture, pine, and native forest catchments. NZ J Mar Freshwat Res 35:751–762

Schiff S, Aravena R, Mewhinney E et al (1998) Precambrian shield wetlands: hydrologic control of the sources and export of dissolved organic matter. Clim Change 40:167–188

Schmidt TS, Clements WH, Wanty RB et al (2012) Geologic processes influence the effects of mining on aquatic ecosystems. Ecol Appl 22:870–879. https://doi.org/10.1890/11-0806.1

Schofield KA, Pringle CM, Meyer JL et al (2001) The importance of crayfish in the breakdown of rhododendron leaf litter. Freshw Biol 46:1191–1204

Shang P, Lu YH, Du YX et al (2018) Climatic and watershed controls of dissolved organic matter variation in streams across a gradient of agricultural land use. Sci Total Environ 612:1442–1453. https://doi.org/10.1016/j.scitotenv.2017.08.322

Siders AC, Compson ZG, Hungate BA et al (2018) Litter identity affects assimilation of carbon and nitrogen by a shredding caddisfly. Ecosphere 9:e02340

Singer GA, Battin TJ (2007) Anthropogenic subsidies alter stream consumer-resource stoichiometry, biodiversity, and food chains. Ecol Appl 17:376–389

Sinsabaugh R, Weiland T, Linkins AE (1992) Enzymic and molecular analysis of microbial communities associated with lotic particulate organic matter. Freshw Biol 28:393–404

Smith RM, Kaushal SS (2015) Carbon cycle of an urban watershed: exports, sources, and metabolism. Biogeochemistry 126:173–195. https://doi.org/10.1007/s10533-015-0151-y

Smock L (1990) Spatial and temporal variation in organic matter storage in low-gradient, headwater streams. Arch Hydrobiol 118:169–184

Sobczak WV, Findlay S (2002) Variation in bioavailability of dissolved organic carbon among stream hyporheic flowpaths. Ecology 83:3194–3209

Sollins P, Glassman CA, Dahm CN (1985) Composition and possible origin of detrital material in streams. Ecology 66:297–299

Spanhoff B, Meyer EI (2004) Breakdown rates of wood in streams. J N Am Benthol Soc 23:189–197

Spencer RG, Hernes PJ, Dinga B et al (2016) Origins, seasonality, and fluxes of organic matter in the Congo River. Glob Biogeochem Cycle 30:1105–1121

Spencer RGM, Butler KD, Aiken GR (2012) Dissolved organic carbon and chromophoric dissolved organic matter properties of rivers in the USA. J Geophys Res-Biogeosci 117:14. https://doi.org/10.1029/2011jg001928

Spencer RGM, Kellerman AM, Podgorski DC et al (2019) Identifying the Molecular Signatures of Agricultural Expansion in Amazonian Headwater Streams. J Geophys Res-Biogeosci 124:1637–1650. https://doi.org/10.1029/2018jg004910

Sponseller R, Benfield E (2001) Influences of land use on leaf breakdown in southern Appalachian headwater streams: a multiple-scale analysis. J N Am Benthol Soc 20:44–59

Stanley EH, Powers SM, Lottig NR et al (2012) Contemporary changes in dissolved organic carbon (DOC) in human-dominated rivers: is there a role for DOC management? Freshw Biol 57:26–42. https://doi.org/10.1111/j.1365-2427.2011.02613.x

Stiling P, Cornelissen T (2007) How does elevated carbon dioxide (CO2) affect plant–herbivore interactions? A field experiment and meta-analysis of CO2-mediated changes on plant chemistry and herbivore performance. Glob Chan Biol 13:1823–1842

Subalusky AL, Dutton CL, Njoroge L et al (2018) Organic matter and nutrient inputs from large wildlife influence ecosystem function in the Mara River, Africa. Ecology 99:2558–2574

Subalusky AL, Dutton CL, Rosi EJ et al (2017) Annual mass drownings of the Serengeti wildebeest migration influence nutrient cycling and storage in the Mara River. Proc Natl Acad Sci U S A 114:7647–7652. https://doi.org/10.1073/pnas.1614778114

Subalusky AL, Post DM (2019) Context dependency of animal resource subsidies. Biol Rev 94:517–538. https://doi.org/10.1111/brv.12465

Suberkropp K, Godshalk G, Klug M (1976) Changes in the chemical composition of leaves during processing in a woodland stream. Ecology 57:720–727

Suberkropp K, Klug M (1976) Fungi and bacteria associated with leaves during processing in a woodland stream. Ecology 57:707–719

Sutfin NA, Wohl EE, Dwire KA (2016) Banking carbon: a review of organic carbon storage and physical factors influencing retention in floodplains and riparian ecosystems. Earth Surf Process Landf 41:38–60

Tank JL, Rosi-Marshall EJ, Griffiths NA et al (2010) A review of allochthonous organic matter dynamics and metabolism in streams. J N Am Benthol Soc 29:118–146

Tant CJ, Rosemond AD, First MR (2013) Stream nutrient enrichment has a greater effect on coarse than on fine benthic organic matter. Freshw Sci 32:1111–1121. https://doi.org/10.1899/12-049.1

Thomas SM, Griffiths SW, Ormerod SJ (2016) Beyond cool: adapting upland streams for climate change using riparian woodlands. Glob Chan Biol 22:310–324

Thurman EM (2012) Organic geochemistry of natural waters, vol 2. Springer Science & Business Media, Dordrecht

Tiegs SD, Clapcott JE, Griffiths NA et al (2013) A standardized cotton-strip assay for measuring organic-matter decomposition in streams. Ecol Indic 32:131–139

Tiegs SD, Costello DM, Isken MW et al. (2019) Global patterns and drivers of ecosystem functioning in rivers and riparian zones. Sci Advanc 5:eaav0486

Tiegs SD, Levi PS, Ruegg J et al (2011) Ecological effects of live salmon exceed those of carcasses during an annual spawning migration. Ecosystems 14:598–614. https://doi.org/10.1007/s10021-011-9431-0

Tomczyk NJ, Parr TB, Gray E et al (2018a) Trophic strategies influence metal bioaccumulation in detritus-based, aquatic food webs. Environ Sci Technol 52:11886–11894. https://doi.org/10.1021/acs.est.8b04009

Tomczyk NJ, Parr TB, Wenger SJ et al (2018b) The influence of land cover on the sensitivity of streams to metal pollution. Water Res 144:55–63. https://doi.org/10.1016/j.watres.2018.06.058

Tonin AM, Gonçalves JF, Bambi P et al (2017) Plant litter dynamics in the forest-stream interface: precipitation is a major control across tropical biomes. Sci Repo 7:10799. https://doi.org/10.1038/s41598-017-10576-8

Tonin AM, Pozo J, Monroy S et al (2018) Interactions between large and small detritivores influence how biodiversity impacts litter decomposition. J Animal Ecol 87:1465–1474

Tuchman NC, Wahtera KA, Wetzel RG et al (2003) Elevated atmospheric CO2 alters leaf litter quality for stream ecosystems: an in situ leaf decomposition study. Hydrobiologia 495:203–211. https://doi.org/10.1023/a:1025493018012

Usio N, Townsend CR (2001) The significance of the crayfish Paranephrops zealandicus as shredders in a New Zealand headwater stream. J Crustacean Biol 21:354–359

Vanni MJ, McIntyre PB, Allen D et al (2017) A global database of nitrogen and phosphorus excretion rates of aquatic animals. Ecology 98:1475

Vannote RL, Minshall GW, Cummins KW et al (1980) River continuum concept. Can J Fish Aq Sci 37:130–137. https://doi.org/10.1139/f80-017

Vazquez E, Amalfitano S, Fazi S et al (2011) Dissolved organic matter composition in a fragmented Mediterranean fluvial system under severe drought conditions. Biogeochemistry 102:59–72. https://doi.org/10.1007/s10533-010-9421-x

Venarsky MP, Benstead JP, Huryn AD et al (2018) Experimental detritus manipulations unite surface and cave stream ecosystems along a common energy gradient. Ecosystems 21:629–642

Wallace JB, Cuffney T, Webster J et al (1991) Export of fine organic particles from headwater streams: effects of season, extreme discharges, and invertebrate manipulation. Limnol Oceanog 36:670–682

Wallace JB, Eggert SL, Meyer JL et al (2015) Stream invertebrate productivity linked to forest subsidies: 37 stream-years of reference and experimental data. Ecology 96:1213–1228. https://doi.org/10.1890/14-1589.1

Wallace JB, Ross DH, Meyer JL (1982a) Seston and dissolved organic carbon dynamics in a southern Appalachian stream. Ecology 63:824–838

Wallace JB, Webster JR, Cuffney TF (1982b) Stream detritus dynamics: regulation by invertebrate consumers. Oecologia 53:197–200

Wallace JB, Webster JR, Eggert SL et al (2001) Large woody debris in a headwater stream: long-term legacies of forest disturbance. Int Rev Hydrobio 86:501–513

Wallace JB, Whiles MR, Eggert S et al (1995) Long-term dynamics of coarse particulate organic matter in three Appalachian Mountain streams. J N Am Benthol Soc 14:217–232

Wallace TA, Ganf GG, Brookes JD (2008) A comparison of phosphorus and DOC leachates from different types of leaf litter in an urban environment. Freshw Biol 53:1902–1913

Wallis P, Hynes H, Telang S (1981) The importance of groundwater in the transportation of allochthonous dissolved organic matter to the streams draining a small mountain basin. Hydrobiologia 79:77–90

Wanner S, Ockenfeld K, Brunke M et al (2002) The distribution and turnover of benthic organic matter in a lowland river: Influence of hydrology, seston load and impoundment. Riv Res App 18:107–122

Wanner SC, Pusch M (2001) Analysis of particulate organic matter retention by benthic structural elements in a lowland river (River Spree, Germany). Archiv für Hydrobiologie:475–492

Wantzen KM, Wagner R (2006) Detritus processing by invertebrate shredders: a neotropical-temperate comparison. Journal of the North American Benthological Society 25:216–232

Wantzen KM, Wagner R, Suetfeld R et al (2002) How do plant-herbivore interactions of trees influence coarse detritus processing by shredders in aquatic ecosystems of different latitudes? InterVer für Limnol 28:815–821

Ward GM, Aumen NG (1986) Woody debris as a source of fine particulate organic matter in coniferous forest stream ecosystems. Can J Fish Aq Sci 43:1635–1642

Ward JV (1974) A temperature-stressed stream ecosystem below a hypolimnial release mountain reservoir. Arch Hydrobio 74:247–275

Ward ND, Bianchi TS, Medeiros PM et al (2017) Where carbon goes when water flows: carbon cycling across the aquatic continuum. Front Mar Sci 4:27. https://doi.org/10.3389/fmars.2017.00007

Webster J, Benfield E (1986) Vascular plant breakdown in freshwater ecosystems. Annu Rev Ecol Syst 17:567–594

Webster J, Covich A, Tank J et al (1994) Retention of coarse organic particles in streams in the southern Appalachian Mountains. J N Am Benthol Soc 13:140–150

Webster J, Golladay S, Benfield E et al (1990) Effects of forest disturbance on particulate organic matter budgets of small streams. J N Am Benthol Soc 9:120–140

Webster J, Meyer JL (1997) Organic matter budgets for streams: a synthesis. J N Am Benthol Soc 16:141–161

Webster J, Morkeski K, Wojculewski C et al (2012) Effects of hemlock mortality on streams in the southern Appalachian Mountains. Am Mid Nat 168:112–132

Webster J, Wallace J, Benfield E (1995) Organic processes in streams of the eastern United States. River and Stream Ecosystems-Ecosystems of the World 22:117–187

Webster J, Wallace J, Benfield E (2006) Organic processes in streams of the eastern United States. River and stream ecosystems of the world University of California Press, Berkeley, California, pp 117–187

Webster JR (1983) The role of benthic macroinvertebrates in detritus dynamics of streams: a computer simulation. Ecol Mono 53:383–404

Webster JR, Benfield EF, Ehrman TP et al (1999) What happens to allochthonous material that falls into streams? A synthesis of new and published information from Coweeta. Freshw Biol 41:687–705

Webster JR, Waide JB (1982) Effects of forest clearcutting on leaf breakdown in a southern Appalachian stream. Freshw Biol 12:331–344

Wenger SJ, Subalusky AL, Freeman MC (2019) The missing dead: The lost role of animal remains in nutrient cycling in North American Rivers. Food Webs 18:e00106. https://doi.org/10.1016/j.fooweb.2018.e00106

Wenisch B, Fernández DG, Szöcs E et al (2017) Does the loss of climate sensitive detritivore species alter leaf decomposition? Aq Sci 79:869–879

West A, Lin C-W, Lin T-C et al (2011) Mobilization and transport of coarse woody debris to the oceans triggered by an extreme tropical storm. Limnol Ocean 56:77–85

Westerhoff P, Anning D (2000) Concentrations and characteristics of organic carbon in surface water in Arizona: influence of urbanization. J Hydrol 236:202–222. https://doi.org/10.1016/s0022-1694(00)00292-4

Whiles MR, Hall RO, Dodds WK et al (2013) Disease-driven amphibian declines alter ecosystem processes in a tropical stream. Ecosystems 16:146–157. https://doi.org/10.1007/s10021-012-9602-7

Wiegner TN, Seitzinger SP, Glibert PM et al (2006) Bioavailability of dissolved organic nitrogen and carbon from nine rivers in the eastern United States. Aq Microb Ecol 43:277–287

Wild R, Gücker B, Brauns M (2019) Agricultural land use alters temporal dynamics and the composition of organic matter in temperate headwater streams. Freshw Sci 38

Williams CJ, Yamashita Y, Wilson HF et al (2010) Unraveling the role of land use and microbial activity in shaping dissolved organic matter characteristics in stream ecosystems. Limnol Oceanog 55:1159–1171. https://doi.org/10.4319/lo.2010.55.3.1159

Wilson HF, Saiers JE, Raymond PA et al (2013) Hydrologic drivers and seasonality of dissolved organic carbon concentration, nitrogen content, bioavailability, and export in a forested New England stream. Ecosystems 16:604–616

Wipfli MS, Baxter CV (2010) Linking ecosystems, food webs, and fish production: subsidies in salmonid watersheds. Fisheries 35:373–387. https://doi.org/10.1577/1548-8446-35.8.373

Wise JL, Van Horn DJ, Diefendorf AF et al (2019) Dissolved organic matter dynamics in storm water runoff in a dryland urban region. J Arid Environ 165:55–63. https://doi.org/10.1016/j.jaridenv.2019.03.003

Wohl E (2015) Legacy effects on sediments in river corridors. Earth-Sci Rev 147:30–53. https://doi.org/10.1016/j.earscirev.2015.05.001

Wohl E, Ogden FL (2013) Organic carbon export in the form of wood during an extreme tropical storm, Upper Rio Chagres, Panama. Earth Surf Proc Land 38:1407–1416

Wotton RS, Malmqvist B (2001) Feces in aquatic ecosystems. Bioscience 51:537–544

Wotton RS, Malmqvist B, Muotka T et al (1998) Fecal pellets from a dense aggregation of suspension-feeders in a stream: an example of ecosystem engineering. Limnol Oceanog 43:719–725

Wright MS, Covich AP (2005a) The effect of macroinvertebrate exclusion on leaf breakdown rates in a tropical headwater stream. Biotropica 37:403–408

Wright MS, Covich AP (2005b) Relative importance of bacteria and fungi in a tropical headwater stream: leaf decomposition and invertebrate feeding preference. Microb Ecol 49:536–546

WWPU (2017) The United Nations World Water Development Report (2017) Wastewater: the untapped resource. UNESCO, Paris

Wymore AS, Compson ZG, Liu CM et al (2013) Contrasting rRNA gene abundance patterns for aquatic fungi and bacteria in response to leaf-litter chemistry. Freshw Sci 32:663–672

Wymore AS, Compson ZG, McDowell WH et al (2015) Leaf-litter leachate is distinct in optical properties and bioavailability to stream heterotrophs. Freshw Sci 34:857–866

Wymore AS, Salpas E, Casaburi G et al (2018) Effects of plant species on stream bacterial communities via leachate from leaf litter. Hydrobiologia 807:131–144

Yates CA, Johnes PJ, Spencer RGM (2019) Characterisation of treated effluent from four commonly employed wastewater treatment facilities: a UK case study. J Environ Manage 232:919–927. https://doi.org/10.1016/j.jenvman.2018.12.006

Yoon B, Raymond PA (2012) Dissolved organic matter export from a forested watershed during Hurricane Irene. Geophys Res Lett 39

Yoshimura C, Gessner MO, Tockner K et al (2008) Chemical properties, microbial respiration, and decomposition of coarse and fine particulate organic matter. J N Am Benthol Soc 27:664–673

Young RG, Huryn AD (1999) Effects of land use on stream metabolism and organic matter turnover. Ecol Appl 9:1359–1376

Yule CM (1996) Trophic relationships and food webs of the benthic invertebrate fauna of two aseasonal tropical streams on Bougainville Island, Papua New Guinea. J Trop Ecol 12:517–534

Yule CM, Leong MY, Liew KC et al (2009) Shredders in Malaysia: abundance and richness are higher in cool upland tropical streams. J N Am Benthol Soc 28:404–415

Microbes occur in or on all surfaces in streams, where they play key roles in supporting stream food webs and ecosystem processes (Findlay 2016). Unlike other common nomenclature in stream ecology (e.g., for algae, fishes, etc.), the term, microbe, is used to define a size class of a great diversity of organisms, rather a single taxonomic group (Findlay 2010). Microbial diversity includes taxa from the three great domains of life, the Bacteria, Archaea, and Eukarya (Pace 2006). Originally classified by their physical structure or basic chemical reactions, newer descriptors rely on molecular methods that allow researchers to examine changes in microbial diversity at the species-level.

Microbes are essential in maintaining biogeochemical processes in benthic and pelagic habitats in streams. The greatest microbial biomass is typically found in benthic sediments (Findlay et al. 2002). In smaller systems, microbial biomass on surfaces or in sediments can exceed the biomass of planktonic species, but the relative contribution of benthic and planktonic species to the microbial community is influenced by physical, chemical, and biological factors (Besemer et al. 2012; Mao et al. 2019). Researchers first recognized microbes for their contribution to the decomposition of allochthonous organic matter and mineralization of inorganic nutrients. After "blue-green algae" (Cyanobacteria) were acknowledged as part of the microbial community, photosynthetic production of autochthonous carbon was recognized as an additional function performed by microbes (Pomeroy et al. 2007). The uptake, storage, and transformation of energy and nutrients by microbial communities in the benthos is a critical flux in stream food webs, especially in smaller, headwater systems (Battin et al. 2016; Meyer 1994; Wagner et al. 2017). Research has also demonstrated that bacterial communities in the water column actively contribute to these processes along the river continuum (Graeber et al. 2018).

Until the past few decades, research in stream microbial ecology has been relatively limited. Yet, enhanced technological capabilities, including the use of ergosterol assays to estimate fungal biomass (Gessner and Chauvet 1993), and direct microscopy to estimate bacterial biomass (Findlay and Arsuffi 1989), allow researchers to begin examining changes in microbial abundance through space and time (Findlay 2010). Our ability to account for specific components of dissolved organic matter, and rapid advances in molecular microbial ecology including the ability to assay for key enzymes and to survey microorganisms for the functional genes that encode particular enzymes, have also generated new insights into such processes as leaf breakdown and biofilm activity (Findlay 2016; Findlay and Sinsabaugh 2006; Zak et al. 2006). Additionally, newer molecular techniques (Zeglin 2015) and the development of new statistical methods (Buttigieg and Ramette 2014) have allowed researchers to estimate the abundance and regulation of functional genes in the microbial community, and have begun to generate information about the ways that microbes regulate biogeochemical processes (Zeglin 2015). For instance, our understanding of microbial diversity has substantially expanded using both polymerase chain reaction (PCR) and direct sequencing (direct rRNA-seq) to amplify the 16S rRNA gene and construct bacterial phylogenies (Rosselli et al. 2016). The application of flow cytometry (Dann et al. 2017; Sgier et al. 2016) and quantitative stable isotope probing (Hayer et al. 2016; Koch et al. 2018) have also permitted researchers to study taxon-specific population dynamics and function in stream microbial communities. Because the microbiome is an emerging concept of interest in ecology, and because of the fundamental contribution of microbes to many aspects of the structure and function of streams and rivers, we discuss microbes and important microbial functions throughout this book. However, the

© Springer Nature Switzerland AG 2021
J. D. Allan et al., *Stream Ecology*,
https://doi.org/10.1007/978-3-030-61286-3_8

ubiquity of microbes and their importance to stream ecosystem ecology warrants further discussion. Thus, in this chapter we focus on the structure and function of benthic and planktonic microbial communities in lotic ecosystems.

8.1 The Microbial Loop

Microbes are important components of ecosystem processes in streams, as they convert resources (i.e., energy and nutrients) into forms that can be ingested and assimilated by consumers at higher trophic levels (Findlay 2010). Bacteria and fungi utilize non-living organic matter as a carbon source, fueling microbial production and mineralizing that carbon as CO_2. Bacterial conversion of dissolved organic matter (DOM) into fine particulate organic matter (FPOM) is an essential flux of carbon within stream food webs (Fig. 8.1). The assimilation of the organic carbon into microbial biomass, the subsequent consumption of this heterotrophic production, and the eventual re-mineralization to CO_2 by community respiration are principal biotic processes driving the removal of DOM from lotic systems.

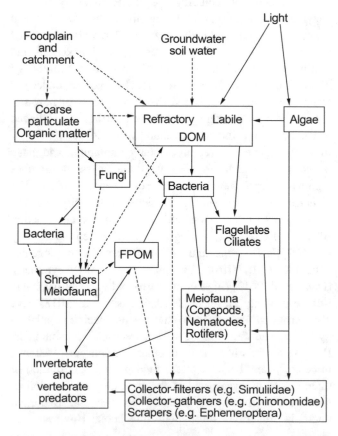

Fig. 8.1 A simplified fluvial food web emphasizing the role of the meiofauna and microbial production. Dotted lines are flows within the microbial web (Reproduced from Schmid-Arraya and Schmid 2000, based on Meyer 1994)

Microbes can also degrade a variety of carbon compounds and use multiple electron acceptors to support their metabolic activities. Autotrophic microbes, such as cyanobacteria, can also fix carbon, and evidence suggests that they may disproportionately contribute to the production of higher trophic levels relative to their biomass (McCutchan and Lewis 2002). Fungi typically respond rapidly to new influxes of coarse particulate organic matter (CPOM) and frequently dominate total microbial biomass on leaves and wood (Hieber and Gessner 2002). Hence, stream reaches with frequent litter inputs can be hot spots of fungal growth and biomass (Findlay 2010). Fungal production on CPOM can be as much as 10% of the total volume of coarse particulates (Methvin and Suberkropp 2003). In contrast, bacterial populations are typically more abundant in deeper sediments and in habitats dominated by fine particulates than in CPOM deposits (Findlay et al. 2002). Microbes mediate the processing of particulate organic matter (POM), as they may make the organic substrate more accessible to large consumers such as leaf-shredding insects by softening the tissues, and the fungi and bacteria are themselves an easily assimilated source of carbon. Allochthonous resources, such as leaf litter, may also lack highly unsaturated omega-3 fatty acids (HUFAs), essential fats for freshwater animals. Many organisms are unable to synthesize HUFAs and must consume microbes capable of generating these fats to meet their physiological requirements (Twining et al. 2016, 2017). The uptake of DOM by microorganisms provides a pathway for an abundant carbon source to enter fluvial food webs. Collectively, the pathways associated with carbon and nutrient flow through microbial food webs in aquatic systems are frequently termed, "the microbial loop" (Azam et al. 1983).

The microbial loop was initially used to describe the relative contribution of microbial food webs to biogeochemical processes in oligotrophic regions of the ocean (Azam et al. 1983; Fenchel 2008; Pomeroy et al. 2007). Studies of both marine and lacustrine systems (Pomeroy et al. 2007; Pomeroy and Wiebe 1988) have documented that carbon flux through the microbial loop is much greater than was previously recognized. Since its inception, the term has also been applied to microbial ecology in lotic systems; yet, the link between microbes and higher-order consumers such as macroinvertebrates in rivers differs from marine food webs in several, potentially significant ways (Hall and Meyer 1998; Meyer 1994). Allochthonous sources of energy and nutrients, and strong links between planktonic and benthic microbial food webs, are characteristics that distinguish the riverine from the marine microbial loop. Additionally, microbial production can be consumed directly by those organisms capable of ingesting individual cells and biofilms, or it can be ingested through its association with larger particles, including FPOM and CPOM. Ingestion of individual bacterial cells implies a food chain of small and then larger micro-consumers before eventually reaching macroinvertebrates and fish, raising the possibility that the original microbial production is largely dissipated by multiple energy transfers within a microbial loop (Meyer 1994). Understanding the environmental conditions and natural history traits that govern the assimilation of microbially-derived energy by higher trophic levels and that control the rates of microbially-mediated nutrient and energy dynamics are essential in predicting how energy and nutrient dynamics will change with anthropogenic activities.

Microbial production forms the base of many stream food webs. Specifically, meiofaua, or protozoans and metazoans less than 0.5 mm in length, are important consumers of microbial production (Schmid-Araya et al. 2016). Meiofauna are an often neglected component of the fauna, yet they may contribute half or more of the diversity and abundance of stream assemblages, particularly due to rotifers and larval midges of the Chironomidae (Robertson 2000). Most of the meiofaunal groups, including protozoans, gastotrichs, rotifers, nematodes, microturbellarians, small oligochaetes, and microcrustaceans, are suspension feeders or browsers feeding on bacteria or on small particles coated with biofilm.

Comparatively few studies have examined trophic interactions of meiofauna, but their impact on the flow of energy and nutrients through the microbial loop and to higher-order consumers may be particularly important. Figure 8.1 is a simplified conceptual model of the relationships among microbial production and the meiofauna within stream food webs (Schmid-Araya and Schmid 2000). This figure emphasizes the potential influence of the physicochemical and biological characteristics of a system on the flow of energy and elements through the microbial loop. However, it does not reflect the temporally dynamic and complex relationships that are most likely present in meiofaunal food webs.

The difficulties of sampling and quantifying meiofauna often require ecologists to aggregate meiofaunal species into a single group when studying riverine food webs. Work by Schmid-Araya et al. (2016) highlights the complex and dynamic interactions that can be found in meiofaunal food webs in streams, and emphasizes that characteristics such as food web size and complexity and predator–prey ratios may be underestimated if meiofauna are considered in aggregate. Their work in the River Lambourn in Berkshire, England, demonstrated that the meiofauna contributed more than one-third of all species in the food web, and that at times, meiofaunal species functioned as top consumers in their study system (Fig. 8.2a). Notably, this study also demonstrated that food web relationships varied with season (Fig. 8.2b–e), indicating there may be corresponding temporal shifts in ecosystem processes within the microbial loop (Schmid-Araya et al. 2016). Clearly, the number of trophic transfers between bacterial production and higher order

consumers can be one or many, with significant consequences for the rate and amount of energy transmitted through food webs.

The number of trophic links in the microbial web is strongly influenced by the small size of its members, and especially the size of a bacterial cell, which is about 0.5 μm. Because few suspension feeders are able to capture particles of this size (Wotton 1994), the ingestion of bacteria by flagellates (ca. 5 μm) and ciliates (ca. 25 μm) provides a micro-consumer pathway for this energy to reach larger consumers. On the other hand, bacteria associated with biofilm in a surface microlayer may be ingested directly by benthic consumers of surface films and by deposit feeders that pass organic matter and associated microbes through their guts. For instance, Perlmutter and Meyer (1991) documented significant consumption of bacteria by a harpacticoid copepod. Some studies have shown that flagellates and ciliates are able to exert significant grazing pressure on bacteria in streambed sediments (Bott and Kaplan 1990) and in the water column (Carlough and Meyer 1991). The effectiveness of direct meiofaunal consumption likely varies with the composition of the assemblage, and has been found to be greater when large ciliates and rotifers are present (Borchardt and Bott 1995). Predator–prey linkages within the meiofauna also are little studied, although many taxa of rotifers, cyclopoid copepods, oligochaetes, microturbellarians, mites, and nematodes are predaceous. The larger members of the meiofauna may be consumed by predaceous invertebrates such as tanypod midges (Schmid and Schmid-Araya 1997), larger predators including *Sialis* and *Plectrocnemia* (Lancaster and Robertson 1995) and even by fish, as Rundle and Hildrew (1992) document for the stone loach *Noemacheilus barbatulus* in English streams. Because members of the meiofauna are small and soft-bodied, assessing their presence in the guts of predators obviously is a challenge, and thus far, the resultant data usually are qualitative rather than quantitative (Schmid-Araya and Schmid 2000).

In benthic food webs, the presence of biofilms and association of bacterial cells with POM in sediments and on substrates also facilitates the direct ingestion of microbial biomass by macro-consumers (Lodge et al. 1988; Meyer 1994). This potentially allows microbial carbon to bypass a lengthy food chain and reach macroinvertebrates in sufficient quantity to contribute significantly to the support of higher trophic levels, as has been nicely demonstrated by a whole stream addition of ^{13}C sodium acetate to label the benthic bacteria of a small stream at the Coweeta Hydrologic Laboratory in the southeastern US (Hall and Meyer 1998). The label soon appeared throughout the food web, including in invertebrate predators. Macroinvertebrates derived from 0 to 100% of their carbon from bacteria, and this amount was significantly related to the percent of amorphous detritus in invertebrate guts (Fig. 8.3), so it is likely that

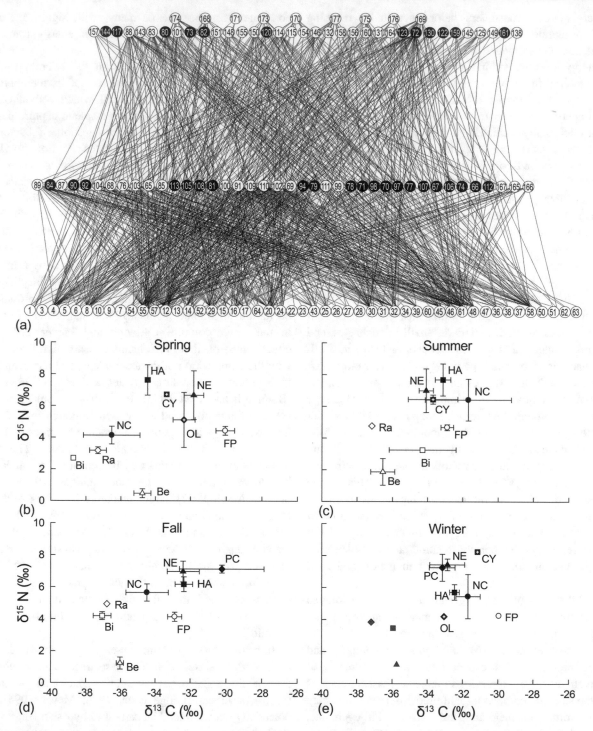

Fig. 8.2 Trophic interactions within the food web of the River Lambourn, United Kingdom. (**a**) Representation of the food web in the summer months. Black-colored bubbles represent meiofaunal-sized taxa. Variation (mean ± SE) of $\delta^{13}C$ and $\delta^{15}N$ for various meiofauna and their basal resources within the benthos of the River Lambourn during (**b**) spring, (**c**) summer, (**d**) fall, and (**e**) winter. Bi, biofilm; Ra, *Ranunculus* spp.; Be, *Berula erecta*; FP, fine particulate organic matter; NC, non-predatory Chironomidae larvae; PC, predatory

Chironomidae larvae; HA, harpacticoid copepods; CY, cyclopoida copepods; NE, nematodes; OL, oligochaetes. The gray symbols in **e** are estimated overall means from other seasons. Greater values of $\delta^{15}N$ are indicative of higher trophic level position. Allochthonous carbon sources have greater $\delta^{13}C$ on average (range: −32.90‰ to −29.99‰) than the autochthonous sources, which are more depleted in $\delta^{13}C$ (range: −38.74‰ and −34.29‰) (Reproduced from Schmid-Arraya et al. 2016)

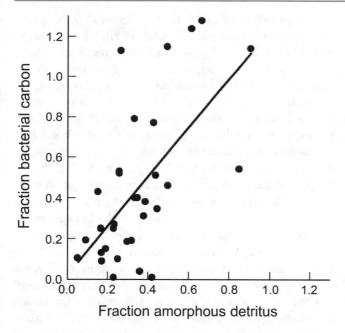

Fig. 8.3 Relationship between the fraction of invertebrate carbon derived from bacteria and the fraction of amorphous detritus in invertebrate guts (Reproduced from Hall and Meyer 1998)

biofilm was ingested incidentally through the consumption of particles of FPOM and CPOM. The assimilation of bacterial carbon may also have resulted from consumption of bacterial exopolymers secreted within the 'slime' and consumed by the grazing of substrate surfaces. Similar work conducted by Collins et al. (2016) in two small streams in upstate New York examined how light availability altered trophic interactions between bacterial and macroconsumers. Researchers employed dual isotope tracer additions (^{13}C-acetate, ^{15}N-ammonium) to examine changes in trophic interactions before and after canopy thinning. They found that the flux of carbon from bacteria to higher trophic levels can be substantial (>70% for some taxa). Increased light tended to reduce the magnitude of carbon fluxes from bacteria to invertebrate consumers. Their results demonstrated that bacterial energy assimilation varied among consumers and within consumer groups under differing environmental conditions.

Microbial response to terrestrially supplied DOC supports expectations that DOC bioavailability varies with catchment vegetation type and flowpaths (Fasching et al. 2019; Freixa et al. 2016; O'Brien et al. 2017). Benthic microbial communities in streamside microcosms along a forested stream responded immediately and positively to increases in terrestrially derived DOC, and more so to DOC extracted from upper soil horizons than from deeper soils (Kreutzweiser and Capell 2003). Findlay et al. (2001) compared water draining from pasture, native forest, and pine plantation in the Whatawhata catchment on the North Island of New Zealand to evaluate the effects of catchment vegetation on the ability of DOC to support bacterial growth. Water from the pasture

surface flowpath supported twice as much bacterial production as pasture groundwater, and differences among the three vegetation types were detected from surface flows but not from groundwater, suggesting less degradation of DOC during the more rapid passage of surface flow.

Enrichment of stream ecosystems with highly labile DOC provides further evidence of a direct connection of bacterial production to higher trophic levels. When the amount of bioavailable carbon was increased in a small stream by the addition of dextrose, the density, growth, and respiration of epilithic bacteria increased in the treatment reach compared to the control, and growth rates of chironomid larvae also increased (Wilcox et al. 2005). It is conceivable that in some habitats the microbial and metazoan webs are linked, and in others the microbial loop is an energy sink, internally dissipating whatever energy is obtained from dissolved and particulate carbon sources. Elucidation of the magnitude of the "link versus sink" role of the microbial loop remains an important challenge in lotic ecology.

Our understanding of the potential influence of anthropogenic change on the structure and function of stream microbial communities is still emerging; however, turnover in microbial composition in response to changing environmental conditions could have profound effects on the physical, biological, and chemical characteristics of lotic systems (Zeglin 2015). Changes in temperature may alter the metabolic rates of microbial species and influence rates of ecosystem respiration and gross primary production. Alterations in stream hydrology associated with the intensification of precipitation and drought events may influence colonization and growth rates of benthic and pelagic species (Besemer 2015). In a Mediterranean river, Freixa et al. (2016) found that fluctuating discharge associated with drought influenced the structure and function of bacterial communities longitudinally from headwaters to higher order streams. The potential uptake rates of more recalcitrant carbon compounds appeared to be greater under base-flow conditions. Their research also suggested that microbial utilization of terrestrially-derived allochthonous carbon was more pronounced in headwater systems, whereas the accumulation and transport of more recalcitrant compounds became more important downstream. Collectively, their findings documented a longitudinal transition of bacterial community reliance on allochthonous (upstream) to autochthonous (downstream) energy sources. Furthermore, it appears that drought forced microbial communities to rely more on autochthonous energy resources, potentially due to a lack of pulses of allochthonous carbon resources that occur during rain events.

Climate and land use change are also enhancing the rate of organic matter leaching from soils, and are making the water darker in many aquatic systems. The implications of these physicochemical changes on food web ecology and biogeochemical processes in streams remains largely unknown (Demars et al. 2018; Demars 2019). As microbial growth and productivity are responsible for a large portion of carbon cycling in inland waters, these changes may collectively influence global carbon dynamics (Milner et al. 2017).

The world is urbanizing at unprecedented rates (Grimm et al. 2008; UNDES 2018), and this has potentially profound implications for stream microbial communities. Hosen et al. (2017) used 16S rRNA sequencing to examine bacterial community composition in streams along a gradient of urbanization. Bacterial community composition in the water column and in the benthos were significantly correlated with the amount of impervious surface (e.g., pavement) in a watershed, but benthic and water column communities differed in their response to environmental factors. Benthic communities were most strongly influenced by instream environmental conditions including discharge, conductivity, and nutrient levels, indicating that environmental filtering may play a substantial role in shaping benthic bacterial communities. In contrast, water column communities were instead directly linked to watershed urbanization. Pollution associated with urban and agricultural runoff may also threaten the structural and functional integrity of microbial communities (LeBrun et al. 2018). In a comparison of microbial colonization of microplastic substrates (polyethylene and polypropylene) and natural substrates (cobblestone and wood), researchers documented that distinct biofilm communities with reduced diversity formed on plastic substrates when compared to natural substrates (Miao et al. 2019).

8.2 The Ecology of Biofilms

The three-dimensional structure and high levels of microbial diversity associated with the bacteria, fungi, and algae growing within biofilms generates complex networks of interactions in biofilm communities (Besemer 2015, 2016). Our understanding of the ecology of biofilms has advanced significantly since early work conducted in the late 1970s and 1980s, which discovered that bacteria attached to surfaces were a prevalent feature of microbial life in streams (Costerton et al. 1978; Geesey et al. 1978; Lock et al. 1984). Early research in biofilm ecology also emphasized the important interactions between microbial autotrophs and heterotrophs within biofilms that support energy, gas, and nutrient transfer. This body of work also highlighted the role biofilms play in the adsorption of DOM (Haack and McFeters 1982; Lock et al. 1984). More recently, Battin et al. (2016) reviewed research advances in our understanding of the ecology of stream biofilms for readers interested in developing a more in-depth understanding of biofilm ecology.

The growth and development of biofilms can be a relatively predictable process (Sobczak 1996), where layers of microbes thicken and grow until cells begin to become limited by access to nutrients, carbon, or oxygen. Biofilm formation involves the conditioning of surfaces by macromolecule sorption, followed by the attachment of bacteria, the production of a polysaccharide matrix, and the development of the biofilm itself, including the growth and later detachment of autotrophic and heterotrophic microorganisms (Fischer 2003). Biofilms are embedded in an exopolysaccharide matrix produced by microorganisms and forming on surfaces of stones, sediments, plants, decaying wood and leaves, and on suspended particles in larger rivers. This matrix binds together algae, bacteria, fungi, protists, detrital particles, various exudates, exoenzymes, and metabolic products in an organic microlayer (Fig. 8.4). Additional organic compounds, including proteins, nucleic acids, and humic compounds also occur in the matrix (Sinsabaugh and Foreman 2003). Energy transformations within biofilms include the conversion of light to chemical energy by algal photosynthesis, adsorption and microbial uptake of heterotrophic carbon, and internal transfers due to extracellular release and cell lysis. Both autotrophs and heterotrophs are likely to benefit from internal fluxes within this highly symbiotic association.

Microbial eukaryotes, such as algae and fungi, can be diverse and abundant in biofilms and can be responsible for much of the biogeochemical activity associated with biofilms (Besemer 2015; Besemer et al. 2012). Proteobacteria, Bacteriodetes, and cyanobacteria are typically the most common bacterial groups in freshwater systems. Fungal species frequently provide needed structural components within biofilms, and they are important decomposers of allochthonous organic matter. Algae, such as Chlorophyta, generate substrates that support biofilm development, and they can be important sources of carbon for their heterotrophic counterparts. Viruses and protists, such as amoebae, ciliates, and flagellates, can influence the growth, diversity, and structure of biofilm communities, subsequently altering biogeochemical processes within biofilms and between biofilms and the surrounding environment. Though they can make up a large proportion of the biofilm community under certain conditions (e.g., in deeper sediments), Archaea are not typically dominant in riverine biofilms (Besemer 2015).

A broad range of higher-level consumers, including ciliates, flagellates, and smaller macroinvertebrates also are found in biofilms (Dopheide et al. 2015; Hakenkamp and Morin 2000). Their feeding behavior can alter biogeochemical dynamics (Risse-Buhl et al. 2012), community structure (Wey et al. 2012), and the physical structure (Bohme et al. 2009; Lawrence et al. 2002) of the biofilm environment (Battin et al. 2016). Grazing organisms, such as heterotrophic protists and small metazoans, can restructure

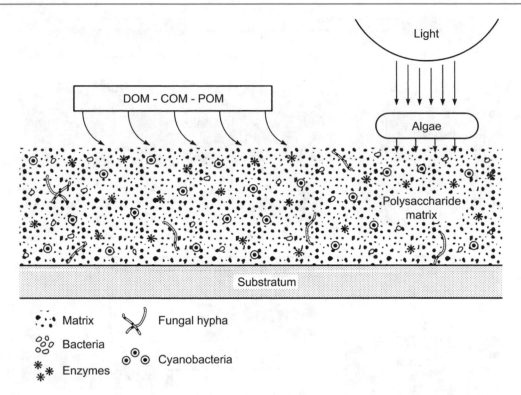

Fig. 8.4 A structural and functional model of the organic microlayer-microbial community found as a surface biofilm on stones and other submerged objects in streams. The matrix of polysaccharide fibrils produced by the microbial community binds together bacteria, algae, and fungi, and is inhabited by protozoans and micro-metazoans that graze on this material. Detrital inputs include dissolved, colloidal, and fine particulate organic matter, while light energy is trapped by algal photosynthesis. Within the matrix, extracellular release and cell death result in enzymes and other molecular products that are retained due to low diffusion rates and thus available for utilization by other microorganisms (Modified from Lock 1981)

biofilm communities by selectively consuming certain species, reducing growth rates via grazing, and altering interspecies interactions. Biofilm-dwelling grazers can import planktonic production into the biofilm food web, thereby coupling planktonic and benthic food webs (Weitere et al. 2018). Protists can influence the biomass of microbes through grazing, and can indirectly affect decomposition rates by excreting potentially limiting nutrients and enhancing microbial activity. Protists also can change the morphology of biofilms and influence organismal diversity within biofilms by enhancing surface roughness and surface area, subsequently promoting nutrient and gas exchange (Marks 2019).

Biofilms grow on inorganic and organic surfaces in benthic and hyporheic zones, and the physical, chemical, and biological characteristics of where they are growing have important implications for their structure and their function (Fig. 8.5; Battin et al. 2008; Besemer 2016). Biofilms growing on boulders, cobbles, pebbles are classified as "epilithic". Epipsammic biofilms grow on grains of sand, and epipelic biofilms grown on sediment surfaces, creating complex matrices that can stabilize finer sediments. Biofilms also grow on macrophytes and submerged wood, referred to as epiphytic and epixylic, respectively. When growing on living tissues, epiphytic and epixylic biofilms can benefit or harm their host. Biofilms can also grow on suspended aggregates of detrital material; these aggregate communities can be considered mobile and they have adapted to substantially different hydrological conditions (Besemer 2015).

A diverse suite of physical, chemical, and biological characteristics that are influenced by climate, hydrology, underlying geology, and land use, governs the growth and structure of biofilms. Physical factors, including disturbance by grazing animals, sediment scouring and high flows, and chemical factors, including nutrient and carbon availability, affect the spatial and temporal heterogeneity biofilm structure and function. Surface topography, flow, and turbulence generate important differences in shear stress and mass transfer among microhabitats, factors that also influence microbial colonization and dispersal (Battin et al. 2007; Woodcock et al. 2013). Interactions between flow and geomorphology control biofilm growth because water flow can also mediate the rate at which biofilms are exposed to oxygen, nutrients, and organic carbon. The high demand of biofilm communities for oxygen, nutrients, and dissolved organic matter can cause microorganisms within the biofilm

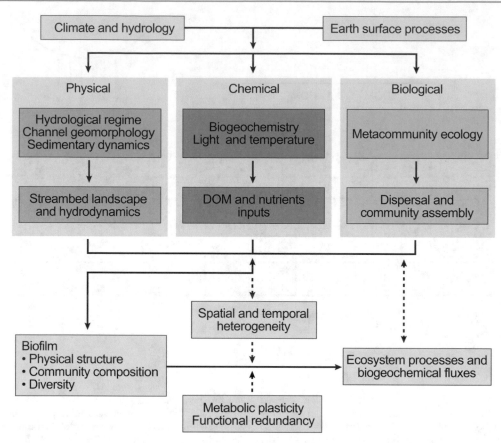

Fig. 8.5 Coupled physical, chemical and biological processes that affect the structure and function of biofilms. These processes are susceptible to global environmental change, including climate change and land use change. The effects of nutrient and organic matter availability, topography, hydrology, and dispersal dynamics of biofilms influence their spatial and temporal dynamics. Temporal and spatial variation in functional redundancy and metabolic plasticity within biofilm communities influence the impact of biofilms on ecosystem processes. Processes can interact across spatial and temporal scales and generate more heterogeneity within and among biofilms. Dashed arrows identify feedbacks and modulating interactions (Reproduced from Battin et al. 2016)

to become nutrient, energy, or oxygen-limited (Battin et al. 2016).

Microbial development varies with the chemical composition of its substrate. McNamara and Leff (2004) tested the response of several bacteria species to leachate from sugar maple leaves at various stages of decomposition using an agar substrate that allowed the leachate to diffuse through filters on which bacteria were enumerated. Species differed in their response to components of leaf leachate such as tolerance to phenolic compounds, demonstrating how the composition of microbial assemblage can influence its ability to utilize the mixture of labile, recalcitrant, and inhibitory compounds. Measurements of uptake of tree-tissue leachate in streambed sediments within recirculating mesocosms found that most DOC was bioavailable, and by extrapolation to bacterial demand for DOC in White Clay Creek, Pennsylvania, could support up to half of community respiration (Wiegner et al. 2005). Tank et al. (1998) reported higher microbial respiration, fungal biomass, and extracellular enzyme activity in wood biofilms in a litter-excluded stream compared with a reference stream, suggesting competition for carbon or nutrients between microorganisms associated with leaf and wood substrates.

Biofilm studies have focused primarily on surfaces that have some contact with current, and less is known about the function of biofilms within the sediments and into the hyporheic zone. Microbial growth can reduce diffusion through the biofilm, thereby constraining the rates of microbial transformation of nutrients and energy and the rate and direction of chemical fluxes through the hyporheic zone (Caruso et al. 2017). Permeability of the sediments and the hydraulic residence time of the infiltrating water will strongly affect how much water column DOC exchanges with pore water. Battin (2000) showed that water velocities influence transport of solutes into surface biofilms, and higher velocities help to overcome diffusional limitation of material exchange. At least in some instances, the availability of POC rather than infiltration of DOC is likely to determine hyporheic microbial metabolism (Brugger et al. 2001). A synthesis of numerous studies of bacterial production documented a significant positive relationship with amount of organic matter in the sediments (Cole et al. 1988). Findlay et al. (1986) reported daily bacterial production to be an order of magnitude higher in sediments of backwater areas than in sandy regions of two blackwater rivers of the southeastern United States, corresponding to differences in organic content between sites. Similarly, bacterial production was higher in an Appalachian mountain spring (Crocker and Meyer 1987) and in a Pennsylvania stream (Kaplan and Bott 1989) than in the two blackwater rivers, and sediment organic matter content also was higher at the former sites.

The relative contributions of POM versus DOM to bacterial production within sediments will vary with the supply of each, and with hyporheic flowpaths and water residence time (Findlay and Sobczak 2000). Within the hyporheic interstices of the Toss River, a gravel-bed stream in Switzerland, greatest bacterial production occurred within interstices dominated by surface water inflow. An attenuation of bacterial abundance was observed in deeper sediment strata (Brunke and Fischer 1999). The abundance of several hyporheic invertebrate taxa, taxon richness, and total invertebrate density correlated with bacterial abundance and production, indicative of a consumer response to resources. In the Spree River, Germany, bacterial production rates in the sediments were nearly three orders of magnitude greater than in the water column, and bacterial respiration was sufficiently high to metabolize a large proportion of the organic carbon retained in the sediments (Fischer and Pusch 2001). To evaluate the influence of hyporheic flowpaths on the availability of stream water DOC within the sediments beneath the stream bed, Sobczak and Findlay (2002) sampled DOC and oxygen concentrations in wells located along transects on gravel bars, where downwelling of surface water was indicated by negative hydraulic gradient (Fig. 8.6).

Between 38–50% of the surface water DOC was removed, and declines in DOC along the flowpath were accompanied by decreases in oxygen concentration, suggesting that DOC removal was due to bacterial metabolism.

In lowland rivers where sediments tend to be finer and water circulates less freely, microbial metabolism within the streambed is likely to show a vertical profile, more so if the sediments are vertically stratified and less so if they are shifting. Based on the rapid turnover of organic carbon in sediments, Fischer et al. (2002) concluded that sediment dynamics significantly foster organic carbon metabolism in lowland rivers and thus strongly influence the metabolism of the whole ecosystem. Comparing biofilms growing on stones and on sandy substrata in a Mediterranean stream in northeastern Spain, Romani and Sabater (2001) observed higher extracellular enzymatic activity in episammic compared with epilithic biofilms, indicating the much greater availability of POC in sandy substrates relative to gravel, and the likely contribution of sand habitats to OM processing.

Spatial and temporal variation in biofilm accumulation rates within and among systems is common (Findlay et al. 1993; Hudson et al. 1992), owing to the diversity of organisms within biofilms (Battin et al. 2003b). Over a two-month period, Veach et al. (2016) studied the development of biofilm communities in a third-order stream in the Konza Prairie in Kansas in the central US. Throughout the observation, microbial abundance and chlorophyll a concentrations continually increased (Fig. 8.7), whereas net primary productivity did not change significantly. Their work suggested that stochastic processes were important in forming biofilm communities in the first two weeks of succession, but deterministic selection driven by environmental conditions played a much larger role as the communities developed. The continual increase in algal biomass and the

Fig. 8.6 Sampling wells positioned along a hyporheic flow path to evaluate DOC transfer from the water column to benthic biofilms. Negative vertical hydraulic gradient (-VHG) indicates down-welling of surface water, while a positive gradient (+VHG) indicates upwelling of hyporheic water (Reproduced from Sobczak and Findlay 2002)

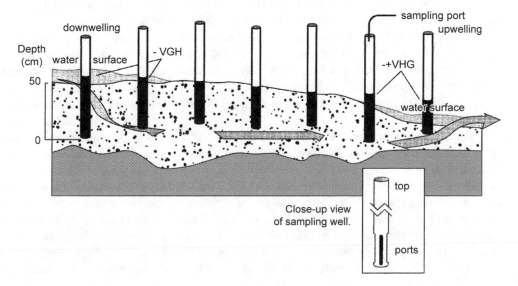

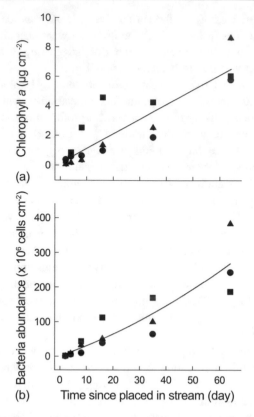

Fig. 8.7 Temporal development of biofilm components in a prairie stream. (**a**) chlorophyll a, a proxy for algal biomass and (**b**) the abundance of bacteria through time. The different symbols represent three sample sites (Reproduced from Veach et al. 2016)

abundance of bacterial cells was unexpected, as other research has reported plateaus in algal biofilm community development within two weeks of flooding (Dodds et al. 1996; Fisher et al. 1982). However, the aforementioned studies examined patterns in secondary, rather than primary succession. Veach et al. (2016) also noted that other research (e.g., Melo and Bott 1997; Siboni et al. 2007) has demonstrated that site-specific factors, such as geochemistry and discharge, could partially explain the different results, as colonization and primary succession may require conditioning prior to biofilm development. Visible grazing scars on the biofilms suggested that consumers may have also reduced biofilm biomass and supported positive increases in microbial abundance and chlorophyll concentrations (Veach et al. 2016).

Using streamside flumes to observe biofilm development under slow and fast current velocities, Battin et al. (2004) found that slow velocities favored thicker biofilms in which internal cycling of carbon was greater, based on DOC uptake and microbial growth rates. Manipulation of substrate heterogeneity resulted in an immediate and significant increase in the respiration of the benthic biofilm of a Virginia piedmont stream, probably due to changes in near-bed flow velocity and turbulence intensity (Cardinale et al. 2002).

Shear stress on biofilms increases under higher flow velocities. Under these conditions, biofilms can form ridges and streamers that oscillate in the stream water (Battin et al. 2003a; Besemer et al. 2007). Biofilms can also mediate the transport of solutes in streams, but the role of biofilms in these processes are still not well understood (Aubeneau et al. 2016).

Algae and bacteria in biofilms produce extracellular material that may help adsorb organic matter from the overlying water (Findlay and Sobczak 1996; Fischer 2003; Freeman and Lock 1995). By reducing diffusion rates, the organic microlayer tends to retain and concentrate compounds, particularly those of higher molecular weight. In addition, the polysaccharide matrix can act as an ion exchange system, attracting and binding charged organic molecules, anions, and cations (Lock 1981; Lock et al. 1984). Dissolved organic matter is adsorbed by the matrix and later diffuses into the biofilm, where it can be used by microorganisms. The presence of high densities of autotrophs can significantly affect carbon transfer between autotrophic and heterotrophic organisms within biofilms (Rier et al. 2007). Fungi may also play important roles in biofilm development, but they are still relatively understudied (Barlocher and Murdoch 1989; Baschien et al. 2008).

Nutrients may limit microbial production when an abundant carbon source exists. Microbial respiration on wood biofilms increased in response to nutrient addition in a reference stream but not in a litter-excluded stream, suggesting lowered nutrient immobilization and thus higher nutrient availability in the absence of decomposing leaf litter and its microbial flora (Tank and Webster 1998). A comparison of nutrient limitation of biofilm algae and fungi in ten streams in North America found that fungi responded more to nutrient addition than did algae, although for both autotrophs and heterotrophs, nitrogen limitation was more common than phosphorus limitation (Tank and Dodds 2003).

Genetic diversity within bacterial communities can be used to infer how species traits influence succession in biofilm communities. Niederdorfer et al. (2017) used the number of copies of the operon rrn on bacterial genomes from several genes encoding ribonucleic acid as a proxy to infer important cellular processes, such as growth rate and sporulation efficiency. The work was based upon a body of research suggesting that bacteria with greater numbers of rrn copies grow rapidly, but use resources inefficiently, in contrast to bacteria with fewer copies of rrn that grow slowly and use resources more efficiently (e.g., Polz and Cordero 2016; Roller et al. 2016; Valdivia-Anistro et al. 2016). This suggests that the number of rrn copies reflected important trade-offs between resource use efficiency and growth rate during biofilm succession, and that flashy systems would favor species characterized by rapid growth rates in spite of relatively limited resources. Niederdorfer et al. (2017) found that in eutrophic systems, biofilms in early stages of development had more taxa with greater numbers of rrn copies when compared to oligotrophic systems. Additionally, biofilms grown in oligotrophic systems had relatively lower growth rates. These findings indicate that early biofilm succession was dominated by slow-growing, but resource-efficient species that have leaky functions, such as producing extracellular polymeric substances (EPS). Initial biofilm formation likely requires energetically costly processes, including cell adhesion, surface conditioning, and EPS production that may compromise the ability for species to grow quickly. Investment in EPS production at early stages of biofilm production is also evolutionary justified, as there are fitness advantages for species living within the EPS matrix. In resource-limited systems, initial colonizers may be able to begin biofilm formation, while more competitive species with comparatively more rrn copies appear when biofilms are more established (Niederdorfer et al. 2017). Though microbial species can be obligate biofilm or obligate benthic species, many species can switch modes, and as such, bacterioplankton may comprise a source of initial bacterial colonization in biofilms.

Interactions among species are essential in maintaining complex structural and functional dynamics in biofilms, and environmental conditions mediate these interactions (Battin et al. 2016). Yet, our understanding of the ecological adaptations and metabolic-tradeoffs that govern biofilm formation and succession under variable environmental conditions is still emerging (Lau et al. 2018; Niederdorfer et al. 2017; Peipoch et al. 2019). Research does suggest that both metabolic performance and biofilm activity can rapidly respond to fluctuating environmental conditions (Besemer 2015), and that community composition may represent an integrated response to the environment (Findlay and Sinsabaugh 2006). To maintain ecosystem processes within biofilms, researchers have argued that complementarity, or niche separation among species, is essential even when certain processes seem to be sustained by functional redundancy (Gamfeldt et al. 2008). In fact, limited functional redundancy was indicated by a negative relationship between taxon diversity and the activity of several extracellular enzymes in biofilm (Besemer 2015; Peter et al. 2011). Relationships between diversity and function in biofilms may be influenced by metabolic plasticity (the ability of a community to respond to environmental change by altering metabolic performance), and by functional redundancy (Besemer 2015; Comte et al. 2013).

In a study using streamside flumes, Singer et al. (2010) created a gradient of physical heterogeneity in artificial streambeds to assess how habitat diversity influenced microbial diversity and function. They found that increasing

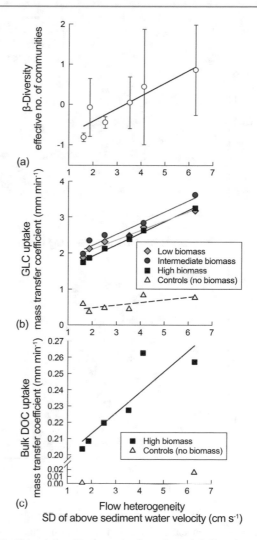

Fig. 8.8 The relationship between flow, bacterial diversity, and the mass transfer of carbon. Flow heterogeneity, as expressed as the length of the 3-dimensional vector velocity, explained patterns in (**a**) beta diversity, (**b**) glucose mass transfer, and (**c**) the bulk mass transfer of dissolved organic carbon. Triangles represent controls with no biofilms in (**b**) and (**c**) (Reproduced from Singer et al. 2010)

flow heterogeneity was significantly related to among-site, or beta diversity, carbon uptake, and the diversity of carbon resources used by the microbial community (Fig. 8.8). Measurement of glucose and DOC uptake in the flumes revealed that glucose uptake was largely driven by physical processes related to flow heterogeneity. However, uptake rates of more recalcitrant forms of DOC were enhanced by increased beta diversity in flumes with more flow heterogeneity, indicating a positive relationship with biodiversity and ecosystem function (Singer et al. 2010).

In well-lit areas, eukaryotic algae and cyanobacteria can dominate biofilm communities, but in light-limited habitats, such as deeper sediments, bacteria and archaea are typically abundant. These shifts in community structure may have important implications for stream biogeochemistry. Wagner

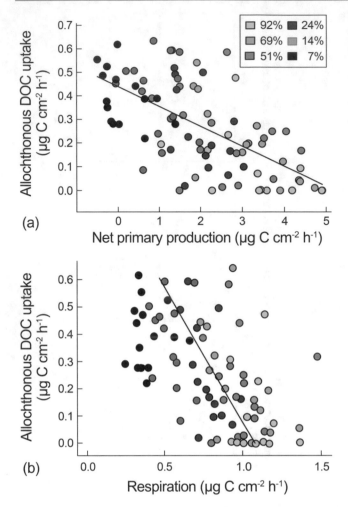

Fig. 8.9 Linear regressions between the uptake of allochthonous dissolved organic carbon (DOC) and (**a**) net primary productions and (**b**) allochthonous DOC uptake and respiration across a light gradient (% transmission of the incident light) during daytime incubations (Reproduced from Wagner et al. 2017)

et al. (2017) documented that biofilms in high light environments have been shown to preferentially use autochthonous DOC, but biofilms in lower-light environments may shift towards the use of allochthonous DOC (Fig. 8.9). An evaluation of 69 streams from sites in Kentucky and Michigan using algal biomass, DOC, and nutrients as possible predictors of bacterial cell density found that algal biomass was the best predictor, but the relationship was not evident at low algal biomass (Rier and Stevenson 2002). As long as they have access to light and inorganic nutrients, older biofilm communities are less dependent on carbon resources from outside the biofilm (Battin et al. 2016). Comparison of a biofilm in an open channel to one inside a dark pipe within the stream channel documented greater DOC uptake in the former (Romani et al. 2004). The light-grown biofilm supported greater biomass and activity of bacteria due to its higher algal biomass, exudates, and

development of a polysaccharide matrix. The dark-grown biofilm, in contrast, was highly dependent on the supply and lability of organic matter and was more efficient in the uptake of labile molecules, indicating that interactions between phototrophs and heterotrophs drive carbon dynamics in biofilms. Labile DOC in stream water may be essential in supporting bacterial growth during initial colonization, while DOC from algae may enhance bacterial growth at later stages of biofilm development (Sobczak 1996). Additionally, research suggests that labile DOC produced within biofilms may stimulate the decomposition of more recalcitrant forms of DOC found in biofilms. This effect, called "priming", has been found to be important in terrestrial ecosystems, but our understanding of the factors governing priming in freshwater systems is limited (Guenet et al. 2010; Halvorson et al. 2019a, b; Wagner et al. 2017).

Our understanding of patterns in diversity of microbial communities over large spatial scales is still relatively limited (Zeglin 2015; Zeglin et al. 2019). Many microbes have resting stages that protect them during long-distance transport, suggesting that regional differences in diversity may be unlikely. Yet, research has shown that landscape-level variation in microbial diversity does exist (Findlay 2010). In a survey of more than 100 streams within a single watershed, Besemer et al. (2013) found that average within-site diversity declined from upstream to downstream reaches. Patterns in species turnover across sites also indicated there was a higher degree of variability in community composition of biofilms among headwater sites than in sites downstream, suggesting that microbial diversity in headwater streams may function as an important source of microbial diversity within larger river networks (Battin et al. 2016). Biofilm diversity can also be governed by environmental conditions, and some studies have shown that land use change and stream water chemistry can have larger impacts on patterns in diversity than geographical distance (Fierer et al. 2007; Lear et al. 2013).

Both regional and local factors influence the initial formation and subsequent development of biofilm communities. Microbes are frequently characterized by short generation times and high dispersal rates that can make geographical distances less important in structuring biofilm communities (Fierer 2008; Fierer et al. 2007). In fact, studies have found that the species composition from the surrounding water column is not sufficient to explain the diversity of biofilms, indicating that local environmental characteristics such as water chemistry act as strong forces of selection in the composition of microbial communities (Besemer 2015). This is not to assert that regional process have no influence on biofilm structure. Lear et al. (2013) found that taxon richness in biofilm bacteria from New Zealand streams was significantly greater at latitudes closer to the equator and reduced at higher elevations. They also observed that patterns in bacterial community similarity were related to both geographic and latitudinal distance, but not to elevational distance. Notably, their work still demonstrated that catchment land use had a larger influence on taxon richness and community similarity than climate or geography (Lear et al. 2013).

Stream microbes are now recognized for their contribution to global elemental cycling. They influence carbon dynamics through their contribution to OM processing and by emitting large volumes of carbon dioxide into the atmosphere (Battin et al. 2008; Raymond et al. 2013). Biofilms also contribute to nitrogen cycling by removing nitrogen from the water via denitrification and emitting nitrous oxide or nitrogen gas into the atmosphere (Beaulieu et al. 2011; Mulholland et al. 2008). Exoenzymes and enzymes derived from cell lysis may be retained and remain active, facilitating the release of molecular products. The accumulation of exoenzymes within the microlayer permits surface film bacteria to divert resources from enzyme synthesis to microbial growth, thereby reducing energy demands on microorganisms for enzyme synthesis. Sampling of biofilm from the surfaces of stones (Sinsabaugh et al. 1991) and organic substrates (Golladay and Sinsabaugh 1991) has documented that exoenzyme accumulation occurs as suggested under the Lock model (Fig. 8.4). This has the beneficial effect that enzyme activity can be spatially distant from the microorganisms that produced the original enzymes, enhancing the availability of organic carbon to microorganisms and helping to maintain growth in an environment where carbon and nutrient sources are variable (Pusch et al. 1998).

8.3 Bacterioplankton

Autotrophic and heterotrophic bacteria found in the water column, or bacterioplankton, can be distinct in form and function from bacterial communities found in benthic habitats and from those found in the surrounding terrestrial environment. Similar to biofilms, the structure and function of bacterioplankton communities are influenced the by physical, chemical, and biological characteristics of a river network. Species that are unique to the water column, soils, or sediments may not persist when they are transported from their preferred habitat. Longitudinal variation in species variation are presumed to reflect changes in the connectivity with soils and sediment. Yet, our knowledge of factors governing bacterioplankton community structure through space and time are still relatively limited (Ann et al. 2019).

Bacterioplankton communities are shaped by microbial dispersal from surrounding environments and in-stream environmental conditions (Niño-García et al. 2016). As noted earlier, headwater streams can be important sources of

bacterial diversity along river networks as they host greater microbial diversity than higher-order rivers. In a study examining community transitions from the headwater tributaries to the mainstem of the River Thames using 16S rRNA gene pyrosequencing, Read et al. (2015) documented ecological succession in the bacterial community composition along the river continuum in a highly urbanized river. Water residence time, rather than site-specific physicochemical parameters, explained observed patterns in community composition. Most likely, these patterns may have also been influenced by a legacy of industrial pollution and other physiochemical changes associated with urbanization. Other studies in watersheds impacted by human activity have documented similar patterns. Teachy et al. (2019) documented declines in within-site diversity of bacterioplankton with increasing stream order in streams in the southeastern US, and Savio et al. (2015) documented declines in both bacterial richness and evenness in both the free-living and particle-associated communities in downstream reaches.

Headwater systems may be dominated by taxa associated with soil and sediments, suggesting that runoff from terrestrial systems seems to be an important source of bacterial communities for rivers (de Oliveira and Margis 2015). When evident, such patterns are likely due to the high surface area to volume ratio in small watersheds. For instance, research in a river network in the southeastern US documented negative longitudinal relationships in the relative abundance of soil- and sediment-associated microbes, and positive longitudinal relationships in the relative abundance of freshwater-associated microbes (Hassell et al. 2018). This work suggested that the importance of bacterial colonization via transport from surrounding terrestrial environments declined with increasing stream order, whereas the influence of instream selection became more important in structuring bacterioplankton communities farther downstream. It should be noted, however, that the study watershed was impacted by human development, and that sample collection was limited to the water column and did not include soil or sediment samples from the watershed.

Trends in bacterioplankton diversity may be related to physicochemical conditions, but our understanding of landscape-level patterns are in a nascent stage (dos Reis et al. 2019; Staley et al. 2015; Zeglin et al. 2019). High precipitation events and resulting increases in discharge may increase soil runoff and the delivery of soil-derived bacterial species and re-suspend sediment that may lead to changes in the plankton community (Hassell et al. 2018). This may be especially true in urbanizing landscapes where watersheds may have their own unique microbial communities and increases in impervious surface can exacerbate the rapid delivery of large volumes of water to streams (Walsh et al. 2005). Phosphorus concentrations and turbidity can also influence the species diversity and functional capacity of instream microbial communities (LeBrun et al. 2018). Fluctuations in water temperature and exposure to UV radiation may act as a force of selection on bacterial communities, filtering out species that exceed their thermal minima or maxima or are sensitive to changes in UV (Tranvik and Bertilsson 2001). Furthermore, increases or reductions in water temperature may also affect the growth and reproduction of microbial communities. Rainfall and water temperature have been found to influence microbial community composition by sediment-associated taxa entering systems through runoff (Staley et al. 2015; Teachey et al. 2019). At four locations in the Rocky Mountains in the western US, Portillo et al. (2012) documented unique communities that experienced significant temporal shifts that appeared to be driven by instream chemical conditions rather than season. Notably, the turnover in community composition was higher than has been documented in many studies of other microbial, plant, or animal communities.

Relative to microbial communities in the benthos, bacterioplankton are thought to make only a limited contribution to instream biogeochemical processes, and this may be especially true in smaller streams (Sinsabaugh et al. 2014). In headwater streams in Peru, where net ecosystem production was estimated using dissolved oxygen sensors and water column metabolism was estimated from light–dark bottle experiments, water column respiration was 25 times less than benthic respiration (Bott and Newbold 2013). Though abundance is not a direct indicator of activity, many studies have documented benthic bacterial densities to be orders of magnitude greater than densities collected in stream water (e.g., Geesey et al. 1978; Kaplan and Bott 1989; Lock 1981).

Environmental variables that are likely to affect planktonic bacterial production and their influence on biogeochemical processes include, but are not limited to, the amount and lability of DOC, nutrients, and pH. Carbon is generally considered to be an important limiting resource for bacterial production; hence, variation in the quantity and lability of DOC and POC sources is considerable in structuring bacterial communities. Bacterioplankton production is expected to be especially important in lowland rivers, where bacterial doubling times potentially can exceed washout rates, provided that DOC is of sufficient lability and quantity and other environmental conditions are favorable. Estimates of production by bacterioplankton vary over at least two orders of magnitude (Fig. 8.10). Values range from 0.14–0.52 μg C L^{-1} h^{-1} in blackwater rivers in the southeastern US and the Amazon and Orinoco Basins (Benner et al. 1995; Castillo et al. 2004), to 40–75 μg C L^{-1} h^{-1} in anthropogenically enriched rivers like the Maumee and the Ottawa in Ohio (Sinsabaugh et al. 1997). Values for planktonic bacterial abundance are less variable than bacterial production, ranging between 0.05 × 10^6 cells mL^{-1} in the Kuparuk River in Alaska to 5 × 10^6 cells mL^{-1} in more disturbed systems like the Hudson River in the northeastern US and the River Rhine in central Europe (Admiraal et al. 1994; Findlay et al. 1991).

Greater availability of nutrients can directly enhance bacterial production whenever the organic matter that serves as a carbon source has a low nutrient content, forcing bacteria to obtain nutrients from elsewhere (Findlay and Sinsabaugh 2003). Nutrient supply can also benefit bacterial production indirectly, by enhancing primary production and thus the carbon supply. The addition of various combinations of nutrients to bacterioplankton samples cultured in situ

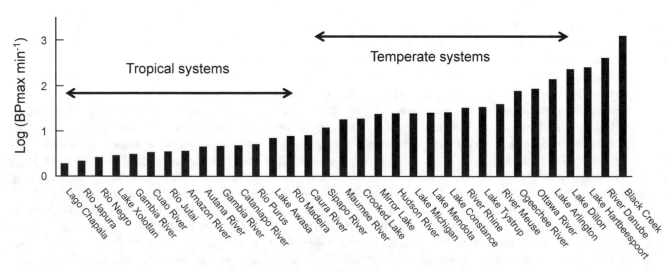

Fig. 8.10 Annual variation in bacterial production measured as the maximum: minimum for several freshwater systems. A logarithmic transformation was applied to the vertical axis to reduce the scale (Reproduced from Castillo et al. 2004)

has shown phosphorus limitation in clearwater rivers of the Amazon basin (Farjalla et al. 2002) and in clear and blackwater rivers draining the Guayana Shield in the Orinoco Basin (Castillo et al. 2003). In the Rio Negro, a blackwater tributary of the Amazon River, bacterial production was enhanced by the simultaneous addition of glucose, ammonium, and phosphorus (Benner et al. 1995), which suggests that C, N, and P were available in approximately the stoichiometric ratio needed by bacteria. The influence of temperature is reflected in seasonal variation in bacterial metabolism, with higher values during the warmer months (Edwards and Meyer 1987; Findlay et al. 1991). The effect of pH is ambiguous, as some studies suggest that bacterial extracellular enzymes can be negatively affected by acidification (e.g., Simon et al. 2009); however, some exoenzymes exhibit an optimum at low pH (e.g., Münster et al. 1992). Low pH may also affect the availability of DOC to bacteria because the cell membrane is more permeable to hydrophilic compounds, and at very low pH, humic compounds become more hydrophobic and thus are less permeable substrates (Edling and Tranvik 1996).

The lability and composition of DOC can limit the production of bacterioplankton (Kaplan et al. 2008; Ruiz-

González et al. 2015; Ward et al. 2017). In fact, correlations of bacterial production with total DOC are rare (Findlay 2003), probably because bacterial production reflects the amount of labile DOC rather than the total (deMelo et al. 2020; Findlay et al. 1998). For instance, in Lake Janauacá, a floodplain lake of the Amazon river, bacterial community composition was associated with shifts in the lability, but not quantity of DOM. Bacteria preferentially used labile and freshly produced sources of DOM (Fig. 8.11; deMelo et al. 2020). The evidence suggests that some constituents of the heterogeneous mix of molecules that comprise DOC are more available than others, and preferential removal of low molecular weight DOC is commonly reported (Kaplan and Bott 1982, 1989; Meyer et al. 1987; Ruiz-González et al. 2015). Peptides and sugars support a large fraction of bacterial production, evidence of the high bioavailability of these compounds, which can be found free or in complexes with humic molecules (Findlay and Sinsabaugh 1999; Foreman et al. 1998). Bioavailability also is related to the proportion of aliphatic compounds, which are more abundant in algal and macrophyte leachate than in leachate from woody plants (Sun et al. 1997). However, bacteria also can use humic substances, which are rich in aromatic components (Moran and Hodson 1990; Tranvik 1990).

As previously mentioned, research often suggests that benthic communities play much larger roles in ecosystem dynamics than do planktonic bacterial communities; however, this may be context dependent. For example, using a short-term laboratory experiment to examine patterns in DOC uptake between planktonic and biofilm bacteria, Graeber et al. (2018) found that planktonic bacteria can respond quickly to pulses of labile DOC, and may have uptake rates similar to bacteria in biofilms. These findings are similar to those documented by Kamjunke et al. (2015), in which planktonic, but not benthic, bacteria responded to changes in the quantity and composition of DOC. Graeber et al. (2018) argued that this response may have been due to the fact that benthic biofilms are typically characterized by a polysaccharide matrix that reduces their connection to the water column and provides alternative carbon sources for microbes. In their study, carbon uptake by planktonic communities was also mediated by phosphorus availability, highlighting the role that coupled biogeochemical cycles most likely play in microbial dynamics in river networks.

Although the supply of organic carbon and nutrients in lotic ecosystems likely is limiting to bacterioplankton growth much of the time, microbial respiration in the water column and sediments of higher-order rivers may nonetheless be sufficient to mineralize large amounts of carbon, significantly contributing to global carbon dynamics through evasion, or out-gassing (Raymond et al. 2013). For instance, the lower Hudson River is almost always super-saturated with CO_2, evidence of an excess of respiration over

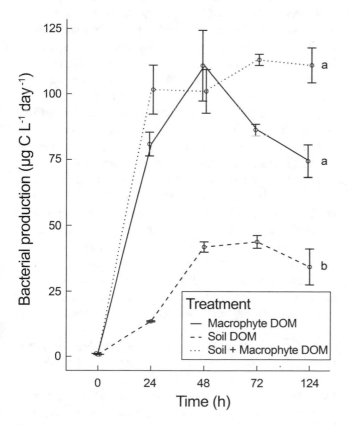

Fig. 8.11 Bacterial production estimates during incubations with dissolved organic matter (DOM) from different sources. The bars are standard error and the letters "a" and "b" indicate significant differences between treatments (Reproduced from DeMelo et al. 2020)

production, and DOC concentrations decline as one proceeds downriver, evidence of its mineralization by microbial activity (Cole and Caraco 2001).

The extent and frequency of drought are expected to increase in drought-prone areas in the near future. In a microcosm study of drought impacts on bacterioplankton, Székely and Langenheder (2017) found that drying altered microbial function, but that drying and subsequent re-wetting altered the structure and function of the community. Changes in the bacterial community and the ability of the system to recover from drought were affected by the initial conditions and the frequency of previous drying–rewetting cycles. This suggests that the history of environmental stress has an important influence, and dispersal may support the recovery of the bacterial community by reintroducing sensitive taxa and supporting the recovery of microbial function.

Land-use change and nutrient pollution may influence planktonic bacterial diversity. Urbanization can influence within- and among-site bacterial diversity within a watershed and the abundance of certain taxa within the planktonic community. Hosen et al. (2017) found that taxa associated with human activities, such as bacteria from the genera *Polynucleobacter*, which has been associated with eutrophic conditions, and *Gallionella*, which is linked to corrosion of water distribution systems, were more common with increasing impervious surfaces. Furthermore, data from their study indicated that urbanization was linked to the loss of keystone taxa from the community. Complex relationships also link microorganisms, ecosystem processes, and water quality. Thus, nutrient pollution may also restructure bacterial communities in rivers. In addition to containing chemical constituents that may influence the microbial community, effluent from wastewater treatment plants (WWTP) may introduce substantial concentrations of microorganisms that act to re-structure instream microbial communities. Mansfeldt et al. (2020) documented that bacterial communities downstream of discharge from a WWTP were a mixture of the communities found in upstream habitats and the effluent. In many of the sites sampled, more than 50% of the communities found downstream of a discharge point were from the effluent.

8.4 Summary

Microbes are diverse assemblages of organisms that occur in or on all surfaces in streams and populate the water column. Both autochthonous and allochthonous sources of energy and nutrients are important to microbial communities, and external sources of carbon can be especially important in fueling microbial production. Bacteria and fungi convert particulate and dissolved organic matter into microbial tissue

available to a wide range of micro-consumers, including protozoans and small metazoans termed meiofauna. Because of the small size of the meiofauna, it has been speculated that microbial production would have to pass through many food chain links before reaching macroinvertebrates and fishes, and so much energy would be dissipated with each step that little energy would reach higher trophic levels. Under this scenario, the microbial food web is a sink for carbon transfer, although still important for mineralizing carbon and nutrients. However, bacterial production has been shown to reach virtually all consumers including fish, almost certainly due to the consumption of microbes occurring on the surface of FPOM and CPOM. The complex linkages between microbial communities and higher trophic levels is known as the microbial loop, and the extent to which it functions as a carbon link or sink is an active topic of research.

Diverse assemblages of microorganisms growing on the streambed and other surfaces are called biofilms. Found on virtually all surfaces in streams, they include algae, bacteria, fungi, protists, detrital particles, various exudates, exoenzymes, and metabolic products in an organic microlayer. The three-dimensional structure and high taxonomic and functional diversity associated within biofilms create complex networks of multi-species interactions and multicellular behaviors. As such, biofilms are important sites of ecosystem functions, including organic matter processing and nutrient cycling. Research on the ecological adaptations and metabolic tradeoffs that govern biofilm formation and succession under variable environmental conditions suggests that metabolic performance and biofilm activity can rapidly respond to fluctuating environmental conditions. In biofilms, complementarity in functional roles among different organisms seems to be essential, even when certain processes appear to be sustained by functional redundancy.

Bacterioplankton communities result from microbial dispersal from surrounding environments and in-stream environmental conditions. Notably, headwater streams seem to function as important sources of bacterial diversity along river networks, and they have been shown to host greater microbial diversity than reaches downstream. Many environmental variables have been shown to influence the growth and function of planktonic bacteria including the amount and lability of DOC, availability of nutrients, and pH.

New molecular tools have advanced our ability to assess microbial diversity and microbial activity in freshwater systems, along with advances in statistical methods, sensor technology, and the ability to process and analyze large amounts of diverse data. Research comparing biogeochemical processes in planktonic and benthic bacterial communities often indicates that benthic communities have the greater influence in system-wide biogeochemistry. However, this may be context dependent. Studies have repeatedly shown that bacterioplankton communities can respond relatively quickly to changing environmental conditions. This may be due in part to the fact that benthic biofilms are typically separated from constituents in the water column by a polysaccharide matrix that reduces diffusion from the water column, while also providing alternative carbon sources for microbes due to the presence of autotrophs within the matrix.

References

Admiraal W, Breebaart L, Tubbing G et al (1994) Seasonal variation in composition and production of planktonic communities in the lower River Rhine. Freshw Biol 32:519–531

Ann V, Freixa A, Butturini A, Romani AM (2019) Interplay between sediment properties and stream flow conditions influences surface sediment organic matter and microbial biomass in a Mediterranean river. Hydrobiologia 828:199–212

Aubeneau A, Hanrahan B, Bolster D et al (2016) Biofilm growth in gravel bed streams controls solute residence time distributions. J Geophys Res Biogeo 121:1840–1850

Azam F, Fenchel T, Field J et al (1983) The ecological role of water-column microbes. Mar Ecol Prog Ser 10:257–263

Barlöcher F, Murdoch JH (1989) Hyporheic biofilms-a potential food source for interstitial animals. Hydrobiologia 184:61–67

Baschien C, Manz W, Neu TR et al (2008) In situ detection of freshwater fungi in an alpine stream by new taxon-specific fluorescence in situ hybridization probes. App Environ Micro 74:6427–6436

Battin TJ, Besemer K, Bengtsson MM, Romani AM, Packmann AI (2016) The ecology and biogeochemistry of stream biofilms. Nat Rev Microbio 14:251

Battin TJ, Kaplan LA, Findlay S et al (2008) Biophysical controls on organic carbon fluxes in fluvial networks. Nat Geosci 1:95–100

Battin TJ, Kaplan LA, Newbold JD et al (2003) Effects of current velocity on the nascent architecture of stream microbial biofilms. App Environ Micro 69:5443–5452

Battin TJ, Kaplan LA, Newbold JD et al (2003) Contributions of microbial biofilms to ecosystem processes in stream mesocosms. Nature 426:439–442

Battin TJ, Sloan WT, Kjelleberg S et al (2007) Microbial landscapes: new paths to biofilm research. Nat Rev Microbio 5:76–81

Battin TJ, Wille A, Psenner R, et al. (2004) Large-scale environmental controls on microbial biofilms in high-alpine streams.Biogeosciences 1:159–171

Beaulieu JJ, Tank JL, Hamilton SK et al (2011) Nitrous oxide emission from denitrification in stream and river networks. Proc Natl Acad Sci U S a 108:214–219

Benner R, Opsahl S, Chin-Leo G et al (1995) Bacterial carbon metabolism in the Amazon River system. Limnol Oceanogr 40:1262–1270

Besemer K (2015) Biodiversity, community structure and function of biofilms in stream ecosystems. Res Microbiol 166:774–781

Besemer K (2016) Microbial biodiversity in natural biofilms. In: Romani AM, Guasch H, Balaguer MD (eds) Aquatic Biofilms: Ecology, Water Quality and Wastewater Treatment. Caister Academic Press, Wymondham, pp 63–88

Besemer K, Peter H, Logue JB et al (2012) Unraveling assembly of stream biofilm communities. ISME J 6:1459–1468

Besemer K, Singer G, Limberger R et al (2007) Biophysical controls on community succession in stream biofilms. App Environ Microbio 73:4966–4974

Besemer K, Singer G, Quince C et al (2013) Headwaters are critical reservoirs of microbial diversity for fluvial networks. Proc R Soc B-Biol Sci 280:8

Bohme A, Risse-Buhl U, Kusel K (2009) Protists with different feeding modes change biofilm morphology. Fems Microbio Ecol 69:158–169

Borchardt MA, Bott TL (1995) Meiofaunal grazing of bacteria and algae in a piedmont stream. J N Am Benthol Soc 14:278–298

Bott TL, Kaplan LA (1990) Potential for protozoan grazing of bacteria in streambed sediments. J N Am Benthol Soc 9:336–345

Bott TL, Newbold JD (2013) Ecosystem metabolism and nutrient uptake in Peruvian headwater streams. Int Rev Hydrobiol 98:117–131

Brugger A, Wett B, Kolar I, Reitner B, Herndl GJ (2001) Immobilization and bacterial utilization of dissolved organic carbon entering the riparian zone of the alpine Enns River. Austria Aqua Microb Eco 24:129–142

Brunke M, Fischer H (1999) Hyporheic bacteria - relationships to environmental gradients and invertebrates in a prealpine stream. Arch Hydrobiol 146:189–217

Buttigieg PL, Ramette A (2014) A guide to statistical analysis in microbial ecology: a community-focused, living review of multivariate data analyses. Fems Microbio Ecol 90:543–550

Cardinale BJ, Palmer MA, Swan CM, Brooks S, Poff NL (2002) The influence of substrate heterogeneity on biofilm metabolism in a stream ecosystem. Ecology 83:412–422

Carlough LA, Meyer JL (1991) Bacterivory by sestonic protists in a southeastern blackwater river. Limnol Oceanogr 36:873–883

Caruso A, Boano F, Ridolfi L, Chopp DL, Packman A (2017) Biofilm-induced bioclogging produces sharp interfaces in hyporheic flow, redox conditions, and microbial community structure. Geophys Res Lett 44:4917–4925

Castillo MM, Allan JD, Sinsabaugh RL et al (2004) Seasonal and interannual variation of bacterial production in lowland rivers of the Orinoco basin. Freshw Biol 49:1400–1414

Castillo MM, Kling GW, Allan JD (2003) Bottom-up controls on bacterial production in tropical lowland rivers. Limnol Oceanogr 48:1466–1475

Cole JJ, Caraco NF (2001) Carbon in catchments: connecting terrestrial carbon losses with aquatic metabolism. Mar Freshwat Res 52:101–110

Cole JJ, Findlay S, Pace ML (1988) Bacterial production in fresh and saltwater ecosystems: a cross-system overview. Mar Ecol Prog Ser 43:1–10

Collins SM, Sparks JP, Thomas SA et al (2016) Increased light availability reduces the importance of bacterial carbon in headwater stream food webs. Ecosystems 19:396–410

Comte J, Fauteux L, del Giorgio PA (2013) Links between metabolic plasticity and functional redundancy in freshwater bacterioplankton communities. Front in Micro 4:11

Costerton JW, Geesey GG, Cheng KJ (1978) How Bacteria Stick. Sci Amer 238:86–95

Crocker MT, Meyer JL (1987) Interstitial dissolved organic carbon in sediments of a southern Appalachian headwater stream. J N Am Benthol Soc 6:159–167

Dann LM, Smith RJ, Jeffries TC et al (2017) Persistence, loss and appearance of bacteria upstream and downstream of a river system. Mar Freshw Res 68:851–862

deMelo ML, Kothawala DN, Bertilsson S et al (2020) Linking dissolved organic matter composition and bacterioplankton communities in an Amazon floodplain system. Limnol Oceanogr 65:63–76 de Oliveira LFV, Margis RJPo, (2015) The source of the river as a nursery for microbial diversity. PLoSO One 10:e0120608

Demars B, Friberg N, Kemp J, et al. (2018) Reciprocal carbon subsidies between autotrophs and bacteria in stream food webs under stoichiometric constraints. bioRxiv:447987

Demars BO (2019) Hydrological pulses and burning of dissolved organic carbon by stream respiration. Limnol Oceanogr 64:406–421

Dodds WK, Hutson RE, Eichem AC et al (1996) The relationship of floods, drying, and light to primary production and producer biomass in a prairie stream. Hydrobiologia 333:151–159

Dopheide A, Lear G, He ZL et al (2015) Functional gene composition, diversity and redundancy in microbial stream biofilm communities. PLoS ONE 10:21

dos Reis MC, Bagatini IL, de Oliveira VL et al (2019) Spatial heterogeneity and hydrological fluctuations drive bacterioplankton community composition in an Amazon floodplain system. PLoS ONE 14:e0220695

Edling H, Tranvik LJ (1996) Effects of pH on glucosidase activity and availability of DOC to bacteria in lakes. Arch. Hydrobiol. Spec. Issues Adv. Limnol 48:123–132

Edwards R, Meyer J (1987) Metabolism of a sub-tropical low gradient black water river. Freshw Biol 71:251–263

Farjalla VF, Esteves FA, Bozelli RL, Roland FJ (2002) Nutrient limitation of bacterial production in clear water Amazonian ecosystems. Hydrobiologia 489:197–205

Fasching C, Wilson HF, D'Amario SC, Xenopoulos MA (2019) Natural land cover in agricultural catchments alters flood effects on DOM composition and decreases nutrient levels in streams. Ecosystems 22:1530–1545

Fenchel T (2008) The microbial loop–25 years later. J Exp Mar Biol Ecol 366:99–103

Fierer N (2008) Microbial biogeography: patterns in microbial diversity across space and time. In: Zengler K (ed) Accessing Uncultivated Microorganisms. American Society of Microbiology, Washington DC, pp 95–115

Fierer N, Morse JL, Berthrong ST, Bernhardt ES, Jackson RB (2007) Environmental controls on the landscape-scale biogeography of stream bacterial communities. Ecology 88:2162–2173

Findlay S (2003) Bacterial response to variation in dissolved organic matter. In: Findlay SEG, Sinsabaugh R (eds) Aquatic Ecosystems. Elsevier, pp 363–379

Findlay S (2010) Stream microbial ecology. J N Am Benthol Soc 29:170–181

Findlay S (2016) Stream microbial ecology in a changing environment. In: Jones J, Stanely E (eds) Stream Ecosystems in a Changing Environment. Elsevier, Amsterdam, pp 135–150

Findlay S, Howe K, Fontvielle D (1993) Bacterial-algal relationships in streams of the Hubbard Brook experimental forest. Ecology 74:2326–2336

Findlay S, Meyer JL, Risley R (1986) Benthic bacterial biomass and production in two blackwater rivers. Can J Fish 43:1271–1276

Findlay S, Pace ML, Lints D et al (1991) Weak coupling of bacterial and algal production in a heterotrophic ecosystem: the Hudson River estuary. Limnol Oceanogr 36:268–278

Findlay S, Quinn JM, Hickey CW et al (2001) Effects of land use and riparian flowpath on delivery of dissolved organic carbon to streams. Limnol Oceanogr 46:345–355

Findlay S, Sinsabaugh R (2006) Large-scale variation in subsurface stream biofilms: a cross-regional comparison of metabolic function and community similarity. Microb Ecol 52:491–500

Findlay S, Sinsabaugh RL (1999) Unravelling the sources and bioavailability of dissolved organic matter in lotic aquatic ecosystems. Mar Freshw Res 50:781–790

Findlay S, Sinsabaugh RL (2003) Response of hyporheic biofilm metabolism and community structure to nitrogen amendments. Aquat Microb Ecol 33:127–136

Findlay S, Sinsabaugh RL, Fischer DT et al (1998) Sources of dissolved organic carbon supporting planktonic bacterial production in the tidal freshwater Hudson River. Ecosystems 1:227–239

Findlay S, Sobczak W (2000) Microbial communities in hyporheic sediments. In: Jones JB, Mullholland PJ (eds) Streams and Ground Waters. Academic Press, San Diego, pp 287–306

Findlay S, Sobczak WV (1996) Variability in removal of dissolved organic carbon in hyporheic sediments. J N Am Benthol Soc 15:35–41. https://doi.org/10.2307/1467431

Findlay S, Tank J, Dye S et al (2002) A cross-system comparison of bacterial and fungal biomass in detritus pools of headwater streams. Microb Ecol 43:55–66

Findlay SEG, Arsuffi TL (1989) Microbial-growth and detritus transformations during decomposition of leaf litter in a stream. Freshw Biol 21:261–269

Fischer H (2003) The role of biofilms in the uptake and transformation of dissolved organic matter. In: Findlay SEG, Sinsabaugh R (eds) Aquatic Ecosystems. Elsevier, pp 285–313

Fischer H, Pusch M (2001) Comparison of bacterial production in sediments, epiphyton and the pelagic zone of a lowland river. Freshw Biol 46:1335–1348

Fischer H, Wanner SC, Pusch M (2002) Bacterial abundance and production in river sediments as related to the biochemical composition of particulate organic matter (POM). Biogeochemistry 61:37–55

Fisher SG, Gray LJ, Grimm NB et al (1982) Temporal succession in a desert stream ecosystem following flash flooding. Ecol Monogr 52:93–110

Foreman C, Franchini P, Foreman R (1998) The trophic dynamics of riverine bacterioplankton: relationships among substrate availability, ectoenzyme kinetics, and growth. Limnol Oceanogr 43:1344–1352

Freeman C, Lock MA (1995) The biofilm polysaccharide matrix-a buffer against changing organic substrate supply. Limnol Oceanogr 40:273–278

Freixa A, Ejarque E, Crognale S et al (2016) Sediment microbial communities rely on different dissolved organic matter sources along a Mediterranean river continuum. Limnol Oceanogr 61:1389–1405

Gamfeldt L, Hillebrand H, Jonsson PR (2008) Multiple functions increase the importance of biodiversity for overall ecosystem functioning. Ecology 89:1223–1231

Geesey G, Mutch R, Jt C et al (1978) Sessile bacteria: An important component of the microbial population in small mountain streams. Limnol Oceanogr 23:1214–1223

Gessner MO, Chauvet E (1993) Ergosterol to biomass conversion factors for aquatic hyphyomycetes. App Environ Microbio 59:502–507

Golladay SW, Sinsabaugh RL (1991) Biofilm development on leaf and wood surfaces in a boreal river. Freshw Biol 25:437–450

Graeber D, Poulson JR, Heinz M et al (2018) Going with the flow: Planktonic processing of dissolved organic carbon in streams. Sci Total Environ 625:519–530

Grimm NB, Faeth SH, Golubiewski NE et al (2008) Global change and the ecology of cities. Science 319:756–760

Guenet B, Danger M, Abbadie L et al (2010) Priming effect: bridging the gap between terrestrial and aquatic ecology. Ecology 91:2850–2861

Haack TK, McFeters GA (1982) Nutritional relationships among relationships among microorganisms in an epilithing biofilm community. Microb Ecol 8:115–126

Hakenkamp CC, Morin A (2000) The importance of meiofauna to lotic ecosystem functioning. Freshw Biol 44:165–175

Hall RO, Meyer JL (1998) The trophic significance of bacteria in a detritus-based stream food web. Ecology 79:1995–2012

Halvorson HM, Barry JR, Lodato MB et al (2019a) Periphytic algae decouple fungal activity from leaf litter decomposition via negative priming. Funct Ecol 33:188–201

Halvorson HM, Francoeur SN, Findlay RH et al (2019b) Algal-mediated priming effects on the ecological stoichiometry of leaf litter decomposition: a meta-analysis. Front Earth Sci 7:76

Hassell N, Tinker KA, Moore T, Ottesen EA (2018) Temporal and spatial dynamics in microbial community composition within a temperate stream network. Environ Microb 20:3560–3572

Hayer M, Schwartz E, Marks JC et al (2016) Identification of growing bacteria during litter decomposition in freshwater through quantitative stable isotope probing. Environ Microb Rep 8:975–982

Hieber M, Gessner MO (2002) Contribution of stream detrivores, fungi, and bacteria to leaf breakdown based on biomass estimates. Ecology 83:1026–1038

Hosen JD, Febria CM, Crump BC et al (2017) Watershed urbanization linked to differences in stream bacterial community composition. Front Microbiol 8:17

Hudson JJ, Roff JC, Burnison BK (1992) Bacterial productivity in forested and open streams in southern Ontario. Can J Fish 49:2412–2422

Kamjunke N, Herzsprung P, Neu TR (2015) Quality of dissolved organic matter affects planktonic but not biofilm bacterial production in streams. Sci Total Environ 506:353–360

Kaplan L, Cory R (2016) Dissolved organic matter in stream ecosystems: forms, functions, and fluxes of watershed tea. In: Jones J, Stanely E (eds) Stream Ecosystems in a Changing Environment. Elsevier, Amsterdam, pp 241–320

Kaplan LA, Bott TL (1982) Diel fluctuations of DOC generated by algae in a piedmont stream. Limnol Oceanogr 27:1091–1100

Kaplan LA, Bott TL (1989) Diel fluctuations in bacterial activity on streambed substrata during vernal algal blooms: effects of temperature, water chemistry, and habitat. Limnol Oceanogr 34:718–733

Kaplan LA, Wiegner TN, Newbold J et al (2008) Untangling the complex issue of dissolved organic carbon uptake: A stable isotope approach. Freshw Biol 53:855–864

Koch BJ, McHugh TA, Hayer M et al (2018) Estimating taxon-specific population dynamics in diverse microbial communities. Ecosphere 9:e02090

Kreutzweiser DP, Capell SS (2003) Benthic microbial utilization of differential dissolved organic matter sources in a forest headwater stream. Can J For Res 33:1444–1451

Lancaster J, Robertson A (1995) Microcrustacean prey and macroinvertebrate predators in a stream food web. Freshw Biol 34:123–134

Lau MP, Niederdorfer R, Sepulveda-Jauregui A et al (2018) Synthesizing redox biogeochemistry at aquatic interfaces. Limnologica 68:59–70

Lawrence JR, Scharf B, Packroff G et al (2002) Microscale evaluation of the effects of grazing by invertebrates with contrasting feeding modes on river biofilm architecture and composition. Microb Ecol 44:199–207

Lear G, Washington V, Neale M et al (2013) The biogeography of stream bacteria. Glob Ecol Biogeogr 22:544–554

LeBrun ES, King RS, Back JA et al (2018) Microbial community structure and function decoupling across a phosphorus gradient in streams. Microb Ecol 75:64–73

Lock M (1981) River epilithon—a light and organic energy transducer. In: Lock MA, Williams DD (eds) Perspectives in Running Water Ecology. Springer, Berlin, pp 3–40

Lock M, Wallace R, Costerton J, Ventullo R, Charlton S (1984) River epilithon: toward a structural-functional model. Oikos:10–22

Lodge DM, Barker JW, Strayer D et al (1988) Spatial heterogeneity and habitat interactions in lake communities. In: Carpenter SR (ed) Complex Interactions in Lake Communities. Springer, Berlin, pp 181–208

Mansfeldt C, Deiner K, Mächler E et al (2020) Microbial community shifts in streams receiving treated wastewater effluent. Sci Tot Environ 709:135727

Mao YF, Liu Y, Li H, He Q, Ai HN, Gu WK, Yang GF (2019) Distinct responses of planktonic and sedimentary bacterial communities to anthropogenic activities: case study of a tributary of the Three Gorges Reservoir, China. Sci Total Environ 682:324–332

Marks JC (2019) Revisiting the fates of dead leaves that fall into streams. Ann Rev Ecol Evol Syst 50:547–568

McCutchan JH, Lewis WM (2002) Relative importance of carbon sources for macroinvertebrates in a Rocky Mountain stream. Limnol Oceanogr 47:742–752

Mcnamara CJ, Leff LG (2004) Response of biofilm bacteria to dissolved organic matter from decomposing maple leaves. Microb Ecol 48:324–330

Melo LF, Bott TR (1997) Biofouling in water systems. Exp Therm Fluid Sci 14:375–381

Methvin BR, Suberkropp K (2003) Annual production of leaf-decaying fungi in 2 streams. J N Am Benthol Soc 22:554–564

Meyer J (1994) The microbial loop in flowing waters. Microb Ecol 28:195–199

Meyer JL, Edwards RT, Risley R (1987) Bacterial growth on dissolved organic carbon from a blackwater river. Microb Ecol 13:13–29

Miao L, Wang P, Hou J, a. (2019) Distinct community structure and microbial functions of biofilms colonizing microplastics. Sci Tot Environ 650:2395–2402

Milner AM, Khamis K, Battin TJ et al (2017) Glacier shrinkage driving global changes in downstream systems. Proc Nat Acad Sci 114:9770–9778

Moran MA, Hodson RE (1990) Bacterial production on humic and nonhumic components of dissolved organic carbon. Limnol Oceanogr 35:1744–1756

Mulholland PJ, Helton AM, Poole GC et al (2008) Stream denitrification across biomes and its response to anthropogenic nitrate loading. Nature 452:202–246

Münster U, Einiö P, Nurminen J et al (1992) Extracellular enzymes in a polyhumic lake: important regulators in detritus processing. Hydrobiologia 229:225–238

Niederdorfer R, Besemer K, Battin TJ et al (2017) Ecological strategies and metabolic trade-offs of complex environmental biofilms. NPJ Biofilms and Microbiomes 3:21

Niño-García JP, Ruiz-González C, del Giorgio PA (2016) Interactions between hydrology and water chemistry shape bacterioplankton biogeography across boreal freshwater networks. the ISME Journal 10:1755

O'Brien JM, Warburton HJ, Graham SE et al (2017) Leaf litter additions enhance stream metabolism, denitrification, and restoration prospects for agricultural catchments. Ecosphere 8:e02018

Pace NR (2006) Time for a change. Nature 441:289–289

Peipoch M, Miller SR, Antao TR, Valett HM (2019) Niche partitioning of microbial communities in riverine floodplains. Sci Rep 9:13

Perlmutter DG, Meyer JL (1991) The impact of a stream-dwelling harpacticoid copepod upon detritally associated bacteria. Ecology 72:2170–2180

Peter H, Ylla I, Gudasz C et al (2011) Multifunctionality and diversity in bacterial biofilms. PLoS ONE 6:8

Polz MF, Cordero OX (2016) Genomics of Metabolic Trade-Offs. Nat Microbiol 1:2

Pomeroy LR, Williams PJ, Azam F et al (2007) The microbial loop. Oceanography 20:28–33

Pomeroy LR, Wiebe WJ (1988) Energetics of microbial food webs. Hydrobiologia 159:7–18

Portillo MC, Anderson SP, Fierer N (2012) Temporal variability in the diversity and composition of stream bacterioplankton communities. Environ Microbio 14:2417–2428

Pusch M, Fiebig D, Brettar I et al (1998) The role of micro-organisms in the ecological connectivity of running waters. Freshwat Biol 40:453–495

Raymond PA, Hartman J, Lauerwald R et al (2013) Global carbon dioxide emissions from inland waters. Nature 503:355

Read DS, Gweon HS, Bowes MJ et al (2015) Catchment-scale biogeography of riverine bacterioplankton. the ISME Journal 9:516

Rier ST, Kuehn KA, Francoeur SN (2007) Algal regulation of extracellular enzyme activity in stream microbial communities associated with inert substrata and detritus. J N Am Benthol Soc 26:439–449

Rier ST, Stevenson RJ (2002) Effects of light, dissolved organic carbon, and inorganic nutrients on the relationship between algae and heterotrophic bacteria in stream periphyton. Hydrobiologia 489:179–184

Risse-Buhl U, Trefzger N, Seifert AG et al (2012) Tracking the autochthonous carbon transfer in stream biofilm food webs. Fems Microbio Ecol 79:118–131

Robertson AL (2000) Lotic meiofaunal community dynamics: colonisation, resilience and persistence in a spatially and temporally heterogeneous environment. Freshwat Biol 44:135–147

Roller BRK, Stoddard SF, Schmidt TM (2016) Exploiting rRNA operon copy number to investigate bacterial reproductive strategies. Nat Microbio 1:7

Romani A, Guasch H, Munoz I et al (2004) Biofilm structure and function and possible implications for riverine DOC dynamics. Microb Ecol 47:316–328

Romani AM, Sabater S (2001) Structure and activity of rock and sand biofilms in a Mediterranean stream. Ecology 82:3232–3245

Rosselli R, Romoli O, Vitulo N et al (2016) Direct 16S rRNA-seq from bacterial communities: a PCR-independent approach to simultaneously assess microbial diversity and functional activity potential of each taxon. Sci Rep 6:32165

Ruiz-González C, Niño-García JP, Lapierre JF et al (2015) The quality of organic matter shapes the functional biogeography of bacterioplankton across boreal freshwater ecosystems. Glob Ecol Biogeogr 24:1487–1498

Rundle S, Hildrew A (1992) Small fish and small prey in the food webs of some southern English streams. Archiv Fuer Hydrobiologie 125:25–35

Savio D, Sinclair L, Ijaz UZ et al (2015) Bacterial diversity along a 2600 km river continuum. Environ Microbiol 17:4994–5007

Schmid-Araya J, Schmid P (2000) Trophic relationships: integrating meiofauna into a realistic benthic food web. Freshw Biol 44:149–163

Schmid-Araya JM, Schmid PE, Tod SP et al (2016) Trophic positioning of meiofauna revealed by stable isotopes and food web analyses. Ecology 97:3099–3109

Schmid P, Schmid-Araya J (1997) Predation on meiobenthic assemblages: resource use of a tanypod guild (Chironomidae, Diptera) in a gravel stream. Freshw Biol 38:67–91

Sgier L, Freimann R, Zupanic A et al (2016) Flow cytometry combined with viSNE for the analysis of microbial biofilms and detection of microplastics. Nat Commun 7:10

Siboni N, Lidor M, Kramarsky-Winter E et al (2007) Conditioning film and initial biofilm formation on ceramics tiles in the marine environment. FEMS Microbiol Lett 274:24–29

Simon KS, Simon MA, Benfield EF (2009) Variation in ecosystem function in Appalachian streams along an acidity gradient. Ecol Appl 19:1147–1160

Singer G, Besemer K, Schmitt-Kopplin P, Hodl I, Battin TJ (2010) Physical heterogeneity increases biofilm resource use and its molecular diversity in stream mesocosms. PLoS ONE 5:11

Sinsabaugh R, Foreman C (2003) Integrating dissolved organic matter metabolism and microbial diversity: an overview of conceptual models. In: Findlay SEG, Sinsabaugh RL (eds) Aquatic Ecosystems. Elsevier, Amsterdam, pp 425–454

Sinsabaugh RL, Belnap J, Findlay SG et al (2014) Extracellular enzyme kinetics scale with resource availability. Biogeochemistry 121:287–304

Sinsabaugh RL, Findlay S, Franchini P et al (1997) Enzymatic analysis of riverine bacterioplankton production. Limnol Oceanogr 42:29–38

Sinsabaugh RL, Repert D, Weiland T et al (1991) Exoenzyme accumulation in epilithic biofilms. Hydrobiologia 222:29–37

Sobczak WV (1996) Epilithic bacterial responses to variations in algal biomass and labile dissolved organic carbon during biofilm colonization. J N Am Benthol Soc 15:143–154

Sobczak WV, Findlay S (2002) Variation in bioavailability of dissolved organic carbon among stream hyporheic flowpaths. Ecology 83:3194–3209

Staley C, Gould TJ, Wang P et al (2015) Species sorting and seasonal dynamics primarily shape bacterial communities in the Upper Mississippi River. Sci Total Environ 505:435–445

Sun L, Perdue E, Meyer J et al (1997) Use of elemental composition to predict bioavailability of dissolved organic matter in a Georgia river. Limnol Oceanogr 42:714–721

Székely AJ, Langenheder S (2017) Dispersal timing and drought history influence the response of bacterioplankton to drying–rewetting stress. the ISME Journal 11:1764

Tank JL, Dodds WK (2003) Nutrient limitation of epilithic and epixylic biofilms in ten North American streams. Freshw Biol 48:1031–1049

Tank JL, Webster JR (1998) Interaction of substrate and nutrient availability on wood biofilm processes in streams. Ecology 79:2168–2179

Tank JL, Webster JR, Benfield EF, Sinsabaugh RL (1998) Effect of leaf litter exclusion on microbial enzyme activity associated with wood biofilms in streams. J N Am Benthol Soc 17:95–103

Teachey ME, McDonald JM, Ottesen EA (2019) Rapid and stable microbial community assembly in the headwaters of a third-order stream. App Environ Microbio 85:15

Tranvik LJ, Bertilsson S (2001) Contrasting effects of solar UV radiation on dissolved organic sources for bacterial growth. Ecol Lett 4:458–463

Tranvik LJ (1990) Bacterioplankton growth on fractions of dissolved organic carbon of different molecular weights from humic and clear waters Applied. Environ Microbiol 56:1672–1677

Twining CW, Brenna JT, Hairston NG Jr et al (2016) Highly unsaturated fatty acids in nature: what we know and what we need to learn. Oikos 125:749–760

Twining CW, Josephson DC, Kraft CE et al (2017) Limited seasonal variation in food quality and foodweb structure in an Adirondack stream: insights from fatty acids. Freshw Sci 36:877–892

UNDES (2018) World Urbanization Prospects 2018. United Nations, New York

Valdivia-Anistro JA, Eguiarte-Fruns D-S et al (2016) Variability of rRNA operon copy number and growth rate dynamics of *Bacillus* isolated from an extremely oligotrophic aquatic ecosystem. Front Microbio 6:15

Veach AM, Stegen JC, Brown SP et al (2016) Spatial and successional dynamics of microbial biofilm communities in a grassland stream ecosystem. Molec Ecol 25:4674–4688

Wagner K, Bengtsson MM, Findlay RH et al (2017) High light intensity mediates a shift from allochthonous to autochthonous carbon use in phototrophic stream biofilms. Biogeosciences 122:1806–1820

Walsh CJ, Roy AH, Feminella JW et al (2005) The urban stream syndrome: current knowledge and the search for a cure. J N Am Benthol Soc 24:706–723

Ward CP, Nalven SG, Crump BC et al (2017) Photochemical alteration of organic carbon draining permafrost soils shifts microbial metabolic pathways and stimulates respiration. Nat Comm 8:1–8

Weitere M, Erken M, Majdi N et al (2018) The food web perspective on aquatic biofilms. Ecol Monogr 88:543–559

Wey JK, Jurgens K, Weitere M (2012) Seasonal and successional influences on bacterial community composition exceed that of protozoan grazing in river biofilms. App Environ Microbiol 78:2013–2024

Wiegner TN, Kaplan LA, Newbold JD et al (2005) Contribution of dissolved organic C to stream metabolism: a mesocosm study using 13C-enriched tree-tissue leachate. J N Am Benthol Soc 24:48–67

Wilcox HS, Wallace JB, Meyer JL et al (2005) Effects of labile carbon addition on a headwater stream food web. Limnol Oceanogr 50:1300–1312

Woodcock S, Besemer K, Battin TJ et al (2013) Modelling the effects of dispersal mechanisms and hydrodynamic regimes upon the structure of microbial communities within fluvial biofilms. Environ Microbiol 15:1216–1225

Wotton R (1994) Particulate and dissolved organic matter as food. In: Wooton RS (ed) The Biology of Particles in Aquatic Systems. Lewis Publishers, Ann Arbo, pp 235–288

Zak DR, Blackwood CB, Waldrop MP (2006) A molecular dawn for biogeochemistry. Trends Ecol Evol 21:288–295

Zeglin LH (2015) Stream microbial diversity in response to environmental changes: review and synthesis of existing research. Front Microbiol 6:454

Zeglin LH, Crenshaw CL, Dahm CN et al (2019) Watershed hydrology and salinity, but not nutrient chemistry, are associated with arid-land stream microbial diversity. Freshw Sci 38:77–91

Trophic Relationships

The network of consumers and resources that comprise fluvial food webs is supported by a diverse mix of energy supplies that originate within the stream and beyond its banks. These include the living resources of algae and higher plants, and the non-living resources of particulate and dissolved organic matter, especially material derived from dead and decaying plant matter. Microorganisms are important mediators of organic matter availability, and there is increasing evidence of their importance as a resource to both small and large consumers. Additionally, energy subsidies in the form of falling terrestrial arthropods and the eggs and carcasses of migrating fish contribute to the support of many stream-dwellers. Energy sources within a stream reach are not necessarily consumed within that part of the stream, as downstream export, insect emergence, and fish movements can supply energy to distant ecosystems.

Trophic organization in river ecosystems can be complex and indistinct. Many consumers are polyphagous rather than monophagous, and exhibit considerable overlap with one another in their diets. The gut contents of invertebrates can be difficult to identify with great confidence, so these consumers often are characterized by the unspecific term of herbivore-detritivore. In many systems, especially those in the temperate zone, the vast majority of fishes eat invertebrates. As a consequence, while a particular species may be classified solely on the basis of what it eats—herbivore, predator, detritivore, and so on—the resulting categories can be of limited usefulness because they offer too few distinctions among feeding roles. Further resolution of trophic status can be achieved by distinguishing among feeding roles on the basis of how the food is obtained, rather than solely in terms of what food is eaten, and by consideration of morphological adaptations for food capture.

When several species consume a common resource and acquire it in similar fashion, they are considered members of the same feeding group, commonly referred to as a trophic category or guild. Thus, a fish species that captures invertebrate prey directly from the bottom would occupy a different trophic category from another species that consumes the same prey, but captures them from the water column. Invertebrates typically are divided into functional feeding groups on the basis of resource category, where or how the resource is obtained, and morphological adaptations for food capture (Cummins 1973). It is important to note that members of different invertebrate functional groups may consume the same resource: for example, fine particulate organic matter can be captured from the water column or collected from depositional areas within a stream. Hence, the main difference between consumers is not the resource, but the organism's method of acquiring it. Compared to macroinvertebrate classifications, fish trophic categories often rely primarily on what resources is consumed, but also may consider feeding location and morphology.

Traditional methods for investigating the feeding roles of invertebrate species include gut-content analysis, fecal analysis, and behavioral observations. Such studies can provide detailed information on primary feeding mode for many if not most species, and have long been been the basis for trophic classification. In addition, studies of feeding can be complemented with morphological analysis of feeding mechanism, chemical analyses of resource and consumer tissue, and experimental studies (Gelwick and McIntyre 2017). Recent years have seen broader use of powerful new tools that provide fresh insight into the energy sources that consumers assimilate into their tissue (Post 2002; Finlay et al. 2010). In many instances, studies provide evidence that less abundant food sources actually contribute disproportionately to consumer energy intake: in other words, what consumers eat and what they assimilate can be quite different. Stable isotopes of C, N, and H are becoming widely used as food web tracers in aquatic ecosystems, because isotopic ratios of consumer proteins reflect the proteins of their food sources, and some isotopic ratios change with each trophic transfer. The ratio of carbon isotopes ($^{13}C/^{12}C$ abbreviated as $\delta^{13}C$) relies on isotopic differences between algae and plant matter of terrestrial origin. Algae generally

© Springer Nature Switzerland AG 2021
J. D. Allan et al., *Stream Ecology*,
https://doi.org/10.1007/978-3-030-61286-3_9

have a higher $^{13}C/^{12}C$ ratio (are "enriched" in ^{13}C) relative to terrestrially-derived C, although variability among algal taxa associated with environmental conditions, including water velocity, may limit insights from this tracer alone (Finlay et al. 2010). The isotopes of nitrogen are most useful for estimating trophic position because the ratio ($^{15}N/^{14}N$ abbreviated as $\delta^{15}N$) of a consumer is typically enriched by 3–4% relative to its diet. In contrast, the $\delta^{13}C$ changes little as carbon moves through food webs, allowing it to be used to evaluate the ultimate sources of carbon for an organism when the isotopic signature of the sources are different. The stable isotope of hydrogen (δ^2H, or δD for Deuterium) can be used to evaluate allochthony in aquatic systems because terrestrially-derived resources in streams are substantially enriched in δ^2H relative to autochthonous material (Doucett et al. 2007). Many animal consumers are dependent on their food supply to obtain lipids important to their growth. Certain fatty acids are biomarkers for cyanobacteria and diatoms, while others are more characteristic of allochthonous resources, and so can distinguish the relative importance of autochthony versus allochthony to consumer assimilation. Although not yet in wide use, recent development of microbial biomarkers holds promise for improved assessment of the microbial compartment of food webs (Middelburg 2014).

Historically the study of trophic relationships has emphasized the larger animal consumers, but a growing appreciation of the contribution of bacteria and fungi to the diet of macro-consumers, as well as the potential importance of protists, micro-metazoans, and early instars of macroinvertebrates in energy transfers, have led to a greater focus on

the importance of microbial food webs. Biofilms in particular have been shown to be important energy complexes where algae and microorganisms living in close association are able to capitalize on the energy obtained from sunlight and from organic matter, and so autotrophic and heterotrophic pathways can be closely linked. The trophic ecology of micro-consumers and energy supplied through microbial production are explored in Chap. 8; in this chapter we focus on the macro-consumers groups of invertebrates, fishes, and other vertebrates.

9.1 Invertebrate Feeding Roles

Macroinvertebrates are major components of riverine food webs, serving as important links between basal resources and higher trophic levels. However, traditional categories of food consumption, such as herbivore, detritivore, predator, etc., are of limited use for stream invertebrates. Most aquatic invertebrates are omnivorous, at least in their early life stages, and many retain this flexibility throughout their lives. Accidental ingestion of diverse food items wherever detritus, algal films, and small invertebrates are intermixed, and the often amorphous nature of gut contents, further compounds the difficulty of portraying stream food webs. The establishment of functional feeding groups (FFGs, Cummins 1973; Cummins and Klug 1979), and their association with most North American taxa (Merritt et al. 2019), was a major advance that has found wide use in stream ecology.

Table 9.1 lists the main FFG categories, their dominant food, feeding mechanism, and typical size range of particles

Table 9.1 Functional feeding groups (FFGs) of aquatic invertebrates, based on Merritt et al. (2017, 2019). Categories and size ranges are generalizations, and main FFG categories can be further subdivided based on feeding mechanism

Functional feeding group	Dominant food	Feeding mechanism	General particle size range of food (mm)
Shredders	Decomposing vascular plant tissue (CPOM) Living vascular plant tissue Wood	Detritivores-chewers of CPOM Herbivore-chewers and miners of live macrophytes Borers and gougers	>1
Collectors	Decomposing FPOM and associated microflora and— microfauna	Detritivore-filterers (suspension feeders) Detritivore-gatherers (deposit feeders)	<1
Scrapers	Periphyton—attached algae and associated microflora and— microfauna	Herbivores scraping, grazing, and browsing on mineral and organic substrates	<1
Macrophyte piercers	Living vascular plant cell tissues and fluids	Herbivores pierce tissues and suck fluids	>1
Predators	Living animal tissue	Carnivores that ingest entire animal or parts Carnivores that pierce tissues and suck fluids	>1
Parasites	Living animal tissue	Internal and external parasites of all life stages	>1

consumed. Scrapers consume non-filamentous attached algae from substrates (coarse sediments, wood, or stems of rooted aquatic vascular plants). Detrital shredders primarily feed on leaves or needles of terrestrial plant litter (coarse particulate organic matter, CPOM) entrained in the stream and colonized by microbes. Gathering collectors have very generalized adaptations to feed on fine particulate organic matter (FPOM) from depositional areas or crevices. Filtering collectors capture FPOM in suspension in streams using morphological structures or silk capture nets. Herbivore shredders are adapted to feed on live rooted aquatic plants, primarily the leaves. Herbivore piercers are adapted to pierce individual filamentous algal cells and suck out the cell contents. Predators are adapted to catch and consume live prey by engulfing the prey or piercing and extracting the prey hemolymph.

Within a given FFG one finds different taxonomic groups and a variety of adaptations with regard to mouthparts and food-gathering mechanisms. Mouthparts adapted for scraping benthic algae from hard surfaces are broadly similar amongst caddis larvae from different families (Glossosomatidae, Helicopsychidae) and a beetle larvae (the water penny, Psephenidae). Similarly, the mandibles of three genera of wood gougers, including *Heteroplectron* (Trichoptera), *Lara* (Coleoptera), and *Lipsothrix* (Diptera) all have three teeth and are scoop-shaped with basal setae that aid in passage into the mouth of the removed wood fragments and contained microbes (Cummins 2018). Aquatic insects specialized for suspension feeding may have different morphologies that nonetheless filter small particles from the water column, such as setae along the forelegs of some mayfly nymphs, and the nets of hydropsychid caddis larvae.

The FFG system assumes a direct correspondence between basal resources available at a stream location and the populations of macroinvertebrates that are adapted to efficiently harvest those food resources (Merritt et al. 2017). Periphyton, coarse and fine particulate organic matter, and other animals are the main categories of basal resources, described in Chaps. 6 and 7 (although the biofilms discussed in Chap. 8, containing microbes, algae, and meiofauna, occur widely and complicate this neat partitioning). Differences in the size of the food items, such as CPOM vs FPOM; its location, either suspended or attached to or deposited on the stream bed; and whether or not resources are living or dead, are further correlates of FFG designation. Thus, the use of FFGs to characterize the macroinvertebrate assemblage at a stream location provides insight into functional roles, basal resource availability, and the importance of allochthony vs autochthony. In essence, functional feeding group classification tells us that shredders feed on CPOM, collectors feed on deposited (gatherers) and suspended (filterers) FPOM, scrapers consume periphyton, piercers feed on live algal cell contents in periphyton, and predators ingest

prey. Parasites and pathogens seldom are considered explicitly, but have much in common with predators.

9.1.1 Consumers of CPOM

Figure 9.1 depicts the shredder:CPOM linkage typical of a small stream in the temperate zone. The consumption of autumn-shed leaves in woodland streams by the invertebrates termed shredders is the most extensively investigated trophic pathway involving CPOM (Cummins and Klug 1979), and shall serve as our model here. Invertebrates that feed on decaying leaves include crustaceans (especially amphipods, isopods, crayfish, and freshwater shrimp), snails, and several groups of insect larvae. The latter includes crane fly larvae (Tipulidae), and several families of the Trichoptera (Limnephilidae, Lepidostomatidae, Sericostomatidae, Oeconesidae), and Plecoptera (Peltoperlidae, Pteronarcidae, Nemouridae). The leaf-shredding activities of insect larvae and gammarid amphipods are particularly well studied (Table 9.2). *Tipula* and many limnephilid caddis larvae eat all parts of the leaf including mesophyll and venation, whereas peltoperlid stonefly nymphs avoid venation and concentrate mainly on mesophyll, cuticle, and epidermal cells (Ward and Woods 1986). The radula of snails and mouthparts of *Gammarus* are most effective at scraping softer tissues, and the bigger crustaceans are able to tear and engulf larger leaf fragments (Anderson and Sedell 1979). Different feeding modes among CPOM detritivores also have consequences for the production of FPOM. The caddis *Limnephilis* produced small particles at a higher rate than two crustacean shredders, and both the size fractions and C: N ratios of fine particles generated varied with the mix of detritivore species (Patrick 2013).

Selection of food by shredders is based on several characteristics of leaves such as toughness, nutrient content, the presence of plant chemical defenses, and the degree of conditioning by microorganisms (Graça 2001). The nutritional quality of leaves is intimately linked with the microorganisms that contribute so greatly to leaf breakdown. Much effort has been directed at determining how microorganisms directly (as food) and indirectly (by modifying the substrate) contribute to the nourishment of CPOM consumers, and what capabilities these detritivores possess to digest the various components of their diet. Invertebrate detritus-feeders unquestionably prefer leaves that have been "conditioned" by microbial colonization in comparison to uncolonized leaves. When presented with elm leaves that were either autoclaved or cultured with antibiotics to inhibit microbial growth, versus normal colonized leaves, *Gammarus* consumed far more of the latter (Kaushik and Hynes 1971). Subsequent work has confirmed that preference is greatest for leaves at the stage of conditioning that

Fig. 9.1 The shredder:CPOM linkages for a small stream within a temperate deciduous forest. Physical abrasion, microbial activity (especially by fungi), and invertebrate shredders reduce much of the CPOM to smaller particles. Chemical leaching and microbial excretion and respiration release DOM and CO_2, but much of the original carbon enters other detrital pools as feces and fragmented material (Reproduced from Cummins and Klug 1979)

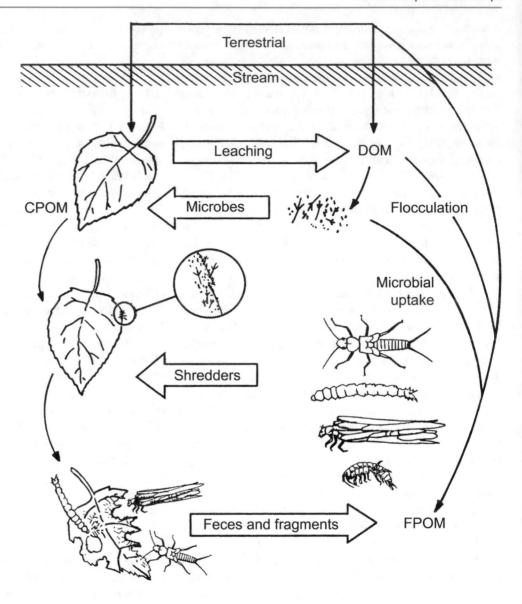

Table 9.2 The contrasting feeding strategies of two CPOM detritivores. Based on Barlöcher (1983)

	Gammarus fossarum	*Tipula abdominalis*
Feeding mechanism	Scrapes at leaf surface	Chews entire leaf
Gut pH and digestive biochemistry	Anterior gut slightly acid Its own enzymes and fungal exoenzymes attack leaf carbohydrates Posterior gut is alkaline, can digest microbial proteins and some leaf proteins	Foregut and midgut highly alkaline (up to 11.6) Results in high proteolic activity but inactivation of fungal exoenzymes, thus little activity toward leaf carbohydrates
Efficiency	Highly efficient at processing conditioned leaves at low metabolic cost	Less dependent upon stage of conditioning, probably good at extracting protein, but at high metabolic cost
Other attributes of feeding ecology	Highly mobile	Low mobility
	Polyphagous	Obligate detritivore

corresponds to the period of greatest microbial growth (Arsuffi and Suberkropp 1984; Suberkropp and Arsuffi 1984). The benefits to the consumer include greater efficiency in converting ingested leaf biomass into consumer biomass and a higher individual growth rate.

Preference trials that compared shredders from tropical and temperate locations provided with conditioned (leaves submerged in the stream for two weeks) and unconditioned leaves from a temperate and a tropical tree found that all shredders preferred conditioned over unconditioned leaves regardless of the region of origin of either the shredders or the leaves (Fig. 9.2). In addition, all grew faster when provided with the conditioned leaves (Graça 2001). Considering that the leaves of the temperate tree (alder, *Alnus glutinosa*) have been shown to be a preferred food in a number of studies, whereas the tropical tree (*Hura crepitans*, Euphorbiaceae) produces a milky juice used by Amerindians to make poison darts, a general preference for alder leaves is no surprise. More surprising is the observation that *Gammarus* showed no preference among conditioned leaves, and shredders grown on *Alnus* and *Hura* did not differ in survival and growth rates, indicating that leaf conditioning was more important than leaf type.

Green and senescent leaves differ in their phenol, lignin, and nutrient content, and thus in their quality as food. Larvae of the caddis *Lepidostoma complicatum* grew more slowly on green than senescent leaves and none reached maturity, whereas 70% of larvae fed senescent leaves reached the adult stage (Kochi and Kagaya 2005). However, larvae that were given both senescent and green leaves had a faster growth rate than those provided with senescent leaves only, probably due to the higher nitrogen content of green leaves. The freshwater shrimp *Xiphocaris elongate* was found to prefer leaves of *Dacryodes excelsa* over *Cecropia scheberiana*, despite their higher secondary compound content and firmness, apparently because of the lower lignin content of *Dacryodes* leaves (Wright and Covich 2005).

Microorganisms may enhance the palatability and nutritional quality of leaves in at least two distinct ways (Barlöcher 1985). One, termed microbial production, refers to the addition of microbial tissue, substances, or excretions to the substrate. Because assimilation efficiencies on fungal mycelia and mixed microflora have been shown to exceed 60%, while values for conditioned and unconditioned leaves average near 20% (Martin and Kukor 1984; Barlöcher 1985), indications are that the nutrient content per unit mass in microorganisms can be several-fold greater than that of the leaf substrate. The second potential role for microorganisms is microbial catalysis, which encompasses all of the changes that render the leaf more digestible. This includes partial digestion of the substrate into sub-units that detritivores are capable of assimilating, and production of exoenzymes that remain active after ingestion. As support for this proposition, Barlöcher (1985) pointed out that structural carbohydrates (cellulose, hemicellulose, pectin) may be partially digested by microorganisms into intermediate products which the gut fluids of invertebrates are then able to degrade. Indeed, leaves subjected to partial hydrolysis with hot HCl were preferred by *Gammarus pseudolimnaeus* over untreated leaves (Barlöcher and Kendrick 1975). Barlöcher (1982) also showed that fungal exoenzymes extracted from decomposing leaves remained active in the presence of gut enzymes of *G. fossarum* for up to four hours at the foregut's pH, indicating that ingested exoenzymes can aid in the digestion of polysaccharides.

Some shredders may be able to actively discriminate between fungi and leaf material. In feeding trials with the freshwater detritivores *Gammarus. pulex* and *Asellus aquaticus*, Graca et al. (1993) found that both species discriminated between fungal mycelia, fungally colonized leaf material, and uncolonized leaf material. Individuals of *A. aquaticus* selectively consumed fungal mycelia whereas *G. pulex* fed preferentially on leaf material, and for the latter species fungi appeared to be more important as modifiers of leaf material. Using radio-labelled food sources and inhibitors of DNA synthesis, Findlay et al. (1984, 1986) demonstrated that only 15% of the respired carbon in the freshwater isopod *Lirceus* and 25% in the stonefly *Peltoperla* was met by consumption of microbes, primarily fungi. In addition, while insect larvae may lack the ability to synthesize cellulolytic enzymes, Sinsabaugh et al. (1985) demonstrated using radio-labelled cellulose substrate that leaf-shredding insects indeed were able to digest and assimilate plant cell wall polysaccharides. The authors inferred that digestion was aided by ingested exoenzymes in the case of *Pteronarcys*, and by endosymbionts in the distinctive rectal lobe of the hindgut of *Tipula*. Leaf-shredding crustaceans produce enzymes that enhance their ability to digest leaf litter of terrestrial origin. The amphipod *G. pulex* produces phenol oxidase and cellulase activity in the hepatopancreas, whereas in the isopod *A. aquaticus* these enzymes are produced by endosymbiotic bacteria (Zimmer and Bartholmé 2003).

Algae and bacteria of biofilms associated with leaf litter may contribute substantially to shredder nutrition. The exclusion of leaf litter from experimental stream reaches forced greater reliance on biofilms, and the shredders *Tallaperla* and *Tipula* derived on average 32 and 14% of their carbon from bacteria respectively, probably in the form of bacterial exopolymers (Hall and Meyer 1998). Shredders can also obtain carbon from algae growing on leaf biofilms, where the algae can increase the food quality of leaf biofilms and also stimulate microbial production by the release of exudates, and thereby enhance the growth of shredders (Franken et al. 2005). Algae attached to leaf litter may also influence the fatty acid composition of shredder diet, and thereby enhance shredder growth.

Fig. 9.2 Preference of tropical (*Nectopsyche argentata, Phylloicus priapulus*) and temperate (*Sericostoma vittatum, Gammarus pulex*) zone shredders for tropical (*Hura crepitans*) and temperate (*Alnus glutinosa*) conditioned and unconditioned leaves. Mean and one standard deviation are shown. (* = P < 0.05; **<0.01; ***<0.001) (Reproduced from Graca et al. 2001)

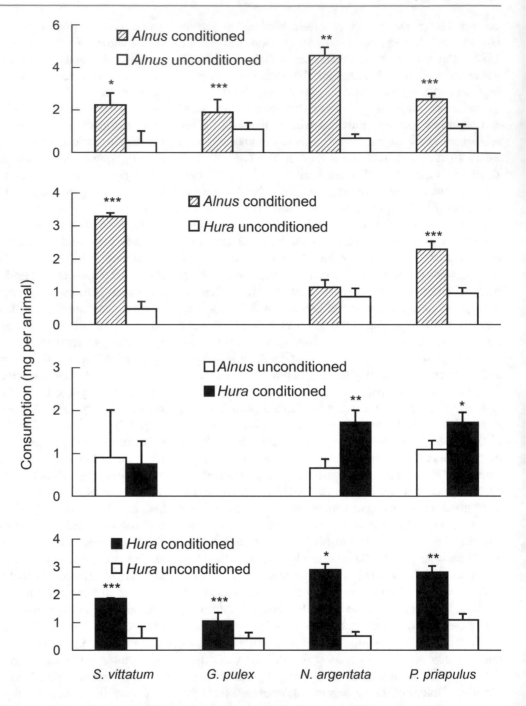

The importance of wood as a geomorphic agent in stream channels, altering flows and increasing habitat diversity, was discussed in Chap. 3. Wood can contribute 15–50% of total litter inputs in small, deciduous forest streams and even more in coniferous regions (Anderson and Sedell 1979). Wood is considered to be a minor energy resource because few invertebrates feed on it directly and wood appears to be a poor food. Although utilized only very slowly (a residence time of at least years to decades, in comparison to weeks to months for leaves), wood provides food and habitat for a number of species. Anderson et al. (1978) found some 40

taxa associated with this resource in wood-rich Oregon streams. Prominent aquatic xylophages included a midge (*Brilla*) which was an early colonizer of phloem on newly fallen branches, two species that gouged the microbially conditioned surface of waterlogged wood (the elmid, *Lara*, and the caddis, *Heteroplectron*), and a cranefly (*Lipsothrix*) that consumed nearly decomposed woody material. Invertebrate standing crop biomass on wood was about two orders of magnitude lower per kg of substrate than on leaf litter. The beetle *Lara avara* possesses robust mandibles capable of slicing away thin strips of wood, but apparently lacks digestive enzymes or gut symbionts to aid digestion. Microscopic inspection of material progressing through the gut indicated no change to the wood (Steedman and Anderson 1985); presumably the larva is nourished by microbiota and their exudates occurring on the wood surface. Not surprisingly, *L. avara* grows very slowly and requires 4–6 yrs to attain maturity. Wood fibers represented a high fraction (63%) of the gut contents of the caddis *Pycnocentria funerea* in streams draining a pine forest in New Zealand, and stable isotope analysis also indicated that most of its nourishment was derived from pine wood (Collier and Halliday 2000).

9.1.2 Consumers of FPOM

The collector:FPOM linkage (Fig. 9.3) depends on fine particulate organic matter captured from suspension or from substrates. Morphological and behavioral specializations for suspension feeding including setae, mouthbrushes, and fans are diverse and well studied (Wallace and Merritt 1980), whereas the mechanisms of deposit feeding appear to be less elaborate (Wotton 1994). FPOM originates in a number of ways. Categories considered to be richest in quality include sloughed periphyton and biofilm, and particles produced in the breakdown of CPOM. Collector-gatherers almost always are a major component of stream food webs and often are reported to be the dominant group present. In ten Hong Kong streams, FPOM was the major dietary component of macroinvertebrates in both shaded and unshaded streams, contributing over 50% of the diets of all primary consumers with the exception of obligate shredders and some scrapers (Li and Dudgeon 2008). Stable isotope signatures of FPOM was intermediate between CPOM and algae or cyanobacteria, indicating its mixed origin.

Caddis larvae in the superfamily Hydropsychoidea (comprised of the Philopotamidae, Psychomyiidae, Polycentropodidae, and Hydropsychidae) spin silken capture nets in a variety of elegant and intricate designs. Most net spinning caddis are passive filter feeders, constructing nets in exposed locations, but some nets act as snares (*Plectrocnemia*) or as depositional traps where undulations by the larvae create current (*Phylocentropus*) (Wallace and Malas 1976). Filter-feeding hydropsychids vary considerably in mesh size and microhabitat placement of their nets. There is evidence that catch nets of larger mesh tend to be found at higher velocities and capture larger prey, whereas fine mesh nets occur in microhabitats of low velocity and retain smaller particles (Wiggins and Mackay 1978). Members of the Arctopsychinae spin coarse nets, capture a good deal of animal prey and larger detritus, and tend to occur in headwaters. The Macronematinae occur in larger rivers, spin fine nets, and capture small particles. The Hydropsychinae are intermediate in net mesh size, more widely distributed, and perhaps because of the broad range of resources utilized, also are richer in genera.

Edler and Georgian (2004) examined the efficiency of particle capture in *Ceratopsyche morosa* (net mesh size

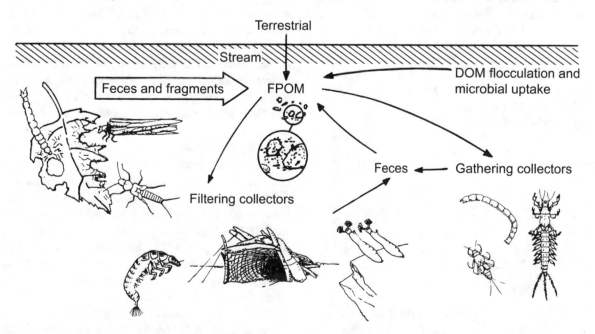

Fig. 9.3 The collector:FPOM linkages for a small stream within a temperate deciduous forest. Sources of detrital particles <1 mm include CPOM fragments, terrestrial inputs, animal feces, and sloughed algal cells and biofilm material. FPOM and associated microorganisms are ingested from the water column by filter feeders and from the streambed by collector-gatherers (Reproduced from Cummins and Klug 1979)

160 × 229 µm) and *C. sparna* (150 × 207 µm) by releasing food items of different sizes including *Artemia* nauplii (mean length 528 mm), and pollen of corn (*Zea mays*, mean diameter 84 mm) and paper mulberry (*Broussonetia papyrifera*, 12.5 mm). Both caddis species ingested more of the largest particles despite the greater availability of smaller particles in suspension (Fig. 9.4), but particles smaller than mesh openings are retained as well. Selective capture of larger particles might be expected to be energetically rewarding, and this is supported by the finding that *H. siltalai* nets retained a larger range of particles size (1–40 mm) than those observed in the water column (1–25 mm) (Brown et al. 2005). Because some captured particles were smaller than the mesh size of *H. siltalai's* net, adherence of particles to the silk apparently has some role in overall particle retention.

The impressive nets of caddis larvae are but one of the many specialized adaptations for capturing particles from suspension that have arisen repeatedly among aquatic invertebrates (Wallace and Merritt 1980). Larval black flies (Diptera: Simuliidae) are highly specialized suspension feeders (Fig. 9.5). Larvae attach to the substrate in rapid, often shallow water and extend their paired cephalic fans into the current (Currie and Craig 1988). Particles apparently are snared by sticky material on the primary fans, which are the main suspension-feeding organs, while secondary and medial fans act to slow and deflect the passage of particles. Food items are removed by the combing action of mandibular brushes and labral bristles, further adaptations to a filtering existence and lacking in some blackfly species that scrape substrates instead. Fans are opened when feeding and closed at other times (Crosskey 1990). The four species

studied by Chance (1970) ingested particles from <1 µm to >350 µm. Field studies generally report the majority of ingested particles to be <10 µm in diameter (Merritt et al. 1982). Visualization of the fields of flow surrounding individual simuliid larvae indicates that they position their fans for maximum filtering effectiveness, and may be able to manipulate flow vortices to enhance feeding (Chance and Craig 1986; Lacoursiere and Craig 1993). Palmer and Craig (2000) suggest that black fly larvae occurring in fast-flowing, particle-rich water will tend to have strong fans with a porous ray structure, whereas larvae found in slow-flowing, particle-poor water will tend to have weak fans with a complex structure.

Despite the evident elegance of the adaptations of larval simuliids for suspension feeding, this is by no means the only feeding mode employed. Currie and Craig (1988) state that scraping the substrate using mandibles and labrum is the second most important method of larval feeding, not including species that lack cephalic fans and are obligate scrapers. In addition, black fly larvae occasionally ingest animal prey, and Ciborowski et al. (1997) demonstrated that black fly larva grow when supplied only with dissolved organic matter. This diversity is a useful reminder that even those taxa displaying great specialization for a particular trophic role also may be capable of great versatility.

Larval black flies are important not only for their ability to filter very fine particles, but also for their production of fecal pellets. In northern rivers and particularly at lake outlets where very dense black fly aggregation occur, fecal pellet loads of several tons of carbon per day have been reported (Wotton and Malmqvist 2001). These pellets are available to filter feeders when in suspension, and to deposit feeders after they have sedimented. When Wotton et al. (1998) induced black fly larvae to produce labeled fecal pellets by adding paint to a lake-outlet stream, the guts of midge larvae, oligochaetes, and black fly larvae contained abundant label, and lesser amounts were found in baetid mayfly nymphs and the isopod *Asellus*.

Fecal pellets likely are an under-appreciated source of FPOM. Feces usually contain undigested food items and often are bound into discrete pellets, although some are diffuse (Wotton and Malmqvist 2001). Pellet size varies with the size of the animal that produced them, and can be as small as 6 × 9 µm in protozoans. Although most organisms produce fecal pellets that are smaller than the food they consume, some suspension feeders such as larval black flies can ingest very small food items and so produce fecal pellets larger than the food they ingest.

Other dipteran families with representatives adapted to a suspension-feeding existence in running waters include the Culicidae, Dixidae, and Chironomidae (Wallace and Merritt 1980). Some Chironominae construct tubes or burrows with catchments and create current by body undulations; others

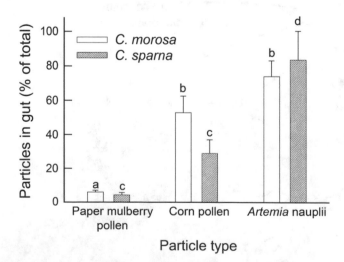

Fig. 9.4 Particles found in the guts of 5th-instar larvae of *Ceratopsyche morosa* and *C. sparna* as fractions of total. For each species, bars marked with the same letter are not significantly different (Reproduced from Edler and Georgian 2004)

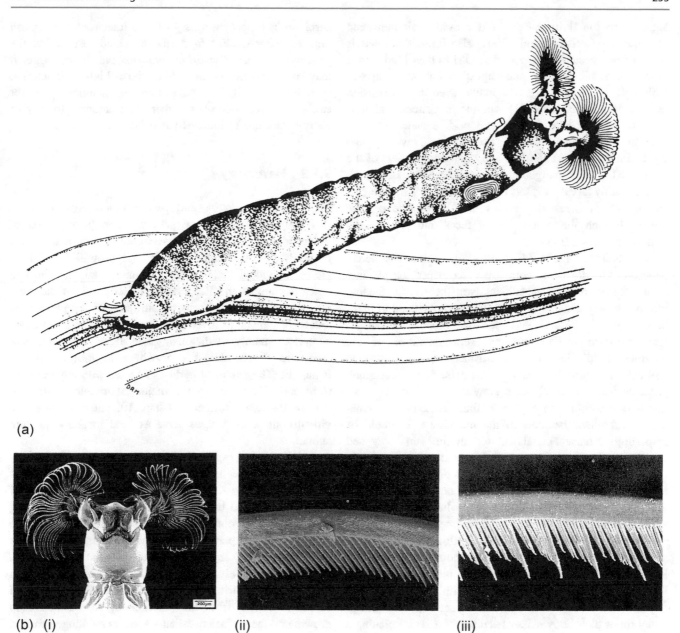

(a)

(b) (i)   (ii)   (iii)

Fig. 9.5 (**a**) The typical filtering stance of a black fly larva (*Simulium vittatum* complex). The larval body extends downstream at progressively greater deflection from vertical with increasing current velocity, and is rotated 90–180° longitudinally as can be seen by following the line of the ventral nerve cord. The position of the paired cephalic fans is upper and lower, rather than side by side. The boundary layer (depth where Ū falls below 90% of mainstream flow) begins at roughly the height of the upper fan (Chance and Craig 1986). (**b**) Details of cephalic fans: left: head of a normal larvae seen from beneath, with cephalic fans fully open; middle: *Simulium atlanticum* with uniform fringe of microtrichia; right: *S. manense* with long and short microtrichia (Reproduced from Crosskey (1990) and SEM photographs of D.A. Craig)

such as *Rheotanytarsus* passively suspension-feed by means of a sticky secretion supported by rib-like structures on the anterior end of the case.

Bivalved mollusks are effective filter feeders, capable of removing very small particles (10 μm and smaller) from their respiratory water current using sieve-like modified gills and mucus to filter and trap particles. Bivalves can remove large amounts of FPOM from the water column, including detritus, bacterioplankton, phytoplankton, and zooplankton (Strayer 1999). Roditi et al. (1996) reported that zebra mussels removed phytoplankton and nonfood particles at the same rate, but other studies suggest that mussels can be selective within the FPOM pool. Based on stable isotope analyses, Nichols and Garling (2000) determined that unionids, which are the dominant group of freshwater mollusks, used bacteria as their main carbon source, although algae were found in the gut and provided vitamins and phytosterols. Christian et al. (2004) also found that mussels were using a bacterial fraction of FPOM as their food source based on stable isotope and digestive enzyme analyses. Although bivalves are traditionally seen as suspension feeders, Raikow et al. (2001) reported that stream unionids obtained 80% of their food from deposited material versus 20% from suspended material. These unionids were probably assimilating the microbial and algal components of the suspended or benthic organic matter rather the bulk material. Like black fly larvae, mussels can consume very small algae, and transform ingested particles into larger size organic matter through the production of feces and pseudofeces (Atkinson et al. 2011).

Mechanisms of FPOM feeding by collector-gatherers either are less diverse in comparison to suspension feeding, or less is known about the subject. Nonetheless, this feeding role is well-represented in most stream ecosystems in numbers of both individuals and species. Among the macroinvertebrates in swifter streams, representatives of the mayflies, caddisflies, midges, crustaceans, and gastropod mollusks are prominent deposit feeders consuming small particles from the benthos. In slow currents and fine sediments one would also expect to find oligochaetes, nematodes, and other members of the meiofauna. It would be surprising if these animals all fed in the same way and consumed the same food. In addition to their particular food-gathering morphologies, these taxa differ in their ability to produce mucus, in mobility and body size, in their digestive capabilities, and in whether they are surface-dwellers or live within the sediments.

Browsing on easily assimilated biofilms may allow consumers to meet their energy needs without having to ingest large quantities of material. This is not the case for animals that ingest low-quality POM mixed with sediments. Many deposit feeders "bulk-feed", processing each day from one to many times their body mass of sediments and assimilating a low fraction of what they ingest. The burrowing mayfly *Hexagenia limbata* ingests more than 100% of its dry mass daily (Zimmerman and Wissing 1978). The assimilation efficiency of FPOM collectors in Sycamore Creek, Arizona, was estimated at 7–15%, and they consumed the equivalent of their body weight every 4–6 h (Fisher and Gray 1983). High quality foods that can be absorbed rapidly should favor high feeding rates and short retention times, whereas feeding should slow to allow longer digestion of poor quality foods. Calow (1975a) demonstrated an inverse relation between ingestion rate and absorption efficiency in two freshwater gastropods. When starved, snails slowed the rate of passage of food through the hepatopancreas, the main site of absorption and digestion. The effect of changing food quality on gut retention time apparently varies with the quality of the food, however. Calow (1975b) found that the

herbivorous limpet *Ancylus fluviatilis* increased its retention time for poor quality food (the expected result), but the detritivorous snail *Planorbis contortus* did the opposite. It may be that whenever the food carrier is highly recalcitrant, as in the case of lignin, it pays to process material rapidly for easily removed microbes rather than attempt to extract energy from nearly indigestible substrate.

9.1.3 Herbivory

The *grazer:periphyton* and *piercer:macrophyte* linkages (Fig. 9.6) are the principal pathways for the ingestion of living primary producers by invertebrates. The latter refers primarily to the microcaddisflies (Hydroptilidae), which pierce individual cells of algal filaments and imbibe cell fluids (Cummins and Klug 1979). Descriptions of the grazing pathway typically focus on attributes of the periphyton mat and the mode of invertebrate herbivory. The periphyton, comprised mainly of diatoms, green algae, and cyanobacteria, are found almost everywhere in running waters (Chap. 6). The extent of herbivory varies with algal growth form and differs among the major taxonomic groups for reasons discussed further in Chap. 10, but it appears that virtually all benthic algae serve as food for some grazing animal.

Morphological specialization of grazing invertebrates includes the blade–like mandibles of glossosomatid caddis larvae, the rasping radula of snails, chewing mouthparts of some mayflies and brush–like structures of others, piercing mandibles of hydroptilid caddis larvae, and so on. These are described as scrapers, grazers, and piercers, respectively. Other FFGs likely ingest plant matter occasionally. Collector–gatherers surely consume loose algae along with microbes and detritus (Lamberti and Moore 1984), and shredders benefit from the presence of an attached flora growing on the surface of fallen leaves, as mentioned earlier. Drifting diatoms and algae also are captured by suspension feeders, especially those taxa possessing fine sieving devices (philopotomid caddisflies, some chironomid and black fly larvae), and even the relatively coarse meshes of most hydropsychids retain some diatom and algal cells. Indeed, within the North American insect fauna, consumption of algae has been noted in at least six orders and 38 families (Merritt et al. 2019). Moreover, the composition of an herbivore's diet changes with many factors, including age, season, food availability, and location.

Just as animals differ in their mode of feeding, members of the periphyton differ in a number of ways that affect their overall vulnerability to particular herbivores. Benthic algae vary markedly in growth form and mode of attachment as well as in overall size (Fig. 6.1), and this must affect their availability to particular kinds of grazers. For example, field

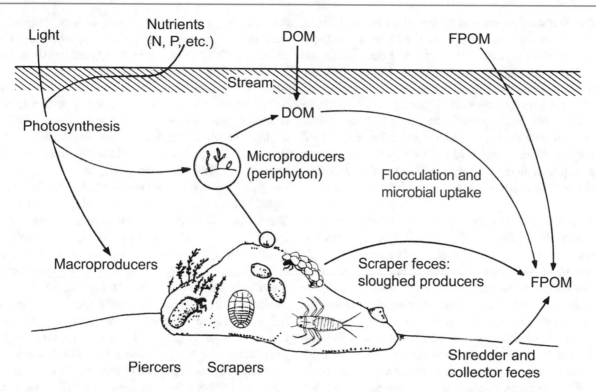

Fig. 9.6 The grazer:periphyton and piercer:macrophyte linkages for a temperate stream. The periphyton-biofilm organic layer on substrate surfaces is scraped or browsed depending on the consumer's mode of feeding. Diatoms and other algae are important constituent of this basal resource, but consumers also may ingest detritus, microorganisms, and occasional very small invertebrates. Piercers such as caddis larvae (Hydroptilidae) imbibe cell fluids through the cell walls of macroalgae (Reproduced from Cummins and Klug 1979)

manipulations of grazer densities in a California stream established that the mayfly *Ameletus* with collector-gatherer mouthparts was most effective with loosely attached diatoms. In contrast, the stout, heavily sclerotized mandibles of the caddis *Neophylax* were effective against tightly adherent diatoms (Hill and Knight 1988). Filamentous algae apparently are difficult for grazing insects to harvest or digest, and so they are consumed principally as new growths (Lamberti and Resh 1983). To the snail *Lymnaea*, however, possessing both a radula for their harvest and a gizzard for their mechanical breakdown, filamentous green algae provide a very satisfactory diet.

The assimilation efficiencies of herbivore-detritivores fed different diets are a useful measure of the wide range of nutritional value of various foods. Based on a review of 45 published values for 20 species of aquatic insects, assimilation efficiencies range from 70–95% on a diet of animal prey, 30–60% for a variety of algal and periphyton diets, and from 5–30% on a diet of detritus (Pandian and Peter 1986). Considerable variation can occur even for a single species feeding on periphyton. Assimilation efficiencies for the snail *Juga silicula* were as high as 70–80% when first added to laboratory streams, but values declined during the course of the study to as low as 40% (Lamberti et al. 1987). This coincided with a shift in composition of the periphyton from diatoms and unicellular green algae to filamentous green algae and cyanobacteria. The decline in assimilation efficiency could be the result of cell senescence and other changes in physiological condition, or a decline in nutritional value owing to successional changes in the periphyton assemblage.

The wide range of assimilation efficiencies observed with periphyton diets is at least partly due to their structural and biochemical characteristics. Variation in protein and lipid content and in cell wall thickness likely is responsible for differences among autotrophs in their nutritional value and palatability. Too high a ratio of carbon to nitrogen makes for a poor diet, indicating a high cellulose and lignin content and a low protein content; in general C:N ratios should be less than 17:1 for animal utilization. On this basis, members of the periphyton appear to be generally suitable (C:N ranges from 4-8:1), whereas aquatic vascular macrophytes appear to be nutritionally less adequate (C:N from 13-69:1) (Gregory 1983). Variation in the nitrogen content of diets was an extremely effective predictor of assimilation efficiency for twenty taxa of aquatic insects reviewed by Pandian and Peter (1986).

Lipid content is another variable likely to influence the nutrition and development of herbivores. Most insects are unable to synthesize polyunsaturated fatty acids and sterols, indicating that the lipid content of their diets is important to food quality. Intense grazing by a snail and a larval caddisfly in laboratory streams altered the fatty acid composition of the periphyton, suggesting that grazing may have been responsive to this aspect of diet quality (Steinman et al. 1987). Cargill et al. (1985) showed that specific fatty acids were critical dietary components to a detritivorous caddis larva, *Clistoronia magnifica*. More generally, the higher algal polyunsaturated fatty-acid (PUFA) content of stream algae is a principal reason why algae are a higher quality food source than dead leaves or even microbes (Guo et al. 2016).

Cyanobacteria are considered to be a poor food supply for freshwater plankton feeders (Wetzel 2001) and possibly for periphyton grazers as well. Cyanobacteria may have a high protein content, but other attributes, including a polymucosaccharide sheath rendering cell walls resistant to digestion, perhaps toxins, and a filamentous growth form all detract from their value as food. However, the evidence from lotic grazers is mixed. For example, in laboratory feeding trials the mayfly *Tricorythodes minutus* ate and assimilated two cyanobacteria, *Anabaena* and *Lyngbya* (McCullough et al. 1979), whereas *Asellus* and *Gammarus* would not consume *Phormidium* (Moore 1975). The nutritional inadequacy of cyanobacteria for gammarids and potentially other benthic invertebrates appears to be at least partially due to a deficiency in certain lipids, as a cyanobacterial diet supplemented with certain lipids markedly improved gammarid growth and survival (Gergs et al. 2014).

Macrophytes of rivers and streams traditionally have been thought to enter food webs primarily as detritus, as their tough cell walls and high lignin content that provide structural support are barriers to herbivore consumption of living plants. More recent syntheses of a large number of experimental studies across all types of freshwater ecosystems suggest this view is incorrect, and indicate that herbivores can remove up to half of all macrophyte biomass as living tissue (Bakker et al. 2016; Wood et al. 2017). Submerged macrophytes require less support tissue than emergent forms, and experience higher rates of herbivory, as one would expect. Herbivorous taxa include some aquatic insects, crayfish, and snails among the invertebrates; some fishes including Asian carps, and a number of leaf, fruit, and seed eating fishes of the tropics; ducks and other waterfowl; and semi-aquatic mammals such as the muskrat of North America and capybara of South America (discussed further in Sect. 9.3). Macrophytes typically are most abundant in large rivers with associated floodplains, lakes, and backwaters, and these habitats are likely to also support the greatest abundance of their consumers.

Intriguingly, the most dramatic effects of invertebrate grazing on living aquatic macrophytes involve herbivores derived mainly from terrestrial insect lineages. These include chrysomelid and curculionid beetles, aquatic and semi-aquatic lepidopterans, and specialized dipterans (Newman 1991). At a site in the Ogeechee River, Georgia, infested with the waterlily leaf beetle *Pyrrhalta nymphacaeae* (Chrysomelidae), leaves of the waterlily *Nuphar luteum* lasted only 17 days compared to more than 6 weeks at another site where the beetle was absent (Wallace and O'Hop 1985). In Brazilian rivers, the native apple snail (*Pomacea canaliculata*) significantly reduced biomass accrual of *Hydrilla verticillate*, a submerged, rooted macrophyte native to Asia and Australia, but did not appear capable of fully suppressing its establishment (Calvo et al. 2019). Some free-floating macrophytes, including the water hyacinth *Eichhornia crassipes* and the giant salvinia, *Salvinia molesta*, can become so abundant that they present serious weed-control problems worldwide, particularly in the sub-tropics and tropics. The salvinia weevil (*Cyrtobagous salviniae*, Curculionidae), a natural enemy of giant salvinia in South America, has been introduced around the world as a biological control agent, and in most instances has reduced plant abundance to acceptable levels. Giant salvinia occurs in at least at least 12 states in the southern US, and where weevils have been released, plant coverage has been dramatically reduced (Tipping et al. 2008).

9.1.4 Predaceous Invertebrates

The *predator:prey linkage* (Fig. 9.7) is ubiquitous. All animals are prey at some stage of their life cycle, and predaceous invertebrates occur in all sizes, from protozoans that engulf other protozoans, to insects and crustaceans capable of ingesting large invertebrates and small fish. Most predators engulf their prey entire or in pieces, but snipe flies (Diptera: Athericidae) and some hemipterans have piercing mouthparts (Cummins 1973). Other distinctions can be made between hunting by ambush versus searching (Peckarsky 1984), and whether prey are obtained from suspension, as in some hydropsychids, or strictly from the substratum, as in flatworms. Occasional predation probably is widespread, particularly the ingestion of micrometazoans, protozoans, and early life history stages of macroinvertebrates. Such unpremeditated carnivory may provide high quality protein needed by many invertebrates to complete their life cycles, and also may form an important link between microbial and macro-consumer food webs.

Mechanical detection is a widespread and varied modality for sensing prey. In many instances this means actual contact, for instance with antennae and setal fringes of limbs as in the stonefly *Dinocras cephalotes* (Sjostrom 1985). Vibrations in the water or of capture nets also serve as signals, as in the hemipteran *Notonecta* (Lang 1980), which captures prey on the water surface, and net-spinning caddis

Fig. 9.7 Two predaceous invertebrates common in streams of the Rocky Mountains, US. (**a**) The stonefly *Megarcys signata*, (**b**) the mayfly *Drunella* devouring *Baetis*, also a mayfly. Photos courtesy of Angus McIntosh

larvae that detect vibrations of prey in their nets (Tachet 1977). Visual cues likely are less important to invertebrate predators, because eyes are not well developed and many species dwell in crevices or are not active by day, but odonates, some heteropterans, and gyrinid beetles rely more on vision (Peckarsky 1984). Larval *Libellula depressa* (Odonata) were observed to strike at a mayfly nymph in response to either mechanical or visual cues, but mechanical cues were primary and did not require contact, and chemical cues apparently were ineffective (Rebora et al. 2004). Indeed, chemical detection of prey is important only in a few predaceous insects in the Hydrometridae and Dytiscidae, but it may be important in other invertebrates. Lake-dwelling triclads exhibit a chemosensory response to their isopod prey (Bellamy and Reynoldson 1974), and presumably stream-dwelling triclads do so also. The water mite *Unionicola crassipes* locates prey primarily by mechanoreception and vision, but it also becomes more sedentary in prey-conditioned water, suggesting that chemical detection promotes area-restricted search behaviors that presumably enhance encounter rates (Proctor and Pritchard 1990).

Sit-and-wait predators include those that simply remain motionless until the prey approaches within striking range, and those that trap their prey using nets (e.g., caddis larvae) or mucus trails (e.g., flatworms). Odonates that usually ambush also will stalk prey, a behavior that may be influenced by hunger level or their own risk of predation. Sjostrom (1985) reported that *D. cephalotes* searched in darkness, but was primarily a sit-and-wait predator in very low light. Risk from its own predators is the most likely explanation, although ability of prey to escape may be an additional factor. Predators often are undiscriminating in their diets, capturing whatever they encounter that is small enough to subdue. Aspects of the predator that bias it towards consuming more of some prey than others include sensory capabilities, foraging mode, and behavioral mechanism of prey capture. For prey, many aspects of body plan, life style, and behavior influence their vulnerability. These traits of predator and prey are not easily separated. From the

many studies of the diet of predaceous invertebrates, usually based on gut analyses and behavioral observations, body size, prey availability, and prey vulnerability are of particular importance in determining what is eaten.

Size relationships between predators and their prey as well as within a guild of invertebrate predators are of critical importance to food web relationships, a topic discussed in greater detail in Sect. 10.2. Typically, the average size of ingested prey increases with size of predator, as does the variety of prey items consumed. Predaceous stoneflies tend to ingest diatoms and other non-animal items when very small. Diet changes gradually over development, often consisting primarily of chironomids in early instars, and then broadening to include a menu in which mayflies, simuliids, and trichopterans supplement and may eventually replace midge larvae as prey. Although some differences are reported among species and study locales, presumably reflecting differing availability of prey, any two stoneflies of about the same size, when in similar habitats, consume diets of similar species composition. By measuring head widths of ingested prey and converting those values to dry mass, Allan (1982) showed a very similar positive relationship between prey size and predator size for several species of predaceous stoneflies and the two most common prey, *Baetis* and Chironomidae (Fig. 9.8). With an increase in the size of prey ingested there usually is an increase in diet breadth as well. Small predators tend to have less diverse diets because they don't reach sufficient size to capture prey larger and more agile than midge larvae.

Analysis of gut contents typically reveals a good correlation between what is eaten and what is available. The rank order of prey taxa in the diet of large *Hesperoperla pacifica* was similar to the rank order of prey in the benthos (Allan and Flecker 1988). There is some evidence that prey availability is such a decisive factor that it may override differences between predators in foraging mode. The net-spinning *P. conspersa* and the more mobile *S. fuliginosa* exhibited considerable overlap in habitat use and diet, although the former consumed more terrestrial items, large stoneflies, and

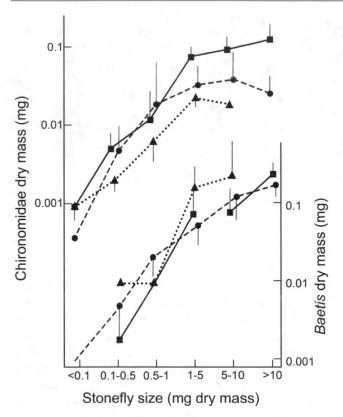

Fig. 9.8 Average dry mass of prey found in the foreguts of three species of predaceous stoneflies, as a function of size groupings of predators. Stoneflies of a particular size consumed prey of the same size for both prey species. Means and 95% confidence limits are shown for *Megarcys signata* (■), *Kogotus modestus* (▲), and *Hesperoperla pacifica* (●) (Reproduced from Allan 1982)

small chironomids, which apparently were more easily trapped in the net of *P. conspersa* (Strategies et al. 1979).

9.1.5 Patterns in FFG Composition

A number of studies have examined the correspondence between FFG composition and basal resources, often within the framework of the River Continuum Concept (Fig. 1.1), which describes changes in basal resources with stream size and order. Terrestrial leaf litter should dominate in shaded headwaters, benthic algae should be most abundant in the wide but relatively shallow mid-order river sections, and FPOM derived from upstream and floodplain sources should be most important in deeper and more turbid lower river sections. The composition of FFG is expected to mirror these changes. In addition, FFGs should respond to differences in terrestrial vegetation: headwaters in open meadows should support more benthic algae and more scrapers, as should stream sections where forest harvest has opened the canopy.

A survey of invertebrate FFGs in streams ranging from first to seventh order in the Cascade Range of Oregon, US, found

reasonable correspondence with the river continuum model (Hawkins and Sedell 1981). Shredders dominated upper, shaded reaches, scrapers were most important in intermediate-sized sections, collectors increased in importance downstream, and predators were nearly constant in relative abundance at all sections. In addition, shredder numerical abundance was significantly correlated with CPOM biomass, scraper abundance correlated with chlorophyll *a* on cobbles, and the abundance of invertebrate predators correlated with the abundance of invertebrate prey. Sampling eleven sites along the river continuum of the Little Tennessee River, North Carolina, US, from a first-order headwater stream to a seventh-order site, classifying taxa into FFGs, and estimating proportional representation using biomass, Grubaugh et al. (1996) found FFG abundance was strongly related to habitat diversity. Collector-filterers generally dominated in cobble and bedrock areas, collector-gatherers in pebble-gravel, and shredders and collector-gatherers in depositional habitats. After weighting FFG biomass according to relative habitat availability among sites, benthic community composition was consistent with predictions of the river continuum concept (Fig. 9.9). In a subsequent study conducted at 5th through 7th-order sites in the same system, employing detailed gut content analyses, leaf detritus consumption decreased with increasing stream order, and consumption of amorphous fine particles increased (Rosi-Marshall and Wallace 2002), again broadly consistent with expectations outlined by the river continuum model. Similarly, FFG distribution showed partial agreement with expected longitudinal changes from headwaters to mid-order in a Puerto Rican stream (Greathouse and Pringle 2006). Shredders decreased, scrapers increased, and predators remained unchanged, as expected. The downstream decrease in filterers normally would be unexpected, but in this system may be explained by the high abundance of filtering atyid shrimp in the headwaters.

In contrast, a number of studies have found poor or no correspondence between FFGs and expectations of trophic diversity based on riparian forest cover. In a comparison of open and closed canopy streams in the western Cascade Mountains of Oregon, US, Hawkins et al. (1982) found that open streams had higher abundances of invertebrates, including of collector-gatherer, filter feeder, herbivore shredder and piercer, and predator FFGs. Moreover, neither shredders nor scrapers exhibited a marked difference in density among canopy types. The distribution of feeding groups along the LaTrobe River, Victoria, Australia, showed some similarity with predictions of the river continuum concept, although habitat played a role (scrapers were most abundant at cobble sites, and filterers avoided sand), and shredders did not decrease downstream as expected (Marchant et al. 1985).

There is much evidence that the presence and importance of shredders vary regionally, being generally scarcer in

Fig. 9.9 Relationships between relative dominance of feeding groups and catchment area in the Ball Creek—Coweeta Creek—Little Tennessee River continuum. Dominance is expressed as percentages of total habitat-weighted biomass at each sampling station (Reproduced from Grubaugh et al. 1996)

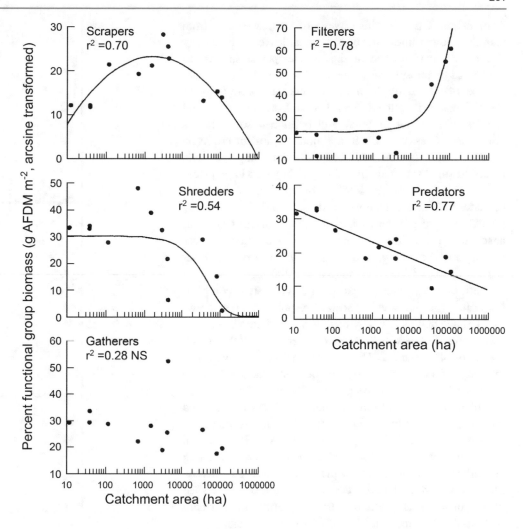

tropical systems. Important shredders of the Northern Hemisphere, including the caddisfly family Limnephilidae, certain plecopteran stoneflies, and certain crustaceans (amphipods and isopods), are weakly represented in the Neotropics. A global analysis of 129 stream sites from 14 regions on six continents did indeed show shredders to be more abundant and diverse in temperate than in tropical streams, with an inverse correlation with temperature, but no relationship with leaf toughness, a measure of leaf palatability (Boyero et al. 2011). The number of shredder taxa was found to be low in some tropical regions (Central America and the Caribbean) but not in others in South America, Asia, and Australia, indicating that shredder distribution is related to biogeography as well as to latitude.

Scarcity of shredding insects implies that microbes are the primary agents of leaf breakdown in those locations. Fresh leaves of several tree species underwent rapid decomposition in a tropical stream in in Mato Grosso, Brazil, but no shredder activity was observed, indicating that biological breakdown was primarily microbial (Wantzen and Wagner 2006). Where they occur, it may be that tropical shredders

must be flexible in their diet, due to a less predictable supply of leaf litter. A gut content analysis of *Phylloicus* larvae (Trichoptera: Calamoceratidae) in sites in southeastern Brazil with differing amounts of riparian vegetation demonstrated that this insect, which is commonly classified as a shredder, had greater amounts of CPOM in their guts at sites with riparian vegetation, but contents were dominated by FPOM at sites without extensive riparian habitat (Ferreira et al. 2015). It should also be noted that while insect shredders of terrestrial leaf litter often are rare in tropical streams, decapod shrimp and crabs, large omnivores able to tear apart tough leaves, can be abundant.

Several explanations have been proposed for the scarcity of shredders in the tropics. Chief among these are the ideas that shredders are from lineages that evolved in cool waters, and so are physiologically poorly adapted to the warmer waters of the tropics (Dobson et al. 2002); and that leaves in tropical streams are both scarce and less palatable, due to better defenses against herbivory (Wantzen et al. 2002), greater toughness, and reduced nutrient concentrations. Indeed, a comparison of leaves of three tropical tree species from French Guyana with leaves of four tree species from Germany found litter quality of tropical leaves to be clearly lower than that of three of the four temperate species, as evidenced by phosphorus content and leaf toughness (Bruder et al. 2014). Similarly, leaves from Venezuela were found to be of greater toughness than leaves from Portugal, and species of shredder preferentially consumed softer leaves (Graça and Cressa 2010). In addition, tropical macroinvertebrates tend to be smaller in body size than temperate zone taxa, and small size may limit ability to break up large pieces of CPOM. The presence of a diverse and specialized shredder FFG in north temperate streams likely is favored by predictable leaf fall during autumn, more synchronized leaf abscission, potentially less mechanical or chemical protection against herbivory, and, relative to the tropics, fewer species of trees (Wantzen and Wagner 2006).

Anthropogenic change can shift FFG relative composition, often (but not always) seen in response to altered land use. Urban streams typically have fewer individuals and less taxonomic diversity than rural counterparts (Walsh et al. 2005; Booth et al. 2016). Depending on the type and intensity of urban stressors, invertebrate functional group representation may also be affected. Taxon richness, diversity metrics, and pollution-intolerant EPT taxa all declined with increasing urbanization across 43 streams in southeastern Wisconsin, US (Stepenuck et al. 2002). Proportional representation of collectors and gatherers increased along the urbanization gradient, while proportions of filterers, scrapers, and shredders decreased with increased watershed imperviousness (Fig. 9.10). Urban streams in northern Maryland, US, experienced a 50% loss of predator taxa and up to 70% loss of collector taxa relative to streams in agricultural settings (Moore and Palmer 2005). Aquatic insect species were fewer and smaller at downstream urban compared with upstream rural sites in five streams in New York, US (Lundquist and Zhu 2018). Urban sites supported significantly less biomass of shredders and predators; collector-gatherers were the most abundant and diverse group overall, and were not markedly less diverse at the urban sites. Studies frequently find a significant increase in the proportion of collector–gatherers with urbanization (Stepenuck et al. 2002; Compin and Cereghino 2007; Sterling et al. 2016), and a decline in predator representation is also common (Smith and Lamp 2008; Sterling et al. 2016). Using percent imperviousness to compare urban, suburban, mixed-use, and rural watersheds in the upper Oconee River basin, Georgia, US, Sterling et al. (2016) observed a decline in biomass of predators, scrapers, and shredders with increasing impervious cover, while dominance by collector-gatherers increased. Oligochaetes and non-predatory chironomids comprised 60–90% of macroinvertebrate biomass at highly urbanized sites.

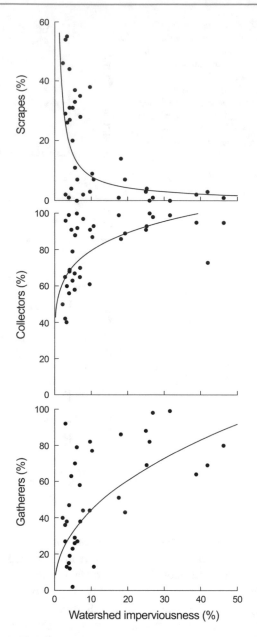

Fig. 9.10 The relative abundance of scrapers, collectors, and gatherers in relation to percent watershed imperviousness at riffle habitats in 43 streams from southeastern Wisconsin, US (Reproduced from Stepenuck et al. 2002)

9.1.6 Assimilation-Based Analyses of Feeding Roles

Traditional methods for investigating the feeding roles of invertebrate species include gut-content analysis, fecal analysis, and behavioral observations. Such information can provide important confirmation of FFG classification, as well as examples of flexible and opportunistic feeding that call into question the utility of FFG classification. Examples exist of algal consumption by predatory (Lancaster et al.

2005) and leaf-shredding (Dangles 2002) invertebrates. Within the scraper category, some *Glossosoma* caddis larvae were found to feed selectively on the algal components of biofilms while heptageniid mayflies consumed bulk biofilm (McNeely et al. 2006). Typically classified as a shredder feeding on leaf litter and associated microbes, the many species of *Gammarus* are actually able to exploit a far wider food base as a facultative herbivore and predator (MacNeil et al. 1997). Comparing the FFG assignment for fifty-six benthic macroinvertebrates where growth or isotope studies provided additional information about resource use, Mihuc (1997) found that half or more of gatherers, scrapers, and shredders consumed resources not indicated by FFG identity, suggesting that they should be classified as generalists. If flexible feeding is indeed widespread, then FFG classification may have limited success in identifying the relative importance of different basal resources, and the extent to which stream energy pathways are predominantly allochthonous or autochthonous.

Recent years have seen broader use of powerful new tools that provide fresh insight into the energy sources that consumers assimilate into their tissue (Post 2002; Finlay et al. 2010). As described earlier, stable isotopes of C, N, and H are becoming widely used as food web tracers in aquatic ecosystems, providing evidence on trophic position and the relative importance of autochthonous versus allochthonous basal resources. Fatty acid analysis can also reveal the importance of algae to a consumer's assimilation and growth, and even in apex predators can reveal the food web pathway of primary importance.

From assimilation-based analyses, a number of studies have found that algae are more important as a basal resource than might traditionally have been expected. In tropical Hong Kong streams, isotope and fatty acid analysis showed that algal sources contributed more than terrestrial sources to the biomass of a snail and two species of shrimp, despite the predominance of terrestrial detritus inputs (Lau et al. 2009). Fine particulates were a more important energy source than leaf litter in most comparisons, attributed to the former's lower C:N ratio and higher palatability. In the Eel River, California, where previous studies have found shredders to consume both terrestrial detritus and algae, shredders spanned a range of $\delta^{13}C$ and δD values from those consistent with consumption of terrestrial detritus to values enriched in ^{13}C and depleted in ^2H, indicating near-complete reliance on algae growing in pool habitats where algal $\delta^{13}C$ was highly enriched (Finlay et al. 2010).

Evidence that the contribution of algal resources to lotic food webs often has been underestimated, especially in shaded streams receiving abundant leaf fall, is one of the most consistent and intriguing outcomes of stable isotope analyses. Using δ^2H enrichment as a measure of consumer reliance on allochthonous resources, Collins et al. (2016)

found considerable flexibility in macroinvertebrate feeding across a range of stream conditions in New York, US, and on the island of Trinidad. Their results were not greatly inconsistent with FFG expectations: scrapers showed the highest reliance on autochthonous energy, predators and shredders made the most use of allochthonous resources, and collector gatherers and filterers spanned a broad range of energy use. However, even shredders relied partly on autochthonous energy, depending on the resource base available for consumption, and grazers consumed mostly allochthonous material in some circumstances. Strong reliance on allochthony by predators presumably reflected the food resources of their prey, and for fishes likely indicated their reliance on terrestrial invertebrates that fell into the stream. Flexibility in feeding, as seen in degree of reliance on allochthonous energy, was significantly related to canopy cover for most taxonomic groups (Fig. 9.11). In the tropical streams of Trinidad, reliance on autochthonous energy sources was even more pronounced, and even more strongly related to a gradient in canopy cover.

Interestingly, the argument that autochthony may be under-estimated has been advocated both recently (Vadeboncoeur and Power 2017) and much earlier. In a counterpoint to the then-widely accepted paradigm that allochthonous energy primarily in the form of leaf fall was the main energy source for streams, Minshall (1978) argued that autotrophic production often is the major or sole source of fixed carbon supplied to stream ecosystems, and can be important in streams considered to be primarily heterotrophic. His compilation of then-available data for a number of streams indicated that heavily forested headwater streams did tend to derive most of their fixed carbon from outside the system, whereas high values for instream production were associated with grassland and desert regions, and the larger, more open forest streams of the deciduous and semi-arid Rocky Mountain areas. As more evidence from assimilation-based studies accumulates, it is apparent that benthic algae can be an important basal resource even in settings where their role has previously been thought to be modest.

In closing, it is worth noting that despite the limitations of FFG classification, this approach remains in wide use. FFG tables in Merritt et al. (2019) recognize the challenge of accurately classifying taxa into feeding groups by providing both primary and secondary designations, and mention of facultative feeding where known, thereby helping to distinguish obligate (specialist) from facultative (generalist) taxa. The ease with which an investigator can translate information on the relative abundance of taxa at a site into a depiction of likely energy pathways within the food web makes this approach extremely useful. Shifts in the proportional representation of FFGs along, for example, an urban gradient, can reveal how anthropogenic change likely is

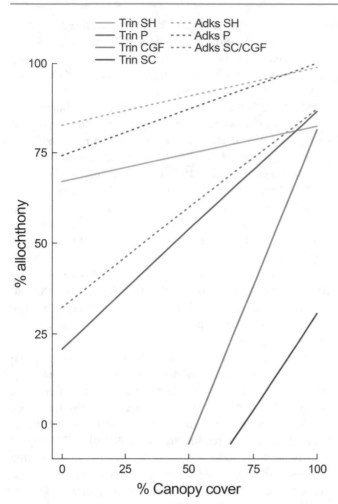

Fig. 9.11 The relationship between reliance on allochthonous energy inputs, determined from hydrogen isotope ratios of consumer organisms, and percent canopy cover differs between functional groups and between temperate and tropical sites. Tropical streams are more autochthonous than temperate streams, and tropical sites become autochthonous more rapidly with decreasing canopy cover. Functional group codes are as follows: SH = shredders, P = predators, CGF = collector gatherer/filterers, SC = scrapers. Solid lines represent data from tropical streams in Trinidad (Trin) and dashed lines represent data from temperate streams of the Adirondacks (Adks) in New York, US (Reproduced from Collins et al. 2016)

affecting basal resources. However, new tools based on energy assimilation make clear that flexible feeding habits, intermixing of energy sources of different origin, and differential assimilation can be important complications, not revealed using traditional methods solely relying on FFG classification and gut content analysis. Stable isotopes, fatty acid analyses, biomarkers and other emerging techniques will increasingly be the approaches most useful in revealing energy pathways in lotic ecosystems.

9.2   Trophic Roles of Lotic Fishes

Viewed across an assemblage of species, stream fishes consume virtually every resource available, and many individual species consume a wide range of resources, often changing diet over ontogeny and across environmental setting (Matthews 1998; Gelwick and McIntyre 2017). This makes trophic categorization understandably difficult, and there does not appear to be universal agreement on any single trophic classification in the literature. Regardless of these challenges, in virtually any study of fishes at a locale where one wishes to understand food web relationships and what determines the composition and relative abundance of species present, trophic categorization provides an important perspective. For convenience, we will begin with a discussion of these categories, and later review the caveats. Trophic categories can be as detailed as needed if one's objective is to describe differences in food consumed and feeding mode within a particular assemblage of species. To provide a general framework, however, the following categories have broad applicability: herbivores, detritivores, planktivores, omnivores, benthic invertivores, midwater-surface feeders (largely on insects), and piscivores. Where appropriate, one can add specialized categories such as snail-eaters, scale and fin eaters, fruit-eaters, parasites such as lampreys, and so on.

9.2.1   Fish Trophic Categories

Most fish species of temperate streams feed on invertebrates, which frequently make up much or virtually all of their diet. Aquatic invertebrate prey can be captured from the benthos, as individuals suspended in the water column, referred to as "drift", and as terrestrial infall on the water surface or entrained in the current. Invertivores thus can be separated into benthic feeders, mid-water feeders, and surface feeders. Benthic invertivores consume mainly aquatic insects, although crustaceans, mollusks, worms, and other invertebrate taxa also are eaten. Invertebrates in the water column are mostly aquatic insects dislodged from the benthos or present due to the phenomenon of drift, discussed further in Chap. 10. Some terrestrial invertebrates may be included in the diet of mid-water fishes, and may be an even more substantial fraction of the diet of surface feeders. Some studies have found it impractical to distinguish by feeding position, and simply refer to generalized invertivores. Aquatic insects and other invertebrates also comprise a significant portion of the diet of tropical stream fishes across most habitats, and terrestrial invertebrates often are important in smaller headwater streams of the tropics (Wolff et al. 2013; Ramírez et al. 2015).

Herbivory is not common among the stream fishes of North America, where only about 55 of over 700 total fish species are primarily herbivorous. Most macrophyte-eating fish are native to tropical waters, or to rivers in Asia or Europe (Matthews 1998), and some, including the grass carp *Ctenopharyngodon idella*, have been introduced into other regions for weed control. Grass carp daily rations (in wet mass of macrophyte tissue) range from 50% to over 100% of their body mass per day, indicating that this feeding strategy is based on processing a high volume of material. Grass carp also are known to have a low metabolic rate and assimilation efficiency relative to other fishes, and to require animal protein for proper growth. *Campostoma*, a cyprinid known as the stoneroller due to male nest-building behavior, is an important herbivore in many regions of the eastern to central United States, scraping algae from rocks and logs with the cartilaginous ridge on its lower jaw. In a small stream in Oklahoma, US, where the presence or absence of *Campostoma* was determined by either the distribution of its predator or by experimental manipulation, dense standing crops of attached algae (predominantly *Spirogyra* and *Rhizoclonium*) accumulated in areas lacking *Campostoma* but were scarce where the herbivore was present (Power et al. 1985).

Herbivorous fish species often make up a larger proportion of the total fish assemblage in tropical compared to temperate streams, where they can have strong effects on plant and animal abundances. Plant matter was a significant percentage (2–25%) of the diet of 77% of the 17 fish species in a Costa Rican stream (Wootton and Oemke 1992). Fish consumed a native grass, *Panicum*, leaves of two riparian plants *Ficus insipida* (a fig) and *Monstera* (a large-leaf member of the Araceae, often seen as a house plant), and periphyton, demonstrated by comparing loss of plant matter in locations exposed to fish, versus cage exclusions. Armored catfish (Loricariidae: Siluriformes) are an important group of grazing fishes, with 92 genera and at least 680 species occupying freshwater habitats of tropical and subtropical Central and South America (Delariva and Agostinho 2001). Noted for their dorsoventrally flattened body form, bony plates covering their bodies, and sucker-like mouths, loracariids are common algivores and detritivores in Neotropical streams, and also popular aquaria fish. Various species feed by scraping attached algae and diatoms from the substrate or by vacuuming up organic detritus, often including associated microorganisms. Armored catfish species also can differ in habitat use and substrates grazed. In streams of Panama, the most common loricariid, *Ancistrus*, grazed periphyton from flat surfaces on wood, bedrock, and clay substrates in pools, whereas others specialized in grazing on substrates such as pebbles in riffles (Power 1983).

Herbivorous and detritivorous fish generally have large gut lengths relative to their body lengths (Delariva and Agostinho 2001; Ward-Campbell et al. 2005), facilitating prolonged digestive action by enzymes and microflora. Indeed, relative gut length is a good indicator of trophic position, distinguishing carnivores, with short guts, from herbivores and detritivores, with their elongated, coiled guts. A review of body length and gut length relationships in 71 fish species found that, for the same body length, species that include plant material in their diet, either exclusively (pure herbivores) or in significant proportions (omnivores with preference for plant material) had greater gut lengths than fishes that prey on other animals, including omnivores with preference for animal material, and carnivores (Karachle and Stergiou 2010).

As with herbivores, relatively few riverine fishes of the temperate zone are detritivores. In North America, some fishes that are often or predominantly detritivorous are abundant and widespread, include the river carpsucker *Carpiodes carpio* and white sucker *Catostomus commersonii* (Cypriniformes: Catostomidae), as well as the introduced European carp *Cyprinus carpio* (Cypriniformes: Cyprinidae). Special adaptations include a muscular stomach to grind food, an intestine with greatly increased absorptive surface due to elongation (up to 20 times body length) or elaborate mucosal folding, and protrusible jaws that allow the fish to suck in fine, flocculant detritus. Although the number of detritivorous fish species may be few, in some circumstances, including larger rivers and reservoirs, they can dominate assemblage biomass (Miranda et al. 2019). At least for catostomids, animal prey can be an important dietary component, and so whether they should be considered detritivores or omnivores may vary from study to study.

In the Neotropics, detritivorous fishes in the families Prochilodontidae and Curimatidae are important components of many South American river systems, comprising over 50% of community biomass in some regions (Bowen 1983). In Africa, detritus-feeding fishes are found in the Citharinidae, the Cyprinidae, and some of the Cichlidae. Unsurprisingly, detritivorous species are most abundant in habitats where detritus is a major resource, often the downstream depositional reaches of larger rivers, in floodplain lakes and backwaters, and during dry seasons as the availability of other food items decrease. Because of their high contribution to biomass, detritivorous fishes play an important role in food webs, linking carbon originating in detritus both to piscivorous fishes (Winemiller 2004) and human fishers (Bowen 1983).

Omnivory is common in riverine ecosystems, more so if all life stages of organisms are considered, and if occasional ingestion of a wider range of food items meets whatever threshold is deemed sufficient to be classified as an omnivore. For example, many invertivores occasionally consume larval fish, and many piscivores also consume invertebrate prey. Some primarily herbivorous fishes can be considered omnivores even though the majority of their diet is plant matter, because animal prey, detritus, and organic-rich sediments frequently are consumed as well; a similar statement could be made for many detritivorous fishes. The central stoneroller minnow *Campostoma anomalum*, which has been shown to strongly limit benthic algae (Power et al. 1985), derived the majority of its growth in a tallgrass prairie stream from consumption of algae (47%), followed by amorphous detritus (30%), animal matter (21%), and leaves (2%) (Evans-White et al. 2003). Owing to differential digestibility, an omnivore may derive more of its growth from the animal portion of its diet, even if it is a lower fraction by mass. For many species, omnivory is a manifestation of what is often referred to as flexible or generalist feeding, topics we shall discuss further below.

Planktivores are found primarily in larger rivers, backwaters, floodplain lakes, and reservoirs, where both phytoplankton and populations of mid-water animal prey such as small crustaceans and rotifers can develop. Zooplanktivores include filter feeders such as paddlefish (family Polyodontidae), a primitive fish of large rivers, and gizzard shad (*Dorosoma cepedianum*), a member of the herring family, that do not use vision; sight-feeding filter feeders such as the alewife (*Alosa pseudoharengus*); and sight-feeding particle feeders such as yellow perch (*Perca flavescens*). Many small-bodied fishes and younger stages of larger fishes are sight-feeding particle feeders. Prey selection is influenced by gill raker spacing in filter feeders, by prey visibility (positively influenced by prey size and negatively influenced by turbidity), and prey ability to escape buccal suction (Matthews 1998).

Two planktivorous species of Asian carp introduced into North America for aquaculture and control of algal blooms now are widespread throughout much of the Mississippi River system. Larvae of the silver carp (*Hypophthalmichthys molitrix*) feed principally on zooplankton, and adults feed primarily on phytoplankton and to a lesser extent also consume zooplankton and detritus. A highly specialized filter feeder, the silver carp exerts a strong current with its buccal cavity, has fused gill rakers capable of filtering particles as small as 4 μm, and an epibranchial organ that secretes mucus that assists in trapping small particles. The related bighead carp (*H. nobilis*) consumes larger particles than silver carp, including a greater proportion of zooplankton. In backwater lakes of the Illinois and Mississippi rivers, planktivorous fishes included three native species: bigmouth buffalo (*Ictiobus cyprinellus*), gizzard shad, and paddlefish (*Polyodon spathula*); and non-native bighead and silver carp (Sampson et al. 2009). The gizzard shad and both carps consumed mainly planktonic rotifers, crustacean zooplankton were the preferred prey of paddlefish, and bigmouth buffalo consumed both rotifers and crustacean zooplankton. Planktivores can be abundant in floodplain lakes of the tropics, where the larvae in the genus *Hypophthalmus* (Siluriformes:Pimilodidae) and *Plagioscion* (Perciformes:Sciaenidae) feed on zooplankton. Both are piscivorous as adults, and important food fish. In a limnetic region of the Paraná River, Brazil, larva of *H. oremaculatus* fed on small cladocerans and rotifers, and larger larvae of *P. squamosissimus* also consumed calanoid copepods (Da Silva and Bialetzki 2019). More surprisingly, some species of the genus of *Rhabdolichops* (Gymnotiformes, New World electric or knife fishes) are highly specialized planktivores occurring in deep, swift waters of the Orinoco River main channel, where they consume large numbers of very small planktonic Crustacea and insect larva (Lundberg et al. 1987). The terminal mouth, relatively large eye, and elongate and bony gill rakers of *R. zareti* are adaptations to planktivory that differentiate it from congeners.

Piscivores have a diet primarily or exclusively of other fish. For such a diet, the piscivore must have a size advantage over its potential prey (usually other young-of-the year fishes), either by being born earlier, being born at a larger size, or growing faster. Some species are piscivorous in their first year, some after age one, and some after more years of growth. Summarizing findings for 27 species of freshwater piscivores from Europe and North America. Mittelbach and Persson (1998) found that species that were born larger and had larger mouth gapes became piscivorous at younger ages and at smaller sizes. The size of prey eaten increased with predator size in all species, and prey sizes in the diets were remarkable similar for piscivores of similar body length despite morphological differences among piscivore species. Evidently, most of the variation in the sizes of prey consumed is due to differences in piscivore body size rather than among species. The number of predator species in any local assemblage varies, but systems with more prey fish species tend to have more piscivorous fish species, often by a factor of three to four (Matthews 1998).

The brown trout (*Salmo trutta*) is a good example of a facultative piscivore that transitions to a primarily piscivorous feeding mode at approximately 30 cm in length (Jensen et al. 2012), although it may consume fish before reaching this size. A detailed energy budget for brown trout found that energy intake, growth, and the optimum temperature for growth all increased markedly when trout changed their diet from invertebrates to fish, indicating significant benefits of shifting to a more energy-rich diet (Elliott and Hurley 2000).

Finally, some fish species exhibit unusual and highly specialized adaptations for feeding, although these specialists may be restricted to certain environments. The redear sunfish *Lepomis microlophus* (Centrarchidae), also called the shellcracker, has thick pharyngeal teeth that allow it to crush snails, its preferred food (Keast 1978). At least 200 species of frugivorous fishes are known from tropical South America, consuming fruit from flooded forest habitats during the high water season (Goulding 1980; Correa et al. 2007). Frugivores contribute a substantial fraction of the fishery harvest, and not incidentally serve as important seed dispersal agents (Anderson et al. 2009). Fruit-eating species of the Characidae possess multicuspid, molariform teeth able to crush hard seeds, whereas catfish swallow fruits and seeds whole. Several South American species are lepidophagous, or scale-eaters, including species of piranha (Serrasalmidae), and *Probolodus*, *Roeboides*, and *Roeboexodon* species of the Characiformes. Some scale-eaters will also feed on fins of other fishes, and many omnivorous or predatory fish may on occasion nip the fins of other fishes. A few African species in the family Distichodontidae are specialized fin-eaters, or pterygophagous. The neotropical knifefish *Hypostomus oculeus* (Gymnotiformes) has spoon-shaped teeth adapted for feeding on wood. Parasitic species include some lampreys, jawless fish of the order Petromyzontiformes that attach and bore into the flesh of other fishes to suck their blood and body fluids; and the candiru (*Vandellia cirrhosa*) in the catfish family Trichomycteridae. Native to the Amazon Basin and known as the vampire fish, the candiru feeds on blood and is commonly found in the gill cavities of other fishes; very occasionally, it invades human orifices of unclothed bathers.

9.2.2 Patterns in Fish Trophic Composition

Given this list of trophic categories, what can be said about the relative number of species occupying each feeding role? As we shall see, some patterns are rather general, but the trophic composition of fish assemblages varies with longitudinal position from headwaters to lowland river, between the temperate zone and the tropics, and with local differences in resource availability and habitat. Common changes in the species composition of stream fish assemblages along a river's length include replacement of species, an increase in overall species richness, and a preponderance of larger species downstream (Horwitz 1978; Schlosser 1987). In both temperate and tropical rivers, the trophic composition of fish

assemblages also changes from headwaters to river mouth in accord with changing resource availability and habitat, approximately as posited by the river continuum concept. Goldstein and Meador (2004) classified 359 species of North American lotic freshwater fish into trophic categories based on mouth position, teeth, pharyngeal accessories, the ratio of gut length to body length, peritoneum color, and stomach morphology together with reported stomach contents. Cross-sorted by stream size, it is evident that invertivores comprise over half of the species list (54-75%) except in large rivers, per cent herbivores increases downstream, and detritivores are uncommon (Table 9.3). These results are broadly consistent with an earlier study of 15 river systems in the US (Horwitz 1978), in which over half of the species present were invertivores/insectivores, and the number of piscivorous species averaged about one-third of the number of invertivore species and less than one-fifth of the total. Planktivores were absent from headwaters, piscivores increased downstream, and fewer than 20% of the species subsisted on a diet of plant and detrital material. Similar findings were obtained from studies of trophic representation along several near-natural and regulated large rivers in Europe, the River Doubs in France and the Rivers Rhine and Meuse in the Netherlands (Aarts and Nienhuis 2003). The proportion of species feeding on benthic invertebrates and periphyton decreased towards the river mouth, while the proportion of zooplanktivorous and phytivorous species increased, and detritivorous species did not show a clear trend. The percentage of piscivorous species was fairly constant at around 15% in all zones. Comparing the freshwater fish assemblages of headwater streams from four continents (Europe, North America, Africa and South America), Ibañez et al. (2009) noted similarities in longitudinal patterns, including an increase in species richness, a decline in invertivores, and an increase in omnivores.

From these studies, it is evident that fish trophic composition in temperate streams indeed changes along a river's length. In addition, it is apparent that most species are insectivore/invertivores, omnivores often are the second most abundant group (but without a strict threshold for omnivory, this designation can vary among studies), piscivores are third, and only a few species occupy remaining

Table 9.3 Percent frequencies of fish species by trophic category and stream size for 359 North American fish species and life history stages. Some species have more than one trophic and stream size preference. Reproduced from Goldstein and Meador (2004)

Trophic category	Stream size category				
	Small streams	Small rivers	Medium rivers	Large rivers	Variable
Herbivore	9.6	9.3	11.7	18.8	6.6
Planktivore	11.5	3.1	8.5	21.9	7.7
Detritivore	5.8	4.3	9.6	3.4	5.5
Invertivore	67.3	75.2	54.3	40.6	56.4
Carnivore	5.8	8.1	16.0	15.6	24.2

trophic categories. As further evidence, a trait analysis constructed for 88 species of riverine fishes in Europe (Logez et al. 2013) showed herbivory, planktivory, and parasitic feeding each represented by two species, detritivory by five, and piscivory by nine, but 24 species were considered omnivorous and 44 were considered insectivorous species.

One might expect a wider range of trophic categories in tropical than in temperate river systems, owing to the much higher species richness found in tropical rivers (Albert et al. 2011). From studies of Atlantic Rain Forest streams of Brazil, Abilhoa et al. (2011) reported that the fish fauna included 269 species belonging to 89 genera and 21 families, in which characins (tetras and relatives), loricariids (armored catfishes and relatives), trichomycterids (candirus), rivulids (killifishes), and poeciliids (guppies) were strongly dominant. Species richness at a given locale can be much higher in tropical than in temperate streams. Combining data from more than 800 stream localities in the temperate zone, Matthews (1998) concluded that few reported 30 or more species at a locality, 20 or more species was not uncommon for eastern North America, and fewer than ten often was reported for the more depauperate streams of the western US. In contrast, Flecker (1992) collected 55 species of fishes in a 500-m study reach of Río Las Marias, an Andean piedmont stream of Venezuela. From seine collections in various habitats (sand, rock, wood; river vs lagoon) of the Cinaruco River, a species rich, backwater and floodplain river located in Venezuela's plains region, Arrington et al. (2005) estimate that as many as 50 to 80 fish species occur per habitat type, while over 280 fish species are known from the system.

Unquestionably one finds a wider range of trophic categories in tropical relative to temperate streams, including more species of herbivores and detritivores, and more unusual specialists. This is especially apparent in large tropical rivers with extensive lateral floodplains, where lake and backwater habitats connect to the river during high water, and where much fish production occurs in seasonally inundated habitats (Welcomme 1979). Plant production and detritus of both autochthonous and allochthonous origin are of great importance in these systems; consequently there is a greater role for mud and detritus feeding (which often supports the greatest biomass of fish), and for predation (which often dominates species richness). The extensive flooded forests of the Amazon, known as várzea forest, make available a wide variety of food items including seeds, nuts, fruits, flowers, leaves, monkey feces, numerous terrestrial invertebrates and the occasional vertebrate (Goulding 1980).

However, invertivores tend to dominate in tropical streams just as they do in their temperate counterparts, especially in upper reaches. Additionally, differences between lower-order, forested streams and larger,

higher-order rivers also have some patterns in common with temperate locations. In a study of forested tropical streams of the Bolivian Amazon, Ibañez et al. (2007) distinguished eight trophic guilds based on statistical clustering of gut contents. Eighteen of the 30 fish species consumed invertebrates, and were further sub-divided by aquatic, terrestrial, and generalist feeding habit. Diet specialization was observed at almost all trophic levels, except for the omnivore and piscivore feeding guilds, which the authors considered to be generalists. Similar results were obtained from a study of small forested streams of the Amazon basin, Bolivia (Pouilly et al. 2006). Diet analysis for 28 fish species identified seven detritivores, four algivores, two piscivores. and 15 invertivores (further divided into 6 generalist, three benthic, and six aquatic specialists) (Fig. 9.12). Invertivores dominated or co-dominated with detritivores at higher elevations, and the trophic composition was more diverse at lower elevation sites owing to an increase in the relative number of detritivore, algivore and piscivore species. A diet dendogram for 48 species from floodplain lakes in Bolivian Amazon provide an interesting comparison (Pouilly et al. 2003) (Fig. 9.13). More species of zooplanktivores and mud feeders were reported; piscivores also were more numerous, and sub-divided into carnivores to represent more exclusive consumption of other fish. Mud feeder, algivore, and piscivore species were considered to exhibit the most dietary and morphological specialization, relative to omnivores, invertivores, and zooplanktivores. Clearly, tropical rivers harbor more trophic diversity, the majority of fish species in most systems feed on invertebrates, and feeding roles that are uncommon in the temperate zone can be well represented in tropical river systems, especially in larger rivers with floodplain lakes and backwaters. Longitudinal changes in trophic representation in tropical river systems are thus broadly similar to what is reported from temperate rivers, with the important qualification that lowland rivers with their floodplain lakes usually have considerably more species of herbivores and detritivores, both uncommon in the temperate zone, as well as more species of piscivores. However, these patterns describe species richness, and can be quite different if expressed as biomass (e.g., Wolff et al. 2013). Owing to their large body size and the abundance of their resource base, detritivores and herbivores can make up a disproportionate share of the biomass at lowland river sites.

9.2.3 Feeding Mode and Morphology

The dominant mode of prey capture in teleost fishes is suction feeding, accomplished by expansion of the buccal cavity, causing water and prey items to flow into the predator's mouth (Lauder 1980; Liem 1980). Some species have evolved more complex skull linkages capable of greater

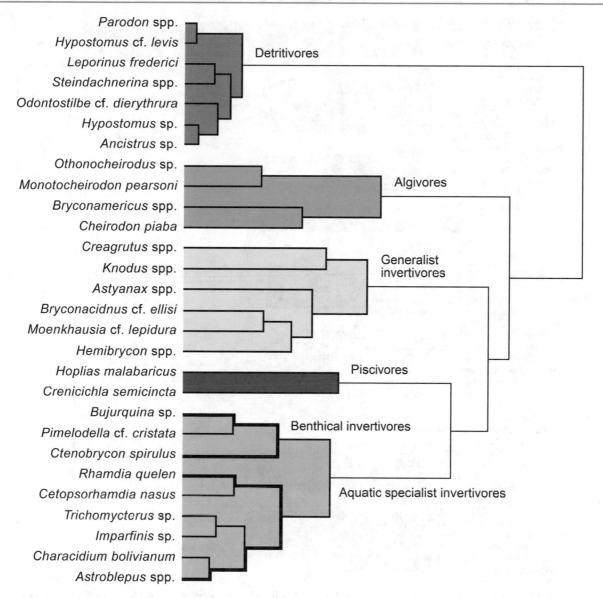

Fig. 9.12 Diet dendogram depicting trophic roles for 28 species of fishes from forested headwater streams in Bolivia (Reproduced from Pouilly et al. 2006)

suction force, and some combine suction feeding with forward body movement in a rapid strike during prey capture, referred to as ram suction. Some species grip prey in their jaws during capture, to manipulate, shred, or crush their prey. Wainwright and Richard (1995) showed that a functional analysis of lever distances and their ratios involved in opening versus closing the jaw could discriminate species that use ram-suction feeding versus biting or manipulation of prey. The jaw-lever systems for mouth opening and closing represent direct trade-offs for speed and force of jaw movement.

Various aspects of fish morphology have been found useful in understanding habitat and feeding preferences of fishes, as described in listings of fish traits (Frimpong and Angermeier 2010). Variation in body size, mouth gape and position, dentition, and gut length are frequently found to be strongly associated with feeding role, as well as visual and chemosensory adaptations. Traits associated with speed and maneuverability, including body shape and fin position, may influence both habitat use and prey capture. In a pioneering study of the relationship between fishes' ecological role and their morphological adaptations, Gatz (1979) examined 56 morphological features of 44 species seined from North Carolina piedmont streams, calculated 3,080 pairwise correlation coefficients among characters, and then used factor analysis to look for associations among characters in the correlation matrix. The first factor separated "lie-in-wait" biting predators from cruising suction feeders, the second

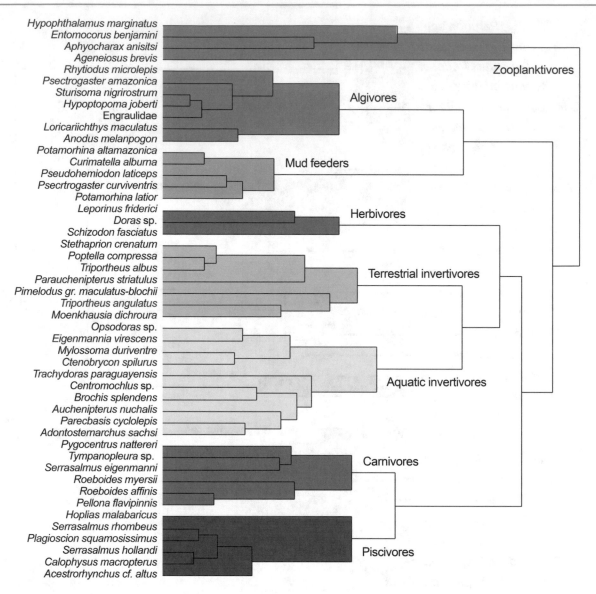

Fig. 9.13 Diet dendogram depicting trophic roles for 42 species of fishes from a low elevation floodplain site in Bolivia. (Reproduced from Pouilly et al. 2003)

reflected the differences in body shape and proportions associated with habitat use, the third factor separated a benthic from mid-water lifestyle, and the fourth separated small insectivores with short guts from other fishes. Fishes with flat, deep bodies were associated with slow water habitats. Fishes with ventral mouths obtained relatively more food from the bottom, those with terminal or anterior mouths did not. Fishes that dwell on or near the bottom in fast water regions had reduced swim bladder volume, and relative gut length was greatest in mud feeders, to list some principal findings.

Many subsequent investigations of what is known as the study of ecomorphology provide further evidence that fish trophic position often accords with their morphological adaptations. Among eleven species of characid fishes

collected from streams in Rio Grande do Sul, Brazil, those whose mouth positions were sub-terminal fed mainly on benthic items such as detritus, organic matter, and benthic aquatic insects (Bonato et al. 2017). *Bryconamericus* is an example, feeding in a head-down position, applying lips and teeth to the substrate to feed by rasping and suction. In contrast, species of *Astyanax* have the mouth in a more superior position, permitting them to forage from the surface and water column on floating terrestrial insects. The angle of their teeth is best suited for biting and tearing plants and ingesting terrestrial insects. Piscivores often are distinguished by their greater mouth gape, longer lower jaws and tooth size, an upturned mouth, and greater snout length. These features describe the characin genus *Oligosarcus*, which feeds primarily on other species of fish. Armored catfish are another example of highly specialized feeding morphology. A study of six species occurring in the Upper Paraná River, Brazil, revealed species adapted for both suction and scraping feeding modes (Delariva and Agostinho 2001) (Fig. 9.14). Species that feed on fine grained detritus obtain food by suction, and possess a well-developed respiratory membrane, long gill rakers, rudimentary labial and pharyngeal teeth, a thin stomach wall, and a long intestine. In comparison, species feeding on periphyton by scraping the substratum have large, strong, spatulate teeth, short gill rakers, a well-developed stomach, and a shorter intestine. Of the six species studied, one (*Rhinelepis aspera*) was clearly a suction feeder, two (*Hypostomus microstomus* and *Megalancistrus aculeatus*) were scrapers, and others (*Hypostomus regani*, *H. ternetzi*, *H. margaritifer*) were intermediate.

An important sub-theme in the study of ecomorphology concerns the extent of morphological convergence between unrelated species that occupy essentially the same trophic role. Unquestionably this occurs, as attested to by similarities in body form and jaw structure among unrelated

piscivores, in the greater gut length of herbivores and detritivores compared with invertivores and piscivores, and so on. A comparative study of 30 ecomorphological traits assessed for the dominant fish species from lowland stream and backwater sites in Alaska, temperate North America, Central America, South America, and Africa found numerous ecomorphological convergences and identified several cases of ecologically equivalent species, despite dominance by different orders of fish within the different biotic regions (Winemiller 1991) (Fig. 9.15). When an ecomorphological analysis is applied to a local fish assemblage, however, one frequently encounters multiple species within the same genus or family, with the consequence that when certain similarities are found in morphology and feeding habit, common ancestry may be the more robust explanation. In an early demonstration of the confounding influence of phylogenetic relatedness, Douglas and Matthews (1992) found that ecomorphological analysis of 17 species of fish from the Roanoke River and its tributaries simply confirmed that trophic ecology frequently conformed to family-level taxonomy. However, morphological variation within eight

Fig. 9.14 Ventral view of the position and form of mouth for six loricariids collected from the Upper Paraná River, southern Brazil. (**a**) *Rhinelepis aspera*; (**b**) *Hypostomus regani*; (**c**) *H. ternetzi*; (**d**) *H. margaritifer*; (**e**) *H. microstomus*; (**f**) *Megalancistrus aculeatus*. *R. aspera* feeds on fine grained detritus using suction to obtain food, and has rudimentary labial and pharyngeal teeth. *M. aculeatus* and *H. microstomus* feed on coarser material such as periphyton by scraping the substrate, and possess large, strong, spatulate teeth. The remaining species are intermediate (Reproduced from Delariva and Agostinho 2001)

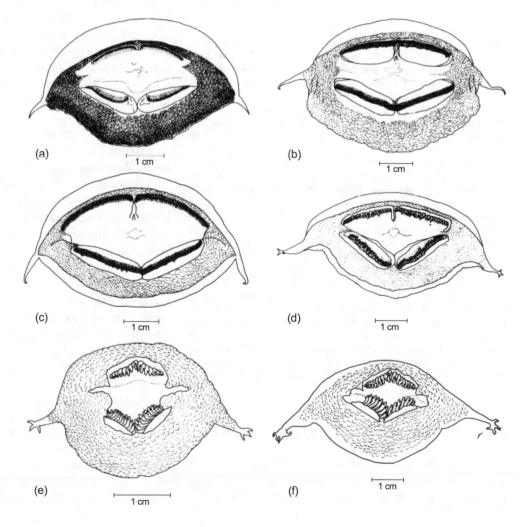

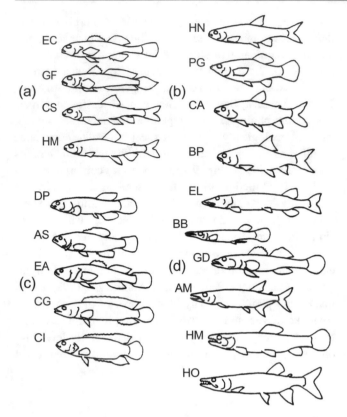

Fig. 9.15 Examples of ecomorphological convergences among fishes of five study regions. Note that fishes are not drawn to the same scale. (**a**) Small, benthic invertebrate-feeders: ED = *Etheostoma chlorosomum* (Perciformes, Percidae; North America), GF = *Gobionellus fasciatus* (Perciformes, Gobiidae; coastal Central America), CS = *Characidium* sp. (Characiformes, Characidiidae; South America), HM = *Hemigrammocharax multifaciatus* (Characiformes, Citharinidae; Africa). (**b**) Small, epibenthic algivores/detritivores with long, coiled guts: HN = *Hybognathus nuchalis* (Cypriniformes, Cyprinidae; North America), PG = *Poecilia gilli* (Cyprinodontiformes, Poeciliidae; Central America), CA = *Steindachnerina* (= *Curimata*) *argentea* (Characiformes, Curimatidae; South America), BP = *Barbus poechi* (Cyrpiniformes, Cyprinidae; Africa). (**c**) Small, cylindrical, vegetation-dwelling invertebrate-feeders: DP = *Dallia pectoralis* (Salmoniformes, Umbridae; Alaska), AS = *Asphredoderus sayanus* (Percopsiformes, Asphredoderidae; North America), EA = *Eleotris amblyopsis* (Perciformes, Eleotridae; Central America), CG = *Crenicichla geayi* (Perciformes, Cichlidae; South America), CI = *Ctenopoma intermedium* (Perciformes, Anabantidae; Africa). (**d**) Fusiform, sit-and-wait/stealth piscivores: EL = *Esox lucius* (Salmoniformes, Esocidae; Alaska), BB = *Belonesox belizanus* (Cyrpinodontiformes, Poeciiidae; Central America), GD = *Gobiomorus dormitor* (Perciformes, Eleotridae; Central America), AM = *Acestrorhynchus microlepis* (Characiformes, Characidae; South America), HM = *Hoplias malabaricus* (Characiformes, Erythrinidae; South America), HO = *Hepsetus odoe* (Characiformes, Hepsetidae; Africa) (Reproduced from Winemiller 1991)

species of minnow (Cyprinidae) did have some power in predicting microhabitat use.

Because similarities between diet and morphology within a fish assemblage can be simply the consequence of shared phylogeny, some authors have found ways to address this.

Using a statistic of taxonomic relatedness to factor out this influence, Ibañez et al. (2007) were able to show that relative intestinal length, standard length, and mouth orientation indeed discriminated among some trophic guilds within an assemblage of 30 fish species from forested tropical streams of the Bolivian Amazon. Fishes belonging to the algivorous and detritivorous guilds had large relative gut lengths. Benthic fishes from the algivorous and mud feeder guilds also exhibited relatively narrow heads and a ventral (Loricariidae) or oblique (Curimatidae) mouth orientation. Small size characterized most fishes of the aquatic invertivorous guild, whereas members of herbivorous and piscivorous guilds tended to be larger. However, although species having similar diet showed some similarity in morphological attributes, Ibañez et al. (2007) point out that this link was rather weak, discriminating among the three trophic guilds just mentioned but not others. A study of fishes inhabiting small streams of the Colombian Amazon also provided mixed evidence for a correspondence between morphology and diet (Ramírez et al. 2015). Several fish species feeding on aquatic invertebrates had multicuspid teeth, but so did a seed-eating species. Two species of knifefishes both had a long coiled intestine, but one was an omnivore, feeding mostly on fish and aquatic invertebrates, and the other an herbivore. However, tooth adaptations did relate to diet: the former has bi-cuspid teeth for invertebrate feeding, the latter has spoon-shaped teeth adapted for feeding on wood.

Although it is generally true that predators are larger than their prey, the largest species in an assemblage may not be predators. Because primary consumer fish species (algivores and detritivores) exhibited a wide size range in a diverse tropical food web in a savannah tributary of the Orinoco River, Venezuela, predatory fishes of all body sizes are able to exploit taxa low on the food web, resulting in relatively short, size-structured food chains for individual components of the overall web (Layman et al. 2005).

In addition to the morphological specializations just described, the sensory systems of fishes can be finely attuned to environmental constraints on food acquisition. Light level differs between day and night, with depth, and with the dissolved and suspended load, creating markedly different visual environments for seeking and capturing prey. River water varies in clarity, perhaps nowhere more evidently than in the Amazon basin. Whitewater streams are heavily colored by their alluvial loads, while blackwaters carry little silt but are darkly stained with dissolved material. Typical Secchi disk (water clarity) readings are less than 0.2 m in the former, 1–1.5 m in the latter (Muntz 1982). Clearwater rivers carry comparatively little silt or dissolved organics, and light penetration often equals or exceeds 4 m. Absorbance of short-wavelength light is relatively great in fresh water, and more so as light penetration is reduced. Levine and MacNichol (1979) examined 43 species of mostly tropical freshwater fishes, dividing them into four groups on the basis of visual pigments. Species with strongly "short-wave-shifted" visual pigments were primarily diurnal, and fed from the surface or in shallow waters. Several species exhibiting the typical behavioral and morphological characteristics of catfishes lay at the other extreme. Their visual pigments were the most long-wave sensitive; in addition they were primarily benthic and probably foraged either nocturnally or in very turbid waters.

Of several environmental variables considered to influence fish assemblage structure in floodplain lakes of the Orinoco River, Venezuela, transparency was a remarkably reliable predictor of species composition, including the numerical density of piscivorous species (Rodríguez and Lewis 1997). Fish with sensory adaptations to low light were dominant in turbid conditions (Secchi transparency <20 cm), whereas visually oriented fishes predominated in clear lakes (Secchi transparency >20 cm) but declined seasonally, concomitant with a decline in transparency. Species of characiforms, cichlids, and clupeomorphs that usually are diurnal and rely on vision were most abundant in clear floodplain lakes, including the peacock bass, *Cichla orinocensis*, and pike-like characiforms species *Acestrorhynchus microlepis*, *A. nasutus*, and *Boulengerella lucia* that lunge and engulf their prey. Catfishes, knifefishes, and a piscivorous sciaenid were most abundant in turbid lakes. Catfishes and knifefishes are primarily nocturnal, foraging efficiently in turbid waters using tactile and chemical sensors in catfishes, and electric sensors in knifefishes, while also gaining refuge from visual predators.

9.2.4 Challenges of Fish Trophic Categorization

As mentioned earlier, the use of fish trophic categories requires important caveats. Many species change their diet as they transition through life stages and grow in size. Such ontogenetic changes are especially well known in many piscivores, which may be invertivores as larvae, generalists or omnivores during the first one or more years of growth, and primarily piscivorous once a certain size threshold is reached. This challenge can be met at least in part by restricting trophic classification to adult individuals, or through the use of "ecological species", in which different life stages of the same taxonomic species are assigned to different trophic categories. A second issue involves habitat use, which can change, for example, with water level between wet and dry seasons, or when risk of predation forces an organism into less preferred locations with possibly different resource availability. Species that can occupy a wide range of habitats likely encounter a wider range of resources. Confronted with a variable and fluctuating resource base, the ability to feed opportunistically on different types of food is advantageous. While it is difficult to operationally define a generalist versus a specialist, one can intuitively grasp that for some species it may be advantageous to be able to eat a broad diet, and others may be more successful by being the best at eating just one thing. Even in the tropics, where some wonderful examples of ecological specialization can be found, many observers have opined that a generalist or opportunistic feeding strategy seems to characterize the majority of species (Lowe-McConnell 1987; Ibañez et al. 2007; Mortillaro et al. 2015). It is not uncommon for the same species to be found to feed quite differently in separate studies. In the Apure and Arauca rivers, two tributaries of the Orinoco River, the catfish *Pseudoplatystoma hemioliopterus* is primarily piscivorous, while in the Amazon basin, fruits and seeds have been found in their stomachs (Barbarino Duque and Winemiller 2003). Goulding (1980) reported that piranhas ingested mostly seeds and fruit during the flooded period.

Traditional studies of dietary analysis of gut contents are increasingly being supplemented by recently developed methods relying on analyses of stable isotopes and fatty acids. These hold considerable promise for identifying not only what species eat, but what food resources are important to assimilation and growth, both for individual species and the entire food web. Employing these methods to study food source utilization by nine fish species from two Amazon floodplains near the confluence of the Solimões and Negro rivers, Mortillaro et al. (2015) reported wide-ranging diets and feeding flexibility. Detritivores were positioned at the base of the food chain as expected, but fatty acid analysis pointed to inclusion of a high-quality food source, such as microalgae, in their diets. Both omnivores and insectivores exhibited opportunist feeding behavior, consuming a wide range of food resources. Piscivores had the most ^{15}N-enriched signature, consistent with their position at the top of the trophic chain. Only one herbivore, *Schizodon fasciatus*, consumed C4 macrophytes, which suggested some digestive specialization to cope with their low digestibility. While the diet of most species was broadly consistent with their trophic designation, results showed considerable dietary flexibility, differences in diet between the same species at different locations, and the ability of species to adapt their feeding behavior to changes in resource availability driven by hydrologic seasonality.

In summary, trophic categorization of fishes has its uses and its limitations. Different feeding roles can be identified within an assemblage of fishes based on differences in diet, feeding behavior, and morphology. What is also evident, however, is that across life stages, habitats, and seasons, many species are flexible in their diet, benefiting from their ability to exploit different food items based on changing availability of resources. Thus, studies often refer to the high frequency of dietary plasticity, or characterize many of the species present as omnivorous, generalists, or opportunists. It also is important to recognize that habitat, enemies, disturbance frequency, and many other environmental variables influence where a species occurs, and very likely the range of food resources. In practice, trophic categories are a useful way to summarize the broad similarities in the feeding ecology of taxa that have similar feeding roles, as long as their use doesn't obscure the individual differences and great flexibility of which fish are capable.

9.3 Other Vertebrates

Although fishes are the principal vertebrate component of most riverine food webs, all vertebrate classes have representatives in running waters. In small headwater streams, salamanders and snakes may be important top predators; there are many species of fish-eating birds, and some that consume aquatic invertebrates; a few mammals feed primarily or exclusively on aquatic prey, and a surprising diversity of mammals do so at least occasionally. Although many of these species consume animal prey, algal, higher plant, and detrital resources are consumed as well. Especially in larger rivers, the littoral zone and floodplain habitats provide feeding opportunities across all trophic levels for vertebrate consumers.

Larval amphibians in the temperate zone are found primarily in standing waters, but they can be diverse and abundant in tropical streams. The majority rasp algae and detritus from substrates and so are herbivores and detritivores, although unknown amounts of microbes and small organisms likely are ingested as well, suggesting a more omnivorous diet. Based on fatty acid analysis, anuran tadpoles sampled from ponds in Illinois, US, consumed larval insects and phytoplankton at one site and mainly periphyton along with sediments at another site, apparently reflecting predominant resource availability (Whiles et al. 2010). Larvae of the web-footed frog, *Rana palmipes*, are widely distributed in Neotropical streams, where they are epibenthic consumers of algae and sediments. Using cage enclosures with a range of densities, Flecker et al. (1999) observed rapid accumulation of benthic sediments when tadpole density was low, but rapid removal of sediments at higher tadpole densities. In addition, tadpole growth was strongly related to sediment supply. Herbivorous and detritivorous amphibian larvae likely obtain some nutrition from occasional ingestion of animal prey, as Schiesari et al. (2009) reported from stable isotope analysis of four ranid species from US wetlands. Amphibian declines have been reported throughout the world in recent decades, often attributed to *Batrachochytrium dendrobatidis*, an amphibian-specific aquatic fungus. Significant declines of larval amphibians in upland streams of the Neotropics have resulted in increased

biomass and production of both algae and invertebrate algal grazers (Colón-Gaud et al. 2009), strongly indicting that tadpoles at their pre-decline abundance reduced the amount of algal primary production available to other consumers.

Salamanders prey upon invertebrates, other amphibians, and fish, and may be the principal vertebrate predators in some headwater streams. Petranka (1984) concluded that the larval two-lined salamander *Eurycea bislineata* was an opportunistic generalist, consuming a variety of insect larvae and crustaceans. Salamanders can attain large size, including *Megalobatrachus* of the Orient, *Cryptobranchus* (the hellbender) and *Necturus* (the mud puppy) of eastern North America, and *Dicamptodon ensatus* of the Pacific Northwest of North America. Using suction feeding, adult hellbenders consume primarily crayfish and fish, but larvae consume mainly invertebrate prey such as mayflies and caddisflies, shifting to larger prey at later life stage (Hecht et al. 2017).

Reptiles that feed in rivers include many families of snakes but especially the Colubridae (water snakes), turtles, alligators, and crocodiles. Turtles are, for the most part, omnivores of sluggish streams and rivers, consuming substantial amounts of invertebrate and fish prey, but some are more specialized as herbivores or carnivores. Snakes and members of the Crocodilia are predators of fish and invertebrates in aquatic environments, but often consume other vertebrates as well. Size of prey relative to size of predator is a common constraint, and many predators increase the size and breadth of their diet as they grow.

Piscivory in snakes has evolved independently in multiple lineages, attracting study of the different behavioral and mechanical solutions to the problem of feeding in water. In a number of species, prey capture is accomplished by a sideways head-sweeping motion to minimize drag on the skull and/or to avoid pushing prey items away from the mouth. However, some aquatic species of garter snakes (*Thamnophis*) have been reported to use fast forward strikes to capture fish and amphibians. Filmed attacks by two species of aquatic piscivorous garter snakes established that they oriented visually toward prey items and struck forward rapidly with peak head velocities that approached speeds attained by fast striking species on land (Alfaro 2002). The dice snake *Natrix tessellate* (Colubridae), widely distributed in Europe and Asia, also uses frontal strikes to capture prey underwater. Aquatic species in the natricine sub-family include primarily piscivorous species with narrow, streamlined heads, and species that prey mainly on frogs and have broader heads, evidently reflecting the antagonistic design requirements of fast underwater striking versus the consumption of bulky prey (Brecko et al. 2011).

Diet studies indicate that aquatic snakes are generalist predators, consuming a wide range of fish, amphibian, and other prey, with some specializing more on fish and others on amphibians. An aquatic population of the Oregon garter snake *T. atratus* fed on small prey along the stream margin as juveniles, but as adults they consumed a wider variety of prey types and sizes, especially concentrating on larvae of the Pacific giant salamander in mid-stream substrates (Lind and Welsh 1994). By palpating the abdomen of *N. tessellata* captured from streams of the Tolfa Mountains, Italy, to induce regurgitation of ingested food, Luiselli et al. (2007) observed that fishes accounted for over 90% of the diet, and anurans the remainder. Larger individuals had a broader diet, including prey species that were never consumed by juveniles. Prey were primarily diurnal, as is this sit-and-wait predator, and the most abundant prey species were also the most frequently consumed. A review of the feeding habits of this well-studied snake reported a total of 113 prey taxa, mostly fish, but invertebrates, amphibians, reptiles and mammals were also eaten (Weiperth et al. 2014). Environmental setting plays a role, as non-fish prey were especially important in deserts, high mountains, and in dry Mediterranean areas. Reports for other species of aquatic snakes also suggest generalist and opportunistic feeding. From stomach analysis of museum specimens of three species of Amazonian water snakes in the genus *Helicops* (Colubridae), de Carvalho et al. (2017) recorded 36 species of fishes and 11 species of anurans. All fishes found in snake guts occupy middle and upper layers of water column and are primarily nocturnal, as are these snakes.

Members of four families of turtles in the sub-order Cryptodira may be found in aquatic habitats, including species of the Emydidae (e.g., Blanding's and painted turtles), Kinosternidae (mud and musk turtles), Trionychidae (soft-shell turtles) and Chelydridae (snapping turtles). Most can be described as carnivorous or omnivorous, although some also consume plant matter. In addition, two families in the sub-order Pleurodira, the Podocnemidae and the Chelidae (side-necked turtles) occur in freshwater habitats in the Southern Hemisphere. The Podocnemidae are also primarily carnivorous. The Chelidae include snake-necked species that mainly are predators of fish, invertebrates, and gastropods; and short-necked forms that are largely herbivorous or molluscivorous, but include several species that primarily eat fruits.

North American species of Emydidae have been reported to have a substantial component of plant matter in their diet. The Ouachita map turtle, *Graptemys ouachitensis,* of the Mississippi River consumes mostly animal prey when small, but plant matter becomes more important with increase in size, including fruits, seeds, leaves, and grasses (Moll 1976). The painted turtle *Chrysemys picta* is considered an omnivorous generalist whose diet includes algae, vascular plants, aquatic invertebrates, insects, and vertebrates such as fish and frogs (Hofmeister et al. 2013). Two widely distributed South American turtles in the Podocnemidae, the yellow-spotted river turtle (*Podocnemis unifilis*) and the

South American river turtle (*P. expansa*) both are primarily herbivorous, although accidental consumption of small animals may contribute important nutrition as well (Lara et al. 2012). Large-bodied species are thought to become increasingly herbivorous as they grow, presumably due to reduced prey capture ability as size increases.

As mentioned above, snake-necked and short-necked species of the Chelidae exhibit markedly different morphological and dietary specializations. Using stomach lavage, Tucker et al. (2011) compared the diets of three chelid species from free-flowing and impounded rivers in southeastern Queensland, Australia. The white-throated snapping turtle *Elseya albagula* was primarily herbivorous, consuming fruits and plant matter; the saw-shelled turtle *Myuchelys latisternum* was a carnivore, consuming insects and crustaceans, but also some fruit and plant matter; and the Australian short-necked turtle *Emydura krefftii* was an omnivore, consuming algae, sponges, fruit and plant matter, as well as insects, snails, and crustaceans.

While some turtles clearly are primarily herbivorous and others, like the snapping turtle *Chelydra serpentina* (Chelydridae) of North America and its massive relative, the alligator snapping turtle *Macrochelys temminckii* of the southeastern US, are strongly carnivorous, omnivory is widespread in most freshwater turtle lineages. Even *C. serpentina* with its powerful jaws consumes plant as well as animal matter, and both scavenges and actively preys upon anything it can capture and swallow. However, the alligator snapping turtle is almost entirely carnivorous, often feeding by opening its mouth to reveal its tongue as a lure to an unwary fish.

American alligators (*Alligator mississippiensis*), the dominant apex predator across many aquatic ecosystems of the southeast US, are considered to be generalist predators. A compilation of data from a large number of studies of alligator stomach contents noted that prey included crustaceans, mollusks, fishes, amphibians, reptiles, mammals, birds, aquatic and terrestrial insects, and seeds (Rosenblatt et al. 2015). Analyzing the diet of over 200 individuals, Rosenblatt et al. established that diet contents varied with alligator size, capture location, and season. Although alligators are dietary generalists as a species, individual animals showed considerable specialization. Likely causes are habitat heterogeneity, with associated differences in prey composition, and the relative abundance and ease of capture of prey present.

At least 11 orders of birds make use of rivers and streams as feeding habitat (Hynes 1970). Many are fish predators but some feed directly on invertebrates (e.g. the Cinclidae or dippers, Ormerod 1985). The American dipper *Cinclus mexicanus*, found in western North America, forages mostly in fast-flowing streams where it feeds on aquatic insects, as well as small fish, fish eggs, and flying insects. Dippers plunge underwater to capture prey, returning to the surface to swallow their prey (Kingery and Willson 2019). Other species of birds are underwater hunters. The snake bird *Anhinga anhinga* stalks its prey underwater, spearing it with a rapid strike before returning to the surface; prey are then swallowed head first. The anhinga's diet includes many different species of small to medium-size fishes as well as invertebrates, snakes, and small turtles (Frederick and Siegel-Causey 2000). The common merganser *Mergus merganser* is an underwater pursuit predator with a slender and serrated bill for grasping prey. Its diet is primarily small fish, but also includes aquatic invertebrates, frogs, small mammals, birds, and plants (Pearce et al. 2015). Examples of aerial hunters include kingfishers, terns, and the osprey. The belted kingfisher *Megaceryle alcyon* preys on fishes near the surface to a maximum depth of about 60 cm, favoring clear water locations for best visibility, and capturing most prey at the surface without submerging (Kelly et al. 2009). Returning to its perch, the bird pounds its prey to stun it before swallowing head first. The least tern *Sternula antillarum* is found along major interior rivers of North America, feeding primarily on small fishes, with over 50 fish species listed as prey, and occasionally on invertebrates (Thompson et al. 2011). Terns fly or hover 1–10 m above water while searching for prey, then plunge-dive and grasp prey with open mandibles without fully submerging. The osprey *Pandion haliaetus* is the only North American raptor that consumes live fish as its main prey source, diving feet first to capture prey from within only about the top meter of water (Bierregaard et al. 2016). Prey typically are 10–30% but may be over 50% of the osprey's body mass, requiring powerful wing strokes for the partially submerged osprey to regain the air. The great blue heron *Ardea herodias* is one of the most prominent wading birds of North and Central America (Vennesland and Butler 2011). Individuals hunt by slowly wading or standing in wait in shallow water, but also will dive feet first after prey. Prey are caught by a rapid forward thrust of the neck and head, and most prey are swallowed whole. They have a broad diet, including fish, insects, mammals, amphibians, birds, and crustaceans. Aerial life stages of aquatic insects are prey for a number of insectivorous birds. The bank swallow *Riparia riparia* is an aerial feeder over aquatic and meadow habitats, consuming a wide variety of aquatic and terrestrial insects including mayflies, dipterans, and odonates (Garrison 1999). Diving ducks consume plant matter and a variety of invertebrates, and some such as the mergansers are specialized piscivores. Non-diving (dabbling) ducks primarily consume aquatic vegetation, but invertebrates such as snails attached to vegetation are also eaten.

While there is some evidence that avian piscivores do not have a major impact on fish populations except when fish are easily captured, such as during low water conditions (Draulans 1988), in some instances bird predation can significantly influence behavior or abundance of their prey. Steinmetz et al. (2003) altered the abundance of great blue herons and belted kingfishers along an Illinois prairie stream by suspending plastic bird netting along an exclusion reach and adding kingfisher perches along an augmentation reach. The mean sizes of two abundant prey, striped shiners and central stonerollers, decreased under normal and elevated predation but increased in the reduced predation reach, in accord with preferred prey sizes of the two predators, apparently due to a combination of direct mortality and prey emigration. Armored catfish in Panamanian streams experience significant predation risk from fishing birds (Power 1984), and this causes larger individuals to avoid shallow waters. Because these fish are effective herbivores, the depth distribution of periphyton inversely mirrors the distribution of fish.

A variety of mammals feed within river systems. Taxa ranging from shrews to racoons to bears occasionally or frequently consume invertebrates and fish, while muskrats, beaver, and the South American capybara consume aquatic and riparian vegetation. The European river otter *Lutra lutra* is primarily piscivorous in temperate European localities, whereas Mediterranean otters behave as more generalist predators, relying less on fish and more on aquatic invertebrates and reptiles (Clavero et al. 2003). Sedges and grasses are important dietary items for muskrats *Ondatra zibethicus* of North America and the capybara (*Hydrochoerus hydrochaeris*) of South America, two rodents capable of digesting this high-fiber diet. The capybara, the world's largest rodent, processes its high-fiber diet by a combination of hind-gut fermentation and re-ingesting its feces. Very large river-dwelling mammals include the plant-eating manatees of Central and South America and West Africa, and dolphins, which feed on invertebrates and fish. River dolphins are top predators and those from the Amazon have been found to eat at least 50 fish species from 19 families, including individuals up to 0.8 m in length (Best and da Silva 1984); in addition, they occasionally consume mollusks, crustaceans, and turtles. Fish consumption by dolphins is perceived as a conflict by at least some indigenous fishers, and there is evidence of intentional killings of the freshwater dolphins *Inia geoffrensis* and *Sotalia fluviatilis* in areas of the Western Brazilian Amazon (Loch et al. 2009).

Seasonal fluxes of anadromous fishes into rivers provide nourishment for a great many mammal species (Willson and Halupka 1995). Stable-isotope analysis of hair samples from 13 brown bear (*Ursus arctos*) populations located over a wide area of western North America identified populations ranging from largely vegetarian to largely carnivorous, and food resources ranged from mostly terrestrial to mostly salmon (Hilderbrand et al. 1999). The proportion of meat in the diet was strongly correlated with female body mass and litter size, and salmon was the most important source of meat for the most productive populations.

A few species of bats have become aquatic carnivores despite their ancestry as terrestrial invertivores. The greater bulldog bat *Noctilio leporinus* (Noctilionidae) of Central and South America is the bat species most specialized for piscivory. Using echolocation, *N. leporinus* captures small fish when they jump out of the water as well as by dragging its enlarged, clawed feet just below the surface in areas where surface disturbance has been detected (Schnitzler et al. 1994). The tail membrane between its legs further assists in transferring prey to the mouth. A few other bat species, including the fish-eating bat *Myotis vivesi* (Vespertilionidae) also consume fish. Piscivorous bats have distinct cranial shapes that enable high bite force at narrow gapes, necessary for processing fish prey (Santana and Cheung 2016). Aerial adults of aquatic insects can be important diet constituents for insectivorous bats. Although it consumed a broad diet of invertebrate species, the little brown bat *Myotis lucifugus* relied heavily on the mass-emerging mayfly genus *Caenis*, shown by molecular analysis of bat fecal pellets collected under roosts in southwestern Ontario, Canada (Clare et al. 2011).

Finally, it is worth noting that the hunting tactics of different vertebrate predators largely determine the size range of prey captured and the habitat, especially water depth range, where prey are encountered. Wading birds typically fish in water no deeper than 20–30 cm. Leg length and striking distance limit their success at greater depths. Diving and skimming predators such as kingfishers, osprey, and bats usually fish very close to the surface. Swimming predators typically feed at greater depth, either to minimize their own risk of predation or, especially if they are of large body size, to have more room to maneuver. The need to capture and swallow prey generally results in a rough correspondence between prey size and predator size, even in species able to extend their gapes or rend prey into pieces. The combination of a predator's depth range and size range may significantly affect the size and depth distribution of fishes in streams, and perhaps affect other members of the biota as well. Indeed, many vertebrate predators may have their impact on riverine communities by influencing the foraging location of their prey. As we shall see in subsequent chapters, the consequences can ramify widely through the food web.

9.4 Summary

Invertebrate and vertebrate animals in riverine food webs occupy a variety of trophic roles as consumers of algae, higher plants, non-living organic matter, and other animals. Their resources include those produced within the river and its floodplain, and external inputs derived mainly from the riparian zone. Microorganisms are important mediators of resource quality, especially the fungi that colonize leaf litter and biofilms of bacteria and other organisms that coat most wetted surfaces. Trophic organization in river ecosystems can be complex and indistinct, as many consumers are generalist feeders and often overlap broadly in their diets. Nonetheless, the classification of consumers into trophic categories provides useful insight into the variety of consumer roles and range of resources available. Initially, trophic categories were based primarily on the resource consumed, hence common categories included invertivore, herbivore, piscivore, and so on. Because so many consumers in riverine food webs are flexible feeders, it may seem that most are simply omnivores. However, trophic categorization can be made more robust by also specifying where the food is obtained (e.g., from the benthos or the water column) and morphological adaptations for capture and digestion.

Invertebrates typically are divided into functional feeding groups (FFGs) on the basis of resource category, where or how the resource is obtained, and morphological adaptations for food capture. Scrapers consume non-filamentous attached algae from substrates. Detrital shredders primarily feed on leaves and other products of terrestrial plants that fall into the stream and are colonized by microbes. Gathering collectors feed on fine particulate organic matter from depositional areas or crevices. Filtering collectors capture FPOM in suspension in streams using morphological structures or silk capture nets. Herbivore piercers are adapted to pierce individual filamentous algal cells and suck out the cell contents. Predators are adapted to catch and consume live prey by engulfing the prey or piercing and extracting the prey hemolymph. It should be noted that members of different invertebrate functional groups may consume the same resource: for example, fine particulate organic matter can be captured from the water column or collected from depositional locales. The main difference is not the resource, but the organism's method of acquiring it.

The FFG system assumes a direct correspondence between FFG composition and basal resources available at a stream location; thus, the use of FFGs to characterize the macroinvertebrate assemblage at a stream location provides insight into functional roles, basal resource availability, and the importance of allochthony vs autochthony. Studies that examine changing basal resource availability longitudinally in a river system or between shaded and open-canopy locations generally, but not invariably, find that FFG composition varies accordingly. Anthropogenic change can shift FFG relative composition, often in response to altered land use. Recent studies employing stable isotope and fatty acid analyses provide new insights into consumer diet, often showing that members of a particular FFG are consuming and assimilating energy from foods not usually attributed to them. Results generally reveal more flexibility in feeding and a greater degree of omnivory than expected.

Stream fishes consume virtually every resource available, and many individual species consume a wide range of resources, often changing diet over ontogeny and across environmental setting. Fish trophic categories rely primarily on what resources are consumed, but also may consider feeding location and morphology. There does not appear to be a single, generally accepted set of trophic categories, but the following are commonly used: herbivores, detritivores, planktivores, omnivores, benthic invertivores, midwater-surface feeders (largely on insects), and piscivores. Where appropriate, one can add specialized categories such as snail-eaters, scale and fin eaters, fruit-eaters, parasites such as lampreys, and so on. In the majority of stream settings, it usually is the case that invertivores comprise over half of the species list, except in large rivers, per cent herbivores increases downstream, and detritivores are uncommon.

From headwaters to river mouth, fish diversity generally increases, and trophic diversity does as well. Unquestionably, one finds a wider range of trophic categories in tropical than temperate streams, including more species of herbivores and detritivores, and more unusual specialists. This is especially apparent in large tropical rivers with extensive lateral floodplains where lake and backwater habitats connect to the river during high water. Plant production and detritus of both autochthonous and allochthonous origin are of great importance in these systems; consequently there is a greater role for mud and detritus feeding (which often supports the greatest biomass of fish), and for predation (which often dominates species richness).

Various aspects of fish morphology are useful in understanding habitat and feeding preferences of fishes. Variation in body size, mouth gape and position, dentition, and gut length are frequently found to be strongly associated with feeding role, as well as visual and chemosensory adaptations. The study of ecomorphology is based on correspondence between the trophic role of a species of fish and its morphological adaptations for capture and digestion. Persuasive examples include similarities in body form and jaw structure among unrelated piscivores, the greater gut length of herbivores and detritivores compared with invertivores and piscivores, and so on. Ecomorphological analysis also has its limitations, especially due to relatedness, as closely related species may have similar morphology and diet due to common ancestry.

Trophic categorization of fishes has its uses and its limitations. Different feeding roles can be identified within an assemblage of fishes based on differences in diet, feeding behavior, and morphology. What is also evident, however, is that across life stages, habitats, and seasons, many species are flexible in their diet, benefiting from their ability to exploit different food items based on changing availability of resources. Thus, studies often refer to the high frequency of dietary plasticity, or characterize many of the species present as omnivorous, generalists, or opportunists. Even in the tropics, where some wonderful examples of ecological specialization can be found, many observers have opined that a generalist or opportunistic feeding strategy seems to characterize the majority of species.

Although fishes are the principal vertebrate component of most riverine food webs, all vertebrate classes have representatives in running waters. In small headwater streams, salamanders and snakes may be important top predators; there are many species of fish-eating birds, and some that consume aquatic invertebrates; a few mammals feed primarily or exclusively on aquatic prey, and a surprising diversity of mammals do so at least occasionally. Other animals frequently are the main prey, but algal, higher plant, and detrital resources are consumed as well. Especially in larger rivers, the littoral zone and floodplain habitats provide feeding opportunities across all trophic levels for vertebrate consumers.

Larval amphibians in the temperate zone are found primarily in standing waters, but they can be diverse and abundant in tropical streams. The majority rasp algae and detritus from substrates and so are herbivores and detritivores, although unknown amounts of microbes and micro-organisms likely are ingested as well. Snakes and members of the Crocodilia are predators of fish and invertebrates in aquatic environments, but often consume other vertebrates. Diet studies indicate that aquatic snakes are generalist predators, consuming a wide range of fishes, amphibians, and other prey. Turtles are for the most part omnivores of sluggish streams and rivers, consuming substantial amounts of invertebrate and fish prey, but some are more specialized as herbivores or carnivores. Many orders of birds make use of rivers and streams as feeding habitat. Many are fish predators but some feed directly on invertebrates. Piscivorous birds capture fish and invertebrates by a wide variety of feeding modes, including underwater pursuers, waders, aerial plunge-divers, and aerial insectivores. Mammals ranging from shrews to raccoons to bears occasionally or frequently consume invertebrates and fish, while muskrats, beaver, and the South American capybara consume aquatic and riparian vegetation. Vertebrate predators have evolved a variety of hunting tactics that largely determine the size range of prey captured and the habitat, especially water depth range, where prey are encountered. The combination of a predator's depth range and size range may significantly affect the size and depth distribution of fishes in streams, and many vertebrate predators may have their greatest impact on riverine communities by influencing the foraging location of their prey.

References

Aarts BGW, Nienhuis PH (2003) Fish zonations and guilds as the basis for assessment of ecological integrity of large rivers. Hydrobiol 500:157–178

Abilhoa V, Braga RR, Bornatowski H, Vitule JRS (2011) Fishes of the Atlantic rain forest streams: Ecological patterns and conservation. In: Grillo O. (Ed.) Changing Diversity in Changing Environment. InTechopen

Albert JS, Petry P, Reis RE (2011) Major biogeographic and phylogenetic patterns. In: Albert JS, Reis RE (eds) Historical biogeography of neotropical freshwater fishes. University of California Press, Berkeley, CA, pp 21–57

Alfaro ME (2002) Forward attack modes of aquatic feeding garter snakes. Funct Ecol 16:204–215

Allan JD (1982) Feeding habits and prey consumption of three setipalpian stoneflies (Plecoptera) in a mountain stream. Ecology 63:26–34

Allan JD, Flecker AS (1988) Prey preference in stoneflies: a comparative analysis of prey vulnerability. Oecologia 76:496–503

Anderson JT, Rojas SJ, Flecker AS (2009) High-quality seed dispersal by fruit-eating fishes in Amazonian floodplain habitats. Oecologia 161:279–290

Anderson NH, Sedell JR (1979) Detritus processing by macroinvertebrates in stream ecosystems. Annu Rev Entomol 24:351–377

Anderson NH, Sedell JR, Roberts LM, Triska FJ (1978) The role of aquatic invertebrates in processing of wood debris in coniferous forest. Am Midl Nat 100:64–82

Arrington DA, Winemiller KO, Layman CA (2005) Community assembly at the patch scale in a species rich tropical river. Oecologia 144:157–167

Arsuffi TL, Suberkropp K (1984) Leaf processing capabilities of aquatic Hyphomycetes: interspecific differences and influence on shredder feeding preferences. Oikos 42:144–154

Atkinson CL, First MR, Covich AP et al (2011) Suspended material availability and filtration-biodeposition processes performed by a native and invasive bivalve species in streams. Hydrobiologia 667:191–204

Bakker ES, Wood KA, Pagès JF et al (2016) Herbivory on freshwater and marine macrophytes: a review and perspective. Aquat Bot 135:18–36. https://doi.org/10.1016/j.aquabot.2016.04.008

Barbarino Duque A, Winemiller KO (2003) Dietary segregation among large catfishes of the Apure and Arauca Rivers, Venezuela. J Fish Biol 63:410–427. https://doi.org/10.1046/j.1095-8649.2003.00163.x

Barlöcher F (1985) The role of fungi in the nutrition of stream invertebrates. Bot J Linn Soc 91:83–94

Barlöcher F, Kendrick B (1975) Leaf-conditioning by microorganisms. Oecologia 20:359–362

Barlöcher FB (1982) The contribution of fungal enzymes to the digestion of leaves by Gammarus fossarum Koch (Amphipoda). Oecologia 52:1–4

Barlöcher FB (1983) Seasonal variation of standing crop and digestibility of CPOM in a Swiss Jura stream. Ecology 64:1266–1272

Bellamy LS, Reynoldson TB (1974) Behaviour in competition for food amongst lake-dwelling Triclads. Oikos 25:356–364

Best RC, da Silva VMF (1984) Amazon river dolphin, Boto Inia geoffrensis (de Blainville 1817). In: Ridgway SH, Harrison RJ (eds) Handbook of marine mammals. Academic Press, London, pp 1–23

Bierregaard RO, Poole AF, Martell MS et al (2016) Osprey (Pandion haliaetus). In: Rodewold PG (ed) The Birds of North America. Cornell Lab of Ornithology, Ithaca, NY

Bonato KO, Burress ED, Fialho CB (2017) Dietary differentiation in relation to mouth and tooth morphology of a neotropical characid fish community. Zool Anz 267:31–40

Booth DB, Roy AH, Smith B, Capps KA (2016) Global perspectives on the urban stream syndrome. Freshw Sci 35:412–420

Bowen SH (1983) Detritivory in neotropical fish communities. Dr W. Junk Publishers, The Hague

Boyero L, Perason RG, Dudgeon D et al (2011) Global distribution of a key trophic guild contrasts with common latitudinal diversity patterns. Ecology 92:1839–1848

Brecko J, Vervust B, Herrel A, Van Damme R (2011) Head morphology and diet in the dice snake, Natrix tessellata. Mertensiella 18:20–29

Brown SA, Ruxton GD, Pickup RW, Humphries S (2005) Seston capture by Hydropsyche siltalai and the accuracy of capture efficiency estimates. Freshw Biol 50:113–126

Bruder A, Schindler MH, Moretti MS (2014) Litter decomposition in a temperate and a tropical stream : the effects of species mixing, litter quality and shredders, 438–449

Calow P (1975a) The feeding strategies of two freshwater gastropods, Ancylus fluviatilis Mill. and Planorbis contortus Linn. (Pulmonata), in terms of ingestion rates and absorption efficiencies. Oecologia 20:33–49

Calow P (1975b) Defaecation strategies of two freshwater gastropods, Ancylus fluviatilis Mill. and Planorbis contortus Linn. (Pulmonata) with a comparison of field and laboratory estimates of food absorption rate. Oecologia 20:51–63

Calvo C, Mormul RP, Figueiredo BRS et al (2019) Herbivory can mitigate, but not counteract, the positive effects of warming on the establishment of the invasive macrophyte Hydrilla verticillata. Biol Invasions 21:59–66

Cargill AS, Cummins KW, Hanson BJ, Lowry RR (1985) The role of lipids as feeding stimulants for shredding aquatic insects. Freshw Biol 15:455–464

Chance M (1970) The functional morphology of the mouthparts of black fly larvae (Diptera: Simuliidae). Quaest Entomol 6:245–284

Chance M, Craig DA (1986) Hydrodynamics and behavior of Simuliidae larvae (Diptera. Can J Zool 64:1295–1309

Christian AD, Smith BN, Berg DJ et al (2004) Trophic position and potential food sources of 2 species of unionid bivalves (Mollusca: Unionidae) in 2 small Ohio streams. J North Am Benthol Soc 23:101–113

Ciborowski JJH, Craig DA, Fry KM (1997) Dissolved organic matter as food for black fly larvae (Diptera:Simuliidae). J North Am Benthol Soc 16:771–780

Clare EL, Barber BR, Sweeney BW et al (2011) Eating local: influences of habitat on the diet of little brown bats (Myotis lucifugus). Mol Ecol 20:1772–1780

Clavero M, Prenda J, Delibes M (2003) Trophic diversity of the otter (Lutra lutra L.) in temperate and Mediterranean freshwater habitats. J Biogeogr 30:761–769

Collier KJ, Halliday JN (2000) Macroinvertebrate-wood associations during decay of plantation pine in New Zealand pumice-bed streams: stable habitat or trophic subsidy? J North Am Benthol Soc 19:94–111

Collins SM, Kohler TJ, Thomas SA et al (2016) The importance of terrestrial subsidies in stream food webs varies along a stream size gradient. Oikos 125:674–685

Colón-Gaud C, Whiles MR, Kilham SS et al (2009) Assessing ecological responses to catastrophic amphibian declines: Patterns of macroinvertebrate production and food web structure in upland Panamanian streams. Limnol Oceanogr 54(1):331–343

Compin A, Cereghino R (2007) Spatial patterns of macroinvertebrate functional feeding groups in streams in relation to physical variables and land-cover in Southwestern France. Landsc Ecol 22:1215–1225

Correa SB, Winemiller KO, Lopez-Fernandez H, Galetti M (2007) Evolutionary perspectives on seed consumption and dispersal by fishes. Bioscience 57:748–756

Crosskey RW (1990) The Natural History of Black Flies. Wiley, New York

Cummins KW (2018) Functional analysis of stream macroinvertebrates. InTech Open. https://doi.org/10.5772/intechopen.79913

Cummins KW (1973) Trophic relations of aquatic insects. Annu Rev Entomol 18:183–206

Cummins KW, Klug MJ (1979) Feeding ecology of stream invertebrates. Ann Rev Ecol Syst 10:147–172

Currie DC, Craig DA (1988) Feeding strategies of larval black flies. In: Kim KC, Merritt RW (eds) Black flies: ecology, population management and annotated world list. Pennsylvania State University, University Park, PA, pp 155–170

Da Silva JC, Bialetzki A (2019) Early life history of fishes and zooplankton availability in a Neotropical floodplain: predator-prey functional relationships. J Plankton Res 41:63–75

Dangles O (2002) Functional plasticity of benthic macroinvertebrates: Implications for trophic dynamics in acid streams. Can J Fish Aquat Sci 59:1563–1573

de Carvalho TC, de Assis Montag LF, dos Santos-Costa MC (2017) Diet composition and foraging habitat use by three species of water snakes, *Helicops* Wagler, 1830, (Serpentes: Dipsadidae) in eastern Brazilian Amazonia. J Herpetol 51:215–222. https://doi.org/10.1670/15-161

Delariva RL, Agostinho AA (2001) Relationship between morphology and diets of six neotropical loricariids. J Fish Biol 58:832–847. https://doi.org/10.1111/j.1095-8649.2001.tb00534.x

Dobson M, Magana A, Mathooko JM, Ndegwa FK (2002) Detritivores in Kenyan highland streams: more evidence for the paucity of shredders in the tropics? Freshw Biol 47:909–919. https://doi.org/10.1046/j.1365-2427.2002.00818.x

Doucett RR, Marks JC, Blinn DW et al (2007) Measuring terrestrial subsidies to aquatic food webs using stable isotopes of hydrogen. Ecology 88:1587–1592

Douglas ME, Matthews WJ (1992) Does morphology predict ecology? Hypothesis testing within a freshwater stream fish assemblage. Oikos 65:213. https://doi.org/10.2307/3545012

Draulans D (1988) Effects of fish-eating birds on freshwater fish stocks: an evaluation. BiolConserv 44:251–263

Edler C, Georgian T (2004) Field measurements of particle-capture efficiency and size selection by caddisfly nets and larvae. J North Am Benthol Soc 23:756–770. Doi: https://doi.org/10.1899/0887-3593(2004)023<0756:fmopea>2.0.co;2

Elliott JM, Hurley MA (2000) Daily energy intake and growth of piscivorous brown trout, *Salmo trutta*. Freshw Biol 44:237–245

Evans-White MA, Dodds WK, Whiles MR (2003) Ecosystem significance of crayfishes and stonerollers in a prairie stream: functional differences between co-occurring omnivores. J North Am Benthol Soc 22:423–441. https://doi.org/10.2307/1468272

Ferreira WR, Ligeiro R, Macedo DR et al (2015) Is the diet of a typical shredder related to the physical habitat of headwater streams in the Brazilian Cerrado? 51:115–124. https://doi.org/10.1051/limn/2015004

Findlay S, Meyer JL, Smith PJ, Smith PJ (1984) Significance of bacterial biomass in the nutrition of a freshwater isopod (*Lirceus* sp.). Oecologia 63:38–42

Findlay S, Meyer JL, Snith PJ, Smith PJ (1986) Contribution of fungal biomass to the diet of a freshwater isopod (*Lirceus* sp.). Freshw Biol 16:377–385

Finlay JC, Doucett RR, McNeely C (2010) Tracing energy flow in stream food webs using stable isotopes of hydrogen. Freshw Biol 55:941–951. https://doi.org/10.1111/j.1365-2427.2009.02327.x

Fisher SG, Gray LJ (1983) Secondary production and organic matter processing by collector macroinvetebrates in a desert stream. Ecology 64:1217–1224

Flecker AS (1992) Fish trophic guilds and the stucture of a tropical stream: weak direct versus strong indirect effects. Ecology 73:927-940 https://doi.org/10.2307/1940169

Flecker AS, Feifarek BP, Taylor BW et al (1999) Ecosystem engineering by a tropical tadpole: density-dependent effects on habitat structure and larval growth rates. Copeia 1999:495–500

Franken RJM, Waluto B, Peeters ETHM et al (2005) Growth of shredders on leaf litter biofilms: the effect of light intensity. Freshw Biol 50:459–466. https://doi.org/10.1111/j.1365-2427.2005.01333.x

Frederick PC, Siegel-Causey D (2000) Anhinga (*Anhinga anhinga*). In: Poole AF, Gill FB (eds) The Birds of North America. Cornell Lab of Ornithology, Ithaca, NY. https://doi.org/10.2173/bna.522

Frimpong EA, Angermeier PL (2010) Trait-based approaches in the analysis of stream fish communities. Am Fish Soc Symp 73:109–136

Garrison BA (1999) Bank Swallow (*Riparia riparia*). In: Poole AF, Gill FB (eds) The Birds of North America. Cornell Lab of Ornithology, Ithaca, NY. https://doi.org/10.2173/bna.414

Gatz AJ (1979) Morphologically inferred niche differentiation in stream fishes. Am Midl Nat 106:10. https://doi.org/10.2307/2425131

Gelwick FP, McIntyre PB (2017) Trophic relations of stream fishes. In: Hauer FR, Lamberti GA (eds) Methods in stream ecology, third. Elsevier Inc., New York, pp 457–479

Gergs R, Steinberger N, Basen T, Martin-Creuzburg D (2014) Dietary supply with essential lipids affects growth and survival of the amphipod Gammarus roeselii. Limnologica 46:109–115. https://doi.org/10.1016/j.limno.2014.01.003

Goldstein RM, Meador MR (2004) Comparisons of fish species traits from small streams to large rivers. Trans Am Fish Soc 133:971–983. https://doi.org/10.1577/t03-080.1

Goulding M (1980) The fishes and the forest. Princeton University Press, Berkeley, CA

Graca M, Maltby L, Calow P (1993) Importance of fungi in the diet of *Gammarus pulex* and *Asellus aquaticus* II. Effects on growth, reproduction and physiology. Oecologia 96:304–309

Graca MAS, Cressa C, Gessner MO, et al. (2001) Food quality, feeding preferences, survival and growth of shredders from temperate and tropical streams. Freshwat Biol 46:947–957

Graça MAS (2001) The role of invertebrates on leaf litter decomposition in streams-a review. Int Rev Hydrobiol 86:383–393

Graça MAS, Cressa C (2010) Leaf quality of some tropical and temperate tree species as food resource for stream shredders. Int Rev Hydrobiol 95:27–41. https://doi.org/10.1002/iroh.200911173

Greathouse EA, Pringle CM (2006) Does the river continuum concept apply on a tropical island? Longitudinal variation in a Puerto Rican stream. 152:134–152. https://doi.org/10.1139/F05-201

Gregory SV (1983) Plant-herbivore interactions in stream ecosystems. In: Barnes JR, Minshall G (eds) Stream Ecology. Plenum Press, New York, pp 157–190

Grubaugh JW, Wallace JB, Houston ES (1996) Longitudinal changes of macroinvertebrate communiries along an appalachian stream continuum. Can J Fish Aquat Sci 909:896–909

Guo F, Lunz WC, Kainz M et al (2016) The importance of high-quality algal food sources in stream food webs—current status and future perspectives. Freshw Biol 61:1411–1422. https://doi.org/10.1111/fwb.12755

Hall RO, Meyer JL (1998) The trophic significance of bacteria in a detritus-based stream food web. Ecology 79:1995–2012

Hawkins CP, Murphy ML, Anderson NH (1982) Effects of canopy, substrate composition, and gradient on the structure of macroinvertebrate communities in Cascade Range streams of Oregon. Ecology 63:1840–1856

Hawkins CP, Sedell JR (1981) Longitudinal and seasonal changes in functional organization of macroinvertebrate communities in four Oregon streams. Ecology 62:387–397

Hecht KA, Nickerson MA, Colclough PB (2017) Hellbenders (*Cryptobranchus alleganiensis*) may exhibit an ontogenetic dietary shift. Southeast Nat 16:157–162. https://doi.org/10.1656/058.016.0204

Hilderbrand GV, Hanley TA, Robbins CT, Schwartz CC (1999) Role of brown bears *Ursus arctos* in the flow of marine nitrogen into a terrestrial ecosystem. Oecologia 121:546–560

Hill WR, Knight AW (1988) Grazing effects of two stream insects on periphyton. Limnologica 33:15–26

Hofmeister NR, Welk M, Freedberg S (2013) Elevated levels of δ15 N in riverine Painted Turtles (*Chrysemys picta*): trophic enrichment or anthropogenic input? Can J Zool 91:899–905. https://doi.org/10.1139/cjz-2013-0121

Horwitz RJ (1978) Temporal variability patterns and the distributional patterns of stream fishes. Ecol Monogr 48:307–321. https://doi.org/10.2307/2937233

Hynes H (1970) The ecology of running waters. University of Toronto Press, Toronto

Ibañez C, Belliard J, Hughes RM et al (2009) Convergence of temperate and tropical stream fish assemblages. Ecography (Cop) 32:658–670. https://doi.org/10.1111/j.1600-0587.2008.05591.x

Ibañez C, Tedesco PA, Bigorne R et al (2007) Dietary-morphological relationships in fish assemblages of small forested streams in the Bolivian Amazon. Aquat Living Resour 20:131–142. https://doi.org/10.1051/alr:2007024

Jensen H, Kiljunen M, Amundsen PA (2012) Dietary ontogeny and niche shift to piscivory in lacustrine brown trout *Salmo trutta* revealed by stomach content and stable isotope analyses. J Fish Biol 80:2448–2462. https://doi.org/10.1111/j.1095-8649.2012.03294.x

Karachle PK, Stergiou KI (2010) Intestine morphometrics of fishes: a compilation and analysis of bibliographic data. Acta Ichthyol Piscat 40:45–54. https://doi.org/10.3750/AIP2010.40.1.06

Kaushik NK, Hynes HBN (1971) The fate of autumn-shed leaves that fall into streams. Arch für Hydrobiol 68:465–515

Keast A (1978) Feeding interrelations between age groups of pumpkinseed (*Lepomis gibbosus*) and comparisons with Bluegill (*L. macrochirus*). J Fish Res Board Canada 35:12–27

Kelly JF, Bridge ES, Hamas MJ (2009) Belted Kingfisher (*Megaceryle alcyon*). In: Poole AF (ed) The Birds of North America. Cornell Lab of Ornithology, Ithaca, NY https://doi.org/10.2173/bna.84

Kingery HE, Willson MF (2019) American Dipper (*Cinclus mexicanus*). In: Rodewald PG (ed) The Birds of North America. Cornell Lab of Ornithology, Ithaca, NY https://doi.org/10.2173/bna.amedip.03

Kochi K, Kagaya T (2005) Green leaves enhance the growth and development of a stream macroinvertebrate shredder when senescent leaves are available. Freshw Biol 50:656–667. https://doi.org/10.1111/j.1365-2427.2005.01353.x

Lacoursiere JO, Craig DA (1993) Fluid transmission and filtartion efficiency of the labral fans of black fly larvae (Diptera: Simuliidae): hydrodynamic, morphological, and behavioural aspects. Can J Zool 71:148–162

Lamberti GA, Ashkenas LR, Gregory SV, Steinman AD (1987) Effects of three herbivores on periphyton communities in laboratory streams. J North Am Benthol Soc 6:92–104

Lamberti GA, Moore JW (1984) Aquatic insects as primary consumers. In: Resh VH, Rosenberg DM (eds) The ecology of aquatic insects. Praeger, New York, pp 164–195

Lamberti GA, Resh V (1983) Stream periphyton and insect herbivores: an experimental study of grazing by a caddisfly population. Ecology 64:1124–1135

Lancaster J, Bradley DC, Hogan A, Waldron S (2005) Intraguild omnivory in predatory stream insects. J Anim Ecol 74:619–629. https://doi.org/10.1111/j.1365-2656.2005.00957.x

Lang HH (1980) Surface wave discrimination between prey and nonprey by the back swimmer *Notonecta glauca* L. (Hemiptera, Heteroptera). Behav Ecol Sociobiol 6:233–246

Lara NRF, Marques TS, Montelo KM et al (2012) A trophic study of the sympatric Amazonian freshwater turtles *Podocnemis unifilis* and *Podocnemis expansa* (testudines, podocnemidae) using carbon and nitrogen stable isotope analyses. Can J Zool 90:1394–1401. https://doi.org/10.1139/cjz-2012-0143

Lau DCP, Leung KMY, Dudgeon D (2009) Are autochthonous foods more important than allochthonous resources to benthic consumers in tropical headwater streams? J North Am Benthol 28:426–439. https://doi.org/10.1899/07-079.1

Lauder GV (1980) The suction feeding mechanism in sunfishes (*Lepomis*): an experimental analysis. J Exp Biol 88:49–72

Layman CA, Winemiller KO, Arrington DA, Jepsen DB (2005) Body size and trophic position in a diverse tropical food web. Ecology 86:2530–2535

Levine JS, MacNichol EF (1979) Visual pigments in teleost fishes: effects of habitat, microhabitat, and behavior on visual system evolution. Sens Processes 3:95–131

Li AOY, Dudgeon D (2008) Food resources of shredders and other benthic macroinvertebrates in relation to shading conditions in tropical Hong Kong streams. Freshw Biol 53:2011–2025. https://doi.org/10.1111/j.1365-2427.2008.02022.x

Liem KF (1980) Adaptive significance of intra-and interspecific differences in the feeding repertoires of cichlid fishes. Am Zool 20:295–314

Lind AJ, Welsh HH (1994) Ontogenetic changes in foraging behaviour and habitat use by the Oregon garter snake, *Thamnophis atratus hydrophilus*. Anim Behav 48:1261–1273

Loch C, Marmontel M, Simões-Lopes PC (2009) Conflicts with fisheries and intentional killing of freshwater dolphins (Cetacea: Odontoceti) in the Western Brazilian Amazon. Biodivers Conserv 18:3979–3988. https://doi.org/10.1007/s10531-009-9693-4

Logez M, Bady P, Melcher A, Pont D (2013) A continental-scale analysis of fish assemblage functional structure in European rivers. Ecography (Cop) 36:80–91. https://doi.org/10.1111/j.1600-0587.2012.07447.x

Lowe-McConnell R (1987) Ecological studies in tropical fish communities. Cambridge University Press, London

Luiselli L, Capizzi D, Filippi E et al (2007) Comparative diets of three populations of an aquatic snake (*Natrix tessellata*, Colubridae) from Mediterranean streams with different hydric regimes. Copeia 426–435. https://doi.org/10.1643/0045-8511

Lundberg JG, Lewis WM, Saunders JF, Mago F (1987) A major food web component in the Orinoco River channel: Evidence from planktivorous electric fishes. Science (80-) 237:81–83

Lundquist MJ, Zhu W (2018) Aquatic insect functional diversity and nutrient content in urban streams in a medium-sized city. Ecosphere 9:1–11

MacNeil BC, Dick JTA, Elwood RW (1997) The trophic ecology of freshwater *Gammarus* spp. (Crustacea:Amphipoda): problems and perspectives concerning the functional feeding group concept. Biol Rev 72:349–364

Marchant R, Metzeling L, Graesser A, Suter P (1985) The organization of macroinvertebrate communities in the major tributaries of the LaTrobe River, Victoria, Australia. Freshw Biol 15:315–331

Martin M, Kukor J (1984) Role of micophagy and bacteriophagy in invertebrate nutrition. In: Klug M, Reddy C (eds) Current perspectives in microbial ecology. American Society of Microbiology, Washington, DC, pp 257–263

Matthews WJ (1998) Patterns in freshwater fish ecology. Kluwer Academic Publishers

McCullough DA, Minshall GW, Cushing CE (1979) Bioenergetics of a stream "collector" organism, *Tricorythodes minutus* (Insecta: Ephemeroptera). Limnol Oceanogr 24:45–58

McNeely C, Clinton SM, Erbe JM (2006) Landscape variation in C sources of scraping primary concsumers in streams. J North Am Benthol Soc 25:787–799. https://doi.org/10.1899/0887-3593(2006) 025

Merritt RW, Cummins KW, Berg MB (2019) Aquatic Insects of North America, 5th edn. Kendall/Hunt, Dubuque Iowa

Merritt RW, Cummins KW, Berg MB (2017) Trophic relationships of macroinvertebrates. In: Hauer FR, Lamberti GA (eds) Methods in Stream Ecology, Third. Elsevier Inc., New York, pp 413–433

Merritt RW, Ross DH, Larson GJ (1982) Influence of stream temperature and seston on the growth and production of overwintering larval black flies (Simuliidae, Diptera). Ecology 63:1322–1331

Middelburg JJ (2014) Stable isotopes dissect aquatic food webs from the top to the bottom. Biogeosciences 11:2357–2371. https://doi.org/10.5194/bg-11-2357-2014

Mihuc TB (1997) The functional trophic role of lotic primary consumers: generalist versus specialist strategies. Freshw Biol 37:455–462

Minshall GW (1978) Autotrophy in stream ecosystems. Bioscience 28:767–771

Miranda LE, Granzotti RV, Dembkowski DJ (2019) Gradients in fish feeding guilds along a reservoir cascade. Aquat Sci 81:15. https://doi.org/10.1007/s00027-018-0615-y

Mittelbach G, Persson L (1998) The ontogeny of piscivory and its ecological consequences. Can J Fish Aquat Sci 55:1454–1465. https://doi.org/10.1139/cjfas-55-6-1454

Moll D (1976) Food and feeding strategies of the Ouachita map turtle (*Graptemys pseudogeographica*). Am Midl Nat 96:478–482

Moore AA, Palmer MA (2005) Invertebrate biodiversity in agricultural and urban headwater streams: implications for conservation and management. Ecol Appl 15:1169–1177

Moore JW (1975) The role of algae in the diet of *Asellus aquaticus*. J Anim Ecol 44:719–730

Mortillaro JM, Pouilly M, Wach M et al (2015) Trophic opportunism of central Amazon floodplain fish. Freshw Biol 60:1659–1670. https://doi.org/10.1111/fwb.12598

Muntz WR (1982) Visual adaptations to different light environments in Amazonian fishes. Rev Can Biol Exp 41:35–46

Newman RM (1991) Herbivory and detritivory on freshwater macrophytes by invertebrates: a review. J North Am Benthol Soc 10:89–114

Nichols SJ, Garling D (2000) Food-web dynamics and trophic-level interactions in a multispecies community of freshwater unionids. Can J Zool 78:871–882

Ormerod SJ (1985) The diet of breeding Dippers *Cinclus cinclus* and their nestlings in the catchment of the River Wye, mid-Wales: a preliminary study by faecal analysis. Ibis (Lond 1859) 127:316–331

Palmer RW, Craig DA (2000) An ecological classification of primary labral fans of filter-feeding black fly (Diptera: Simuliidae) larvae. Can J Zool 78:199–218

Pandian TJ, Peter M (1986) An indirect procedure for the estimation of assimilation efficiency of aquatic insects. Fresh water Biol 16:93–98

Patrick CJ (2013) The effect of shredder community composition on the production and quality of fine particulate organic matter. Freshw Sci 32:1026–1035. https://doi.org/10.1899/12-090.1

Pearce J, Mallory ML, Metz K (2015) Common Merganser (*Mergus merganser*). In: Poole AF (ed) The Birds of North America. Cornell Lab of Ornithology, Ithaca, NY

Peckarsky BL (1984) Predator-prey interactions among aquatic insects. In: Resh VH, Rosenberg DM (eds) The ecology of aquatic insects. Praeger, New York, pp 196–254

Petranka JW (1984) Sources of interpopulational variation in growth responses of larval salamanders. Ecology 65:1857–1865

Post DM (2002) Using stable isotopes to estimate trophic position: models, methods, and assumptions. Ecology 83:703–718

Pouilly M, Barrera S, Rosales DCAN (2006) Changes of taxonomic and trophic structure of fish assemblages along an environmental gradient in the Upper Beni watershed (Bolivia). J Fish 68:137–156. https://doi.org/10.1111/j.1095-8649.2005.00883.x

Pouilly M, Lino F, Bretenoux JG, Rosales C (2003) Dietary-morphological relationships in a fish assemblage of the Bolivian Amazonian floodplain. J Fish Biol 62:1137–1158. https://doi.org/10.1046/j.1095-8649.2003.00108.x

Power ME (1983) Grazing responses of tropical freshwater fishes to different scales of variation in their food. Environ Biol Fishes 9:103–115

Power ME (1984) Depth Distributions of armored catfish: Predator-induced resource avoidance? Ecology 65:523–528

Power ME, Matthews WJ, Stewart AJ (1985) Grazing minnows, piscivorous bass, and stream algae: dynamics of a strong interaction. Ecology 66:1448–1456

Proctor HC, Pritchard G (1990) Prey detection by the water mite *Unionicola crassipes* (Acarai: Unionicolidae). Freshw Biol 23:271–279

Raikow DF, Hamilton SK, Kellogg WK (2001) Bivalve diets in a midwestern U.S. stream: A stable isotope enrichment study. Limnol Ocean 46:514–522

Ramírez F, Davenport TL, Mojica JI (2015) Dietary-morphological relationships of nineteen fish species from an Amazonian terra firme blackwater stream in Colombia. Limnologica 52:89–102. https://doi.org/10.1016/j.limno.2015.04.002

Rebora M, Piersanti S, Gaino E (2004) Visual and mechanical cues used for prey detection by the larva of *Libellula depressa* (Odonata Libellulidae). Ethol Ecol Evol 16:133–144. https://doi.org/10.1080/08927014.2004.9522642

Roditi HA, Camco NF, Cole JJ, Strayer DL (1996) Filtration of Hudson River water by the Zebra Mussel (*Dreissena polymorpha*). Estuaries 19:824–832

Rodríguez MA, Lewis WM (1997) Structure of fish assemblages along environmental gradients of floodplain lakes in the Orinoco River. Ecol Monogr 67:109–128

Rosenblatt AE, Nifong JC, Heithaus MR et al (2015) Factors affecting individual foraging specialization and temporal diet stability across the range of a large "generalist" apex predator. Oecologia 178:5–16. https://doi.org/10.1007/s00442-014-3201-6

Rosi-Marshall EJ, Wallace JB (2002) Invertebrate food webs along a stream resource gradient. Freshw Biol 47:129–141

Sampson SJ, Chick JH, Pegg MA (2009) Diet overlap among two Asian carp and three native fishes in backwater lakes on the Illinois and Mississippi rivers. Biol Invasions 11:483–496. https://doi.org/10.1007/s10530-008-9265-7

Santana SE, Cheung E (2016) Go big or go fish: morphological specializations in carnivorous bats. Proc R Soc B Biol Sci 283:283:20160615. https://doi.org/10.1098/rspb.2016.0615

Schiesari L, Werner EE, Kling GW (2009) Carnivory and resource-based niche differentiation in anuran larvae: Implications for food web and experimental ecology. Freshw Biol 54:572–586. https://doi.org/10.1111/j.1365-2427.2008.02134.x

Schlosser IJ (1987) The role of predation in age-and size-related habitat use by stream fishes. Ecology 68:651–659

Schnitzler HU, Kalko EK, Kaipf I, Grinell AD (1994) Fishing and echolocation behavior of the greater bulldog bat, *Noctilio leporinus*, in the field. Behav Ecol Sociobiol 35:327–345

Sinsabaugh RL, Linkins AE, Benfield EF, Benfield EF (1985) Cellulose digestion and assimilation by three leaf-shredding aquatic insects. Ecology 66:1464–1471

Sjostrom O (1985) Hunting behaviour of the perlid stonefly nymph *Dinocras cephalotes* (Plecoptera) under different light conditions. Anim Behav 33:534–540

Smith RF, Lamp WO (2008) Comparison of insect communities between adjacent headwater and main-stem streams in urban and rural watersheds. J North Am Benthol Soc 27:161–175. https://doi.org/10.1899/07-071.1

Steedman RJ, Anderson NH (1985) Life history and ecological role of the xylophagous aquatic beetle, *Lara avara* LeConte (Dryopoidea: Elmidae). Freshw Biol 15:535–546

Steinman AD, Mcintire D, Lowry RR (1987) Effects of herbivore type and density on chemical composition of algal assemblages in. J North Am Benthol Soc 6:189–197

Steinmetz J, Kohler SL, Soluk DA (2003) Birds are overlooked top predators in aquatic food webs. Ecology 84:1324–1328

Stepenuck KF, Crunkilton RL, Wang L (2002) Impacts of urban landuse on macroinvertebrate communities in southeastern Wisconsin streams. J Am Water Resour Assoc 38:1041–1051

Sterling JL, Rosemond AD, Wenger SJ (2016) Watershed urbanization affects macroinvertebrate community structure and reduces biomass through similar pathways in Piedmont streams, Georgia, USA. Freshw Sci 35:676–688. https://doi.org/10.1086/686614

Strategies F, Townsend CR, Hildrew AG (1979) Resource partitioning by two freshwater invertebrate predators with contrasting foraging strategies. J Anim Ecol 48:909–920

Strayer DL (1999) Effects of alien species on freshwater mollusks in North America. J North Am Benthol Soc 18:74–98

Suberkropp K, Arsuffi TL (1984) Degradation, growth, and changes in palatability of leaves colonized by six aquatic hyphomycetes: interspecific differences and influence on shredder feeding preferences. Mycologia 76:398–407

Tachet H (1977) Vibrations and predatory behavior of *Plectronemia conspersa* larvae (Trichoptera). Zeitschrift fur Tierpsycholgie 45.61–74

Thompson BC, Jackson JA, Burger J et al (2011) Least Tern (*Sternula antillarum*). In: Poole A, Gill FB (eds) The Birds of North America. Cornell Lab of Ornithology, Ithaca, NY

Tipping PW, Martin MR, Center TD, Davern TM (2008) Suppression of *Salvinia molesta* Mitchell in Texas and Louisiana by *Cyrtobagous salviniae* Calder and Sands. Aquat Bot 88:196–202. https://doi.org/10.1016/j.aquabot.2007.10.010

Tucker AD, Guarino F, Priest TE (2011) Where lakes were once rivers: contrasts of freshwater turtle diets in dams and rivers of Southeastern Queensland. Chelonian Conserv Biol 11:12–21

Vadeboncoeur Y, Power ME (2017) Attached algae: the cryptic base of inverted trophic pyramids in freshwaters. Annu Rev Ecol Evol Syst 48:255–279

Vennesland RG, Butler RW (2011) Great Blue Heron (*Ardea herodias*). In: Poole AF, Gill FB (eds) The Birds of North America. Cornell Lab of Ornithology, Ithaca, NY

Wainwright PC, Richard BA (1995) Predicting patterns of prey use from morphology of fishes. Environ Biol Fishes 44:97–113. https://doi.org/10.1007/BF00005909

Wallace JB, Malas D (1976) The fine structuure of capture nets of larval Philopotamidae (Trichoptera) with special emphasis on *Dolophilodes distinctus*. Can J Zool 54:178801802

Wallace JB, Merritt RW (1980) Filter-feeding ecology of aquatic insects. Ann Rev Entomol 25:103–135

Wallace JB, O'Hop J (1985) Life on a fast pad: waterlily leaf beetle impact on water lilies. Ecol 66:1534–1544

Walsh CJ, Roy AH, Feminella JW et al (2005) The urban stream syndrome: current knowledge and the search for a cure. J North Am Benthol Soc 24:706–723. https://doi.org/10.1899/04-028.1

Wantzen KM, Wagner R (2006) Detritus processing by shredders: a tropical-temperate comparison. J North Am Benthol Soc 25:216–232. https://doi.org/10.1899/0887-3593(2006)25

Wantzen KM, Wagner R, Suetfeld R, Junk WJ (2002) How do plant-herbivore interactions of trees influence coarse detritus processing by shredders in aquatic ecosystems of different latitudes? Verhandlungen der Int Vereinigung für Theor und Angew Limnol 28:815–821. https://doi.org/10.1080/03680770.2001.11901827

Ward-Campbell BMS, Beamish FWH, Kongchaiya C (2005) Morphological characteristics in relation to diet in five coexisting Thai fish species. J Fish Biol 67:1266–1279. https://doi.org/10.1111/j.1095-8649.2005.00821.x

Ward GM, Woods DR (1986) Lignin and fiber content of FPOM generated by the shredders *Tipula abdominalis* (Diptra: Tipulidae) and *Taloperla cornelia* Needham & Smith (Plecoptera: Peltoperlidae). Arch fur Hydrobiol 107:545–562

Weiperth A, Gaebele T, Potyó I, Puky M (2014) A global overview on the diet of the dice snake (*Natrix tessellata*) from a geographical perspective: foraging in atypical habitats and feeding spectrum widening helps colonisation and survival under suboptimal conditions for a piscivorous snake. Zool Stud 53. https://doi.org/10.1186/s40555-014-0042-2

Welcomme RL (1979) Fisheries Ecology of Floodplain Rivers. Longman, London

Wetzel RG (2001) Limnology, 3rd edn. Academic Press, San Diego

Whiles MR, Gladyshev MI, Sushchik NN et al (2010) Fatty acid analyses reveal high degrees of omnivory and dietary plasticity in pond-dwelling tadpoles. Freshw Biol 55:1533–1547. https://doi.org/10.1111/j.1365-2427.2009.02364.x

Wiggins GB, Mackay RJ (1978) Some relationships between systematics and trophic ecology in Nearctic aquatic insects, with special reference to Trichoptera. Ecology 59:1211–1220

Willson MF, Halupka KC (1995) Anadromous fish as keystone species in vertebrate communities. Conserv Biol 9:489–497

Winemiller KO (2004) Floodplain river food webs: generalizations and implications for fisheries management. In: Welcomme RL, Petr T (eds) Proceedings of the second international symposium on the management of large rivers for fisheries, vol 2. FAO and the Mekong River Commission, pp 285–307

Winemiller KO (1991) Ecomorphological diversification in lowland freshwater fish assemblages from five biotic regions. Ecol Indic 61:343–365

Wolff LL, Carniatto N, Hahn NS (2013) Longitudinal use of feeding resources and distribution of fish trophic guilds in a coastal Atlantic stream, southern Brazil. Neotrop Ichthyol 11:375–386. https://doi.org/10.1590/S1679-62252013005000005

Wood KA, Hare MTO, Mcdonald C et al (2017) Herbivore regulation of plant abundance in aquatic ecosystems. Biol Rev 92:1128–1141. https://doi.org/10.1111/brv.12272

Wootton JT, Oemke MP (1992) Latitudinal differences in fish community trophic structure, and the role of fish herbivory in a Costa Rican stream. Environ Biol Fishes 35:311–319. https://doi.org/10.1007/BF00001899

Wotton RS (1994) Particulate and dissolved organic matter as food. In: Wotton RS (ed) The biology of particles in aquatic Systems. Lewis, Boca Raton, FL, pp 235–288

Wotton RS, Malmqvist B (2001) Feces in aquatic ecosystems. Bioscience 51:537–544

Wotton RS, Malmqvist B, Muotka T, Larsson K (1998) Fecal pellets from a dense aggregations of suspension feeders in a stream: An example of ecosystem engineering. Limnol Oceanogr 43:719–725

Wright MS, Covich AP (2005) The effect of macroinvertebrate exclusion on leaf breakdown rates in a tropical headwater stream. Biotropica 37:403–408. https://doi.org/10.1111/j.1744-7429.2005.00053.x

Zimmer M, Bartholmé S (2003) Bacterial endosymbionts in *Asellus aquaticus* (Isopoda) and *Gammarus pulex* (Amphipoda) and their contribution to digestion. Limnol Ocean 48:2208–2213

Zimmerman MC, Wissing TE (1978) Effects of temperature on gut-loading and gut-clearing times of the burrowing mayfly, *Hexagenia limbata*. Freshw Biol 8:269–277

The assemblage of species within a stream reach forms a network of linkages and interactions that vary in strength and the number of species affected. The basal resources of algae and detritus, together with associated microorganisms and external energy subsidies, sustain higher consumers including herbivores, predators, the omnivores that blur trophic classification, and the pathogens and parasites that affect all. The availability of resources can limit the abundance of consumers (called bottom-up control), and consumers can regulate the abundance of their prey at lower trophic levels (called top-down control). Species often compete for the same limiting resources such as food or space, and successful competitors are able to retain their place in the assemblage while less aggressive or efficient species may be excluded. The interactions between grazers and algae, predators and their animal prey, parasites and their hosts, and competing species with one another constitute the primary linkages that collectively bind species together within food webs. In this chapter we focus on the ecological consequences and complexity of such linkages, and in the following chapter we explore the forces that ultimately determine which species are found together and the multiple food web interactions that structure lotic communities.

10.1 Herbivory

Primary producers, including algae, cyanobacteria, bryophytes, and vascular plants, are important basal resources in lotic food webs. Macrophytes also can be consumed directly by terrestrial and aquatic organisms, although much macrophyte production is consumed after entering the detrital pool, and phytoplankton play a relatively minor role in most free-flowing rivers. Grazing on benthic algae by invertebrates, some fishes, and a few amphibian larvae is the most important pathway of herbivory in streams, and has received by far the most study. Benthic algae vary in their distribution, growth form, and nutritional value, and grazers differ in their means of scraping and browsing this food supply. Thus, the species of algae that are consumed reflect their vulnerability to particular grazers, and possibly aspects of grazer preference as well. Grazers can have a number of impacts on algae, reducing their abundance, altering assemblage composition, and even stimulating algal growth and overall productivity through the removal of senescent cells and the recycling of nutrients.

10.1.1 Direct Interactions Between Consumers and Producers

The interaction between primary consumers and their food, of which benthic algae are of primary importance, is strongly dependent upon feeding modality of the consumer species, and traits, including growth form and chemical constituents, of the producer species (Steinman 1996; Holomuzki et al. 2013). Benthic algae are comprised of various algal groups and cyanobacteria, ranging in size from small individual cells to large filaments and colonies (Fig. 6.1), and intermixed with other surface-layer organisms including heterotrophic bacteria and micro-consumers such as protists and meiofauna, as well as organic material and extracellular compounds. Consumers of this mixture, often referred to interchangeably as benthic algae, periphyton, or epilithon, are many species of invertebrates, amphibian larvae, and fishes, categorized as grazers or herbivores.

As described in Chap. 9, some herbivores are able to scrape periphyton from hard surfaces, including snails with their rasping radulae, glossosomatid caddisflies with their mandibles, and some fishes such as the stoneroller *Campostoma* and loricariid catfish with sucker-like mouthparts that can remove all but the most tightly adherent and crevice-dwelling algae. Various mayflies, including in the Heptageniidae and Leptophlebiidae, can better be described as browsers and gatherers of upright forms of the algal overstory. As herbivorous species exhibit a range of feeding modalities, so also do algal species vary in size and growth forms that affects their vulnerability. Stalked, erect, and filamentous growth forms often are vulnerable to most herbivores, although filamentous algae can become too large for some. Prostrate forms often are vulnerable only to raspers and scrapers, and species of these feeding modes can have especially strong effects on most periphyton biomass and assemblages. Invertebrates are the dominant grazers in temperate latitudes, but freshwater fishes and amphibian larvae are important in some locations. Algae and organic matter are major dietary components for tropical fishes, which have been shown to strongly reduce attached algal biomass (Power 1984a; Flecker et al. 2002).

10.1.1.1 Grazer Impacts on Periphyton Assemblages

The impact of grazers on periphyton have been studied in both lab and field settings with a wide variety of innovative apparatus and experimental designs. In-depth reviews (Feminella and Hawkins 1995; Steinman 1996; Liess and Hillebrand 2004 , Liess and Kahlert 2007) agree that grazers exert strong control over periphyton biomass. Where herbivorous fish are plentiful, they have been found to wield considerable control over benthic primary producers.

Exclusion of the stoneroller minnow *Campostoma* (due to the presence of a piscivorous fish) resulted in growth of filamentous algae, whereas introduced *Campostoma* caused rapid declines in algal biomass (Power and Matthews 1983). Tropical streams often contain numerous species of grazing fish, as well as algivorous insects, mollusks, crustaceans, and larval amphibians. Armored catfish (Loricariidae), popular for their diligence in cleaning algae from the walls of aquaria, have reached high abundances in southern Mexico, outside their native range. Exclusion experiments by Capps and Flecker (2015) documented very substantial decreases in algal biomass in the presence of grazing catfish, compared to where they were excluded (Fig. 10.1).

Snails and caddisfly larvae have proven to be highly effective grazers of periphyton in a number of studies, and share the traits of being individually large, relatively slow-moving, and well equipped with scraping mandibles or a radula. Periphyton biomass increased some five- to 20-fold relative to control substrates when the caddis larva *Helicopsyche* was excluded using tiles raised above the bed of a California stream (Lamberti and Resh 1983). Barriers of petroleum jelly effectively excluded another caddis larva, *Glossosoma*, from stone surfaces in a Montana stream, resulting in a fivefold increase in algal cell counts (McAuliffe 1984a). Lamberti et al. (1987) compared the mayfly

Fig. 10.1 A field experiment using cages to study the grazing impact of the armored catfish *Pterygoplichthys* (Siluriformes:Loricariidae) in a stream in southern Mexico. (**a**) An experiment with large cages (1.5 × 1.5 × 1.0 m length × width × height, constructed of poultry wire ~2.5-cm diameter). From left to right: *Pterygoplichthys* enclosure (EN), stream reference (SR), cage control (CC), and *Pterygoplichthys* exclosure (EX). (**b**) a cage control treatment from a small cage experiment (24 × 48 × 10 cm). (**c**) a small cage exclosure (Reproduced from Capps and Flecker 2015)

Centroptilum elsa, the caddis *Dicosmoecus gilvipes*, and the snail *Juga silicula*, which they characterized as a browser, scraper, and rasper, respectively. Laboratory streams were inoculated with algal scrapings, consumers were added at approximately natural densities nine days later, and development of the periphyton mat was monitored for 48 days. The effect of the mayfly was slight and confined to small (<2 cm diameter) patches, but *Juga* had a substantial impact and *Dicosmoecus* even more so. However, other studies have found grazing mayflies able to limit benthic algae. An experiment that enclosed the mayfly *Ameletus* at realistic densities in Plexiglas chambers containing natural stream bed material resulted in marked reductions in periphyton standing crops, even at densities of 0.5x ambient (Hill and Knight 1987) (Fig. 10.2).

As one might expect, suppression of grazers results in greater algal biomass, and suppression of periphyton results in fewer grazers. When all herbivorous invertebrates in a 50-m² reach of a Colorado mountain stream were reduced by daily electroshocking with a portable apparatus typically used to collect fish, algal biomass increased substantially in comparison to a reference stream (Taylor et al. 2002). Algal biomass can differ markedly at small spatial scale when the presence of predators restricts the local distribution of grazers. Rings of attached filamentous green algae developed along the margins of a Panama stream where wading birds effectively excluded grazing catfish, while algal biomass was much reduced in deeper water where catfish were safe to feed (Power 1984b). At the larger spatial scale of stream sections or entire streams, the abundance and growth of grazers varies with their food supply. In a small stream draining an area that had recently been clearcut, production of the mayfly *Baetis* was roughly 18 times higher than at a reference site (Wallace and Gurtz 1986). Mayfly guts contained mainly diatoms, and estimates of gut fullness from the open canopy stream were up to double those from the forested stream. Although algal cell densities varied little among sites, periphyton production (based on *Baetis* production and projected food consumption) was estimated to be nearly 30 times greater at the open site. Subsequent forest regrowth resulted in canopy closure, and after six years *Baetis* was much rarer and periphyton production had dropped tenfold. Because clearcutting affected the entire stream, recruitment rather than redistribution is the presumed mechanism.

In a meta-analysis of 865 experimental studies reported in 178 publications, Hillebrand (2009) evaluated grazer control of periphyton biomass in all kinds of aquatic habitats to explore how environmental variables affected the degree of grazer control, and whether experimental results depended on aspects of experimental design. Overall, grazers removed an average of 59% of periphyton biomass, evidence that periphyton is indeed strongly controlled by herbivory. Different consumer groups had significantly different impacts. Two crustacean groups (isopods and amphipods) and trichopteran larvae had strongest effects, and dipterans (mainly Chironomidae) the weakest. The size of the grazer effect, which measured the proportional removal of periphyton, was positively associated with increasing algal biomass, temperature, and scarce resource availability. In lotic experiments in particular, laboratory experiments exhibited stronger effects than field studies, and within field studies, enclosures yielded significantly stronger grazing effects than exclosures. This is best explained by the fact that lab experiments confine grazers and their food without allowing consumer emigration, as do field enclosure experiments. In addition, lab experiments usually were stocked with higher grazer biomass and ran at higher temperatures. Length of experiment also appears to influence outcome, as grazing effects become more pronounced over time, especially when

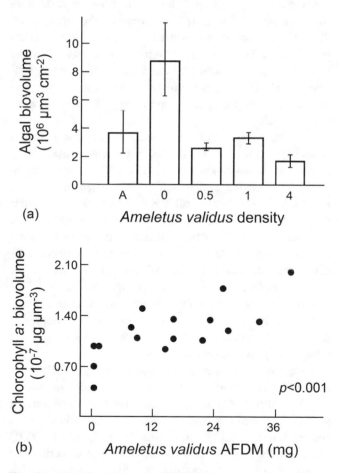

Fig. 10.2 The effect of density of the mayfly *Ameletus* on periphyton standing crop and quality. (**a**) Periphyton abundance under various grazing conditions. A: ambient densities on streambed; 0: cages with zero density; 0.5, 1 and 4: cages with 0.5x, 1x and 4x natural densities, respectively. Results were similar for chlorophyll *a* and AFDW. Note that even low densities of grazers reduced algal biovolume. (**b**) The ratio of chlorophyll *a* per unit biovolume increased significantly with *Ameletus* biomass (Reproduced from Hill and Knight 1987)

spatially confined. This highlights the challenge of reproducing realistic lotic environments in lab settings compared to field experiments.

Species and trait composition usually change in response to grazing, which tends to reduce species richness as well as dominance by a few species, thereby increasing the evenness of species representation in the assemblage (Hillebrand 2009). Upright, overstory, and loosely attached algal taxa are vulnerable to a wide range of grazers, whose presence often shifts benthic assemblages toward prostrate, understory forms. In mesocosm experiments using various combinations of grazing by a snail, a mayfly, and a caddisfly, several prostrate diatom species were dominant on grazed clay tiles, erect growth forms were more abundant in ungrazed controls and in snail alone treatments, and two species of green filamentous algae and a colonial species of cyanobacteria were present only in ungrazed controls (Holomuzki and Biggs 2006). A shift towards basal cells of the grazer-resistant chlorophyte *Stigeoclonium* occurred in response to grazing fishes in a stream in the Andean foothills of Venezuela (Flecker et al. 2002) and to snails in flow-through channels within a stream in Tennessee, US (Rosemond et al. 2000).

The components of the periphyton mat that are most affected by herbivory can vary with the species of grazer and with their feeding mode. Grazing by a snail (*Juga*) and a caddisfly larva (*Dicosmoesus*) caused broadly similar changes to benthic algal assemblages, reducing algal assemblages to representation mainly by the adnate diatom *Achnanthes* and basal cells of the filamentous green alga *Stigeoclonium* (Lamberti et al. 1987). However, it also is possible for two herbivores to have quite different effects, as Hill and Knight (1987, 1988) demonstrated in their comparison of the caddis *Neophylax* and the mayfly *Ameletus*. Loose and adnate layers were sampled separately and *Ameletus* affected principally the former, causing declines in motile diatoms including *Surirella spiralis* and several species of *Nitzschia*. *Neophylax* affected both layers but its major impact was through reducing the abundance of a particularly large, adnate diatom that comprised the bulk of total periphyton biovolume.

Environmental conditions have a strong influence on algal growth form, which in turn affects what herbivore feeding modes are likely to most effective (Fig. 10.3, Vadeboncoeur and Power 2017). Where the scouring effect of storms and grazing pressures are strong, one finds thin films <0.1 mm thick of small, tightly appressed or motile diatoms. In high light and low nutrient environments, tightly attached felts 0.1–2 mm thick of nitrogen-fixing cyanobacteria usually develop, with diatom overgrowth, providing forage for algivorous fishes. When grazing pressure is low or floods are few, mats of filamentous chlorophytes such as *Stigeoclonium* and *Cladophora* develop, usually with an epiphytic layer of diatoms. The first two assemblage types persist under intense grazing, whereas the third occurs when green algae have escaped grazing for sufficient time to become too large and unwieldy for most consumers (but epiphytic diatoms growing on macroalgal filaments may be heavily grazed by small invertebrates).

10.1.1.2 Food Quantity and Quality

The high efficiency of grazers at reducing biomass of benthic algae has several possible explanations (Hillebrand 2009). Microalgae are high-quality food in terms of P and N content. Grazers are large relative to their food items, and rasping or scraping periphyton from hard substrate surfaces allows high proportional removal of periphyton biomass, with different taxa of grazers preferentially using different layers of the periphyton (Steinman 1996). As reviewed above, grazers consume substantial quantities of benthic algae, which supports substantial grazer biomass and in turn contributes to higher trophic levels. However, food quantity alone, commonly measured as biomass or chlorophyll a, likely is an insufficient measure of the consumer's resource base. Food quality, assessed from nutrient and fatty acid content of the resource, and assimilation and growth efficiencies of the consumer, provides important insights into the resource basis of grazers and energy transfers to upper trophic levels.

A grazer's food supply may be limiting because nutrient needs of the consumer are not met by the relative availability of nutrients in the producer, which instead is more likely to reflect their relative availability in the environment (Sterner and Elser 2002). Ecological stoichiometry theory predicts that the relative imbalance between particulate nutrients in consumer and producer biomass determines nutrient limitation of consumer growth, and also the rates at which consumers recycle nutrients. The stoichiometry, or the ratio of elements, of food resources is an important indicator of food quality as it affects the growth and reproduction of primary consumers. High food quality corresponds to low periphyton carbon (C) content relative to other nutrients, such as nitrogen (N) and phosphorus (P).

Research shows that primary producers generally have high and more variable carbon to nutrient ratios relative to aquatic consumers. Incubating periphyton in once-through streamside flumes provided with different stream water concentrations and N:P ratios, Stelzer and Lamberti (2001) found that the N and P content and N:P ratios of periphyton tracked stream water concentrations. Further, growth of the snail *Elimia livescens* was greater on a diet of periphyton grown at higher streamwater P concentrations, but only when food was limited in quantity, suggesting that streamwater dissolved P influenced *Elimia* growth through its influence on periphyton chemical composition (Stelzer and Lamberti 2002). By manipulating nutrients, light, and

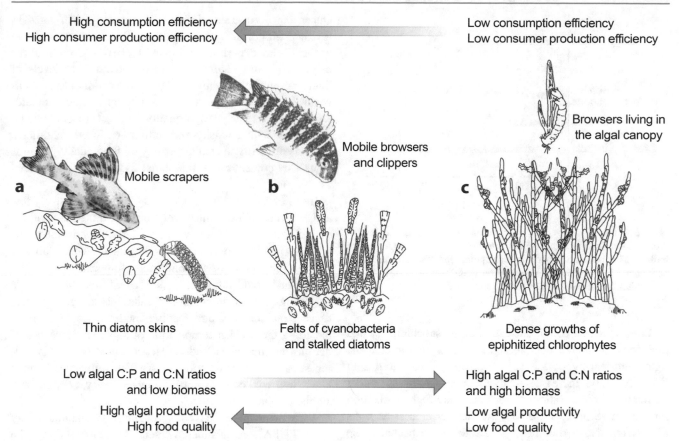

High consumption efficiency
High consumer production efficiency

Low consumption efficiency
Low consumer production efficiency

a Mobile scrapers

b Mobile browsers and clippers

c Browsers living in the algal canopy

Thin diatom skins

Felts of cyanobacteria and stalked diatoms

Dense growths of epiphitized chlorophytes

Low algal C:P and C:N ratios and low biomass

High algal C:P and C:N ratios and high biomass

High algal productivity
High food quality

Low algal productivity
Low food quality

Fig. 10.3 Schematic of the grazer-benthic algae interaction between different grazing modes and algal growth forms. (**a**) Mobile grazers with scraping mouth parts can maintain barely perceptible, thin films of rapidly growing, nutritious diatoms. Although biomass is low, algal quality is high due to high concentrations of nitrogen and phosphorus relative to carbon (low C:N). (**b**) Cyanobacteria in the family Rivulariaceae and stalked diatoms persist in the presence of grazers that consume loose algae from the low-growing felts that coat rocks. (**c**) When grazing pressure is relaxed for periods of weeks, green algae such as *Cladophora* form dense mats that provide habitat for small grazers. As biomass accumulates, algal quality declines because there are high concentrations of carbon in the algae relative to nitrogen and phosphorus. Note: organisms are not to scale (Reproduced from Vadeboncoeur and Power 2017)

grazer abundance in flow-through channels in a Tennessee, US, stream, Rosemond et al. (2000) found that periphyton exhibited less P limitation and snails had higher growth rates when nutrient supply was elevated. And in a field survey of 41 southern New Zealand streams, Liess et al. (2012) reported that periphyton C:N correlated with variation in water column dissolved inorganic N; in addition, the abundance of grazers, and of invertebrates in general, was higher in streams with higher food quality as indicated by lower C:N ratios.

In contrast to the relatively high carbon to nutrient ratios of primary producers, aquatic consumers often have lower and more constant C: nutrient ratios, reflective of their greater need for N and P in their tissues and relatively homeostatic regulation of their body chemistry. Benthic macroinvertebrates collected from streams in the Midwestern US exhibited little variation in body C, N, and P concentrations and ratios within taxa, consistent with the view that many animals exhibit relative homeostasis in elemental composition (Evans-White and Lamberti 2005). However, elemental composition differed among taxonomic groups, as insects, mollusks (soft body tissue only), and crustaceans exhibited declining C:P and N:P ratios, in that order (Fig. 10.4). Interestingly, the range of N:P for benthic insects was greater than in other taxonomic groups for which data were available.

Because animals often have relatively constant tissue stoichiometry and lower C:nutrient ratios when compared to primary producers, food quality as reflected in its elemental composition is typically much more variable for herbivores than it is for fish and invertebrates that consume animal prey (Sterner and Elser 2002). Thus herbivores, and especially herbivorous fish with their boney skeletons, have greater potential for their growth and reproduction to be limited by nutrients rather than energy or other factors. Nutrient demand of the consumer relative to nutrient composition of its resource also has important consequences for nutrient excretion and recycling, discussed further in a later section.

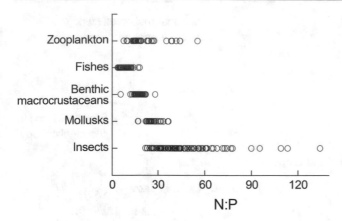

Fig. 10.4 Molar nitrogen to phosphorus ratios of different animal groups. Insect, mollusk, and benthic crustacean data are from streams in Indiana–Michigan and Wisconsin. Each point represents a taxon from a particular stream. Fish and zooplankton data from published sources (Reproduced from Evans-White and Lamberti 2005)

Environmental conditions can alter the stoichiometry and thus food quality of benthic algae to their consumers. The light : nutrient hypothesis (LNH) posits that increased light intensity may increase algal C: nutrient ratios under nutrient-limited conditions because autotroph C-fixation rates increase in response to light (Sterner et al. 1997). As a consequence, herbivore growth rates are expected to be greatest at intermediate light-to-nutrient ratios, where high-quality food is expected to be most abundant. Herbivores are hypothesized to be C limited at low light-to-nutrient ratios and nutrient limited at high light-to-nutrient ratios. In support of this expectation, Fanta et al. (2010) found that periphyton P increased in laboratory and natural streams as water column P increased, and decreased as light increased. Using fast-growing juvenile snails (*Gyraulus chinensis*) in lab streams, Ohta et al. (2011) observed results consistent with the LNH. As light intensity increased, so did periphyton biomass as well as C:N and C:P ratios in periphyton tissue. Snail growth rate and the phosphorus content of its gonadal tissue were maximized at an intermediate light intensity and were most responsive to periphyton C:P ratios under the oligotrophic conditions of the experiment. The light-nutrient interaction occurs because light can only reduce periphyton nutrient content in oligotrophic environments (Fanta et al. 2010), and so the results of experimental studies are most likely to depend on the range of light levels and nutrient concentrations employed. Using three levels of light and two concentrations of P in large, flow-through experimental streams, Hill et al. (2011) found that periphyton C:P and C:N ratios increased with light augmentation and decreased with P enrichment, consistent with the LNH. Manipulation of light (open vs. shaded) and nutrients (N + P in a slow release mixture) in cobble-bed riffles of three headwater streams in southeastern

Queensland, Australia, influenced grazer growth rates as revealed from head capsule widths at the beginning and end of the 42-day experiment (Guo et al. 2016a). Growth of the mayfly *Austrophlebioides* declined under higher levels of light and nutrients, while growth of the caddis larvae *Helicopsyche* responded positively to nutrient additions regardless of light intensity. High quality food was most abundant under low light intensity and nutrient-enriched conditions, and was primarily related to periphyton food quality in terms of its C:N content rather than algal food quantity measured as chlorophyll *a*.

Tests of the LNH in stream ecosystems are relatively few and, when examined collectively, provide mixed support for the LNH. Most likely, the variation in outcome is due to the wide variety of environmental contexts in which lab and field studies have been conducted. Important contextual factors that could influence experimental outcomes include but are not limited to consumer traits, including body size, tissue stoichiometry, and feeding mode; periphyton traits, including the species composition of algae and presence of detritus and microbes; and physicochemical factors, including stream nutrient composition (which influences benthic algal nutrient composition), flow velocity, depth, and ambient light.

Fatty acids, particularly long-chain polyunsaturated fatty acids (PUFA), are considered essential for animal diets. The higher PUFA content of algae relative to plant detritus is a key reason for their higher quality as food, and because invertebrates and fish have limited ability to synthesize fatty acids, their dietary acquisition is important to freshwater animals (Guo et al. 2016b). In an experimental test of the LNH hypothesis, described above (Hill et al. 2011), light also influenced the composition of fatty acids in periphyton, and effects were strongest in phosphorus-poor streams (Fig. 10.5). Experimental manipulation of light and nutrients in a subtropical stream in Queensland, Australia, altered the fatty acid composition of stream periphyton, which in turn influenced fatty acid content of stream herbivores (Guo et al. 2016c). The combined effects of shading and added nutrients led to increased levels of highly unsaturated fatty acids in periphyton, and increased similarity in fatty acid content between stream grazers (the mayfly *Austrophlebioides* and caddisfly *Helicopsyche*) and periphyton. Large instars of both grazers showed higher growth in response to these changes, indicating that the concentration of highly unsaturated fatty acids in periphyton is an indicator of periphyton food quality. There also is evidence that algal biofilms on the surfaces of leaf litter improve the nutritional quality of terrestrial inputs for invertebrate shredders (Guo et al. 2016d)). The stream invertebrate shredder (*Anisocentropus bicoloratus*, Trichoptera) reached larger sizes on leaves that developed an algal biofilm under enriched nutrient conditions, evidently by increasing leaf PUFA content.

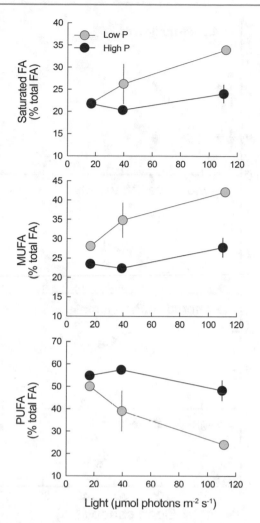

Fig. 10.5 Influence of three levels of light and two concentrations of phosphorus on the fatty acid content of periphyton in large flow through streams (22 m long by 0.3 m wide), supplied with unfiltered, low-phosphorus water from a nearby spring-fed stream. Symbols represent mean ±1 SE (n = 2). SAFA, saturated fatty acids; MUFA, monounsaturated fatty acids; PUFA, polyunsaturated fatty acids (Reproduced from Hill et al. 2011)

10.1.1.3 Behavioral Responses

Algae are patchily distributed, from the smallest scale of the surface of an individual substrate, to an intermediate scale such as from stone to stone, through larger scales such as open versus canopied sections of streams. Although some herbivores might feed essentially at random, an ability to perceive and respond to this patchiness ought to be advantageous. Grazers can concentrate in food-rich locations through behavioral mechanisms at small and even relatively large scales, and such non-random foraging has been established in both vertebrate and invertebrate grazers of periphyton. Richards and Winshall (1988) studied grazer distribution at small scales in an alpine stream, using natural stones that were selected based on visual assessment of

periphyton abundance and in some instances scraped to produce patches of various widths. Stones were replaced in the stream under glass viewing boxes, and insect presence then was determined by photography. Within one to two days, grazing mayflies of the genus *Baetis* were concentrated in patches rich in periphyton. In laboratory microcosms containing rocks with algae from a nearby stream, *Baetis* distributions were unselective when algal biomass was homogeneously distributed, but clustered on high-food rocks when it was heterogeneously distributed (Alvarez and Peckarsky 2005). Detailed analyses of foraging in the caddis larvae *Dicosmoecus* (Hart 1981) and the mayfly nymph *Baetis* (Kohler 1984) document that these insects spend much more time in periphyton-rich patches than would be expected under a model of random movement. When individual *Dicosmoecus* entered an area with abundant periphyton, gathering movements of the forelegs and the rate of mandibular scraping both increased. In addition, overall movement rate slowed, and individuals tended to turn back upon reaching a patch boundary. As a result, time spent in rich patches was two to three times what would be expected by chance alone.

The ability to perceive spatial heterogeneity in food supply and respond by simple movement rules that tend to concentrate foraging in regions of high reward is termed area-restricted search. When the periphyton attached to an artificial substrate were scraped to create a checkerboard design that covered only 20% of the substrate surface, *Baetis* spent up to 80% of its time in food patches (Kohler 1984). By comparing the area searched to the smallest area that circumscribed the sequence of movements, Kohler determined that these mayflies searched food-rich patches very thoroughly. Moreover, search behavior upon departure from a patch was influenced by patch quality. Search intensity was much greater just after departing a high quality patch, as evidenced by high thoroughness and low movement rates in comparison to movements following departure from patches of lower quality (Fig. 10.6).

Whenever highly mobile herbivores concentrate where algal resources are rich, crowding can reduce the rate of return for an individual grazer to approximately what it would experience in a less productive but less crowded region. A likely consequence is for the abundance and biomass of grazers to increase proportionally with algal productivity, but for foraging gain per individual to be roughly constant. Power (1983) observed just this pattern in the distribution of armored catfish among pools in a Panamanian stream. Shaded pools were less productive and supported a lower abundance and biomass of herbivorous fish compared to open pools. However, individual growth rates were similar across this resource gradient (Fig. 10.7). Movements of individuals among pools in a manner similar to the finer

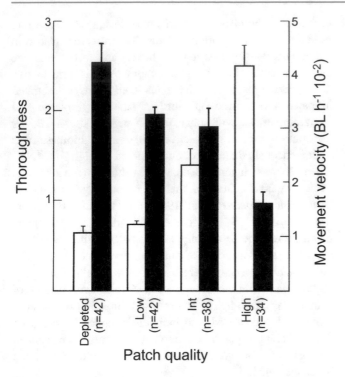

Fig. 10.6 The influence of patch quality (periphyton cell density) on *Baetis* search behavior immediately after leaving a patch. Thoroughness of searching (open bars) increases and movement rate (solid bars) decreases with increasing patch quality (Reproduced from Kohler 1984)

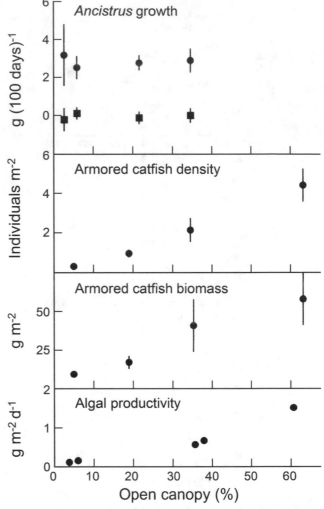

Fig. 10.7 Evidence that the loricariid catfish *Ancistrus* conforms to an ideal-free distribution. Algal productivity increases in relation to openness of canopy. Density and biomass of catfish increase proportionately with algal productivity, but growth rate (●, rainy season; ■, dry season) is constant. Two standard errors are shown (Reproduced from Power 1983)

scale foraging behaviors of *Baetis* and *Dicosmoecus* presumably result in this pattern, referred to as the ideal-free distribution.

In sum, grazing animals respond to locations of high periphyton abundance, both by shifts in distribution and, if conditions persist for long enough, by population recruitment. These concentrations of grazers can either reduce or enhance variation in the distribution and abundance of periphyton, and as we shall see in the next section, influence the composition and physiognomic structure of the periphyton assemblage.

10.1.2 Indirect Effects of Grazer-Resource Interaction

Grazers affect the periphyton by direct consumption, by physical disruption of algal mats, and through indirect pathways, especially by nutrient regeneration. The previous section summarized direct effects of consumption, including reduction in periphyton biomass, compositional changes including a reduction in the overstory component, and marked differences in effects depending on the identity of the grazer. In addition to these direct effects, grazing can indirectly influence algal nutrient content, productivity, diversity

and heterogeneity (Fig. 10.8). The meta-analysis by Liess and Hillebrand (2004) that found a strong, negative relationship between grazing and algal biomass, also revealed that on average, grazing altered algal nutrient ratios, produced algae with relatively higher P concentrations, reduced the area-specific productivity of algae but enhanced algal productivity per unit biomass, increased the spatial heterogeneity of algae, and reduced algal diversity. In general, these effects were significant but of lesser magnitude than direct consumption, and effect sizes correlated with the magnitude of biomass reduction.

Several additional pathways can account for the diverse effects grazers can have on the quality, quantity, and productivity of periphyton (Fig. 10.9). First, periphyton grazing that dislodges or consumes substantial amounts of detritus is

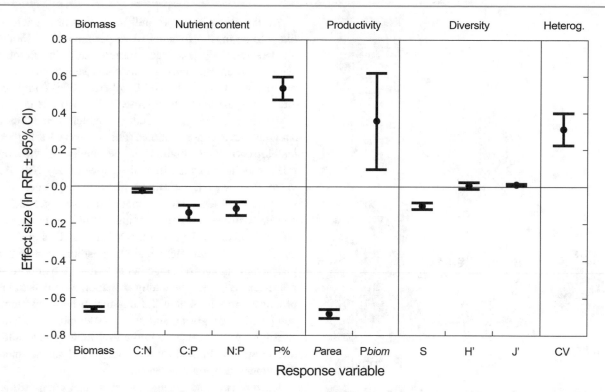

Fig. 10.8 The mean effect size of grazing based on a meta-analysis of 495 experiments reported in 116 studies, based on the log of the response ratio between grazer and control treatments. Periphyton response variables included biomass, nutrient content as percent phosphorus and molar ratios, area- and biomass-specific productivity, taxon diversity metrics and spatial heterogeneity estimated from measurement variance within grazer and control treatments. Diversity measures include number of taxa (S), Shannon diversity (H'), and relative evenness of abundance among taxa. Effect and sample size vary among response measures (Reproduced from Liess and Hillebrand 2004)

likely to result in lower C:N and C:P ratios of the remaining periphyton. Second, by reducing algal biomass, herbivores can indirectly reduce overall nutrient demand and increase nutrient uptake per unit biomass, thereby alleviating nutrient limitation of periphyton mats and increasing particulate nutrients in autotroph tissues. Increased diffusion to the periphyton layer and changes to species composition may also affect nutrient uptake rates. Third, grazer excretion and/or egestion of nutrients can directly increase the supply of nutrients available to the periphyton. Any of these effects can influence C:N:P ratios of the periphyton, and hence food quality for grazers.

Another meta-analysis by Hillebrand et al. (2008) focused on the ecological stoichiometry of grazer–periphyton studies found that grazers generally lowered C:P and C:N ratios, indicating that grazed periphyton had higher P and N content. Grazer effects on periphyton nutrient ratios varied with the nutrient content of grazers and their food, as well as grazer biomass, the amount of biomass removal, and water column nutrients. In addition, Hillebrand et al. (2008) found a trend towards increased periphyton particulate N relative to particulate P, such that periphyton N:P increased with grazer density.

Subsequent studies have more directly examined the importance of nutrient recycling due to consumer excretion

and egestion by explicitly comparing elemental composition of grazer and resource. Ecological stoichiometry, or the study of the balance of elements in ecological processes, predicts that an herbivore with a high body particulate P and thus a low body N:P ratio would be more likely to experience P-limitation than would another herbivore with lower body P and a higher body N:P ratio (Sterner and Elser 2002). One would also expect that P-rich species would selectively retain P and excrete waste with a high N:P ratio. A study of nutrient excretion by snails (body N:P = 28) compared with crayfish (body N:P = 18) supports this prediction (Evans-White and Lamberti 2005). As expected from its body elemental ratio, crayfish excretion had a significantly higher ratio of ammonium: soluble reactive phosphorus than did snails that had higher body N:P ratio. Additionally, the N content of periphyton in lab streams was higher in the presence of crayfish than snails, and the P content was lower in the presence of either grazer, suggesting that benthic grazers can alter nutrient composition and limitation of periphyton via nutrient excretion.

Comparison of two grazers of a Neotropical stream that represent extremes of body stoichiometry in vertebrates provide further insights. Due to their extensive, P-rich bony-plated armor, the armored catfish *Ancistrus triradiatus*

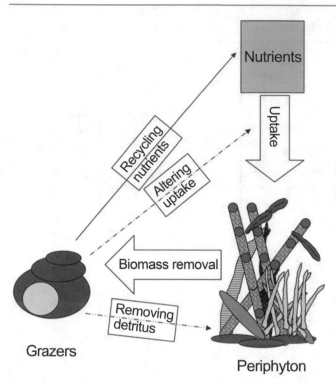

Fig. 10.9 Illustration of the most important direct and indirect effects of grazers on nutrient stoichiometry. Grazers can influence periphyton carbon to nitrogen to phosphorus ratios by removing detritus, which reduces the carbon content of periphyton; by reducing detritus and senescent cells by grazing and dislodgment, thereby increasing nutrient and light availability to actively growing cells; and by recycling nutrients via excretion and egestion (Reproduced from Hillebrand et al. 2008)

has a high body P, and thus low body N:P, whereas early tadpoles of the frog *Rana palmipes* have low body P and thus a high body N:P. As would be expected, catfish retain P and excrete at a high N:P ratio, and tadpoles do the opposite (Vanni et al. 2002). Using a grazer exclusion design in which periphyton was exposed to both grazing and excretion or protected from grazing but exposed to consumer excretion, Knoll et al. (2009) demonstrated that grazer identity can affect periphyton nutrient stoichiometry through both grazing and excretion. Water nutrient concentrations of soluble reactive phosphorus were highest in the tadpole treatment, while nitrate-N concentrations and N:P ratios were highest in the catfish treatment. Periphyton N:P increased in the presence of catfish and decreased in the presence of tadpoles, indicating that the elemental composition of grazers had an effect on periphyton N:P ratios mediated through grazer excretion and streamwater chemistry. Catfish stimulated the growth of periphyton protected from grazing but exposed to grazer excretion, but tadpoles did not, consistent with other evidence that periphyton in this Neotropical stream are primarily N limited.

As the above example indicates, consumer excretion is likely to be most influential when periphyton are limited by the preferentially excreted nutrient, and in oligotrophic systems when nutrients are in short supply. To test this prediction, Evans-White and Lamberti (2006) compared the effects of grazing by stoichiometrically disparate consumers —snails and crayfish—under ambient and elevated streamwater P concentrations. They observed a higher C:P and lower P concentrations in periphyton in the crayfish treatments, as expected from their lower body N:P. At elevated P, however, consumer identity no longer affected periphyton elemental composition, suggesting that consumer-driven nutrient dynamics and consumer identity are more likely to be important when nutrients are in short supply. Clearly, the elemental composition of streamwater, periphyton, and grazers together determine the importance of grazer excretion in alleviating stoichiometric constraints on producer growth. Where the mis-match between elemental composition of grazer and its food supply is substantial, consumer growth may be limited by food quality, and producers may be limited by the volume and stoichiometry of consumer excretion and egestion.

Grazers can also change the elemental composition and the nutrient demand of the periphyton community by shifting community composition toward grazer-resistant taxa that might differ in their C:N:P stoichiometry. This was shown in a multi-factorial lab experiment that assessed periphyton nutrient stoichiometry, algal taxonomy and biomass, and dissolved nutrients in response to grazing by the gastropod *Viviparus viviparus* (Liess and Kahlert 2007). Grazing resulted in strong dominance by the mucilage-producing algae *Chaetophora* and had a pronounced effect on periphyton nutrient stoichiometry because mucilage has high C and N content, but low P content.

10.1.3   Disturbance and Herbivory

The interaction between grazers and periphyton generally is considered to be strongest under stable environmental conditions and diminished when environmental conditions are extreme or highly variable, often referred to as the "harsh-benign" hypothesis. Most field studies of grazing have been conducted under low flow conditions, and thus do not adequately represent interactions during the more physically stressful conditions associated with environmental extremes that can occur seasonally or episodically (Feminella and Hawkins 1995). Such extremes are commonly referred to as disturbances. Often, disturbances are generated by abiotic factors, such as changes in flow or temperature; however, they can also be generated by biotic agents such as invasive species. Many of the best documented examples of disturbance influencing herbivory involve extremes of current.

Grazing may result in an algal assemblage that is less vulnerable to scouring by floods, presumably by reducing mat build-up. When the periphyton in laboratory streams subjected to variation in grazing pressure by the snail *Elimia clavaeformis* experienced a common scour disturbance, the structural characteristics of the periphyton exposed to snail grazing were more resistant to change than periphyton communities with no previous exposure to snails (Mulholland et al. 1991). Because stream current influences the architecture and taxonomic composition of periphyton, it creates conditions under which some species can forage more effectively than others, and facilitates conditions whereby interactions between a given species and flow can influence populations of other grazing species. For example, in a mesocosm experiment with *Glossosoma verdona* (Trichoptera) and *Drunella grandis* (Ephemeroptera) (Wellnitz and Poff 2012), senescent filaments of *Ulothrix* (a green alga) became abundant under low flow conditions, entangling *Glossosoma* and causing weight loss and mortality. However, *Drunella* was able to reduce senescent filaments across all experimental treatments, and *Glossosoma* survivorship and weight gain in slow current was positively correlated with *Drunella* density.

Experimental reduction of grazers at locations of differing current velocity revealed an interaction between grazing and current in a Colorado stream (Opsahl et al. 2003). After 45 days, tiles that were experimentally electrified to reduce grazer populations had significantly fewer grazers and more than twice the algal biomass compared with control tiles. Greater algal abundance on control tiles in slow currents suggested that grazers differed in their ability to regulate algae across the current velocity gradient. Hintz and Wellnitz (2013) observed a subsidy-stress effect of increasing current velocity on accumulation of algal biomass in artificial streams, as increasing current first facilitated biomass accumulation (subsidy) but at higher velocities removed it (stress). Determinants of either the facilitation or reduction of biomass varied with the identity of three grazing mayflies, presumably the result of differences in tolerance of species to variations in current. Algal biomass and assemblage composition showed some variation between wet and dry seasons in Hong Kong streams that was attributable to spate-induced disturbance caused by monsoonal rains during the wet season (Yang et al. 2009). Averaged across four streams, algal biomass was only modestly greater in the dry season, possibly due to strong grazing by the algivorous fish *Pseudogastromyzon myersi*.

In summary, grazing by invertebrates, amphibian larvae, and fishes is an important energy pathway to higher trophic levels. Grazing has strong direct effects on the abundance,

physiognomy, and species composition of benthic algae, and can influence algal nutrient availability and productivity. In this chapter we have limited our focus to the interaction between just these two trophic levels. However, benthic algae are influenced by a number of different environmental variables (Chap. 6), grazers may be as influenced by their predators as their food supply (Sect. 10.2), and these two-way interactions are embedded in a much more complex food web with interactions across multiple trophic levels (Chap. 12). As we broaden our perspective to include all of the environmental variables that affect periphyton, including seasonal and episodic disturbance as well as the predators and parasites that can regulate grazer abundance, it becomes increasingly apparent that biological assemblages are complex entities subject to multiple, interacting controls.

10.2 Predation

Predation is ubiquitous. All hetcrotrophic organisms are prey for others at some stage of their life cycles, and many species encounter predation risk throughout their lives. The potential effects of predation are diverse, and include reduction in abundance or even the elimination of a species from a region, restrictions on behavior, habitat use, and foraging efficiency that affect growth rates and reduce fitness, and adaptation via natural selection to persistent predation risk. Top predators can cause a potential cascade of interactions through the food web, directly affecting prey by reducing their abundance and changing their foraging behavior, and indirectly influencing additional species to which the prey are linked as food or competitors. Furthermore, changes in energy pathways and species composition may have consequences for nutrient utilization and regeneration. In this chapter we consider the predator-prey linkage as an interaction that has effects on populations, directly through consumption and mortality, and indirectly through behavioral and morphological adaptations that may entail some fitness cost to the prey in order to survive. In Chap. 12 we will examine how predation can trigger trophic cascades that have consequences for the entire ecosystem.

10.2.1 The Predator-Prey Interaction

All predators show some degree of preference, feeding mainly on certain species, size classes, or types of prey. A large literature documents the diet of vertebrate and invertebrate predators in running waters. It is evident that predation is complex and its influence depends in part on aspects of the predators, including their morphology, foraging mode, means of prey detection, and size relative to the size of prey. Prey characteristics also play a major role in predation, as prey abundance, activity, visibility, and size strongly influence detection probability, attack, and capture rates. Predator and prey behavior and population size can vary seasonally and across habitats, adding spatial and temporal complexity in predator-prey interactions. Further, most predators must also contend with their own predation risk from larger predators and from aggressive competitors.

10.2.1.1 Prey Selection by Fishes

Most fishes of temperate steams, including the younger life stages of more piscivorous fish, feed on invertebrates. Aquatic invertebrate prey can be captured from the benthos and as individuals suspended in the water column, referred to as "drift". Terrestrial invertebrates falling onto the water surface also can be an important diet supplement for aquatic predators. By comparing prey abundance, average size, and species composition collected in fish stomach contents with the characteristics of the aquatic invertebrates collected in stream habitats, numerous studies have found prey abundance and size to be strong predictors of fish diet. In general, the number of prey eaten increases with prey abundance for all types of predators, at a decelerating rate due to the time limitation imposed by the handling and ingestion of individual food items. This relationship is known as a functional response curve. Whenever more than one type of prey is present, gut analyses generally find that prey that are abundant in the environment are also common in the diet (Allan 1981). However, the correspondence often is not 1:1, indicating some degree of predator selectivity. Prey choice can be strongly influenced by contrast, motion, and size, all of which serve to make certain prey more conspicuous. Larger prey items are expected to be preferred both because they offer a greater energy reward and because they are more readily detected.

Feeding trials using juvenile coho salmon *Oncorhynchus kisutch* in lab streams nicely illustrate the variables influencing fish reaction to prey floating on the water surface (Dunbrack and Dill 1983). Both reaction distance (Fig. 10.10) and attack distance (Fig. 10.11) were increasing functions of food width and fish size. Attack distance declined in satiated fish and the probability of ingestion following an attack declined with largest prey. The predicted size composition of a fish's diet based on these relationships was tested with fish captured from the wild, held for 24 h without food to ensure their guts were empty, and then released into cages placed in a stream. Predicted and actual diet corresponded closely, and showed a strong bias towards larger prey.

Feeding behavior often changes with experience and learning in vertebrate predators. Searching often improves via greater reactive distances, higher swimming speeds, and greater path efficiency, while attack latency may decrease and capture success may increase (Dill 1983). The result is a

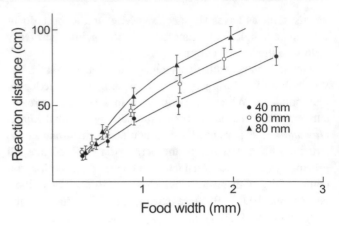

Fig. 10.10 Reaction distance (cm) as a function of prey (stonefly) width (mm) and fish size (40, 60 and 80 mm length) for young coho salmon (*Oncorhynchus kisutch*). Lines fitted by eye. Vertical bars represent 1 SE (Reproduced from Dunbrack and Dill 1983)

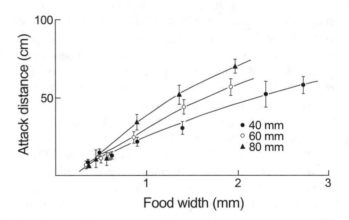

Fig. 10.11 Attack distance (cm) (distance swum by fish from its station to point of capture) as a function of food width and fish length for young coho salmon (*Oncorhynchus kisutch*). Lines fitted by eye. Vertical bars represent 1 SE (Reproduced from Dunbrack and Dill 1983)

tendency to specialize on the prey that the predator has consumed most frequently in its recent feeding history, with an accompanying increase in foraging efficiency. Hunger can influence predation rate by modifying any of several aspects of predatory behavior. As hunger declines, search behavior also declines owing to changes in movement speed and reactive distance. In addition, the probability that an attack will follow an encounter declines, and handling time of prey tends to increase (Ware 1972). Capture rate consequently varies with hunger level.

Prey of terrestrial origin falling onto the water surface can be an important part of the diet of some stream fishes, especially day-active, visual predators. Based on stomach content analysis and floating surface traps that captured terrestrial infall into small streams of southeastern Alaska, US, juvenile coho salmon consumed approximately equal fractions of terrestrial and aquatic prey (Allan et al. 2003).

Terrestrial inputs may be especially important during summer, when warmer temperatures result in greater energetic demands, and aquatic invertebrates are primarily of small size following adult emergence and reproduction. In the Horonai stream of Japan during July-August, terrestrial prey comprised over 70% of daily biomass consumption by rainbow trout *O. mykiss* (Nakano et al. 1999b; Nakano and Murakami 2001). Trout foraging was greatest at dusk and dawn, when light may limit drift-feeding effectiveness, and terrestrial input peaked near dusk, whereas drift by benthic invertebrates peaked near midnight (Fig. 10.12). Thus, diel periodicity in the availability of terrestrial and aquatic invertebrates, as well as timing of foraging by trout, explain the dominance of terrestrial prey in trout diet. Again, prey size played an important role, as terrestrial invertebrates were larger than aquatic items, and were selectively consumed.

Environmental variables, especially those that affect prey visibility, can significantly modify predation rates. Although visually-dependent predators can feed under quite dim light, prey capture success declines with falling light levels. A light intensity of 0.1 lx, corresponding to late dusk or a full moon, often is the lower threshold for effective visual location of prey (Hyatt 1979). Even within the range we consider daylight, however, gradation in light level can be influential on rates of predation. Wilzbach et al. (1986) compared the feeding of cutthroat trout in pools from forested sections of streams with pools from open (logged) sections. Prey were captured at higher rates in open pools, and artificial shading lowered the capture rate to that observed in shaded pools. Under varying light conditions

corresponding to twilight, moonlight, and overcast night conditions, the foraging efficiency of young Atlantic salmon *Salmo salar* in the laboratory was unaffected by current velocity until light levels fell below 0.1 lx, at which point the fish were more efficient at prey capture in slower currents (Metcalfe et al. 1997). When provided a choice of foraging location, juvenile salmon shifted towards slower velocity position as light level declined.

While studies of prey capture by fish feeding on surface or water column prey nicely illustrate how size and visibility influence prey capture, predators are opportunistic, shifting their foraging behavior between sit-and-wait and active search. They move between habitats, and exploit prey from the surface, the water column, and the benthos. In a study of prey selection by brook trout, *Salvelinus fontinalis*, Forrester et al. (1994) established trout densities in replicate 35-m long stream sections at either medium or high levels relative to natural densities. Trout fed selectively on larger prey during the daytime, preferring cased caddis larvae and several species of mayflies and stoneflies, but showed no size selection at night. Because cased caddis rarely drift, their consumption presumably reflects benthic foraging. Other prey may have been captured by drift-feeding, and reduced visibility and the resulting smaller visual field at night may explain the absence of size selection at night.

Although less well studied than the salmonids, research into foraging by other stream fishes illustrates some very different feeding modes. Fishes in the family Galaxiidae (Fig. 10.13) are the dominant and most species-rich group of freshwater fishes throughout the cool southern hemisphere, with >50 species occurring in many parts of Australasia, Patagonian South America, and South Africa (McDowall 2006). Galaxiids actively pursue prey at night, apparently detecting prey by disturbing them or by contact, although they also capture prey during the day from a fixed position (McIntosh and Townsend 1995). In Central Europe, the gudgeon (*Gobio gobio*) and stone loach (*Barbatula barbatula*) are benthivorous fish that feed primarily at night and non-visually (Worischka et al. 2015). Principal prey were small bodied, active, and abundant aquatic invertebrates, especially chirononomids and simuliids (Diptera). Visual detection may play a small role in prey selection in the gudgeon, but not in the loach; however, these two predators were broadly similar in diet. They differed in habitat occupation, however. Field video revealed that the gudgeon foraged almost exclusively in pools and was more active at night than during the day, whereas the stone loach used both riffles and pools, and foraged only at night (Worischka et al. 2012).

Feeding from a fixed location on prey items delivered by the current is common in many stream fishes, and is the basis for pioneering efforts to model foraging efficiency based on energy gain from prey consumption and energy expenditure

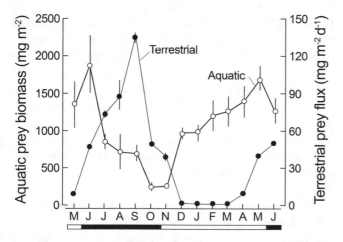

Fig. 10.12 Seasonal variation in prey availability as aquatic invertebrates versus terrestrial infall to the Horonai Stream, Hokkaido, Japan. Aquatic prey biomass estimated from substrate sampling, terrestrial prey inputs from pan traps on the stream surface. Black and white portions of horizontal bars at bottom of figures indicate leafing and defoliation periods, respectively (Reproduced from Nakano and Murakami 2001) © 2001 by The National Academy of Sciences of the USA

Fig. 10.13 (**a**) a species of galaxias (*Galaxias vulgaris*) from New Zealand. Fishes in the family Galaxiidae are the dominant and most species-rich group of freshwater fishes (with >50 species) throughout the cool southern hemisphere, found in many parts of Australasia, Patagonian South America, and South Africa. (**b**) The brown trout, *Salmo trutta*, native to northern Europe and now common in many streams in the southern hemisphere, frequently restricting or eliminating galaxiids. Photos by Angus McIntosh

in prey capture (Fausch 1984; Hughes and Dill 1990). Foraging theory holds that individuals attempt to optimize their net energy intake (NEI), and thus their growth and fitness, by strategies that provide high energy intake while minimizing energy expenditures. For drift-feeding fish, this translates to selecting locations that maximize prey delivery while minimizing swimming costs, and it further implies that more dominant individuals will successfully compete for preferred locations. On the premise that a drift-feeding fish could optimize its net energy intake by selecting locations in low water velocity near faster currents that deliver abundant drifting invertebrates, Fausch (1984) developed a simple model of positions held by salmonids. Tested in laboratory streams, the growth rate of juvenile trout and salmon increased with NEI, and the rank of NEI at positions held by coho salmon correlated nearly perfectly with their rank in the dominance hierarchy.

Subsequently, Hughes and Dill (1990) developed a more detailed model of position choice by drift-feeding fish in which the number of prey a fish encounters varies with its reaction distance to prey, water depth, and water velocity, while the proportion captured declines with water velocity. Net energy intake is derived from food consumption minus the swimming cost calculated by using water velocity at the fish's focal point. Positions chosen by solitary Arctic grayling (*Thymailus arcticus*) in the pools of a mountain stream in Alaska closely matched Hughes and Dill (1990) model predictions and proved superior to predictions from the Fausch (1984) model, due to more realistic assumptions about the number of prey the fish detected and the influence of water velocity on the prey capture abilities of the fish.

As elegant as this model is, it likely needs further refinement to accurately describe fish foraging behavior. A test of the Hughes and Dill model with brown trout in a New Zealand stream found that reaction distance equations predicted prey size selection well, but the model over-estimated prey capture rates by a factor of two, prey detection averaged only half the expected value, and capture probability decreased rapidly with distance from fish's focal

point (Hughes et al. 2003). In addition, trout captured about two-thirds of their prey downstream of the focal point, rather than upstream. Water temperature also may play a role. In experiments with juvenile brown trout at different temperatures, Watz et al. (2012) observed station-holding at temperatures below 10 °C, and a positive relationship between the proportion of time a fish spent holding a foraging station and the fish's capture success when feeding on drifting prey. Above 10 °C, however, trout shifted from a sit-and-wait foraging mode to cruise foraging, achieving nearly equal capture success regardless of foraging mode, and capturing most of the prey encountered. At low temperatures, fish may lack the swimming capacity to pursue prey detected further downstream, compared with their ability in warmer water. The shift from drift to cruise feeding may also compensate for low prey densities because cruising fish search both the water column and the stream bed for food (Fausch et al. 1997; Nakano et al. 1999a). This suggests that foraging models based mainly on drift feeding may need to be broadened to incorporate both feeding modes. An additional energetic cost experienced by drift-feeding fishes is related to pursuit of false food items, such as drifting debris. Using high-definition video to measure the reactions of drift-feeding juvenile Chinook salmon (*O. tshawytscha*) to natural debris, Neuswanger et al. (2014) found that up to 25% of an individual's foraging time was expended in capturing and rejecting inedible particles. This likely represents an energetic cost that varies with environmental conditions, but is not yet built into models of foraging time and energy return.

At present, fish habitat needs are often estimated using physical habitat models based on statistical associations of fish assemblages with simple hydraulic variables such as width and depth, known as habitat suitability curves. As discussed in Chap. 5, there is a growing sense among fisheries managers that such methods fail to represent a mechanistic understanding of fish habitat requirements (Railsback 2016). Further refinement of foraging models has the potential to predict fish habitat requirements based on food supply and net energy intake, thereby generating an improved understanding of larger-scale patterns of stream fish distribution, growth, and abundance.

Discussion of fish foraging behavior would be incomplete without acknowledgement that predators need to be wary of their own enemies. Exclusion of great blue herons (*Ardea herodias*) and belted kingfishers (*Ceryle alcyon*) from sections of two small prairie streams in Illinois, US, resulted in significant increases in medium size classes of two common prey, striped shiners (*Luxilus chrysocephalus*) and central stonerollers (*Campostoma anomalum*) (Steinmetz et al. 2003). As with most predators, fish-eating birds have size preferences that vary among species, avoiding prey that are either too small or too large, but this topic appears understudied. Riparian predators can also present challenges for fishes. Photographic monitoring of rainbow and cutthroat (*O. clarkii*) trout tethered in shallow microhabitats that lacked cover recorded successful captures by riparian predators, including raccoons and eight species of birds (Harvey and Nakamoto 2013). While capture rates likely were higher in this study than is the case with free-swimming fish, they demonstrate that fish in streams are at significant chronic risk from a variety of predators. Of course, habitat choice, including use of cover and locations of greater depth, will influence predation risk. As Power (1984b) showed in streams in Panama, herbivorous fishes avoided shallow-water areas around the margins of pools, resulting in "bathtub rings" of algae.

What a predator eats is often determined simply by an analysis of stomach contents, either by dissection after capture of both invertebrate and vertebrate predators, or non-lethally with fish by forcing water into the stomach with a syringe, flushing out prey. Because the diet of fish often reflects prey availability, which can change seasonally and with habitat, stomach analysis provides only a snapshot of what a particular fish is eating at a particular time. An alternative approach uses stable isotopes of carbon and nitrogen as indirect tracers of fish diets, and because the isotopes are from body tissues, they reflect an integration over time. Compared with stomach analyses, stable isotopes provide time- and space-integrated insights into trophic relationships that can be used to develop models of trophic structure (Layman et al. 2012). The most commonly analyzed elements in food-web analyses are nitrogen (N) and carbon (C). The ratio of ^{15}N to ^{14}N (expressed relative to a standard, $\delta^{15}N$) increases (becomes "enriched") with each trophic transfer, and thus is an estimator of trophic position. The ratio of ^{13}C to ^{12}C ($\delta^{13}C$) differs among primary producers with different photosynthetic pathways (e.g. C3 versus C4 photosynthetic pathways in plants), but changes little with trophic transfers. Because algae primarily use the C3 pathways and many terrestrial plants (and hence allochthonous detritus) are C4, a plot of δ15 N against δ13C for organisms in a food web reveals both trophic position and the relative importance of autochthonous vs allochthonous production within a food web.

The usefulness of stable isotope analysis to identify predator feeding habits is nicely illustrated by a study of four crocodilian species in the central Amazon (Villamarin et al. 2017). Most crocodilians are considered generalist opportunistic predators that feed on any source of protein available, with any diet differences attributable to habitat use. However, the $\delta^{13}C$ signature, determined from a claw and small piece of dorsal tail muscle from captured animals, revealed differences in diet resulting not only from habitat selection but also from prey preferences. Mean $\delta^{13}C$ values were highest in the headwater species, intermediate in two species of flooded-forest streams and lowest in the species occupying floodplains, reflecting an increasing downstream reliance on aquatic over terrestrial resources. Significant differences were also observed between two co-occurring species of dwarf caiman, *Paleosuchus trigonatus* and *P. palpebrosus*, indicating different prey bases despite habitat overlap.

10.2.1.2 Invertebrate Predators

Relative body size of species within a food web strongly influences trophic relationships, with consequences for resource partitioning, diet breadth, and predator-prey interactions. Most predators consume prey that are smaller than themselves, and at least for invertebrate predators in running waters, on average the mass of their prey is roughly two orders of magnitude less, or about one percent, of the mass of the predator (Woodward and Warren 2007). While ease of subduing prey provides an obvious mechanism, it is instructive to consider each component of a successful predation event. Likelihood of detection may increase with size, and very small prey may not elicit an attack. Attack probability is often thought of as unimodal, around an optimal prey size, although this may not be universal. Capture success may decrease with largest prey, while handling time (and thus the overall rate of obtaining prey) increases for larger prey. Larger species may outgrow predation risk, entering a 'size refuge' at some stage of their life cycle, whereas smaller species may never reach a size where they escape predation.

Although these statements may apply to all predator-prey interactions, they are especially true within invertebrate systems, as is illustrated by the extensive mutual predation and cannibalism seen within the predator guild of Broadstone Stream in southern England (Woodward and Hildrew 2002). The six species (three predaceous midges, a caddisfly, an alderfly, and a dragonfly) exhibited marked size differences, but relative size relationships changed seasonally due to growth (Fig. 10.14). Small predators had the narrowest diets. The overlap in the size of prey consumed was greater when predator sizes overlapped strongly, but declined as predator size diverged (Fig. 10.15). The largest predator, *Cordulegaster boltonii*, was preyed upon only by larger conspecifics, and the smallest, *Zavrelimyia barbatipes*, was eaten by all five of the larger species and by conspecifics. The direction of intra-guild predation could be reversed whenever early instars of large species coexisted with late instars of small species. In this system, clearly, food web structure was influenced mainly by body size relationships, although encounter probabilities and foraging mode also were influential.

Predator foraging mode affects prey vulnerability, interacting with aspects of prey movement to influence localized encounter rates and departures. Mobile prey are likely to flee if able to detect the approach of large, actively searching predators, and so predator impact may be greatest with least mobile prey. This is a complication for cage experiments, which have the potential to overestimate predator impact when predator and prey are confined, and to underestimate whenever prey can escape or colonize from the surrounding environment (Wooster and Sih 1995). For sit-and-wait predators, prey mobility may increase their mortality as a consequence of increased encounter rates. In the Broadstone Stream, predation by the dragonfly *C. boltonii*, a sit-and-wait predator, fell most heavily on mobile mayflies, which were

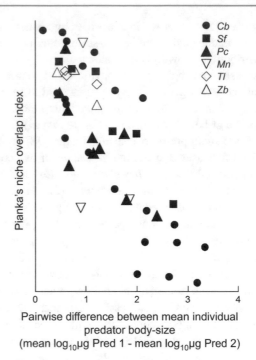

Fig. 10.15 Pairwise dietary overlap among invertebrate predators as a function of differences in individual predator body-size using mean log dry mass of pairs of predators among size classes within each species. See Fig. 10.14 for species codes (Reproduced from Woodward and Hildrew 2002)

Fig. 10.14 Relative abundance size-spectra of benthic macroinvertebrates in the Broadstone Stream, U.K, on six sampling occasions in 1996–97. The double-headed arrows indicate the size ranges of the six predator species. From largest to smallest the predators include the dragonfly *Cordulegaster boltonii*, the alderfly *Sialis fulginosa*, the caddisfly *Plectrocnemia conspersa*, and three tanypod midges *Macropelopia nebulosa*, *Trissopelopia longimana*, and *Zavrelimyia barbatipes* (Reproduced from Woodward and Hildrew 2002)

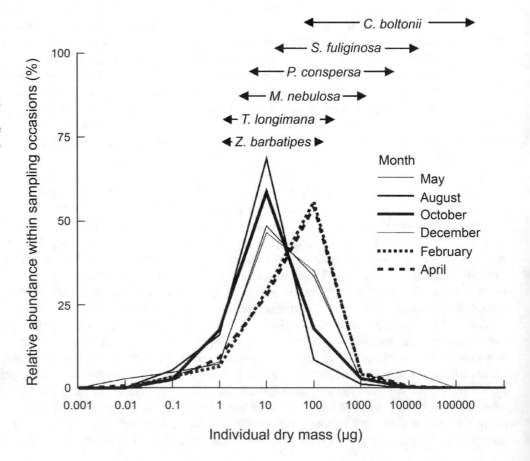

not greatly depleted due to high prey exchange rates, but their losses were indeed attributable to consumption rather than flight (Woodward and Hildrew 2002). In the same system, the net-spinning caddis *Plectrocnemia conspersa* also was reported to have the greatest impact on mobile prey (Lancaster et al. 1991). Prey abundance, movement by crawling or drifting, and speed of prey movement and predator attack likely are additional variables affecting encounter rate and capture success with sit-and-wait predators.

The foraging behavior of predaceous invertebrates does not appear to be much influenced by prey availability or prior experience, although it has been suggested that predators aggregate in areas of high prey density (Malmqvist and Sjostrom 1980). However, Peckarsky and Dodson (1980) found that predaceous stoneflies were no more likely to colonize cages containing high prey densities than cages with few prey. Peckarsky (1985) argued that the absence of any aggregative behavior in these predators is explained by the ephemeral nature of prey patches, since highly mobile potential victims like *Baetis* can rapidly disperse. Hunger level did influence which prey were consumed by the stonefly *Hesperoperla pacifica* offered a choice between the soft-bodied, agile mayfly *Baetis bicaudatus*, and the slow and clumsy *Ephemerella altana*, which has a spiny and rigid exoskeleton (Molles and Pietruszka 1983). Starved stoneflies ate mostly *E. altana*, while satiated stoneflies ate both prey in about equal numbers. When freshly-killed prey were offered to starved predators, however, a preference for *Baetis* was evident. The proposed explanation was that starved predators attacked both prey about equally, but with increasing satiation began to restrict their attack only to *Baetis*.

Habitat complexity and the availability of refuges can markedly alter predation rates. Refuges may be absolute, rendering the prey unavailable, but more commonly they serve to reduce the likelihood of encounter and capture. In laboratory trials with two invertebrate predators, four invertebrate prey, and various substrate conditions, Fuller and Rand (1990) showed that all variables affected prey capture rates. *Baetis* was more vulnerable than the other prey (an ephemerellid mayfly, a blackfly larva, and several hydropsychids), probably because its mobility led to high encounter rates. The predators, a stonefly and an alderfly, differed in their predation rates on various substrates due to differences in their sensing of prey with their antennae and with pursuit success. The substrates, which included sand, gravel mixed with pebbles, and artificial turf, resulted in differential capture success via its effects on encounter rates and by facilitating the construction of stronger retreats by some caddis larvae. Although the particular outcomes may be influenced by specifics of the experimental design, such effects of habitat complexity on prey capture probably are common.

10.2.2 Effects of Predation on Prey Populations

10.2.2.1 Direct Effects on Prey Populations

Predators may consume enough prey to reduce prey populations. Often referred to as a direct effect or consumptive effect, predator regulation of prey populations has been investigated by comparisons of predator and prey spatial distributions, by comparisons of prey production with predator consumption, and with field experiments using enclosure/exclosure designs from small cages to large stream sections. Experimental results vary widely, from strong effects of predators on their prey, including increases in benthic algae due to control of grazers; to modest impact on certain, usually large, prey; to no discernible influence at all. Some of this variation appears to be associated with the environmental context, and some may be due to details of experimental design. Most field manipulations are of fish presence/absence, although a few involve invertebrate predators, and the extent of prey reduction attributable to consumption or emigration is not always resolved.

Size classes and species of fishes that are vulnerable to piscivores frequently show an inverse relationship between predator and prey abundances or exhibit non-overlapping distribution patterns. Surveys of fish assemblages at 86 pool sites in tropical streams in Trinidad provided cases where the widely distributed killifish *Rivulus hartii* occurred alone, as well as in various combinations with other species (Gilliam 1993). Its distribution was largely complementary to the piscivorous fish *Hoplias malabaricus*, and its abundance at sites with other species was only about one-third of that predicted from expectations based on *Rivulus*-only pools. In a comparison of 18 streams with trout and six streams without trout in central Finland, *Baetis* densities were five-fold higher in troutless streams, midge larvae showed a non-significant trend towards greater abundance in trout streams, and cased caddis larvae did not differ between the two types of systems (Meissner and Muotka 2006).

Field manipulations have also demonstrated predator impact on prey populations. In a tropical river in northern Australia, macroinvertebrates were roughly twice as abundant in large fish exclusion cages that were approximately 20 m in length and 2–7 m wide (Garcia et al. 2015). Using enclosures with open cobble/gravel bottoms and large-mesh netting that allowed invertebrates to move freely, Dahl (1998) introduced brown trout (*Salmo trutta*) and leeches (*Erpobdella octoculata*) separately and together to compare the influence of a predator selective for large prey, the trout, and one selective for small prey, the leech. Invertebrate size distribution and biomass were reduced in the presence of trout due to a strong effect of trout on the amphipod *Gammarus*. This pattern was due to direct consumption of the amphipod as well as emigration of the amphipod out of the enclosures. Although leeches alone did not reduce prey biomass, their presence skewed prey size distribution towards larger individuals, mainly due to emigration by smaller invertebrates from the enclosures. Whether invertebrate predators can have as strong an impact on prey as vertebrate predators is subject to some controversy, but clearly the influence of the former can be substantial, although the species and size classes affected most likely differ.

Several studies of secondary production of predators and prey have concluded that predators consume a large proportion of the available prey production. For instance, brown trout (*Salmo trutta*) consumed essentially all macroinvertebrate production in a New Zealand stream (Huryn 1996), and macroinvertebrate predators likewise consumed all prey production in a southeastern US stream (Wallace et al. 1997). Although the demonstration that the majority of energy produced at one trophic level is consumed by higher trophic levels is not definitive evidence of either bottom-up or top-down control, it certainly indicates that consumption by predators is the principal fate of the trophic level in question. Estimated prey consumption by stoneflies, the most abundant invertebrate predators present at several sites in a Rocky Mountain stream, was roughly half that attributed to trout, suggesting that the influence of invertebrate predators was less than that of fish (Allan 1983). On the other hand, when fish are absent it seems plausible that invertebrate predators consume all secondary production at lower trophic levels. Indeed, predaceous invertebrates consumed nearly all production by detritivorous invertebrates in a coastal stream (Smith and Smock 1992), and consumption by invertebrate predators also was high in small, fishless streams in the southeastern US (Hall et al. 2000).

As a counterpoint to the above, several predator manipulations have reported only modest or no effects on species composition. Removal of the top predator *Abedus herberti* (Hemiptera: Belostomatidae, the giant water bug) from mesocosms placed in arid-land stream pools in southeastern Arizona, US, had no overall effect on species richness or abundance of invertebrate prey but consistently affected large-bodied species (Boersma et al. 2014). Reaching lengths up to 4 cm, the giant water bug primarily reduced abundances of mid-sized (> 10 mm) predators such as dragonfly nymphs. Similarly, the overall diversity and abundance of aquatic insects colonizing substrate did not differ between cages that were exposed to fish predation, compared with exclusion cages, but the diversity of large (>8 mm) invertebrates increased in the absence of predators (Flecker and Allan 1984). Exclusion of trout from 100-m reaches of a small stream in Finland resulted in significant benefits to large prey, particularly predaceous invertebrates and cased caddis, but *Baetis* mayflies and chironomid larvae were unaffected (Meissner and Muotka 2006). Removal of an

entire macroconsumer assemblage of fishes and predatory shrimp from pools in a tropical stream in Hong Kong resulted in negligible impact on benthic invertebrates, although grazing mayflies increased modestly (Ho and Dudgeon 2016). Several studies manipulating trout abundance at both large (Allan 1982) and small (Ruetz et al. 2004; Zimmerman and Vondracek 2007) experimental scale have failed to detect any change in prey abundance, possibly because of high dispersal movements of abundant invertebrate prey.

Whether direct, top-down control of stream invertebrates by predators is strong or weak most likely depends on natural context and experimental design. When predator populations are limited by available habitat or their own predators, they are unlikely to consume enough prey to limit prey populations, and prey dispersal can quickly compensate for local losses. In a trout removal study in a Rocky Mountain stream, US, high drift rates suggested that dispersal may have been sufficient to mask any impact of predation (Allan 1982). The availability of prey external to a coupled predator-prey interaction, as when terrestrial invertebrates supplement fish diet, can reduce predation pressure on benthic invertebrates, and any reduction in that subsidy may result in greater predation pressure on benthic prey. In a forest stream in northern Japan, predation by Dolly Varden charr reduced the biomass of herbivorous aquatic arthropods when terrestrial arthropod input was experimentally reduced by placing greenhouse covers over sections of streams. However, no predation effect was evident on the aquatic arthropods when terrestrial arthropod supply occurred naturally (Nakano et al. 1999b).

Details of experimental design also may influence the outcome of predation experiments, since studies typically use a mesh deemed coarse enough to contain fish yet allow the free flow of water to maintain stream conditions as close to natural as possible and minimize sedimentation within cages. When mesh size is large enough to allow dispersal of aquatic invertebrates and cages themselves are small (i.e., on the order of a few square meters or less) prey dispersal may overwhelm prey consumption. Based on a literature review and several specific studies, Cooper et al. (1990) argued that the magnitude of prey exchange (both immigration and emigration) among substrate patches strongly influences the perceived effects of predators on prey populations. Reported predator effects are strongest in cages using small mesh that confined predator and prey to an enclosed area. When Dahl and Greenberg (1999) measured prey exchange rates and predation effects in a Swedish stream using small enclosures (1.5 × 0.5 × 0.5 m) with either 3 mm or 6 mm mesh, with and without brown trout, predation effects were strongest in the 3 mm cages, which had the lower exchange rate. Beyond revealing a possible bias of experimental design, these findings imply stronger predator effects in pools, where prey replenishment rates are low, compared with fast-flowing stream sections, where prey dispersal and colonization tend to be higher. These studies provide further evidence that predation can reduce prey within cages directly, by consumption, and indirectly by inducing emigration and predator avoidance. The importance of these two mechanisms appeared to vary among taxa, likely reflecting taxon-specific differences in mobility, vulnerability, and response to predator cues. The non-consumptive effects of predation are seen in a number of predator-avoidance responses of prey that act to reduce the direct, consumptive influence of predators on their prey, often exacting a cost in reduced energy intake in both predator and prey populations.

10.2.2.2 Non-consumptive Effects of Predation

Exposure to risk of predation is a common consequence of foraging and other activities, and thus individuals will benefit from adaptations that minimize predation risk while maintaining energy intake. Prey species may respond to a predator's presence by escape behaviors, by becoming increasingly nocturnal, or by altering growth and development. Referred to as non-consumptive effects or indirect effects, they provide the obvious benefit of immediate survival, but may result in foregone foraging and thus exact a cost in future growth and fecundity. Some avoidance mechanisms operate regardless of the physical proximity of predator and prey and serve to reduce the likelihood of initiation of an attack, whereas others function to foil attack and capture. Anti-predatory traits can be fixed, such as protective armor or invariant nocturnal activity, or induced by the presence of the predator, such as fleeing after tactile or visual contact or perceiving the 'scent of death' when water-born chemicals reveal a predator's presence (Kats and Dill 1998; Preisser et al. 2005). Flexible predator avoidance requires some ability to assess risk and make an escape, and so fixed responses may be favored when those conditions are not met or the cost in lost foraging opportunity is modest. Because predation can rarely be entirely ignored or eliminated, threat-sensitive predator avoidance behaviors should be a widespread solution to the tradeoff between foraging and avoiding being eaten.

Many characteristics of prey can reduce their vulnerability to predators, including small size, nocturnal or crepuscular behavior, use of interstitial spaces, body shape and armor, and so on, although it may not be certain whether those traits have evolved as anti-predator defenses. Mayfly larvae exposed to predaceous stoneflies (*Megarcys signata* and *Kogotus modestus*) in microcosms differed in their vulnerability and responses to predator-prey encounters (Peckarsky 1996). Soft-bodied *Baetis*, an abundant and favored prey, escaped by entering the drift, whereas *Ephemerella*, morphologically defended by a thick exoskeleton,

shows no avoidance behavior. Anti-predator responses are also common in fish and amphibians of running waters. Young fish may rely especially on predator avoidance, while older fish may experience reduced predation risk due to size, defensive armor and spines, olfactory cues, and learning (Fuiman and Magurran 1994; Kelley and Magurran 2003). A review of 43 studies of predator–prey fish interactions in lotic systems found that predators caused prey fish to move to habitats of reduced predation risk, such as shallower margins within a given macrohabitat, and in the opposite direction to avoid predation by terrestrial predators. In addition, prey reduced their overall activity and foraging in the presence of predators, resulting in slower growth rates. In laboratory trials, young coho salmon *Oncorhychus kisutch* offered houseflies as prey reduced reaction and attack distances and shortened attack time in the presence of a model rainbow trout, compared to young salmon foraging in the absence of threat (Dill and Fraser 1984). Moreover, the responsiveness of young coho salmon to the model was reduced by higher hunger levels and the presence of a competitor. Anti-predator defenses are found in some amphibians, usually based on palatability or use of chemical cues for avoidance (Kats et al. 1988). Those lacking these defenses are largely restricted to breeding in temporary pools that lack fish.

The downstream transport of benthic invertebrates and its pronounced nocturnal periodicity have fascinated stream ecologists since early researchers, impressed by the quantities of invertebrates drifting downstream, wondered how benthic populations could persist in the face of constant unidirectional losses. Known as the "drift paradox" (Muller 1963; Waters 1972), explanations centered on whether compensatory upstream movement sufficed for population persistence or, alternatively, if drift represented surplus production in excess of carrying capacity. It is now apparent that compensatory upstream movements are modest and depletion of upstream populations is not observed, suggesting instead that drift largely represents small-scale movements important for individual habitat selection and distribution, but with little discernible influence on overall population dynamics of stream invertebrates (Naman et al. 2016).

The nocturnal periodicity of invertebrate drift is best interpreted as an evolved response to risk from predators whose visual field is greatest in daylight. Evidence that nocturnal periodicity varied among size classes and taxa in accordance with predation risk, so that smaller, less vulnerable individuals were nearly aperiodic while larger, more vulnerable size classes and taxa were nocturnal, strongly indicates that nocturnal drift is an adaptation to reduce predation (Allan 1978). Patterns in drift behavior of mayflies in a series of Andean streams provides additional evidence that nocturnal periodicity is a relatively fixed behavioral response to risk of predation by drift-feeding fishes (Flecker 1992). Mayfly drift was primarily nocturnal in piedmont streams with natural populations of visually hunting predators, but was aperiodic in mountain streams that historically lacked drift-feeding fishes. However, where rainbow trout have been introduced to naturally fishless Andean streams, the mayfly *Baetis* exhibited strong nocturnal peaks in drift, suggesting a rapid evolutionary response to an exotic predator. Indeed, nocturnal drift became more pronounced along a gradient of predation regimes (Fig. 10.16), and was observed even when fish were experimentally excluded, suggesting that nocturnal activity has evolved as a fixed behavioral response to predation. In a comparison of fishless and fish-bearing streams in the Rocky Mountains, Colorado, US, the ratio of night:day drift numbers for all mayfly taxa was near 1:1 in fishless streams and near 10:1 for several mayfly taxa in streams with trout (McIntosh et al. 2002).

Why aquatic insects enter the drift is not fully resolved, because entry could be due to accidental dislodgment resulting from foraging movements and turbulence, referred to as passive drift (or catastrophic drift at substrate-mobilizing flows); or entry could be deliberate, to depart from an unprofitable foraging location, avoid benthic predators, or depart from unfavorable abiotic conditions. The

strong nocturnal periodicity of drift suggests that at least the nighttime component reflects a trade-off where daytime foraging is restricted to reduce predation risk from drift-feeding fish. Nocturnally foraging benthic predators such as stoneflies may also induce drift entry, as Hammock et al. (2012) observed when the perlid stonefly *Doroneuria* was added to 2-m long channels placed within a fishless stream, relative to control channels that lacked the stonefly. Hydraulic conditions, including turbulence and back eddies, may contribute to dislodgment at any time (Oldmeadow et al. 2010).

Chemical cues (kairomones) are chemicals emitted by an organism that alert prey to the presence of a predator, and predators to the presence of prey. Many freshwater fishes release chemical cues into the water following mechanical damage to epidermal tissues, which serve as a warning to conspecifics and an attractant to predators. In streams of Trinidad, the pike cichlid *Crenicichla alta*, a solitary, visually foraging ambush predator, increased its presence in the vicinity of damage-released chemical cues from its preferred prey, the guppie *Poecilia reticulata* (Elvidge and Brown 2012). Illustrating that cues influence prey as well as predator, lab-reared guppies exposed continuously to chemical cues emitted by *C. alta* strongly reduced food intake in the presence of predator cues, relative to

Fig. 10.16 Night:day drift ratio of mayfly drift densities from a series of streams in the Venezuelan Andes representing a gradient from low to high predation. Note that drift is greater by day in high elevation streams lacking drift-feeding fish (Rio Albarregas [ALB] and Quebrada La Fria [FRI]) compared to nearby streams containing introduced trout (Quebradas Coromoto [COR] and Mucunutan [MUC]) (Reproduced from Flecker 1992)

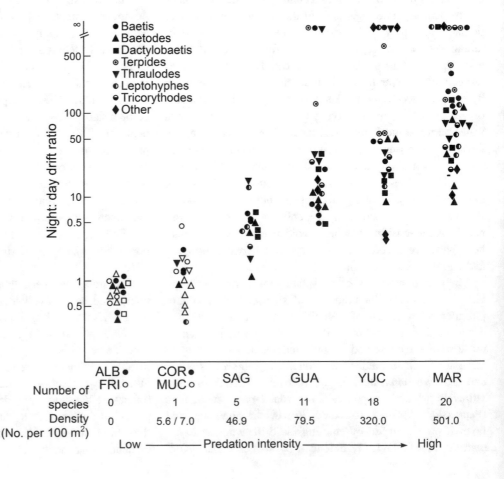

individuals reared in the absence of cues, but assimilated nutrients more efficiently, indicating an induced physiological shift (Dalton and Flecker 2014).

Studies show that mayfly larvae can respond to trout chemical cues by altering behavior and development. When water from holding tanks containing trout was added to microcosms with *Baetis* larvae, nocturnal drift was substantially depressed relative to controls (McIntosh and Peckarsky 2004). In a Rocky Mountain stream containing brook trout and presumably fish odor, release of water containing brook trout immediately upstream of drift nets resulted in a rapid decline in the drift of larger individuals of the mayfly *Baetis*, and an increase in numbers of smaller individuals (Mcintosh and Taylor 1999). The increase in drift by smaller size classes is unexplained, but could reflect an evolved behavior to exit locations of high predation risk prior to reaching a more vulnerable size. Chemical cues from the nocturnal and benthivorous gudgeon *Gobio gobio* suppressed activity of the mayfly larvae *B. rhodani* (Schaffer et al. 2013) and the amphipod *Gammarus pulex*, but amphipods did not react to stone loach *Barbatula barbatula*, which poses less predation risk (Szokoli et al. 2015).

Faster development and maturation at a smaller size appears also to be a response induced by fish "odor". *Baetis* mayflies just prior to emergence are easily identified from their enlarged, darkened wing pads. A survey of size of mature larvae in streams in the Rocky Mountains of Colorado, US, revealed that size at metamorphosis of summer generation *Baetis* was smaller in fish streams than in fishless streams (Peckarsky et al. 2001). This was interpreted as evidence that larval development was accelerated where predation risk was greater, resulting in metamorphosis of younger and smaller individuals. Because fecundity increases with body size, accelerated development likely incurs a cost in future reproduction relative to extended development and larger size at metamorphosis in safe environments. Similarly, field surveys of streams in central New York, US, found that *Ephemerella* mayflies emerged earlier and at smaller sizes in streams with high relative densities of fish, compared to streams with low fish abundance (Dahl and Peckarsky 2003). Laboratory rearing experiments showed that *Ephemerella* larvae exposed to fish chemical cues likewise exhibited faster larval development and smaller size at maturity compared to no such cue, indicating an inducible change in development associated with perceived predation risk.

In closing, predation in stream ecosystems unquestionably exerts a direct influence upon prey populations. Predator foraging mode and prey vulnerability result in differential predation rates that constitute a strong force driving adaptations to minimize predation risk. In some circumstances, predators clearly are able to exclude prey species from certain habitats or markedly reduce their abundance. The immediate effects are lost foraging opportunity and reduced growth rates, or direct mortality, which affects some size classes and species more than others. Because the prey themselves are consumers of other animal or plant resources, the potential exists for predation to create top-down trophic cascades and indirectly affect other species in the food web. In fact, some of the most dramatic effects of predation in lotic ecosystems are revealed in far-reaching cascades, offering convincing evidence that predation is a strong force shaping biological communities.

10.3 Competition

Competition occurs when members of the same or different species utilize shared resources that are in limited supply, thereby reducing one another's individual fitness and population abundance through the depletion of those resources. This definition encompasses two mechanisms of competitive interaction. Exploitation (also termed indirect) competition involves the depletion of shared resources such that another individual is disadvantaged. Interference or direct competition usually is of an aggressive nature, for instance when one individual excludes another from a preferred habitat and foraging location. Competition has long been viewed as a challenge to species coexistence, requiring sufficient differences between species to prevent competitive exclusion. Thus, niche specialization becomes a key consideration in community assembly, a topic we return to in Chap. 12. Competition has been demonstrated in many different settings, however, and when it occurs it often is asymmetrical, with one species able to exclude a second species, which persists by occupying habitats or using resources largely unutilized by the superior species (Begon et al. 2005). Competition may be less evident a force than predation and herbivory, perhaps because it acts more gradually. In addition, competition may often be diffuse, emanating from many species rather than just a pair-wise interaction.

A rigorous demonstration of competition generally requires evidence of an adverse effect of numbers of one population upon the abundance, growth, or survival of individuals of another population under reasonably natural conditions, and also some insight into the mechanism driving the competitive interaction. However, many studies either document some overlap in resource use, from which competition is inferred, or some differences in resource use, from which niche partitioning is inferred. Although such investigations must be viewed as weak evidence for competition, they make up a large portion of the existing literature. We shall first consider the evidence in support of resource partitioning, and then look to other lines of evidence including experimental and natural comparisons. Finally, since unrestrained competition ultimately should result in the elimination of all but the best competitors, it is necessary to ask how commonly and under what conditions this situation occurs. Harsh conditions imposed by abiotic factors, floods in particular, appear to be important in counteracting strong competition in a number of instances, and biotic interactions are likely to be primary only under environmentally benign conditions.

10.3.1   Resource Partitioning

Resource overlap typically is evaluated based on similarities between individuals along three major axes: food, habitat, and time (season or time of day) when the organism is active. The evidence from many studies of resource partitioning, encompassing a variety of taxa in both aquatic and terrestrial settings, indicates that habitat segregation occurs more commonly than dietary segregation, which in turn is more common than temporal segregation (Schoener 1974). Schoener also reported a tendency for trophic separation to be of relatively greater importance among aquatic organisms. Evidence of food specialization historically has been obtained from inspection of gut contents; thus, it matters a great deal whether food items fall into easily distinguished categories. Not surprisingly, food partitioning is reported more commonly from studies of grazers and predators than of detritivores. However, stable isotope analysis of food items and consumer tissue is providing new insights not only into food consumed but also assimilation and growth (Aberle et al. 2005; McNeely et al. 2007). Fish and invertebrates both have been studied extensively from a resource partitioning perspective, but the literature for aquatic plants and benthic algae is scant.

10.3.1.1   Algae

Few studies explicitly address competition between species of algae in lotic ecosystems. It is well established that algal abundance and species composition in artificial streams change in response to adjustments in nutrient, light, or current regime, and assemblages also undergo succession under a particular environmental regime. Many field studies describe shifts in algal dominance associated with changing environmental conditions. The tendency for filamentous green algae to dominate under high light levels is suggestive of a competitive advantage, while their scarcity under low light regimes may be due to the reduced pigment diversity of chlorophytes relative to other common stream algae (Steinman and McIntyre 1987). In comparing the influence of nutrient availability in the water column versus nutrient diffusing substrates, Pringle (1990) observed the diatoms *Navicula* and *Nitzschia* to dominate the overstory and

interfere with the establishment of understory taxa *Achnanthes* and *Cocconeis*. Because nuisance blooms of the invasive benthic alga *Didymosphenia geminate* showed a negative correlation with densities of native algal taxa, predominantly of other diatoms that formed thick filamentous mats, Bray et al. (2017) interpreted this as interference competition. Expression of this apparent effect of competition with native taxa varied with nutrient treatments and water velocity. Considering the diversity of periphyton species and variety of growth forms (Fig. 6.1), and the likely competition for substrate space and access to light, future studies of individual algal species may be expected to provide stronger evidence of the importance of competitive interactions within assemblages of benthic algae.

Algae potentially may compete with heterotrophic bacteria within biofilms for access to nutrients, although co-occurrence potentially is mutually beneficial, as when bacteria utilize dissolved organic carbon (DOC) released by photosynthesizing algae, and bacterial breakdown of organic matter releases nutrients important to algae. However, when other sources of labile DOC are present or when algae are light- or nutrient-limited, bacteria are less likely to utilize algal-generated DOC and may instead compete with algae for inorganic nutrients. Rier and Stevenson (2002) tested the latter possibility by experimentally manipulating light levels and using nutrient diffusing substrates to release inorganic nutrients and/or glucose in an oligotrophic stream located in Kentucky, US. They found no evidence that algae were negatively affected by competition with bacteria for nitrogen and phosphorus, nor that bacteria were using algae as a carbon source. They speculated that algae serve as a substrate for bacterial colonization, and that is the basis for their positive association.

10.3.1.2   Invertebrates

Many studies document habitat partitioning among stream-dwelling invertebrates. Temporal separation of life cycles over seasons is frequent among the univoltine (one generation per year) insects of temperate streams, and differences in diet are reported principally in animals that consume easily categorized food items. The distribution and abundance of filter-feeding caddisfly larvae provide an attractive system for the study of resource partitioning, as they utilize the common resource of FPOM, require space to attach their nets, and have seemingly ample opportunity for resource partitioning via differences in the mesh size of capture nets and location of attachment. Indeed, differences in food particle size consumed (Wallace et al. 1977), microhabitat distribution (Hildrew and Edington 1979), longitudinal distribution (Lowe and Hauer 1999), and life cycle (Mackay 1977) have each been demonstrated. The instars of a species also differ in habitat use, and their preferred current velocity typically increases over their development (Osborne and Herricks 1981). On the other hand, it does not appear that either food or space commonly is limiting to co-occurring caddis larvae. After estimating the size fraction captured by six filter-feeding caddis larvae as well as total availability of organic particles, Georgian and Wallace (1981) found no evidence that food was limiting or that resource partitioning occurred. The size fractions captured showed very high overlap, and amounted to only about 0.1% of available FPOM.

Although it remains unclear whether competitive interactions among species of filter-feeding caddis larvae frequently limit their distribution and abundance, this guild nonetheless illustrates resource partitioning along multiple dimensions. Some species clearly differ in net dimensions, current, and other microhabitat preferences, and in temperature adaptations that determine larger-scale spatial segregation. Detailed analysis of microhabitat use by filter-feeding caddis larvae in lake outlet streams in northern Finland revealed differences among instars and species, particularly in their association with moss and with a hydraulic metric, the Froude number (Muotka 1990). *Polycentropus falvomaculatus* was a microhabitat generalist but three species of *Hydropsyche* (*H. augustipennis*, *H. pellucidula*, and *H. saxonica*) were more specialized. It has been suggested that larger species with their larger mesh sizes are suited to higher current velocities and also to capture larger food items, and smaller mesh sizes might function best in slow currents (Alstad 1987), but Muotka (1990) found only partial agreement with this expectation, and in one study the species with the smallest mesh net was most abundant at the highest velocities (Wallace et al. 1977).

Although temporal segregation apparently is less common than habitat or diet partitioning across diverse taxa (Schoener 1974), numerous examples from temperate running waters illustrate a distinct seasonal succession among closely related species. Co-occurring species of stoneflies in the genus *Leuctra* and family Nemouridae in small streams of the English Lake District display staggered life cycles that minimize the temporal overlap in their resource demands (Elliott 1987a, b). The periods of maximum larval growth were sufficiently out of phase among five species of riffle-dwelling *Ephemerella* mayflies to ensure at least ten-fold size differences on any given date between the most closely related species (Fig. 5.14). Highly synchronized, non-overlapping life cycles among presumed competitors are not always the rule, however. Of six leptophlebeid mayflies in a New Zealand stream, only two had reasonably well-defined growth periods, and overlap of life histories was pronounced (Towns 1983). As with resource partitioning along microhabitat and food axes, temporal specialization likely is offset by the advantages of flexible habits and life cycles.

To evaluate whether temporal partitioning should be attributed to competition, Tokeshi (1986) developed a null model of expected overlap by assuming that species' life cycles were distributed independently of one another throughout the year, with the constraint that most growth should occur during favorable seasons. For nine species of chironomid larvae living epiphytically on spiked water-milfoil and consuming a similar diet of diatoms, actual overlap of life cycles was greater than expected by chance alone. Since this result is the opposite of that expected in temporal partitioning, it appears that all nine species were tracking seasonal peaks in resource abundance. Competition, if it occurred, was not manifested in temporal partitioning.

10.3.1.3 Fishes

Resource partitioning within fish assemblages has received a great deal of study, and as with macroinvertebrates, extensive segregation can be documented along the axes of diet, habitat, and time. In a review of some 116 such studies conducted primarily with salmonids of cool streams or small, warmwater fishes of temperate regions, Ross (1986) found that segregation along habitat and food axes was about equally frequent while temporal separation was less important. However, even among similar faunas the importance of space, food, and time axes varied considerably. Although the resource partitioning perspective is upheld in many studies, other researchers report greater overlap and attribute co-occurrence to a combination of individual specializations and the importance of environmental variation in mitigating competitive interactions.

Studies of darters and minnows in North America provide examples of low overlap in species distributions or in microhabitat and feeding position at a single locale. Eight species of cyprinids that co-occurred in a Mississippi stream showed considerable microhabitat segregation with respect to vertical position in the water column and association with aquatic vegetation (Baker and Ross 1981). Only two species failed to separate on these two axes, and one was the only nocturnal feeder in the assemblage. Direct observation by snorkeling in streams of West Virginia, US, found evidence of habitat partitioning by depth, substrate size, and water velocity for 10 darter species. *Percina* typically occurred in the water column, whereas species of *Etheostoma* were benthic and segregated by occurring under, between, and on top of rocks (Welsh and Perry 1998). Moyle and Senanayake's (1984) study of an even more diverse group of fish in a small rain forest stream describes a highly structured assemblage with minimal overlap based on fish morphology, habitat use, and diet.

The extent of resource overlap versus partitioning is likely to vary with food availability owing to opportunistic feeding when certain prey are very abundant. Seasonal changes in diet overlap are well illustrated by Winemiller's (1989) study of nine species of piscivorous fish that were abundant in a lowland stream and marsh habitat in western

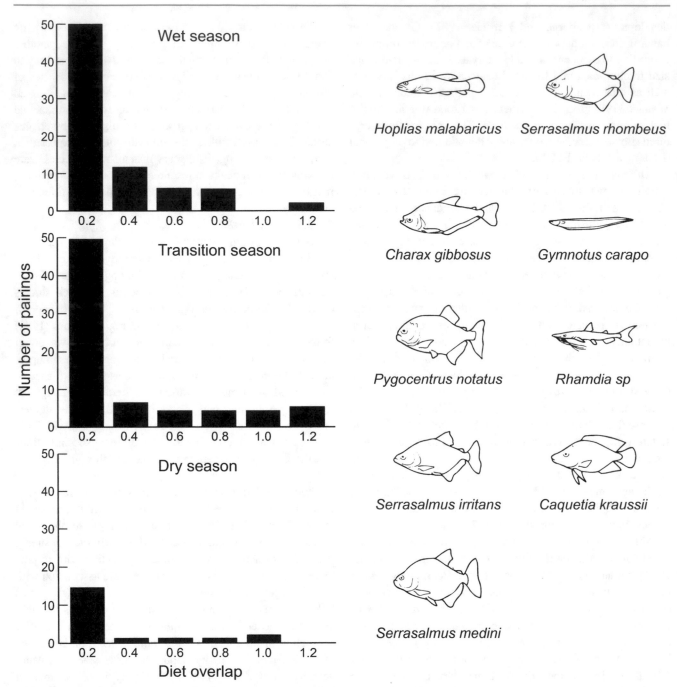

Fig. 10.17 Frequency histograms of dietary overlap exhibited by each of nine piscivorous fish during different seasons at a lowland creek-and-marsh site in Venezuela. Wet season lasts from May to August, transition season from September to December, and dry season from January to April. Diet overlap was computed from pairwise comparisons of ingested prey after converting prey abundance to volume as an approximation of biomass. Dry season data are less extensive because not all species were present and many had empty guts. Over half of overlap estimates were <0.10 (Reproduced from Winemiller 1989)

Venezuela. Members of this guild exhibited substantial resource partitioning in food type, food size, and habitat. Of the possible 72 species combinations among the nine piscivores, only one pair of fin-nipping piranhas exhibited substantial overlap on all three niche dimensions. For the most part, diet overlap of pairs of piscivore species within their feeding guild was low (Fig. 10.17). Highest overlap occurred during the wet season when prey were abundant and lowest overlap occurred during the transition season when prey were least available. Thus, despite the opportunities for competition in this species-rich tropical system, food resource partitioning was widespread. Winemiller (1991) concluded that the higher species diversity of tropical fish assemblages relative to temperate assemblages was paralleled by higher ecomorphological diversity (body and mouth shape, dentition, etc.), which facilitates niche partitioning and reduces competition.

Some studies of habitat partitioning among stream fishes have reported segregation between groups of species occupying distinct microhabitat guilds but considerable remaining overlap at the species level. In an observational field study of six minnows in an Ozark stream, using the habitat variables of water depth, current, substrate, vertical and lateral position of the fish, and their use of pools, riffles, and glides, Gorman (1988) found a clear separation between species occupying higher versus lower water column position but considerable overlap among species within those two categories. The fishes of Coweeta Creek, North Carolina, were separable into three microhabitat guilds: benthic, lower, and mid-water column, but differences in microhabitat use between species within these guilds were not easily distinguished (Grossman et al. 1998). This is interpreted as evidence that environmental variation is more influential than resource availability in limiting population densities of stream fishes, as several authors have argued (Gorman 1988; Baltz and Moyle 1993). In this view it is advantageous that fishes exhibit flexibility and overlap in their use of resources, and assemblage structure reflects the combined influence of environmental variation, particularly in hydrology, together with differences among species in their individual ecology.

Further evidence of competitive interactions among stream fishes can be found in the many examples of novel species combinations that result from invasions, range expansions, and intentional introductions. The many introductions of salmonids to benefit recreational fishing provide numerous examples where a non-native has displaced ecologically similar species over wide regions. Size advantage, aggressiveness, and possibly differences in foraging adaptations as well as predation by the invader on the native species are underlying mechanisms. Relative to native consumers, invaders may possess a novel feeding mechanism, a broad feeding niche, greater energetic efficiency in transforming prey resources into consumer biomass, and greater resilience in the face of natural disturbances (Simon and Townsend 2003).

The widespread replacement of southern hemisphere native galaxiids, which are mostly benthic foragers that use mechanical cues to feed both day and night, with day-feeding, northern hemisphere trout, illustrates this well (Fig. 10.13). Fishes in the family Galaxiidae are the dominant and most species-rich group of freshwater fishes across the cool southern hemisphere, with >50 species in Australasia, Patagonian South America, and South Africa (McDowall 2006). Invasion of cold temperate, southern hemisphere, fresh waters by rainbow trout and brown trout is associated with widespread decline and extirpation of native

fish species. Brown and rainbow trout have been widely introduced in Australia and New Zealand, where they are the mainstay of a highly successful sport-fishing industry (Crowl et al. 1992). In a survey of 198 sites of the Taieri River Drainage in the South Island of New Zealand, both species co-occurred in only nine sites, and then at significantly lower densities (Townsend and Crowl 2006). A survey of 54 small streams on an island in Chilean Patagonia found rainbow trout (*Oncorhynchus mykiss*) to be more widely established than brown trout (*Salmo trutta*), and the two invaders have had different impacts (Young et al. 2010). *Aplochiton*, a trout-like, drift-feeding native species of the Galaxiidae, inhabits uninvaded streams and co-occurs with rainbow trout, but has been nearly eliminated where brown trout occur. Rainbow trout may be more widespread because of greater dispersal tendencies or broader environmental tolerances, but brown trout may have the greater impact where they occur due to a greater reliance on piscivory. The eventual outcome for *Aplochiton* is uncertain: they may persist where habitat or other aspects of the invader's niche limits its abundance, or gradually become extirpated as non-native trout continue to colonize remaining locations.

In many streams of western North America, native bull trout (*Salvelinus confluentus*) and cutthroat trout (*Oncorhynchus clarkii*) have been replaced by the now widespread brook trout, *S. fontinalis*, introduced from eastern North America, and brown trout from Europe. And in mountain streams of the Pyrenees in southwest France, rainbow trout native to western North America have negatively affected habitat use, growth, and survival of native brown trout (Blanchet et al. 2007). Aggressive encounters among salmonid species and between size classes are well known, providing the mechanism for direct or interference competition. Fausch and White (1981) recorded the daytime positions of brook trout in the presence of brown trout, which is the behaviorally dominant species, and then removed brown trout from a section of a Michigan stream. Brook trout subsequently shifted to resting positions that afforded more favorable water velocity characteristics and greater shade, and this habitat shift was greatest in the larger individuals. Such interactions are not limited to the salmonids, of course. A study of interactions between a native fish, the rosyside dace (*Clinostomus funduloides*), and an invasive, the yellowfin shiner (*Notropis lutipinnis*), collected from an Appalachian stream in the southeastern US, found that larger fish dominated in intraspecific interactions, and the invasive species in interspecific interactions (Hazelton and Grossman 2009). However, velocity and turbidity influenced habitat position and aggressive encounters, illustrating the role of environmental context in mediating species interactions.

A common distribution pattern among many now-rare native salmonids is to be restricted to high-elevation headwaters, with replacement by nonnative fishes in lower-elevation reaches. This has led to an hypothesis of temperature-mediated competition, whereby high-elevation species are considered to be superior competitors at cold temperatures, excluding low elevation species, while the reverse is thought to be true at low-elevation, warmer sites. In a laboratory study of growth, feeding, and aggression of bull and brook trout at different temperatures, McMahon et al. (2007) reported a clear metabolic advantage for brook over bull trout at warmer temperatures, but bull trout, increasingly restricted to cool headwaters, did not gain a competitive, size, or survival advantage over brook trout at colder temperatures. However, the competitive advantage of brook trout is reduced at low temperatures, and that, along with complex habitat and proximity to nearby bull trout populations, may explain bull trout persistence. Similarly, in a field study measuring growth and condition (a measure of body mass at a given length) of cutthroat and brown trout at six locations along the thermal gradient of the Logan River, Utah, US, McHugh and Budy (2005) found that brown trout negatively affected cutthroat trout growth and condition, but the converse was not supported. Temperature had little effect on the outcome when both species co-occurred, but precluded brown trout invasion of upper elevations. These studies lend weight to the view that abiotic factors and physiological limits of the downstream species thus determine the upstream boundary of its distribution, and reversals in dominance in which the upstream species is demonstrably the better competitor at colder temperatures rarely occur.

Environmental conditions can strongly influence not only the local distribution of a non-native species, but also its ability to establish populations in a novel environment. The invasion success of rainbow trout, one of the most widely introduced species worldwide, is strongly influenced by a match between timing of fry emergence and months of low flood probability. It has been most successful where hydrologic regimes are most similar to its native range, and only moderately successful or has failed where spring or summer flooding has hampered recruitment (Fausch et al. 2001).

In closing, the extensive literature on resource partitioning clearly demonstrates that species differ in their habitat use, food capture abilities, and timing of activity or growth. These are in effect descriptions of the specializations that constitute a species' niche. Such differences constitute weak evidence of competition, as they do not resolve whether species adversely affect one another, or whether these examples of niche segregation reflect ecological specializations acquired and fixed over the species' evolutionary history. It is for these reasons that recent work has focused on more rigorous tests, typically involving experiments under fairly natural conditions, and we now consider such studies.

10.3.2   Experimental Studies of Competition

Competitive interactions can be manipulated in the laboratory or in the field, conditional on the ingenuity of the investigator to construct an experiment that allows realistic interactions and reveals the mechanisms involved. Laboratory experiments usually offer the greatest experimental control and can be particularly useful in demonstrating the potential for competition and in identifying mechanisms. Field experiments offer greater realism—at least in principle —although they too have artifacts and their outcomes may be influenced by the relatively small scale and short timeframe of the typical field study. Investigators often have used a combination of laboratory and field experiments, and in some cases have exploited natural comparisons or unusual environmental events to provide evidence of competitive effects at large scale and under natural conditions.

Interference competition is well documented in space-limited taxa of invertebrates. Large size almost invariably conveys an advantage, and the loser may be injured or cannibalized. In the latter instance the line blurs between competition and predation. Larvae of the netwinged midge (*Blepharicera*) and black flies (Simuliidae) compete for space on stone surfaces in swift-flowing small streams, even though the former feed on attached periphyton and the latter are primarily suspension feeders. Their densities were inversely correlated in a California stream, and behavioral observations of interactions revealed that larval black flies "nipped" at *Blepharicera* within reach, disrupting their feeding (Dudley et al. 1990). *Blepharicera* spent significantly less time feeding in the presence of black fly larvae, compared to when the investigators removed all simuliids within a 5 cm radius. Aggressive competitors such as caddis larvae affect other species by multiple mechanisms including interference, predation, behavioral avoidance, and by modifying flow patterns (Hemphill 1988). Small-scale density manipulations of *Hydropsyche siltalai* illustrate the multiple pathways of its influence in a lake outlet stream in north Sweden (Englund 1993). The presence of *H. siltalai*, a net-spinning filter feeder that aggressively monopolizes space, resulted in reduced numbers of the mayfly *Ephemerella ignita* and the black fly *Simulium truncatum*, and increases in the free-living, predatory caddis *Rhyacophila nubila* and chironomid larvae. Direct mortality due to predation was the primary cause of declines, although hydropsychid nets likely also interfere with the attachment and feeding of simuliid larvae. Increased abundance of *R. nubila* and chironomid larvae is somewhat surprising since they also are consumed by *H. siltalai*, but apparently hydropsychid nets enhanced food availability to these taxa, and so the presence of *H. siltalai* resulted in a positive facilitation.

Strong competitive interactions involving stream macroinvertebrates have been demonstrated with sessile or slow-moving grazers, due to a combination of exploitation and interference competition. Snails of the family Pleuroceridae appear to be competitive dominants in those headwater streams where they reach high abundances. Both *Juga silicula* in the northwestern US and *Elimia clavaeformis* in the southeast have been reported to reach high densities and to make up more than 90% of the invertebrate biomass (Hawkins and Furnish 1987; Hill 1992). Snails can graze periphyton to very low levels, and because of their large size, individual snails also may harm other species by "bulldozing" over substrate surfaces. The interaction between *Elimia clavaeformis* and the caddis *Neophylax etnieri* in a headwater stream in Tennessee makes a strong case that the snails' influence is via exploitative competition (Hill 1992). High dietary overlap determined from gut analysis suggested that these two grazers were competing for periphyton. Both species substantially increased their growth rates and condition (ash free dry mass per unit wet mass) when transferred from the stream to a high quality diet in the laboratory, suggesting food limitation in nature. In a natural experiment, Hill examined periphyton abundance and *Neophylax* condition in six streams lacking *Elimia*, and in six streams where the snail was abundant. Periphyton biomass was three times greater and caddis larvae at diapause roughly twice as large in the absence of snails (Fig. 10.18).

Manipulation of a large herbivorous snail (*Sulcospira hainanensis*) in Hong Kong streams demonstrated a competitive effect on grazing insects, due to reduction of algal abundance by grazing snails, but only during the dry season (Yeung and Dudgeon 2013). During the dry season, algal biomass was greater and mayfly densities markedly higher on tiles where wire mesh barriers reduced snail densities by about 30–50% relative to controls (Fig. 10.19). Differences

were minor during the wet season, when spate-induced disturbance overrode the biotic interaction. This is in agreement with predictions of the harsh-benign hypothesis, which asserts that biotic interactions are most important under relatively benign environmental conditions—often low and constant flows—and less so under unfavorable conditions such as frequent scouring flows.

Armored grazers, including snails, stone-cased caddisfly larvae, and neotropical armored catfishes, whose protection from predators provides an advantage over less defended grazers such as mayflies, can become very abundant and are capable of reducing algal levels sufficiently that other grazers are disadvantaged. The caddis larva *Glossosoma* has been shown to be an effective grazer of periphyton and competitor with other macroinvertebrates in several studies. Its slow rate of movement and efficient grazing reduce periphyton to low levels, allowing it to maintain high population densities at relatively low resource levels, and its stone case deters predation. By erecting barriers of petroleum jelly in a Montana stream, McAuliffe (1984a, b) was able to achieve approximately a fivefold reduction in *Glossosoma* densities, which resulted in a twofold increase in algal cell density. Mobile grazers such as *Baetis* were significantly more abundant in areas where *Glossosoma* was excluded. At normal densities *Glossosoma* appears able to reduce algal densities to levels where *Baetis* experiences resource limitation, and thus exploitation rather than interference is the primary mechanism. A 10-month exclusion experiment in a Michigan springbrook of very constant flow provides additional evidence of the community-wide effects of *Glossosoma*. Periphyton biomass increased substantially, as did the densities of most grazers (Kohler 1992). Following a 60-day experiment in a headwater stream in northern California, US, in which *G. penitum* was removed from 4-m² stream sections, reducing its biomass by 80–90%, chlorophyll a

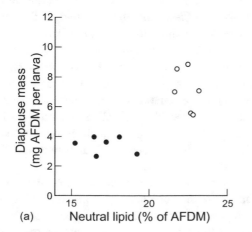
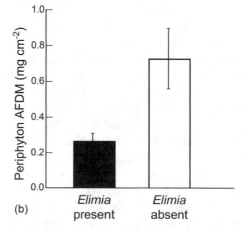

Fig. 10.18 Comparison of six streams in the southwestern US lacking the snail *Elimia clavaeformis* versus six streams where the snail was extremely abundant. (**a**) Average mass of diapausing larvae of the caddis *Neophylax* were higher in the absence of snails. (**b**) Periphyton mass also was higher in the absence of snails (Reproduced from Hill 1992)

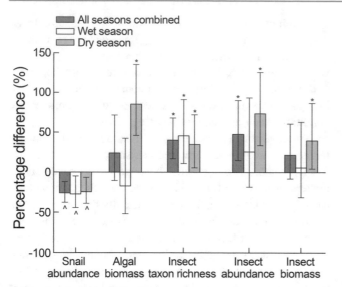

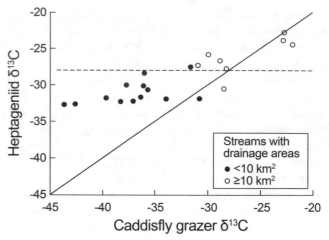

Fig. 10.19 Experimental reduction of an abundant grazer, the snail *Sulcospira hainanensis*, resulted in much more algal biomass, and greater abundance and biomass of grazing insects in Hong Kong streams, but effects were strong only in the dry season. Spate-induced disturbance during the wet season appeared to override this biotic interaction, an example of what is often referred to as the harsh-benign hypothesis. Values shown are mean percentage difference and associated 95% confidence intervals (Reproduced from Yeung and Dudgeon 2013)

Fig. 10.20 Stable carbon isotope ratios of heptageniid mayfly larvae (*Cinygma*, *Cinygmula*, *Nixe*, and *Epeorus*) from the South Fork Eel River watershed (California, US) plotted against those of grazing caddisflies (*Glossosoma penitum*, *Glossosoma califica*, and *Neophylax rickeri*). Each point represents isotope ratios of caddisflies and mayflies collected from the same stream and habitat on the same date. Streams range from drainage areas of 0.8–145 km². Algal ¹³C values generally increase with stream size in the watershed (Finlay 2004); the most ¹³C-depleted values correspond to the smallest and least productive streams. Solid circles represent streams with drainage areas < 10 km². In small unproductive streams, mayflies' carbon isotope ratios are intermediate between those of detritus (dotted line) and scraping caddisflies. As stream size increases, the stable carbon isotope ratios of mayflies and caddisflies converge (Reproduced from McNeeley et al. 2007)

increased twofold and algal consumption by heptageniid mayfly larvae increased as well (McNeeley et al. 2007). While this result supports the view that mayflies experienced competitive release due to an increased food supply, less algal carbon passed through mayfly grazers than expected, possibly because scraping caddisflies consume primarily algae whereas mayfly diets include algae and detrital carbon in variable proportions. Indeed, comparison of stable carbon isotope ratios of caddis and mayfly larvae from multiple stream sites suggests that their diets converge at larger and more productive sites where algae are more abundant and comprise the primary diet of mayflies (Fig. 10.20). This provides yet another example that environmental context can determine whether competition has noticeable effects.

Collectively, these studies demonstrate that competitive interactions indeed take place among macroinvertebrate assemblages of streams, but the extent and magnitude of competitive effects in natural systems are not well resolved. However, observations that accompanied the collapse of *Glossosoma nigrior* populations in Michigan trout streams due to outbreaks of the microsporidian *Cougourdella* provide impressive documentation of the ecosystem-wide influence of a dominant grazer (Kohler and Wiley 1997). The pathogen-induced decline of *Glossosoma* resulted in marked increases in the biomass of periphyton (Fig. 10.21) and in the abundance of most grazers and filter-feeders (Fig. 10.22). Remarkably, these changes dovetailed with

results from Kohler's (1992) prior laboratory and field experiments in a stable, spring-fed stream where *Glossosoma* was the dominant grazer. Long-term exclusion of *Glossosoma* resulted in increases in periphyton biomass and in the abundance and growth of a number of other species of grazers including midge larvae, indicating that the caddis larva might influence other members of the assemblage through diffuse competition. Two sessile filter-feeders, the black fly *Simulium* and the midge *Rheotanytarsus* also increased in the *Glossosoma* exclusion, which may reflect interference competition from physical encounters with the more robust *Glossosoma*. Inspection of Fig. 10.22 leaves little doubt that, in this system, competition from a dominant grazer has dramatic, system-wide consequences. It also is instructive that although all responses were in accord with results from previous, smaller-scale experiments, those studies underestimated the extent and magnitude of *Glossosoma*'s influence.

Finally, as several examples show, environmental factors influence the outcome of competitive interactions. Resource suppression by a dominant grazer resulting in indirect competitive effects on grazing insects was limited to the dry season (Fig. 10.19, Yeung and Dudgeon 2013) when conditions were relatively benign. Invasion success and extent

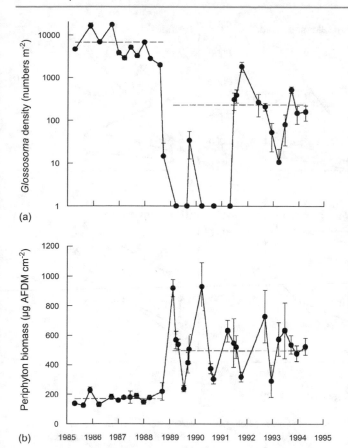

(a)

(b)

Fig. 10.21 Density of *Glossosma nigrior* (**a**) and biomass of periphyton (**b**) in Spring Brook, Michigan. Horizontal dashed lines are the overall mean density or biomass for the periods before and after *Glossosoma's* collapse in 1988. Values are means +1 SE (Reproduced from Kohler and Wiley 1997)

of spatial overlap between competitively dominant non-natives and native species depends on both temperature and hydrologic regime (Fausch et al. 2001; McHugh and Budy 2005). For all of the above reasons, it is likely that the influence of competition within stream communities will vary among locations, over seasons, and between different species assemblages.

10.4 Parasitism

Organisms that infect or parasitize their hosts typically result in the reduced health and fitness of the host, and sometimes parasitism results in death. Microparasites, also called pathogens, are small organisms capable of multiplying within a host, including viruses, bacteria, fungi, and protists. Macroparasites typically are larger and do not multiply directly within individual hosts; this includes flatworms (trematodes), roundworms (nematodes), tapeworms (cestodes), spinyheaded worms (acanthocephalans) and ectoparasitic arthropods (ticks, copepods, mites, and water

lice) (Johnson and Paull 2011). Parasites can infect a single host; however, some pass through one or more intermediate hosts before reaching their definitive (terminal) host.

Parasites can have profound effects on the behavior and abundance of their hosts, including ripple effects that affect other species that interact with the host. The immediate effects of a pathogen on its host include reduced growth, increased susceptibility to predators, and reduced fitness. Beyond these effects on individuals, pathogens can profoundly influence host population dynamics, and when a susceptible species encounters a virulent, often novel pathogen, the results can be catastrophic. Pathogen populations typically are assessed based on their prevalence (the proportion of the host population that is infected), and abundance (also called intensity, the number of individual pathogens per infected host) (Bush et al. 1997). In a host-pathogen interaction where long exposure presumably has allowed the host to adapt to infection, host and parasite abundance may remain relatively stable. However, just as a new infection can sweep through a human population, the arrival of a novel pathogen can result in such a rapid decline of its host that its populations virtually disappear.

Parasites and pathogens potentially may also produce effects that spread indirectly to other members of the community. By influencing host abundance, they may have top-down effects similar to strongly interacting predators and grazers. In addition, parasites and pathogens can influence host traits including behavior and morphology, thus having trait-mediated indirect effects on other species. When the host species is itself a strong interactor with other members of the food web, the effects of its decline can ramify widely.

Freshwater ecosystems harbor pathogens and parasites responsible for a number of well-known animal diseases including salmonid whirling disease, amphibian chytridiomycosis, and crayfish plague. Diseases affecting human populations, such as malaria, cholera, and river blindness, among others, are also linked to fresh waters (Johnson and Paull 2011). Fishes are affected by many types of pathogens including viruses causing anemia and kidney failure, myxozoans causing whirling disease and proliferative kidney disease, as well as by bacteria, fungi, helminths, protists, and arthropod parasites. In contrast, host-pathogen interactions among aquatic insects have received little study, and the majority of research has focused on the effects on host fitness, with only a small number of studies examining population or community-level effects (Kohler 2008). In all likelihood, the influence of pathogens on aquatic invertebrates is under-appreciated. From DNA extraction of 10 species of aquatic insect larvae and two amphipods of the genus *Gammarus* from a low mountain stream in Germany, a total of 26 parasite species, including 12 microsporidians and 10 helminths, were detected in 12 host species, most

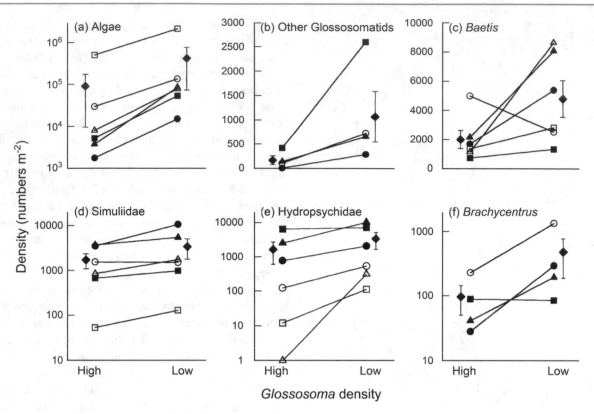

Fig. 10.22 Mean abundances of periphyton (as algal cells cm^{-2}) (**a**), periphyton-grazing insects (**b–c**), and filter-feeders (**d–f**) as a function of *Glossosoma* density (high, low = prior to or during recurrent pathogen outbreaks, respectively) in six streams from southwest and northern lower Michigan. Invertebrate densities are expressed as number of individuals m^{-2}. Symbols denote the six streams (Reproduced from Kohler and Wiley 1997)

with surprisingly high prevalence (Grabner 2017). Microsporidians were present in all host species, and trematodes were present in all but one caddis larva. Nematodes were detected in five host species, and three acanthocephalan species were detected in the amphipods. However, specific host-parasite associations were relatively rare.

Nematodes in the family Mermithidae are frequent internal parasites of mayflies. Infections often are visible as coiled juveniles within the abdomen of hosts, and have been reported in Ephemeroptera, Plecoptera, Trichoptera, and Diptera. In high elevation streams of the Rocky Mountains, US, the number of individuals of larval mayflies infected by the genus *Gasteromermis* varied widely, but was estimated at 22% of the summer generation of the genus *Baetis* (Vance and Peckarsky 1996). Half of the infected larvae were estimated to die before emerging as adults, and because mermithids castrate their hosts, even those that survive to adulthood fail to reproduce, indicating that parasitism can be an important source of population mortality. However, parasites may be overlooked for a variety of reasons. While mermithids often are readily seen through the integument of a larval insect, this may not always be the case. Lancaster and Bovill (2017) observed well-developed parasitic juvenile worms in the abdominal cavities of six species of the

caddis *Ecnomus* from a stream of central Victoria, Australia, but only in adults, not in larvae. Presumably the infection occurred during the larval stage, and the parasite underwent rapid growth during or shortly after pupation.

Some diseases that have been known for many decades have re-emerged in recent years, and others that have only recently emerged are becoming widespread. Emerging infectious diseases are those that appear for the first time in a population, or suddenly increase in incidence or geographic range (Daszak et al. 2000). The basis for disease emergence can be difficult to determine, as it may involve newly identified species or newly evolved pathogen traits interacting with human-induced environmental change, species invasions, and evolutionary adaptations by pathogen and host. Some dramatic examples are discussed below, each of which meets the definition of an emerging disease.

Amphibian populations have declined dramatically and globally in recent decades, mostly since 1980 (Lips et al. 2006). While a number of causal factors have been implicated and some declines are enigmatic, a virulent fungal pathogen of amphibians, *Batrachochytrium dendrobatidis*, is thought to be responsible in many instances (Collins and Storfer 2003). First described in the late 1990s, Chytridiomycosis or "chytrid" is now known from six continents.

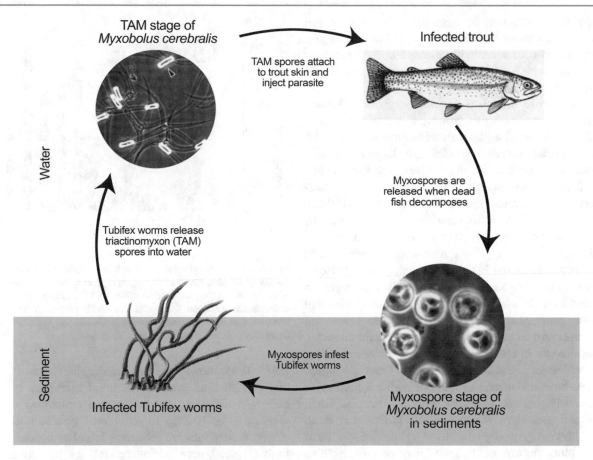

Fig. 10.23 Life cycle of *Myxobolus cerebralis*, the agent of whirling disease

A cutaneous fungal infection that manifests only in amphibians, its free living zoospore stage infects the skin, causing some species to die within weeks, while others are little affected. Mortality apparently is due to disrupted epidermal function and osmotic imbalance. The only known host is larval and adult amphibians, and chytrid is thought to be amongst the greatest wildlife epizootic events ever documented, causing a wave of species loss in the New World tropics. Many affected species are stream-associated frogs at medium to high altitude, in protected forested sites in the tropics of Central and South America and northern Australia (Skerratt et al. 2007). North American species also have been severely impacted, including the mountain yellow-legged frog (a species complex consisting *of Rana muscosa* and *Rana sierrae*) in California's Sierra Nevada mountains (Vredenburg et al. 2010).

Myxobolus cerebralis, a myxozoan fish pathogen known as whirling disease, was first detected in 1893 in Europe, in non-native rainbow trout, but is suspected of originating in brown trout in Eurasia (Bartholomew and Reno 2002; Sarker et al. 2015). It spread worldwide with the cultivation of salmonids in fish hatcheries, appearing in wild populations in the western US after the 1980s. *Myxobolus cerebralis* has a complex, 2-host life cycle alternating between salmonid

fish and the oligochaete *Tubifex tubifex*, producing waterborne spores within each host (Fig. 10.23). The only species of aquatic oligochaete known to be a suitable host, *T. tubifex* is a cosmopolitan and hardy species able to survive drought and food shortages by secreting a protective cyst and lowering its metabolic rate. Spores attach to mucous cells of the trout's epidermis, the gill respiratory epithelium, and the buccal cavity, penetrating epidermal tissues and migrating along nervous tissue to cartilage. Common symptoms are skeletal deformities and tail-chasing swimming behavior that gives this disease its name. The parasite produces myxospores inside infected fish that are released into the sediment upon death of the fish, and then ingested by tubifex worms. Species of salmonids differ in susceptibility, as rainbow and cutthroat trout are rated as highly susceptible, whereas brown trout are resistant and only develop the disease when exposed to very high parasite doses. Whirling disease has been effectively managed in hatchery-reared salmonids by rearing young fish in well water to prevent or reduce exposure of young susceptible fish, and also by chemical treatment and exposure to UV radiation. Initially thought to be primarily a problem within hatcheries, whirling disease has since been implicated in the decline of wild cutthroat trout populations in the western US. Greatest infection risk appears to be associated with habitat and temperature conditions that favor *T. tubifex* and spore production, and co-occurrence with diseased trout, particularly rainbow trout, which are both highly susceptible and widely stocked (Ayre et al. 2014).

The fungus-like organism *Aphanomyces astaci* (Oomycetes) causes crayfish plague, growing in melanized areas of the cuticle and spreading by asexual swimming zoospores. North American crayfish species are largely immune, apparently as a result of having evolved with the parasite, but the disease is devastating to European crayfishes, causing stress to the immune system and rapid death (Holdich et al. 2009). First introduced into Europe in the mid-nineteenth century, the late 20th century arrival of several, highly invasive North American crayfish species that are natural hosts has led to further spread of the disease and declines in European species. Availability of migration routes and suitable habitat for invasive crayfish species will determine geographic spread. However, some argue that *Aphanomyces astaci* may be over-diagnosed as the causative agent of die-offs in crayfish, at least until the recent, wider use of molecular identification methods. Other pathogens may have been undetected as crayfish harbor a number of disease-causing agents including viruses, bacteria, fungi, protists and metazoans (Longshaw 2011; Edgerton et al. 2019). Although generally considered a pathogen specific to crayfish, other freshwater decapods may be vulnerable as well (Svoboda et al. 2014).

Invasive species may be important hosts and vectors of pathogens brought to a new region where, much as introduced predators can devastate naïve prey, novel pathogens can quickly cause epizootics (i.e., outbreaks of infectious diseases in non-human animals) (Poulin et al. 2011). The introduction of crayfish plague from North America to Europe is one example, and the spread of whirling disease and its subsequent transfer from hatchery-reared trout to wild trout is another. When a pathogen is of high virulence to a native species but low virulence to an invading species, 'spill-over' of the parasite can cause high mortality to the native species, or reduce its ability to compete with the invader. This is sometimes referred to as parasite-mediated competition, and is likely to be most severe when the invading species is sufficiently abundant to cause high exposure of the native species to the pathogen. The timeline of the decline of the noble crayfish (*Astacus astacus*) in Sweden due to crayfish plague illustrates this well (Bohman et al. 2006). Due to declines of noble crayfish following first arrival of plague in the early 1900s, and market demand for freshwater crayfish, beginning in 1969 the government began a large-scale introduction of the American signal crayfish (*Pacifastacus leniusculus*), a chronic carrier of *Aphanomyces astaci*. Comparison of the number of outbreaks for the 20-year period before (1940–1960) and after

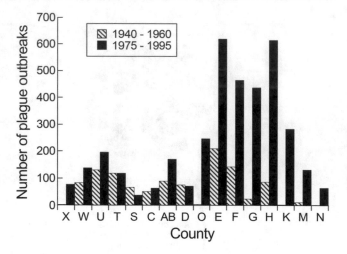

Fig. 10.24 A comparison of crayfish plague outbreaks for each county in Sweden occurred during two periods, before (1940–60) and after (1975–94) the introduction of the signal crayfish in Sweden. The counties are arranged from north to south (letters X to M). Only counties that have reported crayfish outbreaks are represented in the chart (Reproduced from Bohman et al. 2006)

(1975–1995) shows the further geographic spread of crayfish plague with time, and the increased frequency of disease outbreaks following the widespread cultivation of signal crayfish (Fig. 10.24). In addition, once introduced, a pathogen may take up residence in native species, and its further spread may be independent of the original host species. The intestinal nematode *Camallanus cotti* is thought to have invaded Hawaii via the introduction of poeciliid fishes used in mosquito control, and now infects native Hawaiian fishes including the goby *Awaous stamineus*. Surveys by Gagne et al. (2015) show that the parasite has become decoupled from its original host, as it has spread beyond the range of its introduced host to streams that have no record of poeciliid presence, although the mechanism of its dispersal is not yet established.

One final example of parasitism in stream-dwelling organisms deserves mention, that involving the larval stage of certain mussels, because it is so fascinating and unique. Unionid mussels have a larval stage that is temporarily parasitic on fish. The benefit to mussels is dispersal, including upstream transport, and the energetic cost to the fish of an immune response generally is low. The variety of ways that glochidia are released, their attachment mechanisms, and in some mussels the ability to attract fish hosts by mimicking prey, make fascinating natural history (Barnhart et al. 2008). Mature glochidia are expelled individually, in loosely bound mucous masses, or bound together in discrete packages by a mucoid or gelatinous matrix, and can be distinguished by the number and shape of attachment hooks. Some aggregates have the appearance of prey and are ingested by fish, settling in the gills. Several species of Lampsilinae have evolved mantle margins that undulate to

Fig. 10.25 A greenfin darter (*Etheostoma chlorobranchium*) freshly captured by a female Oyster Mussel (*Epioblasma capsaeformis*). The mussel will release her parasitic larvae to attach onto fins and soft tissues, and then release the fish to disperse these larvae. Clinch River, Tennessee. Photo by David Herasimtshschuk, Freshwaters Illustrated

attract fish seeking prey, and the genus *Epioblasma* captures its fish host in its valves to facilitate transfer of glochidia (Fig. 10.25). After a brooding time that varies among taxa, larvae detach and fall to the substrate, taking on the typical form of a juvenile mussel. Mussels show varying degrees of host specificity from use of a single host species to many, since as temporary parasites, mussels must circumvent the immune responses of the host to complete this life stage. However, the preferred hosts are known for only some of the 300 or so species of North American Unionidae.

10.4.1 Direct and Indirect Effects

The most obvious direct effect of a pathogen on its host is a reduction in growth and reproductive fitness, or simply death of the host. Changes in the morphology or behavior of an infected host also may occur, possibly as an adaptation that benefits either host or parasite, depending on which of the two appears to gain some advantage. Changes in host behavior as a consequence of parasitism could be simply a pathological side effect of parasite infection, such as reduced mobility, a host-adaptive strategy such as behavioral fevers, or a parasite-adaptive strategy, such as making a secondary host more vulnerable to predation by a definitive (terminal) host. Seeking warm water when infected has been documented in a number of fish species, including the Trinidadian guppy *Poecilia reticulata* when infected with a common helminth ectoparasite that feeds on skin and fin tissue. Infected guppies showed a preference for temperatures above 30 °C, which are known to impede trematode growth (Mohammed et al. 2016).

In two-host systems where it benefits the parasite to induce the intermediate host to 'commit suicide" so as to reach its definitive host, some impressive examples of behavior modification have been documented. The life cycle of parasitic hairworms (Nematomorpha: Gordiida) requires two critical transitions across habitats, first from aquatic hosts such as insect larvae to terrestrial definitive hosts (crickets and grasshoppers) and second from the definitive hosts back to water (Schmidt-Rhaesa and Ehrmann 2001). In a well-known example of parasite manipulation of host behavior, crickets infected by mature hairworms seek and jump into water, at which time adult worms emerge and free-living individuals begin searching for sexual partners. The crickets, now wriggling at the water surface, constitute a terrestrial prey subsidy to fish (Ponton et al. 2006; Sato et al. 2008). Amphipods in the genus *Gammarus* can be infected with the acanthocephalan parasite *Pomphorhynchus laevis*, which completes its life cycle when the amphipod is eaten by a fish predator capable of serving as definitive host. In the River Teme, England, approximately 20% of adult *G. pulex* were infected with *P. laevis*, and the drift of parasitized amphipods was significantly greater than that of unparasitized individuals (McCahon et al. 1991). Because the acanthocephalan must complete its development in the gut of a trout or other drift-feeding fish, altering amphipod behavior to make it more likely to drift is clearly to the parasite's advantage.

Mermithid nematodes also have been reported to alter host behavior. Infected nymphs of mayflies in the genus *Baetis* were consumed significantly more by a predaceous stonefly than unparasitized nymphs, apparently because they were less prone to escape the approaching predator

(Peckarsky and Vance 1997). Similarly, mermithid-infected nymphs of mayflies in the genus *Deleatidium* were disproportionately represented in drift samples compared with benthic samples, suggesting that infection by mermithids results in an increased tendency to drift, where they would be more susceptible to fish predation (Williams et al. 2019). However, in neither case is there an apparent advantage to the parasite, which usually emerges when the adult mayfly enters the water to lay its eggs. These two studies of the increased susceptibility of infected mayfly hosts to predation suggest a third possibility, namely that the change in behavior is a non-adaptive, pathological side effect of infection. In neither case is there an advantage to the parasite, such as transference to a final host, as the insect is the only host to the nematode; in addition, the life cycle of most common mermithids found in stream insects is completed when adult mayflies enter the water to oviposit on substrates. Thus, the most apparent effect of mermithid parasitism is to harm host reproductive fitness, and it is in the parasite's interests that its host not be eaten during its larval life.

Interestingly, the mermithid *Gasteromermis* does alter the behavior of adult *Baetis* males (Vance 1996). In uninfected individuals, adult females crawl down the side of a rock into the water to oviposit, but males of course do not. When an infected female enters the water, the mermithid escapes through a puncture wound in the mayfly's abdomen. Infected males are feminized, meaning they behave like females, with the same ovipositing behavior that favors mermithid escape. Further illustrating the complexity of host-parasite interactions, the mermithid *Pheromermis* requires two hosts: an aquatic insect as an intermediate aquatic host and a terrestrial definitive host, vespid wasps, which acquire the infection by consuming the adult of the aquatic insect (Poinar 1976). Surveys of the Guare and Emilia rivers in northern Venezuela found *Pheromermis* juveniles coiled inside stonefly nymphs collected at multiple sites (Gamboa et al. 2012). Prevalence was over 90% in both in *Anacroneuria blanca* and *A. caraca*, indicating that larvae of this stonefly to be important hosts of this mermithid in these rivers. However, the terminal host remains unknown.

When the prevalence and abundance of a pathogen is great enough, it may regulate its host's population or, in extreme cases, drive it to extinction as appears to be the case with chytrid and some anurans. A 15-year monitoring study of a caddisfly (*Brachycentrus americanus*) and the prevalence of a microsporidian disease provided convincing evidence that the disease was largely responsible for driving observed *Brachycentrus* population dynamics in a stream in Michigan, US (Kohler and Holland 2001). Both the host and its pathogen exhibited cyclical dynamics with a lag of one generation, evidence of delayed density-dependent parasitism as disease prevalence tracked increases and declines in the caddis population. The caddisfly *Glossosoma nigrior*

occurs with the microsporidian *Cougourdella* at some locations but not others, causing strikingly different *Glossosoma* dynamics depending on the presence of the microsporidian (Kohler 1992). *Glossosoma* populations did not fluctuate greatly in the absence of *Cougourdella*. In contrast, where *Cougourdella* infections had been detected, *Glossosoma* populations existed either at high densities where *Cougourdella* prevalence was consistently low, or at low densities due to recurrent *Cougourdella* epizootics. *Glossosoma* is a particularly strong interactor capable of reducing attached algae to very low levels and relatively invulnerable to predators due to its case of small stones (Kohler and Wiley 1997). As described earlier, since this system provided strong evidence of competition, pathogen-induced reductions in *Glossosoma* abundance resulted in increased abundance of attached algae (Fig. 10.21), and population sizes of most other algal consumers increased, evidence of competitive dominance by the caddis larva and the indirect benefit to other grazers resulting from *Cougourdella* outbreaks (Fig. 10.22). As further evidence of the community-wide influence of this pathogen-host interaction, invertebrate predators doubled in abundance following *Glossosoma* population collapse.

Because the tadpoles of many species of amphibian feed on algae and detritus, and chytrid-induced declines of amphibian populations can be extreme, ripple effects throughout food webs can be substantial (Whiles et al. 2006). Manipulation of tadpole access to experimental tiles in a Panamanian stream using an electric field showed that tadpoles significantly reduced organic particles and diatom abundance and diversity, while tadpole exclusion enhanced abundance of grazing mayflies (Ranvestel et al. 2004). The dramatic decline in amphibian tadpole populations driven by a chytrid outbreak had profound, ecosystem-level consequences in a relatively undisturbed wet-forest stream in Panama (Whiles et al. 2013). Tadpole biomass in 2008 was 2% of the value in 2006, resulting in a doubling of algal and fine detrital biomass. The uptake and cycling of nitrogen, measured with a tracer addition of ^{15}N, was markedly lower, apparently because of changes to mineralization rates (less ingestion and excretion by tadpoles) and lower export of fine particulates (reduced bioturbation). Surprisingly, unlike the study of Kohler and Wiley (1997), other grazers did not increase, indicating that other grazers were not able to compensate for the loss of tadpoles.

10.5 Summary

Species are interconnected through the proximate food chain linkages of herbivory, predation, competition, and parasitism. The supply of resources is potentially limiting to consumers, and an abundance of consumers may in turn deplete resource levels. The importance of these bottom-up and top-down effects is seen in foraging and risk-avoidance adaptations, in the size of populations, and through indirect effects on species several steps removed from the initial interaction. Consumers of the same resource are competitors whenever resource sharing is mutually detrimental, but various mechanisms of resource partitioning may sufficiently reduce overlap to permit coexistence. Parasites and pathogens can reduce host growth and fitness, alter host behavior, and in some cases cause direct mortality. The strength of species interactions is most evident when the abiotic environment is moderate, and may be reduced or undetected whenever environmental variation is extreme.

The study of herbivory in stream ecosystems has focused mainly on the grazing of benthic algae by invertebrates, some fishes, and a few amphibian species. Benthic algae vary in their distribution, growth form, and nutritional value, and grazers differ in their means of scraping and browsing this food supply. Grazers have numerous impacts on algae, reducing their abundance, altering assemblage composition, and even stimulating algal growth and overall productivity through the removal of senescent cells and the recycling of nutrients. Studies comparing the balance of carbon, phosphorus, and nitrogen between herbivores and their food resources provide insight into the nutritional quality of periphyton to herbivores, and the role of consumer egestion and excretion in recycling nutrients to the benefit of primary producers. Strong effects of herbivory in streams have been documented with a number of invertebrates, including snails and some caddisfly and mayfly larvae, and in fishes such as the stoneroller in North America and armored catfishes in the Neotropics. Under the usually moderate environments in which most grazing studies are carried out, top-down control of algae by grazing appears to be at least as strong as bottom-up control by nutrient supply. Disturbance, particularly due to extremes of flow, can alter the grazer-algal dynamic by reducing grazer abundance, and heavy grazing pressure can reduce algal biomass to a level where it is less vulnerable to scouring during high flows.

Predation affects all organisms at some stage of the life cycle, and many species encounter predation risk throughout their lives. It affects individuals and populations directly through consumption and mortality, and also can result in behavioral and morphological adaptations that may entail some fitness cost to the prey. The many fascinating examples of the foraging behavior of predators and risk-avoidance tactics of prey attest to the importance of this interaction to both parties. Prey species depart from risky environments, restrict the time of day and location of foraging, and evolve defensive morphologies that may exact a cost in growth or subsequent reproduction. Because the prey are themselves consumers of other resources, these responses help us to understand how the indirect effects of top predators can extend throughout the food web. Top predators have often but not invariably been demonstrated to limit the abundance of prey populations, to confine the prey's distribution to habitats where the predator is absent or ineffective, and in some instances to trigger an elaborate cascade of interactions with consequences for whole ecosystems. Habitat conditions, the identity of the top predator, the magnitude of external subsidies, and environmental disturbance can act as switches that turn a cascade into a trickle, or the reverse.

Competition between consumers for a shared resource either through its mutual exploitation or by aggressive interference depends on the extent of niche overlap versus niche segregation. Estimates of overlap in diet, habitat, or temporal activity of groups of species that share a common resource are often used to infer competition. Field observations such as the different mesh sizes and locations of the nets of hydropsychid caddis larvae or use of stream habitat and time of day for foraging by stream fishes suggest how interactions within groups of potentially competing species can be ameliorated through the partitioning of diet, space, or time. The large literature on resource partitioning among stream-dwelling invertebrates and fishes provides much insight into the specialization of individual species, but because the extent to which resources actually are limiting often is unknown, this is weak evidence for the importance of competition. Experimental studies with invertebrates have documented numerous cases of aggressive interference, mainly involving space limitation, and in some cases the interaction is as much predation as competition. Evidence to date may reflect challenges of experimental design and scale, as is suggested by the system-wide effects that followed the decline of an abundant grazing caddis affected by a parasite outbreak. It seems that competition can be an important interaction in stream assemblages, but the extent of its influence is not well understood.

Freshwater ecosystems harbor pathogens and parasites responsible for a number of well-known animal diseases including salmonid whirling disease, amphibian chytridiomycosis, and crayfish plague, as well as diseases affecting humans, such as malaria, cholera, and river blindness. Some infect a single host, while others pass through one or more intermediate hosts before reaching their definitive (terminal) host. In the latter case, parasites may alter host behavior, making it more vulnerable to consumption by the definitive host. Host-parasite relationships involving fishes have received considerable study, relative to those involving aquatic invertebrates. Where long exposure presumably has allowed the host to adapt to infection, host and parasite abundance may remain relatively stable. However, the arrival of a novel pathogen can result in such a rapid decline that the host population virtually disappears, as evidenced by many species of neotropical amphibians infected by a virulent fungal pathogen. Severe infections may arise when brought to a new region by an invasive species that hosts but is relatively resistant to the parasite or pathogen, or when some change to environmental conditions favor its emergence.

References

Aberle N, Hillebrand H, Grey J, Wiltshire KH (2005) Selectivity and competitive interactions between two benthic invertebrate grazers (*Asellus aquaticus* and *Potamopyrgus antipodarum*): an experimental study using. Freshw Biol 50:369–379. https://doi.org/10.1111/j.1365-2427.2004.01325.x

Allan JD (1978) Trout predation and the size composition of stream drift. Limnol Oceanogr 23:1231-1237. https://doi.org/10.4319/lo.1978.23.6.1231

Allan JD (1981) Determinants of diet of brook trout (*Salvelinus fontinalis*) in a mountain stream. Can J Fish Aquat Sci 38:184–192

Allan JD (1982) The effects of reduction in trout density on the invertebrate community of a mountain stream. Ecology 63:1444–1455

Allan JD (1983) Food consumption by trout and stoneflies in a Rocky Mountain stream, with comparison to prey standing crop. In: Fontaine TD, Bartell SM (eds) Dynamics of lotic ecosystems. Ann Arbor Science Publishers, Ann Arbor, MI, pp 371–390

Allan JD, Wipfli MS, Caouette JP et al (2003) Influence of streamside vegetation on inputs of terrestrial invertebrates to salmonid food webs. Can J Fish Aquat Sci 60. https://doi.org/10.1139/f03-019

Alstad DN (1987) Particle size, resource concentration, and the distribution of net-spinning caddisflies. Oecologia 71:525–531

Alvarez M, Peckarsky BL (2005) How do grazers affect periphyton heterogeneity in streams? Oecologia 142:576–587. https://doi.org/10.1007/s00442-004-1759-0

Ayre KK, Caldwell CA, Stinson J, Landis WG (2014) Analysis of regional scale risk of whirling disease in populations of Colorado and Rio Grande Cutthroat Trout using a Bayesian Belief Network model. Risk Anal 34:1589–1605. https://doi.org/10.1111/risa.12189

Baker JA, Ross ST (1981) Spatial and temporal resource utilization by Southeastern Cyprinids. Copeia 1981:178–189

Baltz DM, Moyle PB (1993) Invasion resistance to introduced species by a native assemblage of California stream fishes. Ecol Appl 3:246–255

Barnhart MC, Haag WR, Roston WN (2008) Adaptations to host infection and larval parasitism in Unionoida. J North Am Benthol Soc 27:370–394. https://doi.org/10.1899/07-093.1

Bartholomew JL, Reno PW (2002) The history and dissemination of whirling disease. Am Fish Soc Symp 26:1–22

Begon M, Townsend CR, Harper J (2005) Ecology: from individuals to ecosystems. Blackwell, Oxford

Blanchet S, Loot G, Grenouillet G, Competitive BS (2007) Competitive interactions between native and exotic salmonids: a combined field and laboratory demonstration. Ecol Freshw Fish 16:133–143. https://doi.org/10.1111/j.1600-0633.2006.00205.x

Boersma KS, Bogan MT, Henrichs BA, Lytle DA (2014) Top predator removals have consistent effects on large species despite high environmental variability. Oikos 123:807–816. https://doi.org/10.1111/oik.00925

Bohman P, Nordwall F, Edsman L (2006) The effect of the large-scale introduction of signal crayfish on the spread of crayfish plague in Sweden. Bull Français la Pêche la Piscic 1291–1302. https://doi.org/10.1051/kmae:2006026

Bray J, Kilroy C, Gerbeaux P, Harding JS (2017) Ecological eustress? Nutrient supply, bloom stimulation and competition determine dominance of the diatom *Didymosphenia geminata*. Freshw Biol 62:1433–1442. https://doi.org/10.1111/fwb.12958

Bush AO, Lafferty KD, Lotz JM, Shostak AW (1997) Parasitology meets ecology on its own terms: Margolis et al. revisited. J Parasitol 83:575–583

Capps KA, Flecker AS (2015) High impact of low-trophic-position invaders: nonnative grazers alter the quality and quantity of basal food resources. Freshw Sci 34:784–796. https://doi.org/10.1086/681527

Collins JP, Storfer A (2003) Global amphibian declines: sorting the hypotheses. Divers Distrib 9:89–98

Cooper SD, Walde SJ, Peckarsky BL (1990) Prey exchange rates and the impact of predators on prey populations in streams. Ecology 71:1503–1514

Crowl TA, Townsend CR, Mcintosh AR (1992) The impact of introduced brown and rainbow trout on native fish: the case of Australasia. Rev Fish Biol Fisheries 2:217–241

Dahl J (1998) The impact of vertebrate and invertebrate predators on a stream benthic community. Oecologia 117:217–226. https://doi.org/10.1007/s004420050651

Dahl J, Greenberg L (1999) Effects of prey dispersal on predator - prey interactions in streams. Freshw Biol 41:771–780

Dahl J, Peckarsky BL (2003) Developmental responses to predation risk in morphologically defended mayflies. Oecologia 137:188–194. https://doi.org/10.1007/s00442-003-1326-0

Dalton CM, Flecker AS (2014) Metabolic stoichiometry and the ecology of fear in Trinidadian guppies: consequences for life histories and stream ecosystems. Oecologia 176:691–701. https://doi.org/10.1007/s00442-014-3084-6

Daszak P, Cunningham AA, Hyatt AD (2000) Emerging infectious diseases of wildlife—threats to biodiversity and human health. Science (80-) 287:443–449. https://doi.org/10.1126/science.287.5452.443

Dill LM (1983) Adaptive flexibility in the foraging behavior of fishes. Can J Fish Aquat Sci 40:398–408

Dill LM, Fraser AHG (1984) Risk of predation and the feeding behavior of juvenile eoho salmon (*Oncorhynchus kisutch*). Behav Ecol Sociobiol 16:65–71

Dudley Bytoml, Antonio CMD, Cooper SD (1990) Mechanisms and consequences of interspecific competition between two stream insects. J Anim Ecol 59:849–866

Dunbrack RL, Dill LM (1983) A model of size dependent surface feeding in a stream dwelling salmonid. Environ Biol Fishes 8:203–216. https://doi.org/10.1007/BF00001086

Edgerton BF, Henttonen P, Jussila J et al (2019) Understanding the causes of disease in European freshwater crayfish. Conserv Biol 18:1466–1474

Elliott JM (1987a) Egg hatching and resource partitioning in stoneflies: The six British *Leuctra* spp. (Plecoptera: Leuctridae). J Anim Ecol 56:415–426

Elliott JM (1987b) Temperature-induced changes in the life cycle of Leuctra nigra (Plecoptera: Leuctridae) from a Lake District stream. Freshw Biol 18:177–184

Elvidge CK, Brown GE (2012) Visual and chemical prey cues as complementary predator attractants in a tropical stream fish assemblage. Int J Zool. https://doi.org/10.1155/2012/510920

Englund G (1993) Effects of density and food availability on habitat selection in a net-spinning caddis larva, *Hydropsyche siltalai*. Oikos 68:473–480

Evans-White MA, Lamberti GA (2006) Stoichiometry of consumer-driven nutrient recycling across nutrient regimes in streams. Ecol Lett 9:1186–1197. https://doi.org/10.1111/j.1461-0248.2006.00971.x

Evans-White MA, Lamberti GA (2005) Grazer species effects on epilithon nutrient composition. Freshw Biol 50:1853–1863. https://doi.org/10.1111/j.1365-2427.2005.01452.x

Fanta SE, Hill WR, Smith TB, Roberts BJ (2010) Applying the light: nutrient hypothesis to stream periphyton. Freshw Biol 55:931–940. https://doi.org/10.1111/j.1365-2427.2009.02309.x

Fausch KD (1984) Profitable stream positions for salmonids: relating specific growth rate to net energy gain. Can J Zool 62:441–451. https://doi.org/10.1139/z84-067

Fausch KD, Nakano S, Khano S (1997) Experimentally induced foraging mode shift by sympatric chairs in a Japanese mountain stream. Behav Ecol 8:414–420

Fausch KD, Taniguchi Y, Nakano S et al (2001) Flood disturbance regimes influence rainbow trout invasion success among five holarctic regions. Ecol Appl 11:1438–1455

Fausch KD, White RJ (1981) Competition between Brook Trout (*Salvelinus fontinalis*) and Brown Trout (*Salmo trutta*) for positions in a Michigan Stream. Can J Fish Aquat Sci 38:1220–1227

Feminella JW, Hawkins CP (1995) Interactions between stream herbivores and periphyton: a quantitative analysis of past experiments. J North Am Benthol Soc 14:465–509

Flecker AS (1992) Fish predation and the evolution of invertebrate drift periodicity: evidence from neotropical streams. Ecology 73:438–448

Flecker AS, Allan JD (1984) The importance of predation, substrate and spatial refugia in determining lotic insect distributions. Oecologia 64:306–313. https://doi.org/10.1007/BF00379126

Flecker AS, Taylor BW, Bernhardt ES et al (2002) Interactions between herbivorous fishes and limiting nutrients in a tropical stream ecosystem. Ecology 83:1831–1844

Forrester GE, Chace JG, McCarthy W (1994) Diel and density-related changes in food consumption and prey selection by brook charr in a New Hampshire stream. Environ Biol Fishes 39:301–311. https://doi.org/10.1007/BF00005131

Fuiman LA, Magurran AE (1994) Development of predator defences in fishes. Rev Fish Biol Fish 4:145–183

Fuller RL, Rand PS (1990) Influence of substrate type on vulnerability of prey to predacious aquatic insects. J Am Water Resour Assoc Resour 9:1–8

Gamboa M, Castillo MM, Guerrero R (2012) Anacroneuria spp. (Insecta: Plecoptera: Perlidae) as paratenic hosts of Pheromermis sp. (Nematoda: Mermithidae) in Venezuela. Nematology 14:185–190. https://doi.org/10.1163/138855411X584133

Gagne RB, Hogan JD, Pracheil BM et al (2015) Spread of an introduced parasite across the Hawaiian archipelago independent of its introduced host. Freshw Biol 60:311–322. https://doi.org/10.1111/fwb.12491

Garcia EA, Townsend SA, Douglas MM (2015) Context dependency of top-down and bottom-up effects in a Northern Australian tropical river. Freshw Sci 34:679–690. https://doi.org/10.1086/681106

Georgian TJ, Wallace JB (1981) A model of seston capture by net-spinning caddisflies. Oikos 36:147–157

Gilliam JF (1993) Structure of a tropical stream fish community: a role for biotic interactions. Ecology 74:1856–1870. https://doi.org/10.2307/1939943

Gorman OT (1988) An experimental study of habitat use in an assemblage of Ozark minnows. Ecology 69:1239–1250. https://doi.org/10.2307/1941279

Grabner DS (2017) Hidden diversity: parasites of stream arthropods. Freshw Biol 62:52–64. https://doi.org/10.1111/fwb.12848

Grossman GD, Ratajczak REJ, Crawford M, Freeman MC (1998) Assemblage organization in stream fishes: effects of environmental variation and interspecific interactions. Ecol Monogr 68:395–420

Guo F, Kainz MJ, Valdez D et al (2016a) The effect of light and nutrients on algal food quality and their consequent effect on grazer growth in subtropical streams. Freshw Biol 35: https://doi.org/10.1086/688092

Guo F, Kainz MJ, Sheldon F, Bunn SE (2016b) The importance of high quality algal food sources in stream food webs—current status and future perspectives. Freshw Biol 61:815–31

Guo F, Kainz MJ, Sheldon F, Bunn SE (2016c) Effects of light and nutrients on periphyton and the fatty acid composition and somatic growth of invertebrate grazers in subtropical streams. Oecologia 181:449–462. https://doi.org/10.1007/s00442-016-3573-x

Guo F, Kainz MJ, Valdez D et al (2016d) High-quality algae attached to leaf litter boost invertebrate shredder growth. Freshw Sci 35:1213–1221. https://doi.org/10.1086/688667

Hall RO, Wallace JB, Eggert SL (2000) Organic matter flow in stream food webs with reduced detrital resource base. Ecology 81:3445–3463

Hammock BG, Krigbaum NY, Johnson ML (2012) Incorporating invertebrate predators into theory regarding the timing of invertebrate drift. Aquat Ecol 46:153–163. https://doi.org/10.1007/s10452-012-9388-x

Hart DD (1981) Foraging and resource patchiness: field experiments with a grazing stream insect. Oikos 37:46–52

Harvey BC, Nakamoto RJ (2013) Seasonal and among-stream variation in predator encounter rates for fish prey. Trans Am Fish Soc 142:621–627. https://doi.org/10.1080/00028487.2012.760485

Hawkins CP, Furnish JK (1987) Are snails important competitors in stream ecosystems? Oikos 49:209–220

Hazelton PD, Grossman GD (2009) Turbidity, velocity and interspecific interactions affect foraging behaviour of rosyside dace (*Clinostomus funduloides*) and yellowfin shiners (*Notropis lutippinis*). Ecol Freshw Fish 18:427–436. https://doi.org/10.1111/j.1600-0633.2009.00359.x

Hemphill N (1988) Competition between two stream dwelling filter-feeders, *Hydropsyche oslari* and *Simulium virgatum*. Oecologia 77:73–80

Hildrew AG, Edington JM (1979) Factors facilitating the coexistence of Hydropsychid caddis larvae (Trichoptera) in the same river system. J Anim Ecol 48:557–576

Hill WR (1992) Food limitation and interspecific competition in snail-dominated streams. Can Entomol 49:1257–1267

Hill WR, Knight AW (1987) Experimental analysis of the grazing interaction between a mayfly and stream algae. Ecology 68:1955–1965

Hill WR, Knight AW (1988) grazing effects of two stream insects on periphyton. Limnologica 33:15–26

Hill WR, Rinchard J, Czesny S (2011) Light, nutrients and the fatty acid composition of stream periphyton. Freshw Biol 56:1825–1836. https://doi.org/10.1111/j.1365-2427.2011.02622.x

Hillebrand H (2009) Meta-analysis of grazer control of periphyton biomass across aquatic ecosystems. J Phycol 45:798–806. https://doi.org/10.1111/j.1529-8817.2009.00702.x

Hillebrand H, Frost P, Liess A (2008) Ecological stoichiometry of indirect grazer effects on periphyton nutrient content. Oecologia 155:619–630. https://doi.org/10.1007/s00442-007-0930-9

Hintz WD, Wellnitz T (2013) Current velocity influences the facilitation and removal of algae by stream grazers. Aquat Ecol 47:235–244. https://doi.org/10.1007/s10452-013-9438-z

Ho BSK, Dudgeon D (2016) Are high densities of fishes and shrimp associated with top-down control of tropical benthic communities? A test in three Hong Kong streams. Freshw Biol 61:57–68. https://doi.org/10.1111/fwb.12678

Holdich DM, Reynolds JD, Souty-Grosset C, Sibley PJ (2009) A review of the ever increasing threat to European crayfish from non-indigenous crayfish species. Knowl Manag Aquat Ecosyst 11. https://doi.org/10.1051/kmae/2009025

Holomuzki JR, Biggs BJF (2006) Food limitation affects algivory and grazer performance for New Zealand stream macroinvertebrates. Hydrobiologia 561:83–84. https://doi.org/10.1007/s10750-005-1606-2

Holomuzki JR, Feminella JW, Power ME (2013) Biotic interactions in freshwater benthic habitats. J North Am Benthol Soc 29:220–244. https://doi.org/10.1899/08-044.1

Hughes NF, Dill LM (1990) Position choice by drift-feeding salmonids: Model and test for Arctic Grayling (*Thymallus arcticus*) in subarctic mountain streams, interior Alaska. Can J Fish Aquat Sci 47:2039–2048. https://doi.org/10.1139/f90-228

Hughes NF, Hayes JW, Shearer KA, Young RG (2003) Testing a model of drift-feeding using three-dimensional videography of wild brown trout, *Salmo trutta*, in a New Zealand river. Can J Fish Aquat Sci 1476:1462–1476. https://doi.org/10.1139/F03-126

Huryn AD (1996) An appraisal of the Allen paradox in a New Zealand trout stream. Limnol Oceanogr 41:217–241

Hyatt KD (1979) Feeding strategy. In: Hoar WS, Randall DJ, Brett JR (eds) Fish Physiology, vol 8. Academic Press, New York, pp 71–119

Johnson PTJ, Paull SH (2011) The ecology and emergence of diseases in fresh waters. Freshw Biol 56:638–657. https://doi.org/10.1111/j.1365-2427.2010.02546.x

Kats LB, Dill LM (1998) The scent of death: Chemosensory assessment of predation risk by prey animals. Ecoscience 5:361–394. https://doi.org/10.1080/11956860.1998.11682468

Kats LB, Petranka JW, Sih A (1988) Antipredator defenses and the persistence of amphibian larvae with fishes. Ecology 69:1865–1870

Kelley JL, Magurran AE (2003) Learned predator recognition and antipredator responses in fishes. Fish Fish 4:216–226

Kohler SL (1984) Search mechanism of a stream grazer in patchy environments: the role of food abundance. Oecologia 62:209–218

Kohler SL (1992) Competition and the structure of a benthic stream community. Ecol Monogr 62:165–188

Kohler SL (2008) The ecology of host-parasite interactions in aquatic insects. In: Lancaster J, Briers RA (eds) Aquatic Insects: Challenges to Populations. CAB International, pp 55–80

Kohler SL, Holland WK (2001) Population regulation in an aquatic insect: the role of disease. Ecology 82:2294–2305

Kohler SL, Wiley MJ (1997) Pathogen outbreaks reveal large-scale effects of competition in stream communities. Ecology 78:2164–2176

Knoll LB, Mcintyre PB, Vanni MJ, Flecker AS (2009) Feedbacks of consumer nutrient recycling on producer biomass and stoichiometry: separating direct and indirect effects. Oikos 118:1732–1742

Lamberti GA, Ashkenas LR, Gregory SV, Steinman AD (1987) Effects of three herbivores on periphyton communities in laboratory streams. J North Am Benthol Soc 6:92–104

Lamberti GA, Resh V (1983) Stream periphyton and insect herbivores: An experimental study of grazing by a caddisfly population. Ecology 64:1124–1135

Lancaster J, Bovill WD (2017) Species-specific prevalence of mermithid parasites in populations of six congeneric host caddisflies of *Ecnomus* McLachlan, 1864 (Trichoptera: Ecnomidae). Aquat Insects 38:67–78. https://doi.org/10.1080/01650424.2017.1299866

Lancaster J, Hildrew AG, Townsend CR (1991) Invertebrate predation on patchy and mobile prey in streams. J Anim Ecol 60:625–641

Layman CA, Araujo MS, Boucek R et al (2012) Applying stable isotopes to examine food-web structure: An overview of analytical tools. Biol Rev 87:545–562

Liess A, Le Gros A, Wagenhoff A et al (2012) Landuse intensity in stream catchments affects the benthic food web: consequences for nutrient supply, periphyton C: nutrient ratios, and invertebrate richness and abundance. Freshw Sci 31:813–824. https://doi.org/10.1899/11-019.1

Liess A, Hillebrand H (2004) Invited review: direct and indirect effects in herbivore—periphyton interactions. Arch für Hydrobiol 159:433–453. https://doi.org/10.1127/0003-9136/2004/0159-0433

Liess A, Kahlert M (2007) Gastropod grazers and nutrients, but not light, interact in determining periphytic algal diversity. Oecologia 152:101–111. https://doi.org/10.1007/s00442-006-0636-4

Lips KR, Brem F, Brenes R et al (2006) merging infectious disease and the loss of biodiversity in a Neotropical amphibian community. Proc Natl Acad Sci 103:3165–3170

Longshaw M (2011) Diseases of crayfish: a review. J Invertebr Pathol 106:54–70. https://doi.org/10.1016/j.jip.2010.09.013

Lowe WH, Hauer FR (1999) Ecology of two large, net-spinning caddisfly species in a mountain stream: distribution, abundance,

and metabolic response to a thermal gradient. Can J Zool 77:1637–1644

Mackay RJ (1977) Behavior of *Pycnopsyche* (Trichoptera: Limnephilidae) on mineral substrates in laboratory streams. Ecology 58:191–195

Malmqvist B, Sjostrom P (1980) Prey size and feeding patterns in *Dinocras cephalotes* (Plecoptera. Oikos 35:311–316

McAuliffe JR (1984a) Resource depression by a stream herbivore: effects on distributions and abundances of other grazers. Oikos 42:327–333

McAuliffe JR (1984b) Competition for space, disturbance, and the structure of a benthic stream community. Ecology 65:894–908

McCahon CP, Maund SJ, Poulton MJ (1991) The effect of the acanthocephalan parasite *Pomphorhynchus laevis* on the drift of its intermediate host *Gammarus pulex*. Freshw Biol 25:507–513. https://doi.org/10.1111/j.1365-2427.1991.tb01393.x

McDowall RM (2006) Crying wolf, crying foul, or crying shame: Alien salmonids and a biodiversity crisis in the southern cool-temperate galaxioid fishes? Rev Fish Biol Fish 16:233–422. https://doi.org/10.1007/s11160-006-9017-7

McHugh P, Budy P (2005) An experimental evaluation of competitive and thermal effects on brown trout (*Salmo trutta*) and Bonneville cutthroat trout (*Oncorhynchus clarkii* utah) performance along an altitudinal gradient. Can J Fish Aquat Sci 62:2784–2795. https://doi.org/10.1139/f05-184

McIntosh AR, Peckarsky BL (2004) Are mayfly anti-predator responses to fish odour proportional to risk? Arch für Hydrobiol 160:145–151. https://doi.org/10.1127/0003-9136/2004/0160-0145

McIntosh AR, Peckarsky BL, Taylor BW (2002) The influence of predatory fish on mayfly drift: extrapolating from experiments to nature. Freshw Biol 47:1497–1513. https://doi.org/10.1046/j.1365-2427.2002.00889.x

McIntosh AR, Taylor BW (1999) Rapid size-specific changes in the drift of *Baetis bicaudatus* (Ephemeroptera) caused by alterations in fish odour concentration. Oecologia 118:256–264

McMahon TE, Danehy RJ, Ecology CA (2007) Temperature and competition between Bull Trout and Brook Trout: a test of the elevation refuge hypothesis. Trans Am Fish Soc 136:1313–1326. https://doi.org/10.1577/T06-217.1

McNeely C, Finlay JC, Power ME (2007) Grazer traits, competition, and carbon sources to a headwater-stream food web. Ecology 88:391–401

Meissner K, Muotka T (2006) The role of trout in stream food webs: integrating evidence from field surveys and experiments. J Anim Ecol 75:421–433. https://doi.org/10.1111/j.1365-2656.2006.01063.x

Metcalfe NB, Valdimarsson SK, Fraser NHC (1997) Habitat profitability and choice in a sit-and-wait predator: Juvenile salmon prefer slower currents on darker nights. J Anim Ecol 66:866–875

Mohammed RS, Reynolds M, James J et al (2016) Getting into hot water: sick guppies frequent warmer thermal conditions. Oecologia 181:911–917. https://doi.org/10.1007/s00442-016-3598-1

Molles MC, Pietruszka RD (1983) Mechanisms of prey selection by predaceous stoneflies: roles of prey morphology, behavior and predator hunger. Oecologia 57:25–31

Moyle PB, Senanayake FR (1984) Resource partitioning among the fishes of rainforest streams in Sri Lanka. J Zool London 202:195–223

Mulholland PJ, Steinman AD, Palumbo AV et al (1991) Role of nutrient cycling and herbivory in regulating periphyton communities in laboratory streams. Ecology 72:966–982. https://doi.org/10.2307/1940597

Muller K (1963) Diurnal rhythm in 'Organic Drift' of *Gammarus pulex*. Nature 198:806–807

Muotka T (1990) Coexistence in a guild of filter feeding caddis larvae: do different instars act as different species? Oecologia 85:281–292

Nakano S, Fausch KD, Kitano S (1999a) Flexible niche partitioning via a foraging mode shift: a proposed mechanism for coexistence in stream-dwelling charrs. J Anim Ecol 68:1079–1092

Nakano S, Kawaguchi Y, Taniguchi Y, Miyasaka H (1999b) Selective foraging on terrestrial invertebrates by rainbow trout in a forested headwater stream in northern Japan. Ecol Res 14:351–360

Nakano S, Murakami M (2001) Reciprocal subsidies: dynamic interdependence between terrestrial and aquatic food webs. Proc Natl Acad Sci U S A 98:166–170

Naman SM, Rosenfeld JS, Richardson JS (2016) Causes and consequences of invertebrate drift in running waters: from individuals to populations and trophic fluxes. Can J Fish Aquat Sci 73:1292–1305. https://doi.org/10.1139/cjfas-2015-0363

Neuswanger J, Wipfli MS, Rosenberger AE, Hughes NF (2014) Mechanisms of drift-feeding behavior in juvenile Chinook salmon and the role of inedible debris in a clear-water Alaskan stream. Environ Biol Fishes 97:489–503. https://doi.org/10.1007/s10641-014-0227-x

Ohta T, Miyake Y, Hiura T (2011) Light intensity regulates growth and reproduction of a snail grazer (Gyraulus chinensis) through changes in the quality and biomass of stream periphyton. Freshw Biol 56:2260–2271. https://doi.org/10.1111/j.1365-2427.2011.02653.x

Oldmeadow DF, Lancaster J, Rice SP (2010) Drift and settlement of stream insects in a complex hydraulic environment. Freshw Biol 55:1020–1035. https://doi.org/10.1111/j.1365-2427.2009.02338.x

Opsahl RW, Wellnitz T, Poff NL (2003) Current velocity and invertebrate grazing regulate stream algae: results of an in situ electrical exclusion. Hydrobiologia 499:135–145

Osborne LL, Herricks EE (1981) Microhabitat characteristics of Hydropsyche (Trichoptera: Hydropsychidae) and the importance of body size. J North Am Benthol Soc 6:115–124

Peckarsky BL (1996) Alternative predator avoidance syndromes of stream-dwelling mayfly larvae. Ecology 77:1888–1905

Peckarsky BL, Dodson SI (1980) An experimental analysis of biological factors contributing to stream community structure. Ecology 61:1283–1290

Peckarsky BL, Taylor BW, Mcintosh AR et al (2001) Variation in mayfly size at metamorphosis as a developmental response to risk of predation. Ecology 82:740–757

Peckarsky BL, Vance SA (1997) The effect of mermithid parasitism on predation of nymphal Baetis bicaudatus (Ephemeroptera) by invertebrates. Oecologia 110:147–152

Peckarsky L (1985) Do predaceous stoneflies and siltation affect the structure of stream insect communities colonizing enclosures? Can J Zool 63:1519–1530

Poinar GOJ (1976) Presence of Mermithidae (Nematoda) in invertebrate paratenic hosts. J Parasitol 62:843–844

Ponton F, Lebarbenchon C, Lefèvre T et al (2006) Hairworm anti-predator strategy: a study of causes and consequences. Parasitology 133:631–638. https://doi.org/10.1017/S0031182006000904

Poulin R, Paterson RA, Townsend CR et al (2011) Biological invasions and the dynamics of endemic diseases in freshwater ecosystems. Freshw Biol 56:676–688. https://doi.org/10.1111/j.1365-2427.2010.02425.x

Power M (1984a) Habitat quality and the distribution of algae-grazing catfish in a Panamanian stream. J Anim Ecol 53:357–374

Power ME (1983) Grazing responses of tropical freshwater fishes to different scales of variation in their food. Environ Biol Fishes 9:103–115

Power ME (1984b) Depth distributions of armored catfish: predator-induced resource avoidance? Ecology 65:523–528

Power ME, Matthews WJ (1983) Algae-grazing minnows (Campostoma anomalum), piscivorous bass (Micropterus spp.) and the distribution of attached algae in a small prairie-margin stream. Oecologia 60:328–332

Preisser EL, Bolnick DI, Benard ME (2005) Scared to death? The effects of intimidation and consumption in predator-prey interactions. Ecology 86:501–509

Railsback SF (2016) Why it is time to put PHABSIM out to pasture. Fisheries 41:720–725. https://doi.org/10.1080/03632415.2016.1245991

Ranvestel AW, Lips KR, Pringle CM et al (2004) Neotropical tadpoles influence stream benthos: evidence for the ecological consequences of decline in amphibian populations. Freshw Biol 49:274–285. https://doi.org/10.1111/j.1365-2427.2004.01184.x

Richards C, Winshall GW (1988) The influence of periphyton abundance on Baetis bicaudatus distribution and colonization in a small stream. J North Am Benthol Soc 7:77–86

Rier ST, Stevenson RJ (2002) Effects of light, dissolved organic carbon, and inorganic nutrients on the relationship between algae and heterotrophic bacteria in stream periphyton. Hydrobiol 489:179–184

Rosemond AD, Mulholland PJ, Brawley SH (2000) Seasonally shifting limitation of stream periphyton: response of algal populations and assemblage biomass and productivity to variation in light, nutrients, and herbivores. Can J Fish Aquat Sci 57:66–75

Ross ST (1986) Resource partitioning in fish assemblages: a review of field studies. Copeia 1986:352–388

Ruetz CRI, Vondracek B, Newman RM (2004) Weak top-down control of grazers and periphyton by slimy sculpins in a coldwater stream. J North Am Benthol Soc 23:271–286

Sarker S, Kallert DM, Hedrick RP, El-Matbouli M (2015) Whirling disease revisited: pathogenesis, parasite biology and disease intervention. Dis Aquat Organ 114:155–175. https://doi.org/10.3354/dao02856

Sato T, Arizono M, Sone R, Harada Y (2008) Parasite-mediated allochthonous input: do horsehair worms enhance subsidized predation of stream salmonids on crickets? Can J Zool Zool 86:231–235

Schaffer M, Winkelmann C, Hellmann C, Benndorf J (2013) Reduced drift activity of two benthic invertebrate species is mediated by infochemicals of benthic fish. Aquat Ecol 47:99–107

Schmidt-Rhaesa A, Ehrmann R (2001) Horsehair worms (Nematomorpha) as parasites of praying mantids with a discussion of their life cycle. Zool Anz 240:167–179. https://doi.org/10.1078/0044-5231-00014

Schoener TW (1974) Resource partitioning in ecological communities. Science 185:27–39

Simon KS, Townsend CR (2003) Impacts of freshwater invaders at different levels of ecological organisation, with emphasis on salmonids and ecosystem consequences. Freshw Biol 48:982–994. https://doi.org/10.1046/j.1365-2427.2003.01069.x

Skerratt LF, Berger L, Speare R et al (2007) Spread of chytridiomycosis has caused the rapid global decline and extinction of frogs. EcoHealth 4:125–134. https://doi.org/10.1007/s10393-007-0093-5

Smith LC, Smock LA (1992) Ecology of invertebrate predators in a Coastal Plain stream. Freshw Biol 28:319–329

Steinman AD (1996) Effects of grazers on benthic freshwater algae. In: Stevenson RJ, Bothwell ML, Lowe RL (eds) Algal Ecology. Academic Press, San Diego, pp 341–373

Steinman AD, McIntyre CD (1987) Effects of irradiance on the community structure and biomass of algal assemblages in laboratory streams. Can J Fish Aquat Sci 44:1640–1648

Steinmetz J, Kohler SL, Soluk DA (2003) Birds are overlooked top predators in aquatic food webs. Ecology 84:1324–1328

Stelzer R, Lamberti GA (2002) Ecological stoichiometry in running waters: periphyton chemical composition and snail growth. Ecology 83:1039–1051. https://doi.org/10.2307/3071912

Stelzer RS, Lamberti GA (2001) Effects of N: P ratio and total nutrient concentration on stream periphyton community structure, biomass, and elemental composition. Limnol Oceanogr 46:356–367

Sterner RW, Elser JJ (2002) Ecological stoichiometry. Princeton University Press, Princeton, NJ

Sterner RW, Elser JJ, Fee EJ et al (1997) The light: nutrient ratio in lakes: the balance of energy and materials affects ecosystem structure and process. Am Nat 150:663–684

Svoboda J, Strand DA, Vrålstad T et al (2014) The crayfish plague pathogen can infect freshwater-inhabiting crabs. Freshw Biol 59:918–929. https://doi.org/10.1111/fwb.12315

Szokoli F, Winkelmann C, Berendonk TU, Worischka S (2015) The effects of fish kairomones and food availability on the predator avoidance behaviour of Gammarus pulex. Fundam Appl Limnol/Arch für Hydrobiol 186:249–258. https://doi.org/10.1127/fal/2015/0633

Taylor BW, Mcintosh AR, Peckarsky BL (2002) Reach-scale manipulations show invertebrate grazers depress algal resources in streams. Limnol Oceanogr 47:893–899

Tokeshi M (1986) Resource utilization, overlap and temporal community dynamics: a null model analysis of an epiphytic chironomid community. J Anim Ecol 55:491–506

Towns DR (1983) Life history patterns of six sympatric species of Leptophlebiidae (Ephemeroptera) in a New Zealand stream and the role of interspecific competition in their evolution. Hydrobiologia 99:37–50

Townsend CR, Crowl TA (2006) Fragmented population structure in a native New Zealand fish: an effect of introduced Brown Trout? Oikos 61:347. https://doi.org/10.2307/3545242

Vadeboncoeur Y, Power ME (2017) Attached algae: the cryptic base of inverted trophic pyramids in freshwaters. Annu Rev Ecol Evol Syst 48:255–279

Vance SA (1996) Morphological and behavioural sex reversal in mermithid-infected mayflies. Proc R Soc B Biol Sci 263:907–912

Vance SA, Peckarsky BL (1996) The infection of nymphal Baetis bicaudatus by the mermithid nematode Gasteromermis sp. Ecol Entomol 21:377–381. https://doi.org/10.1046/j.1365-2311.1996.00009.x

Vanni MJ, Flecker S, Hood JM (2002) Stoichiometry of nutrient recycling by vertebrates in a tropical stream: linking species identity and ecosystem processes. Ecol Lett 5:285–293

Villamarin F, Jardine TD, Bunn SE et al (2017) Opportunistic top predators partition food resources in a tropical freshwater ecosystem. Freshw Biol 62:1389–1400. https://doi.org/10.1111/fwb.12952

Vredenburg VT, Knapp RA, Tunstall TS, Briggs CJ (2010) Dynamics of an emerging disease drive large-scale amphibian population extinctions. Proc Natl Acad Sci 107:9689–9694. https://doi.org/10.1073/pnas.0914111107

Wallace JB, Eggert SL, Meyer JL, Webster JR (1997) Multiple trophic levels of a forest stream linked to terrestrial litter inputs. Science (80-) 277:102–105

Wallace JB, Gurtz M (1986) Response of Baetis mayflies (Ephemeroptera) to catchment logging. Am Midl Nat 115:25–41

Wallace JB, Webster JR, Woodall WR (1977) The role of filter feeders in flowing waters. Arch fur Hydrobiol 79:506–532

Ware DM (1972) Predation by Rainbow Trout (Salmo gairdneri): the influence of hunger, prey density, and prey size. J Fish Res Board Canada 29:1193–1201

Waters TF (1972) The drift of stream insects. Annu Rev Entomol 17:253–272

Watz J, Piccolo JJ, Greenberg L, Bergman E (2012) Temperature-dependent prey capture efficiency and foraging modes of brown trout Salmo trutta. J Fish Biol 81:345–350. https://doi.org/10.1111/j.1095-8649.2012.03329.x/full

Wellnitz T, Poff NL (2012) Current-mediated periphytic structure modifies grazer interactions and algal removal. Aquat Ecol 46:521–530. https://doi.org/10.1007/s10452-012-9419-7

Welsh SA, Perry SA (1998) Habitat partitioning in a community of darters in the Elk River, West Virginia. Environ Biol Fishes 51:411–419

Whiles MR, Hall RO, Dodds WK et al (2013) Disease-driven amphibian declines alter ecosystem processes in a tropical stream. Ecosystems 16:146–157. https://doi.org/10.1007/s10021-012-9602-7

Whiles MR, Lips KR, Pringle CM et al (2006) The effects of amphibian population declines on the structure and function of Neotropical stream ecosystems. Front Ecol Environ 4:27–34. https://doi.org/10.1890/1540-9295(2006)004%5b0027:teoapd%5d2.0.co;2

Williams JT, Townsend CR, Townsend CR (2019) Mermithid nematode infections and drift in the mayfly Deleatidium spp. (Ephemeroptera). J Parasitol 87:1225–1227

Wilzbach MA, Cummins KW, Hall JD (1986) Influence of habitat manipulations on interactions between cutthroat trout and invertebrate drift. Ecology 67:898–911

Winemiller K (1989) Ontogenetic diet shifts and resource partitioning among piscivorous fishes in the Venezuelan llanos. Environ Biol Fishes 26:177–199

Winemiller KO (1991) Ecomorphological diversification in lowland freshwater fish assemblages from five biotic regions. Ecol Indic 61:343–365

Woodward G, Warren P (2007) Body size and predatory interactions in freshwaters: scaling from individuals to communities. In: Raffaelli DG, Edmonds-brown R (eds) Hildrew AG. The structure and function of aquatic ecosystems. Cambridge University Press, Body size, pp 98–117

Woodward G, Hildrew AG (2002) Body-size determinants of niche overlap and intraguild predation within a complex food web. J Anim Ecol 71:1063–1074

Wooster D, Sih A (1995) A review of the drift and activity responses of stream prey to predator presence. Oikos 73:3–8

Worischka S, Koebsch C, Hellmann C, Winkelmann C (2012) Habitat overlap between predatory benthic fish and their invertebrate prey in streams: The relative influence of spatial and temporal factors on predation risk. Freshw Biol 57:2247–2261. https://doi.org/10.1111/j.1365-2427.2012.02868.x

Worischka S, Schmidt SI, Hellmann C, Winkelmann C (2015) Selective predation by benthivorous fish on stream macroinvertebrates—the role of prey traits and prey abundance. Limnologica 52:41–50. https://doi.org/10.1016/j.limno.2015.03.004

Yang GY, Tang T, Dudgeon D (2009) Spatial and seasonal variations in benthic algal assemblages in streams in monsoonal Hong Kong. Hydrobiologia 632:189–200. https://doi.org/10.1007/s10750-009-9838-1

Yeung ACY, Dudgeon D (2013) A manipulative study of macroinvertebrate grazers in Hong Kong streams: do snails compete with insects? Freshw Biol 58:2299–2309. https://doi.org/10.1111/fwb.12210

Young KA, Dunham JB, Stephenson JF et al (2010) A trial of two trouts: Comparing the impacts of rainbow and brown trout on a native galaxiid. Anim Conserv 13:399–410. https://doi.org/10.1111/j.1469-1795.2010.00354.x

Zimmerman JKH, Vondracek B (2007) Brown trout and food web interactions in a Minnesota stream. Freshw Biol 52:123–136. https://doi.org/10.1111/j.1365-2427.2006.01681.x

The forces that shape community structure are those that determine which species and how many species occur together, which are common and which are rare, and the interactions amongst them. The number of coexisting species in ecological communities is a consequence of processes operating at both local and regional scales (Rosenzweig 1995; Gaston 2000). Regional diversity is determined by species formation and loss over earth history, and opportunities for dispersal. Local diversity is determined by species adaptations to the abiotic and biotic factors encountered at a site, and by each species' dispersal ability and life history characteristics. Thus, the composition of an assemblage of species at a location is the product both of large-scale spatial patterns and the forces responsible for them, and local conditions that influence which species are best adapted to that environment, or best able to colonize it.

Community structure refers to the organization of a biological community based on numbers of individuals within different taxonomic groups and functional roles, and the underlying processes that maintain that organization. Explanations for patterns in species diversity and community structure frequently invoke niche-based models (MacArthur 1972; Chase and Leibold 2003), in which the presence and abundance of individual species is a reflection of their fit to habitat conditions and success in interspecific interactions, the subjects of Chaps. 5 and 10 respectively. In relatively stable or moderate environments, biological interactions are considered to be particularly influential in the assembly and maintenance of communities. For communities to exhibit predictable structure requires that their assembly is the outcome of non-random processes that result in repeatable patterns. This leads us to expect that the same species, in roughly the same abundances, will be found in the same locale as long as environmental conditions do not change greatly, and that similar communities should occur wherever environmental circumstances are comparable. In addition, many ecosystems experience periodic disturbances, and streams are no exception. Environmental disturbances such as floods and droughts, when sufficiently extreme or frequent, are likely to prevent biotic interactions from acting with the strength and regularity required to result in consistent community patterns. Very harsh environments or frequent disturbances may severely restrict the number of species that can survive those conditions and thus reduce diversity, whereas a moderate level of disturbance may enhance diversity by counteracting the tendency of a few superior species to become dominant.

The countering view to niche models asserts that assemblages are an unstructured sample of whichever species are able to survive and reproduce under local environmental conditions, changing as conditions change, and by chance. A more formalized version, the neutral model of Hubbell (2001, 2005), considers all species of a particular trophic level to be essentially interchangeable, and so random replacement following stochastic colonization and extinction determines the momentary composition of short-lived assemblages. Under this model, species are ecologically equivalent and substitutable, to be replaced from a regional species pool whenever a chance local extinction depletes site diversity. The recolonization of a lost population requires dispersal, and so distance, life-history traits, and other factors, such as terrain, can determine whether dispersal limits the opportunities for a particular species to re-establish.

Studies of local assemblages often assume that communities are determined solely by environmental conditions and species interactions at the local scale, without regard for such larger scale processes as dispersal, speciation, and historical biogeography. However, regional species pools and factors that influence dispersal at large spatial scales influence local diversity and assemblage structure by controlling the pool of species available to colonize a location (Ricklefs and Schluter 1993; Rosenzweig 1995). Processes occurring at large spatial scales, such as the size and composition of the regional species pool, opportunities for dispersal, and climate can influence local communities (Leibold and Chase

2017). This is especially true in local communities if species interactions are not the primary force governing community structure, including systems with frequent disturbance events such as floods and droughts, or when communities are dominated by rare species with limited dispersal capacities (Cornell and Harrison 2014).

Species sorting describes how regional diversity affects local diversity through the action of environmental factors acting at progressively smaller scales, often visualized as a hierarchical series of filters (Poff 1997). It is also important to remember that local communities are not isolated from one another. They are connected via dispersal to other local communities, forming a network or metacommunity, that in turn is influenced by spatial heterogeneity and environmental factors that are operating at different scales (Heino and Mykrä 2008; Leibold and Chase 2017). Consideration of regional diversity also reminds us that the long-term persistence of a species usually does not depend solely upon its survival in any one local community. Separate populations of a species may exhibit different trends in different locales, with the consequence that dispersal permits a long-term regularity on a larger scale that is not apparent by detailed investigation on a finer scale. Such a perspective lessens the need for equilibrium-enhancing interactions because regional processes of immigration and emigration may contribute some of the buffering against extinction that otherwise must be attributed to biotic factors.

11.1 Global and Regional Patterns in Species Diversity

Freshwater biodiversity varies at all scales, including among regions of the world, along latitudinal and environmental gradients, among ecoregions, and across habitats. As discussed in Chapter 1, freshwater habitats support more than 125,000 species of freshwater animals, comprising nearly 10% of all known species and approximately one-third of vertebrate species (Balian et al. 2008; Strayer and Dudgeon 2010). Freshwater fishes are better known than invertebrates and microorganisms, but are still far from fully catalogued, especially in tropical latitudes where the diversity of freshwater fishes is known to be much greater than at temperate latitudes (Lévêque et al. 2008). A comprehensive assessment of freshwater biodiversity is presented in a series of articles edited by Balian et al. (2008). These accounts are extremely useful for cataloguing known diversity in all major freshwater groups, providing much information on biogeographic patterns and the current status of many species.

Among freshwater organisms, fishes have been most intensively scrutinized from the perspective of identifying and attempting to explain global and regional patterns,

because species-level distributions are well known for a large fraction of the fauna. The global distribution of freshwater fish species richness (number of species) is shown in Fig. 11.1 (Oberdorff et al. 2011), and interesting patterns are immediately apparent. More species occur at low (tropical) than higher latitudes. Eastern North America has more fish species than western North America or western Europe. Some regions are notably depauperate. Three main hypotheses have been invoked in the study of global patterns: species-area, species-energy, and historical events. Studies find that species richness increases with the area under consideration, with available energy and ecosystem productivity, and as a consequence of historical factors that influence speciation and dispersal. Variation in regional species richness itself is difficult to fully explain, but events in earth history that influence species formation, extinction, and differences in the ability of individual species to disperse clearly influence the regional species pool, and contemporary patterns in abundance and diversity in turn are influenced by overall productivity, habitat diversity, and biological interactions (Rosenzweig 1995; Cornell and Harrison 2014). In an analysis of freshwater fish species richness for 132 West European and North American rivers, Griffiths (2006) concluded that contemporary ecological factors provided the strongest statistical explanation of the variation in freshwater fish species richness for both continents. Historical factors, while also significant, had less explanatory power. We shall briefly consider each explanation, and because studies often infer that each plays a role and researchers usually try to deduce their relative weights, most examples provide evidence for multiple causal mechanisms.

11.1.1 Species-Area Relationships

The species-area hypothesis relates to a well-established relationship between the number of species and the area of study, first established using islands of differing sizes (MacArthur and Wilson 1963, 1967). On a log-log scale, species richness (S) increases with area (A) by an exponent, z, which is empirically determined and can be compared among taxa and locations studied. The relationship typically is log-linear, according to the equation:

$$S = cA^z \tag{11.1}$$

where S = species richness, A = area of habitat, and c and z are parameters determined from the data. The slope parameter z quantifies the rate of increase in number of species with area surveyed, and frequently falls between 0.2 and 0.4.

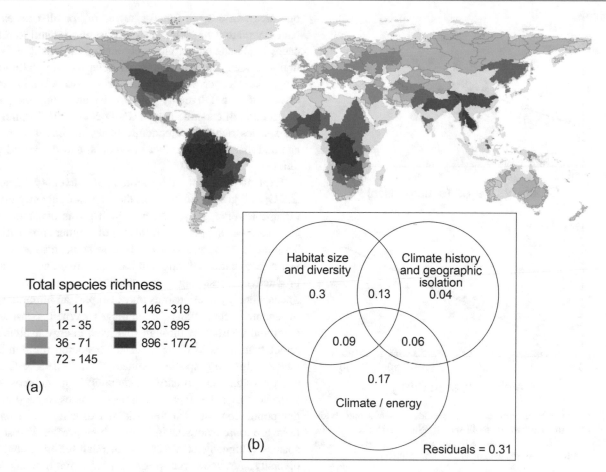

Fig. 11.1 (**a**) Freshwater fish species richness for worldwide drainage basins and (**b**) contribution of area-related (habitat size and diversity), climate-related (climate/energy) and historical variables (climate history and geographic isolation) to explain variation in species richness (Reproduced from Oberdorff et al. 2011)

A number of studies have established that species richness increases log-linearly with either drainage basin area or river discharge (Fig. 11.2), including for the mollusks of North America (Sepkoski and Rex 1974), and fishes of West Africa (Hugueny 1989), South America (De Mérona et al. 2012), and Europe and North America (Oberdorff et al. 1997). Species richness is highly correlated with basin area for rivers worldwide, and stronger statistical relationships are observed when analyzed separately by continent (Amarasinghe and Welcomme 2002). The exponent of the species-area relationship always is greater in tropical than temperate regions (0.25 for Europe, 0.26 for Asia, 0.49 for Africa, 0.51 for South America), demonstrating a trend toward a more rapid increase in species richness with increasing river size at low latitudes.

On a global scale, total drainage area of the watershed and river discharge as a proxy of river size can explain a large proportion of the variation in freshwater fish species richness (Oberdorff et al. 1995). The underlying basis is still debated, but area-determined changes in rates of species formation and loss, and in habitat complexity, are possible explanations. Larger rivers may harbor greater diversity of habitat and food resources because of their size, and thus sustain a greater number of coexisting species. More habitat area may also permit larger population numbers within species, thereby reducing the probability of species extinction. Rates of speciation may be higher in larger systems due to corresponding increases in habitat heterogeneity and the occurrence of geographical barriers, such as waterfalls (Oberdorff et al. 2011). For example, higher speciation in fishes was positively related to the level of natural fragmentation caused by waterfalls in sub-drainages of the Orinoco Basin (Dias et al. 2013). Research also suggests that many of the species found in larger rivers appear to be unable to live in small streams. Longitudinal studies of riverine fishes find that addition, rather than replacement, of species characterizes the change in community assemblages along a river continuum, and a preponderance of larger species is observed downstream (e.g., Horwitz 1978; Schlosser 1987).

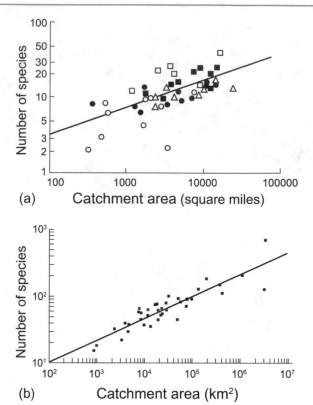

Fig. 11.2 The relationship between species richness and basin drainage area (**a**) freshwater mollusks of North American rivers, (**b**) freshwater fishes of West African rivers (Reproduced from Sepkoski and Rex 1974 and Hugueny 1989, respectively)

11.1.2   Latitudinal Diversity Gradients

An increase in numbers of species as one proceeds from high latitudes towards the tropics is one of the most general of geographic patterns of species richness (Hillebrand 2004; Kinlock et al. 2018). Several hypothesis based on the stability of tropical climates over time, their areal extent and higher biological productivity, and historic rates of speciation have been proposed to explain this pattern (Mittelbach et al. 2007; Heino 2011).

Although stronger in terrestrial than in freshwater environments, this relationship unquestionably holds for the fishes of running waters. Despite the incomplete state of taxonomic knowledge, more than 4,000 species of freshwater fish are estimated to occur in the Neotropics and approximately 3,000 in the Afrotropical region. This greatly exceeds the roughly 1,400 species found in the lakes and rivers of North America and 330 species of Europe (Lévêque et al. 2008). In an analysis of the distribution of freshwater taxa, Tisseuil et al. (2013) found greater species richness and endemism of fishes, aquatic amphibians, birds, and mammals in tropical and subtropical large drainage basins. Endemism refers to the number of species that only occur in

one area and are not found elsewhere. For all taxa except fishes, variation in global distribution was related to differences in climate and productivity among regions. For fishes, drainage area and environmental heterogeneity (altitudinal range, land cover, and climate variability) were the main drivers of global diversity patterns. By analyzing the global distribution of endemic fishes, Hanly et al. (2017) found that the species richness of endemic fishes increased with lake age and area, and with river basin area, and decreased with latitude.

Explanations for the latitudinal diversity gradient (LDG) and greater diversity in the tropics frequently invoke the species-energy hypothesis, which posits that more species can be supported in areas of higher productivity. Tropical environments, with higher solar insolation and warm temperatures throughout the year, are on average more productive, resulting in greater species diversity. Species-energy relationships may also play an important role in shaping community assemblages. When considering community structure through this lens, energy availability, rather than the size of an area, is considered the dominant factor influencing species richness (Wright 1983). Greater energy (e.g., net primary productivity) in an ecosystem suggests that either there are more resources to support larger populations and so the risk of extinction is lower, or there are more resources to be used by a greater diversity of species (Hugueny et al. 2010). In analyses of global fish diversity, net terrestrial primary productivity, along with river size and flow regime, were important variables explaining gradients in diversity (Oberdorff et al. 1995; Guégan et al. 1998). Energy can also be related to climate variables such as temperature, solar radiation, and precipitation that in turn are related to the physiological tolerances of species (Turner et al. 1987; Currie 1991). In a study that included 926 basins worldwide (Fig. 11.1), area and climate-related variables explained 77% of the variation in fish diversity, although past climate and geographic isolation also appeared as important factors in shaping fish communities (Oberdorff et al. 2011). By analyzing diversity patterns of Odonata in Europe and Northern Africa, Keil et al. (2008) found that evapotranspiration, which can be used as a proxy for the energy input to an area, best explained species richness. Although some studies support the species-energy hypothesis, the mechanisms connecting energy and diversity are not clear (Hugueny et al. 2010; Heino 2011; Oberdorff et al. 2011).

The fishes of Europe and North America show a strong latitudinal diversity gradient (Fig. 11.3); however, the gradient is strong for resident but not for potadromous species (fishes that migrate within river systems), evidence that species vagility influences conformity to the LDG relationship (Griffiths 2015). Note also that these analyses extend only to 20° latitude, and so do include the greater species richness found in the tropics. The LDG also may vary with type of freshwater habitat.

Whether a latitudinal diversity gradient exists for aquatic invertebrates, resulting in more species in the tropics, is uncertain because of limited information on geographic patterns of taxonomic richness at the species level. For a very large dataset of mayflies, stoneflies, and caddisflies, Vinson and Hawkins (2003) found there was no simple latitudinal gradient in local genus-level diversity, other than a decline at very high latitudes. Ephemeroptera showed three peaks in richness at 30°S, 10°S and N, and 40°N latitude, with lower values near the poles. Plecoptera peaked at 40°N and 40°S latitude. In contrast, Trichoptera showed less latitudinal variation. Diversity of Ephemeroptera was highest in the Afrotropical realm, Plecoptera in the Nearctic, and Trichoptera in Australia. Some of this variation likely is due to incomplete sampling, and some to areas of radiation and spread. In addition, the results of the study indicated that local environmental factors explained a larger proportion of the variation than historical factors related to biogeographical realms. By comparing aquatic insects sampled in 100 streams in subtropical Brazil and in Finland, Heino et al. (2018) found richness at the genus level was greater in Brazil than in Finland at both the regional and local level, although differences were greater at the regional scale. In contrast, aquatic insect abundance was greater in Finland, likely because of higher nutrient availability.

While comparison of site-scale diversity provides insight into regional patterns, it is limited in that the relationship between site and regional diversity may not be linear. To circumvent this difficulty, Pearson and Boyero (2009) first developed statistical relationships between taxonomic diversity and regional area, and then examined whether latitude explained the remaining variation (the "residuals") for seven freshwater taxa. Diversity was greater at higher latitudes for Ephemeroptera and Plecoptera, at low latitudes

for Odonata, bony fishes, and Anura, and showed no latitudinal trend for Trichoptera and Caudata. For Plecoptera, a group that has radiated primarily in cold Nearctic regions and is found mainly at higher altitudes in the tropics, a positive relationship between latitude and residuals was obtained; in contrast, a negative relationship was found for Odonata, which are highly diverse at tropical latitudes.

Several studies in Europe document that species richness of aquatic insects declines as one moves from mid to high latitudes. Iversen et al. (2016) report that species richness of Ephemeroptera, Plecoptera, and Trichoptera (EPT) fauna declines steadily along a gradient from 35° to 75° latitude. However, the decrease in richness with latitude was most evident for fauna of the headwaters, while higher richness was observed in the middle latitudes for fauna of lowland rivers. Lower species richness at high latitudes can be explained by low productivity resulting from low temperatures and solar insolation for much of the year, and also time since glaciation for colonization and adaptation by species from more temperate regions. While species richness in Ephemeroptera, Odonata, and Plecoptera was negatively related to latitude in Europe, Heino (2009) noted that highest diversity occurred in countries with high mountain ranges such as France, Germany, and Austria, and the lowest levels of diversity occurred in lowland countries such as the Netherlands, Latvia, and Denmark, suggesting that habitat heterogeneity also plays a fundamental role in shaping species diversity.

11.1.3 Historical Explanations

Changes in landforms, climate, and connectivity over earth history provide historical explanations for patterns in species diversity, as they can influence speciation and extinction events, open and close dispersal pathways, and affect time for recolonization following disruptive events such as

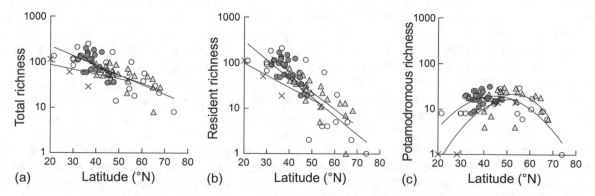

Fig. 11.3 Latitudinal diversity gradients for freshwater fishes in North America and Europe. Regional species richness as a function of latitude for (**a**) total, (**b**) resident, and (**c**) potamodromous species for Atlantic (Mississippi regions filled circles, extra-Mississippi regions open circles), Pacific (crosses) and European (triangles) realms (Reproduced from Griffiths 2015)

glaciation. Comparisons of freshwater fishes of Europe and North America provide evidence that historical events, in particular recent glaciations, have left strong signatures on their respective faunas. The European fish fauna is less diverse than that of North America, and regions within North America differ greatly. All of Canada and Alaska contain some 180 species of fish, considerably fewer than the rich Mississippi basin where most of the major adaptive radiations in North America have occurred. The Tennessee and Cumberland Rivers drainage realm alone includes some 250 species of fishes (Starnes and Etnier 1986). Species richness declines from east to west across the United States (Moyle and Cech 2006), due in part to major differences in extinction rates during the Pleistocene, as well as earlier geologic influences (Smith et al. 2010). As a consequence, the western U.S. contains only about one-fourth as many fish species as does the east.

Comparison of species-area relationships for the fish faunas of Europe and North America provides further evidence that evolutionary and biogeographic history can have a profound influence on regional diversity (Fig. 11.4). Post-glacial recolonization was more restricted in Europe relative to North America because drainage divides in Europe tend to run from east to west and re-establishment from Iberia and the Adriatic was restricted by mountain ranges, therefore glaciated areas were recolonized largely from the Danube Basin. This likely limited both southward retreat and subsequent northward recolonization for the European fauna, whereas the North American fauna had a much larger refugial area free from glaciation, and easier routes for recolonization (Mahon 1984; Oberdorff et al. 1997). In both Europe and North America, the fish faunas of glaciated regions are species-poor in comparison to unglaciated regions, and contain species that are larger, more migratory, and give less parental care compared with the unglaciated regions of the Mississippi and Missouri basins (Moyle and Herbold 1987; Griffiths 2006).

At the last glacial maximum (LGM) some 18,000 years ago, sea levels were lower and tropical forests became fragmented and reduced in extent, the result of more arid climatic conditions. Contemporary fish diversity and distribution in tropical realms bears the signature of these events. In a study including tropical basins of South America, Central America, and Africa, Tedesco et al. (2005) found that present day fish diversity was greater in basins that were connected with rain forest refuges during the LGM, when compared to regions that lacked connectivity. The authors suggested that basins connected to rain forest refuges may have had greater stability in discharge than more isolated basins. In contrast, basins without connectivity most likely experienced drier conditions and greater fish extinction rates.

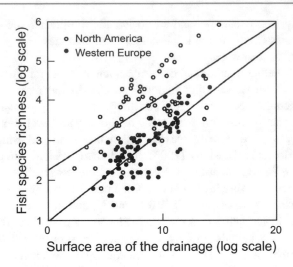

Fig. 11.4 Fish species richness as a function of drainage basin area for West European and North American rivers. Lines represent best fit using a power function (Reproduced from Oberdorff et al. 1997)

Changes in sea level during the Pleistocene due to alternating glacial advances and retreats may have influenced connectivity between drainage basins, depending on basin topography and coastal slope. Increased connectivity among currently isolated coastal river systems in ancient drainage networks (palaeo-drainages) during glaciated periods that lowered sea levels could facilitate fish dispersal among catchments. In a worldwide study analyzing the influence of historical connectivity on freshwater fishes, Dias et al. (2014) found greater diversity and lower endemism in basins that were connected during the LGM, when sea level declined by as much as 120 m, compared to those that remained isolated. These results highlight the role of dispersal processes in the present-day distribution of fishes. Similarly, past connections among coastal catchments during the LGM explained a large fraction (75%) of the genetic variation in the freshwater tetra *Hollandichthys multifasciatus*, endemic to streams of the Atlantic forest in southern Brazil (Thomaz et al. 2015). Present-day distributions of the 26 tetra populations sampled in this study were better explained by associations with 12 palaeo-drainages (Fig. 11.5), rather than by habitat stability, a term estimated by the permanency of Atlantic forest in the present and during the last glacial maximum.

Far more ancient events in earth history also are reflected in present-day diversity patterns, as is illustrated by fish diversity gradients in the Amazon Basin. The modern Amazon was formed by the joining of its western to its eastern basins, which were isolated throughout the Miocene and only began flowing eastward 1–9 million years ago (Oberdorff et al. 2019). Present-day diversity patterns suggest that the main center of fish diversity was located in western areas, progressing eastward due to fish dispersal

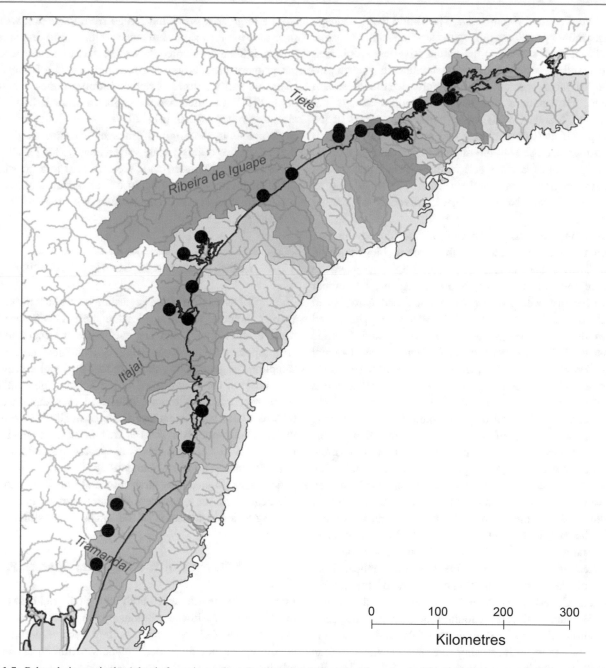

Fig. 11.5 Palaeodrainages in the Atlantic forest in southern Brazil during the last glacial maximum. The black line represents the current shoreline, and the palaeodrainages are shown in blue-green shades. Black dots represent the sampled populations (Reproduced from Thomaz et al. 2015)

after the basins were united and the Amazon River assumed its modern course toward the Atlantic. This historical analysis helps to explain the unexpected finding of a decline in diversity along an upriver-downriver gradient, which is contrary to most findings but may reflect the history of the Amazon drainage network.

Comparative studies of taxonomically related fish groups clearly show that history and biogeography can influence the taxonomic and ecological diversity of a region. The Nida River in south-central Poland and the Grand River in

Ontario, Canada, are two river systems that exhibit similar gradients from the headwaters downstream and occupy similar climates. Hence, they might be expected to support about the same number of species, filling roughly similar ecological roles. There are in fact many similarities (Fig. 11.6), due largely to the abundance of cyprinids in both, and comparable species in the Esocidae, Cottidae, and Gasterosteidae (Mahon 1984). Yet, there also are prominent differences, including the diversity in North America of Centrarchidae and Ictaluridae, and radiation within the genera *Notropis* and *Etheostoma*. The Grand River drainage contains more species overall, especially in smaller streams. In addition, more species in the Grand River are specialized stream dwellers, whereas the Nida includes a greater proportion of large species that are only occasional stream dwellers. Explanations for such differences always are speculative. Mahon (1984) suggests that the success of the lentic specialists (Centrarchidae and possibly Ictaluridae) in North America closed out the migratory option typical of the larger cyprinids of Europe, and favored species that formed resident populations in small streams. This may help to explain the greater North American diversity of smaller species.

Geographic diversity patterns of the Ephemeroptera, Plecoptera, and Trichoptera, differ among biogeographic realms. The greatest number of species of Ephemeroptera are found in the Palearctic and Nearctic regions, but higher genus diversity occurs in the Neotropical and Afrotropical realms (Barber-James et al. 2008). Plecoptera genus and species diversities are greatest in Palearctic and Nearctic realms, while Trichoptera are most diverse in the Oriental region (De Moor and Ivanov 2008; Fochetti and Tierno De Figueroa 2008). Given that the initial evolution of these taxa occurred prior to the splitting of the Pangean supercontinent (Balian et al. 2008), such disparate patterns among aquatic insects suggest either different centers of radiation or differential success in surviving subsequent environmental changes. For example, greater genus richness and endemism of Ephemeroptera in the southern hemisphere can be related to greater diversity in Gondwana than in Laurasia, greater extinction rates in the northern hemisphere, and higher climate heterogeneity in the southern hemisphere (Barber-James et al. 2008). Both recent (glaciation) and older (tectonic) geological influences are implicated by studies of the distribution of Trichoptera of the Iberian Peninsula. Consistent with other studies of Mediterranean basins, Bonada et al. (2005) found that trichopteran species composition shifts from a dominance of Palearctic species in northern basins to an increasing presence of North African and endemic species in southern basins. This reflects the combined influence of glaciations during the Quaternary that shifted northern species to more southern latitudes, and the joining of the Eurasian and African plates during the Tertiary that enabled North-African species to enter the southeast of the Iberian Peninsula.

It is evident that species-area, species-energy, and earth history all are valid explanations for large-scale patterns in species richness. From a number of analyses, Oberdorff et al. (2011) concluded that the macroecological variables of drainage area and system productivity provided the strongest explanations of global patterns in freshwater fish species richness, and history less so, with the caveat that the influence of history is more challenging to establish. Distance

from a large area providing refuge from glaciation was one measure of historical influence with some explanatory power. It should be noted, however, that after river size and net primary productivity are factored out, North American rivers are still 1.7 times as rich as European ones (Oberdorff et al. 2011), and their fish assemblages differ in body form, size, and other aspects (Fig. 11.6) that no doubt have their explanations in the origin of lineages. Clearly, the distribution of species at large scale varies greatly, for reasons that we at least partly understand. We turn next to a question of much importance for the study of ecological communities: how does local species richness vary with regional species richness?

11.2   Local Patterns in Species Diversity

The most basic information regarding a local assemblage of species includes which species are present, what are their relative abundances, and can we expect to encounter the same species in the same relative abundance at the next sampling. In general, we shall see that the local assemblage is a subset of the species pool of the region. The number of species recorded is strongly influenced by sampling effort, thus requiring some standardization of methods in order to make valid comparisons. As is true in all ecological communities, a moderate number of species are common and most are rare. The most abundant species at a site tend to persist over time, unless environmental conditions change. We begin with an assessment of these patterns before proceeding to a consideration of the forces responsible.

11.2.1   Influence of the Regional Species Pool

Local richness is influenced both by regional richness and by local factors. A linear relationship between regional and local species richness suggests that local environmental conditions do not limit local richness, while an asymptotic relationship suggests that local environmental or biotic factors may ultimately limit the number of species at a site (Heino et al. 2009). From an analysis of fish species richness across a large area of the southwestern Iberian Peninsula of Europe, encompassing 436 sites distributed across 23 river basins, Filipe et al. (2010) found that local species richness (LSR) was strongly associated with regional species richness (RSR) for both native and introduced species (Fig. 11.7). Environmental factors including rainfall seasonality and stream slope also influenced local richness, but had less influence. However, considerable variation at the local scale remained unexplained, indicating a role for additional environmental variables and biological interactions. A study of the LSR–RSR relationship in stream diatoms in boreal

Fig. 11.6 Some of the fish species occupying small drainage basins (300 km² or less) in (**a**) the Grand River system, southern Ontario, and (**b**) in the Nida River system, south-central Poland (Reproduced from Mahon 1984)

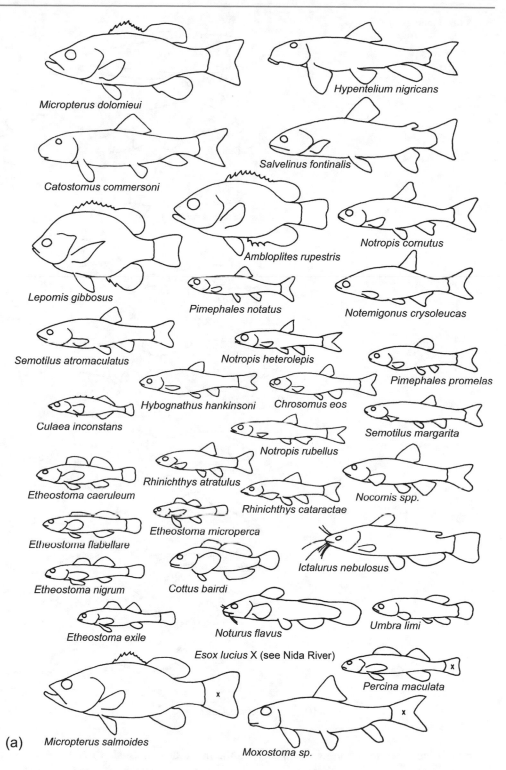

(a)

Micropterus dolomieui

Hypentelium nigricans

Catostomus commersoni

Salvelinus fontinalis

Lepomis gibbosus

Ambloplites rupestris

Notropis cornutus

Pimephales notatus

Notemigonus crysoleucas

Semotilus atromaculatus

Notropis heterolepis

Pimephales promelas

Culaea inconstans

Hybognathus hankinsoni

Chrosomus eos

Semotilus margarita

Notropis rubellus

Etheostoma caeruleum

Rhinichthys atratulus

Nocomis spp.

Etheostoma flabellare

Rhinichthys cataractae

Etheostoma microperca

Ictalurus nebulosus

Etheostoma nigrum

Cottus bairdi

Etheostoma exile

Noturus flavus

Umbra limi

Esox lucius X (see Nida River)

Percina maculata

Micropterus salmoides

Moxostoma sp.

streams in Finland found a relatively strong linear relationship between diatom species richness sampled from stream riffles, and regional species richness across multiple drainages on a 1,100 km north-south latitudinal gradient (Soininen et al. 2009). Roughly speaking, the number of diatoms collected from a single riffle ranged from a maximum of one-third to one-half of the number in the drainage system, to a minimum of 10–25%. Using data on macroinvertebrate assemblages from 705 streams sampled as part of the Swedish national lake and stream survey, Stendera and Johnson (2005) found no simple generalization between LSR and RSR. However, it did appear that

Fig. 11.6 (continued)

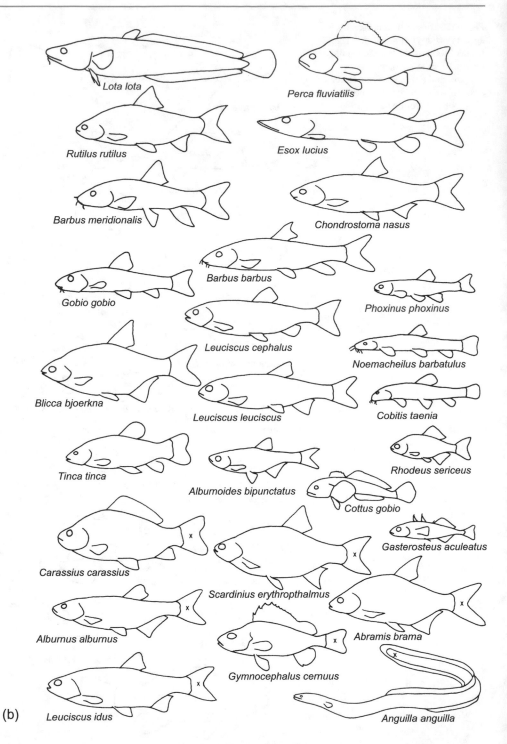

(b)

maximum local species richness was usually in the range of 15–35% of regional species richness.

11.2.2 Influence of Sampling Effort

A complete survey of all the species within a local assemblage is an extremely challenging task. Such studies are unusual partly because the taxonomic knowledge of many groups is inadequate, and partly because the exhaustive compilation of a species list is rarely a priority. It is more usual to find either a detailed study of a single taxon, or an ecological investigation where the focus is on the more common species, while taxa that are difficult to identify are lumped, often at the family or genus levels. Estimated species richness at a locale is highly sensitive to sample size and

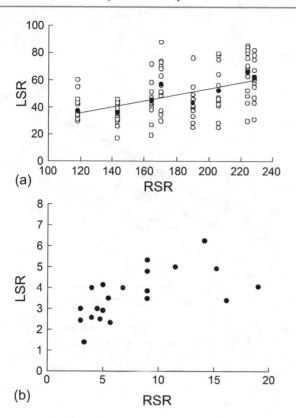

Fig. 11.7 Relationship between local species richness (LSR) and regional species richness (RSR) in (**a**) diatoms in streams in Finland, and (**b**) fishes in Mediterranean streams of the Iberian Peninsula. (Reproduced from Soininen et al. 2009 and Filipe et al. 2010, respectively)

the thoroughness with which all habitats are sampled. This is because many species are rare, hence often undetected, but an increasing number of rare species will be included as sampling effort increases. Two statistical approaches are commonly used to address this issue and make sample estimates more comparable across sites (Chao et al. 2014). Rarefaction uses a plot of number of species against number of samples from a random re-sampling of the data set to estimate species richness for a given sample size. Asymptotic estimators extend the species accumulation curve to derive an estimate of species richness that is relatively independent of additional sampling effort.

The importance of sample size is nicely illustrated by a study of macroinvertebrates collected from individual stones in a large reach of rapids (20–40 m wide, approximately 1 km in length) of the River Lestijoki, Finland (Kuusela 1979). The number of individuals per stone was positively correlated with the number of species per stone (Fig. 11.8a). In addition, the cumulative number of species increased with the logarithm of the cumulative number of stones sampled; that is, at a decelerating rate. The latter relationship has been reported from streams of widely different regions

(Fig. 11.8b), and clearly illustrates the dependence of local species richness on sampling effort. Whether sampling included one or multiple habitats also can influence the number of species collected. For diatoms, Smucker and Vis (2011) found that a multiple habitat approach (pools, riffles, rocks, sand) resulted in higher species richness and evenness than when sampling only epilithic habitat (rocks in riffles).

11.2.3 Common and Rare Species

A general observation from surveys of ecological communities is that a few species are common and most are quite rare. A collection of 52,000 insect specimens that emerged as adults from a stream flowing underneath an 11-m^2 greenhouse near Schlitz, Germany, yielded a total of 148 species, but the 15 most abundant species contributed 80% of the total number of individuals (Illies 1971). Woodward et al. (2002) report a similar finding for the Broadstone Stream, UK, which is relatively species-poor due to the stream's acidity derived from acid deposition. Demonstrating remarkable consistency over 24 years of study, a core community of eight taxa always was present, contributing 75% or more of total individuals (Fig. 11.9). A practical consequence of the tendency for a few species to dominate an assemblage is that collection of a small number of samples will include most of the common species, whereas further sampling effort will continue to produce additional species almost (but not quite) indefinitely. This underlies the relationship between sample size and local species richness (Fig. 11.8), which in turn influences the amount of sampling effort necessary to characterize a system.

Why a few species should be more abundant, widespread, and successful and many species quite rare remains one of the great enigmas of ecology. Species that have wide regional distributions usually are locally abundant as well (Gaston and Blackburn 2000). By comparing stream insects collected at 50 headwater stream sites located in a drainage basin with data collected at 110 stream sites in 5 ecoregions in Finland, Heino (2005) found a positive and significant relationship between the local abundance of species and regional distribution. Although such a relationship can be an artifact of sampling, such as failing to detect rare species, it is likely that the explanation has an ecological basis in organism-niche relationships, dispersal abilities, and population growth rates.

Because most species are minor components of ecological webs, the completeness of the species list may not matter greatly for most ecological analyses. Nonetheless, the few exhaustive inventories are of interest because they give a sense of just how biologically diverse stream communities

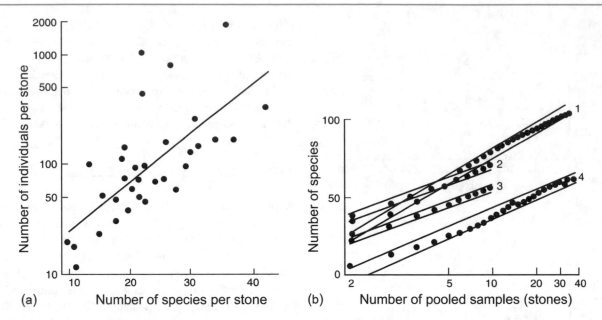

Fig. 11.8 (**a**) The number of species collected increases with the size of the sample, illustrated by Kuusela's (1979) study of the fauna on individual stones in a large Finnish river. (**b**) The cumulative number of species collected increases with the logarithm of cumulative number of stones sampled: 1, River Javavankoski, Finland (Kuusela 1979); 2, Vaal River, South Africa (Chutter and Noble 1966); 3, Lytle Creek, Utah (Gaufin et al. 1956); 4, Rio Java, Costa Rica (Stout and Vandermeer 1975). Solid lines are 95% confidence limits

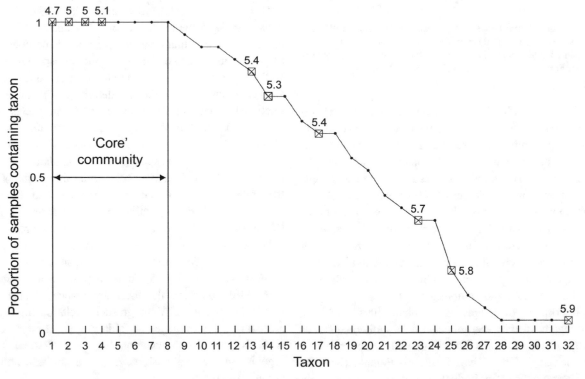

Fig. 11.9 The proportion of sampling occasions that included each taxon over 24 years of study in the Broadstone Stream, U.K. Numbers along the lower axis refer to individual species. The core community consisted of eight taxa that were always present. These included 1, *Nemurella pictetii*; 2, *Leuctra nigra*; 3, *Plectrocnemia conspersa*; 4, *Sialis fuliginosa*; 5, Pentaneurini; 6, Ceratopogonidae; 7, *Heterotrissocladius marcidus/Brillia modesta*; 8, *Polypedilum abicorne*. The pH optima for individual taxa are indicated where known (Reproduced from Woodward and Hildrew 2002)

can be. Perhaps the most complete species lists come from long-term studies of a small German stream, the Breitenbach (Table 11.1), and of the Broadstone Stream in the UK (Schmid-Araya et al. 2002). More than one thousand invertebrate species have been collected from the Breitenbach, and because this list was compiled using aerial as well as aquatic collections it is uncertain what fraction derives from habitats other than the stream, including a small impoundment and other standing water habitats. However, it is thought to be less than one third of the invertebrate species in the stream (P. Zwick, personal communication). This compilation indicates that the greatest invertebrate diversity is located in a few groups, including several minute, interstitial phyla (Nematoda, Rotatoria, Annelida, and Platyhelminthes) and the highly diverse Diptera, especially the midge family Chironomidae. In general, it is the smallest taxa that are most diverse (Palmer et al. 1997). The Broadstone Stream is relatively depauperate because of high acidity; however, its count of 131 invertebrate species likewise reflects high representation by small taxa that are often overlooked (Schmid-Araya et al. 2002).

In summary, there are numerous factors that contribute to some areas being relatively rich in species while other areas are less so. A first level of explanation must take into account regional diversity, which is influenced by history, topography, climate, and geography; intensity of the sampling effort at both within and between-habitat levels; and local environmental conditions. There is also good reason to believe that interactions between species, which likely vary in their intensity depending upon environmental conditions, play a major role in determining local species richness. This provides the link between the topics of species diversity and community structure, and we turn now to the latter.

11.3 Local and Regional Controls of Community Assemblages

The discussion of community structure and the rules that govern community assembly has generated a rich literature in ecology. To enter into this topic, it is useful to distinguish some key ideas, keeping in mind that they are not fully independent from one another. Niche-based models focus on the interplay between biotic interactions (usually predation, herbivory, competition) and abiotic forces (primarily habitat and disturbance) that determine the suitability of a place for a particular species. The habitat template model (Southwood 1988) emphasizes the association of species with habitat features, such that individual species occur where they are best suited and more species are found where habitat conditions are most diverse. In this long-favored explanation of stream community structure, the physiological, morphological, behavioral, and life history attributes of individual species determine which will successfully colonize and maintain populations in a particular environment. Both abiotic and biotic factors can be visualized as a series of filters that determine the subset of the regional species pool that is most likely to successfully colonize and maintain populations. In contrast, explanations emphasizing stochastic variability suggest that local extinctions and dispersal are more influential than local environmental conditions in structuring communities.

Disturbance models emphasize the interplay between species interactions and variation in flow, temperature, and other environmental factors that periodically reduce the abundance of some or all species in an assemblage. Because predation, competition, and herbivory can potentially eliminate local populations, disturbance can serve as a countering

Table 11.1 A total of 1085 species of metazoans reported as of 1989 from a two-km stretch of the Breitenbach, a small stream near Schlitz in northern Germany. From Zwick (1992)

Insecta	Number	Other metazoa	Number
Diptera	468	Nematoda	141
Coleoptera	71	Rotatoria	130
Trichoptera	57	Annelida	56
Ephemeroptera	18	Platyhelminthes	50
Plecoptera	18	Crustacea	24
Hymenoptera	3	Hydrachnellea	22
Megaloptera	2	Mollusca	12
Planipennia	2	Gastrotricha	6
Odonata	1	Vertebrata	3
		Nematomorpha	1
Total	640		445

force, limiting the effectiveness of strongly interacting species, and either facilitate or prevent recolonization by displaced species. A focus on disturbance seems appropriate to fluvial ecosystems because they appear to be highly variable and occasionally harsh environments. In addition, benthic invertebrates and algae are patchily distributed, and this suggests that disturbance, biotic interactions, and recolonization may combine to govern population dynamics at the local scale (Townsend 1989). Because dispersal ability varies among species, individual mobility, propensity to drift, and aerial flight ability all are important traits that may permit the persistence of weak competitors and vulnerable prey in environments where they might otherwise lose out.

11.3.1 Consistency in Assemblage Composition

Evidence that communities have predictable structure, implying primacy of habitat and niche-based explanations, has often been sought by analyzing patterns in species composition. Consistency in assemblage composition over time, and similarity in assemblage structure among locations whose environments are comparable, suggests that underlying processes, rather than randomness, govern community formation. A core community of eight taxa in the aforementioned Broadstone Stream (Fig. 11.9) was present for more than two decades of study, and species turnover generally was low. High persistence of assemblage composition and dominance by the same handful of species in the stream strongly suggests that dominant species in the locale are not simply a random sampling of a larger species pool, but instead are those whose traits allow them to be especially successful in the environmental conditions at that location. This community was persistent until the 1990s when reduction in acidity due to controls in anthropogenic emissions created conditions amenable to colonization by a large, predaceous dragonfly, *Cordulegaster boltoni*. Additionally, an increase pH over time supported an invasion of the stream by brown trout *Salmo trutta* in 2005. The abundance of invertebrates declined in response to predator introduction, and food chains lengthened with the addition of top predators, while modelling indicated that persistence and food web stability decreased. Notably, most species that were observed in the initial Broadstone studies conducted in acidic waters four decades earlier were still present. Other invertebrates typically associated with less acid conditions have not colonized the streams, suggesting that shifts in food web dynamics may be delaying community recovery (Layer et al. 2011).

The degree of community persistence appears to vary with environmental conditions. Two surveys of 27 streams in the same locale as the Broadstone Stream suggested greater persistence of taxa within cold- than warm-water streams (Townsend et al. 1987). Species persistence of fish

assemblages was high and similar in two streams surveyed nine years apart, but the difference between collections was greater in the stream that exhibited higher seasonal and year-to-year variation in flow regime, maximum summer temperatures, and frequency of dewatering (Ross et al. 1985). Other studies reporting less overall persistence in assemblage structure (Grossman et al. 1982) have also attributed this to environmental variability. Based on ten years of sampling in Coweeta Creek, North Carolina, Grossman et al. (1998) concluded that environmental variability in flows rather than habitat or resource availability best explained variation in fish assemblages. Indeed, lack of assemblage persistence or of relationships between species composition and habitat variables may be a frequent finding wherever unpredictable floods and droughts introduce high temporal variation into stream assemblages (Angermeier and Schlosser 1989).

11.3.2 Local Environmental Factors and Spatial Characteristics of River Networks

Spatial comparisons of biological assemblages provide insight into the roles of community structuring mechanisms. Local communities of species that are spatially separate often are partially linked by dispersal in what is referred to as a metacommunity (Wilson 1992; Leibold et al. 2004). This framework highlights the importance of spatial processes, including spatial variability in environmental conditions, in dispersal pathways and site proximity, and in the dispersal abilities of individual species. Although each local community is the result of species sorting, in which 'filtering' by local environmental factors ensures that each species occurs where environmental conditions are suitable, dispersal from neighboring communities also plays a role (Heino et al. 2015). Comparisons among biological communities across space should find greatest differences when local conditions are variable and dispersal is low. Conversely, when dispersal rates are high, communities are typically more similar, especially when among-site environmental differences are modest, and species can often persist in habitats that are less favorable because of frequent recolonization (Leibold et al. 2004; Chase et al. 2005). Populations of organisms with the ability to disperse over long distances are often strongly influenced by local environmental factors, which determine whether a species can persist once it reaches a given locality. In contrast, populations of species with lesser dispersal ability are frequently limited by the relative connectivity of a system, as spatially and/or temporally heterogeneous environments may constrain their ability to colonize new habitats (Heino 2011; Padial et al. 2014).

In a study of macroinvertebrate assemblages from ten grassland streams in the Taieri River catchment on the South

Island of New Zealand, Thompson and Townsend (2006) posited that if assemblage structure was explained best by dispersal, the greatest similarity among community assemblages would be in adjacent sites. However, if more distant assemblages were similar, it would potentially lend support to the argument that similar environmental conditions were more important in structuring communities. Their results indicated that dispersal was the primary factor. Assemblage similarity was best explained by spatial proximity for species with low dispersal ability, a mixed model worked best for species with moderate dispersal abilities, and neither model worked especially well for species of high dispersal ability. It is important to note that there is a tendency for ecological conditions to be spatially correlated; therefore, using comparisons of natural assemblages to test models based on species sorting by environmental conditions versus dispersal of ecologically equivalent species in structuring communities should be done with caution. However, in this study, Thompson and Townsend (2006) did not find a relationship between distance and the similarity of physical and chemical conditions for the ten streams. Thus, they attributed the observed negative relationship between spatial distance and community similarity to the distance limitations of aerial dispersal and the possible influence of chance order of arrival on community assembly.

The species of tropical floodplain rivers provide a natural experiment in community assembly because seasonal fluctuations in discharge result in repeated cycles of extirpation and recolonization of floodplain habitat at the local scale. By experimentally manipulating habitat complexity, Arrington et al. (2005) showed that species-specific differences in dispersal ability of fishes significantly affected the fish assemblage response to changes in habitat availability. Dispersal was most important to community assembly in newly formed habitat patches, whereas in older patches, abundances of individual species increasingly were influenced by habitat characteristics. Similarly, Padial et al. (2014) assessed the relative importance of environmental and spatial variables to aquatic communities of the Paraná River floodplain in Brazil. Spatial variables, primarily those related to distance between sampling sites and connectivity between sites, were more important in explaining variation in the presence of larger organisms with lower dispersal abilities, such as fish and macrophytes, while local environmental variables were more important in explaining the distribution of smaller organisms, such as phytoplankton, that had greater dispersal capacities (Fig. 11.10).

The roles of local conditions and dispersal capabilities in explaining patterns in community assemblage can vary depending on the position of a site within a river network. Headwaters are more isolated than larger streams because the overland distance between headwater streams is greater, and organisms have to move against the current to colonize upstream sites (Heino et al. 2015). Several studies have showed that local processes are more important in headwater communities than in downstream reaches. Brown and Swan (2010) found no correlation in headwater streams between similarity of macroinvertebrate communities and distance between sites (Fig. 11.11a); however, there was a significant correlation between community assemblage and environmental conditions, suggesting that species sorting rather than dispersal was the main factor influencing community structure in headwater streams (Fig. 11.11c). In contrast, in mainstem sites, similarity between communities declined with increasing distance (Fig. 11.11b), and there was a positive relationship between community assemblage and the similarity of environmental conditions (Fig. 11.11d). This suggests that community assemblage in the mainstem is influenced both by dispersal capabilities and local species sorting. However, other studies have not found clear associations in species composition between headwater and mainstem sites. The relative importance of local conditions and spatial variables to community composition differed among taxonomic groups in streams and rivers of the Danube basin in Hungary (Schmera et al. 2018). For macroinvertebrates, local environmental variables had a strong influence on community structure in streams, while both environmental conditions and spatial variables were important in communities in larger systems. For fishes and diatoms, environmental conditions and spatial characteristics influenced community assemblages in smaller streams, but did not explain variation in assemblage structure in the larger sites.

Seasonal variation also can affect the relative influence of local environmental conditions on the composition of aquatic communities along a river network. In northern Sweden, Göthe et al. (2013) found that local variables including water chemistry and frequency of low flows acted as strong filters at upstream sites in all seasons, but were important at downstream sites only during autumn, suggesting that seasonal changes in dispersal can create temporal variation in community assemblages. Comparing the effects of local (e.g., flow regime and habitat characteristics) and regional (e.g., distance among sites) factors in dryland streams of Arizona, US, Cañedo-Argüelles et al. (2015) found that local variables were most important in explaining community assemblages. This is unsurprising, as dryland systems are characterized by distinct seasonality in flow and spatial heterogeneity in environmental conditions, including in both perennial and intermittent reaches. Similar to many of the aforementioned studies, the effect of local environmental conditions on community assemblage varied among taxa with different dispersal capacities. The distributions of invertebrates with low dispersal capacities were most influenced by local habitat conditions, while strong dispersers were present in all suitable habitats. Interestingly, the

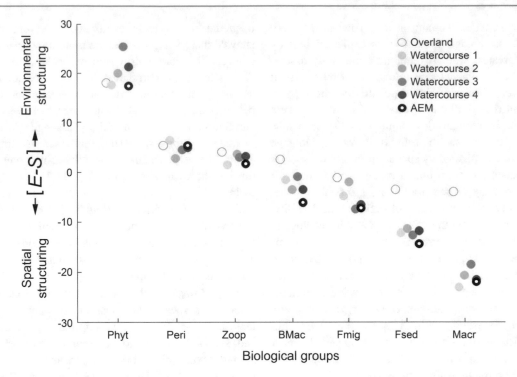

Fig. 11.10 Difference between the contribution of environmental and spatial variables (E-S) for different biological groups ordered based on their dispersal ability, decreasing from left to right. (Phyt, phytoplankton; Peri, periphyton; Zoop, zooplankton; BMac, benthic macroinvertebrates; Fmig, migratory fish; Fsed, sedentary fish; Macr, macrophytes) (Reproduced from Padial et al. 2014)

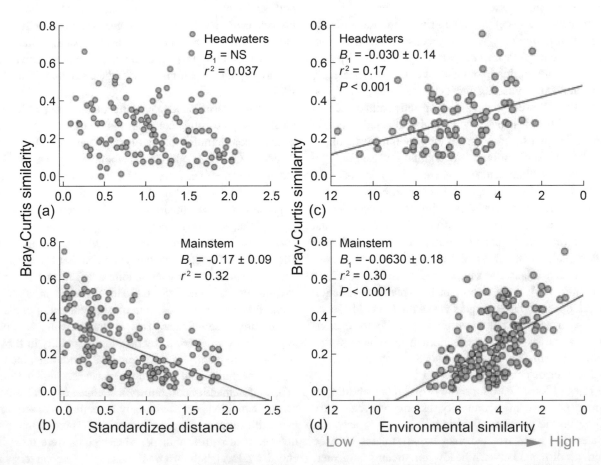

Fig. 11.11 Relationships between community similarity (Bray-Curtis) and geographic distance among sites for benthic macroinvertebrates across 52 sites in (**a**) headwaters and (**b**) mainstems. Relationships between community similarity and environmental similarity for the 52 sites in (**c**) headwaters and (**d**) mainstems (Reproduced from Brown and Swan 2010)

presence of taxa with intermediate dispersal capacities was related to topographic distance, suggesting that these species can escape from most harsh conditions, but do not have the ability to reach all suitable habitats.

How dispersal capabilities and spatial relationships of the river system influence community assemblages is governed by the size of a river network and connectivity within the network (Heino et al. 2015). Depending on connectivity within a river network and dispersal mode of organisms, stream systems can differ in the importance of dispersal in overcoming spatial fragmentation of populations and communities (Tonkin et al. 2018). In low connectivity systems, such as arid-land streams that can be highly fragmented by episodic drying, the river network will play a smaller role as a dispersal route than in systems with continuous connectivity by water (Fig. 11.12). Variation among organisms in dispersal mode also influences how species distributions interact with connectivity and spatial characteristics of a watershed. Freshwater fishes can only disperse along aquatic corridors, but many macroinvertebrates and amphibians are only restricted to aquatic environments for part of their development. In headwater streams of the Central Amazon, Stegmann et al. (2019) found that distance to large-river floodplain systems had a greater influence on fish assemblages than distance among sites or local conditions. Streams located closer to large rivers showed greater taxonomic and functional richness, suggesting that floodplains can influence upstream assemblages, probably through increased dispersal during flood pulses and higher habitat availability during the dry season. In contrast, overland connectivity influences the structure of aquatic invertebrates, as Razeng et al. (2016) found for streams in central Australia. Dispersal through valleys was more important in structuring invertebrate communities than dispersal based on distance between sites or through waterways, likely related to the arid conditions of the study sites. Compared to other studies conducted in arid lands, local environmental conditions had little influence on the community structure in this example, probably because the study sites were relatively homogenous.

Understanding the relative influence of the factors governing community structure can provide insight to management options (Datry et al. 2015). For example, when local environmental variables are most important in shaping local communities, conservation efforts may be more effective if they are focused on local habitat conservation. In contrast, in systems where dispersal is the main factor regulating the distribution of species and community assemblage, efforts to maintain connectivity within a river network and to enhance refuges or areas of the stream that are relatively buffered from disturbance during times of limited connectivity should be prioritized.

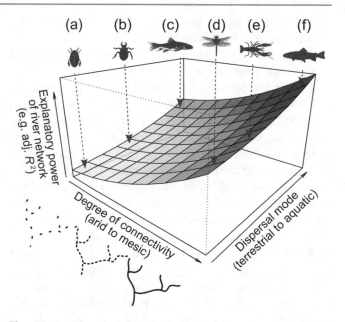

Fig. 11.12 Conceptual model of the explanatory power of river network, degree of connectivity, and dispersal mode to local community composition via overall dispersal. Taxa illustrating differences in dispersal mode: (**a**) diving beetle, *Boreonectes aequinoctialis*; (**b**) giant waterbug, *Abedus herberti*; (**c**) desert sucker, *Catostomus clarki*; (**d**) dragonfly, *Ophiogomphus occidentis*; (**e**) crayfish, *Pacifastacus leniusculus*; (**f**) rainbow trout, *Oncorhynchus mykiss* (Reproduced from Tonkin et al. 2018)

11.3.3 The Habitat Template and Species Traits

Habitat template theory places particular emphasis on the matching of habitat requirements of individual species to the abiotic and biotic conditions of a locale (Southwood 1988; Townsend and Hildrew 1994). Increasingly, such efforts examine the traits of species with the expectation that attributes such as size, body shape, lifespan, and mode of dispersal will help us understand why certain species succeed where others do not, and may also provide clues regarding the environmental factors that are responsible for structuring communities. A conceptual elaboration of this approach connects the regional to the local species pool through a hierarchical series of filters that determines the likelihood that a particular subset of colonists will be successful at a locale (Fig. 11.13). As initially developed by Tonn et al. (1990) and Poff (1997), and further elaborated by Heino (2009) and Cornell and Harrison (2014), the local assemblage of species is filtered from a regional species pool first through dispersal barriers, which are influenced by factors such as topography and climate, but also by differences in species' dispersal capabilities; and secondly by environmental filters that exist at a hierarchy of scales, such as flow and temperature regimes, and macro- and microhabitat variables. We expect systems that experience frequent disturbance and recolonization to be strongly influenced by the species pool, whereas those more strongly structured by species interactions may be less so.

The ability of individual species to reach a location and then persist there in the face of whatever environmental conditions and biological interactions a species encounters will be determined by many aspects of its morphology, physiology, life history, and dispersal ability. There are referred to as species traits, and biological assemblages increasingly are being viewed from the perspective of species traits rather than taxonomic identity (Poff et al. 2006; Menezes et al. 2010). Using a trait-based approach allows scientists to more efficiently make comparisons of species assemblages among regions because traits, rather than species, may be shared more commonly among sites (Verberk et al. 2013). Suites of traits have been defined for a variety of taxonomic groups, including plants, algae, invertebrates, and fishes (Statzner and Bêche 2010). Over the past few decades, traits have been applied in the study of community assembly (Poff 1997; Lamouroux et al. 2004), stream health and biomonitoring (Menezes et al. 2010; Statzner and Bêche 2010), community and ecosystem function (Naeem and Wright 2003; Tolonen et al. 2017), and in predicting species invasions (Olden et al. 2006; Statzner et al. 2008).

Traits are typically divided into biological and ecological categories (Vieira et al. 2006). Biological traits traditionally include morphological, physiological, and life history attributes, while ecological traits are related to habitat

preferences. However, suites of traits are different for different taxonomic groups. In fishes, traits typically reflect life-history strategies (e.g., longevity, migration), body size, reproductive attributes, trophic habits, habitat preference, and tolerance to physicochemical variables (Frimpong and Angermeier 2010). In macroinvertebrates, biological traits typically include attributes such as maximum body size, number of reproduction cycles and descendants per year, method of reproduction, method of dispersal, resistant life stages, respiration mode, locomotion mode, diet, and feeding habits. Ecological traits for macroinvertebrates are frequently related to habitat requirements such as altitude, substrate, lateral and longitudinal distribution, and to physicochemical tolerance (e.g., salinity, temperature, and pH among other variables) (Schmera et al. 2017). Each trait typically includes several subcategories or modalities (Table 11.2) that are assigned specific codes and values that are used for empirical analyses. Specialists have generated databases of traits for fishes and macroinvertebrates that have facilitated the use of species traits to address ecological and management questions (Macroinvetebrates: Usseglio-Polatera et al. 2000; Poff et al. 2006; Tomanova et al. 2006; Statzner et al. 2007; Fishes: Noble et al. 2007; Frimpong and Angermeier 2009). However, further work is still needed to expand these data bases to other genera and regions (Statzner and Bêche 2010). Verberk et al. (2013) emphasize that traits do not occur in isolation, nor does a single trait define the relationship between a species and the environment. Rather, the combination of traits within a species govern the adaptive response of an organism to environmental conditions. Thus, environmental conditions do not act on single traits, but rather on a species, which is an assemblage of traits.

Efforts to demonstrate that species traits match with environmental variables have met with at least moderate success. To test the hypothesis that functional organization of fish communities is related to hydrological variability, Poff and Allan (1995) described habitat, trophic, morphological, and tolerance characteristics using six categories of species traits for each of the 106 fish species present at 34 sites in Wisconsin and Minnesota. Two ecologically-defined assemblages were identified, associated with either hydrologically variable streams (high variation in daily flows, moderate frequency of spates) or hydrologically stable streams (high predictability of daily flows, stable baseflow conditions). Fish assemblages from variable sites had generalized feeding strategies, were associated with silt and general substrate categories, were characterized by slow-velocity species with headwater affinities, and were tolerant of sedimentation. These findings indicate that hydrologic regime acts as an environmental filter, supporting theoretical predictions that variable habitats should harbor

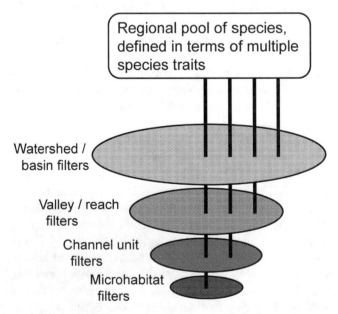

Fig. 11.13 The number and types of species at a site reflect the influence of regional diversity and species traits (trophic, habitat, life history, etc.) that allow them to pass through multiple biotic and abiotic filters at hierarchical spatial scales. The species found within a particular microhabitat possess traits suitable for prevailing watershed/basin, valley bottom/stream reach, and channel unit and habitat conditions (Reproduced from Poff 1997)

Table 11.2 Trait categories and their modalities from an analysis of relationships between species traits and environmental variables for invertebrate assemblages of streams in France. Based on Lamouroux et al. (2004)

Trait	Modalities
Maximum size	7 categories from <2 mm to >80 mm
Body flexibility	None (<10°), low (10–45°), high (>45°)
Body form	Streamlined, flattened, cylindrical, spherical
Life span	<1 year, >1 year
Voltinism	<1, 1, >1 generation/year
Aquatic stages	Egg, larva, nymph, imago
Reproduction	Ovoviviparous, individual eggs,[a] egg masses,[b] asexual
Dispersal	Aquatic active, aquatic passive, aerial active, aerial passive
Resistance form	Eggs, cocoons, cells resists desiccation, diapause/dormancy, none
Respiration	Tegument, gill, plastron, spiracle (aerial)
Locomotion/relation to substrate	Flyer, surface swimmer, swimmer, crawler, borrower, interstitial, temporarily attached, permanently attached
Feeding habits	Absorber, deposit feeder, shredder, scraper, filter feeder, piercer, predator, parasite, parasitoid

[a]Isolated eggs can be free or cemented to substrate. [b]Egg masses can be free, cemented, in vegetation or deposited terrestrially

resource generalists whereas stable habitats should include a higher proportion of specialist species.

The hierarchical filter model implies that associations between traits of the species assemblage and habitat variables should exist at multiple spatial scales. Lamouroux et al. (2004) assessed correlations between traits and environmental variables across spatial scales within two river basins in France, after first summarizing the functional composition of invertebrate communities using 60 categories of 12 biological traits (Table 11.2). Roughly half of the tested relationships between traits and environmental variables were significant at the microhabitat scale, and about one-fourth of the tests were significant at the reach scale. Although a number of invertebrate traits differed between basins, this was not attributable to between-basin habitat differences. In this example, filters at the microhabitat scale clearly were most influential on the suite of community traits. Using 11 ecological traits and 11 biological traits, Usseglio-Polatera et al. (2000) identified gradients in body size, reproductive rate, and feeding ecology within 472 European macroinvertebrate taxa. Because they were able to aggregate taxa into groups with similar traits, the authors speculated that improved resolution of habitat affinities or response to pollution might be attained using a subset of species sharing a similar suite of traits, rather than using the entire assemblage.

Any consideration of species traits must recognize that some traits tend to co-occur and others may rarely, if ever be found, in the same species. For example, large body size, long lifespan, and low reproductive potential form one common suite of attributes; small body size, short lifespan, and high reproductive potential another. These are frequently referred to as slow and fast, or K and r species, respectively (Begon et al. 2006). Mixtures between these two suites of attributes are rare, suggesting a trade-off exists between two alternate life styles, or these two suites of traits may represent two ends of a spectrum.

Three life history strategies for fishes in tropical regions and in North America have been proposed: (i) opportunistic (characterized by small body size, earlier maturation, high reproductive effort, low fecundity, and low investment in individual offspring), (ii) periodic (characterized by large body size, late maturation, moderate reproductive effort, high fecundity, and low investment per offspring), and (iii) equilibrium (characterized by variable body size, moderate to long life span, low reproductive effort, low batch fecundity, and high investment per offspring). In this model, the r strategy is divided into opportunistic and periodic strategies and the equilibrium strategy is related to K species (Winemiller and Rose 1992; Winemiller 2005). This model has been applied to investigate the relationship between fishes and environmental variables, species distribution, the role of local and regional factors on fish assemblages, and the impact of hydrologic alteration in rivers (Hoeinghaus et al. 2007; Frimpong and Angermeier 2010).

The relative contribution of local environmental conditions and spatial characteristics of a river network on community structure, described earlier for species, has also been investigated using species traits. Using both taxonomic and functional approaches, Hoeinghaus et al. (2007) found that the taxonomic make-up of fish assemblages in Texas, US, was related to species' geographic distributions and historical processes, and linked primarily to dispersal, suggesting that regional factors strongly influenced local fish assemblages. However, functional analyses based on trophic and life-history traits indicated that local factors such as habitat stability and predation pressure can have equal influence on local assemblage composition. The functional approach was able to distinguish more stable, deepwater sites dominated by piscivores from sites with more variable habitat where opportunistic groups were dominant. This study suggests that trait analysis may be a more effective approach than species identity in describing relationships between assemblages and the environment. Similar results were obtain by Logez et al. (2013) for fishes in rivers of Europe, where temperature and physical structure (e.g., slope, benthic sediment structure) were the primary environmental variables that explained trait distribution in the community, indicating that similar environmental conditions resulted in similar trait distributions.

Additional studies have found trait-based analyses to provide insight into the relative influence of environmental and spatial factors. An analysis of species and trait composition of macrophytes, macroinvertebrates, and fishes in headwaters and larger streams in Denmark found that environmental variables were more strongly associated with variation in traits than with taxonomic composition of fishes and macroinvertebrates (Göthe et al. 2017). Larger scale spatial factors had a stronger influence on macrophyte and fish communities, likely related to their lower dispersal capacity compared with macroinvertebrates. In the River Teno drainage in Finland and Norway, Tolonen et al. (2016) also found that trait composition of macroinvertebrate communities was more strongly related to local environmental variables than was species composition, indicating that local processes may act as filters to combinations of traits.

Several studies have compared the performance of species traits and taxonomic composition in detecting anthropogenic impacts on riverine systems. An investigation of the impacts of agriculture on macroinvertebrate communities in streams of New Zealand was able to distinguish sites based on land use practices from analyses of both traits and taxonomy. However, differences among the land use categories were more pronounced using traits, such as number of reproductive cycles, life duration of adults, egg laying modes, and parental-care behavior (Dolédec et al. 2006). Similarly, trait analysis was informative in a study of the responses of macroinvertebrates to a gradient of human disturbance in the Han River, the largest tributary of the Yangtze River in China (Li et al. 2019). Multivoltinism, fast development, short life span, and small body size were dominant at more disturbed sites, while univoltine species, characterized by slower development, larger body size, and longer life span, were more abundant at reference sites. Interestingly, species with strong flying dispersal capacity were found at more disturbed sites, most likely because this trait facilitates dispersal and recolonization. Burrowers and sprawlers were also more abundant at more disturbed sites, possibly due to their ability to cope with increased siltation.

Both trait and taxonomic composition differed with extent of human disturbance, but environmental conditions explained a larger proportion of trait than taxonomic variation, as suggested by the habitat template model. Further application of trait analysis to the study of anthropogenic impacts will benefit from identification of additional traits that more precisely indicate the causes of changes to lotic communities (Statzner and Bêche 2010; Verberk et al. 2013).

Although assemblage diversity traditionally has been evaluated based on the number and relative abundance of species or other taxonomic unit, recent attention has turned to the diversity of organisms' functional traits. Functional diversity and richness are determined by converting an assemblage of species into an assemblage of traits, most commonly using feeding habits, body size, substrate preferences, and locomotion. Functional diversity is then calculated from the relative abundance of organisms possessing those traits (Schmera et al. 2017).

Recent evidence suggests that functional diversity can reveal differences among assemblages more effectively than analyses based on species richness and abundance (Cadotte et al. 2011; Gagic et al. 2015). For example, Ephemeroptera, Plecoptera, and Trichoptera exhibited no differences in taxon richness between land use categories in savanna streams of the Cerrado, Brazil (Castro et al. 2018). However, functional richness and functional dispersion (measures of functional diversity) were higher in the less disturbed site, where Shannon's index of species diversity was also greater (Fig. 11.14). This reduction in functional diversity indicates that environmental conditions in the most disturbed habitats could support only a limited number of functional roles, consistent with the habitat template hypothesis. In other cases, however, species richness and functional measures of diversity can show similar patterns in the same system. Pease et al. (2012) observed the expected increase in species richness and functional diversity of fishes from the headwater to the lower reaches in the Grijalva River, Mexico, likely related to greater food and habitat resources in the lowland reaches.

11.4 Disturbance, Diversity, and Community Structure

Hurricanes, fires, floods, and droughts are well-known examples of extreme disturbances that episodically cause high mortality to populations of many species in forest, grassland, coral reef, river, and other ecosystems. Because species vary in their resistance to disturbance, as well as rates of recolonization and recovery, disturbance can ameliorate strong biological interactions and help to maintain populations of species that might otherwise be eliminated by their consumers or competitors. Streamflow is both the most

Fig. 11.14 Difference among three categories of land use pressure in (**a**) genus richness, (**b**) functional richness, (**c**) Shannon diversity, and (**d**) functional dispersion. Medians (lines), quartiles (box end), and maximum and minimum values (whiskers) are represented. Significant differences among the categories are indicated with different letters (Reproduced from Castro et al. 2018)

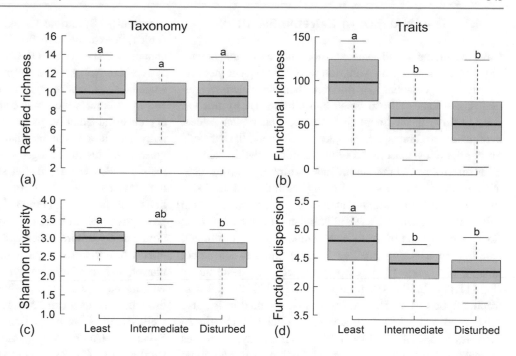

obvious and the most readily quantified disturbance variable in fluvial ecosystems, and flow variability is evident at all scales from the turbulence around a stone to the occasional extremes of major floods and droughts. Thus, small-scale disturbance might act almost continuously, whereas larger disturbances occur less frequently. Seasonal temperature extremes, pathogen outbreaks, sediment pulses from bank failure, and the arrival of a novel species each might constitute a disturbance to a particular system.

Two major axes of environmental variation that affect stream community structure are environmental harshness and disturbance frequency. These axes are independent, as one environment might be uniformly harsh, a second subject to frequent and extreme disturbances (Peckarsky 1983). Both act to restrict the abundance and diversity of species that are found at a location, but they differ in that perennially harsh environments such as those that are very cold or highly acid have an on-going effect, whereas disturbance implies an alternation with periods of more benign conditions. Thus harsh environments would be expected to contain relatively few species and experience less species replacement, whereas frequently disturbed environments would be expected to contain more species and exhibit higher species turnover. Because strong biotic interactions including competition and predation can reduce species diversity, it may be that an intermediate level of disturbance promotes diversity by ameliorating species effects on one another (Connell 1978; Ward and Stanford 1983).

Because the stream benthos is subject to ongoing turbulence and disturbance, it is possible that strong biological interactions may be frequently interrupted, resulting in a

continuously shifting mosaic of habitat conditions and species colonization and replacement. This is the model of patch dynamics (Townsend 1989), and it is consistent with the patchy distribution of algae and invertebrates on the streambed (Downes et al. 1998). A fluctuating environment combined with continual dispersal and colonization permits more species to co-occur than would be true if conditions exhibited greater constancy. It also confers some regularity to pattern, because environmental circumstances are predictable in the aggregate even though they are unpredictable for any given place and time.

Individual species will differ in their vulnerability to a particular disturbance event, reflecting many aspects of morphology, behavior, and lifestyle; thus the impact of the same disturbance may differ among species (Lytle and Poff 2004). In addition, between-habitat differences occur as a consequence of bed and substrate characteristics, influencing the availability of refuges within the substrate (Matthaei et al. 1999) and resulting in microhabitats where disturbance is less pronounced (Lancaster et al. 1991; Matthaei et al. 2000).

Several authors have highlighted the importance of characterizing disturbances in terms of type, intensity, duration, frequency, and spatial extent when studying their effects on ecosystems, as it will aid in comparing events within and among systems (Lake 2000; Winemiller et al. 2010). Peckarsky et al. (2014) describes several methods to assess streambed disturbance in headwater streams, including photographs of the stream bed, hydrological indices, Shields number, and visual characterization of channel stability.

11.4.1 The Influence of Extreme Events

Changes in local and seasonal abundances of the stream biota in response to flow variation received frequent mention in previous chapters. Periphyton, benthic macroinvertebrates, and fishes all can be strongly influenced by fluctuations in flow. In piedmont rivers of the Venezuelan Andes, benthic invertebrates are subject to frequent flash floods during the rainy season and droughts during the dry season. Total macroinvertebrate abundance exhibited a strong negative relation with average monthly rainfall, used as a surrogate of flow because no stream gauges were available (Flecker and Feifarek 1994). Numbers recovered during flood-free periods due to colonization and recruitment, resulting in a strong positive relationship between insect abundances and time elapsed since the last storm (Fig. 11.15). In the Glenfinish River in Ireland, a catastrophic flood (a 1-in-50-year event) that occurred in August, 1986, produced a 70% decline in taxon richness and reduced densities to 5% of pre-flood levels. Interestingly, responses to the flood differed among taxa. The smallest and more abundant chironomids were less affected than Ephemeroptera, Plecoptera, and Trichoptera, which needed 3–5 years to recover. Insects with fast life cycles (r-strategy) recovered faster than slow life cycle, large-bodied species (K-strategy), which in turn showed faster recovery than mollusks and crustaceans (Woodward et al. 2016).

Disturbance has been shown to have dramatic, ecosystem-wide effects. Examples include those where floods and droughts are of sufficient magnitude that habitat is disrupted and many of the organisms present are displaced, comparative studies of rivers basins with different disturbance regimes, and instances where disturbance most severely affects a species that is a strong interactor within a

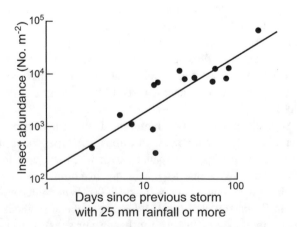

Fig. 11.15 The number of aquatic insects in a river of the Andean foothills subject to a pronounced dry season and frequent floods during the rainy season. Time elapsed since the most recent rainfall event >25 mm is used in lieu of stormflow data (Reproduced from Flecker and Feifarek 1994)

community. Sycamore Creek, Arizona, is a desert stream subject to extreme flash floods that might occur a few times annually. Fisher et al. (1982) describe the recovery of the system following a late summer flood, until another flood some 60 days later re-started the sequence. Biomass of algae and invertebrates were reduced by nearly 100%, but recovery occurred quickly, particularly by the algae, which initially was dominated by diatoms and later by filamentous green and blue-green algae (Fig. 11.16). Macroinvertebrate recovery also was rapid, but slower by several weeks than the algal recovery. Nutrient uptake and community metabolism changed over community succession, and this highly productive system began to export surplus primary production. The Sycamore Creek example nicely demonstrates the effects of a disturbance that occurs with unpredictable regularity, and so ecosystem dynamics can only be understood within the cycle of disturbance and recovery. Periodic drought obviously can severely disrupt stream ecosystems as well. Based on an evaluation of resilience and recovery of Sycamore Creek to a number of spates and droughts, Boulton et al. (1992) concluded that droughts had the greater impact.

Extreme drying can have profound impacts on aquatic communities. Bogan et al. (2015) described temporal shifts in community composition in intermittent, arid-land streams of the Madrean Sky Islands in Arizona, US, during seasonal drying. In these headwater systems, flow is generally intermittent and streams can remain dry for more than 9 months of the year; however, some streams are perennial, and contain macroinvertebrate taxa that are rare in the intermittent streams. As droughts begin, lateral connectivity with riparian areas is lost. This is followed by loss of longitudinal connectivity as flow ceases, leaving only isolated pools (Fig. 11.17). In response, most lotic species disappear, and only relatively tolerant Coleoptera, Hemiptera, and Trichoptera remain. The new habitat is subsequently colonized by other coleopterans and hemipterans that are not typically encountered under flowing conditions. As the drought persists, vertical connectivity is lost and invertebrate richness decreases markedly. Such declines are even evident in the drought-tolerant groups. Some invertebrates may initially migrate to hyporheic habitats, but as this refuge dries, only organisms that have evolved resting stages can remain. As flow returns to the stream, only 10–12 species of resistant taxa typically are present, but recovery is rapid, reaching abundance levels similar to perennial streams in 8–10 weeks. Recovery continues as strong aerial dispersers colonize streams. These species are drought-tolerant taxa that remained dominant in perennial pools during the drought. Total recovery of the macroinvertebrate assemblage is typically observed in streams that are near dry season refuges (e.g. headwater seeps and springs, perennial pools,

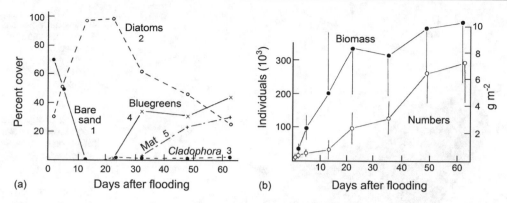

Fig. 11.16 Temporal succession of the biota in Sycamore Creek, Arizona, following flooding. (**a**) Percent cover of algal patch types; (**b**) Mean invertebrate numbers and biomass after flooding. Values are means and 95% confidence intervals (Reproduced from Fisher et al. 1982)

Fig. 11.17 Changes in aquatic invertebrate species richness in streams of Madrean Sky Island as flow decreased (solid line) and recovered (dashed line). A–D: thresholds during the drying period. E: recovery by taxa resistant to drought. F: recovery via aerial recolonisation of resilient taxa. G: full recovery via multiple resistance and resilience (instream and overland) pathways. Processes contributing to species loss and recovery are indicated at the top of the figure (Reproduced from Bogan et al. 2015)

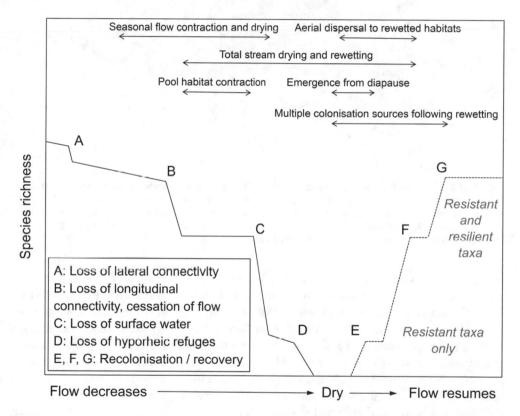

flowing reaches) that maintained flow for at least 4–5 months.

Notably, these patterns were observed during predictable seasonal droughts in conditions under which the macroinvertebrate community has evolved. Extreme droughts can convert perennial to intermittent systems, and researchers have documented changes in expected species composition. In particular, large predators were lost from many habitats. However, macroinvertebrate densities increased and species richness remained similar when compared to typical seasonal drying (Bogan and Lytle 2011). Collectively, these results indicate that drought severity and habitat isolation can influence local and regional invertebrate richness

(Fig. 11.18). The greatest number of species were observed in streams that experienced mild and more predictable droughts and had greater connectivity to perennial refuges, while only the most resistant species are found in more isolated streams under severe drought conditions (Bogan et al. 2015).

11.4.2 Disturbance Frequency and Biotic Responses

Flow variation has been shown to mediate species interactions. Throughout most of Arizona, the introduced

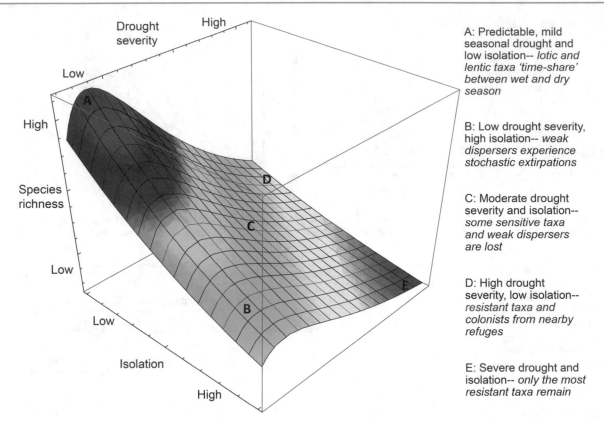

Fig. 11.18 Conceptual model describing the interactive effects of drought severity (based on drying intensity and duration) and habitat isolation (distance to nearest perennial refuge) on stream invertebrate species richness. A: Richness increased by the colonization of lentic taxa during part of the year under moderate drought disturbance. B–D: Richness decreased due to high drought severity or isolation, or a combination of both. E: in highly isolated sites under high drought severity, only highly resistant or resilient taxa remain (Reproduced from Bogan et al. 2015)

mosquitofish *Gambusia affinis* has replaced a native poeciliid, the Sonoran topminnow *Poecilopsis occidentalis*, largely through predation on juveniles (Meffe 1984). In mountainous regions subject to extreme flash floods, however, long-term coexistence of the two species results from the native fish's superior ability to avoid downstream displacement. Hydropsychids (Trichoptera) and simuliids (Diptera) exhibit a similar interaction seasonally in coastal Californian streams (Hemphill and Cooper 1983). On hard substrates in fast-flowing sections, black fly larvae are more abundant in spring and early summer, while caddis larvae predominate thereafter. Winters of high discharge lead to greater numbers of simuliids, and winters of low flow lead to higher densities of hydropsychids. By scrubbing substrate surfaces with a brush, Hemphill and Cooper showed that disturbance benefited simuliids because they were the more rapid colonizers, whereas caddis larvae were superior at monopolizing space on rock surfaces. As time passed since the last disturbance, hydropsychid larvae gradually replaced simuliids due to their aggressive defense of net sites.

Because flow variation is such a pervasive feature of the fluvial environment, many organisms show adaptation to resist or minimize its effects (Lytle and Poff 2004). Timing of life cycle events such as egg-laying or emergence can be effective when flow extremes have a degree of predictability. The emergence of young rainbow trout from spawning gravel in spring appears to be an example of synchronizing a sensitive life cycle stage with the long-term average dynamics of the flow regime. The rainbow trout *Oncorhyncus mykiss*, one of the most widely introduced species world-wide, succeeds in new environments where the flow regime matches its native range and fails in other environments where it does not, apparently because floods harm trout fry (Fausch et al. 2001). Position shifts to protected areas are a common behavioral response to high flows, such as the relocation of fishes from the thalweg to the floodplain during a flood event in an Illinois river (Schwartz and Herricks 2005). In arid-land streams of the southwestern US prone to flash flooding, giant waterbugs (Belostomatidae) crawl out of streams in advance of the flood, using rainfall as the cue (Lytle 1999). Morphological adaptations including streamlining and other adaptations to minimize drag were described in Sect. 5.1.2.

Any pattern of disturbance that recurs with some long-term regularity is referred to as a disturbance regime. Riseng et al. (2004) characterized disturbance in 97 Midwestern US streams that differed in their frequency of spates and droughts using several measures of low and high flow occurrence, substrate movement, and summer water temperature (which can reach stressful levels during low flow periods). In locations where scouring floods reduced consumer biomass, algal biomass was strongly influenced by nutrient supply, whereas in more stable streams grazers depressed algal biomass regardless of nutrient concentration. The different disturbance regimes identified in this study were clearly a product of regional patterns in geology and precipitation, and so exhibited a degree of spatial predictability. In some California rivers, a grazing caddis larva is abundant during flood-free periods and escapes most predation due to its large size and robust case, but is highly vulnerable to high flows and rolling stones during floods (Wootton et al. 1996). During flood-free periods and in dammed rivers, the main energy pathway is from algae to the caddis; however, after flooding occurs in unregulated river segments, more energy flows to smaller grazers and then to young steelhead trout. In essence, disturbance regime acts as a switch, causing one of two possible food web configurations to become dominant.

Disturbance frequency can influence the responses of entire aquatic communities. Many studies have focused on the effects of floods; however, more recent work has been focused on low flows and drought in response to climate change and the intensification of anthropogenic water withdrawals. Using stream mesocosms, Ledger et al. (2008) investigated the effects of low flow frequency on periphyton by conducting short dewatering disturbances (6 days) in high frequency (33-day) and low frequency (99-day) cycles. In the controls, *Gongrosira incrustans*, a crustose green alga, was dominant and was accompanied by small unicellular diatoms (*Rhoicosphenia*) and blue-greens (*Phormidium*). With dewatering events, *Gongrosira* decreased in abundance due to its vulnerability to drying. When the frequency of dewatering was low, *Gongrosira* was able to partially recover, but under high frequency disturbances the species only grew in small patches, leaving space available to other algae, particularly mat-forming diatoms. Contrasting traits between *Gongrosira* and the diatoms likely explain the effects of disturbance on the algal community. *Gongrosira* has the ability to gain and retain space through crust formation, but cannot resist dry conditions, while diatoms can take advantage of available substrate space through rapid growth, but do not have the traits to retain space when competing with other species.

The effects of low water events can be species- and habitat-specific. Haghkerdar et al. (2019) experimentally manipulated the number of daily disturbances in stream mesocosms by simulating the effects of flood disturbance on substrate. The richness of macroinvertebrate taxa declined as disturbance frequency increased. The abundance of Chironomidae, the dominant group of organisms, was not affected by disturbance frequency, but the abundance of the remaining populations decreased as disturbance increased, suggesting that disturbance effects can vary among taxa. Walters and Post (2011) studied the effects of low flows on benthic communities by diverting between 40 and 80% of the water from three streams in the western US. Total biomass of aquatic insects declined with decreasing flows, with greater effects observed in riffles than pools. Some feeding groups such as collectors and scrapers were most affected, but family-level richness did not vary with altered flows.

In tropical headwater streams of Bolivia that experienced a range of environmental conditions related to altitude, climate, and flow regime, environmental variables were considered to be more important than dispersal capabilities in shaping communities (Datry et al. 2016). However, streams that experienced moderate environmental conditions were more influenced by dispersal. Under extreme environmental conditions, neither species sorting nor dispersal explained community structure, suggesting that neutral processes may shape communities in more extreme environments. Datry et al. (2015) speculate that under the dynamic conditions experienced by many streams and rivers, the relative contribution of species sorting and dispersal can vary as the system recovers from a disturbance. Shortly after disturbance, dispersal is an important process that enhances the arrival of organisms, but after a patch has been colonized, processes related to species sorting can shape local communities.

As previously mentioned, the effects of disturbance on communities can be mitigated by the presence of refuges, or regions of the stream buffered from a disturbance. Refuges from one disturbance event sometimes provide protection from multiple kinds of disturbance, but this is not always true. For instance, examples of refuges from increased flow include pools, floodplains, areas downstream of large boulders, and interstitial areas between rocks. Refuges from drought can be found in perennial pools, the hyporheic zone, and leaf litter accumulations. Organisms that persist in refuges after a disturbance can play a fundamental role in the persistence of populations as they can colonize habitat patches in the stream where organisms were lost due to disturbance (Lake 2000).

It is important to note that organisms can also be the source of disturbance, altering the physical, chemical, and biological structure of streams. Such species are considered ecological engineers, and their ability to influence habitat structure has the potential to affect the abundance and distribution of other species (Jones et al. 1994). The grazing fish *Parodon apolineri* has been classified as an ecological engineer, as it creates observable grazing scars and a patchy distribution of algae and sediments on stone surfaces. By quantifying the spatial pattern of scars in a neotropical stream and by experimentally manipulating densities of *Parodon*, Flecker and Taylor (2004) demonstrated that grazing enhanced habitat heterogeneity, although with time or at higher densities grazing tended to reduce algae and sediments to a more uniform condition of low abundance. However, the hypothesized correspondence between habitat heterogeneity and richness of the algal and invertebrate assemblages was not observed, perhaps because the time-scale at which those taxa respond was slower than the loss and renewal dynamics generated by grazing *Parodon*. Sockeye salmon *Oncorhynchus nerka* can affect stream benthic communities through their seasonal spawning migrations. In streams of Alaska, the biomass of algae and insects tends to increase during the spring, but then decreases by 75–85% during salmon nest excavation, and also post spawning, when salmon disturb the streambed to maintain the nest free of sediments (Moore and Schindler 2008). After salmon die, algae quickly reach pre-spawning levels, but invertebrate recovery is slower. Anthropogenic activities have altered the distribution of some species of ecosystem engineers through the introduction of non-native species. Beavers (*Castor canadensis*), notorious for their engineering capabilities, were introduced into the Cape Horn Biosphere Reserve in southern Chile in the mid-1940s (Anderson and Rosemond 2007). They have subsequently altered riparian and in-stream diversity through changes in hydrogeomorphic and biogeochemical processes.

11.5 Summary

The number of coexisting species in ecological communities is a consequence of processes operating at both local and regional scales. The assemblage of species at a location are a sub-set of the species in the region, which are themselves a product of species formation, loss, and dispersal over earth history. The local assemblage is also the result of species sorting that is influenced by local environmental conditions and other species present, and by each species' dispersal ability and life history characteristics.

Patterns in the regional species richness of freshwater organisms are best known for fishes, whose global distribution reveals a number of intriguing patterns. The number of species of freshwater fishes is inversely related to latitude, increases with drainage area and discharge of the river system, and differs among regions at the same latitude in both number of species and the body plans and traits of species present. Greater productivity, more habitat diversity, and more habitat area are amongst the explanations for these patterns. Historical changes in landforms, climate, and

connectivity provide further insight into patterns in species diversity, as they can influence speciation and extinction events, open and close dispersal pathways, and affect time for recolonization following disruptive events such as glaciation. Comparisons of freshwater fishes of Europe and North America provide evidence that historical events, in particular recent glaciations, have left strong signatures on their respective faunas. Post-glacial recolonization was more restricted in Europe relative to North America because drainage divides in Europe tend to run from east to west, and re-establishment from Iberia and the Adriatic was restricted by mountain ranges. In contrast, the North American fauna had a much larger refugial area in unglaciated regions of the Mississippi and Missouri basins, and easier routes for recolonization. Present-day patterns in freshwater fish diversity in tropical rivers also bear the signature of past glacial events. Lower sea levels during the last glacial maximum allowed for connections at their lower termini between some now-isolated river systems, and greater aridity resulted in differential isolation among rivers during periods of tropical forest fragmentation. It appears that macroecological variables of drainage area and system productivity provide the strongest explanations of global patterns in freshwater fish species richness, and history less so, with the caveat that the influence of history is more challenging to establish. Although the historical biogeography of aquatic insects is less well known, diversity in major insect groups likewise shows continental-scale patterns such that number of species and genera differ among biogeographical regions.

Local species richness varies in proportion to regional species richness, indicating that site surveys are likely to find more species in areas with a larger species pool. A comprehensive survey of all species at a site is a challenging task, especially for invertebrates and the smallest organisms, due to incomplete taxonomic knowledge and the fact that most species are rare and escape detection in all but the most extensive sampling. As a consequence, most invertebrate studies focus on the more common species in the assemblage. Community structure refers to the organization of a biological community based on numbers of individuals within different taxonomic groups and functional roles, and the underlying processes that maintain that organization. Niche-based explanations for patterns in species diversity and community structure focus on the fit of individual species and their traits to habitat conditions, and the influence of interspecific interactions. However, environmental disturbances such as floods and droughts, when sufficiently extreme or frequent, are likely to prevent biotic interactions from acting with the strength and regularity required to result in consistent community patterns. In these circumstances, species assemblages are likely to be composed of those species best adapted to environmental extremes and those most capable of dispersal and recolonization of disrupted environments. Studies of species assemblages in streams frequently explore the relative importance of these two contrasting explanations.

All organisms have some capacity to disperse, and local assemblages most likely are connected to other, neighboring assemblages by occasional or frequent movements between them. Differences in dispersal modes, as when fish are limited to connecting waterways but insects are capable of aerial dispersal, and degree of physical connectivity, influence whether each local assemblage is structured more by local environmental conditions or more by proximity to neighbors.

Species traits such as size, body shape, lifespan, and mode of dispersal are increasingly explored as an alternative to species identity to help us understand why certain species succeed where others do not. This perspective intersects well with a conceptual framework in which the local assemblage of species is filtered from a regional species pool first through dispersal barriers, which are influenced by factors such as topography and climate, but also by differences in species' dispersal capabilities; and secondly by environmental filters that exist at a hierarchy of scales, such as flow and temperature regimes, and macro- and microhabitat variables. Supporting evidence exists in trait-based analyses that provide insight into the relative influence of environmental and spatial factors in determining the composition of local assemblages.

The role of disturbance in local community structure has been the focus of much study in lotic ecosystems, especially with regard to flood and drought events. Local conditions can be perennially harsh, as when temperature, pH, or some other environmental variable is continually extreme, or can alternate between moderate and extreme condition, as with episodically recurring floods and droughts. This is seen in desert streams, when scoured communities accrue biomass following a flood, only to be re-set when the next flood occurs; and in tropical streams during the rainy season, when insect numbers correspond to days since the last substantial rain event. Experimental manipulation of the frequency of de-watering shows how species traits determine patterns of recovery and species replacement. Differential dispersal ability among species and connectivity within a river network also are important explanatory variables in the response of stream communities to hydrologic disturbance. Finally, not all disturbance is abiotic, as organisms also are disturbance agents. This is seen when the nest-building activities of large salmon runs disturb the stream bed, and in some tropical streams where very abundant sediment-feeding fishes create a patchwork of feeding scars on benthic substrates.

References

Amarasinghe US, Welcomme RL (2002) An analysis of fish species richness in natural lakes. Environ Biol Fishes 65:327–339. https://doi.org/10.1023/A:1020558820327

Anderson CB, Rosemond AD (2007) Ecosystem engineering by invasive exotic beavers reduces in-stream diversity and enhances ecosystem function in Cape Horn, Chile. Oecologia 154:141–153. https://doi.org/10.1007/s00442-007-0757-4

Angermeier PL, Schlosser IJ (1989) Species-area relationships for stream fishes. Ecology 70:1450–1462. https://doi.org/10.2307/1938204

Arrington DA, Winemiller KO, Layman CA (2005) Community assembly at the patch scale in a species rich tropical river. Oecologia 144:157–167. https://doi.org/10.1007/s00442-005-0014-7

Balian EV, Leveque C, Segers H, Martens K (2008) Freshwater animal diversity assessment. Hydrobiologia 595:1–637. https://doi.org/10.1007/978-1-4020-8259-7

Barber-James HM, Gattolliat JL, Sartori M, Hubbard MD (2008) Global diversity of mayflies (Ephemeroptera, Insecta) in freshwater. Hydrobiologia 595:339–350

Begon M, Harper JL, Townsend CR (2006) Ecology: individuals, populations and communities, 4th edn. Blackwell Publishing, Oxford

Bogan MT, Boersma KS, Lytle DA (2015) Resistance and resilience of invertebrate communities to seasonal and supraseasonal drought in arid-land headwater streams. Freshw Biol 60:2547–2558. https://doi.org/10.1111/fwb.12522

Bogan MT, Lytle DA (2011) Severe drought drives novel community trajectories in desert stream pools. Freshw Biol 56:2070–2081. https://doi.org/10.1111/j.1365-2427.2011.02638.x

Bonada N, Zamora-Muñoz C, Rieradevall M, Prat N (2005) Ecological and historical filters constraining spatial caddisfly distribution in Mediterranean rivers. Freshw Biol 50:781–797. https://doi.org/10.1111/j.1365-2427.2005.01357.x

Boulton AJ, Peterson CG, Grimm NB, Fisher SG (1992) Stability of an aquatic macroinvertebrate community in a multiyear hydrologic disturbance regime. Ecology 73:2192–2207. https://doi.org/10.2307/1941467

Brown BL, Swan CM (2010) Dendritic network structure constrains metacommunity properties in riverine ecosystems. J Anim Ecol 79:571–580. https://doi.org/10.1111/j.1365-2656.2010.01668.x

Cadotte MW, Carscadden K, Mirotchnick N (2011) Beyond species: functional diversity and the maintenance of ecological processes and services. J Appl Ecol 48:1079–1087. https://doi.org/10.1111/j.1365-2664.2011.02048.x

Cañedo-Argüelles M, Boersma KS, Bogan MT et al (2015) Dispersal strength determines meta-community structure in a dendritic riverine network. J Biogeogr 42:778–790. https://doi.org/10.1111/jbi.12457

de Castro DMP, Dolédec S, Callisto M (2018) Land cover disturbance homogenizes aquatic insect functional structure in neotropical savanna streams. Ecol Indic 84:573–582. https://doi.org/10.1016/j.ecolind.2017.09.030

Chao A, Gotelli NJ, Hsieh TC et al (2014) Rarefaction and extrapolation with Hill numbers: a framework for sampling and estimation in species diversity studies. Ecol Monogr 84:45–67. https://doi.org/10.1890/13-0133.1

Chase JM, Amarasekare P, Cottenie K et al (2005) Competing theories for competitive metacommunities. In: Holyoak M, Leibold MA, Holt RD (eds) Metacommunities: spatial dynamics and ecological communities. The University of Chicago Press, Chicago, pp 335–354

Chase JM, Leibold MA (2003) Ecological niches: linking classical and contemporary approaches. University of Chicago Press, Chicago

Chutter FM, Noble R (1966) The reliability of method of sampling stream invertebrates. Arch Für Hydrobiol 62:95–103

Connell JH (1978) Diversity in tropical rain forests and coral reefs. Science 199:1302–1310. https://doi.org/10.1126/science.199.4335.1302

Cornell HV, Harrison SP (2014) What are species pools and when are they important? Annu Rev Ecol Evol Syst 45:45–67. https://doi.org/10.1146/annurev-ecolsys-120213-091759

Currie DJ (1991) Energy and large-scale patterns of animal- and plant-species richness. Am Nat 137:27–49

Datry T, Bonada N, Heino J (2015) Towards understanding the organisation of metacommunities in highly dynamic ecological systems. Oikos 125:149–159. https://doi.org/10.1111/oik.02922

Datry T, Melo AS, Moya N et al (2016) Metacommunity patterns across three Neotropical catchments with varying environmental harshness. Freshw Biol 61:277–292. https://doi.org/10.1111/fwb.12702

De Mérona B, Tejerina-Garro FL, Vigouroux R (2012) Fish-habitat relationships in French Guiana rivers: a review. Cybium 36:7–15

De Moor FC, Ivanov VD (2008) Global diversity of caddisflies (Trichoptera: Insecta) in freshwater. Hydrobiologia 595:393–407

Dias MS, Cornu JF, Oberdorff T et al (2013) Natural fragmentation in river networks as a driver of speciation for freshwater fishes. Ecography (Cop) 36:683–689. https://doi.org/10.1111/j.1600-0587.2012.07724.x

Dias MS, Oberdorff T, Hugueny B et al (2014) Global imprint of historical connectivity on freshwater fish biodiversity. Ecol Lett 17:1130–1140. https://doi.org/10.1111/ele.12319

Dolédec S, Phillips N, Scarsbrook M et al (2006) Comparison of structural and functional approaches to determining landuse effects on grassland stream invertebrate communities. J North Am Benthol Soc 25:44–60. https://doi.org/10.1899/0887-3593(2006)25[44:COSAFA]2.0.CO;2

Downes BJ, Lake PS, Schreiber ESG, Glaister A (1998) Habitat structure and regulation of local species diversity in a stony, upland stream. Ecol Monogr 68:237–257. https://doi.org/10.1890/0012-9615(1998)068[0237:HSAROL]2.0.CO;2

Fausch KD, Taniguchi Y, Nakano S et al (2001) Flood disturbance regimes influence rainbow trout invasion success among five holarctic regions. Ecol Appl 11:1438–1455. https://doi.org/10.1890/1051-0761(2001)011[1438:FDRIRT]2.0.CO;2

Filipe AF, Magalhães MF, Collares-Pereira MJ (2010) Native and introduced fish species richness in Mediterranean streams: the role of multiple landscape influences. Divers Distrib 16:773–785. https://doi.org/10.1111/j.1472-4642.2010.00678.x

Fisher SG, Gray LJ, Grimm NB, Busch DE (1982) Temporal succession in a desert stream ecosystem following flash flooding. Ecol Monogr 52:93–110. https://doi.org/10.2307/2937346

Flecker AS, Feifarek B (1994) Disturbance and the temporal variability of invertebrate assemblages in two Andean streams. Freshw Biol 31:131–142. https://doi.org/10.1111/j.1365-2427.1994.tb00847.x

Flecker AS, Taylor BW (2004) Tropical fishes as biological bulldozers: density effects on resource heterogeneity and species diversity. Ecology 85:2267–2278. https://doi.org/10.1890/03-0194

Fochetti R, Tierno De Figueroa JM (2008) Global diversity of stoneflies (Plecoptera; Insecta) in freshwater. Hydrobiologia 595:365–377. https://doi.org/10.1007/s10750-007-9031-3

Frimpong EA, Angermeier PL (2009) FishTraits: a Database of ecological and life-history traits of freshwater fishes of the United States. Fisheries 34:487–495

Frimpong EA, Angermeier PL (2010) Trait-based approaches in the analysis of stream fish communities. Am Fish Soc Symp 73:109–136

Gagic V, Bartomeus I, Jonsson T et al (2015) Functional identity and diversity of animals predict ecosystem functioning better than species-based indices. Proc R Soc B Biol Sci 282:20142620. https://doi.org/10.1098/rspb.2014.2620

Gaston KJ (2000) Global patterns in biodiversity. Nature 405:220–227. https://doi.org/10.1038/35012228

Gaston KJ, Blackburn TM (2000) Pattern and processes in macroecology. Blackwell Scientific, Oxford

Gaufin AR, Harris EK, Walter J (1956) A statistical evaluation of stream bottom sampling data obtained from three standard samplers. Ecology 37:643–648

Göthe E, Angeler DG, Sandin L (2013) Metacommunity structure in a small boreal stream network. J Anim Ecol 82:449–458. https://doi.org/10.1111/1365-2656.12004

Göthe E, Baattrup-Pedersen A, Wiberg-Larsen P et al (2017) Environmental and spatial controls of taxonomic versus trait composition of stream biota. Freshw Biol 62:397–413. https://doi.org/10.1111/fwb.12875

Griffiths D (2006) Pattern and process in the ecological biogeography of European freshwater fish. J Anim Ecol 75:734–751. https://doi.org/10.1111/j.1365-2656.2006.01094.x

Griffiths D (2015) Connectivity and vagility determine spatial richness gradients and diversification of freshwater fish in North America and Europe. Biol J Linn Soc 116:773–786. https://doi.org/10.1111/bij.12638

Grossman GD, Moyle PB, Whitaker JO (1982) Stochasticity in structural and functional characteristics of an Indiana stream fish assemblage: a test of community theory. Am Nat 120:423–454. https://doi.org/10.1086/284004

Grossman GD, Ratajczak RE, Crawford M, Freeman MC (1998) Assemblage organization in stream fishes: effects of environmental variation and interspecific interactions. Ecol Monogr 68:395–420. https://doi.org/10.1890/0012-9615(1998)068[0395:AOISFE]2.0.CO;2

Guégan JF, Lek S, Oberdorff T (1998) Energy availability and habitat heterogeneity predict global riverine fish diversity. Nature 391:382–384. https://doi.org/10.1038/34899

Haghkerdar JM, McLachlan JR, Ireland A, Greig HS (2019) Repeat disturbances have cumulative impacts on stream communities. Ecol Evol 9:2898–2906. https://doi.org/10.1002/ece3.4968

Hanly PJ, Mittelbach GG, Schemske DW (2017) Speciation and the latitudinal diversity gradient: insights from the global distribution of endemic fish. Am Nat 189:604–615. https://doi.org/10.1086/691535

Heino J (2011) A macroecological perspective of diversity patterns in the freshwater realm. Freshw Biol 56:1703–1722. https://doi.org/10.1111/j.1365-2427.2011.02610.x

Heino J (2009) Biodiversity of aquatic insects: spatial gradients and environmental correlates of assemblage-level measures at large scales. Freshw Rev 2:1–29. https://doi.org/10.1608/FRJ-2.1.1

Heino J (2005) Positive relationship between regional distribution and local abundance in stream insects: a consequence of niche breadth or niche position? Ecography (Cop) 28:345–354. https://doi.org/10.1111/j.0906-7590.2005.04151.x

Heino J, Melo AS, Jyrkänkallio-Mikkola J et al (2018) Subtropical streams harbour higher genus richness and lower abundance of insects compared to boreal streams, but scale matters. J Biogeogr 45:1983–1993. https://doi.org/10.1111/jbi.13400

Heino J, Melo AS, Siqueira T et al (2015) Metacommunity organisation, spatial extent and dispersal in aquatic systems: patterns, processes and prospects. Freshw Biol 60:845–869. https://doi.org/10.1111/fwb.12533

Heino J, Mykrä H (2008) Control of stream insect assemblages: roles of spatial configuration and local environmental factors. Ecol Entomol 33:614–622. https://doi.org/10.1111/j.1365-2311.2008.01012.x

Heino J, Virkkala R, Toivonen H (2009) Climate change and freshwater biodiversity: detected patterns, future trends and adaptations in northern regions. Biol Rev 84:39–54. https://doi.org/10.1111/j.1469-185X.2008.00060.x

Hemphill N, Cooper SD (1983) The effect of physical disturbance on the relative abundances of two filter-feeding insects in a small stream. Oecologia 58:378–382. https://doi.org/10.1007/BF00385239

Hillebrand H (2004) On the generality of the latitudinal diversity gradient. Am Nat 163:192–211. https://doi.org/10.1086/381004

Hoeinghaus DJ, Winemiller KO, Birnbaum JS (2007) Local and regional determinants of stream fish assemblage structure: inferences based on taxonomic vs. functional groups. J Biogeogr 34:324–338. https://doi.org/10.1111/j.1365-2699.2006.01587.x

Horwitz RJ (1978) Temporal variability patterns and the distributional patterns of stream fishes. Ecol Monogr 48:07–321. https://doi.org/10.2307/2937233

Hubbell SP (2005) Neutral theory in community ecology and the hypothesis of functional equivalence. Funct Ecol 19:166–172. https://doi.org/10.1111/j.0269-8463.2005.00965.x

Hubbell SP (2001) The unified neutral theory of biodiversity and biogeography. Princeton University Press, Princeton

Hugueny B (1989) West African rivers as biogeographic islands: species richness of fish communities. Oecologia 79:236–243. https://doi.org/10.1007/BF00388483

Hugueny B, Oberdorff T, Tedescco PA (2010) Community ecology of river fishes: a large-scale perspective. Am Fish Soc Symp 73:29–62

Illies J (1971) Emergenz 1969 im Breitenbach. Arch Für Hydrobiol 69:14–59

Iversen LL, Jacobsen D, Sand-Jensen K (2016) Are latitudinal richness gradients in European freshwater species only structured according to dispersal and time? Ecography (Cop) 39:1247–1249. https://doi.org/10.1111/ecog.02183

Jones CG, Lawton JH, Shachak M (1994) Organisms as ecosystem engineers. Oikos 69:373–386. https://doi.org/10.2307/3545850

Keil P, Simova I, Hawkins BA (2008) Water-energy and the geographical species richness pattern of European and North African dragonflies (Odonata). Insect Conserv Divers 1:142–150. https://doi.org/10.1111/j.1752-4598.2008.00019.x

Kinlock NL, Prowant L, Herstoff EM et al (2018) Explaining global variation in the latitudinal diversity gradient: meta-analysis confirms known patterns and uncovers new ones. Glob Ecol Biogeogr 27:125–141. https://doi.org/10.1111/geb.12665

Kuusela K (1979) Early summer ecology and community structure of the macrozoobenthos on stones in the Javajankoski Rapids on the River Lestijoki, Finland. Acta Univ Ouluensis (Ser A, No. 87, Oulu, Finland)

Lake PS (2000) Disturbance, patchiness, and diversity in streams. J North Am Benthol Soc 19:573–592. https://doi.org/10.2307/1468118

Lamouroux N, Dolédec S, Gayraud S (2004) Biological traits of stream macroinvertebrate communities: effects of microhabitat, reach, and basin filters. J North Am Benthol Soc 23:449–466. https://doi.org/10.1899/0887-3593(2004)023%3c0449:BTOSMC%3e2.0.CO;2

Lancaster J, Hildrew AG, Townsend CR (1991) Invertebrate predation on patchy and mobile prey in streams. J Anim Ecol 60:625–641. https://doi.org/10.2307/5302

Layer K, Hildrew AG, Jenkins GB et al (2011) Long-term dynamics of a well-characterised food web. Four decades of acidification and recovery in the Broadstone Stream model system. Adv Ecol Res 44:69–117. https://doi.org/10.1016/B978-0-12-374794-5.00002-X

Ledger ME, Harris RML, Armitage PD, Milner AM (2008) Disturbance frequency influences patch dynamics in stream benthic algal communities. Oecologia 155:809–819. https://doi.org/10.1007/s00442-007-0950-5

Leibold MA, Chase JM (2017) Metacommunity ecology. Princeton University Press, Princeton

Leibold MA, Holyoak M, Mouquet N et al (2004) The metacommunity concept: a framework for multi-scale community ecology. Ecol Lett 7:601–613. https://doi.org/10.1111/j.1461-0248.2004.00608.x

Lévêque C, Oberdorff T, Paugy D et al (2008) Global diversity of fish (Pisces) in freshwater. Hydrobiologia 595:545–567

Li Z, Wang J, Liu Z et al (2019) Different responses of taxonomic and functional structures of stream macroinvertebrate communities to local stressors and regional factors in a subtropical biodiversity hotspot. Sci Total Environ 655:1288–1300. https://doi.org/10.1016/j.scitotenv.2018.11.222

Logez M, Bady P, Melcher A, Pont D (2013) A continental-scale analysis of fish assemblage functional structure in European rivers. Ecography (Cop) 36:080–091. https://doi.org/10.1111/j.1600-0587.2012.07447.x

Lytle DA (1999) Use of rainfall cues by Abedus herberti (Hemiptera: Belostomatidae): a mechanism for avoiding flash floods. J Insect Behav 12:1–12

Lytle DA, Poff NLR (2004) Adaptation to natural flow regimes. Trends Ecol Evol 19:94–100. https://doi.org/10.1016/j.tree.2003.10.002

MacArthur RH (1972) Geographical ecology: patterns in the distribution of species. Harper and Row, New York

MacArthur RH, Wilson EO (1963) An equilibrium theory of insular zoogeography. Evolution (N Y) 17:373–387

MacArthur RH, Wilson EO (1967) The theory of island biogeography. Princeton University Press, Princeton

Mahon R (1984) Divergent structure in fish taxocenes of north temperate streams. Can J Fish Aquat Sci 41:330–350. https://doi.org/10.1139/f84-037

Matthaei CD, Arbuckle CJ, Townsend CR (2000) Stable surface stones as refugia for invertebrates during disturbance in a New Zealand stream. J North Am Benthol Soc 19:82–93. https://doi.org/10.2307/1468283

Matthaei CD, Peacock KA, Townsend CR (1999) Scour and fill patterns in a New Zealand stream and potential implications for invertebrate refugia. Freshw Biol 42:41–57. https://doi.org/10.1046/j.1365-2427.1999.00456.x

Meffe GK (1984) Effects of abiotic disturbance of coexistence of predator-prey fish species. Ecology 65:1525–1534. https://doi.org/10.2307/1939132

Menezes S, Baird DJ, Soares AMVM (2010) Beyond taxonomy: a review of macroinvertebrate trait-based community descriptors as tools for freshwater biomonitoring. J Appl Ecol 47:711–719. https://doi.org/10.1111/j.1365-2664.2010.01819.x

Mittelbach GG, Schemske DW, Cornell HV et al (2007) Evolution and the latitudinal diversity gradient: speciation, extinction and biogeography. Ecol Lett 10:315–331. https://doi.org/10.1111/j.1461-0248.2007.01020.x

Moore JW, Schindler DE (2008) Biotic disturbance and benthic community dynamics in salmon-bearing streams. J Anim Ecol 77:275–284. https://doi.org/10.1111/j.1365-2656.2007.01336.x

Moyle PB, Cech JJJ (2006) Fishes: an introduction to ichthyology. Prentice-Hall, Englewood Cliffs, NJ

Moyle PB, Herbold B (1987) Life-history patterns and community structure in stream fishes of western North America: comparisons with eastern North America and Europe. In: Matthews WJ, Heins DC (eds) Community and evolutionary ecology of North American stream fishes. University of Oklahoma Press, Norman

Naeem S, Wright JP (2003) Disentangling biodiversity effects on ecosystem functioning: deriving solutions to a seemingly insurmountable problem. Ecol Lett, 567–579

Noble RAA, Cowx IG, Goffaux D, Kestemont P (2007) Assessing the health of European rivers using functional ecological guilds of fish communities: standardising species classification and approaches to metric selection. Fish Manag Ecol 14:381–392. https://doi.org/10.1111/j.1365-2400.2007.00575.x

Oberdorff T, Dias MS, Jézéquel C et al (2019) Unexpected fish diversity gradients in the Amazon basin. Sci Adv 5:1–10. https://doi.org/10.1126/sciadv.aav8681

Oberdorff T, Guégan J-F, Hugueny B (1995) Global scale patterns of fish species richness in rivers. Ecography (Cop) 18:345–352

Oberdorff T, Hugueny B, Guégan J-F (1997) Is there an influence of historical events on contemporary fish species richness in rivers? J Biogeogr 24:461–467

Oberdorff T, Tedesco PA, Hugueny B et al (2011) Global and regional patterns in riverine fish species richness: a review. Int J Ecol 2011:967631. https://doi.org/10.1155/2011/967631

Olden JD, Leroy Poff N, Bestgen KR (2006) Life-history strategies predict fish invasions and extirpations in the Colorado River Basin. Ecol Monogr 76:25–40. https://doi.org/10.1890/05-0330

Padial AA, Ceschin F, Declerck SAJ et al (2014) Dispersal ability determines the role of environmental, spatial and temporal drivers of metacommunity structure. PLoS One 9:1–8. https://doi.org/10.1371/journal.pone.0111227

Palmer MA, Covich AP, Finlay BJ et al (1997) Biodiversity and ecosystem processes in freshwater sediments. Ambio 26:571–577. https://doi.org/10.2307/4314671

Pearson RG, Boyero L (2009) Gradients in regional diversity of freshwater taxa. J North Am Benthol Soc 28:504–514. https://doi.org/10.1899/08-118.1

Pease AA, González-Díaz AA, Rodiles-Hernández R, Winemiller KO (2012) Functional diversity and trait-environment relationships of stream fish assemblages in a large tropical catchment. Freshw Biol 57:1060–1075. https://doi.org/10.1111/j.1365-2427.2012.02768.x

Peckarsky BL (1983) Biotic interactions or abiotic limitations? A model of lotic community structure. In: Fontaine TI, Bartell S (eds) Dynamics of lotic ecosystems. Ann Arbor Science, Ann Arbor, MI, pp 303–323

Peckarsky BL, McIntosh AR, Horn SC et al (2014) Characterizing disturbance regimes of mountain streams. Freshw Sci 33:716–730. https://doi.org/10.1086/677215

Poff NL (1997) Landscape filters and species traits: towards mechanistic understanding and prediction in stream ecology. J North Am Benthol Soc 16:391–409. https://doi.org/10.2307/1468026

Poff NL, Allan JD (1995) Functional organization of stream fish assemblages in relation to hydrological variability. Ecology 76:606–627. https://doi.org/10.2307/1941217

Poff NL, Olden JD, Vieira NKM et al (2006) Functional trait niches of North American lotic insects: traits-based ecological applications in light of phylogenetic relationships. J North Am Benthol Soc 25:730–755

Razeng E, Morán-Ordóñez A, Brim Box J et al (2016) A potential role for overland dispersal in shaping aquatic invertebrate communities in arid regions. Freshw Biol 61:745–757. https://doi.org/10.1111/fwb.12744

Ricklefs RE, Schluter D (1993) Species diversity in ecological communities. Oxford University Press, Oxford

Riseng CM, Wiley MJ, Stevenson RJ (2004) Hydrologic disturbance and nutrient effects on benthic community structure in midwestern US streams: a covariance structure analysis. J North Am Benthol Soc 23:309–326. https://doi.org/10.1899/0887-3593(2004)023%3c0309:HDANEO%3e2.0.CO;2

Rosenzweig ML (1995) Species diversity in space and time. Cambridge University Press, Cambridge

Ross ST, Matthews WJ, Echelle AA (1985) Persistence of stream fish assemblages: effects of environmental change. Am Nat 122:583–601. https://doi.org/10.1086/284393

Schlosser IJ (1987) The role of predation in age- and size-related habitat use by stream fishes. Ecology 68:651–659. https://doi.org/10.2307/1938470

Schmera D, Árva D, Boda P et al (2018) Does isolation influence the relative role of environmental and dispersal-related processes in stream networks? An empirical test of the network position hypothesis using multiple taxa. Freshw Biol 63:74–85. https://doi.org/10.1111/fwb.12973

Schmera D, Heino J, Podani J et al (2017) Functional diversity: a review of methodology and current knowledge in freshwater macroinvertebrate research. Hydrobiologia 787:27–44. https://doi.org/10.1007/s10750-016-2974-5

Schmid-Araya JM, Hildrew AG, Robertson A et al (2002) The importance of meiofauna in food webs: evidence from an acid stream. Ecology 83:1271–1285

Schwartz JS, Herricks EE (2005) Fish use of stage-specific fluvial habitats as refuge patches during a flood in a low-gradient Illinois stream. Can J Fish Aquat Sci 62:1540–1552. https://doi.org/10.1139/f05-060

Sepkoski JJ, Rex MA (1974) Distribution of freshwater mussels: coastal rivers as biogeographic islands. Syst Zool 23:165–188. https://doi.org/10.2307/2412130

Smith GR, Badgley C, Eiting TP, Larson PS (2010) Species diversity gradients in relation to geological history in North American freshwater fishes. Evol Ecol Res 12:693–726

Smucker NJ, Vis ML (2011) Contributions of habitat sampling and alkalinity to diatom diversity and distributional patterns in streams: implications for conservation. Biodivers Conserv 20:643–661. https://doi.org/10.1007/s10531-010-9972-0

Soininen J, Heino J, Kokocinski M, Muotka T (2009) Local-regional diversity relationship varies with spatial scale in lotic diatoms. J Biogeogr 36:720–727. https://doi.org/10.1111/j.1365-2699.2008.02034.x

Southwood TRE (1988) Tactics, strategies and templets. Oikos 52:3–18. https://doi.org/10.2307/3565974

Starnes W, Etnier D (1986) Drainage evolution and fish biogeography of the Tennessee and Cumberland Rivers drainage realm. In: Hocutt CH, Wiley E (eds) The zoogeography of North American Freshwater Fishes. Wiley Interscience, New York, pp 325–361

Statzner B, Bêche LA (2010) Can biological invertebrate traits resolve effects of multiple stressors on running water ecosystems? Freshw Biol 55:80–119. https://doi.org/10.1111/j.1365-2427.2009.02369.x

Statzner B, Bonada N, Dolédec S (2008) Biological attributes discriminating invasive from native European stream macroinvertebrates. Biol Invasions 10:517–530. https://doi.org/10.1007/s10530-007-9148-3

Statzner B, Bonada N, Dolédec S (2007) Conservation of taxonomic and biological trait diversity of European stream macroinvertebrate communities: a case for a collective public database. Biodivers Conserv 16:3609–3632. https://doi.org/10.1007/s10531-007-9150-1

Stegmann LF, Leitão RP, Zuanon J, Magnusson WE (2019) Distance to large rivers affects fish diversity patterns in highly dynamic streams of Central Amazonia. PLoS One 14:1–17. https://doi.org/10.1371/journal.pone.0223880

Stendera SES, Johnson RK (2005) Additive partitioning of aquatic invertebrate species diversity across multiple spatial scales. Freshw Biol 50:1360–1375. https://doi.org/10.1111/j.1365-2427.2005.01403.x

Stout J, Vandermeer J (1975) Comparison of species richness for stream-inhabiting insects in tropical and mid-latitude streams. Am Nat 109:263–280. https://doi.org/10.1086/282996

Strayer DL, Dudgeon D (2010) Freshwater biodiversity conservation: recent progress and future challenges. J North Am Benthol Soc 29:344–358. https://doi.org/10.1899/08-171.1

Tedesco PA, Oberdorff T, Lasso CA et al (2005) Evidence of history in explaining diversity patterns in tropical riverine fish. J Biogeogr 32:1899–1907. https://doi.org/10.1111/j.1365-2699.2005.01345.x

Thomaz AT, Malabarba LR, Bonatto SL, Knowles LL (2015) Testing the effect of palaeodrainages versus habitat stability on genetic divergence in riverine systems: study of a Neotropical fish of the Brazilian coastal Atlantic Forest. J Biogeogr 42:2389–2401. https://doi.org/10.1111/jbi.12597

Thompson R, Townsend C (2006) A truce with neutral theory: local deterministic factors, species traits and dispersal limitation together determine patterns of diversity in stream invertebrates. J Anim Ecol 75:476–484. https://doi.org/10.1111/j.1365-2656.2006.01068.x

Tisseuil C, Cornu JF, Beauchard O et al (2013) Global diversity patterns and cross-taxa convergence in freshwater systems. J Anim Ecol 82:365–376. https://doi.org/10.1111/1365-2656.12018

Tolonen KE, Leinonen K, Marttila H et al (2017) Environmental predictability of taxonomic and functional community composition in high-latitude streams. Freshw Biol 62:1–16. https://doi.org/10.1111/fwb.12832

Tolonen KE, Tokola L, Grönroos M et al (2016) Hierarchical decomposition of trait patterns of macroinvertebrate communities in subarctic streams. Freshw Sci 35:1032–1048. https://doi.org/10.1086/687966

Tomanova S, Goitia E, Helešic J (2006) Trophic levels and functional feeding groups of macroinvertebrates in neotropical streams. Hydrobiologia 556:251–264. https://doi.org/10.1007/s10750-005-1255-5

Tonkin JD, Altermatt F, Finn DS et al (2018) The role of dispersal in river network metacommunities: patterns, processes, and pathways. Freshw Biol 63:141–163. https://doi.org/10.1111/fwb.13037

Tonn WM, Magnuson JJ, Rask M, Toivonen J (1990) Intercontinental comparison of small-lake fish assemblages: the balance between local and regional processes. Am Nat 136:345–375. https://doi.org/10.1086/285102

Townsend CR (1989) The patch dynamics concept of stream community ecology. J North Am Benthol Soc 8:36–50. https://doi.org/10.2307/1467400

Townsend CR, Hildrew AG (1994) Species traits in relation to habitat templet for river systems. Freshw Biol 31:265–276. https://doi.org/10.1111/j.1365-2427.1994.tb01740.x

Townsend CR, Hildrew AG, Schofield K (1987) Persistence of stream invertebrate communities in relation to environmental variability. J Anim Ecol 56:597–613. https://doi.org/10.2307/5071

Turner JRG, Gatehouse CM, Corey CA (1987) Does solar energy control organic diversity? Butterflies, moths and the British climate. Oikos 48:195–205. https://doi.org/10.2307/3565855

Usseglio-Polatera P, Bournaud M, Richoux P, Tachet H (2000) Biological and ecological traits of benthic freshwater macroinvertebrates: relationships and definition of groups with similar traits. Freshw Biol 43:175–205

Verberk WCEP, Van Noordwijk CGE, Hildrew AG (2013) Delivering on a promise: integrating species traits to transform descriptive community ecology into a predictive science. Freshw Sci 32:531–547. https://doi.org/10.1899/12-092.1

Vieira NKM, Poff NL, Carlisle DM et al (2006) A Database of lotic invertebrate traits for North America. U.S. Geological Survey Data Series 187

Vinson MR, Hawkins CP (2003) Broad-scale geographical patterns in local stream insect genera richness. Ecography (Cop) 26:751–767. https://doi.org/10.1111/j.0906-7590.2003.03397.x

Walters AW, Post DM (2011) How low can you go? Impacts of a low-flow disturbance on aquatic insect communities. Ecol Appl 21:163–174. https://doi.org/10.1890/09-2323.1

Ward JV, Stanford JA (1983) The serial discontinuity concept of lotic ecosystems. In: Fontaine TI, Bartell JM (eds) Dynamics of lotic systems. Ann Arbor Science, Ann Arbor, MI, pp 29–42

Wilson DS (1992) Complex interactions in metacommunities, with implications for biodiversity and higher levels of selection. Ecology 73:1984–2000

Winemiller KO (2005) Life history strategies, population regulation, and implications for fisheries management. Can J Fish Aquat Sci 62:872–885. https://doi.org/10.1139/f05-040

Winemiller KO, Flecker AS, Hoeinghaus DJ (2010) Patch dynamics and environmental heterogeneity in lotic ecosystems. J North Am Benthol Soc 29:84–99. https://doi.org/10.1899/08-048.1

Winemiller KO, Rose KA (1992) Patterns of life-history diversification in North American fishes: implications for population regulation. Can J Fish Aquat Sci 49:2196–2218. https://doi.org/10.1139/f92-242

Woodward G, Bonada N, Brown LE et al (2016) The effects of climatic fluctuations and extreme events on running water ecosystems. Philos Trans R Soc B Biol Sci 371:20150274. https://doi.org/10.1098/rstb.2015.0274

Woodward G, Hildrew AG (2002) Food web structure in riverine landscapes. Freshw Biol 47:777–798. https://doi.org/10.1046/j.1365-2427.2002.00908.x

Woodward G, Jones JI, Hildrew AG (2002) Community persistence in Broadstone Stream (U.K.) over three decades. Freshw Biol 47:1419–1435. https://doi.org/10.1046/j.1365-2427.2002.00872.x

Wootton JT, Parker MS, Power ME (1996) Effects of disturbance on river food webs. Science 273:1558–1561. https://doi.org/10.1126/science.273.5281.1558

Wright DH (1983) Species-energy theory: an extension of species-area theory. Oikos 41:496–506. https://doi.org/10.2307/3544109

Zwick P (1992) Stream habitat fragmentation-a threat to biodiversity. Biodivers Conserv 1:80–97. https://doi.org/10.1007/BF00731036

Energy Flow and Nutrient Cycling in Aquatic Communities

Community connections that link species and support the flow of energy and nutrients between abiotic and biotic spheres of the environment arise from many interactions among organisms. For instance, food webs and energy flow diagrams depict the network of vertical and horizontal linkages extending from basal food resources to top consumers in an integrated visualization of a biological community. Although the number of connecting links can be very large, a modest number of species often make up the majority of the biomass and are responsible for most energy flow. Species also can be grouped together based on their functional similarity in regard to feeding ecology, productivity, capacity for elemental mineralization, etc., revealing their common roles in shaping the structure and function of ecosystems. Some species have complementary or overlapping roles, suggesting that they are functionally redundant for a given function, but at least in some cases certain species appear to be functionally irreplaceable. Thus, the potential loss of a species from communities due to over-harvest, habitat degradation, or other human actions implies that simplified biological communities may be functionally distinct from communities observed in their unaltered state.

In this chapter, we examine some of the ways in which scientists study interactions between aquatic communities and the flow of energy and nutrients. Specifically, we will focus on two concepts that are inextricably linked to one another: food web ecology and the functional role of communities in ecosystem processes.

12.1 Food Web Ecology

Food webs describe the network of consumer-resource interactions among a group of organisms (Layman et al. 2015). The network of interactions depicted in a food web provides one of the most complete, yet succinct, visual summaries of a biological community. The community is assembled from a regional species pool, as individual species are matched with available resources and habitats, influenced by disturbance and dispersal, and modified by the internal dynamics of species interactions. The study of food webs has a long history in ecology, incorporating at least two major lines of inquiry: one emphasizing how species are inextricably linked through their interactions with one another, hence focusing on population and species interactions as discussed in Chaps. 10 and 11; the other concerned with the flux of organic matter and energy that we will discuss here. Most examples can be categorized very roughly as connectance food webs, which attempt to identify all possible linkages; energy flux food webs, which quantify organic matter flow along a limited number of major pathways; and trophic-interaction food webs, which emphasize population processes and species interactions.

An impressively detailed food web for the Broadstone Stream in southern England provides an excellent example of a connectivity food web (Fig. 12.1). All species are of equal importance in a connectance web and all lines are of equal weight, because the web is constructed from diet presence-absence data. This food web is amongst the most detailed on record and includes all of the benthic community including macrofauna, meiofauna, protozoa, and algae (Schmid-Arraya et al. 2002; Woodward et al. 2005). It includes 131 consumer species supported by eight basal resources and three additional food sources such as eggs, for a total of 842 links. The meiofauna comprise 70% of the species present, demonstrating the need to include small-bodied organisms in this type of analysis. Food web structure varied seasonally due to changes in species richness, resulting in temporal changes in the proportion of species at the top and the base of the food web. Despite its complexity, relatively simple patterns in food web structure could be found in relation to body size. Meiofaunal and macrofaunal sub-webs were effectively two compartments because large prey were invulnerable to small predators and large predators were not effective in consuming very small prey (Woodward et al. 2005).

© Springer Nature Switzerland AG 2021
J. D. Allan et al., *Stream Ecology*,
https://doi.org/10.1007/978-3-030-61286-3_12

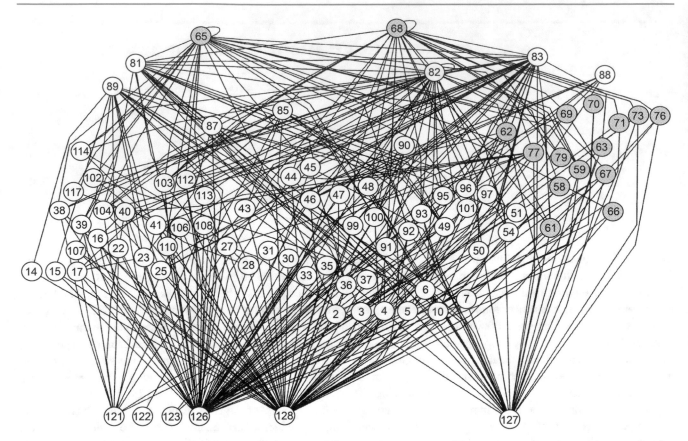

Fig. 12.1 Connectivity food web for the invertebrate community in Broadstone Stream, England in the autumn of 1996. Numbers represent consumers and resources: Protozoa: 1–10. Turbellaria: 11. Rotifera: 12–37. Nematoda: 38–41. Oligochaeta: 42–48. Tardigrada: 49–50. Acari: 51–56. INSECTA: Odonata: 57 *Cordulegaster boltonii*; Plecoptera: 58 *Leuctra nigra*, 59 *Leuctra hippopus*, 60 *Leuctra fusca*, 61 *Nemurella pictetii*, 62 *Siphonoperla torrentium*, 63 Plecoptera larvulae, 64 *Leuctra nigra* adult; Trichoptera: 65 Plectrocnemia conspersa, 66 *Potamophylax cingulatus*, 67 *Adicella reducta*; Megaloptera: 68 *Sialis fuliginosa*; Coleoptera: 69 *Platambus maculatus*, 70 Helodidae sp., 71 Elmidae sp.; Diptera: Ceratopogonidae 72 *Bezzia* sp.; Tipulidae: 73 *Limonia* sp., 74 *Limonia modesta*, 75 *Dicranota* sp., 76 *Pedicia* sp., 77 *Limnophila* sp., 78 *Hexatoma* sp., 79 Limoniinae Gen. sp.; 80 *Rhypholophus* sp.; Chironomidae: 81 *Macropelopia nebulosa*, 82 *Trissopelopia longimana*, 83 *Zavrelymyia barbatipes*, 84 *Conchapelopia viator*, 85 *Apsectrotanypus trifascipennis*, 86 *Zavrelymyia* sp. 2, 87 *Paramerina* sp., 88 *Krenopelopia* sp., 89 *Pentaneura* sp., 90 *Natarsia* sp., 91 *Prodiamesa olivace*, 92 *Brillia modesta*, 93 *Heterotrissocladius marcidus*, 94 *Heterotanytarsus* sp., 95 *Eukiefer-iella* sp., 96 *Georthocladius luteicornis*, 97 *Corynoneura lobata*, 98 Chirononomus/Einfeldia sp., 99 *Polypedilum albicorne*, 100 *Micropsectra bidentata*, 101 *Mectriocnemus* sp. Adult; Simulidae: 102 *Simulium* sp. CRUSTACEA: Ostracoda:103; Cladocera:104–105; Copepoda: Cyclopoida: 106–111; Harpacticoida: 112–116; Algae and Plant material: 118–123. Various: 123 Plecoptera eggs, 124 Turbellaria eggs, 125 Rotifera eggs, 126 Fine particulate organic matter (FPOM), 127 Coarse particulate organic matter (CPOM), 128 *Leptothrix ochracea*. (Reproduced from Schmid-Araya et al. 2002)

Measurement of energy flux provides a quantitative assessment of the strength of linkages along each pathway. Figure 12.2 shows a less detailed food web, but one that quantifies organic matter pathways by converting information from gut analyses into annual ingestion rates for caddisfly larvae dwelling on snag habitat in the Ogeechee River in the southeastern US (Benke and Wallace 1997). The pathways from amorphous detritus and diatoms to several filter-feeding caddis larvae were particularly strong, but *Hydropsyche rossi* derived substantially more energy from consuming animal prey than did the other filter feeders.

12.1.1 Assessing Energy Flow Through Food Webs

As a food web essentially describes who eats what or whom, the construction of food webs requires information on diet. Traditionally, this has been inferred from general knowledge of trophic ecology (Chap. 9) or from detailed studies that combine gut analyses with estimates of assimilation efficiencies. More recently, feeding pathways have been inferred from chemical signatures in consumer tissues. These approaches are reviewed briefly below.

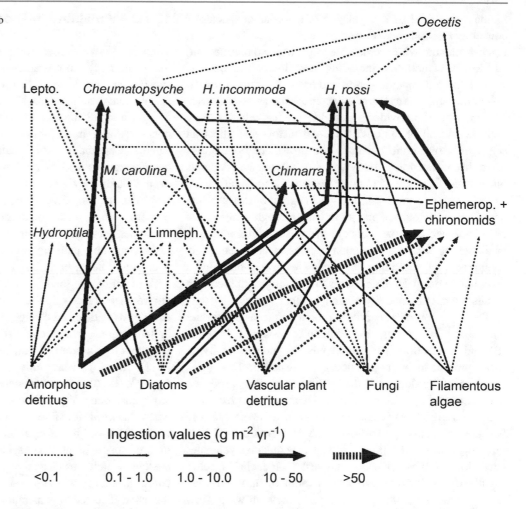

Fig. 12.2 Energy flow food web for caddisfly larvae in the Ogeechee River, Georgia, US. The width of the arrow reflects the magnitude of resources ingested. Abreviations: Ephemerop. = Ephemeroptera, Lepto. = Leptoceridae, Limneph. = Limnephilidae, *M. carolina* = *Macrostemum carolina*. (Reproduced from Benke and Wallace 1997)

12.1.1.1 Stable Isotopes and Other Dietary Indicators

Gut content analysis remains a widely used approach to the estimation of energy fluxes, despite the difficulty of identifying bits of soft-bodied prey and amorphous detritus, and the fact that it is extremely time intensive. To address some of these challenges, stable isotope analysis has become one of the primary methods to assess food resources, trophic position, and energy flow (Layman et al. 2012, 2015; Vander Zanden et al. 2016). Stable isotopes of carbon (δ^{13}C), sulfur (δ^{34}S), nitrogen (δ^{15}N), hydrogen (δ^{2}H) and oxygen (δ^{18}O) can provide relevant information on major food sources, trophic level, and species movements. They are especially useful because they provide information about the type and location of resources that consumers use through time, generating spatial and temporal insights into trophic relationships and supporting the development of models of trophic structure (Layman et al. 2012).

The isotopes of different elements are analyzed to answer different questions. Carbon and sulfur isotopes are useful to determine food sources, because their isotopic ratios vary among primary producers. Nitrogen isotopes are useful to examine trophic relationships among species within a community, because they change with each trophic transfer. Stable hydrogen isotopes can help to determine the relative contribution of allochthonous and autochthonous nutrient sources, and provide additional insights about trophic relationships. In combination with δ^{2}H, δ^{18}O has the potential to generate information about food webs; however, oxygen isotopes are primarily affected by the source of water in a system rather than by diet (Vander Zanden et al. 2016). Stable isotope analysis can provide quantitative information about diet variation within a species, trophic composition, food chain length, and anthropogenically-derived shifts in community structure (Layman et al. 2012). In addition, other methods such as DNA sequencing of stomach contents or feces, fatty acids profiles, biomarkers, and stable isotopes of specific compounds can also be applied to quantify the composition of diet (Nielsen et al. 2018).

12.1.1.2 Secondary Production and Ingestion

Biological production is a measure of biomass produced within some component of an ecosystem. The most basic distinctions are between primary production, microbial production, and the production of all higher trophic levels, termed secondary production. Secondary production is the amount of biomass produced by a population of a consumer species or consumer assemblage over some time period (Benke 1993). It is a net amount, with accrual due to growth and recruitment, and losses to mortality, including predation and disease, as well as emigration, emergence, and tissue loss during starvation or molting. It is a rate, often expressed on an annual basis. The growth of individuals depends on individual energy budgets and all of the environmental factors that affect an organism's metabolic balance. The growth of populations through reproduction and survival of recruits likewise depends on the many factors that influence the dynamics of populations. Thus, secondary production is an integrative measure that describes the amount of energy flowing through a population or assemblage at some location. When such information is available for the most common species in an assemblage, comparison of secondary production estimates provides a quantitative perspective on energy flow through that food web and its overall productivity relative to other studies.

The concept of secondary production can be further understood by noting the difference between biomass (B) and production (P). Biomass is the amount of living tissue present in the population at an instant in time (e.g., grams m^{-2}). Production is the amount of biomass produced over time (e.g., grams m^{-2} yr^{-1}). Their ratio, expressed as P/B, is referred to as biomass turnover or turnover rate, and has the units time^{-1}. Turnover rate is a measure of how rapidly growth and reproduction of a population generates new biomass. Fast growing and fast reproducing species, usually of small body size, may exhibit biomass turnover several to ten or more times in a year. Slow growing, slow reproducing species, usually of large body size, may turnover the population's biomass fewer than once per year (Benke and Huryn 2010).

Secondary production of a population is estimated by measuring population size, typically per unit of area or habitat, at frequent intervals, often monthly for aquatic insects, and converting numbers to biomass using published length–weight regressions (Benke 1993). Changes in monthly biomass are summed to provide an estimate of annual production. Alternatively, estimates of individual growth rates can be combined with biomass data. In an additional analysis, the relative contribution of each food web pathway to the secondary production of each species or taxon (analyses can be at the genus or family level) can be determined by quantification of gut contents and applying an estimate of assimilation efficiency to each food type. By quantifying both the secondary production and the contribution of each resource category to the energy assimilated by each of the most important members of an assemblage, one gains considerable insight into the relative importance of various pathways in food webs (Benke 2018).

12.1.2 Variability in Food Web Structure

Though freshwater food webs vary across space and time, they commonly are characterized by high generalism (each species has many food links) and redundancy (multiple species have similar trophic roles). They also typically contain a small number of larger predators in top trophic positions, and many smaller organisms as primary consumers. Cannibalism and mutual predation are also frequently documented in food webs (Ings et al. 2009). Yet, even though they often share many commonalities, food webs can be very diverse, particularly in their quantitative properties (e.g., abundance and biomass of resources and consumers, strength of interactions among species and resources) rather than in their composition, which can persist over time (Olesen et al. 2010).

Variability in food web structure arises from many environmental factors. In some cases, studies conducted over decades have detected changes in food webs due to variation in temperature, water chemistry, and the addition of top predators to a system (Woodward et al. 2010b; Layer et al. 2011). In an 18-year study of streams in California, US, Power et al. (2008) documented greater biomass of the filamentous green algae *Cladophora* in the spring of years when bankfull flooding occurred during winter months. They attributed greater algal biomass to reduced populations of predator-resistant grazers, such as the caddisfly *Dicosmoecus*, which was negatively impacted by high flows. In contrast, smaller winter floods or spring spates resulted in the short-distance downstream transport of *Dicosmoecus*, rather than increased mortality rates. Therefore, higher grazer abundance reduced biomass of *Cladophora* during the summer in lower-flow years. Food web interactions of fishes were also affected by flooding in this system. During post-flood periods, fishes had a stronger impact on benthic invertebrate populations than during drought years. Thus, flow-mediated changes in fish behavior also indirectly influenced algal populations.

In other cases, researchers have used space-for-time substitutions along a gradient of sites representing potential future conditions (e.g., water temperature gradients, urbanization gradients) to evaluate the responses of food webs to changing environmental conditions. For example, Woodward et al. (2010a) used geothermal streams with a water temperature gradient from 5 to 43 °C to assess the potential impact of climate change on trophic interactions in streams. Brown trout *Salmo trutta* fed on chironomids in the colder streams but fed primarily on the snail *Radix peregra* and the larval black fly *Simulium vitattum* in warmer streams. Trout were larger in warmer streams, suggesting that trout may shift their diet to larger prey in warmer systems to meet higher metabolic demand. Similarly, researcher compared food webs of 20 streams along a pH gradient (5.0 to 8.4) to examine how changes in acidity influenced stream community interactions (Layer et al. 2013). Herbivore-detritivore generalists were found in each stream; however, these taxa were dominant in the more acidic sites. In less acidic (higher pH) streams, the diversity and biomass of algae and the abundance of specialist grazers increased. At higher pH levels, specialist grazers and herbivore-detritivores consumed more algae, indicating shifts in dominant basal food resources along an acidity gradient. This indicates that generalist species were able to shift from a detritus-dominant to an algal-dominant diet along a gradient of pH, and suggests the lack of specialist grazers in acidic streams may contribute to the slow recovery of streams to anthropogenically-derived acid deposition.

As discussed in detail in previous chapters, basal resources, productivity, and habitat conditions are all associated with position along a river continuum; hence, food web structure should also change with landscape position (Woodward and Hildrew 2002). Landscape features govern the physical characteristics of streams, influencing flow regime and habitat features, the relative importance of basal resources in food webs, the magnitude of external subsidies of nutrients, detritus, and prey, and conveying a high degree of individuality to the species assemblage and food web structure among locales (Woodward and Hildrew 2002). Models of stream ecology, particularly the river continuum concept (Fig. 1.1), attempt to explain variation in structure and function of stream communities due to longitudinal patterns and shifts in energy inputs between forested and open sites (Vannote et al. 1980). In brief, the river continuum concept predicts that allochthonous sources support food webs in low-order streams because of their small size and connectivity to terrestrial environments, autochthonous sources increase in importance in middle reaches as the river widens and shading of the river channel declines, and large rivers are strongly dependent on material exported from upstream and lateral connectivity. The influence of the riparian corridor on basal resource availability is well known, especially in small, temperate streams where the extent of summer shade, autumn leaf-fall, and spring leaf-out influence the patterns in allochthony and autochthony. In a study conducted in two Canadian rivers using stable carbon (δ^{13}C) and nitrogen (δ^{15}N) isotopes, Hayden et al. (2016) observed that most invertebrate and fish consumers relied on autochthonous sources along the river network, except for shredder species, which primarily depended upon allochthonous organic matter.

Anthropogenic activities may also shift food web relationships with regard to allochthony and autochthony in river networks. Along the Pecos River, in the southwestern United States, East et al. (2017) found that in middle reaches of the river, fish and macroinvertebrate populations consumed carbon from both autochthonous and riparian sources, and the food webs had few piscivorous species. However, in downstream reaches affected by salinization that was associated with flow alteration, food chains were shorter and consumers were predominantly supported by instream carbon sources. Reduced discharge and habitat availability likely were responsible, because flows became intermittent and reaches were reduced to isolated pools. Downstream, where the river receives perennial fresh inputs from springs, the authors documented decreases in salinity, increases in pool habitats, and a return to longer food chains and dependence on both allochthonous and autochthonous resources, reflecting a discontinuum along the Pecos River.

Urbanization may also affect stream trophic structure (El-Sabaawi 2018). Primary production in urban watersheds can be impacted by increased nutrient loading and by the reduction of shade as riparian forests are removed for development. Increases in the intensity of storm discharge can scour streambeds, removing algae and displacing grazing invertebrate species. Changes in riparian plant composition through the introduction of species, wastewater introduction through aging and obsolete infrastructure, and the loss of sensitive herbivorous and detritivorous species can also change basal food resource availability and consumer diversity and abundance in urban streams. For example, using stable isotope analysis to examine differences in aquatic communities receiving wastewater discharge, Singer and Battin (2007) found that sewage-derived particulate organic matter doubled macroinvertebrate secondary production relative to reaches without discharge.

12.1.3 Factors Influencing Secondary Production

Organisms of small body size typically exhibit rapid growth and development, have short life spans, and have multiple generations per year relative to those of large body size. This is illustrated in Fig. 12.3 for several aquatic insect taxa in a tropical stream where warm temperatures and small body size result in very high growth rates (Hall et al. 2011). It is also well established that ectotherms generally exhibit more rapid growth with increasing temperature. The relationship between body size, body temperature, and metabolic rates, and the tendency for small-bodied organisms to have higher mass-specific metabolic rates than larger-bodied organisms, leads to predictions about the scaling of secondary production measures to body size. Estimates of total invertebrate secondary production ranged from 5 to 25 g m^{-2} yr^{-1} based on a review of a number of studies of low-order streams in North America (Walther and Whiles 2011), although higher values have been recorded (Benke 1993). The turnover rate (P/B) of biomass for a single generation (cohort) of an individual species often is roughly five (Benke 1993), and so a population with an annual life cycle will have an annual P/B of roughly five as well. A population with two cohorts per year will have an annual P/B of about 10, and small-bodied, very fast growing species may have an annual P/B of 20 or more. Indeed, estimates of annual biomass turnover rates for individual species of benthic freshwater macroinvertebrates vary widely, with the majority between two and 20, but both lower and higher values have been reported. The entire macroinvertebrate assemblage at a site will include both slow and fast-growing species, and so assemblage P/B will tend towards intermediate values.

The influence of temperature on the variation in secondary production among populations is clearly evident from estimates of fish assemblage production across cold, cool, and warm water streams of the Appalachian Mountains, US (Myers et al. 2018). Only two streams in this study were considered warm-water, and they had higher biomass and production. Production-to-biomass (P/B) ratios ranged from 0.20 to 1.07, similar to other reported studies, although some higher values have been reported elsewhere. As the mean P/B for all 25 streams combined was 0.65, these fish assemblages turned over more than half of their total biomass annually. A synthesis of a large data base of fish production studies estimated a median P/B for fish assemblages at 0.86 and for individual species of 1.2, indicating that most freshwater fish communities and populations of

single species turn over approximately annually (Rypel and David 2017). Their synthesis of a large number of studies indicated that fish assemblage production showed a decreasing trend with higher latitude. However, much scatter was evident across intermediate (temperate) latitudes, possibly due to local differences in ecosystem productivity and fish assemblage diversity. Individual species may exhibit a counter-gradient trend of production with latitude, suggestive of physiological adaptation for faster individual growth rates during short growing seasons.

In addition to temperature and body size, which exert fundamental control over secondary production and turnover rates, species are likely to be influenced by a wide variety of environmental factors. Reflecting the truism that species are most successful in those environments to which they are best adapted, and do less well in environments that we call marginal based on species' performance, much variation in secondary production estimates no doubt reflects the environmental conditions under which production is measured. By the same token, secondary production serves as a useful measure of suitable conditions, identifying productive habitats for a species. In essence, all of the environmental variables that influence individual growth and reproduction, and the dynamics of populations, will also influence secondary production (Patrick et al. 2019).

Numerous studies have documented variation in secondary production of aquatic invertebrates in response to habitat conditions. Sampling of different habitats allows a weighted-area estimate of total invertebrate production and provides insight into the relative importance of different habitat types for different functional groups. Comparing bedrock-outcrop, riffle, and pool habitats in an Appalachian mountain stream in the southeastern US, Huryn and Wallace (1987) found highest production of collector-filterers in bedrock-outcrop habitats, followed by riffles, and then pools, whereas shredders showed the opposite, and scrapers were most productive in riffle habitats. Annual invertebrate production on a floodplain continuously inundated over nine months was several-fold higher than one flooded only occasionally during storms, and floodplain production was one to two orders of magnitude higher than in channels (Gladden and Smock 1990). Annual production of macroinvertebrate communities along a first- to seventh-order river continuum in the southern Appalachian Mountains, US, displayed large variation among habitats (Fig. 12.4; Grubaugh et al. 1997). Production was relatively low in depositional habitats and some gravel substrates, greater in bedrock habitats, and greatest on plant-covered cobbles, which can stabilize the substrate and enhance access of collector-filtering invertebrates to entrained food resources. After accounting for the proportional availability of habitats along the continuum, estimates of total production increased significantly with stream size. Annual production estimates for sixth- and seventh-order reaches

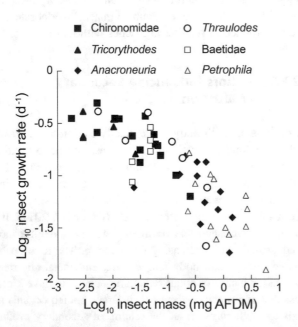

Fig. 12.3 Insect specific growth rates declined with body size in in-stream incubations. Growth rates were estimated using 1–5 individuals that were incubated in a mesh container between 3–7 days. For smaller individuals, 3–4 individuals were incubated together. Each point represents a mean value collected from 3–6 individuals in cages, or from individual organisms (i.e., *Anacroneuria* and *Petrophila*). (Reproduced from Hall et al. 2011)

Fig. 12.4 Estimates of the relative contributions of functional feeding groups to habitat weighted, secondary production in the benthic macroinvertebrate community along the Ball Creek-Coweeta Creek-Little Tennessee River continuum in the southeastern US. The contributions of the feeding groups are depicted as percentages of total habitat-weighted production at each site. The data were arcsine transformed before plotting. The lines represent the best-fit relationships between the catchment area and the contribution of each functional feeding group to production: (**a**) scrapers, (**b**) shredders, (**c**) gatherers, (**d**) filterers, (**e**) predators. (Reproduced from Grubaugh et al. 1997)

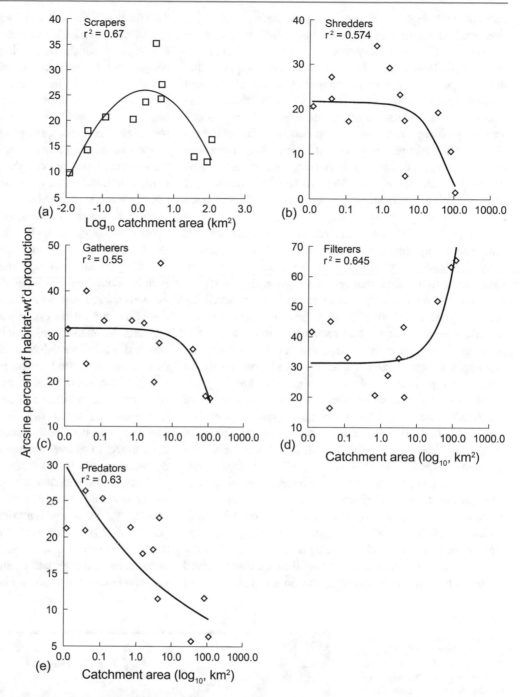

were amongst the highest reported for lotic systems. Moreover, production estimates for individual functional feeding-groups, estimated by proportional habitat weighting, generally supported predictions of the river continuum concept.

12.1.4 Spatial Subsidies and Aquatic Community Response

Evidence from a variety of ecosystems reveals that nutrients and energy flowing across habitat boundaries, or spatial subsidies, can fundamentally alter food webs and biogeochemical cycling in recipient ecosystems (Fig. 12.5; Polis et al. 1997; Polis and Hurd 1996). In its inception, spatial subsidies were defined as a donor-controlled resource (e.g., animal tissue, leaf litter and other plant tissues, etc.) moving from one habitat to another, resulting in greater system productivity and affecting consumer–resource dynamics (Layman et al. 2015). Energy flow across habitat boundaries has been a central point of discussion in stream ecology for many decades. Likens and Bormann (1974) were amongst the first to quantify the influence of terrestrial inputs of nutrient and organic matter on ecosystem function in streams. Foundational concepts in stream ecology, including the river continuum (Vannote et al. 1980) and flood-pulse (Junk et al. 1989), focus on in-stream energy supplies that depend upon the flow of materials between terrestrial and aquatic environments. Food web studies have added significantly to our understanding of the magnitude of various subsidy pathways and how these pathways differ in importance with environmental context, particularly landscape setting. Energy subsidies to fluvial ecosystems that can be extremely important to ecosystem metabolism include leaf litter, the infall of terrestrial invertebrates, and the carcasses and reproductive products of migrating fishes.

Spatial subsidies are expected to have their greatest influence on aquatic communities and ecosystem processes when the donor and recipient systems substantially differ in their productivity, such that resources move from a highly productive system to a system of much lower productivity (Layman et al. 2015). Additionally, the characteristics of the boundary between donor and recipient systems will influence the lability, quantity, timing, and duration of a subsidy entering the new system (Subalusky and Post 2019). In streams, the perimeter: volume ratio may be especially relevant in understanding the potential influence of terrestrially-derived subsidies on rivers and streams. As highlighted in the river continuum concept, energy dynamics in smaller, headwater streams in forested regions of the world are expected to be fueled by allochthonous inputs from riparian vegetation, whereas floodplain subsidies can be important to lowland rivers with intact floodplains.

Allochthonous inputs of leaf litter and other plant products, extensively reviewed in Chap. 7, are an important and well-studied terrestrial subsidy to aquatic ecosystems. However, except in the case of migrating fishes, animal-derived subsidies have received less attention. In a recent review of the effects of animal-derived subsidies, Subalusky and Post (2019) developed a conceptual model (Fig. 7.1) describing the ways in which the net impact of subsidies on stream ecology is influenced by the characteristics of the resources crossing ecosystem boundaries, and the size, species diversity, productivity, resource availability, and seasonality of both the donor and recipient systems. In this model, lability (referring to resource quality), quantity, timing, and duration of material moving from a donor to a recipient system, and the characteristics of both the donor and recipient ecosystems, determine if the material crossing ecosystem boundaries functions as a subsidy, and whether it enhances or reduces productivity. Subsidy flow between systems can occur passively, through wind, atmospheric deposition, overland flow, and downstream flow; or by active processes including animal movement and from anthropogenic sources such as wastewater discharge (Subalusky and Post 2019).

Downstream transport of nutrients and organic matter by river flow and lateral exchanges with the floodplain during seasonal inundation are dominant physical processes by which subsidies are delivered. Increasing evidence suggests that animal-mediated subsides are diverse and support essential biogeochemical processes in lotic environments (Flecker et al. 2010). Subsidies introduced through animal movement are often rich in limiting nutrients, such as nitrogen and phosphorus. Additionally, many species have the ability to move upstream, bringing resources from downstream locations.

Anadromous and catadromous species that move between freshwater and marine environments, such as salmon, transfer marine-derived nutrients into freshwater systems (Levi et al. 2013; Tiegs et al. 2011; Holtgrieve and Schindler 2011). Salmon carcasses even provide a subsidy to riparian areas, seen in the growth of streamside trees, as a result of carcasses moved onto stream banks by predators and floods

Fig. 12.5 Generalized conceptual model of reciprocal flows of spatial subsidies (i.e., invertebrate prey and inputs of plant material) that have direct and indirect effects in both stream and riparian food webs. (Reproduced from Baxter et al. 2005)

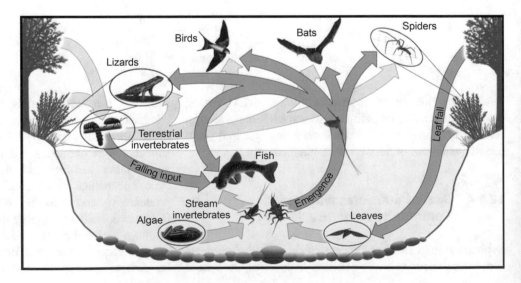

(Hocking and Reynolds 2012). Additionally, animals that move between terrestrial and aquatic environments via shifts in ontogeny, such as amphibians and many aquatic insects, or through daily migrations, such as hippos, actively move energy and nutrients across habitat boundaries [e.g., amphibians (Capps et al. 2015b; Luhring et al. 2017), hippos (Dutton et al. 2018)]. Large herds of migrating animals can also introduce vast quantities of energy and nutrients to river systems via mass drowning events (Wenger et al. 2019; Subalusky et al. 2017).

Inputs of terrestrial insects can be an important energy subsidy to top consumers such as fishes, as Nakano et al. (1999) demonstrated by placing a fine-mesh net over a forest stream in Japan. The exclusion of terrestrial invertebrates resulted in greater fish predation on benthic aquatic invertebrates, triggering a trophic cascade and an increase in periphyton biomass. Further, the biomass of herbivorous invertebrates and periphyton did not differ between treatments with or without fish when terrestrial invertebrate inputs were allowed, suggesting that the supply of arthropods from land normally prevented strong top-down control (Baxter et al. 2005). The complicated consequences for food web interactions due to external subsidies is also evident in a competitive shift in feeding by native Dolly Varden charr *Salvelinus malma* in northern Japan in response to non-native rainbow trout (Baxter et al. 2004). Trout were better able to consume the terrestrial invertebrates that fell into the stream, forcing char to feed on herbivorous invertebrates and resulting in an increase in periphyton biomass. However, the biomass of adult aquatic insects emerging from the stream decreased, and this in turn reduced the density of spiders in the riparian forest.

Subsidies can also travel from streams to the surrounding riparian area, as exemplified by the dependence of spiders and lizards in riparian zones on the emergence of aquatic insects, which can provide 25–100% of the energy required by those animals (Sabo and Power 2002; Baxter et al. 2005). Even the frequency of pollinator visits to riparian plants can be affected by fish predation on larval odonates, which determines the abundance of adult dragonflies and hence predation pressure on pollinating insects (Knight et al. 2005). Some riparian predators, like the fishing spider, lurk at the water's edge (Fig. 12.6), but aquatic subsidies can extend some distance into the riparian zone. Muehlbauer et al. (2014) defined the "stream signature" as the distance from the stream at which terrestrial organisms are receiving aquatic subsidies. The 50% signature refers to the distance where half of the food web energy of organisms within the terrestrial ecosystem comes from stream subsidies. In their analyses, the 50% signature was at 1.5 m, suggesting that most aquatic insects stay near the streams and most recipients are also near the stream. However, the 10% signature was at 550 m, indicating that a fraction of stream subsides extends much farther. Notably, the signatures vary among aquatic and terrestrial taxa, and chironomids, or midges, had the largest stream signature (Fig. 12.7).

For migratory and mobile species, stable isotope analysis has been useful in determining the location where carbon was assimilated, thus providing insight into the magnitude of cross-system subsidies. Differences in the $\delta^{13}C$ of muscle tissue in the migratory tropical fish *Semaprochilodus insignis* established that carbon produced in blackwater systems contributed to fish stocks harvested from whitewater systems of the Amazon (Benedito-Cecilio and Araujo-Lima

Fig. 12.6 A fishing spider (*Dolomedes)* hunts near the water's surface in a Costa Rican stream, as a group of Creek Tetras (*Bryconamericus*) swim below the surface. Photo by David Herasimtshschuk, Freshwaters Illustrated

Fig. 12.7 Estimates of the distance subsides travel into recipient landscapes, or "stream signatures" for different organisms. The values listed for each group of organisms represent the distances at which the abundance of a given group are the 50% and 10% of its near-stream levels. The horizontal lines depict the confidence intervals for each distance. Moving from the bottom to the top of the figure, the organisms are: terrestrial predators (webbed spiders, ground/hunting spiders and predaceous beetles [Coleoptera]), mayflies (Ephemeroptera), stoneflies (Plecoptera), caddisflies (Trichoptera), and midges (Chironomidae). The jagged line documents the furthest distance midges were collected in the study; therefore, distances past this point should be evaluated with caution. (Reproduced from Muehlbauer et al. 2014)

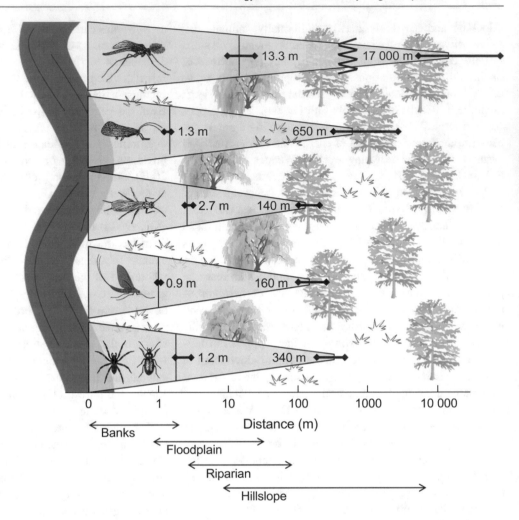

2002). Because epilithic algae, detritus, and algal filaments varied in abundance among benthic habitats and streams in the headwaters of the Eel River of northern California, Finlay et al. (2002) were able to use carbon isotope ratios to assess habitat use by different consumers. The $\delta^{13}C$ values of collector-gatherers and scrapers indicated a reliance on algae from local sources within their riffle or shallow pool habitats, whereas filter feeders derived more carbon from upstream shallow pools. Algal production from shallow pools was the dominant resource base for vertebrate predators in late summer regardless of the habitat where they were collected. The drift of pool insects into riffles, rather than movement of trout among habitats, was the presumed mechanism, illuminating between-habitat subsidies similar to cross-ecosystem subsidies.

12.1.5 Flow Food Webs and Ecosystem Processes

A flow food web constructed from annual production of consumers, proportions of food categories consumed, and

assimilation efficiencies for the different food types not only provides insights into basal resources, as described above, but also allows a detailed description of energy flow between resources and consumers (Marcarelli et al. 2011; Benke 2018). Figure 12.8 illustrates how the magnitude of flows between basal resources and consumers, indicated by thickness of arrows, reveals which consumer taxa have assimilated the greater fraction of energy available from each basal resource. As Hall et al. (2000) point out, their food webs encompassed most (84–91%) of invertebrate secondary production, but less than 30% of the estimated total links, which would require data for a large number of rare taxa. Because they assessed diets of invertebrates responsible for some 80–90% of total production, it seems likely that omission of rare taxa is not a hindrance in food web analysis.

Flow food webs for multiple sites along the Little Tennessee River, North Carolina, US, were structurally similar, because the sites had similar taxa and resources (Rosi-Marshall and Wallace 2002). However, the rates of secondary production of the taxa and the form of the dominant resource differed, and so these structurally similar webs were functionally very different. What was consumed, the

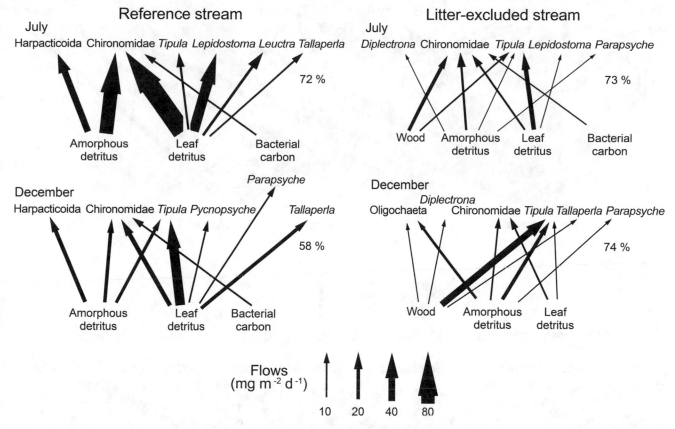

Fig. 12.8 Comparison of energy flows through stream food webs in reference streams and streams where leaf litter was experimentally excluded. The top panels represent the months of December. The percentage values on the right side of each food web are the percentages of organic mater flow represented in the diagram. The width of the arrows is proportional to the magnitude of the flows. (Reproduced from Hall et al. 2000)

rates at which food resources were consumed, and which taxa dominated the consumption of resources all differed along the river gradient.

The magnitude of individual energy flow pathways within a food web can be quantified from estimates of secondary production and food ingestion, as described earlier. Such quantification of energy flow among species and between trophic levels in an assemblage is significantly more informative that food web diagrams that simply show who eats whom, known as connectivity or linkage webs (Fig. 12.1). Although labor intensive, once such data are obtained, they can be used to answer at least three questions (Benke 2018). By comparing the amount of secondary production that is fueled by different basal resources, secondary production reveals the trophic basis for production. By comparing the magnitude of linkages from basal resources through primary consumers to predators, an energy flow food web reveals the dominant energy pathways and dominant actors in the web. Trophic position can be estimated by averaging all flow pathways, which is especially advantageous with omnivores. Finally, by comparing the amount of biomass consumed by a predator to the

production of its prey, one derives a measure of interaction strength, and possible top-down control. In this regard, food web analysis is an alternative to experimental manipulation as a way to detect strong interactors in a food web, and has an advantage in that experimental manipulation of all possible linkages is impractical or infeasible.

Flow food webs can be very detailed. The highly resolved snag web reported by Benke (2018) contained 462 quantified links ≥ 1 mg m^{-2} yr^{-1}. Detailed flow food webs were also constructed for invertebrate assemblages in experimental stream channels in southern England, which were subjected to simulated intermittent drought by 6-days of dewatering per month (Ledger et al. 2013a; Ledger et al. 2013b). Biomass fluxes from detritus were responsible for 96% of all energy transfers, including to predators, and just 5% of links accounted for 90% of biomass flux. As also seen in the Hall et al. (2000) study, the flow food web for just the dominant pathways provides an adequate description of the major linkages (Fig. 12.9). Experimental exposure to drought reduced the biomass flux by more than half, and resulted in substantial losses of species and links, especially among predators.

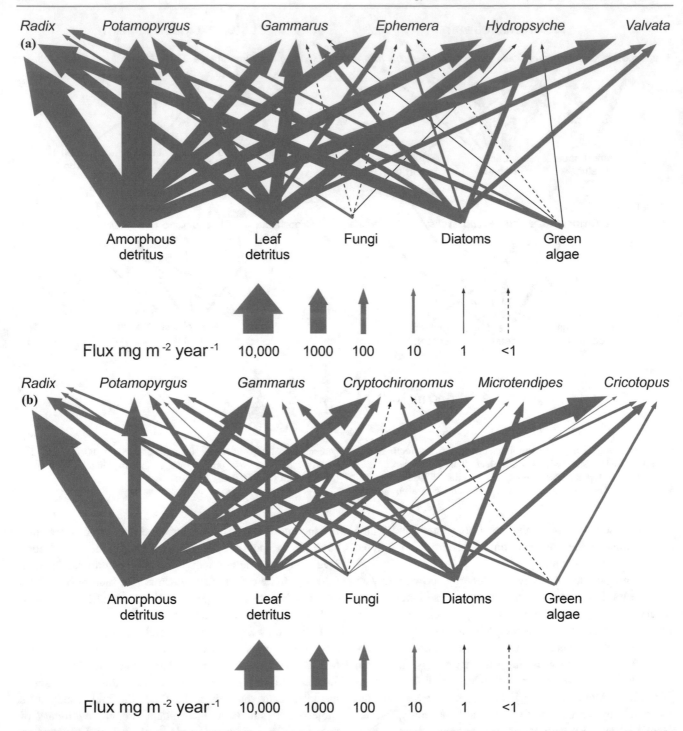

Fig. 12.9 Biomass fluxes from resources to consumers in a experimental manipulation to assess the effects of drought. The fluxes in (**a**) are from control treatments and in (**b**) are from drought-disturbed treatments. The width of arrows are proportional to the size of the annual flux on a per area basis. (Reproduced from Ledger et al. 2013a)

The information in a flow food web also can be used to assign taxa to trophic levels. Owing to omnivory and the utilization of resources at different trophic levels, species often cannot be assigned to an integer trophic level. Rather than using a linear sequence of trophic links, in which the placement of omnivores is problematic, trophic position (TP) can be determined from the trophic positions of all food resources consumed by a particular species, providing a mean estimate of how many links that species is removed from its basal resources. Trophic position above basal resources (assigned level 1) and their primary consumers (level 2) is calculated using the fractional amount of each food type assimilated by each consumer and TPs of all such food items. Using this approach, trophic categories and positions estimated from ingestion and production are fewer than reported for integer-based food chains. The snag food web analyzed by Benke (2018) had seven integer trophic levels based on linkages, but estimated trophic positions ranged only between 1.0 to 3.69. Most link-three species had calculated trophic position only slightly greater than two, and link-five species were only slightly above trophic position three.

Strengths of interactions within a flow food web can be estimated in biomass terms by dividing the rate of organic matter flux of a given food resource to a consumer by the production rate of that resource (Marcarelli et al. 2011; Benke 2018). Ingestion divided by production (I/P, both in units mg m^{-2} yr^{-1}, hence dimensionless) is estimated by dividing a predator's consumption of a given prey by that prey's production, providing a direct measure of predator impact. It thus is a measure of top-down interaction strength, typically used to determine predator impact on their prey base. This serves as an alternative to experimental manipulations, which have long been the traditional approach for estimating per capita interaction strength. A study from rocky intertidal habitat found that interaction strength estimated from secondary production data compared well with experimental findings (Wootton 1997).

Estimates of interaction strength have been used to identify strong and weak interactions within food webs, to assess how interaction strength can change with environmental conditions, and to make inferences about the relative importance of top-down versus bottom-up control over food webs. In Benke's (2018) analysis of the snag web, most impact measures were weak, implying that most predators had little effect on their prey. Other studies have found weak I/P to be common and strong interactions to be less so (Woodward et al. 2005; Cross et al. 2013).

Predator control of prey was stronger when basal resources were reduced in a whole-stream litter exclusion experiment (Hall et al. 2000). Because fewer prey taxa were available in the litter-excluded stream, the dominant flows were from the remaining common prey taxa to predators. In the reference stream with more prey available, predation likely was more diffuse and thus interaction strengths less strong. Interaction strengths were markedly different in food webs of the Colorado River among locations stretching 386 km within the Grand Canyon, Arizona, US (Cross et al. 2013). A few energy pathways dominated at upstream sites (from diatoms to a few invertebrate taxa to rainbow trout), energy efficiencies were low as only about 20% of invertebrate production was consumed by fishes, and the web showed a small number of relatively high interaction strength links. Farther downriver, invertebrate production was much lower, detritus was a more important basal

resource, energy transfer to higher trophic levels were more efficient, and food webs exhibited a higher frequency of weak interactions, defined as consumption/production < 0.1 (< 10%).

In any system where inputs and exports of organic matter and organisms are relatively modest, one expects the energy produced at lower trophic levels to be consumed by the ingestion and secondary production of higher trophic levels. In a leaf-litter exclusion study, predator production varied in accord with non-predator production (Wallace et al. 1997). Unsurprisingly, more energy produced at lower trophic levels generally results in more energy consumed at higher trophic levels. However, studies have reported that predators consume essentially all secondary production, only a small fraction of secondary production, or even that the food supply appears inadequate to support predator production. In a much cited example of the latter, based on study of the trout population of the Horokiwi stream in New Zealand, Allen (1951) noted an apparent paradox: the biomass of the trout population of this stream appeared to be much greater than its food supply could support. This came to be known as the Allen paradox and led to much further research into energy supply to top consumers.

There has to be an answer, of course, and eventually it came in three forms. First, evidence accumulated that aquatic food webs can receive external subsidies in the form of invertebrate prey entering streams from the surrounding riparian area (Baxter et al. 2005). Second, invertebrate production may previously have been underestimated, based on evidence that production and turnover (P/B ratios), especially of smaller invertebrates, can be higher than previously thought (Benke and Huryn 2010). Third, better accounting showed that, as one would expect, the productivity of lower trophic levels does provide a sufficient food supply for higher consumers, who may indeed consume all or nearly all of available secondary production. A comprehensive production budget for a trout stream in the southeastern part of South Island, New Zealand, similar in productivity to the Horokiwi stream, included estimates of primary production, production by brown trout and surficial and hyporheic macroinvertebrates, input of terrestrial invertebrates, and cannibalism by trout (Fig. 12.10; Huryn 1996). Food sources were indeed adequate to support a trout population typical of highly productive streams, but only when all budget compartments, including external inputs, were accounted for in the analysis.

When the highest trophic level consumes almost all of the secondary production of the trophic level below, and the lower trophic level consumes only a modest fraction of its basal resources, top-down control is strongly implicated. This appears likely in Huryn's (1996) study, as secondary production by aquatic invertebrates required only about 20% of total primary production, implying top-down control of

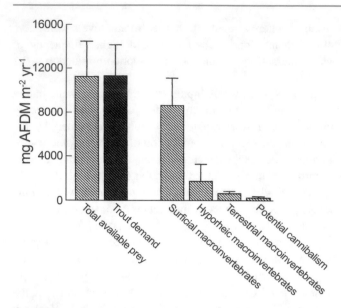

Fig. 12.10 Annual production estimates of the prey available for trout and source-specific prey availability. Trout demand was calculated using published data about the ecological efficiency of trout. The error bars depict 95% confidence intervals. (Reproduced from Huryn 1996)

invertebrates by trout. The inference of strong top-down control in productive New Zealand trout streams gained additional support from studies showing that macroinvertebrate biomass decreased markedly and periphyton biomass increased in experimental enclosures with trout (Flecker and Townsend 1994).

However, it is by no means assured that predators will have strong top-down influence. Production budgets for a New Zealand stream inhabited by native river galaxias (*Galaxias eldoni*) differed from trout stream budgets (Huryn 1998). Fish consumed approximately 18% of available prey production, and invertebrates consumed approximately 75% of primary production. Unlike the trout stream, which provided strong evidence of top-down control and surplus primary production, the galaxias stream likely experienced a mix of top-down and bottom-up controls, in which primary and secondary production were presumably mutually limiting. Experimental studies again were supportive, as the influence of galaxias in enclosures was less strong than the effect of trout (Flecker and Townsend 1994). In the snag food web reported by (Benke and Wallace 2015), invertebrate predators and omnivores consumed about 60% of all invertebrate production, implying that the remainder is available for fish consumption, adult emergence, and other mortality. Most fish predation presumably occurs when invertebrates depart from a snag into the drift, which could indicate that fishes are limited by their food supply rather than the reverse.

The ability of I/P estimates and related calculations to resolve the strength of predator control may be limited to some degree by measurement precision. Huryn's (1996) results suggesting that trout consumed 100% of prey

production raises a different question, namely what insect production remains available for the aerial stage's emergence and reproduction? Studies suggest that as much as 20% of production may be lost to emergence (Jackson and Fisher 1986; Statzner and Resh 1993) although it probably could be quite a bit less. Some mortality may be difficult to capture in budget calculations, including loss to parasitism and disease, burial in sediments, and so on. While comparative inferences among various pathways no doubt reveal which species are the strong interactors, and top-down control is highly likely when predators consume 80% or more of secondary production, measurement error cannot be discounted when making inferences from budget comparisons that have so many terms to estimate.

Anthropogenic activities have been shown to alter the flow of energy and nutrients through food webs. Reductions in secondary production of heptageniid mayflies provided evidence of the effects of chronic metal contamination on insect populations in Rocky Mountain (US) streams (Carlisle and Clements 2005, 2003). Using two measures of urban nonpoint pollution, stream-water conductivity and nutrient concentrations, Johnson et al. (2013) reported a decline in species richness but an increase in total invertebrate production, due to pollution-tolerant non-insect taxa (e.g. Oligochaeta, *Gammarus* sp., *Lirceus* sp., *Physa* sp.), some of which are resistant to macroinvertebrate predation.

12.2 Aquatic Communities and Ecosystem Function

Intact biological assemblages carry out various ecosystem processes including primary production, organic matter decomposition, nutrient cycling, and secondary production. This is thought to be the consequence of the presence of a diverse mix of species with different functional capabilities, and perhaps also with some level of functional redundancy that conveys resiliency in the face of an ever-changing environment. As species are lost from ecosystems due to the relentless pace of human activities, the dependence of system function and resilience on the number and characteristics of species present becomes an issue of considerable concern (Covich et al. 2004; Reid et al. 2019; Strayer and Dudgeon 2010).

12.2.1 Biodiversity and Ecosystem Function

The expectations that biodiversity matters to ecosystem function and also that high biodiversity serves as a buffer against the consequences of species loss have theoretical and empirical support (Loreau et al. 2002). Much of this work has been conducted in terrestrial ecosystems characterized by relatively low levels of diversity (Lecerf and Richardson 2010). Several mechanisms potentially are responsible for relationships between biodiversity and ecosystem function (Giller et al. 2004). When species have complementary (overlapping but not identical) roles, the rate and efficiency of a process should increase when multiple species are present, and especially whenever the activities of one species facilitate those of a second. Thus, any species loss is expected to lower the efficiency of the process in question. When species have redundant roles, the loss of one species may not immediately result in a decline in the rate or efficiency of some process, and the presence of multiple, ecologically similar species provides insurance against a breakdown in ecosystem function should one or more species be adversely affected by environmental change (Toussaint et al. 2016).

Much of the work devoted to understanding biodiversity-ecosystem function relationships has been done in the context of trying to predict functional changes in rivers and streams as native biodiversity is lost from systems. In a review, Vaughn (2010) summarizes some of the primary conclusions that have emerged. First, the assemblage of species traits within a community mediates ecosystem processes through a combination of overlapping and non-overlapping roles. Additionally, taxa that are classified in the same functional group are not always ecologically equivalent, meaning their contribution to ecosystem processes may differ. A decline in the abundance of a common species is comparable to species loss, as a species does not have to be extirpated from a system to lose its functional role in ecosystem processes. Research in biodiversity-ecosystem function relationships in freshwater systems has also emphasized that losses to native biodiversity can affect an entire food web, and that the net effects of biodiversity loss are also dependent on the environmental context and change with shifting spatial and temporal scales. Long-term research is essential to understand how environmental conditions influence the effects of biodiversity on ecosystem processes, and lab and field manipulations allow researchers to control for environmental variables and track energy and nutrients as they move between organisms and the environment. (Vaughn 2010).

Leaf litter breakdown, discussed in detail in Chap. 7, provides evidence that this important ecosystem function is influenced by both species richness and species identity (Kominoski et al. 2010). Using laboratory microcosms, Jonsson and Malmqvist (2000) demonstrated that leaf mass loss increased with number of shredder species present, due either to differences in mode of feeding, facilitation of feeding efficiency, or both. By assembling a dataset of litter breakdown studies from 36 streams of northern Sweden and northeastern France, Dangles and Malmqvist (2004) were able to evaluate the influence of species richness versus

relative abundance components of diversity. The litter decomposition rate increased with the number of species present, but at a lower rate at sites with low dominance. In other words, detrital processing was more rapid in streams that were strongly dominated by one or a few species. Other studies have shown that the presence or absence of a single species, *Gammarus fossorum*, has a disproportionately strong influence on litter breakdown, and it is noteworthy that the highest breakdown rate occurred in a stream where this amphipod was the sole shredder present.

The effects of biodiversity on ecosystem processes can be more pronounced in systems characterized by low diversity, as redundancy in the functional capabilities of taxa may be limited. In a study of 24 high altitude (3200–3900 m), low diversity streams in the tropics, Dangles et al. (2011) employed in-situ experiments and observational data with modelling to assess the influence of shredder diversity on decomposition. They found that decomposition rates were best predicted by a linear model that described rates of decomposition as a function of both the relative abundance of the most effective shredders and increasing shredder diversity. When their approach was tested with datasets from 49 French and Swedish streams with more complex shredder diversity, the effects of richness and identity observed in less diverse systems were lost as the shredder community became more complex.

In contrast to the demonstrated influence of shredder diversity on leaf breakdown, fungal diversity apparently had no effect of leaf mass loss or fungal spore production (a measure of microbial production) in stream microcosms (Dang et al. 2005). This was true with oak and alder leaves, at high and low nutrient levels, and across a range of from one to eight fungal species. There was some suggestion of greater variability in fungal activity at low fungal diversity, consistent with a portfolio effect (the averaging and dampening out of the influence of extreme species as richness increases, just as a diverse stock portfolio is expected to guard against swings in a single stock's value). Although Bärlocher and Corkum (2003) reported a positive effect of fungal diversity on leaf decomposition rates, their result is less convincing because those authors relied on initial inoculates rather than realized communities, which typically have greater unevenness.

The particular suite of traits and functional roles of each species must be carefully considered, as individual species can have unique roles in ecosystems, be disproportionately abundant, dominate energy fluxes, and strongly influence other members of the assemblage. In some instances, individual species have been shown to play such a strong and unique role that species identity rather than overall diversity is of primary importance. This raises the possibility that a positive relationship between biodiversity and ecosystem function may be an artifact of sampling, because these species with unique traits and roles are more likely to be encountered when more of the species pool is included in an experimental study.

If species loss is expected to result in reductions in ecosystem function, then highly diverse communities should be better buffered by their presumed greater levels of complementarity and redundancy, and therefore less affected by the loss of a single species (McIntyre et al. 2007). However, this is not always the case, as evidenced by a particularly striking example of a single species having substantial effects on ecosystem structure and function in a hyperdiverse tropical stream. During the dry season, the flannel-mouth characin *Prochilodus mariae* migrates from floodplain locations into headwater streams in the foothills of the Venezuelan Andes, where it feeds on organic-rich sediments on the streambed, creating visible feeding scars and enhancing sediment transport. *Prochilodus* is an ecosystem engineer, a species that significantly modifies habitat, often influencing habitat heterogeneity and species diversity. Comparisons between open and *Prochilodus*-restricted stream sections (Fig. 12.11) convincingly demonstrated that this one species of detritus-feeding fish uniquely influences carbon flow and ecosystem metabolism (Taylor et al. 2006). In its absence, the amount of particulate organic carbon on the stream bed was higher, the downstream flux of POC declined due to reduced bioturbation and consumption, heterotrophic respiration increased due to greater biofilm growth, and primary production doubled. Because respiration increased more than primary production, net ecosystem metabolism showed a greater deficit. This example is of more than academic interest: fishes of the Prochilodontidae are the most important commercial species of South America and are declining due to the combined influence of dams and over-fishing. Changes to ecosystem function seem all too likely to occur over large areas of the Neotropics due to the loss of species in this family.

12.2.2 Temporal and Spatial Variation in Ecosystem-Level Effects

The relationship between biodiversity and ecosystem function is influenced by changes in community composition and species interactions that vary with both biotic and abiotic conditions (Cardinale and Palmer 2002). Habitat heterogeneity affects ecosystem-level dynamics by influencing how organisms experience environmental conditions and by mediating the colonization and persistence of populations. By varying the size of streambed sediments in riffles while keeping the median substrate size constant, Cardinale et al. (2002) showed that increased heterogeneity in the physical habitat had immediate, stimulatory effects on primary productivity and respiration in benthic communities.

Fig. 12.11 Images of the *Prochilodus mariae* removal experiment. (**a**) The plastic divider in a 210 m section of Río Las Marías in Venezuela. (**b**) Benthic particulate matter after removing *P. mariae* on the right compared with the intact fish assemblage on the left. (Reproduced from Taylor et al. 2006). (**c**) *Prochilodus* feeding on benthic sediments in the Prata river in the region of Bonito, Mato Grosso do Sul, Brazil. Photo courtesy of Ana Carolina O. Neves

Responses were attributed to changes in near-bed flow velocity and turbulence, suggesting disturbance regime may alter ecosystem processes.

Environmental variables, geographic location, and isolation all have been shown to affect biodiversity-ecosystem function relationships. In watersheds that have limited connectivity to other freshwater systems, fewer species are available to colonize habitats and contribute to leaf litter processing. Pacific island streams in Micronesia have exceptionally low insect species richness and density, a complete absence of Ephemeroptera, Trichoptera, and Plecoptera, and an insect community dominated ($\sim 85\%$ biomass) by non-shredding Chironomidae larvae (Benstead et al. 2009). In contrast to many streams in the temperate zone, macroconsumers, including decapods and gastropods, dominate this invertebrate community. Exclusion of macroconsumers by means of experimental electrified quadrats demonstrated that in streams of this region, microbes, rather than invertebrates, were primarily responsible for leaf-litter breakdown (Benstead et al. 2009).

Throughout the world, rivers and streams experience pronounced wet and dry seasons, and this can alter the net effects of species assemblages on productivity, respiration, and nutrient dynamics (Niu and Dudgeon 2011). Flashy, high flows can result in scouring of the benthos, removing primary producers, other microbes, and higher order consumers. Unless organismal biomass increases with increasing discharge, more frequent high flow events are likely to reduce the net influence of consumers on system-wide nutrient dynamics. Additionally, higher flows are typically correlated with more overland flow and more carbon and nutrients washing into streams. This may alter nutrient and carbon availability in river networks and cause changes in biogeochemical cycling. Seasonality is often linked to changing discharge, which can physically and chemically alter stream habitats and can also be linked to changes in community composition as certain species migrate in and out of a given reach.

Anthropogenic activities also may generate changes in species composition that subsequently alter ecosystem processes. For instance, microplastics are now commonly found in freshwater systems (Hoellein et al. 2014, 2017), and create new substrate for biofilm colonization. Though results are just beginning to emerge, studies suggest that biofilms colonizing microplastics are distinct from communities found in the surrounding environment. Using experimental manipulations and high-throughput sequencing of 16S rRNA to examine diversity patterns on biofilm communities on microplastics, Miao et al. (2019) observed that species richness, evenness, and diversity were lower in communities associated with microplastics than on natural substrates. Other research has found that the metabolic pathways of biofilm communities were distinct from those found on natural substrates, suggesting that changes in microbial diversity due to microplastic pollution may result in changes in microbially-mediated ecosystem processes (Aßhauer et al. 2015).

Neotropical streams have lost many of their amphibian species due to the spread of a parasitic fungus, *Batrachochytrium dendrobatidis* (Bd), and this has subsequently altered ecosystem function in affected streams (Rantala et al. 2015; Whiles et al. 2013). To assess the impacts of the loss of tadpoles on leaf-litter decomposition in streams infected with Bd, Rugenski et al. (2012) evaluated the effects of grazing tadpoles, shredding macroinvertebrates, and both together on leaf decomposition and the associated microbial activity. Microbial respiration rates were greatest in treatments with tadpoles, indicating that grazing and nutrient mineralization by tadpoles enhanced microbial activity. Decomposition, measured by leaf area loss, was greatest in treatments with both tadpoles and macroinvertebrates, suggesting tadpole grazing facilitated invertebrate leaf processing by enhancing microbial production (Fig. 12.12). These results indicate that the presence of tadpoles benefits both microbial biofilms and shredding macroinvertebrates, and all play a functional role in the decomposition and generation of POM in these Neotropical headwater streams.

12.2.3 Community-Driven Nutrient Dynamics

The contribution of animals in mediating nutrient cycling, or consumer-driven nutrient dynamics (CND), influences stream ecosystem structure and function, with effects that vary among species and across ecosystems (Atkinson et al. 2017; Vanni 2002; Vanni et al. 2002). Through their consumption and storage of resources in body tissues, and subsequent mineralization of elements by excretion, egestion, and death and decomposition, consumers have the potential to regulate the flux of limiting nutrients, such as nitrogen and phosphorus, in lotic systems (Capps et al. 2015a). Hence, changes in community composition may alter freshwater food webs, as mineralization by consumers is expected to stimulate primary and secondary production, especially in nutrient limited systems (Vaughn 2010; Flecker et al. 2010). Two conceptual frameworks, ecological stoichiometry and the metabolic theory of ecology, help us understand the relative contribution of species to system-wide biogeochemistry through CND.

Ecological stoichiometry (Sterner and Elser 2002) is the study of the balance of elements in ecological processes, linking the body elemental composition of an organism to the resources it consumes and to the waste it produces through excretion, egestion, and decomposition. Because an organism must consume enough of each element that it requires to grow, any element in short supply relative to others (i.e., a limiting nutrient) will be preferentially retained by the organism and not released as waste. Put colloquially, if you eat it and need it, you should not excrete it. Ecological

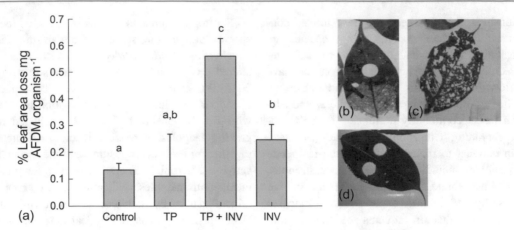

Fig. 12.12 Differences in leaf-litter decomposition rates among experimental treatments in chambers with no decomposers (control), tadpoles only (TP), tadpoles and macroinvertebrate shredders (TP + INV), and macroinvertebrates only (INV). (**a**) Average (±1SD) amount (%) of leaf area lost per unit detritivore (i.e., tadpoles and macroinvertebrates) biomass after 31-day incubations. Photographs of examples of leaves (**b**) with tadpoles only (TP), (**c**) tadpoles and invertebrate shredders (TP + INV) and (**d**) macroinvertebrates only (INV). The circles within the leaves in each of the photographs were sections of the leaves removed to estimate microbial respiration. The different letters in (**a**) reflect significant differences between treatments ($\alpha = 0.05$; one-way analysis of variance and Tukey's multiple comparisons). (Reproduced from Rugenski et al. 2012)

stoichiometry largely rests on what is known as the Law of the Minimum, developed in the mid-nineteenth century, which states that whatever element is in least supply in the environment relative to an organism's needs will limit its growth. Interest in stoichiometric applications grew with the observation that the ratio of carbon to nitrogen to phosphorus in marine phytoplankton is a relatively constant 106:16:1 and similar to dissolved nutrient pools in sea water (Redfield 1958), and with further studies of the elemental mass balance of aquatic organisms (Corner et al. 1976). These early studies set the stage for research into the influence of CND on food web ecology and biogeochemistry (Hessen et al. 2013).

The metabolic theory of ecology (Brown et al. 2004) outlines a theoretical basis for linking first principles of biology, chemistry, and physics to how organisms interact with their environment. Specifically, this theory explores how metabolic rate can be quantitatively linked to body size and temperature. The rates at which an organism uses resources within the environment, and how these resources are allocated to survival, growth, and reproduction, govern ecological processes including but not limited to, the rate of development, population growth rate, species interactions, and system-wide productivity and respiration rates.

When considered together, ecological stoichiometry and the metabolic theory of ecology suggest that nutrient mineralization by animals (i.e., excretion, egestion, and tissue decomposition) should depend on the expression of certain species traits, especially body size, trophic ecology, and ontogeny (Vanni and McIntyre 2016; Atkinson et al. 2017). For instance, amphibians experience large shifts in the ratio of carbon, nitrogen, and phosphorus in their tissues as they develop from cartilage-based larvae into juvenile animals with boney skeletons, a phosphorus-rich tissue (Tiegs et al. 2016; Luhring et al. 2017).

Variation in growth or ingestion among individuals may reflect different patterns in nutrient limitation among organisms and across systems, and this may shift relationships between the stoichiometry of body tissues and the stoichiometry of waste released by an animal (Vanni and McIntyre 2016). If growth rates among or within taxa are not strongly nutrient-dependent, and species are not consistently growing or building new tissues through their lifetime, body stoichiometry may influence excretion stoichiometry, and animals as could release wastes at the same N:P ratio as consumed in their food. However, when a nutrient is in high demand for building tissue and in short supply in the animal's diet, as for example P required for bone elaboration, the ratio of nutrients in excretion can indicate selective retention of a limiting nutrient.

Variation in ingestion rates can also produce relationships contrary to the predictions of ecological stoichiometry, as animals frequently compensate for low-quality diets by consuming more food (Bowen et al. 1995). Hence, the net intake of nutrients by a consumer may be regulated by the quantity rather than the nutritional value of their food. Animal growth is often limited by carbon or energy rather than by nutrients such as phosphorus (Schindler and Eby 1997; Frost et al. 2006), though phosphorus limitation may occur in primary consumers with low phosphorus diets (Sterner 1990; Hood et al. 2005; Benstead et al. 2014). Animals that are limited by energy may consume, and subsequently excrete or egest, excess nutrients in order to acquire enough energy for growth or reproduction, decoupling the relationship between the chemical composition of body tissues and their waste (Vanni and McIntyre 2016; Vanni et al. 2017).

The net contribution of a population or community of organisms to system-wide biogeochemistry depends on their abundance, the rates at which they store and mineralize nutrients, and the physicochemical conditions of the environment, including discharge and ambient nutrient concentrations (McIntyre et al. 2008). For example, Wheeler et al. (2015) and Atkinson and Vaughn (2015) assessed the effects of stream discharge on the roles of fish and mussel excretion on ecosystem processes in streams, respectively. As water volumes declined in drier weather, the ratio of organismal biomass to discharge increased, as did the relative contribution of organism to system-wide nutrient dynamics. Ambient nutrient concentrations can also influence how consumers contribute to nutrient dynamics. In streams along an agricultural gradient, Wilson and Xenopoulos (2011) demonstrated that excretion of nitrogen and phosphorus by fishes was able to fulfill more of the ecosystem-level demand for both elements as the amount of cropland increased. This was especially notable because ambient nutrient concentrations also increased along this gradient, attributed to increases in fish biomass concurrent with increasing ambient nutrient concentrations. In contrast, increasing ambient nutrient concentrations along an agricultural gradient catchment all but eliminated the contribution of mussels to system-wide biogeochemical cycling (Spooner et al. 2013). This work demonstrated that while species can persist in anthropogenically-disturbed environments, their functional roles in ecosystem-level processes may be compromised.

Drought is both a natural and an anthropogenically-mediated disturbance that affects ecosystem processes by altering freshwater community structure. Mussels dominate biomass in some freshwater systems, and although their distribution can be patchy, they can be important contributors to nutrient storage and cycling (Vaughn et al. 2015). Of all groups of freshwater organisms, mussels are often of greatest conservation concern, as many species are exceptionally sensitive to environmental change (Vaughn 2010). In a survey of mussel populations at nine sites within three rivers in the south-central US before and after an exceptional regional drought, Atkinson et al. (2014) documented changes in nutrient storage and mineralization in response to declines in mussel density and biomass. Mussel die-off reduced nitrogen concentrations in the water column and reduced phosphorus storage in mussel tissue, potentially influencing system-wide nutrient dynamics (Fig. 12.13).

As previously discussed, animals can transport elements across habitat boundaries, generating spatial subsidies in rivers and streams. Childress and McIntyre (2015) showed that longnose suckers *Catostomus catostomus* transported elements from the Laurentian Great lakes into tributary streams via excretion and egg deposition (Fig. 12.14), and excretion had large effects on nutrient inputs even in streams influenced by agriculture. Similarly, Subalusky et al. (2015) coupled empirical measurements collected at a zoological park with observed hippopotami (*Hippopotamus amphibius*) densities in the Maasai Mara National Reserve in Kenya to estimate that hippos in the watershed could contribute large proportions of the Mara River's coarse particulate organic matter, total nitrogen, and total phosphorus relative to nutrient loading from the upstream catchment (Fig. 12.15).

Animals often aggregate in space (e.g., mussel beds) and time (e.g., diurnal pool use by hippos), and this can influence when and where they consume and mineralize resources, and thus the heterogeneity of biogeochemical cycling in rivers and streams. In a study of the fish community of a diverse, nitrogen-limited stream in Venezuela, McIntyre et al. (2008) measured the nitrogen and phosphorus excretion rates and population densities of approximately 50 species of fishes. Although excretion rates of both elements varied considerably among species of fishes, collectively the excretion by the fish community could meet more than 75% of the ecosystem demand for dissolved inorganic nitrogen at the reach scale. Fish population densities differed sufficiently among riffles, runs, and pools such that, when controlled for size of the habitat, there was a 47-fold difference in nitrogen excretion and a 14-fold difference in phosphorus excretion among habitat units, suggesting that fishes may be creating hotspots of biogeochemical activity in this system. The introduction of a novel species also may have demonstrable effects on ecosystem processes. In river systems in southern Mexico, diel aggregations of the non-native catfish (Loricariidae: *Pterygoplichthys*; Fig. 12.16), a nocturnal feeder, generated measurable changes in nitrogen and phosphorus dynamics in a nutrient-limited river (Capps and Flecker 2013a, b).

Though not extensively studied, the influence of parasites and disease in freshwater communities may also generate variability in the contribution of organisms to ecosystem function by changing consumer metabolism and behavior. In a study of the elemental content of trematodes and their gastropod hosts, Bernot (2013) found a mismatch of the ratio of nitrogen to phosphorus in the parasite and in most gastropod tissues. Additionally, nutrient mineralization ratios in snails infected by trematodes differed from uninfected snails. Much more remains to be learned about these interactions, but this study suggests that parasite infection should be considered when examining the functional contribution of species to ecosystem-level processes.

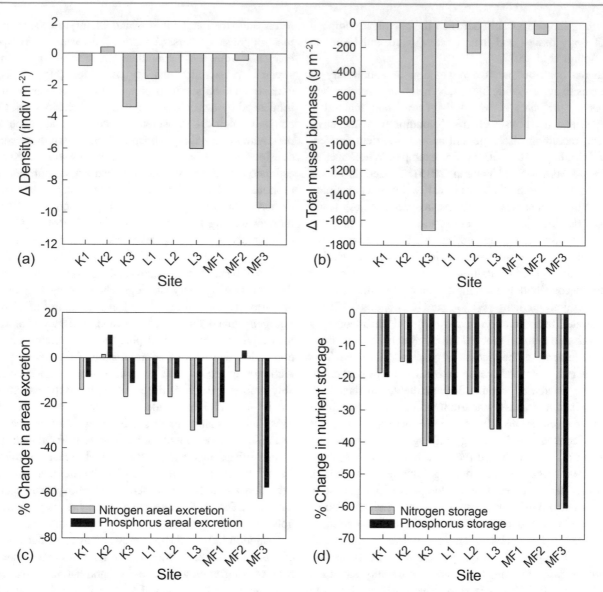

Fig. 12.13 Changes in the density (**a**) and total biomass (dry tissue + shell biomass) (**b**) of freshwater mussels in streams in the south-central US observed in 2010 (pre-drought) and 2012 (post-drought) conditions. The percent change in areal nitrogen and phosphorus (**c**) excretion rates and (**d**) storage in living mussel tissues between the two time periods also is shown. (Reproduced from Atkinson et al. 2014)

12.3   Summary

Species within aquatic communities play essential roles in system-wide energy and nutrient dynamics, connecting the biotic with the abiotic components of rivers and streams. The network of interactions portrayed in a food web provides the most complete yet succinct visual summary of a biological community. Identification of all links, although rarely achieved, serves as a useful reminder of the potential complexity within biological communities. Most energy flows through a subset of species, and typically it is the common species that dominate energy pathways.

Spatial subsidies occur when nutrients and energy flow across habitat boundaries, often with profound influence on recipient ecosystems. These subsidies are diverse across seasons and systems, and include allochthonous litter inputs, the exchange of invertebrates between the stream ecosystem and the terrestrial riparian zone, and upstream–downstream transfers due to water flows, animal migrations, and anthropogenic inputs. Differences in productivity and characteristics of the boundary between donor and recipient systems influence the lability, quantity, timing, and duration of a subsidy entering the new system; thus, the network of interacting species within stream ecosystems is strongly influenced by landscape setting.

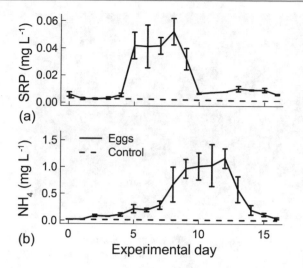

Fig. 12.14 Changes in ambient nutrient concentrations in microcosms through the decomposition of sucker eggs in microcosms with and without eggs: (**a**) soluble reactive phosphorus (SRP), and (**b**) ammonium (NH_4^+). (Reproduced from Childress and McIntyre 2015)

Community structure refers to the organization of a biological assemblage based on numbers of individuals within different taxonomic groups and functional roles, and the underlying processes that maintain that organization. Functional classifications can be based on a variety of species attributes, including but not limited to, ecosystem engineering, feeding ecology, productivity, and the capacity for elemental mineralization.

Ecosystem function relies on the presence of multiple species. Species performing complementary roles should increase the rate and efficiency of a process, and redundancy in the functional role of species increases the resilience of an ecosystem, ensuring that the loss of one species may not immediately result in a decline in the rate or efficiency of some process. Studies have revealed that one or a few species can be especially strong interactors in an ecosystem process, such that the loss of a single species can profoundly alter ecosystem function.

Fig. 12.15 Nutrient and carbon concentrations in a habitat pool for hippopotamus at the Milwaukee County Zoo before (white) and after (grey) approximately one day of use by three adult hippopotami. The bars are ± 1 standard error. (**a**) total nitrogen (TN), (**b**) total phosphorus (TP), (**c**) ammonium-nitrogen (NH_4^+-N), (**d**) soluble reactive phosphorus (SRP), (**e**) nitrate-nitrogen (NO_3^--N), (**f**) dissolved organic carbon (DOC). (Reproduced from Subalusky et al. 2015)

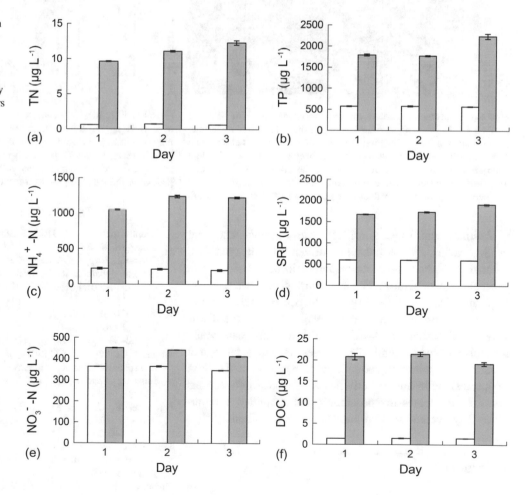

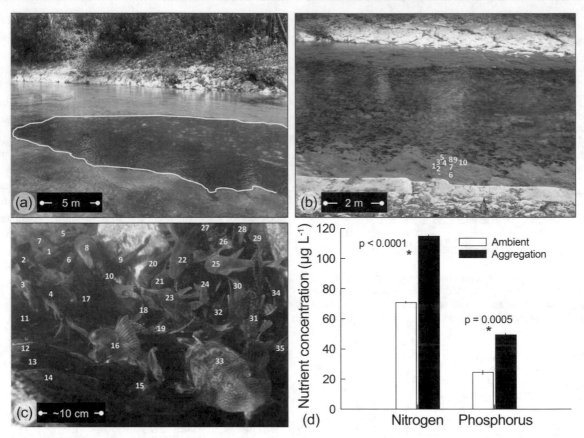

Fig. 12.16 Invasive armored catfish (*Pterygoplichthys*) in southern Mexico. (**a**) Daytime aggregation of catfish outlined by the white line. (**b**) Fish leaving the aggregation at dusk. Each dark spot is at least one fish and a small group of fishes (1–10) has been marked with individual numbers to emphasize the abundance of the catfish. (**c**) Underwater image of aggregation where individual fish are marked with white numbers (1–35). (**d**) Mean values (±1 SE) of nitrogen (N; NH$_4^+$-N) and phosphorus (P; PO$_4^{3-}$-P) collected from paired sites within and outside of catfish aggregations. Aggregations were defined as groups of *Pterygoplichthys* that had an area of at least five square meters with at least 40 *Pterygoplichthys* per m^2. Ambient samples were collected from sites parallel to the aggregations without immediate upstream aggregations of the fish. (Reproduced from Capps and Flecker 2013b)

Animals can play a critical role in biogeochemical cycling, referred to as consumer-driven nutrient dynamics. Animals consume and store resources in body tissues, and subsequently mineralize elements through excretion, egestion, and death and decomposition. Spatial and temporal variation in resource and habitat availability and the distribution and activity of species within systems can create hotspots of biogeochemical processes, including biological production, respiration, and nutrient mineralization. The loss of species due to human activities is an increasingly serious concern, and raises the possibility that simplified communities may become less productive and less resilient.

References

Allen KR (1951) The Horokiwi Stream: a study of a trout population. Bulletin of New Zealand Department of Fishery 10:1–231

Aßhauer KP, Wemheuer B, Daniel R et al (2015) Tax4Fun: predicting functional profiles from metagenomic 16S rRNA data. Bioinformatics 31:2882–2884. https://doi.org/10.1093/bioinformatics/btv287

Atkinson CL, Capps KA, Rugenski AT et al (2017) Consumer-driven nutrient dynamics in freshwater ecosystems: from individuals to ecosystems. Biol Rev 92:2003–2023. https://doi.org/10.1111/brv.12318

Atkinson CL, Julian JP, Vaughn CC (2014) Species and function lost: Role of drought in structuring stream communities. Biol Conserv 176:30–38. https://doi.org/10.1016/j.biocon.2014.04.029

Atkinson CL, Vaughn CC (2015) Biogeochemical hotspots: temporal and spatial scaling of the impact of freshwater mussels on ecosystem function. Freshw Biol 60:563–574. https://doi.org/10.1111/fwb.12498

Bärlocher F, Corkum M (2003) Nutrient enrichment overwhelms diversity effects in leaf decomposition by stream fungi. Oikos 101:247–252

Baxter CV, Fausch KD, Murakami M et al (2004) Fish invasion restructures stream and forest food webs by interrupting reciprocal prey subsidies. Ecology 85:2656–2663

Baxter CV, Fausch KD, Saunders WC (2005) Tangled webs: reciprocal flows of invertebrate prey link streams and riparian zones. Freshw Biol 50:201–220

Benedito-Cecilio E, Araujo-Lima C (2002) Variation in the carbon isotope composition of Semaprochilodus insignis, a detritivorous fish associated with oligotrophic and eutrophic Amazonian rivers. J Fish Biol 60:1603–1607

Benke AC (1993) Concepts and patterns of invertebrate production in running waters. Verh Int Ver Theor Angew Limno l25:15–38

Benke AC (2018) River food webs: an integrative approach to bottom-up flow webs, top-down impact webs, and trophic position. Ecology 99:1370–1381

Benke AC, Huryn AD (2010) Benthic invertebrate production-facilitating answers to ecological riddles in freshwater ecosystems. J N Am Benthol Soc 29:264–285. https://doi.org/10.1899/08-075.1

Benke AC, Wallace JB (1997) Trophic basis of production among riverine caddisflies: implications for food web analysis. Ecology 78:1132–1145

Benke AC, Wallace JB (2015) High secondary production in a Coastal Plain river is dominated by snag invertebrates and fuelled mainly by amorphous detritus. Freshw Biol 60:236–255

Benstead JP, Hood JM, Whelan NV et al (2014) Coupling of dietary phosphorus and growth across diverse fish taxa: a meta-analysis of experimental aquaculture studies. Ecology 95:2768–2777

Benstead JP, March JG, Pringle CM et al (2009) Biodiversity and ecosystem function in species-poor communities: community structure and leaf litter breakdown in a Pacific island stream. J N Am Benthol Soc 28:454–465. https://doi.org/10.1899/07-081.1

Bernot RJ (2013) Parasite-host elemental content and the effects of a parasite on host-consumer-driven nutrient recycling. Freshw Sci 32:299–308. https://doi.org/10.1899/12-060.1

Bowen SH, Lutz EV, Ahlgren MO (1995) Dietary protein and energy as determinants of food quality: trophic strategies compared. Ecology 76:899–907

Brown JH, Gillooly JF, Allen AP et al (2004) Toward a metabolic theory of ecology. Ecology 85:1771–1789. https://doi.org/10.1890/03-9000

Capps KA, Atkinson CL, Rugenski AT (2015a) Implications of species addition and decline for nutrient dynamics in fresh waters. Freshw Sci 34:485–496. https://doi.org/10.1086/681095

Capps KA, Berven KA, Tiegs SD (2015b) Modelling nutrient transport and transformation by pool-breeding amphibians in forested landscapes using a 21-year dataset. Freshw Biol 60:500–511. https://doi.org/10.1111/fwb.12470

Capps KA, Flecker AS (2013a) Invasive aquarium fish transform ecosystem nutrient dynamics. Proc R Soc B-Biol Sci 280. https://doi.org/10.1098/rspb.2013.1520.10.1098/rspb.2013.1520

Capps KA, Flecker AS (2013b) Invasive fishes generate biogeochemical hotspots in a nutrient-limited system. PLoS ONE 8:e54093. https://doi.org/10.1371/journal.pone.0054093

Cardinale BJ, Palmer MA (2002) Disturbance moderates biodiversity-ecosystem function relationships: Experimental evidence from caddisflies in stream mesocosms. Ecology 83:1915–1927

Cardinale BJ, Palmer MA, Swan CM et al (2002) The influence of substrate heterogeneity on biofilm metabolism in a stream ecosystem. Ecology 83:412–422. https://doi.org/10.2307/2680024

Carlisle DM, Clements WH (2003) Growth and secondary production of aquatic insects along a gradient of Zn contamination in Rocky Mountain streams. J N Am Benthol Soc 22:582–597

Carlisle DM, Clements WH (2005) Leaf litter breakdown, microbial respiration and shredder production in metal-polluted streams. Freshw Biol 50:380–390

Childress ES, McIntyre PB (2015) Multiple nutrient subsidy pathways from a spawning migration of iteroparous fish. Freshw Biol 60:490–499

Corner EDS, Head RN, Lindapennycuick CCK (1976) On the nutrition and metabolism of zooplankton X. Quantitative aspects of Calanus helgolandicus feeding as a carnivore. J Mar Biol Assoc UK 56:345–358

Covich AP, Austen MC, Barlocher F et al (2004) The role of Biodiversity in the functioning of freshwater and marine benthic ecosystems. Bioscience 54:767–775. https://doi.org/10.1641/0006-3568(2004)054[0767:Trobit]2.0.Co;2

Cross WF, Baxter CV, Rosi-Marshall EJ et al (2013) Food-web dynamics in a large river discontinuum. Ecol Monogr 83:311–337

Dang CK, Chauvet E, Gessner MO (2005) Magnitude and variability of process rates in fungal diversity-litter decomposition relationships. Ecol Lett 8:1129–1137. https://doi.org/10.1111/j.1461-0248.2005.00815.x

Dangles O, Crespo-Pérez V, Andino P et al (2011) Predicting richness effects on ecosystem function in natural communities: insights from high-elevation streams. Ecology 92:733–743

Dangles O, Malmqvist B (2004) Species richness-decomposition relationships depend on species dominance. Ecol Lett 7:395–402. https://doi.org/10.1111/j.1461-0248.2004.00591.x

Dutton CL, Subalusky AL, Hamilton SK et al (2018) Organic matter loading by hippopotami causes subsidy overload resulting in downstream hypoxia and fish kills. Nat Commun 9:1951

East JL, Wilcut C, Pease AA (2017) Aquatic food-web structure along a salinized dryland river. Freshw Biol 62:681–694. https://doi.org/10.1111/fwb.12893

El-Sabaawi R (2018) Trophic structure in a rapidly urbanizing planct. Funct Ecol 32:1718–1728. https://doi.org/10.1111/1365-2435.13114

Finlay JC, Khandwala S, Power ME (2002) Spatial scales of carbon flow in a river food web. Ecology 83:1845–1859. https://doi.org/10.1890/0012-9658(2002)083[1845:Ssocfi]2.0.Co;2

Flecker AS, McIntyre PB, Moore JW et al (2010) Migratory fishes as material and process subsidies in riverine ecosystems. In: Gido KB, Jackson DA (eds) Community Ecology of Stream Fishes: Concepts, Approaches, and Techniques, vol 73. American Fisheries Society Symposium. Amer Fisheries Soc, Bethesda, pp 559–592

Flecker AS, Townsend CR (1994) Community-wide consequences of trout introduction in New Zealand streams. Ecol Appl 4:798–807

Frost PC, Benstead JP, Cross WF et al (2006) Threshold elemental ratios of carbon and phosphorus in aquatic consumers. Ecol Lett 9:774–779. https://doi.org/10.1111/j.1461-0248.2006.00919.x

Giller PS, Hillebrand H, Berninger UG et al (2004) Biodiversity effects on ecosystem functioning: emerging issues and their experimental test in aquatic environments. Oikos 104:423–436

Gladden JE, Smock LA (1990) Macroinvertebrate distribution and production on the floodplains of two lowland headwater streams. Freshw Biol 24:533–545

Grubaugh J, Wallace B, Houston E (1997) Production of benthic macroinvertebrate communities along a southern Appalachian river continuum. Freshw Biol 37:581–596

Hall RO, Taylor BW, Flecker AS (2011) Detritivorous fish indirectly reduce insect secondary production in a tropical river. Ecosphere 2:1–13

Hall RO, Wallace JB, Eggert SL (2000) Organic matter flow in stream food webs with reduced detrital resource base. Ecology 81:3445–3463

Hayden B, McWilliam-Hughes SM, Cunjak RA (2016) Evidence for limited trophic transfer of allochthonous energy in temperate river food webs. Freshw Sci 35:544–558. https://doi.org/10.1086/686001

Hessen DO, Elser JJ, Sterner RW et al (2013) Ecological stoichiometry: An elementary approach using basic principles. Limnol Oceanog 58:2219–2236

Hocking MD, Reynolds JD (2012) Nitrogen uptake by plants subsidized by Pacific salmon carcasses: a hierarchical experiment. Can J For Res 42:908–917

Hoellein T, Rojas M, Pink A et al. (2014) Anthropogenic litter in urban freshwater ecosystems: distribution and microbial interactions. PLoS One 9

Hoellein TJ, McCormick AR, Hittie J et al (2017) Longitudinal patterns of microplastic concentration and bacterial assemblages in surface and benthic habitats of an urban river. Freshw Sci 36:491–507. https://doi.org/10.1086/693012

Holtgrieve GW, Schindler DE (2011) Marine-derived nutrients, bioturbation, and ecosystem metabolism: reconsidering the role of salmon in streams. Ecology 92:373–385

Hood JM, Vanni MJ, Flecker AS (2005) Nutrient recycling by two phosphorus-rich grazing catfish: the potential for phosphorus-limitation of fish growth. Oecologia 146:247–257

Huryn AD (1996) An appraisal of the Allen paradox in a New Zealand trout stream. Limnol Oceanog 41:243–252

Huryn AD (1998) Ecosystem-level evidence for top-down and bottom-up control of production in a grassland stream system. Oecologia 115:173–183

Huryn AD, Wallace JB (1987) Local geomorphology as a determinant of macrofaunal production in a mountain stream. Ecology 68:1932–1942

Ings TC, Montoya JM, Bascompte J et al (2009) Ecological networks–beyond food webs. J Anim Ecol 78:253–269

Jackson JK, Fisher SG (1986) Secondary production, emergence, and export of aquatic insects of a Sonoran Desert stream. Ecology 67:629–638

Johnson RC, Jin HS, Carreiro MM et al (2013) Macroinvertebrate community structure, secondary production and trophic-level dynamics in urban streams affected by non-point-source pollution. Freshw Biol 58:843–857. https://doi.org/10.1111/fwb.12090

Jonsson M, Malmqvist B (2000) Ecosystem process rate increases with animal species richness: evidence from leaf-eating, aquatic insects. Oikos 89:519–523

Junk W, Bayley P, Sparks R (1989) The flood pulse concept in river-floodplain systems. Canadian Special Publication of Fisheries and Aquatic Sciences 106:110–127

Knight TM, McCoy MW, Chase JM et al (2005) Trophic cascades across ecosystems. Nature 437:880

Kominoski JS, Hoellein TJ, Leroy CJ et al (2010) Beyond species richness: Expanding biodiversity–ecosystem functioning theory in detritus-based streams. River Res Appl 26:67–75

Layer K, Hildrew AG, Jenkins GB et al (2011) Long-term dynamics of a well-characterised food web: four decades of acidification and recovery in the Broadstone Stream model system. In: Woodward G, Ogorman EJ (eds) Advances in Ecological Research, vol 44. Elsevier. San Diego, Elsevier Academic Press Inc, pp 69–117

Layer K, Hildrew AG, Woodward G (2013) Grazing and detritivory in 20 stream food webs across a broad pH gradient. Oecologia 171:459–471

Layman CA, Araujo MS, Boucek R et al (2012) Applying stable isotopes to examine food-web structure: an overview of analytical tools. Biol Rev 87:545–562

Layman CA, Giery ST, Buhler S et al (2015) A primer on the history of food web ecology: fundamental contributions of fourteen researchers. Food Webs 4:14–24

Lecerf A, Richardson JS (2010) Biodiversity-ecosystem function research: Insights gained from streams. River Res Appl 26:45–54

Ledger ME, Brown LE, Edwards FK et al. (2013a) Extreme climatic events alter aquatic food webs: a synthesis of evidence from a mesocosm drought experiment. In: Woodward G, Ogorman EJ (eds) Advances in Ecological Research, Vol 48: Global Change in Multispecies Systems. Elsevier Academic Press Inc, San Diego.

Ledger ME, Brown LE, Edwards FK et al (2013) Drought alters the structure and functioning of complex food webs. Nat Clim Chang 3:223–227

Levi PS, Tank JL, Tiegs SD et al (2013) Biogeochemical transformation of a nutrient subsidy: salmon, streams, and nitrification. Biogeochemistry 113:643–655. https://doi.org/10.1007/s10533-012-9794-0

Likens GE, Bormann FH (1974) Linkages between terrestrial and aquatic ecosystems. Bioscience 24:447–456

Loreau M, Naeem S, Inchausti P (2002) Biodiversity and ecosystem functioning: synthesis and perspectives. Oxford University Press, Oxford

Luhring TM, DeLong JP, Semlitsch RD (2017) Stoichiometry and life-history interact to determine the magnitude of cross-ecosystem element and biomass fluxes. Front Microbiol 8:11. https://doi.org/10.3389/fmicb.2017.00814

Marcarelli AM, Baxter CV, Mineau MM et al (2011) Quantity and quality: unifying food web and ecosystem perspectives on the role of resource subsidies in freshwaters. Ecology 92:1215–1225

McIntyre PB, Flecker AS, Vanni MJ et al (2008) Fish distributions and nutrient cycling in streams: Can fish create biogeochemical hotspots? Ecology 89:2335–2346. https://doi.org/10.1890/07-1552.1

McIntyre PB, Jones LE, Flecker AS et al (2007) Fish extinctions alter nutrient recycling in tropical freshwaters. Proc Natl Acad Sci USA 104:4461–4466. https://doi.org/10.1073/pnas.0608148104

Miao LZ, Wang PF, Hou J et al (2019) Distinct community structure and microbial functions of biofilms colonizing microplastics. Sci Total Environ 650:2395–2402. https://doi.org/10.1016/j.scitotenv.2018.09.378

Muehlbauer JD, Collins SF, Doyle MW et al (2014) How wide is a stream? Spatial extent of the potential "stream signature" in terrestrial food webs using meta-analysis. Ecology 95:44–55

Myers BJE, Dolloff CA, Webster JR et al (2018) Fish assemblage production estimates in Appalachian streams across a latitudinal and temperature gradient. Ecol Freshw Fish 27:363–377

Nakano S, Miyasaka H, Kuhara N (1999) Terrestrial-aquatic linkages: Riparian arthropod inputs alter trophic cascades in a stream food web. Ecology 80:2435–2441

Nielsen JM, Clare EL, Hayden B et al (2018) Diet tracing in ecology: Method comparison and selection. Methods Ecol Evol 9:278–291. https://doi.org/10.1111/2041-210x.12869

Niu SQ, Dudgeon D (2011) The influence of flow and season upon leaf-litter breakdown in monsoonal Hong Kong streams. Hydrobiologia 663:205–215. https://doi.org/10.1007/s10750-010-0573-4

Olesen JM, Dupont YL, O'Gorman E et al (2010) From Broadstone to Zackenberg: space, time and hierarchies in ecological networks. In: Woodward G, Ogorman EJ (eds) Advances in Ecological Research, vol 42. Elsevier Academic Press Inc., San Diego, pp 1–69

Patrick C, McGarvey D, Larson J et al. (2019) Precipitation and temperature drive continental-scale patterns in stream invertebrate production. Sci Adv 5:eaav2348

Polis GA, Anderson WB, Holt RD (1997) Toward an integration of landscape and food web ecology: The dynamics of spatially subsidized food webs. Annu Rev Ecol Syst 28:289–316. https://doi.org/10.1146/annurev.ecolsys.28.1.289

Polis GA, Hurd SD (1996) Linking marine and terrestrial food webs: Allochthonous input from the ocean supports high secondary productivity on small islands and coastal land communities. Am Nat 147:396–423. https://doi.org/10.1086/285858

Power ME, Parker MS, Dietrich WE (2008) Seasonal reassembly of a river food web: Floods, droughts, and impacts of fish. Ecol Monogr 78:263–282. https://doi.org/10.1890/06-0902.1

Rantala HM, Nelson AM, Fulgoni JN et al (2015) Long-term changes in structure and function of a tropical headwater stream following a disease-driven amphibian decline. Freshw Biol 60:575–589. https://doi.org/10.1111/fwb.12505

Redfield AC (1958) The biological control of chemical factors in the environment. Amer Sci 64:205–221

Reid AJ, Carlson AK, Creed IF et al (2019) Emerging threats and persistent conservation challenges for freshwater biodiversity. Biol Rev 94:849–873

Rosi-Marshall EJ, Wallace JB (2002) Invertebrate food webs along a stream resource gradient. Freshw Biol 47:129–141

Rugenski AT, Murria C, Whiles MR (2012) Tadpoles enhance microbial activity and leaf decomposition in a neotropical headwater stream. Freshw Biol 57:1904–1913. https://doi.org/10.1111/j.1365-2427.2012.02853.x

Rypel AL, David SR (2017) Pattern and scale in latitude–production relationships for freshwater fishes. Ecosphere 8:e01660

Sabo JL, Power ME (2002) River-watershed exchange: Effects of riverine subsidies on riparian lizards and their terrestrial prey. Ecology 83:1860–1869. https://doi.org/10.2307/3071770

Schindler DE, Eby LA (1997) Stoichiometry of fishes and their prey: Implications for nutrient recycling. Ecology 78:1816–1831

Schmid-Araya JM, Hildrew AG, Robertson A et al (2002) The importance of meiofauna in food webs: Evidence from an acid stream. Ecology 83:1271–1285. https://doi.org/10.2307/3071942

Singer GA, Battin TJ (2007) Anthropogenic subsidies alter stream consumer-resource stoichiometry, biodiversity, and food chains. Ecol Appl 17:376–389

Spooner DE, Frost PC, Hillebrand H et al (2013) Nutrient loading associated with agriculture land use dampens the importance of consumer-mediated niche construction. Ecol Lett 16:1115–1125. https://doi.org/10.1111/ele.12146

Statzner B, Resh VH (1993) Multiple-site and-year analyses of stream insect emergence: a test of ecological theory. Oecologia 96:65–79

Sterner RW (1990) The ratio of nitrogen to phosphorus resuppied by herbivores: zooplankton and the algal competitive arena. Am Nat 136:209–229

Sterner RW, Elser JJ (2002) Ecological stoichiometry: The biology of elements from molecules to the biosphere. Princeton University Press, Princeton

Strayer DL, Dudgeon D (2010) Freshwater biodiversity conservation: recent progress and future challenges. J N Am Benthol Soc 29:344–358. https://doi.org/10.1899/08-171.1

Subalusky AL, Dutton CL, Rosi-Marshall EJ et al (2015) The hippopotamus conveyor belt: vectors of carbon and nutrients from terrestrial grasslands to aquatic systems in sub-Saharan Africa. Freshw Biol 60:512–525. https://doi.org/10.1111/fwb.12474

Subalusky AL, Dutton CL, Rosi EJ et al (2017) Annual mass drownings of the Serengeti wildebeest migration influence nutrient cycling and storage in the Mara River. Proc Natl Acad Sci U S a 114:7647–7652. https://doi.org/10.1073/pnas.1614778114

Subalusky AL, Post DM (2019) Context dependency of animal resource subsidies. Biol Rev 94:517–538. https://doi.org/10.1111/brv.12465

Taylor BW, Flecker AS, Hall RO (2006) Loss of a harvested fish species disrupts carbon flow in a diverse tropical river. Science 313:833–836. https://doi.org/10.1126/science.1128223

Tiegs SD, Berven KA, Carmack DJ et al (2016) Stoichiometric implications of a biphasic life cycle. Oecologia 180:853–863. https://doi.org/10.1007/s00442-015-3504-2

Tiegs SD, Levi PS, Ruegg J et al (2011) Ecological effects of live salmon exceed those of carcasses during an annual spawning migration. Ecosystems 14:598–614. https://doi.org/10.1007/s10021-011-9431-0

Toussaint A, Charpin N, Brosse S et al (2016) Global functional diversity of freshwater fish is concentrated in the Neotropics while functional vulnerability is widespread. Sci Rep 6:22125

Vander Zanden HB, Soto DX, Bowen GJ et al (2016) Expanding the isotopic toolbox: Applications of hydrogen and oxygen stable isotope ratios to food web studies. Front Ecol Evol 4. https://doi.org/10.3389/fevo.2016.00020

Vanni MJ (2002) Nutrient cycling by animals in freshwater ecosystems. Annu Rev Ecol Syst 33:341–370. https://doi.org/10.1146/annurev.ecolsys.33.010802.150519

Vanni MJ, Flecker AS, Hood JM et al (2002) Stoichiometry of nutrient recycling by vertebrates in a tropical stream: linking species identity and ecosystem processes. Ecol Lett 5:285–293. https://doi.org/10.1046/j.1461-0248.2002.00314.x

Vanni MJ, McIntyre PB (2016) Predicting nutrient excretion of aquatic animals with metabolic ecology and ecological stoichiometry: a global synthesis. Ecology 97:3460–3471. https://doi.org/10.1002/ecy.1582

Vanni MJ, McIntyre PB, Allen D et al (2017) A global database of nitrogen and phosphorus excretion rates of aquatic animals. Ecology 98:1475–1475. https://doi.org/10.1002/ecy.1792

Vannote RL, Minshall GW, Cummins KW et al (1980) River continuum concept. Can J Fish Aquat Sci 37:130–137. https://doi.org/10.1139/f80-017

Vaughn CC (2010) Biodiversity losses and ecosystem function in freshwaters: Emerging conclusions and research directions. Bioscience 60:25–35. https://doi.org/10.1525/bio.2010.60.1.7

Vaughn CC, Atkinson CL, Julian JP (2015) Drought-induced changes in flow regimes lead to long-term losses in mussel-provided ecosystem services. Ecol Evol 5:1291–1305. https://doi.org/10.1002/ece3.1442

Wallace JB, Eggert SL, Meyer JL et al (1997) Multiple trophic levels of a forest stream linked to terrestrial litter inputs. Science 277:102–104

Walther DA, Whiles MR (2011) Secondary production in a southern Illinois headwater stream: relationships between organic matter standing stocks and macroinvertebrate productivity. J N Am Benthol Soc 30:357–373

Wenger SJ, Subalusky AL, Freeman MC (2019) The missing dead: The lost role of animal remains in nutrient cycling in North American Rivers. Food Webs 18:e00106. https://doi.org/10.1016/j.fooweb.2018.e00106

Wheeler K, Miller SW, Crowl TA (2015) Migratory fish excretion as a nutrient subsidy to recipient stream ecosystems. Freshw Biol 60:537–550

Whiles MR, Hall RO, Dodds WK et al (2013) Disease-driven amphibian declines alter ecosystem processes in a tropical stream. Ecosystems 16:146–157. https://doi.org/10.1007/s10021-012-9602-7

Wilson HF, Xenopoulos MA (2011) Nutrient recycling by fish in streams along a gradient of agricultural land use. Glob Change Biol 17:130–139. https://doi.org/10.1111/j.1365-2486.2010.02284.x

Woodward G, Dybkjaer JB, Ólafsson JS et al (2010) Sentinel systems on the razor's edge: Effects of warming on Arctic geothermal stream ecosystems. Glob Change Biol 16:1979–1991

Woodward G, Perkins DM, Brown LE (2010) Climate change and freshwater ecosystems: Impacts across multiple levels of organization. Philos Trans R Soc B-Biol Sci 365:2093–2106. https://doi.org/10.1098/rstb.2010.0055

Woodward G, Speirs DC, Hildrew AG et al (2005) Quantification and resolution of a complex, size-structured food web. Adv Eco Res 36:85–135

Woodward G, Hildrew AG (2002) Food web structure in riverine landscapes. Freshw Biol 47:777–798

Wootton JT (1997) Estimates and tests of per capita interaction strength: diet, abundance, and impact of intertidally foraging birds. Ecol Monogr 67:45–64

Nutrient Dynamics

Various chemical compounds must be acquired by plants and microbes in order for synthesis of new organic matter to take place, and by animals to sustain their growth and metabolism. Although energy exhibits a one-way flow from capture to dissipation via ecosystem metabolism, the chemical constituents of living organisms are continuously re-utilized as they cycle between the biota and the environment. These chemical constituents are referred to as nutrients because they are necessary to sustain life. Nutrient availability can affect genes and the function they encode, cellular processes, individual growth rates, community composition, biological productivity of ecosystems, and the provisioning of ecosystems services (Guignard et al. 2017). Heterotrophs obtain most of their nutrients from their food and by ingesting or absorbing water. Their growth and development are typically limited by energy in the form of organic carbon, rather than nutrients, although this is not always the case. Conversely, the growth and production of autotrophs is often limited by the availability of nutrients, and changes in nutrient availability can produce striking changes in autotrophic communities, such as algal blooms, in aquatic environments.

Elements that are most heavily utilized are referred to as macro-nutrients, including nitrogen, phosphorus, potassium, calcium, sulfur, and magnesium. Other elements including iron, manganese, copper, silica, molybdenum, chloride, and zinc are required in smaller quantities, and are referred to as trace elements or micro-nutrients. The demand for certain nutrients, especially nitrogen and phosphorus, is often much greater than their availability in aquatic systems; as a consequence, the supply of nitrogen and phosphorus often limits biological activity. However, in many systems, human activities have profoundly altered nitrogen and phosphorus dynamics by increasing the available supply in surface and ground waters, resulting in eutrophication of lakes, rivers, and coastal zones worldwide (Smith et al. 2006). Eutrophication is characterized by proliferation of algae or aquatic plants due to nutrient enrichment, and can lead to harmful algal blooms and reduced availability of dissolved oxygen, affecting water quality, ecosystem functioning, and biodiversity in continental and coastal waters (Le Moal et al. 2019).

Dissolved inorganic nutrients can enter a stream reach from many sources: upstream, groundwater, and surface runoff, the waste products of mobile animals, and atmospheric inputs. Nutrients are converted to organic forms through biological uptake and assimilation, subsequently moving through the food webs via consumption, and are returned to inorganic form through mineralization via excretion, egestion, and the decomposition of organic matter (Fig. 13.1). In terrestrial and lacustrine environments, the cycling of nutrients between abiotic and biotic compartments is often thought of as taking place within fixed boundaries. However, the flow of water adds a pronounced spatial dimension to nutrient cycling in streams and rivers. Nutrients mineralized at one location are often transported some distance before subsequent re-utilization. The term "nutrient spiraling" describes the interdependent processes of nutrient cycling and downstream transport (Webster and Patten 1979). Nutrients that are in demand relative to their supply (hence are limiting to biological processes) should be taken up rapidly, resulting in short transport distances and rapid cycling, relative to elements whose supply is less critical. Nutrient dynamics in streams are further complicated by various abiotic uptake and release mechanisms that influence ambient nutrient concentrations. Nitrogen dynamics are even more complex as biotic and abiotic factors influence the transformation of the element among inorganic states.

In this chapter we examine the often complex relationships between availability of inorganic nutrients and their utilization by the biological community. There are, broadly speaking, two perspectives from which we will examine nutrient dynamics: how nutrient supply affects biological productivity, and how processes within the stream ecosystem influence the quantity of nutrients that are transported downstream. As discussed in Chaps. 6 and 7, nutrient supply

© Springer Nature Switzerland AG 2021
J. D. Allan et al., *Stream Ecology*,
https://doi.org/10.1007/978-3-030-61286-3_13

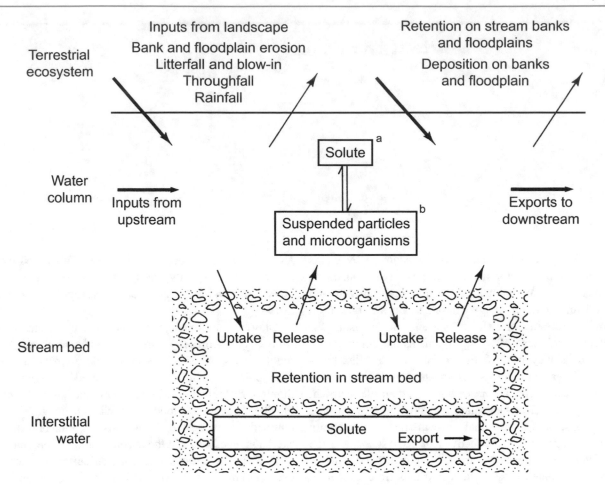

Fig. 13.1 Conceptual diagram of solute processes in streams. Arrow widths indicate approximate magnitude of process. Most materials are transported in dissolved form (**a**), but phosphorus, trace metals, and hydrophobic organics are transported mainly as particulates (**b**) (Modified from Stream Solute Workshop 1990)

can limit rates of photosynthesis and the decomposition of organic matter. Therefore, nutrients often govern the rate at which basal resources are supplied to stream food webs. Rivers transport substantial quantities of dissolved materials to receiving lakes and oceans, and so instream transformation, storage, and removal processes may be significant in large-scale element budgets.

13.1   Sources and Cycling of Nitrogen and Phosphorus

Nitrogen and phosphorus are the primary macronutrients that have been found to influence rates of primary production and the activity of heterotrophic microbes in aquatic and terrestrial environments. Nitrogen is a major component of enzymes that facilitate many biochemical reactions, including cellular respiration; and phosphorus is an important part of genetic material (DNA, deoxyribonucleic acid), energy transfer (ATP, Adenosine triphosphate), and cell membranes

(Schlesinger and Bernhardt 2013). Research has shown that benthic algal productivity can be limited by either nitrogen or phosphorus singly, be co-limited by both, or not be nutrient limited (Chap. 6). Sources and supplies of nitrogen and phosphorus vary considerably with geology, soils, climate, and vegetation, and their concentrations often are substantially elevated owing to anthropogenic inputs. Micronutrients such as silicon and iron, although important for aquatic ecosystems, have received less attention from researchers in running waters.

13.1.1   Nitrogen Sources

Molecular nitrogen (N_2) comprises 78% of the atmosphere and is the primary reservoir of nitrogen on earth. It is a very stable form of nitrogen and unavailable to most organisms; however, some microbes can convert N_2 to reactive nitrogen through the process of biological fixation. Nitrogen bonded to oxygen, hydrogen, or carbon, including ammonium

(NH$_4^+$), nitrate (NO$_3^-$), nitrite (NO$_2^-$), and organic nitrogen, are forms of biologically reactive nitrogen.

Nitrogen occurs in fresh waters in many chemical states (Table 13.1). Dissolved inorganic nitrogen (DIN) includes ammonium, nitrate, and nitrite. Nitrate dominates DIN in streams with adequate dissolved oxygen, while higher ammonium concentrations can indicate inputs of groundwater or pollution, because it is more abundant in anoxic waters (Dodds and Whiles 2010). Nitrate is a highly soluble and mobile ion that can leach through soils, while ammonium is often adsorbed by sediments (Schlesinger and Bernhardt 2013). Dissolved organic nitrogen (DON) consists of amino-nitrogen compounds (e.g., polypeptides, free amino compounds) and other organic molecules. Most particulate organic nitrogen (PON) occurs as bacteria and detritus. Total nitrogen concentration is a measurement that includes all dissolved and particulate forms of organic and inorganic nitrogen. Nitrogen also occurs in gaseous forms as dinitrogen N$_2$ and in association with oxygen as NO$_x$.

Many techniques have been developed to measure nitrogen in aquatic environments. Inorganic nitrogen as ammonium, nitrate, and nitrite generally is measured using colorimetric methods, although nitrate must be reduced to nitrite by passing the sample through a column filled with cadmium granules. Alternatively, sensors have recently been developed to measure nitrate and ammonium concentrations with high accuracy, precision, and temporal resolution, allowing in situ monitoring of nitrogen concentrations (Pellerin et al. 2016). The determination of particulate and dissolved organic forms is more complex because it involves digestion of the sample after separation of dissolved from particulate forms by filtration.

Nitrogen concentrations in rivers are derived from many sources. Atmospheric deposition as precipitation and dry fallout primarily supplies ammonium and nitrate to surface waters, but organic forms of nitrogen can also enter rivers from the atmosphere (Holland et al. 1999; Neff et al. 2002). Atmospheric nitrogen can be transformed by N-fixing microbes, principally cyanobacteria, into forms of nitrogen that can be used by other organisms. Nitrogen fixation also occurs in terrestrial environments by soil microbes and symbionts associated with vegetation roots, and fixed nitrogen can enter streams through surface and groundwater runoff (Hagedorn et al. 2000). During baseflow, most

Table 13.1 Major forms of nitrogen and phosphorus found in natural waters. Nitrogen also is present as dissolved N$_2$ gas (not shown)

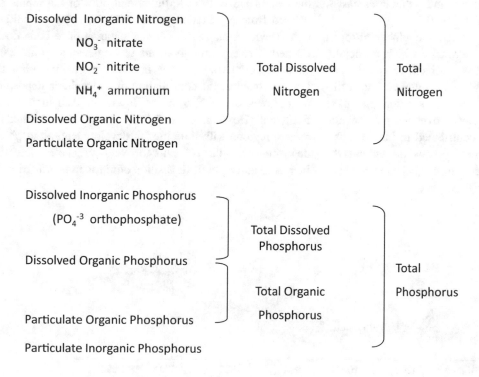

nitrogen inputs from the watershed are derived from subsoil leaching. At the beginning of a rain event, inputs from throughfall (rain that drips from vegetation) can be significant, and as the rain intensifies, nitrogen leached from topsoils rich in DON becomes more important to instream nitrogen concentrations.

Unlike phosphorus, rock weathering has usually been discounted as a source of nitrogen to streams. Yet, recent evidence indicates that some sedimentary rocks contain large amounts of fixed nitrogen and represent a potentially large pool in the global nitrogen budget (Holloway et al. 1998; Houlton et al. 2018). In certain regions, including high northern latitudes and mountainous regions, where N-rich rocks are more prevalent and weathering rates are higher, the weathering of bedrock may provide significant amounts of nitrate to running waters.

Animals may also provide important inputs of nitrogen by transporting energy and elements, often referred to as spatial subsidies, from other ecosystems to rivers (Subalusky and Post 2019). For example, mass migration of animals, such as Pacific salmon (*Oncorhynchus* spp.), can provide important seasonal inputs of marine-derived nitrogen to rivers in North America through their breeding migrations. Research has documented that marine-derived nutrients enter riverine food webs through salmon excretion and through the decomposition of egg masses and salmon carcasses (Janetski et al. 2009; Kohler et al. 2013).

Humans have significantly altered the global nitrogen cycle through the Haber-Bosch Process, which artificially fixes atmospheric nitrogen and is the basis of commercial fertilizer production. The ability to fix nitrogen has facilitated great increases in global food production that have supported human population growth over the past several decades. However, it has also contributed to the eutrophication of freshwater and marine systems throughout the world, as anthropogenically fixed nitrogen enters watersheds

through surface and groundwater runoff polluted with fertilizer and human and animal waste (Boyer et al. 2002; Gächter et al. 2004). Fertilizer production has increased greatly since the 1950's, and now more ammonia is produced through artificial means than is produced through biological fixation from natural environments (Galloway et al. 2013; Schlesinger and Bernhardt 2013). Human influence on the nitrogen cycle is so great that Caraco and Cole (1999) could explain more than 80% of the thousand-fold variation in nitrate export from 35 global rivers using a simple model that incorporated fertilizer use, atmospheric deposition, and human sewage production.

Land management can also alter nutrient cycling in the soils, increasing N leaching into running waters. In the Thames River Basin, UK, a marked increase in nitrate concentration occurred during Second World War, when grasslands were converted to agricultural land, and the effects of ploughing on mobilization of soil nitrate, rather than fertilizer application, increased nitrate export to streams (Fig. 13.2). Further increase was observed in the early 1970s that was related to the use of synthetic fertilizers and land drainage (Howden et al. 2010). The cultivation of N-fixing crops, such as soybeans, has also contributed to increases in instream nitrogen concentrations (Galloway et al. 2004). For example, 52% of nitrogen inputs to the Mississippi River in the central US is from fertilizer and 40% of the inputs are derived from N_2 fixation associated with soy-bean cultivation (David et al. 2010). Similar patterns have been documented in rivers throughout the globe (Castillo et al. 2000; Munn et al. 2010). However, in some regions, including the US and Sweden, the contribution of atmospheric deposition to the transport of N by rivers is greater than inputs from fertilizers (Howarth et al. 2012). Even when located in remote regions with little human development, atmospheric deposition can alter N inputs to rivers (Wilcke et al. 2013). Increased atmospheric deposition of nitrogen is primarily the

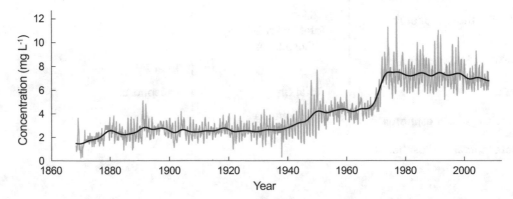

Fig. 13.2 Monthly nitrate concentrations for the Thames River at Hampton, UK, between 1868 and 2008 (grey lines). The dark trend line is based on one year period moving average (Reproduced from Howden et al. 2010)

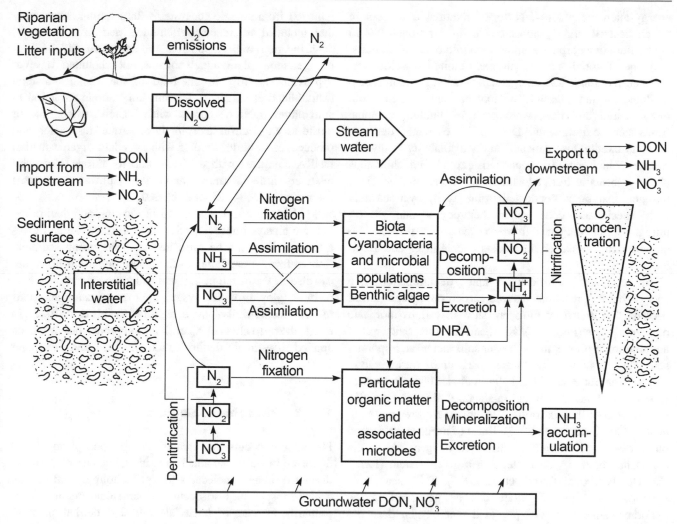

Fig. 13.3 Nitrogen dynamics in a stream ecosystem. Bioavailable inorganic nitrogen consists mainly of nitrate and ammonia, which is immobilized by autotrophs and microbial heterotrophs in biofilms or in suspension, and by higher plants. Assimilatory uptake refers to nutrients that are incorporated into cellular constituents and are potentially available to higher trophic levels. Excretion, decomposition, and production of exudates are the principal pathways by which elements are recycled to an inorganic state. Various dissimilatory transformations of inorganic forms of nitrogen by bacteria add to the complexity of the nitrogen cycle. Cyanobacteria and other microorganisms capable of nitrogen fixation transform N_2 gas into ammonia. Nitrification, which takes place under aerobic conditions, and denitrification, which takes place under anaerobic conditions, further influence the quantities and availability of dissolved inorganic nitrogen

consequence of the burning of fossil fuels, but forest fires can also contribute to N deposition.

13.1.2 Nitrogen Cycling

Nitrogen occurs in several chemical states, and many species of bacteria play specialized roles in converting nitrogen from one form to another (Fig. 13.3). It helps to recognize that a subset of the transformations is used to obtain nitrogen for structural synthesis (assimilatory uptake), while other transformations are energy-yielding reactions (dissimilatory uptake). Nitrogen fixation and the assimilation of dissolved inorganic nitrogen by autotrophs and heterotrophs are used to support structural synthesis, whereas nitrification and denitrification are reactions where bacteria use ammonia as a fuel or nitrate as an oxidizing agent to obtain energy.

Primary producers rely primarily on the surrounding water to supply the nutrients they need to synthesize protein. In comparison, bacteria and fungi can meet much of their nutrient requirements using their carbon substrate, but when they are located on a nutrient-poor substrate, they must also rely on nutrients in the water. Biological uptake and incorporation of nutrients into new tissue is referred to as immobilization, and is completed by both autotrophs and heterotrophs. Ammonium is taken up more readily than nitrate. This is because nitrate must be converted to ammonium prior to assimilation, and this is an energy-intensive process. Nitrogen fixation, a process in which bacteria and cyanobacteria convert nitrogen gas to ammonium to incorporate ammonium into bacterial biomass, may be favored under nitrogen limitation. However, nitrogen-fixation is also energetically costly. Fixation is also inhibited in the presence of oxygen and requires both molybdenum and iron, which can be limiting in some freshwater environments. Due to its energetic demand, nitrogen fixation also depends on the availability of organic carbon, which can be generated through autochthonous photosynthesis or from allochthonous organic matter. Thus, nitrogen fixation is conducted primarily by cyanobacteria with specialized cells called heterocysts that protect nitrogen-fixing enzymes from oxygen, or is restricted to organic-rich, anoxic environments (Schlesinger and Bernhardt 2013).

In contrast, nitrification and denitrification are energy-yielding reactions that are carried out by specialized bacteria that transform nitrogen between various inorganic oxidation states (Fig. 13.3). Nitrification, or the oxidation of ammonium to nitrate, in catchment soils can be an important source of nitrate to stream water, particularly during snowmelt (Bernhardt et al. 2002; Pellerin et al. 2006). In tropical forested catchments, soil and in-stream nitrification are also important contributors to the N exported by streams (Neill et al. 2001; Koenig et al. 2017). Rates of nitrification depend on access to ammonium and dissolved oxygen, and on the abundance of nitrifying bacteria in a given system (Kemp and Dodds 2002a; Bernot et al. 2006). This means that significant amounts of nitrate can be generated in the hyporheic zone of rivers, provided that oxygen does not become limiting due to high rates of bacterial respiration or inadequate water exchange (Holmes et al. 1994; Edwardson et al. 2003).

In denitrification, specialized bacteria use nitrate or nitrite, rather than oxygen, as an electron acceptor to oxidize organic matter anaerobically in energy-yielding reactions that are analogous to aerobic respiration. The complete process of denitrification converts nitrate (NO_3^-) to dinitrogen gas (N_2). However, the process typically is completed through a series of half reactions that are dependent upon specific enzymes, where nitrate is reduced to nitrite (NO_2^-) that is reduced to nitric oxide (NO), converted to nitrous oxide (N_2O), and finally released as dinitrogen gas (N_2). Denitrification depends on the availability of nitrate and organic matter, and requires low oxygen conditions (Findlay et al. 2011). Thus, rates of denitrification are typically greater in anoxic sediments, including in hyporheic and riparian areas (Seitzinger et al. 2006; Zarnetske et al. 2011). Because the end product of denitrification is N_2 gas, which is unavailable to most of the biota and can be released back into the atmosphere, denitrification represents an important pathway by which excess nitrogen can be permanently

removed from aquatic ecosystems. In a cross-system study that included reference, agricultural, and urban streams, denitrification was responsible for approximately 16% of nitrogen removal among all streams, but accounted for up to approximately 40% in one-third of the 72 sites studied (Mulholland et al. 2009). Dissimilatory nitrate reduction to ammonium (DNRA) occurs when bacteria use nitrate or nitrite as an electron acceptor in anaerobic respiration and produce ammonium. In sites with abundant organic matter, DNRA competes with denitrification for nitrate. However, relatively little is known about this pathway in streams (Storey et al. 2004; Ribot et al. 2013).

Methods to estimate rates of nitrification and denitrification are always improving. Sediment nitrification and denitrification rates can be estimated by transporting sediment slurries from stream sites to the laboratory. Both processes can also be estimated in the field using incubation chambers or by employing whole system approaches using additions of ^{15}N tracers (Mulholland et al. 2009; Dodds et al. 2017). In rivers, the open channel N_2 exchange approach has also been applied to estimate denitrification (Laursen and Seitzinger 2002).

13.1.3 Phosphorus Sources

Phosphorus occurs in streams as orthophosphate (PO_4^{3-}) dissolved in water and attached to inorganic particles, and as dissolved organic molecules like phospholipids and nucleic acids. Phosphorus also occurs in particulate organic form, primarily as part of bacterial cells and detrital particles (Table 13.1) (Dodds and Whiles 2010). The various phosphorus fractions are analyzed using filtration and digestion to separate its forms, followed by measurement using colorimetry and additional reactions. Total phosphorus (TP) is determined by digestion of unfiltered samples, and encompasses all forms of phosphorus, including those present in organisms, detritus, and adsorbed to inorganic complexes such as clays and carbonate. Total dissolved phosphorus is measured in filtered water after a digestion procedure, and includes both organic (colloids, esters) and inorganic (orthophosphate and polyphosphates) forms of phosphorus. An operational category known as soluble reactive phosphorus (SRP), based on the reaction of soluble phosphorus with molybdate, is commonly used as a measure of orthophosphate (PO_4^{3-}) in filtered samples. However, there is evidence that the SRP fraction can also include polyphosphates, and therefore may overestimate orthophosphate concentrations (Dodds 2003). In common usage, orthophosphate, phosphate, SRP, and dissolved inorganic phosphorus are interchangeable terms that refer to the form of phosphorus available for biological uptake by organisms. Total reactive phosphorus (TRP) is similar to SRP, but is measured from unfiltered samples. As with nitrogen, sensors have been developed to continuously monitor phosphate in the field (Pellerin et al. 2016), but this technology is still very expensive and requires relatively intensive monitoring when deployed.

Both SRP and TP concentrations in the water column are widely used as indicators of trophic status, but there is some controversy over which is the preferred measure to use for environmental monitoring. Soluble reactive phosphorus is typically considered to be the best indicator of what is immediately available for uptake by aquatic organisms, but because phosphorus can cycle rapidly among its various states, TP may be a better measure of overall availability of phosphorus. For broad comparisons of trophic status and nutrient limitation across stream ecosystems, TP may prove to be more useful as a measure (Dodds 2003). Total dissolved phosphorus is also recommended to be monitored and regulated in streams, because dissolved, rather than particulate phosphorus, is available to algae (Lewis et al. 2011). The usefulness of SRP versus TP in predicting algal productivity most likely varies with the residence time of phosphorus in the ecosystem, because a longer residence time allows for more efficient use of phosphorus by the biota (Edwards et al. 2000). Thus, SRP may be most effective in predicting algal production when water residence time is short, as is frequently true in small streams.

In contrast to nitrogen, which is abundant in the atmosphere and has its cycling driven by microbes, phosphorus does not have a major gaseous component. Rather, the phosphorus cycle is predominantly a geologic cycle where phosphorus is released through the weathering of rocks rich in calcium phosphate (Schlesinger and Bernhardt 2013). Because this typically is a slow process, phosphorus in unpolluted water is often in short supply relative to metabolic demand. In watersheds with limited human development, ambient phosphorus concentrations are typically governed by the composition of underlying bedrock. Phosphorus levels are generally higher in regions draining sedimentary rock deposits, and lower in regions of crystalline (igneous or metamorphic) bedrock. High SRP concentrations in Costa Rican streams influenced by geothermal activity indicate that geothermal groundwater can also be a significant source of phosphorus (Pringle et al. 1993; Small et al. 2016).

Atmospheric deposition of phosphorus in the form of dust, even if limited, can be ecologically significant in drainages where phosphorus is scarce. This is true in the Caura River, a tributary of the Orinoco draining the highly weathered Guyana Shield in Venezuela (Lewis et al. 1987). Similarly, dust transported from Africa represents an important source of phosphorus in the Amazon Basin (Bristow et al. 2010). The forest canopy can also contribute phosphorus to watersheds through leaching as water from precipitation and cloud cover is deposited on vegetative surfaces. For instance, in relatively undisturbed parts of the Upper River Severn in Wales, UK, phosphorus concentrations in throughfall and stemflow were markedly higher than observed in rain water, illustrating the potential strong influence of canopy cover on phosphorus flow into streams (Neal et al. 2003).

Because weathering is slow, phosphorus generated from terrestrial organic matter (e.g., leaf litter, soil organic matter) can be an important source of phosphorus that enters streams through surface runoff and subsurface pathways (McDowell et al. 2001). Orthophosphate readily adsorbs to charged particles such as clay; thus, it is often transported with sediment that erodes during storms. Phosphorus entering streams through sediment runoff can make up a large fraction of the total flux of phosphorus, especially in reaches where slopes are steep and vegetation cover is minimal. The concentrations in runoff vary with the amount of phosphorus in surface soils (Sharpley 1995; Weld et al. 2001) and with the proximity of phosphorus-rich soils to the stream channel (Sharpley et al. 1999). Terrestrial vegetation and microbial communities can also mediate surface- and subsurface-flow of phosphorus, as uptake of phosphorus by plant roots and immobilization by soil microbes can reduce concentrations in mineral soil water (Kaiser et al. 2000; Goller et al. 2006). For example, in a montane forest in Ecuador, the soil organic layer was the main source of organic phosphorus, while runoff from the forest canopy was the main source of inorganic phosphorus. In this system, because phosphorus concentrations in stream water were as low as those observed in rainfall, mineral soil appeared to function as a phosphorus sink, retaining a large proportion of phosphorus (Goller et al. 2006).

Anthropogenic sources of phosphorus are varied. They include municipal and industrial wastewater, termed point source pollution because it enters surface waters at a point (typically a pipe). The concentration of total phosphorus in untreated waste water typically falls in the 4–12 mg L^{-1} range. Although treatment processes remove some phosphorus, and USEPA standards under the 1972 Clean Water act set an effluent limit of 1 mg L^{-1} total phosphorus as a monthly average, total phosphorus concentration in these effluents can be between 0.5 and 10 mg P L^{-1} depending on the technology applied (Carey and Migliaccio 2009). Because dissolved phosphorus is highly bioavailable and the supply is essentially constant throughout the year, including during the growing season when low flows result in less dilution, sewage waste effluent can have a disproportionately large influence on receiving waters. Streams receiving wastewater effluent are generally characterized by high concentrations of nutrients during periods of low flow, and lower concentrations of nutrients during periods of high flow because of a dilution effect. Diel variation in phosphorus concentration due to wastewater effluence can also occur because of predictable changes in daily domestic water use (Withers and Jarvie 2008). Globally, approximately 80% of wastewater is discharged into surface waters without treatment, but this value is influenced by many factors, including dominant land cover (e.g. urban vs. rural), population growth rates, and economic status (WWAP 2017). For example, in Latin America only 20–30% of the urban wastewater undergoes treatment, indicating that many urban streams and rivers are strongly impacted by wastewater inputs. It is important to note, however, that this is an issue of global concern and even in wealthier countries such as the United States, many watersheds are characterized by water quality issues from untreated wastewater entering streams and rivers through aging and obsolete water infrastructure (ASCE 2011, 2017).

Additionally, nonpoint (diffuse) sources of phosphorus, including agricultural, pasture, and urban runoff enter streams and rivers via surface and sub-surface runoff. Small rural wastewater treatment plants, leachate from septic tanks, sediment runoff from construction, and atmospheric deposition are also considered non-point sources of phosphorus in a watershed (Carpenter et al. 1998). Generally speaking, point sources provide more continuous inputs relative to non-point sources of nutrients, which are frequently episodic due to timing of fertilizer application and storm periodicity (Withers and Jarvie 2008). Within a watershed, temporal and spatial variation in agricultural practices, rural wastewater management, and hydrological events can result in heterogeneity in non-point source pollution entering rivers and streams (Withers et al. 2009).

Anthropogenically-derived phosphorus can be conveyed in particulate form in association with sediments, particularly where erosion is high, while phosphorus originating from fertilizer and manure is often transported in dissolved forms (Hatch et al. 1999; Vanni et al. 2001). Sediment erosion is an important process adding phosphorus to lotic systems (Kleinman et al. 2011). The contribution of erosion-derived phosphorus is generally higher during storm events, and this is particularly true in areas that have recently received applications of manure and/or fertilizer (Withers and Jarvie 2008). Phosphorus entering streams through subsurface flow was originally considered a minor flux of non-point source pollution; however, in certain regions it can be important. For example, watersheds with fertilizer-intensive agriculture, especially those with soils with low phosphorus sorption capacity (e.g., sandy soils) or with engineered drainage systems can experience high concentrations of phosphorus in subsurface runoff (King et al. 2015).

Atmospheric deposition can also represent an important anthropogenic input of phosphorus in some catchments. For example, phosphorus emissions in China produced during coal combustion are transported by seasonal winds to Japan, where they are deposited as particulate phosphorus (PP) and phosphate. In the forested Hii River watershed in southwest Japan, PP in precipitation and in river water had similar concentrations during the warm months, indicating that atmospheric deposition was a seasonally important source of phosphorus. In the same river, total phosphorus in precipitation and in river water increased between 2002 and 2012, likely as result of increases in phosphorus deposition (Miyazako et al. 2015). Similarly, increases in urban development and industrial activities between 2007 and 2011 in the Ganga River, India, are also correlated with a two-fold increase in atmospheric deposition of phosphorus, which in turn is highly correlated with phosphorus concentrations in river water and runoff (Pandey et al. 2013).

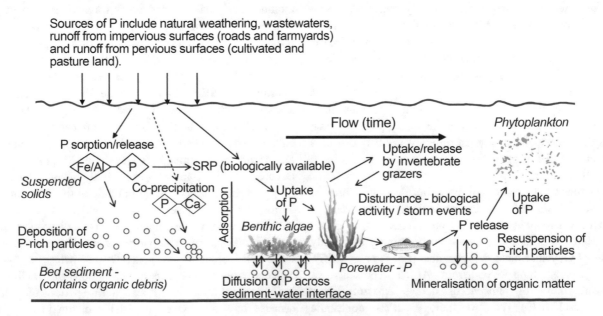

Fig. 13.4 Processes influencing phosphorus cycling in streams (Modified from Withers and Jarvie 2008)

13.1.4 Phosphorus Cycling

Phosphorus dynamics in streams are influenced by physical, chemical, and biological processes (Fig. 13.4). Autotrophic and heterotrophic uptake of dissolved inorganic phosphorus and its assimilation into cellular constituents, the transfer of organic phosphorus through the food chain, and its eventual release and mineralization by excretion and the decomposition of egested material are the biological activities that dominate phosphorus cycling. Algae and microbes in biofilms likely obtain the majority of their phosphorus from the water column, but rooted macrophytes and benthic algae can also remove phosphorus from the sediments. Phosphorus can be released following cell lysis directly as dissolved inorganic phosphorus, or released as dissolved organic phosphorus, which is subsequently mineralized to orthophosphate through bacterial activity. The decomposition of organic matter, including feces, dead organisms, and leaf litter, also releases phosphorus into the water column and sediment pore water (Mainstone and Parr 2002). Phosphorus availability is influenced by physical-chemical transformations. For instance, sorption-desorption reactions can, in part, regulate dissolved phosphorus concentrations in aquatic environments. Sorption of orthophosphate onto charged clays and organic particles occurs at relatively high phosphorus concentrations, while desorption of phosphorus from charged particles is favored under low phosphorus concentrations. In addition, changes in oxygen concentrations can regulate phosphorus dynamics. Under aerobic conditions, dissolved inorganic and organic phosphorus can complex with metal oxides and hydroxides (such as $Fe(OH)_3$) to form insoluble precipitates. When anaerobic conditions occur, the phosphate is released back into the water column. Since the extent of the anaerobic zone tends to vary seasonally with organic matter loading (more organic loading promotes aerobic respiration and leads to decreased oxygen concentrations), the availability of dissolved phosphate tends to covary with the relative volume of organic matter. The magnitude of these processes interacts with other activities mediating phosphorus dynamics and contribute to the retention or downstream export of phosphorus.

13.2 Transport and Spiraling

Nutrients are most available to freshwater autotrophs and microbial heterotrophs when present as inorganic solutes in stream water. Biotic and abiotic uptake, almost exclusively associated with the stream bed in smaller rivers and streams, transforms and temporarily retains dissolved nutrients as they move downstream with flow, but eventually these nutrients that are bound will return to the water column in mineral form. Thus, uptake and retention slows the

downstream passage of dissolved materials. Coupled with flow, uptake retention dynamics stretch nutrient cycles into spirals as they move downstream (Fig. 13.5). To account for this unique feature of nutrient cycling in lotic systems, stream ecologists have developed models that quantify transport distance and uptake rates, by comparing the transport dynamics of a dye or other non-reactive solute with the transport dynamics of reactive nutrients, such as nitrogen and phosphorus (Newbold et al. 1982; Baker and Webster 2017). Since biotic and abiotic uptake occurs primarily at sediment surfaces, exchange of water between the channel and interstitial areas can greatly influence nutrient dynamics.

Solutes such as lithium and bromide that are not readily utilized by the biota or otherwise transformed from state to state by physical and chemical processes are referred to as conservative, and pass unaltered by stream physicochemical or biological conditions through the stream ecosystem. In contrast, the downstream passage of reactive solutes, especially those that regulate metabolic processes such as nitrogen and phosphorus, is delayed by uptake and temporary storage. This distinction between conservative and reactive solutes, while useful, is not absolute, and may depend upon the relationship between supply and demand for a particular element at a particular time and place. For example, the concentration of chloride, a solute that is reactive under some circumstances, is not significantly altered by biological or physicochemical processes because it is typically in excess in running waters and therefore can be considered a conservative solute (Baker and Webster 2017).

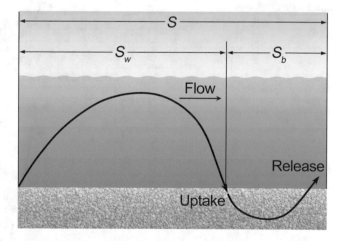

Fig. 13.5 Nutrient uptake and release in streams is coupled with downstream transport, stretching cycles into spirals. Spiraling length (*S*, in meters) is the sum of the distance traveled by a nutrient atom in dissolved inorganic form in the water column, called the uptake length (*S_w*, in meters), and the distance traveled within the biota before being mineralized and returned to the water column, called the turnover length (*S_B*, in meters). Arrows show uptake and release of nutrient retained within the streambed

13.2.1 Physical Transport

Solute dynamics are closely coupled with the physical movement of water, and the net flux of solutes under most conditions is downstream. Models describing the transport of conservative solutes are limited to physical processes, whereas models depicting reactive solute transport must incorporate physical, chemical, and biological dynamics. Stream ecologists can empirically measure the flux of conservative tracers by releasing the tracer at a known point and monitoring changes in tracer concentration at some point downstream. Its concentration will initially rise, reach a peak or plateau, and then decline as the pulse passes through the point of monitoring. The resulting concentration pattern is called a breakthrough curve (Fig. 13.6). A first approximation of this curve can be achieved with a basic equation describing advection and dispersion that takes into account stream dimensions and water velocity. The solute is transported from its point of release due to the unidirectional force of current (advection), and disperses due to molecular

diffusion primarily by turbulent mixing. Under certain simple conditions, which include a uniform channel, constant discharge, and no subsurface flow, the change in solute concentration (C) over time (t) is described as follows:

$$\frac{\partial C}{\partial t} = -V\frac{\partial C}{\partial x} + D\frac{\partial^2 C}{\partial x^2} \qquad (13.1)$$

The first term describes advection in the downstream direction (x, distance) and is proportional to water velocity, v. The second term describes mixing of the solute randomly throughout the water mass according to a dispersion coefficient, D.

More complicated models are needed to account for additional variables such as groundwater and tributary inputs, channel storage, and subsurface flow (Stream Solute Workshop 1990). Indeed, the study of transient storage and the hydrological interactions between the surface stream and subsurface compartments, such as the hyporheic zone, and also the influence of in-channel structures, such as pools, debris, and algal mats, has become increasingly complex and

Fig. 13.6 Concentrations of (**a**) ammonium and (**b**) nitrate and conductivity at 1330 m (Site 1), 1430 m (Site 2), 1750 m (Site 3), and 2610 m (Site 4) downstream from a pulse release point at the Snake River, Wyoming, US. A conservative tracer (NaCl) was also added and its travel time was followed with a conductivity meter. The rise and decrease in concentration after a pulse of a solute is released in a stream is called a breakthrough curve (Reproduced from Tank et al. 2008)

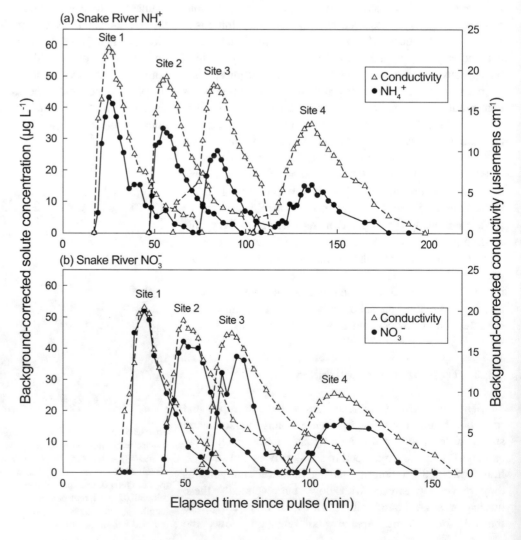

sophisticated (Boano et al. 2014; Harvey and Gooseff 2015). Small streams tend to have substantial exchange of surface water with interstitial water, back eddies behind obstructions, and areas of slow-moving water that all influence solute transport (Bencala and Walters 1983). The net effect is that water and solutes tend to move downstream more slowly than would be expected based on estimates of flow alone. This can be demonstrated by releasing a conservative solute such as a sodium chloride and recording its passage at successive distances downstream. Less is known about transient storage in large rivers, in part because methodological limitations of tracer additions. Although solute dynamics may appear less complex in large rivers compared to small streams, channel elements such as islands, bars, and meanders can influence exchange of water with the hyporheic zone (Boano et al. 2014).

One can model the complex effects of surface-subsurface exchange and back eddies by assuming that solutes are temporarily retained in a "transient storage zone" of slowly moving or even stationary water. Solutes diffuse into the storage zone during the initial passage of the pulse, and they are released back into the stream as the pulse passes and stream concentrations decline. Groundwater inputs can also affect solute concentration depending on the amount of groundwater entering the stream and the concentration of the solute in the inflow. The equations to describe the temporal and spatial changes in the concentration of a conservative solute, including transient storage and dilution from groundwater are:

$$\frac{\partial C}{\partial t} = -\frac{Q}{A}\frac{\partial C}{\partial x} + \frac{1}{A}\frac{\partial}{\partial x}\left[AD\frac{\partial C}{\partial x}\right] + \frac{Q_L}{A}(C_L - C) + \alpha(C_\theta - C)$$

(13.2)

$$\frac{\partial C_S}{\partial t} = -\alpha\frac{A}{A_S}(C_S - C)$$

(13.3)

where A (m^2) is the main channel cross-sectional area, A_S is the cross-sectional area of the modeled storage zone, Q_L is the lateral inflow of groundwater, and C_L is the solute concentration of the lateral inflow. Equation 13.2 includes the advection-dispersion terms (first two terms) and the effects of lateral inputs (third term) and transient storage (fourth term). Equation 13.3 describes the rate of dispersion of solute in or out of the transient storage zone, which is proportional to the transient storage exchange coefficient (α), the ratio between the cross-sections of the stream and the transient storage, and the difference in solute concentration in the storage zone (C_S) and in the water column (C). Adding a term for transient storage permits the model to account for significant features of the observed passage of a solute pulse and the effect of dilution from lateral inputs that Eq. 13.1 is unable to model. Specifically, incorporating this term into the model typically produces a more gradual rise and a more prolonged decline of the solute relative to the simpler model (Baker and Webster 2017).

It should be recognized that these models are empirically useful descriptions of observed dynamics in which transient storage clearly takes place. However, the storage component of the model is just an abstraction. In contrast to the cross-sectional area of the stream channel, which can be measured directly, the cross-sectional model of the storage zone (A_S) is determined by fitting the model to observed solute dynamics. Nonetheless, instream storage zones exist and can be numerous in natural and modified systems. For example, Bencala and Walters (1983) identified five types of storage zones in their study of solute transport in small mountain streams. They included turbulent eddies generated by large-scale bottom irregularities, large but slowly moving recirculating zones along the sides of pools, small but rapidly recirculating zones behind flow obstructions, side pockets, and flow in and out of beds of coarse substrate (hyporheic flow). There are limitations in assuming that a single term representing transient storage accounts for the variability among all types of storage zones, as each zone may uniquely influence solute transport (Bencala et al. 2011). In response, other models have been developed to include additional transient storage components that separate surface and hyporheic storage, and that use multiple solutes to improve the understanding of the transport processes (e.g., Marion et al. 2008; Briggs et al. 2010; Kelleher et al. 2019). Researchers can also employ "smart" tracers, such as resazurin, which are sensitive to aerobic respiration. Such tracers can provide more detailed information about the functioning of transient storage by estimating the fraction of the storage that is metabolically active and can complement the physical characterization of storage (Haggerty et al. 2009; Hanrahan et al. 2018)

13.2.2 Nutrient Spiraling

The processes of advection, dispersion, and transient storage included in the above models describe only the influences of hydrology and the channel upon the downstream transport of a solute. However, reactive solutes experience additional processes that influence the rate of their downstream passage. By comparing the transport of reactive and conservative solutes, the magnitude of these additional processes can be quantified. Some reactive solutes may have their dynamics governed solely by physical-chemical processes such as sorption-desorption, whereas nutrients, such as nitrogen and phosphorus, are also strongly influenced by biological processes. To account for the processes of uptake and mineralization (i.e., the release) of a reactive solute, the hydrologic models of the previous section must be modified to include an uptake rate and a release or mineralization rate. Adding terms representing these processes to Eq. 13.1 converts the relationship to:

$$\frac{\partial C}{\partial t} = -V\frac{\partial C}{\partial x} + D\frac{\partial^2 C}{\partial x^2} - \lambda_C C + \frac{1}{Z}\lambda_b C_b \qquad (13.4)$$

And the rate of change in the immobilized nutrient over time is represented by:

$$\frac{\partial C_b}{\partial t} = Z\lambda_c C - \lambda_b C_b \qquad (13.5)$$

where depth (z), the dynamic uptake rate (λ_c), the mineralization rate (λ_b), and the mass per unit of area of immobilized nutrient in the stream bed (C_b) have been added to the formula.

The complete cycle of a nutrient atom as it is transported downstream includes its transformation from inorganic to organic form by biological uptake, and its subsequent release and mineralization (Newbold et al. 1981). Therefore, spiraling length (S, in meters) is the sum of the distance traveled in dissolved inorganic form in the water column, called the uptake length (S_W), and the distance traveled within the biota before being mineralized and returned to the water column, called the turnover length (S_B) (Fig. 13.5).

$$S = S_w + S_B \qquad (13.6)$$

Uptake length (S_W) is a measure of both nutrient limitation and the efficiency of nutrient use in streams. Short travel distances indicate high demand relative to supply, and greater retentiveness for a nutrient by the stream ecosystem. Turnover length (S_B) is a measure of the distance traveled by an atom within biota until its eventual mineralization. The biological component is generally associated with the stream bed, in the form of attached microorganisms, periphyton, and benthic invertebrates. Typically, an atom will travel the greatest distance in the water column; thus, the expectation is that S_W will be much greater than S_B. Field studies have demonstrated that S_W often represents the greatest fraction of total spiraling distance and uptake is much easier to measure than S_B, so most studies focus on uptake length and related metrics. For example, uptake length for phosphorus in a forest stream was 165 m while turnover length associated to particulate organic matter was 25 m, thus uptake length represent almost 90% of the total spiraling length (Newbold et al. 1983).

Uptake length is estimated using plateau values of the concentration of a reactive solute and a conservative tracer at successive points downstream from its release (Fig. 13.6). When concentrations reach a stable maximum, or plateau at the injection site, water samples are collected at locations downstream of the injection site, and nutrient uptake over the length of the reach from the injection site to the final downstream sampling location can be calculated. The conservative tracer accounts for dilution along the reach. To estimate uptake length, the reactive solute concentrations are divided by the background-corrected conservative tracer concentrations at each sampling site (Tank et al. 2017). This relationship will form a straight line plot against distance on a logarithmic scale. The slope (k_w) is $1/S_W$. The term k_w, also known as the longitudinal uptake rate, is the dynamic uptake rate (λ_c) divided by velocity (v):

$$k_w = \frac{\lambda_C}{v} \qquad (13.7)$$

Uptake length depends strongly on discharge and velocity; hence, it is often desirable to standardize S_w by converting it to a measure of the uptake velocity (v_f) of the solute. Also referred to as the mass transfer coefficient, v_f quantifies the velocity at which a molecule moves from the water column to the stream bottom as a result of biotic or abiotic processes. It is calculated using the relationship:

$$v_f = \frac{vz}{s_w} \qquad (13.8)$$

Because it includes velocity (v) and depth (z) in its calculation, v_f standardizes S_w for discharge; thus, this measure is best suited to comparisons across stream ecosystems.

Uptake velocity quantifies the benthic demand for a nutrient in relation to its supply (Davis and Minshall 1999; Baker and Webster 2017). By combining v_f and the nutrient concentrations of stream water (C), the quantity of inorganic nutrients immobilized by the stream bed per unit of time can be expressed as an areal uptake rate, U, in the units mass per area per time, as:

$$U = v_f C \qquad (13.9)$$

In nutrient spiraling studies, large enough quantities of nutrient must be added to the stream to increase concentrations above ambient levels and achieve measurable changes in nutrient concentrations. However, nutrient addition is likely to saturate biological uptake, and so this method may overestimate S_w and U, and underestimate v_f. The release of trace amounts of stable or radioactive isotopes is an alternative method that can be used to avoid this problem because much smaller quantities are needed, thus more accurate estimates of ambient uptake rates can be obtained (Trentman et al. 2015). In a comparison of methods in a small woodland stream using an isotope-labeled nitrate release with and without simultaneous nitrate enrichment, transport distance was estimated to be 36 m using the tracer alone and 100 m when the stream was enriched (Mulholland et al. 2004). In addition to providing more reliable estimates of uptake lengths and rates, tracers provide the opportunity to quantify transformations among nutrient forms and the role of different biotic compartments in nutrient uptake and retention (Mulholland et al. 2000; Webster et al. 2003). For example, by following the movement of ^{15}N-labelled ammonium in a Kansas prairie stream, Dodds et al. (2000) were able to measure not only uptake of ammonium but also its rate of transformation to nitrate, and trace the amount of N assimilated by primary producers and microbes into primary consumers and predators. In contrast, the only isotope of phosphorus (^{32}P) that is readily available is radioactive; hence, its release into the environment is heavily restricted. Although releasing isotopes to characterize solute dynamics has obvious advantages, it is still an expensive process, especially for larger streams that would require greater quantities of isotope tracers.

A more recent technique called Tracer Additions for Spiraling Curve Characterization (TASCC) calculates spiraling metrics using concentrations measured at different time intervals at the end of the study reach, rather than at several sites downstream from the release point (Fig. 13.7).

This method estimates ambient (background), added (from nutrient release), and total (ambient plus added) uptake rates (Covino et al. 2010). In a study comparing the use of stable isotope tracers and TASCC to estimate nutrient uptake, Trentman et al. (2015) concluded that each method has its advantages and disadvantages and the selection must consider the objective of the study, stream size, and field logistics. For example, using isotope tracers generates more accurate estimates of ambient uptake rates, while TASCC does a better job of characterizing the relationship between uptake rate and nutrient concentration.

Recently developed nutrient sensors are being used to estimate uptake in rivers. High frequency nutrient time series obtained from measurements at a single site or two sites, and from longitudinal profiles, can provide data to assess nutrient uptake in rivers (Heffernan and Cohen 2010; Hensley et al. 2014; Kunz et al. 2017). Continuous solute sensors are still relatively costly, but as prices decline and technology improves, these kinds of data will become more common, and will enhance our ability to model potential changes in instream nutrient dynamics.

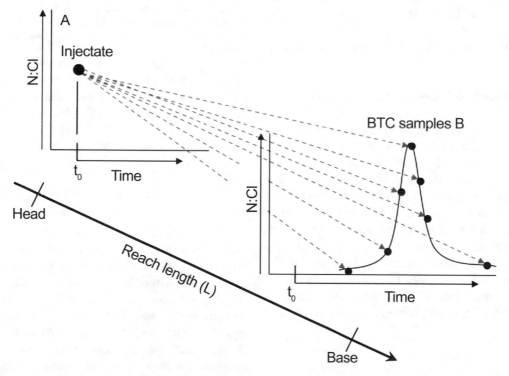

Fig. 13.7 Illustration of method for calculating spiraling metrics using concentrations measured at different time intervals at the end of the study reach. The instantaneous release of a conservative tracer (Cl^{-1}) and nutrient (Nitrogen, N) at time 0 at the head of a reach (**a**) is accompanied by the collection of grab samples at different times at the downstream sampling site (**b**) (base), allowing construction of a tracer breakthrough curve (BTC). Conductivity measurements guide the collection of water samples through time. N:Cl are the nitrogen concentrations (N) divided by the concentration of the conservative tracer (Cl) to correct for any dilution effect from lateral input (Reproduced from Covino et al. 2010)

13.3 Factors Influencing Nutrient Dynamics

Processes in streams that affect nutrient retention and export can affect ecosystem production and the flow of nutrients to downstream reaches; in addition, they have the potential to buffer downstream reaches from the impacts of anthropogenically-derived nutrient additions (Withers and Jarvie 2008). Environmental factors, including discharge, disturbance regimes, temperature, and light penetration, influence abiotic and biotic processes that govern nutrient uptake rates and transformations. Even under base flow conditions (under which most nutrient studies are conducted), rates of nutrient uptake and transformation can be extremely variable. For instance, a ^{15}N tracer study in headwater streams from biomes throughout North America reported wide variation in rates of ammonium uptake and its nitrification to nitrate (Peterson et al. 2001). Similar variability was observed among multiple streams within the Hubbard Brook Experimental Forest, New Hampshire (Bernhardt et al. 2002). Some of the variation can be explained by stream size, as smaller streams have a greater area of stream bed relative to water volume. Abiotic exchange mechanisms are also influenced by sediment characteristics, pH, and ambient nutrient concentrations in stream water. Biotic uptake and mineralization rates vary with overall biological productivity and nutrient availability.

13.3.1 Ambient Nutrient Concentrations

Nutrient uptake rates can respond to ambient nutrient concentrations in a number of ways, and multiple models have been proposed to describe variable responses to ambient nutrients. We will discuss three of them here, but it should be noted that there are datasets relating uptake rates and ambient nutrient concentrations that do not adhere to these relationships. By characterizing the form of the relationship between nutrient concentration and uptake rates, these models help to distinguish among the influences of mass transport and biological demand relative to supply. First, areal uptake (U) can increase as nutrient concentration (C) in the stream increases (Fig. 13.8a), following a first order or linear relationship:

$$U = kC \qquad (13.10)$$

where k is a constant. In this case, the uptake velocity (v_f) is constant (Fig. 13.8b). This relationship most likely occurs in streams characterized by low to moderate ambient nutrient concentrations, and it suggests that uptake is limited by mass transport (Dodds et al. 2002). Second, nutrient uptake can increase with increases in nutrient concentrations until

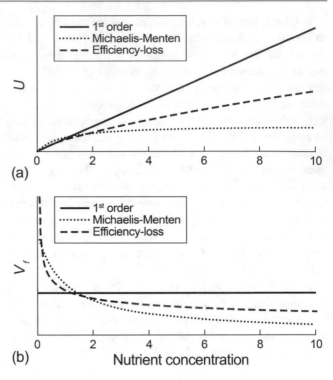

Fig. 13.8 Responses of areal (U) (**a**) and uptake velocity (v_f) (**b**) to variation in nutrient concentration under the three models proposed to explain the relationship (First order or Linear model, Michaelis-Menten, and Efficiency-Loss) (Modified from O'Brien et al. 2007)

supply exceeds biotic demand, at which point saturation of nutrient uptake occurs and the relationship between concentration and uptake reaches a plateau (Fig. 13.8a) (Dodds et al. 2002; Simon et al. 2005). This response is best described by a Michaelis Menten relationship:

$$U = \frac{U_{max}C}{C_{half} + C} \qquad (13.11)$$

where U is areal uptake, C is nutrient concentration, U_{max} is maximum uptake, and C_{half} is the concentration at which uptake equals half of U_{max}. Under this model, v_f will show a non-linear decline with C (Fig. 13.8b). This relationship suggests that the supply of a nutrient is in excess of the biological demand (Dodds et al. 2002; Mulholland et al. 2002; Demars 2008). The Efficiency Loss Model, a third model, proposes that areal uptake (U) can increase with nutrient concentration but at a reduced efficiency (a decrease in v_f) (Fig. 13.8a and b) that is described by a power relationship:

$$U = kC^b \qquad (13.12)$$

where b is less than one (O'Brien et al. 2007; Baker and Webster 2017).

The response of uptake to variation in ambient nutrient concentrations can also change with the forms of nutrients found in a given system. For instance, in agricultural streams of the Midwestern US where nutrient concentrations were ten to one hundred times greater than undisturbed sites in the same region, nitrate uptake appeared to be saturated, as uptake rates were lower and v_f declined in sites with higher nitrate concentrations. In contrast, uptake rates of phosphorus and ammonium increased with higher water column concentrations, indicating the sites were not saturated with respect to these two nutrients. However, declining v_f at higher ambient concentrations suggests that sites may have been approaching saturation for both phosphorus and ammonium (Bernot et al. 2006). Variation in the relationships between ambient nutrient concentrations and uptake rates has also been documented in streams of Central Siberia, Russia. In these systems, Diemer et al. (2015) found that the relationships between areal uptake and the concentration of both nitrate and phosphate followed a first order model, whereas uptake rate of ammonium was best described by the Efficiency Loss model.

Relationships between ambient nutrient concentrations and uptake rates can also provide insights regarding in-stream nutrient availability and limitation. In a stream in Scotland, UK, Demars (2008) found that Michaelis-Menten relationships described phosphorus dynamics, suggesting that phosphorus was probably limiting because phosphorus uptake length was very short. In addition, soluble reactive phosphorus concentrations were very low (4.7 µg L^{-1}), especially when compared to the estimated C_{half} (123 µg L^{-1}), and phosphorus uptake rate was low when compared to U_{max}. In a stream in Catalonia, Spain, uptake length for ammonium was five times shorter than for nitrate, which had concentrations many times greater than those of ammonium. In these systems, v_f for ammonium was also greater than for nitrate (Ribot et al. 2017). In oligotrophic mountain streams of Montana, US, Piper et al. (2017) conducted dual nitrate and phosphate additions to examine the coupling between nitrogen and phosphorus uptake dynamics. In some streams, the response to additions was similar and was not influenced by the order in which the elements were added, suggesting that the nutrients were co-limiting. However, in other cases small increases in phosphorus under low ambient phosphate (<10 µg L^{-1}) and high nitrate availability (>400 µg L^{-1}) resulted in greater uptake of nitrate, while increases in nitrate concentration did not stimulate phosphate uptake, indicating that the systems were limited by phosphorus.

The response of areal uptake to variation in ambient nutrient concentration is important for predicting how stream ecosystems may respond to nutrient loading from anthropogenic activities. In 72 streams in the continental United States and Puerto Rico, higher rates of areal uptake of nitrate were measured in agricultural and urban streams across biomes, where nitrate concentrations were also higher. However, uptake velocity decreased with ambient concentrations, suggesting that nitrate removal efficiency also decreased as instream concentrations increased—a finding that indicated a decrease in benthic demand for nitrate as anthropogenic sources of nitrate increased in a reach, resulting in greater downstream export (Mulholland et al. 2008).

13.3.2 Physical-Chemical Controls

Physical-chemical processes, such as sorption onto and precipitation from sediments, can greatly impact phosphate and, to a lesser extent, ammonium dynamics in streams. Comparatively, physical-chemical processes are not typically considered to influence nitrate concentrations in the water column. Sorption of phosphate ions onto charged clays and charged organic particles occurs when concentrations of soluble reactive P in stream or sediment pore water are high relative to an equilibrium SRP value of no net exchange. Desorption, or precipitation of phosphate from charged particles, occurs when SRP concentrations are lower than the equilibrium value (House 2003). Phosphorus sorption to suspended sediments is rapid, and occurs more readily with smaller particles, such as silts. Because of the association of phosphorus with sediments, amounts of both particulate and total phosphorus in stream water often vary temporally in parallel with concentrations of suspended sediments (Jordan et al. 1997; Ekholm et al. 2000). The affinity of orthophosphate for sediments is also responsible for the link between sediment and phosphorus concentrations in surface runoff. Once this particulate inorganic phosphorus enters the river, desorption often occurs, increasing the concentrations of bioavailable phosphorus.

Sorption-desorption processes can act as a buffer on streamwater nutrient concentrations by removing nutrients from solution when concentrations are high and releasing them back into solution when concentrations are low. Nutrients sorbed onto sediments can be released back into the water column after weeks or months (Peterson et al. 2001). After an 8-year phosphorus addition to a stream in Costa Rica, 10% of the total phosphorus added to the stream over the study period was bonded to aluminum and iron in the sediments (Small et al. 2016). This represented 99% of total phosphorus storage in the study reach. Four years after the experiment had ended, sediment phosphorus returned to pre-addition levels by slowly releasing phosphorus back into the water. Adsorption of ammonium on sediments can also represent an important fraction of the total uptake rate of the ion. For example, Ribot et al. (2017) estimated that sediment adsorption represented 22% of the uptake in a stream in Catalonia in northeastern Spain. In a meta-analysis conducted by Tank et al. (2008), ammonium uptake length estimated using isotope tracers were shorter than those made using ammonium releases (Fig. 13.9). This pattern was not observed for nitrate and probably was due to abiotic sorption of ammonium onto sediments. During spiraling experiments, sorption of phosphate or ammonium on sediments can overestimate uptake and the degree of nutrient limitation, particularly when the pulse release technique is used, due to the experimental increase of concentrations above ambient. Leaving an appropriate distance between release point and sampling sites, estimating the optimum distance to reduce the uptake effect, or modelling of the abiotic uptake can help

to reduce the effect of sorption on the calculation of spiraling metrics (Mulholland et al. 2002; Powers et al. 2009)

In addition to sorption-desorption processes, oxygen dynamics may also play an important role in nutrient exchanges. For example, under aerobic conditions both dissolved inorganic and dissolved organic phosphorus can complex with metal oxides and hydroxides to form insoluble precipitates. Under anaerobic conditions this phosphate can be released back into the water column (Fig. 13.10). The extent of the anaerobic zone varies seasonally and spatially with the amount of organic matter in the sediments, and the availability of dissolved phosphate mirrors these patterns. In a stream in England, SRP in pore water in the first 3 cm of the sediments was several times higher than in stream water, particularly at sites affected by sewage inflow. This pattern was likely related to low oxygen levels and organic matter decomposition in the sediments; however, there was no SRP diffusion across the sediment-water interface upstream from the sewage inputs.

Fig. 13.9 Meta-analysis of nutrient uptake studies showing the variation of (**a**) nitrate (NO_3) and (**b**) ammonium (NH_4) uptake lengths with discharge for releases using ^{15}N tracer and short-term nutrient releases. Grey dots represent those studies than increase nutrient concentration above 50 μg NH_4-N L^{-1} or above 100 μg NO_3-N L^{-1} (Modified from Tank et al. 2008)

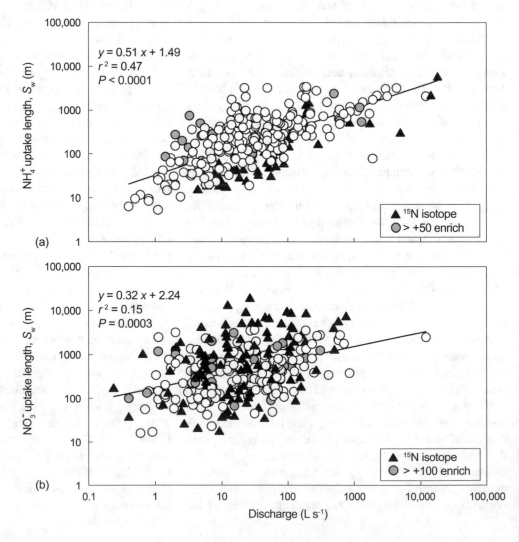

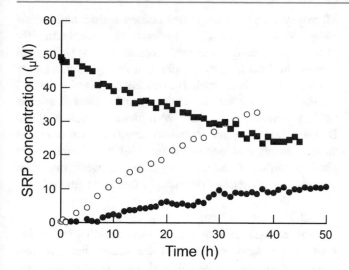

Fig. 13.10 Net uptake and release of SRP by river sediments. Uptake was measured after spiking the overlying water with orthophosphate. The greater release of SRP under anaerobic compared with aerobic conditions is attributed to the dissolution of phosphate that had precipitated with Fe hydroxide minerals (Reproduced from House 2003)

This may have been due to the precipitation of phosphorus with iron (the "oxic cap") in the well-oxygenated waters. Although the oxic cap was also present at sites impacted by sewage discharge, diffusion of phosphorus into stream water was observed under the reducing conditions that resulted from the decomposition of organic matter above the sediment surface (Palmer-Felgate et al. 2010).

13.3.3 Hydrologic Controls

13.3.3.1 Transient Storage
Storage zone mechanisms constitute an important hydrologic influence over nutrient dynamics. Transient storage typically takes place over short distances and time scales, and includes retention in both hyporheic and surface areas (Harvey et al. 1996; Brunke and Gonser 1997). Transient storage has the potential to increase nutrient processing by slowing the rate of downstream transport and increasing the exposure of dissolved nutrients to locations of high nutrient uptake. This effect is likely to be greatest when storage is the result of surface-subsurface exchange (Fig. 13.11) because water entering sediments will come in close contact with biofilms; storage in pools and eddies is less likely to promote this type of interaction (Hall et al. 2002).

Transient storage varies with numerous stream features including channel geomorphology, stream size, discharge, and flow obstructions (D'Angelo et al. 1993; Harvey et al. 1996; Morrice et al. 1997). It is greatest in headwater streams and declines with increasing stream size and discharge, most likely because of reductions in channel complexity, and the increased cross-sectional area of the wetted channel. A common metric to compare the relative size of the storage zone among streams is the ratio As/A, where As is the cross-sectional area of the storage zone and A is the channel cross-sectional area (Runkel 2002). The size of the transient storage zone varied from 0.16 to 0.71 among 13 forested headwater streams in New Hampshire (Hall et al. 2002). In the eighth-order Willamette River, Oregon, which has extensive gravel beds promoting hyporheic flow, A_S/A averaged 0.28, and was highest in two unconstrained reaches where the river was able to rework its gravel bed (Fernald et al. 2001). Similarly, in arctic streams of Siberia, mean A_S/A was 0.24, and showed a negative relationship with discharge, most likely related to limited hyporheic flow due to frozen soil and permafrost (Schade et al. 2016). Values of A_S/A greater than 1 have been recorded in streams with highly permeable bed material (Valett et al. 1996; Martí et al. 1997; Butturini and Sabater 1999). For example, in an Antarctic stream fed by glacial meltwater, A_S/A exceeded 1 in all stream reaches due to high gradient and porous alluvial materials (Runkel et al. 1998). By using the tracer resazurin to compare transient storage among substrates, Argerich et al. (2011) found metabolically active transient storage (MATS) represented 37% of transient storage in a bedrock reach dominated by surface storage, but was 100% in a reach with deep alluvial deposits and dominated by hyporheic storage. Even though bedrock was 16-times more metabolically active than alluvial

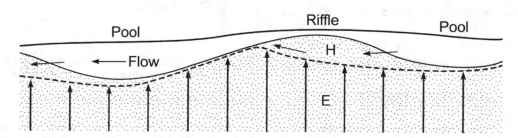

Fig. 13.11 Postulated distribution of hyporheic zones (H) and groundwater zones (E) beneath a pool-riffle-pool sequence in a Michigan river, as inferred from temperature profiles (Reproduced from White et al.1987)

substrate, its contribution was smaller because MATS had a very small surface area (0.002 m^2) relative to the alluvial deposits (0.292 m^2).

Transient storage is highly influenced by obstructions within the stream channel, such as boulders and wood. When vegetation and wood were removed from a vegetated agricultural stream and a forested blackwater stream, transient storage area decreased by 61% and 43%, respectively (Ensign and Doyle 2005). In the same system, experimental baffles were subsequently added to create in-channel transient storage, and A_S increased more than threefold in the agricultural stream and more than doubled in the blackwater stream. Uptake velocities were also enhanced by the addition of the baffles. Both ammonium and phosphate uptake increased markedly (Ensign and Doyle 2005). In sand-bed streams of Mississippi, US, transient storage increased with flow obstruction by wood and artificial devices, also indicating that storage was dominated by surface processes rather than by interactions in the hyporheic zone (Stofleth et al. 2008).

Although transient storage is expected to increase uptake rates because of longer exposure of dissolved nutrients to sediments and biofilms, empirical evidence is equivocal. For instance, as predicted, phosphorus uptake measured with the radioisotope of phosphorus (^{33}P) in two forested streams was greater in the stream that had deeper sediments, and so presumably also had a larger transient storage zone (Mulholland et al. 1997). Valett et al. (1996) also documented a positive relationship between transient storage and uptake in a stream in New Mexico with a large storage zone, but both were also correlated with discharge, and so the causal relationship was unclear. In contrast, Martí et al. (1997) documented no relationship between nutrient uptake length and size of the transient storage zone, even though the latter varied greatly following a flood disturbance in Sycamore Creek in the southwestern US. Similarly, in a comparison of 13 forested headwater streams in New Hampshire, transient storage explained only 35% of between-stream variation in ammonium v_f during summer months and 14% on an annual basis. This may have been because most of the transient storage occurred in the shallow hyporheic zones in pools of these bedrock streams (Hall et al. 2002). Similarly, an interbiome study comparing 10 undisturbed streams reported a low correlation between ammonium uptake and transient storage (Webster et al. 2003).

Under some circumstances, such as systems where the hyporheic zone is minimal and surface algae and biofilms are productive, transient storage associated with the bed surface may be of primary importance. In streams in Wisconsin in the north-central US, transient storage was positive correlated to macrophyte biomass and phosphate uptake velocity (Bohrman and Strauss 2018). Macrophyte biomass explained 63% of the variation in transient storage, suggesting that a large proportion of storage was surficial.

Macrophyte beds can increase residence time and reduce current velocity, which can increase the opportunities for phosphorus uptake. In a study conducted in 12 tropical streams in Brazil, reaches with greater A_S/A and longer water residence times, such as those that contained pools and marshes, had higher ammonium uptake rates than did reaches dominated by runs and meanders (Gücker and Boëchat 2004). Because transient storage occurred mostly in surface waters and accounted for 52–85% of ammonium uptake, surface transient storage zones were important locations of ammonium retention in these systems.

13.3.3.2 Stream Size

As stream size and discharge increase, the interaction between the water column and the stream bed declines, potentially influencing benthic and water column processes that mediate nutrient dynamics. For example, uptake length, the distance traveled by a nutrient atom before immobilization, increases with discharge and water velocity, as seen in a cross-biome comparison of ammonium S_w in 11 North American rivers (Fig. 13.12) (Peterson et al. 2001; Webster et al. 2003). Similar results were obtained by Tank et al. (2008), where a meta-analyses of nitrogen uptake studies and pulse addition experiments in the Upper Snake River, Wyoming, US, indicated a significant relationship between discharge and ammonium and nitrate S_w (Fig. 13.9). In two New Zealand streams, temporal variation in uptake lengths of ammonium, nitrate, and phosphate was also largely explained by changes in water velocity and depth (Simon et al. 2005). An increase in uptake length at higher discharge likely is the consequence of the reduced opportunity for solutes in stream water to interact with sediment surfaces and biofilms (Peterson et al. 2001).

Small streams have been described as "hotspots" for biogeochemical cycling because their shallow depths promote water-sediment interactions. As indicated in an isotope

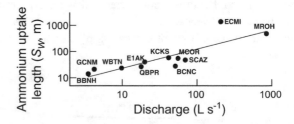

Fig. 13.12 Relationship between uptake length and stream discharge from ^{15}N-ammonium tracer additions at eleven stream sites in the US: Ball Creek, North Carolina (BCNC); West Fork Walker Branch, Tennessee(WBTN); Sycamore Creek, Arizona (SCAZ); Bear Brook, New Hampshire (BBNH); Gallina Creek, New Mexico(GCNM); Quebrada Bisley, Puerto Rico(QBPR); Kings Creek, Kansas (KCKS); Eagle Creek, Michigan (ECMI); Mack Creek, Oregon (MCOR); El Outlet, Alaska (E1AK); East Fork Little Miami River, Ohio(MROH) (Reproduced from Webster et al. 2003)

tracer study in headwater streams throughout North America, more than half of the dissolved inorganic nitrogen received from the catchment can be retained or transformed in small streams (Peterson et al. 2001). Similarly, in a tropical stream network in Puerto Rico, ammonium removal was greater in smaller streams, which received larger N inputs from land and showed higher uptake velocities. This resulted in reduced downstream export of ammonium, which likely limits ammonium uptake in lower reaches of river networks (Koenig et al. 2017). However, recent evidence suggests that larger streams also can play important roles in nutrient uptake (Ensign and Doyle 2006; Ye et al. 2017). For example, in the aforementioned study by Tank et al. (2008), uptake velocities in larger streams were similar to those in small streams. Similarly, Hall et al. (2013) estimated that ammonium uptake velocity was constant along the river network. In contrast, v_f for nitrate and soluble reactive phosphorus tend to decrease with increasing stream size. This could be due to enhanced abiotic sorption of phosphorus and denitrification rates in smaller streams. Despite efforts to understand nutrient cycling at the river network scale, data for large rivers are still relatively scarce, and this is especially true in the tropics. This is primarily due to methodological limitations, and the cost and complexity associated with completing biogeochemical research in regions where research funding is limited. Additionally, as both Tank et al. (2008) and Hall et al. (2013) have noted, the processes operating in smaller streams, such as benthic algal uptake and hyporheic exchange, will probably not apply to larger rivers, where planktonic demand and interaction with the floodplain may influence nutrient cycling.

The relatively recent development of high frequency nutrient sensors, in combination with new developments in modelling, will contribute to a broader understanding of nutrient dynamics at the network scale. For example, Yang et al. (2019) compared autotrophic uptake between forested and agricultural streams using continuous nitrate and dissolved oxygen data in the Selke River in north-central Germany. Greater nitrate uptake occurred in the agricultural reach (Fig. 13.13), and seasonal patterns were pronounced in both streams. Using a nitrate transport and removal model to scale results from the reach to the network level indicated that nitrate uptake was greater in higher order streams (3rd–5th order), while uptake efficiency was greater in smaller streams (1st and 2nd order). These patterns likely were related to changes in riparian shading and nitrate loading along the river network.

Though most studies have focused on the role of benthic processes in nutrient cycling, processes in the water column can also influence nutrient uptake rates. By comparing three watersheds in Wyoming, Indiana, and Michigan, US, including streams ranging from 1st to 5th order, Reisinger et al. (2015) found that water column uptake velocity (v_f) of ammonium increased with stream size, suggesting that ammonium removal increases with greater depth and with less canopy cover. In contrast, the uptake velocity of nitrate or soluble reactive phosphorus in the water column did not predictably vary with stream size. In the same study, water column nutrient demand estimates ranged between 4 and 19% of published stream estimates from other reach-scale studies, suggesting that nutrient removal in the water column may be relatively small in some cases.

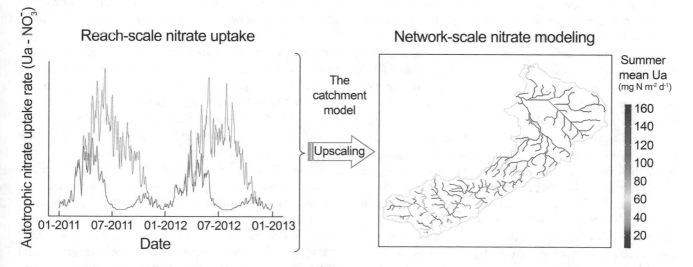

Fig. 13.13 Autotrophic nitrate uptake for a forested upstream (green line) reach and an agricultural (yellow line) downstream reach of the Selke River, Germany. Measurements were conducted using high-frequency sensors, which were upscaled from the reach to the network scale, showing higher uptake rates in larger streams (Modified from Yang et al. 2019)

13.3.4 Biotic Controls of Nutrient Cycling

The efficiency with which stream ecosystems are able to utilize, retain, and recycle nutrients is largely determined by environmental factors that influence uptake and assimilation by the biota and, in the case of nitrogen, the various biologically-mediated transformations that influence its chemical form. Nutrient uptake and cycling is expected to vary directly in response to biotic demand by both primary producers (benthic algae, macrophytes, bryophytes) and heterotrophic microorganisms (bacteria and fungi). Productive systems should cycle nutrients at higher rates, owing to higher uptake rates, and higher rates of regeneration through consumption, excretion, and egestion and subsequent mineralization. Animal consumers such as herbivores can stimulate rates of production by increasing the turnover rate of the producer community and by excreting and egesting nutrients that can support producer demand. The dissimilatory activities of nitrifying and denitrifying bacteria change the concentrations of various forms of inorganic nitrogen, and either enhance or limit the bioavailability of nitrogen.

13.3.4.1 Assimilatory Uptake

Assimilation of nutrients by aquatic organisms can represent an important fraction of total nutrient uptake rates in stream ecosystems. Researchers have used ^{15}N tracer releases to estimate relative importance of assimilatory and dissimilatory processes in nitrogen cycling. In a comparative study of 11 headwater streams from biomes throughout North America, assimilation by autotrophs, bacteria, and fungi and via sorption to sediments were primarily responsible for ammonium uptake rates (70–80% of total uptake), and nitrification also played an important role (20–30%; Peterson et al. 2001; Webster et al. 2003). The importance of assimilatory uptake by autotrophs is well illustrated by a study of 72 streams in the United States and Puerto Rico, where autotrophs were responsible for more than 80% of total nitrate uptake, and nitrification was responsible for 16% (Hall et al. 2009). Biotic uptake only removes nitrogen from the system if species are mobile or are harvested from a stream, as most is recycled within a relatively short time-frame and travel distance. However, biotic uptake can temporarily store large fractions of nitrogen, and enhance the possibility that denitrification can occur and remove the nitrogen prior to downstream export.

Nitrogen uptake by autotrophs often exceeds uptake by heterotrophs. For example, ^{15}N additions conducted in multiple bioclimatic regions around the world suggest that mass-specific ammonium uptake by autotrophs including epilithic biofilm, filamentous algae, and macrophytes was greater and more variable than uptake by heterotrophs associated with leaf litter, wood biofilm, and fine benthic

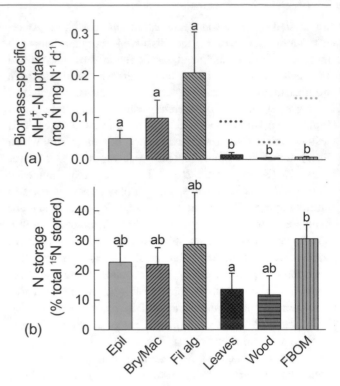

Fig. 13.14 Contribution of different nutrient uptake compartments to (**a**) biomass-specific NH₄-N uptake, and (**b**) N storage at the reach scale. Values shown are mean plus one standard deviation of the study streams. Different letters indicate significant differences among uptake compartments (p < 0.05). Dashed lines above bars in panel A indicate recalculated biomass-specific NH₄-N based on microbial biomass (Epil, epilithic biofilm; Bry/Mac, bryophytes and macrophytes; Fil alg, filamentous algae; Leaves, biofilm on decomposing leaves; Wood, biofilm on small wood; FBOM, fine benthic organic matter) (Reproduced from Tank et al. 2018)

organic matter (Fig. 13.14; Tank et al. 2018). Autotrophic uptake was especially important in reaches with open canopies. In contrast to observed differences in uptake rates, the fraction of nitrogen stored in autotrophic and heterotrophic compartments was similar. Assimilatory uptake can vary between forms of nitrogen. The aforementioned study by Ribot et al. (2017) estimated that assimilatory uptake comprised only 16% of nitrate uptake but 61% of ammonium uptake in forested Catalonian stream in northeastern Spain. In this study, biotic processes associated with microbial activity dominated nitrate uptake rates, but nitrification was responsible for a large proportion of ammonium uptake (39%), indicating that much of the ammonium is being converted into nitrate that can be transported to downstream reaches.

Biotic demand influences phosphate uptake, particularly during periods when streams receive fresh inputs of leaf litter. For instance, phosphorus uptake lengths were negatively correlated with amount of leaf litter found in a woodland stream, reflecting increased microbial demand for phosphorus associated with increased leaf-litter decomposition (Mulholland et al. 1985). Similarly, when leaf litter and then wood was removed for an Appalachian headwater stream, nutrient uptake lengths increased substantially in comparison to a reference stream (Webster et al. 2000). A comparison of two streams that differed in community respiration found a higher demand for phosphorus in the stream with the higher community respiration (Mulholland et al. 1997).

Hydrologic regime influences the abundance of primary producers and the retention of particulate organic matter, as nicely illustrated by Grimm's (1987) study of successional events in Sycamore Creek, Arizona, in the southwestern United States. Following a flood that eliminated virtually all of the biota, biomass initially accumulated rapidly and then plateaued as the system acquired a thick periphyton mat and increased densities of invertebrates. Measurements of hydrologic inputs, storage, and outputs of nitrogen in the biota documented substantial retention of inorganic nitrogen within the 90-m reach, attributed to nitrogen uptake and accumulation in living tissue. In a subsequent study in the same system that employed nitrate and chloride additions to measure uptake length, Martí et al. (1997) showed that nitrate S_w was short relative to other published studies, consistent with evidence that nitrogen was limiting in this desert stream. Nitrate uptake length in Sycamore Creek doubled after a modest midsummer flood.

Similar processes have been documented in streams in other biomes. Periods of low flow and scouring high flows are associated with changes in nutrient retention and the reduction of algal mats dominated by filamentous cyanobacteria in Antarctic streams (McKnight et al. 2004). In an agricultural catchment in Denmark, periods of drought resulted in reductions in ammonium uptake and increases in phosphorus uptake that were attributed to reductions in autotrophic activity and increased heterotrophic demand, respectively (Riis et al. 2017). In the same study, a ^{15}N ammonium release indicated that reduced flow altered the dominant locations where assimilatory uptake occurred in the stream and caused a four-fold decrease in the reach-scale rate of assimilatory update. Under low flows, assimilatory nitrogen uptake primarily occurred in locations of fine benthic organic matter; however, uptake by macrophytes and epiphytic biofilm dominated assimilatory uptake under higher flows.

Other environmental variables such as light can limit primary production (Sect. 6.1.1), thereby influencing nutrient uptake. In a comparison of forested versus logged reaches of a Mediterranean stream, Sabater et al. (2000) found greater algal biomass in open relative to shaded reaches, and uptake lengths for ammonium and phosphorus also were shorter in the open reaches. The uptake velocity

for phosphorus was correlated to primary production in both reaches, suggesting that phosphorus retention in the system were strongly governed by algal productivity. In contrast, v_f for ammonium was poorly correlated with primary production, suggesting that microbial heterotrophic processes or abiotic mechanisms were primarily responsible for ammonium uptake. Similarly, Burrows et al. (2013) documented enhanced phosphate uptake in reaches that had been affected by logging when compared to old-growth forest streams. The authors attributed the increased uptake rates to enhanced light availability, which supported algal growth, and larger quantities of woody debris in streams that may have promoted the growth of microbial biofilms. In addition, abiotic sorption of phosphorus on sediments and charred woody debris likely played a role in enhanced phosphorus uptake in logged reaches.

Aquatic macrophytes and bryophytes are capable of removing substantial amounts of nutrients from flowing water. Meyer (1979) recorded significant removal of phosphorus as a pulse passed over a bryophyte bed in a forested stream in New Hampshire. Likewise, in Walker Branch, Tennessee, ammonium uptake by the bryophyte *Porella* represented 41% of total N retention at the end of a six-week ^{15}N addition experiment (Mulholland et al. 2000). In the River Lillcaa in Denmark, uptake of ammonium was four times higher in areas with macrophytes (*Ranunculus aquatilis* and *Callitriche* sp.) than in non-macrophyte habitats. Epiphytes on macrophytes accounted for 30% of the uptake in macrophyte reaches, while in non-macrophyte habitat, fine benthic organic matter accounted for 98% of the ammonium uptake. Turnover rates of nitrogen in macrophytes was also slower than in other autotrophic compartments, suggesting that macrophytes can play an important role in long term retention of nitrogen in streams (Riis et al. 2012). Laboratory studies of the impact of plants on porewater nutrient pools indicated that plants reduce nutrients in porewater through metabolic uptake; however, field studies do not always support these findings. For instance, research in the middle Hudson River found that in some cases, porewater influenced by submerged vegetation had enriched nutrient concentrations (Wigand et al. 2001). The authors suggested the apparent contradiction was due to the accumulation of allochthonous particulate organic matter in macrophyte beds, and the corresponding microbial mineralization of nutrients. In effect, this work demonstrated the uptake and sequestration of nutrients by macrophyte beds can be masked by processes that promote replenishment of porewater nutrients (Wigand et al. 2001).

Environmental factors that influence algal and microbial production often vary seasonally, leading to temporal changes in nutrient uptake and retention. In a forested headwater stream in Tennessee, water column nitrate and phosphate concentrations were lowest in autumn and early spring, corresponding to periods of highest heterotrophic and autotrophic activity, respectively (Mulholland et al. 2004). In the temperate zone, high uptake rates are seasonally driven by microbes colonizing fresh leaf litter in the fall and the photosynthetic demands of algae and bryophytes in the spring before leaf-out. Uptake rates tend to decline with seasonal increases in discharge and with increased shading from foliage. Marked seasonal patterns in autotrophic nitrate uptake in forested and agricultural streams in Germany in the study by Yang et al. (2019), described previously (Fig. 13.13), were also attributed to changes in light availability due to riparian shading. The closed canopy of the forested streams resulted in lower nitrate uptake rates than in the agricultural streams, which have no gallery trees (Yang et al. 2019).

13.3.4.2 Dissimilatory Transformations

By transforming ammonium to nitrate, nitrifying bacteria can influence the concentrations of both ions in stream water, and thus influence the rate at which nitrogen is removed from the stream through denitrification. The process of nitrification contributes to ammonium uptake and to the production of nitrate in streams; however, this varies among systems. For instance, nitrification accounted for approximately half of total ammonium removal in streams in Michigan (Hamilton et al. 2001) and Puerto Rico (Koenig et al. 2017), but for less than 20% of ammonium uptake in Walker Branch, Tennessee (Mulholland et al. 2000). In the Puerto Rican streams, nitrification declined with stream order as a result of lower ammonium availability (Koenig et al. 2017). Similar variation can occur at smaller spatial scales as variation in stream hydraulic features and in the availability of organic carbon can influence nitrification rates (Bernhardt et al. 2002).

After nitrification occurs, the resulting nitrate can be immobilized by the biota, exported downstream, or transformed into N_2 gas by denitrifying bacteria. Nitrification was implicated as a source of nitrate in a Mojave Desert stream, where streamwater nitrate concentrations were two times greater than expected from groundwater inputs (Jones 2002). Researchers attributed the high nitrate values to the mineralization of organic nitrogen to ammonium that supported high rates of nitrification. Nitrifying bacteria require aerobic conditions; hence, nitrification rates should be greater where subsurface flows provide oxygenated water, such as in shallow, disturbed sediments. Nitrification rates within the hyporheos of Sycamore Creek were highest in regions of downwelling, which presumably supplied organic carbon and oxygenated water (Jones et al. 1995). In this system, most of the nitrogen demand of surface algae was met by hyporheic mineralization of ammonium and subsequent nitrification. In a third-order stream in Oregon, in the

western US, a transition between nitrification and denitrification was measured in hyporheic zone along a gravel bar. Net nitrification occurred in the upstream end of the bar, where oxic conditions and shorter residence times dominated. In contrast, net denitrification occurred in the downstream reaches of the gravel bar where lower oxygen levels and longer residence times were common (Zarnetske et al. 2011). Both of these studies demonstrate the importance of coupled surface and subsurface habitats in supporting stream biogeochemical processes.

The reduction of nitrate to nonreactive N_2 and N_2O gases by denitrifying bacteria is an important pathway in the nitrogen cycle because it represents the only process that permanently removes reactive nitrogen from the stream network. Denitrification rates are enhanced under low oxygen conditions where nitrate and organic matter supplies are rich. Denitrifying bacteria can be found in association with fine benthic organic matter and within senescing mats of *Cladophora* and cyanobacteria, where anoxic zones are common (Triska and Oremland 1981; Kemp and Dodds 2002a, b). For example, sediment-rich streams in agricultural watersheds are frequently characterized by high denitrification rates (David and Gentry 2000; Royer et al. 2004). However, nitrate levels can remain high in these systems due to anthropogenic inputs of nitrogen-rich fertilizer.

In a review of nitrogen loss from streams in the northeastern US, Seitzinger et al. (2002) concluded that the collective amount of nitrogen lost from all streams within a catchment is much greater than would be estimated based on nitrogen loss from a single reach, because there is a cumulative effect from continued removal as nitrogen exported from one reach can be removed in downstream sections of the river network. In their model, between 37 and 76% of nitrogen inputs were lost through denitrification along the river network. Approximately half of the loss occurred in first through fourth-order streams, and the remaining nitrogen was lost in higher-order rivers within the network. This finding may be regional and tied to dominant land cover within a watershed. For example, Royer et al. (2004) suggested that floodplains and wetlands may be more important than headwater streams as locations of denitrification in watersheds in the agricultural midwest of the US. There is still considerable uncertainty about the magnitude of denitrification in river systems. For example, estimated denitrification rates for the Elbe, a 8th order river in Europe, were relatively high, particularly during the spring and summer, resulting in the removal of 10% of the annual nitrogen inputs in the study reach (Ritz et al. 2018). These estimates are double those estimated by Seitzinger et al. (2002) for nitrogen removal in large rivers.

Further insights into the factors influencing denitrification rates are provided by a large, cross-biome ^{15}N tracer addition experiment conducted in 72 streams in the US and Puerto

Rico (Mulholland et al. 2008, 2009). Among systems, denitrification was highly variable, representing between 0.5 and 100% (median of 16%) of total nitrate uptake. In one-third of the streams, denitrification was responsible for more than 40% of total uptake. Areal denitrification (U_{den}) increased with nitrate concentrations among study streams. However, uptake velocity (v_f) decreased as nitrate concentration increased, indicating that uptake efficiency declined with increasing nitrate concentrations. This pattern was not dependent upon dominant land use in the watershed. Denitrification rates were positively correlated with estimates of ecosystem respiration, which is not surprising as increased microbial activity, and subsequent increases in respiration, often depend on an abundance of organic carbon and result in lower oxygen levels, conditions that favor denitrification.

Notably, denitrification can produce N_2O, which is a greenhouse gas. Results from the cross-biome ^{15}N tracer addition experiment indicate that higher instream nitrate concentrations enhanced N_2O production, but did not influence the amount of N_2O produced relative to the amount of N_2 released during denitrification. Though the proportion of N_2O produced through stream denitrification is small (<1% of the total N denitrified), global estimates suggest that the N_2O produced in streams could equal 10% of total amount of N_2O produced through anthropogenic activities (Beaulieu et al. 2010). Nitrification also can contribute to N_2O production. In the Chao Lake basin in China, Yang and Lei (2018) found a strong correlation between N_2O and ammonium in agricultural rivers and with nitrate concentrations in rivers draining populated rural areas, suggesting that nitrification and denitrification both influence N_2O production.

Enhancing stream denitrification is a potential management option to remove nitrogen from aquatic ecosystems, and decrease nitrogen loading to downstream and coastal areas. Restoration actions such as reducing stream incision and reshaping banks to reconnect the stream channel with its floodplain can increase denitrification rates in urban streams (Kaushal et al. 2008). Creating areas that enhance organic matter accumulation, such as debris dams, can also increase denitrification (Craig et al. 2008). On a smaller scale, experimental addition of leaf packs in New Zealand streams increased denitrification rates compared to control streams by increasing the availability of labile organic carbon and microbial activity (O'Brien et al. 2017). Although denitrification can remove nitrogen from aquatic ecosystems, the efficiency of nitrogen removal declines as instream nitrogen concentrations increase. Furthermore, increased instream nitrogen concentrations also are linked to increased greenhouse gas production through denitrification. Therefore, the best management practice seems to be to reduce nitrogen loading to the watershed (Beaulieu et al. 2010).

13.3.4.3 Role of Consumers in Nutrient Dynamics

Animals influence nutrient dynamics by consuming, metabolizing, and releasing organic and inorganic forms of nutrients through excretion, egestion, and the decomposition of tissue (Vanni 2002). These processes, also termed consumer-driven nutrient dynamics (Capps et al. 2015a), can be an important part of biogeochemical cycling in riverine systems (Atkinson et al. 2017; Subalusky and Post 2019). Nutrients are often excreted in inorganic forms such as ammonium and phosphate, and thus are available to primary producers, but other forms of waste, such as urea and feces, are organic in nature and require additional processing by microbes to be available for uptake and assimilation. Because nutrient availability can strongly limit the overall productivity of a stream ecosystem, high rates of animal consumption and growth can stimulate rates of nutrient cycling. Furthermore, because many species are mobile, animals can play important roles in transporting nutrients between ecosystems, and these spatial subsidies can, in turn, alter the community structure and ecosystem dynamics in recipient systems (Subalusky and Post 2019).

Environmental factors such as temperature, discharge, and nutrient limitation can influence the impact of consumers on system-wide nutrient cycling (Vanni 2002). For example, in an early examination of consumer-driven nutrient dynamics in a desert stream, Sycamore Creek, Arizona, in the southwestern US, animal growth and production were observed to be very high, despite strong nitrogen limitation (Grimm and Fisher 1986). Based on laboratory estimates of mass-specific excretion and egestion rates, Grimm (1988) estimated that approximately half of all ingested nitrogen was egested as fecal material and presumably was recycled through coprophagy (re-ingestion) by consumers, or was leached from feces and became available for uptake by autotrophs following leaching or microbial breakdown.

Organismal traits such as tissue nutrient content, body size, and trophic ecology mediate species-specific nutrient consumption, retention, and mineralization. Ecological stoichiometry uses a mass-balance approach to examine relationships between the stoichiometry, or the ratio of elements, among an organisms' diet, tissue nutrient content, and waste products from excretion and egestion (Sterner and Elser 2002). This framework predicts that consumers will preferentially retain elements in the shortest supply relative to their demand (similar to Liebig's Law of the Minimum Sect. 6.1.1), and they will release elements that are consumed in excess of their needs. For example, in a phosphorus-limited system, an organism with high phosphorus demand for its internal structures should preferentially retain phosphorus and excrete and egest less phosphorus relative to other nutrients. In a comparison of excretion rates and ratios for N and P for 28 species of fishes

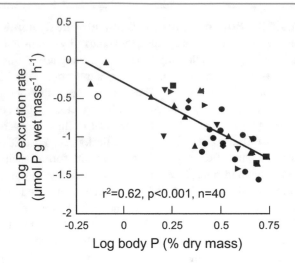

Fig. 13.15 Relationship between P body content and P excretion rate in fishes and amphibians in Río Las Marias, Venezuela (Reproduced from Vanni et al. 2002)

and amphibians in Río Las Marias, an Andean piedmont stream in Venezuela, Vanni et al. (2002) demonstrated that phosphorus excretion rates and N:P excretion ratios were negatively correlated with body phosphorus content and body N:P ratios, respectively (Fig. 13.15). Total excretion by the assemblage of consumers in this stream was estimated to meet 49% of algal demand for nitrogen and 126% of algal demand for phosphorus. Interestingly, armored catfish (Siluriformes, Loricariidae) have a very high body content of phosphorus, consume a stoichiometrically imbalanced diet that is low in phosphorus, and produce excretions of low phosphorus content and high N:P ratios (Hood et al. 2005). As a result, these grazing fishes appear to act as P sinks, decreasing the availability of phosphorus for algae, which in turn may result in lowering the quality of food available to consumers.

The amount of nutrients excreted per unit body mass and time usually decreases with increasing body mass as a consequence of the scaling of metabolism with body size. If larger prey are preferentially ingested, the excretion rate of the prey assemblage will increase (Vanni 2002). The metabolic theory of ecology asserts that metabolic rates in an organism should vary predictably as a function of temperature and body size. According to this theory, small-bodied consumers exposed to high temperatures should have the highest mass-specific excretion rates (Brown et al. 2004).

Animal consumers also affect nutrient cycling indirectly, through their influence on benthic algal biomass, organic matter dynamics, and prey assemblages (Vanni 2002). Whenever grazing sharply reduces algal and biofilm biomass, uptake rates are expected to decrease and spiraling distance to lengthen, whereas moderate grazing that stimulates primary and microbial production should have the opposite effect. Consumption of algae and leaf litter by the snail *Goniobasis*

clavaeformes in artificial channels stimulated mass-specific metabolic rates of microbial populations, but periphyton and microbial biomass were so reduced that overall biotic uptake of phosphorus declined (Mulholland et al. 1983, 1985). As a result, spiraling distance was shortest in the absence of snails. Low grazing pressure by snails allowed the accumulation of an algal mat that created transient storage effects, which in turn enhanced nutrient recycling within the mat (Mulholland et al. 1994). The conversion of CPOM into FPOM by detritivores is likely to enhance transport of particulate N and P because fine particulates are more easily exported during storms, but also may favor retention whenever fine particles are consumed (Vanni 2002).

Animals also transport nutrients among habitats and across ecosystems by their movements and migrations (Subalusky and Post 2019). Migratory fishes such as Pacific salmon can add significant amounts of phosphorus to rivers via excretion and the decomposition of carcasses (Janetski et al. 2009), where the net import of nutrients depends on both the abundance of adult fish migrating from marine systems, and the average size of the juvenile fish emigrating to the ocean (Kohler et al. 2013). Emergence of the adult stages of aquatic insects is one such process. Various authors agree that it is a small fraction of overall nutrient transport (<1%: Meyer et al. 1981; Grimm 1987), although potentially important to animal consumers of the riparian zone as a supply of organic carbon. In contrast, spawning runs of anadromous fish may import substantial amounts of marine-derived nutrients to streams and lakes by their excretion, release of gametes, and their own mortality, especially if many or all die after reproducing. Spawning salmon provide an important nutrient subsidy to freshwater ecosystems of the Pacific Northwest that are generally nutrient-poor (Naiman et al. 2002). Where salmon are abundant, a large fraction of the nitrogen in the stream biota likely is derived from spawning fish (Bilby et al. 2001), and significant quantities appear in riparian vegetation and a host of animal consumers. In Sashin Creek, Alaska, isotope analysis showed that nitrogen and carbon derived from a spawning run of Pacific salmon were incorporated into periphyton, macroinvertebrates, and fish (Kline et al. 1990). A comparison of an Alaskan stream in which downstream reaches contained salmon, but upstream reaches did not, found that salmon-supporting reaches had higher soluble reactive phosphorus concentrations, epilithon abundance, and chironomid biomass, but mayfly biomass was lower (Chaloner et al. 2004). Because of declines in salmon populations, it is estimated that only 6–7% of the marine-derived nitrogen and phosphorous that historically was transported into the rivers of the Pacific Northwest by salmon is currently reaching those rivers (Gresh et al. 2000). Similar estimates have been made for other historical fish migrations (Flecker et al. 2010).

Recent work calls attention to the impact that terrestrial mammals can have on instream nutrient dynamics. In the Mara River, Kenya, mass drownings of wildebeests (*Connochaetes taurinus*) during their annual migration introduce approximately 1,100 tons of nutrient-rich biomass that can contribute between 31–451% of the total phosphorus transported by the river during the 28-day period carcasses persist in the river (Subalusky et al. 2017). Large herding mammals once were much more common globally, but have been extirpated from many regions of the world due to habitat loss and exploitation. This suggests that river networks may have lost potentially important sources of allochthonous subsidies through time. For example, modelling efforts suggest a single mass drowning event of bison on the Assiniboine River in western Canada could have contributed half of the annual load of total phosphorus to the river (Wenger et al. 2019).

Anthropogenic activities can also influence consumer-driven nutrient dynamics in lotic environments. Wilson and Xenopoulos (2011) demonstrated that the volume of nitrogen and phosphorus excreted by fish communities in streams in Ontario, Canada, was positively correlated with the amount of cropland in the riparian zone. Species loss due to extirpation or extinction or the addition of species through introductions can also alter the contribution of consumers to instream nutrient dynamics (Capps et al. 2015b). Atkinson et al. (2014) documented how decreases in density and biomass of mussels in response to extreme drought in rivers in Oklahoma, US, resulted in lower availability of N and storage of phosphorus. Similarly, research focused on the ecosystem-level impacts of amphibian declines highlighted the profound impact stream tadpoles had on nitrogen cycling in Central American streams (Whiles et al. 2013). Amphibian decline was related to lower ammonium demand due to reduced algal turnover, as well as reduced nitrogen mineralization due to lower ingestion and excretion by tadpoles and lower bioturbation. Illustrating the impact of an invasive species, Hall et al. (2003) estimated that the exotic snail *Potamopyrgus antipodarum* consumed 75% of algal production, and its excretion supplied two-thirds of the algal mat's ammonium demand, in a highly productive geothermal spring stream in Yellowstone National Park, in the western US. High snail biomass, rather than high rates of consumption and excretion per individual, explained the snail's dominant role in nutrient flux in this system. Similar work in the phosphorus-limited Chacamax River in Chiapas, Mexico, demonstrated that high densities of an invasive population of phosphorus-rich armored catfish (mentioned above; *Pterygoplichthys* spp) reduced the distance required for fish excretion to turn over the nutrient ambient pools more than twenty times for ammonium (21 to 0.8 km) and for phosphorus (102 to 4.1 km) when compared with native fishes (Capps and Flecker 2013).

13.4 Global Trends in Stream Nutrient Dynamics

In the first global assessment of nutrient concentrations and fluxes from the rivers to the world's oceans, Meybeck (1982) derived estimates based on data from approximately 30 of the world's rivers. Subsequent analysis of approximately 165 sites (Smith et al. 2003) were in agreement with Meybeck's estimates, but higher, indicating that nutrient loads of world rivers had increased in the span of a few decades. Modeling studies support the view that riverine nutrient export has increased over time (Fig. 13.16) (Beusen et al. 2016; Vilmin et al. 2018). The proportion of inorganic forms in total N and P inputs has increased, attributable mainly to runoff from fertilized agricultural lands and increased sewage inputs of DIN, and particle-bound inorganic P associated with soil loss from agricultural lands. Regional differences are pronounced. Losses of P from agricultural lands were a dominant source as early as 1900 in large areas of the US, Europe, India, China and in the Andes region. In contrast, natural sources remain dominant in large areas of the Amazon basin and in tropical Africa over the entire 20th century.

Spatial and temporal patterns in nutrient loads for a region or watershed can be understood from changes in nutrient inputs and exports, which are usually accounted for in nutrient budgets. Budgets have been constructed for small and large catchments, and for large river networks at regional and global scales. Robust budgets are important because they provide estimates of the quantity of riverine nutrients discharged to lakes and oceans, and they provide critical information for the management of eutrophication. By quantifying changes in input, standing stocks, and export rates, we have enhanced our understanding of the ways in which humans have modified stream biogeochemistry.

A nutrient budget is constructed from estimates of all known inputs and outputs (also termed fluxes) of a given element from a defined area. For catchments, the output is the total mass of nutrient exported (also known as load) by the river per unit time ($kg\ yr^{-1}$), which can be divided by catchment area so that it is a yield (often reported as $kg\ ha^{-1}\ yr^{-1}$). Output estimates can be much less than input estimates in budget calculations due to storage and, in the case of nitrogen, loss of the element to the atmosphere through denitrification. As previously mentioned, estimation of nutrient export from watersheds is of particularly interest because of their impacts on eutrophication processes in lakes and coastal waters. Export can be estimated using empirical data and models. Direct measurements of instream nutrient concentration coupled with water discharge data from stream gages frequently are used to estimate load. Model estimates range from those using simple relationships based on

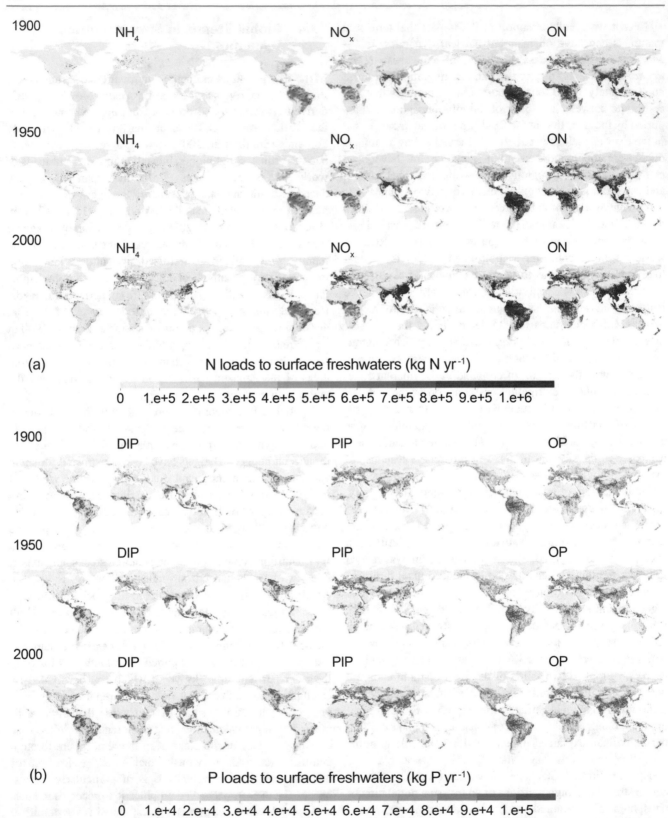

Fig. 13.16 Average inputs of nitrogen (**a**) and phosphorus (**b**) to rivers in 1900, 1950, and 2000 (NOx, nitrate plus nitrite; ON, organic nitrogen; DIP, dissolved inorganic phosphorus; PIP, particulate inorganic phosphorus; OP, organic phosphorus) (Reproduced from Vilmin et al. 2018)

watershed nutrient inputs and fluxes to more complex approaches that integrate biophysical, hydrological, chemical, and socioeconomic information to estimate nutrient export (Seitzinger et al. 2010; Howarth et al. 2012).

13.4.1 Nitrogen

Fluxes of riverine nitrogen from large regions are commonly estimated using the "net anthropogenic nitrogen inputs" (NANI) budget approach (Fig. 13.17). In a NANI budget, anthropogenic inputs include fertilizers, atmospheric deposition of NOy, nitrogen fixation by agricultural crops, and nutrients contained in imported human food and animal feed (Boyer et al. 2002). The final term is a net term, as a catchment can either import more food and feed than it produces, or export more food and feed than is consumed within the catchment. Nitrogen budgets reveal strong human influence on inputs, as humans have more than doubled the global supply of reactive nitrogen. Inputs of N to catchments have increased dramatically due to increased use of agricultural fertilizers, some of which enters rivers via runoff; the burning of fossil fuels, which release NOy gases to the atmosphere and result in wet and dry deposition; and the global trade in food, which can result in large imbalances in both agricultural (nutrient exporting) and urban (nutrient receiving) watersheds. Because of dispersion of air masses,

even the most remote watersheds receive some atmospheric deposition. Undisturbed tropical regions are thought to be minimally affected by atmospheric deposition from distant sources, although fires from savannahs and deforested locations can act as a more regional source to tropical forests (Chen et al. 2010). In the absence of disturbance or significant anthropogenic inputs, much of the nitrogen exported by rivers is in organic form, and evidence indicates that average TN yields are higher from undisturbed tropical rivers than from near-pristine temperate rivers (Lewis et al. 1999, Clark et al. 2000). Brookshire et al. (2012a) estimated more than ten-fold higher annual exports of nitrate from relatively pristine tropical streams (5 kg ha^{-1} yr^{-1}) compared to similar systems in the temperate zone (0.3 kg ha^{-1} yr^{-1}). Larger nitrogen yields in tropical systems are attributed to greater rates of N-fixation, and to a relative absence of nitrogen limitation in tropical terrestrial systems that favors the export of nitrogen from terrestrial to aquatic environments (Downing et al. 1999; Brookshire et al. 2012b). Where anthropogenic inputs are substantial, dissolved inorganic N, mainly as nitrate, predominates.

Riverine exports of nitrogen from human-dominated watersheds are highly correlated with estimated nitrogen inputs. For example, by summing all anthropogenic nitrogen inputs, Howarth et al. (1996) demonstrated that the quantities of nitrogen exported by major rivers of Europe and the

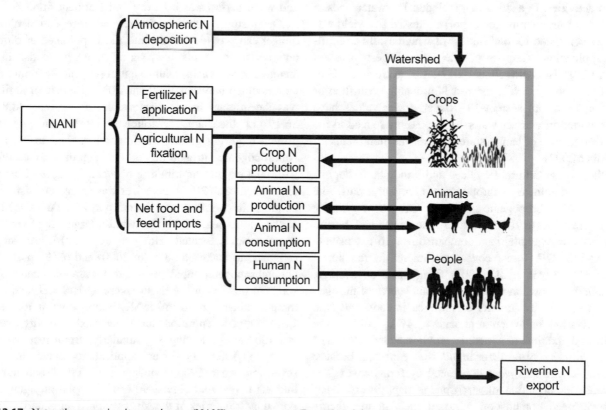

Fig. 13.17 Net anthropogenic nitrogen inputs (NANI) compartments (Reproduced from Hong et al. 2011)

Fig. 13.18 Relationship between riverine total nitrogen (TN) flux and (**a**) NANI and (**b**) synthetic N fertilizer inputs across 154 watersheds in the US and Europe (Reproduced from Howarth et al. 2012)

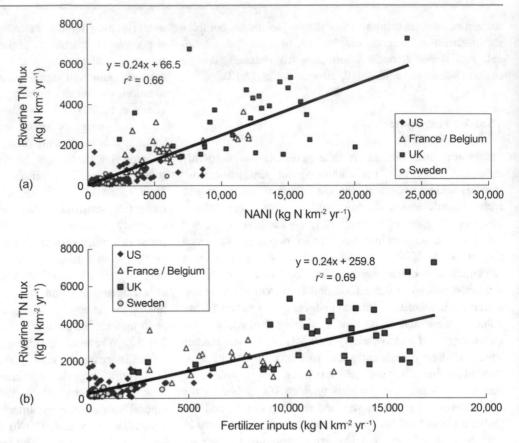

US into the North Atlantic Ocean are largely controlled by human activities. In a study that included 154 watersheds in the US and Europe, nitrogen export expressed as a yield was strongly correlated to total inputs, of which synthetic fertilizer application frequently was the largest fraction (Fig. 13.18) (Howarth et al. 2012). On average 25% of NANI is exported, while the rest is probably denitrified or stored in terrestrial systems. However, lower values have been reported in warmer watersheds, possibly related to the positive influence of temperature on denitrification (Schaefer and Alber 2007).

Although agricultural practices and atmospheric deposition often dominate anthropogenic nitrogen inputs, the magnitude of anthropogenic sources of nitrogen vary greatly among sites due to differences in land cover and human activities. For example, in a comparison of 16 extensively forested (48–87% forest cover) catchments in the northeastern US, Boyer et al. (2002) found that, on average, atmospheric deposition was the dominant source of nitrogen (31%), followed by imports of nitrogen in food and feed (25%), fixation in agricultural lands (24%), fertilizer use (15%), and fixation in terrestrial environments (5%). The smaller influence of fertilizer inputs is as expected, because the study catchments were dominated by forest cover. The substantial contribution of atmospheric deposition reflects the widespread combustion of fossil fuels in the energy sector and by vehicles, releasing NO_x gases that reach land and water surfaces as both dry and wet deposition.

Other studies provide further evidence that a nitrogen budget can provide insight into the importance of different anthropogenic inputs to riverine nitrogen loads. In the Yangtze River basin, China, anthropogenic N inputs more than doubled between 1980 and 2012. The use of fertilizers was the primary driver, responsible for between 40.8 and 56.1% of the increase, followed by human population growth requiring greater net imports of food to the basin. Anthropogenic nitrogen inputs were significantly correlated with DIN export, contributing between 37–66% of the river load (Chen et al. 2016). Fertilizer was again demonstrated to be the main contributor to river export in the drainage basin of the Baltic Sea, where the world´s largest hypoxic "dead zone" zone is located (Hong et al. 2017). Fertilizer was estimated to make up 58% (in 2000) and 68% (in 2010) of net anthropogenic inputs, with greater amounts entering the southern portion of the basin where agricultural land use is more dominant. In the lower Mississippi River in the central US, nitrate concentrations have increased markedly over the last 100 years, showing a particularly strong increase from 1970 to 1983. Fertilizer applications have increased seven-fold since 1960 (Goolsby et al. 1999), and nitrogen budgets reveal that almost 90% of the total nitrogen transported by tributaries of the Mississippi River is derived from diffuse sources. Fertilizer and soil organic matter contribute 50% of total nitrogen, atmospheric deposition, groundwater, and soil inputs supply 24%, and the application of animal manure introduces 15% of total inputs. The remaining 11% is generated through wastewater discharge from urban and industrial areas.

Several studies analyzing temporal trends in nitrogen exports have documented marked changes over time. At a global scale, Seitzinger et al. (2010) estimated a 35.2% increase in transport of dissolved inorganic nitrogen and 17.9% in total nitrogen between 1970 and 2000, with much of the increase attributable to changes taking place in Asia. This is consistent with studies conducted in rivers of China that have documented large increases in nitrogen export over the last few decades (Liu et al. 2009; Ti and Yan 2013; Strokal et al. 2015). Similarly, Beusen et al. (2016) estimated that nitrogen export to the oceans almost doubled during the 20th century despite the marked increase in nitrogen retention behind dams. Again, this study indicated that changing agricultural practices were responsible for this shift, contributing only 19% of the flux in 1900, but 51% of the total N exported in 2000.

It is important to mention that some regions of the world have experienced reductions in nitrogen exports from rivers to the coast. For instance, Oelsner and Stets (2019) analyzed nutrient trends in US rivers draining to the Great Lakes or coasts between 2002 and 2012, and found that both nitrate and total nitrogen load decreased in at least half of the sites. Total nitrogen tended to decline in urban watersheds, and this was most likely related to enhanced waste water management during the time period. In contrast, agricultural watersheds did not show consistent trends in N export. A notable exception was nitrogen export to Lake Erie, one of the Great Lakes, where nitrogen loading from rivers declined. Nitrogen loading declined in most watersheds draining into the Chesapeake Bay in the eastern US, and the nitrogen load delivered by the Mississippi River remained stable over recent years. Similar patterns have been documented in Europe. For example, for rivers across Europe, Bouraoui and Grizzetti (2011) documented declines in nitrate concentrations between 1991 and 2004 in 30% of the 39 rivers analyzed. Declines were particularly evident in rivers draining to the North Sea, including the Elbe, Rhine, and Weser. However, nitrate load increased in 10% of the rivers, such as the Mersey in England, which doubled its nitrate concentration through the study period. Work in Danish streams indicates that point source inputs of nitrogen were reduced by 75% between 1980 and 2013, while loading from non-point, or diffuse sources, declined by 43%. In the same study, total nitrogen concentrations were reduced by more than half, which was associated with a decline in phytoplankton biomass, an increase in water transparency, and an increase in benthic macroalgae in coastal ecosystems. The Danish government achieved these environmental outcomes by initiating a national action plan that included enhanced waste water treatment, fertilizer and crop rotation plans, manure application management, and wetland and forest restoration (Kronvang et al. 2008; Riemann et al. 2016).

13.4.2 Phosphorus

As with nitrogen, one can estimate net anthropogenic inputs of phosphorus in order to examine trends in inputs and their relationship with riverine export. Phosphorus inputs to watersheds include fertilizer application, the net trade of human food and animal feed, and phosphorus-based detergents (Russell et al. 2008). Unlike nitrogen cycling, there is no loss that is analogous to denitrification or fixation, and atmospheric deposition of phosphorus is typically minor (however, deposition through dust can be important in certain regions of the world). In the absence of anthropogenic disturbance, natural weathering is the primary natural input of phosphorus to streams, underscoring the influence of variation in underlying geology. Natural weathering rates are difficult to measure, and usually are small relative to anthropogenic sources of phosphorus.

Estimates of natural background yields of total phosphorus from relatively undisturbed streams are also highly variable and are typically related to geology. In the US, phosphorus loads from streams have been reported to range up to 0.82 kg ha^{-1} yr^{-1} (Clark et al. 2000). Somewhat lower P loads were estimated for two tributaries of the Orinoco River in Venezuela. The annual P yield for the Caura River, a blackwater tributary, was 0.46 kg ha^{-1} yr^{-1}. In the Apure, a moderately disturbed white water tributary, the P yield was 0.68 kg ha^{-1} yr^{-1} (Lewis and Saunders 1990). Phosphorus export from agricultural catchments can be much higher. For instance, load estimates for rivers of the Mississippi-Atchafalaya River basin were up to 1.9 kg ha^{-1} yr^{-1}, between 0.7 and 1.1 kg ha^{-1} yr^{-1} in highly developed agricultural catchments in Illinois (Goolsby et al. 1999), and approximately 1.4 kg ha^{-1} yr^{-1} in the Sandusky and Maumee rivers, two highly agricultural watersheds in Ohio (Baker and Richards 2002). Phosphorus yields of 2 kg ha^{-1} yr^{-1} were documented under intensive livestock farming in British watersheds, compared to intensive agriculture (0.5 kg ha^{-1} yr^{-1}) (Jarvie et al. 2010). Urban dominated catchments can also exhibit high P yields, with values from 0.25 to 1.3 kg ha^{-1} yr^{-1} (Duan et al. 2012; Hobbie et al. 2017).

Estimation of P inputs to watersheds strongly influenced by human activities generally shows, as expected, the strong influence of fertilizer in agricultural watersheds, and of the import of food (and resulting export of human waste) in urban watersheds. Anthropogenic budget estimates and other modeling approaches are useful to identify primary sources and sinks and to compare spatial and temporal variation in phosphorus loading among watersheds. By constructing P budgets for headwater catchments of the Huai River Basin, an important agricultural area that is characterized by the highest human population densities in China, Zhang et al. (2015) documented that fertilizers were the dominant P source (70%), followed by phosphorus imported as food and feed (24%), and phosphorus derived from non-food items (6%). An average of 3.2% of P inputs were exported by the river, although this was quite variable, ranging from 1.5 to 19.2%. Regardless, more than 80% of anthropogenic P inputs presumably were stored in the watershed, likely due to the strong association of phosphorus with particles, which can enhance its retention in soils and river channels. Similarly, fertilizer was the dominant P input to Lake Erie and Lake Michigan watersheds in the north-central US, with the exception of a highly urbanized region where detergents were the main source of phosphorus (Han et al. 2011). P riverine export represented an average of 5% of net anthropogenic P inputs to Lake Michigan watersheds and 10% of P inputs to Lake Erie watersheds, which contrasts to the higher and relatively constant input-export relationship reported for nitrogen (25%) (Howarth et al. 1996, 2012). The small fraction of net anthropogenic P inputs reflected in riverine P export probably results from the capacity of phosphorus to adsorb to soil particles, and its storage in the landscape (Russell et al. 2008). Greater variability among regions can be related to soil type, land use, and precipitation. For example, poorly drained soils, higher precipitation, and fertilizer application can enhance riverine phosphorus export (Han et al. 2011).

When budget calculations are made over an extended period of time, they can help detect long-term trends in inputs, and estimate changes in residence time as nutrients work their way through the catchment and into riverine export. Phosphorus inputs exceeded outputs for a number of catchments in Illinois for the years between 1965 and 1990, but were in balance thereafter, due to a reduction in fertilizer applications (David and Gentry 2000). Phosphorus that accumulated in the soil during years of surplus loading constitutes an important storage component, termed legacy phosphorus, that may contribute to river exports for years to decades. As a consequence, riverine phosphorus exports may remain elevated for some time despite reductions in inputs. This stored phosphorus can delay responses to management actions employed to reduce P concentrations in surface waters (Kleinman et al. 2011).

As previously mentioned, dominant land use influences the source of phosphorus inputs into watersheds. Urban watersheds typically receive most of their phosphorus inputs from wastewater, whereas agricultural watersheds are typically affected more through nonpoint sources carried in runoff. In British catchments studied by Bowes et al. (2010), phosphorus inputs from sewage were dominant in highly urbanized catchments; however, annual loads were controlled by nonpoint sources in most rivers, particularly in rural catchments. Interestingly, sewage inputs were the major contributor to total phosphorus inputs during most times of the year in most rivers, indicating that effective phosphorus management efforts should include the enhancement of wastewater treatment infrastructure. Similar work to construct a phosphorus budget in Great Britain suggested that urban areas contribute 60% of the total phosphorus export of the region, and exports have declined eight-fold between 1974 and 2011 due to improvement in wastewater treatment plants (Worrall et al. 2016).

Efforts to improve wastewater treatment infrastructure may be a very effective management tool to reduce the flow of anthropogenically-derived phosphorus into rivers and streams. In the United Kingdom, Neal et al. (2010) found SRP concentrations decreased by fivefold in the Thames River and threefold in Thame, a tributary of the Thames, after the application of a tertiary phosphorus stripping method that removed soluble reactive phosphorus in effluent from waste water treatment plants. It is important to mention that further decreases are still necessary to meet environmental targets, which have likely been delayed by P desorption from river sediments. This is another example of legacy phosphorus, which can delay and mask management efforts (Sharpley et al. 2013).

In addition to enhanced wastewater treatment and alterations in land management practices, the retention of phosphorus in dams and impoundments can also affect phosphorus export to downstream reaches. A comparison of two rivers in southeastern Michigan, US, found that the more impounded river exported a lower proportion of phosphorus inputs than a river with few dams (Bosch and Allan 2008). By modelling nutrient export at the global scale, Seitzinger et al. (2010) estimated that export of particulate forms of P should decrease over the next several decades due to increased retention by dams. Similar modeling efforts by Maavara et al. (2017) estimated that 12% of the total phosphorus load carried globally by rivers was detained by dams in 2000, and this value is expected to increase to 17% of the total load in 2030. Dams can retain approximately 40% of phosphorus carried by rivers, and dams are still being constructed throughout the globe, strongly suggesting that the proportion of total phosphorus load retained by dams will continue to increase. Although a decline in phosphorus has been detected in North America and Europe, at a global scale Seitzinger et al. (2010) calculated that dissolved inorganic phosphorus export increased 29% between 1970 and 2000, particularly in South Asia and South America. Increases in sewage discharge due to increases in population accounts for a major proportion of phosphorus inputs. This trend indicates that the risk of eutrophication likely will increase in some regions, as anthropogenic sources of phosphorus continue to threaten the integrity of aquatic ecosystems.

13.5 Summary

Nutrients are inorganic materials necessary for life, and whose supply is potentially limiting to biological activity within lotic ecosystems. Although many macro- and micro-nutrients are required for enzymatic activity and protein synthesis, phosphorus and nitrogen are the primary nutrients that limit biological activity. In addition, the supply of silica can be important to diatoms because of the high silica content of their cell walls. Phosphorus and nitrogen occur in numerous forms including dissolved and particulate, and inorganic and organic. They are most bioavailable in their dissolved inorganic forms, as phosphate, nitrate, and ammonium. Their concentrations are typically very low in unpolluted waters, but are greatly elevated in many areas, including most large temperate rivers, due to human inputs of agricultural fertilizers, human and animal waste, atmospheric deposition, and industrial pollution. Nitrogen and phosphorus concentrations in rivers often exceed the levels that cause eutrophication in standing water, and consequently the load of nutrients delivered to lakes and coastal waters by river export is a serious management concern.

There are, broadly speaking, two perspectives on nutrient cycling in lotic ecosystems: how nutrient supply affects biological productivity, and how processes within the ecosystem influence the quantity of nutrients that are transported downstream. Because rivers export substantial quantities of nutrients to receiving lakes and oceans, instream storage and removal processes have the potential to influence large-scale element budgets, and reduce the quantity of nutrients delivered to receiving waters.

Nutrient cycling describes the passage of an atom or element from dissolved inorganic nutrient through its incorporation into living tissue to its eventual re-mineralization by excretion or egestion and decomposition. In many ecosystems nutrients largely cycle in place, but in lotic ecosystems, downstream flow stretches cycles into spirals. The distance traveled by an atom as inorganic solute before its immobilization in the stream bed is called the uptake length, S_w. Because uptake length depends strongly on discharge and velocity it is desirable to standardize nutrient uptake by estimating an uptake velocity, v_f, which quantifies the velocity at which a molecule moves from the water column to retention sites on and within the stream bed. Areal uptake rate, U, is the quantity of inorganic nutrients immobilized by the stream bed per unit of time.

A number of abiotic and biotic processes influence nutrient spiraling. Ambient nutrient concentrations influence nutrient uptake rates and several models have been proposed to describe this relationship. Sorption-desorption reactions, in which both inorganic and organic molecules are bound to the surfaces of sediments, can help to regulate nutrient availability by serving as temporary storage sites when a nutrient is present in stream water at high concentrations. Under aerobic conditions, dissolved inorganic and organic phosphorus can complex with metal oxides and hydroxides to form insoluble precipitates, which are released back to the water column when anaerobic conditions prevail. Transient storage capacity, which accounts for the slow passage of a conservative tracer relative to water column flow, is a useful descriptor of the extent to which channel complexity affects downstream passage. Transient storage can enhance nutrient processing and retention. Stream size influences nutrient cycling through the interaction of the water column with the stream bed, the relative importance of benthic and planktonic processes, and interaction with the floodplain.

Both nitrogen and phosphorus are removed from stream water by biotic processes. The immobilization of nutrients by autotrophs involves assimilatory uptake for the incorporation of nutrients into new tissue. Heterotrophs also require nutrients for synthesis of new structural compounds. Nitrification and denitrification are dissimilatory reactions where bacteria obtain energy by using ammonia as a fuel or nitrate as an oxidizing agent. The capacity of lotic ecosystems to influence the dynamics of nutrients during their downstream passage depends on the factors that influence biotic and abiotic uptake. Biological demand for nutrients varies seasonally with environmental conditions that favor high rates of primary and heterotrophic production. During periods of high discharge, biotic uptake declines, and streams tend to export more nutrients than under lower flow conditions. Because the stream bed and its interstices are locations of biofilm development and organic matter accumulation, sub-surface flowpaths can retard the downstream passage of nutrients and increase their exposure to sites of uptake, thereby contributing to nutrient retention and utilization. Once nutrients are assimilated by primary producers and heterotrophic microbes, they can pass through multiple links within food webs before their eventual mineralization to a bioavailable state. In some systems, recycling by consumers makes a significant contribution to nutrient availability, and in some circumstances, selective retention of nutrients in consumer biomass may significantly contribute to nutrient storage within a reach.

Nutrient budgets provide an accounting of all inputs, exports, and internal stores for some delineated spatial unit, such as a stream reach, catchment, or large river basin or region. Outputs are typically much less than inputs in budget calculations because storage can be important and, in the case of nitrogen, denitrification can contribute to loss of instream nitrogen to the atmosphere. Nutrient budgets have proven to be especially useful in revealing the magnitude of anthropogenic inputs to aquatic systems. Over the last decades nitrogen loads have increased in many watersheds worldwide primarily due the application of fertilizers and combustion of fossil fuels, although in some rivers inputs of nitrogen have decreased due to enhanced wastewater treatment and land management. Phosphorus inputs have decreased in many rivers due to advances in wastewater treatment and reduced fertilizer application, but at the global scale dissolved inputs still show an increasing trend.

References

Argerich A, Haggerty R, Martí E et al (2011) Quantification of metabolically active transient storage (MATS) in two reaches with contrasting transient storage and ecosystem respiration. J Geophys Res Biogeosciences 116:3034. https://doi.org/10.1029/2010JG001379

ASCE (2017) America's infrastructure report card. Washington D.C

ASCE (2011) Failure to act: the economic impact of current investment trends in water and wastewater treatment infrastructure. Washington D.C

Atkinson CL, Capps KA, Rugenski AT, Vanni MJ (2017) Consumer-driven nutrient dynamics in freshwater ecosystems: from individuals to ecosystems. Biol Rev 92:2003–2023. https://doi.org/10.1111/brv.12318

Atkinson CL, Julian JP, Vaughn CC (2014) Species and function lost: role of drought in structuring stream communities. Biol Conserv 176:30–38. https://doi.org/10.1016/j.biocon.2014.04.029

Baker DB, Richards RP (2002) Phosphorus budgets and riverine phosphorus export in Northwestern Ohio watersheds. J Environ Qual 31:96–108

Baker MA, Webster JR (2017) Conservative and reactive solute dynamics. In: Lamberti GA, Hauer FR (eds) Methods in stream ecology, vol 2: ecosystem function, 3rd edn. Academic Press, Boston, pp 129–145

Beaulieu JJ, Tank JL, Hamilton SK et al (2010) Nitrous oxide emission from denitrification in stream and river networks. Proc Natl Acad Sci 108:214–219. https://doi.org/10.1073/pnas.1011464108

Bencala KE, Gooseff MN, Kimball BA (2011) Rethinking hyporheic flow and transient storage to advance understanding of stream-catchment connections. Water Resour Res 47:1–9. https://doi.org/10.1029/2010WR010066

Bencala KE, Walters RA (1983) Simulation of solute transport in a mountain pool-and-riffle stream: a transient storage model. Water Resour Res 19:718–724. https://doi.org/10.1029/WR019i003p00718

Bernhardt ES, Hall RO, Likens GE (2002) Whole-system estimates of nitrification and nitrate uptake in streams of the Hubbard Brook Experimental Forest. Ecosystems 5:419–430. https://doi.org/10.1007/s10021-002-0179-4

Bernot MJ, Tank JL, Royer TV, David MB (2006) Nutrient uptake in streams draining agricultural catchments of the midwestern United States. Freshw Biol 51:499–509. https://doi.org/10.1111/j.1365-2427.2006.01508.x

Beusen AHW, Bouwman AF, Van Beek LPH et al (2016) Global riverine N and P transport to ocean increased during the 20th century despite increased retention along the aquatic continuum. Biogeosciences 13:2441–2451. https://doi.org/10.5194/bg-13-2441-2016

Bilby RE, Fransen BR, Walter JK et al (2001) Preliminary evaluation of the use of nitrogen stable isotope ratios to establish escapement levels for Pacific salmon. Fisheries 26:6–14. https://doi.org/10.1577/1548-8446(2001)026%3c0006:PEOTUO%3e2.0.CO;2

Boano F, Harvey JW, Marion A et al (2014) Hyporheic flow and transport processes: mechanisms, models, and biogeochemical implications. Rev Geophys 52:603–679. https://doi.org/10.1002/2012RG000417

Bohrman KJ, Strauss EA (2018) Macrophyte-driven transient storage and phosphorus uptake in a western Wisconsin stream. Hydrol Process 32:253–263. https://doi.org/10.1002/hyp.11411

Bosch NS, Allan JD (2008) The influence of impoundments on nutrient budgets in two catchments of Southeastern Michigan. Biogeochemistry 87:325–338. https://doi.org/10.1007/s10533-008-9187-6

Bouraoui F, Grizzetti B (2011) Long term change of nutrient concentrations of rivers discharging in European seas. Sci Total Environ 409:4899–4916. https://doi.org/10.1016/j.scitotenv.2011.08.015

Bowes MJ, Neal C, Jarvie HP et al (2010) Predicting phosphorus concentrations in British rivers resulting from the introduction of improved phosphorus removal from sewage effluent. Sci Total Environ 408:4239–4250. https://doi.org/10.1016/j.scitotenv.2010.05.016

Boyer EW, Goodale CL, Jaworski NA et al (2002) Anthropogenic nitrogen sources and relationships to riverine nitrogen export in the northeastern USA. Biogeochemistry 57:137–169

Briggs MA, Gooseff MN, Arp CD, Baker MA (2010) A method for estimating surface transient storage parameters for streams with concurrent hyporheic storage. Water Resour Res 46:0–27. https://doi.org/10.1029/2008WR006959

Bristow CS, Hudson-Edwards KA, Chappell A (2010) Fertilizing the Amazon and equatorial Atlantic with West African dust. Geophys Res Lett 37:3–7. https://doi.org/10.1029/2010GL043486

Brookshire ENJ, Gerber S, Menge DNL, Hedin LO (2012a) Large losses of inorganic nitrogen from tropical rainforests suggest a lack of nitrogen limitation. Ecol Lett 15:9–16. https://doi.org/10.1111/j.1461-0248.2011.01701.x

Brookshire ENJ, Hedin LO, Newbold JD et al (2012b) Sustained losses of bioavailable nitrogen from montane tropical forests. Nat Geosci 5:123–126. https://doi.org/10.1038/ngeo1372

Brown JH, Gillooly JF, Allen AP et al (2004) Toward a metabolic theory of ecology. Ecology 85:1771–1789. https://doi.org/10.1890/03-9000

Brunke M, Gonser T (1997) The ecological significance of exchange processes between rivers and groundwater. Freshw Biol 37:1–33. https://doi.org/10.1046/j.1365-2427.1997.00143.x

Burrows RM, Fellman JB, Magierowski RH, Barmuta LA (2013) Greater phosphorus uptake in forested headwater streams modified by clearfell forestry. Hydrobiologia 703:1–14. https://doi.org/10.1007/s10750-012-1332-5

Butturini A, Sabater F (1999) Importance of transient storage zones for ammonium and phosphate retention in a sandy-bottom Mediterranean stream. Freshw Biol 41:593–603. https://doi.org/10.1046/j.1365-2427.1999.00406.x

Capps KA, Atkinson CL, Rugenski AT (2015a) Consumer-driven nutrient dynamics in freshwater ecosystems: an introduction. Freshw Biol 60:439–442. https://doi.org/10.1111/fwb.12517

Capps KA, Atkinson CL, Rugenski AT (2015b) Implications of species addition and decline for nutrient dynamics in fresh waters. Freshw Sci 34:485–496. https://doi.org/10.1086/681095

Capps KA, Flecker AS (2013) Invasive aquarium fish transform ecosystem nutrient dynamics. Proc R Soc B Biol Sci 280:20131520. https://doi.org/10.1098/rspb.2013.1520

Caraco NF, Cole JJ (1999) Human impact on nitrate export: an analysis using major world rivers. Ambio 28:167–170

Carey RO, Migliaccio KW (2009) Contribution of wastewater treatment plant effluents to nutrient dynamics in aquatic systems. Environ Manage 44:205–217. https://doi.org/10.1007/s00267-009-9309-5

Carpenter SR, Caraco NFN, Correl DL et al (1998) Nonpoint pollution of surface waters with phosphorus and nitrogen. Ecol Appl 8:559–568

Castillo MM, Allan JD, Brunzell S (2000) Nutrient concentrations and discharges in a midwestern agricultural catchment. J Environ Qual 29:1142–1151. https://doi.org/10.2134/jeq2000.00472425002900040015x

Chaloner DT, Lamberti GA, Merritt RW et al (2004) Variation in responses to spawning Pacific salmon among three south-eastern Alaska streams. Freshw Biol 49:587–599. https://doi.org/10.1111/j.1365-2427.2004.01213.x

Chen F, Hou L, Liu M et al (2016) Net anthropogenic nitrogen inputs (NANI) into the Yangtze River basin and the relationship with riverine nitrogen export. J Geophys Res Biogeosciences 121:451–465. https://doi.org/10.1002/2015JG003186

Chen Y, Randerson JT, Van Der Werf GR et al (2010) Nitrogen deposition in tropical forests from savanna and deforestation fires. Glob Chang Biol 16:2024–2038. https://doi.org/10.1111/j.1365-2486.2009.02156.x

Clark GM, Mueller DK, Mast MA (2000) Nutrient concentrations and yields in undeveloped stream basins of the United States. J Am Water Resour Assoc 36:849–860. https://doi.org/10.1111/j.1752-1688.2000.tb04311.x

Covino TP, McGlynn BL, McNamara RA (2010) Tracer additions for spiraling curve characterization (TASCC): quantifying stream nutrient uptake kinetics from ambient to saturation. Limnol Oceanogr Methods 8:484–498. https://doi.org/10.4319/lom.2010.8.484

Craig LS, Palmer MA, Richardson DC et al (2008) Stream restoration strategies for reducing river nitrogen loads. Front Ecol Environ 6:529–538. https://doi.org/10.1890/070080

D'Angelo DJ, Webster JR, Gregory SV, Meyer JL (1993) Transient storage in Appalachian and Cascade Mountain streams as related to hydraulic characteristics. J North Am Benthol Soc 12:223–235. https://doi.org/10.2307/1467457

David MB, Drinkwater LE, McIsaac GF (2010) Sources of nitrate yields in the Mississippi River Basin. J Environ Qual 39:1657–1667. https://doi.org/10.2134/jeq2010.0115

David MB, Gentry LE (2000) Anthropogenic inputs of nitrogen and phosphorus and riverine export for Illinois, USA. J Environ Qual 29:494–508. https://doi.org/10.2134/jeq2000.00472425002900020018x

Davis JC, Minshall GW (1999) Nitrogen and phosphorus uptake in two Idaho (USA) headwater wilderness streams. Oecologia 119:247–255. https://doi.org/10.1007/s004420050783

Demars BOL (2008) Whole-stream phosphorus cycling: testing methods to assess the effect of saturation of sorption capacity on nutrient uptake length measurements. Water Res 42:2507–2516. https://doi.org/10.1016/j.watres.2008.02.010

Diemer LA, McDowell WH, Wymore AS, Prokushkin AS (2015) Nutrient uptake along a fire gradient in boreal streams of Central Siberia. Freshw Sci 34:1443–1456. https://doi.org/10.1086/683481

Dodds WK (2003) Misuse of inorganic N and soluble reactive P concentrations to indicate nutrient status of surface waters. J North Am Benthol Soc 22:171–181. https://doi.org/10.2307/1467990

Dodds WK, Burgin AJ, Marcarelli AM, Strauss EA (2017) Nitrogen transformations. In: Lamberti GA, Hauer FR (eds) Methods in stream ecology, vol 2: ecosystem function, 3rd edn. Academic Press, pp 173–196

Dodds WK, Evans-White MA, Gerlanc NM et al (2000) Quantification of the nitrogen cycle in a prairie stream. Ecosystems 3:574–589. https://doi.org/10.1007/s100210000050

Dodds WK, López AJ, Bowden WB et al (2002) N uptake as a function of concentration in streams. J North Am Benthol Soc 21:206–220. https://doi.org/10.2307/1468410

Dodds WK, Whiles MR (2010) Freshwater ecology: concepts and environmental applications of limnology, 2nd edn. Academic Press, San Diego

Downing JA, McClain M, Twilley R et al (1999) The impact of accelerating land-use change on the N-Cycle of tropical aquatic ecosystems: Current conditions and projected changes. Biogeochemistry 46:109–148. https://doi.org/10.1007/BF01007576

Duan S, Kaushal SS, Groffman PM et al (2012) Phosphorus export across an urban to rural gradient in the Chesapeake Bay watershed. J Geophys Res Biogeosciences 117:G01025. https://doi.org/10.1029/2011JG001782

Edwards AC, Twist H, Codd GA (2000) Assessing the impact of terrestrially derived phosphorus on flowing water systems. J Environ Qual 29:117–124. https://doi.org/10.2134/jeq2000.00472425002900010015x

Edwardson KJ, Bowden WB, Dahm C, Morrice J (2003) The hydraulic characteristics and geochemistry of hyporheic and parafluvial zones in Arctic tundra streams, north slope, Alaska. Adv Water Resour 26:907–923. https://doi.org/10.1016/S0309-1708(03)00078-2

Ekholm P, Kallio K, Salo S et al (2000) Relationship between catchment characteristics and nutrient concentrations in an agricultural river system. Water Res 34:3709–3716. https://doi.org/10.1016/S0043-1354(00)00126-3

Ensign SH, Doyle MW (2005) In-channel transient storage and associated nutrient retention: evidence from experimental manipulations. Limnol Oceanogr 50:1740–1751. https://doi.org/10.4319/lo.2005.50.6.1740

Ensign SH, Doyle MW (2006) Nutrient spiraling in streams and river networks. J Geophys Res Biogeosciences 111. https://doi.org/10.1029/2005JG000114

Fernald AG, Wigington PJ, Landers DH (2001) Transient storage and hyporheic flow along the Willamette River, Oregon: field measurements and model estimates. Water Resour Res 37:1681–1694. https://doi.org/10.1029/2000WR900338

Findlay SEG, Mulholland PJ, Hamilton SK et al (2011) Cross-stream comparison of substrate-specific denitrification potential. Biogeochemistry 104:381–392. https://doi.org/10.1007/s10533-010-9512-8

Flecker AS, Moore JW, Anderson JT et al (2010) Migratory fishes as material and process subsidies in riverine ecosystems. Am Fish Soc Symp 73:559–592

Gächter R, Steingruber SM, Reinhardt M, Wehrli B (2004) Nutrient transfer from soil to surface waters: differences between nitrate and phosphate. Aquat Sci 66:117–122. https://doi.org/10.1007/s00027-003-0661-x

Galloway JN, Dentener FJ, Capone DG et al (2004) Nitrogen cycles: past, present, and future. Biogeochemistry 70:153–226. https://doi.org/10.1007/s10533-004-0370-0

Galloway JN, Leach AM, Bleeker A, Erisman JW (2013) A chronology of human understanding of the nitrogen cycle. Philos Trans R Soc B Biol Sci 368:20130120. https://doi.org/10.1098/rstb.2013.0120

Goller R, Wilcke W, Fleischbein K et al (2006) Dissolved nitrogen, phosphorus, and sulfur forms in the ecosystem fluxes of a montane forest in Ecuador. Biogeochemistry 77:57–89. https://doi.org/10.1007/s10533-005-1061-1

Goolsby DA, Battaglin WA, Lawrence GB, et al (1999) Flux and sources of nutrients in the Mississippi–Atchafalaya River Basin: topic 3 report for the integrated assessment on hypoxia in the Gulf of Mexico. NOAA Coastal Ocean Program Decision Analysis Series No. 17. NOAA Coastal Ocean Program. Silver Spring

Gresh T, Lichatowich J, Schoonmaker P (2000) An estimation of historic and current levels of salmon production in the northeast Pacific ecosystem: evidence of a nutrient deficit in the freshwater systems of the Pacific Northwest. Fisheries 25:15–21. https://doi.org/10.1577/1548-8446(2000)025%3c0015:AEOHAC%3e2.0.CO;2

Grimm NB (1987) Nitrogen dynamics during succession in a desert stream. Ecology 68:1157–1170. https://doi.org/10.2307/1939200

Grimm NB (1988) Role of macroinvertebrates in nitrogen dynamics of a desert stream. Ecology 69:1884–1893. https://doi.org/10.2307/1941165

Grimm NB, Fisher SG (1986) Nitrogen limitation in a Sonoran Desert Stream. J North Am Benthol Soc 5:2–15. https://doi.org/10.2307/1467743

Gücker B, Boëchat IG (2004) Stream morphology controls ammonium retention in tropical headwaters. Ecology 85:2818–2827. https://doi.org/10.1890/04-0171

Guignard MS, Leitch AR, Acquisti C et al (2017) Impacts of nitrogen and phosphorus: from genomes to natural ecosystems and agriculture. Front Ecol Evol 5:70. https://doi.org/10.3389/fevo.2017.00070

Hagedorn F, Schleppi P, Waldner P, Hannes F (2000) Export of dissolved organic carbon and nitrogen from Gleysol dominated catchments-the significance of water flow paths. Biogeochemistry 50:137–161. https://doi.org/10.1023/A:1006398105953

Haggerty R, Martí E, Argerich A et al (2009) Resazurin as a "smart" tracer for quantifying metabolically active transient storage in stream ecosystems. J Geophys Res Biogeosciences 114:1–14. https://doi.org/10.1029/2008JG000942

Hall RO, Baker MA, Rosi-Marshall EJ et al (2013) Solute-specific scaling of inorganic nitrogen and phosphorus uptake in streams. Biogeosciences 10:7323–7331. https://doi.org/10.5194/bg-10-7323-2013

Hall RO, Bernhardt ES, Likens GE (2002) Relating nutrient uptake with transient storage in forested mountain streams. Limnol Oceanogr 47:255–265. https://doi.org/10.4319/lo.2002.47.1.0255

Hall RO, Tank JL, Dybdahl MF (2003) Exotic snails dominate nitrogen and carbon cycling in a highly productive stream. Front Ecol Environ 1:407–411. https://doi.org/10.1890/1540-9295(2003)001%5b0407:esdnac%5d2.0.co;2

Hall RO, Tank JL, Sobota DJ et al (2009) Nitrate removal in stream ecosystems measured by ^{15}N addition experiments: total uptake. Limnol Ocean 54:653–665

Hamilton SK, Tank JL, Raikow DF et al (2001) Nitrogen uptake and transformation in a midwestern U.S. stream: a stable isotope enrichment study. Biogeochemistry 54:297–340. https://doi.org/10.1023/A:1010635524108

Han H, Bosch N, Allan JD (2011) Spatial and temporal variation in phosphorus budgets for 24 watersheds in the Lake Erie and Lake Michigan basins. Biogeochemistry 102:45–58. https://doi.org/10.1007/s10533-010-9420-y

Hanrahan BR, Tank JL, Shogren AJ, Rosi EJ (2018) Using the raz-rru method to examine linkages between substrate, biofilm colonisation and stream metabolism in open-canopy streams. Freshw Biol 63:1610–1624. https://doi.org/10.1111/fwb.13190

Harvey JW, Gooseff M (2015) River corridor science: hydrologic exchange and ecological consequences from bedforms to basins. Water Resour Res 51:6893–6922. https://doi.org/10.1002/2015WR017617.Received

Harvey JW, Wagner BJ, Bencala KE (1996) Evaluating the reliability of the stream tracer approach to characterize stream-subsurface water exchange. Water Resour Res 32:2441–2451. https://doi.org/10.1029/96WR01268

Hatch LK, Reuter JE, Goldman CR (1999) Daily phosphorus variation in a mountain stream. Water Resour Res 35:3783–3791. https://doi.org/10.1029/1999WR900256

Heffernan JB, Cohen MJ (2010) Direct and indirect coupling of primary production and diel nitrate dynamics in a subtropical spring-fed river. Limnol Oceanogr 55:677–688. https://doi.org/10.4319/lo.2009.55.2.0677

Hensley RT, Cohen MJ, Korhnak LV (2014) Inferring nitrogen removal in large rivers from high-resolution longitudinal profiling. Limnol Oceanogr 59:1152–1170. https://doi.org/10.4319/lo.2014.59.4.1152

Hobbie SE, Finlay JC, Benjamin D et al (2017) Contrasting nitrogen and phosphorus budgets in urban watersheds and implications for managing urban water pollution. Proc Natl Acad Sci 114:4177–4182. https://doi.org/10.1073/pnas.1706049114

Holland EA, Dentener FJ, Braswell BH, Sulzman JM (1999) Contemporary and pre-industrial global reactive nitrogen budgets. Biogeochemistry 46:7–43. https://doi.org/10.1007/BF01007572

Holloway JM, Dahlgren RA, Hansen B, Casey WH (1998) Contribution of bedrock nitrogen to high nitrate concentrations in stream water. Nature 395:785–788. https://doi.org/10.1038/27410

Holmes RM, Fisher SG, Grimm NB (1994) Parafluvial nitrogen dynamics in a desert stream ecosystem. J North Am Benthol Soc 13:468–478. https://doi.org/10.2307/1467844

Hong B, Swaney DP, Howarth RW (2011) A toolbox for calculating net anthropogenic nitrogen inputs (NANI). Environ Model Softw 26:623–633. https://doi.org/10.1016/j.envsoft.2010.11.012

Hong B, Swaney DP, McCrackin M et al (2017) Advances in NANI and NAPI accounting for the Baltic drainage basin: spatial and temporal trends and relationships to watershed TN and TP fluxes. Biogeochemistry 133:245–261. https://doi.org/10.1007/s10533-017-0330-0

Hood JM, Vanni MJ, Flecker AS (2005) Nutrient recycling by two phosphorus-rich grazing catfish: the potential for phosphorus-limitation of fish growth. Oecologia 146:247–257. https://doi.org/10.1007/s00442-005-0202-5

Houlton BZ, Morford SL, Dahlgren RA (2018) Convergent evidence for widespread rock nitrogen sources in Earth's surface environment. Science (80-) 360:58–62. https://doi.org/10.1126/science.aan4399

House WA (2003) Geochemical cycling of phosphorus in rivers. Appl Geochem 18:739–748. https://doi.org/10.1016/S0883-2927(02)00158-0

Howarth R, Swaney D, Billen G et al (2012) Nitrogen fluxes from the landscape are controlled by net anthropogenic nitrogen inputs and by climate. Front Ecol Environ 10:37–43. https://doi.org/10.1890/100178

Howarth RW, Billen G, Swaney D et al (1996) Regional nitrogen budgets and riverine N & P fluxes for the drainages to the North Atlantic Ocean: natural and human influences. Biogeochemistry 35:75–139. https://doi.org/10.1007/BF02179825

Howden NJK, Burt TP, Worrall F et al (2010) Nitrate concentrations and fluxes in the River Thames over 140 years (1868–2008): are increases irreversible? Hydrol Process 24:2657–2662. https://doi.org/10.1002/hyp.7835

Janetski DJ, Chaloner DT, Tiegs SD, Lamberti GA (2009) Pacific salmon effects on stream ecosystems: a quantitative synthesis. Oecologia 159:583–595. https://doi.org/10.1007/s00442-008-1249-x

Jarvie HP, Withers PJA, Bowes MJ et al (2010) Streamwater phosphorus and nitrogen across a gradient in rural–agricultural land use intensity. Agric Ecosyst Environ 135:238–252. https://doi.org/10.1016/j.agee.2009.10.002

Jones JB (2002) Groundwater controls on nutrient cycling in a Mojave desert stream. Freshw Biol 47:971–983. https://doi.org/10.1046/j.1365-2427.2002.00828.x

Jones JB, Fisher SG, Grimm NB (1995) Nitrification in the hyporheic zone of a desert stream ecosystem. J North Am Benthol Soc 14:249–258. https://doi.org/10.2307/1467777

Jordan TE, Correll DL, Weller DE (1997) Nonpoint source discharges of nutrients from piedmont watersheds of Chesapeake Bay. J Am Water Resour Assoc 33:631–645. https://doi.org/10.1111/j.1752-1688.1997.tb03538.x

Kaiser K, Guggenberger G, Zech W (2000) Organically bound nutrients in dissolved organic matter fractions in seepage and pore water of weakly developed forest soils. Acta Hydrochim Hydrobiol 28:411–419. https://doi.org/10.1002/1521-401X(20017)28:7%3c411:AID-AHEH411%3e3.0.CO;2-D

Kaushal SS, Groffman PM, Mayer PM et al (2008) Effects of stream restoration on denitrification in an urbanizing watershed. Ecol Appl 18:789–804

Kelleher C, Ward A, Knapp JLA et al (2019) Exploring tracer information and model framework trade-offs to improve estimation of stream transient storage processes. Water Resour Res 55:3481–3501. https://doi.org/10.1029/2018WR023585

Kemp MJ, Dodds WK (2002a) Comparisons of nitrification and denitrification in prairie and agriculturally influenced streams. Ecol Appl 12:998–1009. https://doi.org/10.1890/1051-0761(2002)012%5b0998:conadi%5d2.0.co;2

Kemp MJ, Dodds WK (2002b) The influence of ammonium, nitrate, and dissolved oxygen concentrations on uptake, nitrification, and denitrification rates associated with prairie stream substrata. Limnol Oceanogr 47:1380–1393. https://doi.org/10.4319/lo.2002.47.5.1380

King KW, Williams MR, Macrae ML et al (2015) Phosphorus transport in agricultural subsurface drainage: a review. J Environ Qual 44:467. https://doi.org/10.2134/jeq2014.04.0163

Kleinman P, Sharpley A, Buda A et al (2011) Soil controls of phosphorus in runoff: management barriers and opportunities. Can J Soil Sci 91:329–338. https://doi.org/10.4141/cjss09106

Kline TC, Goering JJ, Mathisen OA et al (1990) Recycling of elements transported upstream by runs of Pacific salmon: I, δ^{15}N and δ^{13}C evidence in Sashin Creek, Southeastern Alaska. Can J Fish Aquat Sci 47:136–144. https://doi.org/10.1139/f90-014

Koenig LE, Song C, Wollheim WM et al (2017) Nitrification increases nitrogen export from a tropical river network. Freshw Sci 36:698–712. https://doi.org/10.1086/694906

Kohler AE, Kusnierz PC, Copeland T et al (2013) Salmon-mediated nutrient flux in selected streams of the Columbia River basin, USA. Can J Fish Aquat Sci 70:502–512. https://doi.org/10.1139/cjfas-2012-0347

Kronvang B, Andersen HE, Børgesen C et al (2008) Effects of policy measures implemented in Denmark on nitrogen pollution of the aquatic environment. Environ Sci Policy 11:144–152. https://doi.org/10.1016/j.envsci.2007.10.007

Kunz JV, Hensley R, Brase L et al (2017) High frequency measurements of reach scale nitrogen uptake in a fourth order river with contrasting hydromorphology and variable water chemistry (Weiße Elster, Germany). Water Resour Res 53:328–343. https://doi.org/10.1002/2016WR019355

Laursen AE, Seitzinger SP (2002) Measurement of denitrification in rivers: an integrated, whole reach approach. Hydrobiologia 485:67–81. https://doi.org/10.1023/A:1021398431995

Le Moal M, Gascuel-Odoux C, Ménesguen A et al (2019) Eutrophication: a new wine in an old bottle? Sci Total Environ 651:1–11. https://doi.org/10.1016/j.scitotenv.2018.09.139

Lewis WM, Wurtsbaugh WA, Paerl HW (2011) Rationale for control of anthropogenic nitrogen and phosphorus to reduce eutrophication of inland waters. Environ Sci Technol 45:10300–10305. https://doi.org/10.1021/es202401p

Lewis WM Jr, Saunders JF (1990) Chemistry and element export by the Orinoco main stem and lower tributaries. In: Weibezahn FH, Alvarez H, Lewis WM (eds) The Orinoco River as an Ecosystem. Galac, Caracas, pp 211–239

Lewis WM Jr, Hamilton SK, Jones SL, Runnels DD (1987) Major element chemistry, weathering and element yields for the Caura River drainage, Venezuela. Biogeochem 4:159–181

Lewis WM Jr, Melack JM, McDowell WH, et al. (1999) Nitrogen yields from undisturbed watersheds in the Americas. Biogeochem 46:149–162

Liu SM, Hong G, Ye XW, Xiang XL (2009) Nutrient budgets for large Chinese estuaries and embayment. Biogeosciences 6:2245–2263. https://doi.org/10.5194/bg-6-2245-2009

Maavara T, Lauerwald R, Regnier P, Van Cappellen P (2017) Global perturbation of organic carbon cycling by river damming. Nat Commun 8:1–10. https://doi.org/10.1038/ncomms15347

Mainstone CP, Parr W (2002) Phosphorus in rivers-ecology and management. Sci Total Environ 282–283:25–47. https://doi.org/10.1016/S0048-9697(01)00937-8

Marion A, Zaramella M, Bottacin-Busolin A (2008) Solute transport in rivers with multiple storage zones: the STIR model. Water Resour Res 44:10406. https://doi.org/10.1029/2008WR007037

Martí E, Grimm NB, Fisher SG (1997) Pre- and post-flood retention efficiency of nitrogen in a Sonoran Desert stream. J North Am Benthol Soc 16:805–819. https://doi.org/10.2307/1468173

McDowell R, Sharpley A, Folmar G (2001) Phosphorus export from an agricultural watershed: linking source and transport mechanisms. J Environ Qual 30:1587–1595. https://doi.org/10.2134/jeq2001.3051587x

McKnight DM, Runkel RL, Tate CM et al (2004) Inorganic N and P dynamics of Antarctic glacial meltwater streams as controlled by hyporheic exchange and benthic autotrophic communities. J North Am Benthol Soc 23:171–188. https://doi.org/10.1899/0887-3593(2004)023%3c0171:INAPDO%3e2.0.CO;2

Meybeck M (1982) Carbon, nitrogen, and phosphorus transport by world rivers. Am J Sci 282:401–450. https://doi.org/10.2475/ajs.282.4.401

Meyer J, Likens G, Sloane J (1981) Phosphorus, nitrogen and organic carbon in a headwater stream. Arch für Hydrobiol 91:28–44

Meyer JL (1979) The role of sediments and bryophytes in phosphorus dynamics in a headwater stream ecosystem. Limnol Oceanogr 24:365–375. https://doi.org/10.4319/lo.1979.24.2.0365

Miyazako T, Kamiya H, Godo T et al (2015) Long-term trends in nitrogen and phosphorus concentrations in the Hii River as influenced by atmospheric deposition from East Asia. Limnol Oceanogr 60:629–640. https://doi.org/10.1002/lno.10051

Morrice JA, Valett HM, Dahm CN, Campana ME (1997) Alluvial characteristics, groundwater-surface water exchange and hydrological retention in headwater streams. Hydrol Process 11:253–267. https://doi.org/10.1002/(SICI)1099-1085(19970315)11:3%3c253:AID-HYP439%3e3.0.CO;2-J

Mulholland PJ, Elwood JW, Newbold JD, Ferren LA (1985) Effect of a leaf-shredding invertebrate on organic matter dynamics and phosphorus spiralling in heterotrophic laboratory streams. Oecologia 66:199–206. https://doi.org/10.1007/BF00379855

Mulholland PJ, Hall RO, Sobota DJ et al (2009) Nitrate removal in stream ecosystems measured by 15 N addition experiments: denitrification. Limnol Oceanogr 54:666–680. https://doi.org/10.4319/lo.2009.54.3.0666

Mulholland PJ, Helton AM, Poole GC et al (2008) Stream denitrification across biomes and its response to anthropogenic nitrate loading. Nature 452:202–205. https://doi.org/10.1038/nature06686

Mulholland PJ, Marzolf ER, Webster JR, Hart DR (1997) Evidence that hyporheic zones increase heterotrophic metabolism and phosphorus uptake in forest streams. Limnol Oceanogr 42:443–451. https://doi.org/10.4319/lo.1997.42.3.0443

Mulholland PJ, Newbold JD, Elwood JW, Hom CL (1983) The effect of grazing intensity on phosphorus spiralling in autotropic streams. Oecologia 58:358–366

Mulholland PJ, Steinman AD, Marzolf ER et al (1994) Effect of periphyton biomass on hydraulic characteristics and nutrient cycling in streams. Oecologia 98:40–47. https://doi.org/10.1007/BF00326088

Mulholland PJ, Tank JL, Sanzone DM et al (2000) Nitrogen cycling in a forest stream determined by a ^{15}N tracer addition. Ecol Monogr 3:471–493

Mulholland PJ, Tank JL, Webster JR et al (2002) Can uptake length in streams be determined by nutrient addition experiments? Results from an interbiome comparison study. J North Am Benthol Soc 21:544–560. https://doi.org/10.2307/1468429

Mulholland PJ, Valett HM, Webster JR et al (2004) Stream denitrification and total nitrate uptake rates measured using a field 15 N tracer addition approach. Limnol Oceanogr 49:809–820. https://doi.org/10.4319/lo.2004.49.3.0809

Munn M, Frey J, Tesoriero A (2010) The influence of nutrients and physical habitat in regulating algal biomass in agricultural streams. Environ Manag 45:603–615. https://doi.org/10.1007/s00267-010-9435-0

Naiman RJ, Bilby RE, Schindler DE, Helfield JM (2002) Pacific salmon, nutrients, and the dynamics of freshwater and riparian ecosystems. Ecosystems 5:399–417

Neal C, Reynolds B, Neal M, et al. (2003) Soluble reactive phosphorus levels in rainfall, cloud water, throughfall, stemflow, soil waters, stream waters and groundwaters for the Upper River Severn area, Plynlimon, mid-Wales. Sci Total Environ 314:99–120 doi: 10.1016/s0048-9697(03)00099-8

Neal C, Jarvie HP, Williams R et al (2010) Declines in phosphorus concentration in the upper River Thames (UK): links to sewage effluent cleanup and extended end-member mixing analysis. Sci Total Environ 408:1315–1330. https://doi.org/10.1016/j.scitotenv.2009.10.055

Neff JC, Holland EA, Dentener FJ et al (2002) The origin, composition and rates of organic nitrogen deposition: a missing piece of the nitrogen cycle? Biogeochemistry 57(58):99–136

Neill C, Deegan LA, Thomas SM, Cerri CC (2001) Deforestation for pasture alters nitrogen and phosphorus in small Amazonian streams. Ecol Appl 11:1817–1828. https://doi.org/10.1890/1051-0761(2001)011%5b1817:dfpana%5d2.0.co;2

Newbold JD, Elwood JW, O'Neill RV, Van Winkle W (1981) Measuring nutrient spiralling in streams. Can J Fish Aquat Sci 38:860–863. https://doi.org/10.1139/f81-114

Newbold JD, O'Neill RV, Elwood JW, Van Winkle W (1982) Nutrient spiralling in streams: implications for nutrient limitation and invertebrate activity. Am Nat 120:628–652. https://doi.org/10.1086/284017

Newbold JW, Elwood JW, O'Neill RV, Sheldon AL (1983) Phosphorus dynamics in a woodland stream ecosystem: a study of nutrient spiralling. Ecology 64:1249–1265. https://doi.org/10.2307/1937833

O'Brien JM, Dodds WK, Wilson KC et al (2007) The saturation of N cycling in Central Plains streams: ^{15}N experiments across a broad gradient of nitrate concentrations. Biogeochemistry 84:31–49. https://doi.org/10.1007/s10533-007-9073-7

O'Brien JM, Warburton HJ, Elizabeth Graham S, et al (2017) Leaf litter additions enhance stream metabolism, denitrification, and restoration prospects for agricultural catchments. Ecosphere 8:e02018. https://doi.org/10.1002/ecs2.2018

Oelsner GP, Stets EG (2019) Recent trends in nutrient and sediment loading to coastal areas of the conterminous US: insights and global context. Sci Total Environ 654:1225–1240. https://doi.org/10.1016/j.scitotenv.2018.10.437

Palmer-Felgate EJ, Mortimer RJG, Krom MD, Jarvie HP (2010) Impact of point-source pollution on phosphorus and nitrogen cycling in stream-bed sediments. Environ Sci Technol 44:908–914. https://doi.org/10.1021/es902706r

Pandey J, Singh AV, Singh A, Singh R (2013) Impacts of changing atmospheric deposition chemistry on nitrogen and phosphorus loading to Ganga River (India). Bull Environ Contam Toxicol 91:184–190. https://doi.org/10.1007/s00128-013-1016-5

Pellerin BA, Kaushal SS, McDowell WH (2006) Does anthropogenic nitrogen enrichment increase organic nitrogen concentrations in runoff from forested and human-dominated watersheds? Ecosystems 9:852–864. https://doi.org/10.1007/s10021-006-0076-3

Pellerin BA, Stauffer BA, Young DA et al (2016) Emerging tools for continuous nutrient monitoring networks: sensors advancing science and water resources protection. J Am Water Resour Assoc 52:993–1008. https://doi.org/10.1111/1752-1688.12386

Peterson BJ, Wollheim WM, Mulholland PJ et al (2001) Control of nitrogen export from watersheds by headwater streams. Science 292:86–90. https://doi.org/10.1126/science.1056874

Piper LR, Cross WF, McGlynn BL (2017) Colimitation and the coupling of N and P uptake kinetics in oligotrophic mountain streams. Biogeochemistry 132:165–184. https://doi.org/10.1007/s10533-017-0294-0

Powers SM, Stanley EH, Lottig NR (2009) Quantifying phosphorus uptake using pulse and steady-state approaches in streams. Limnol Oceanogr Methods 7:498–508. https://doi.org/10.4319/lom.2009.7.498

Pringle CM, Rowe GL, Triska FJ et al (1993) Landscape linkages between geothermal activity and solute composition and ecological response in surface waters draining the Atlantic slope of Costa Rica. Limnol Oceanogr 38:753–774. https://doi.org/10.4319/lo.1993.38.4.0753

Reisinger AJ, Tank JL, Rosi-Marshall EJ et al (2015) The varying role of water column nutrient uptake along river continua in contrasting landscapes. Biogeochemistry 125:115–131. https://doi.org/10.1007/s10533-015-0118-z

Ribot M, von Schiller D, Martí E (2017) Understanding pathways of dissimilatory and assimilatory dissolved inorganic nitrogen uptake in streams. Limnol Oceanogr 62:1166–1183. https://doi.org/10.1002/lno.10493

Ribot M, von Schiller D, Peipoch M et al (2013) Influence of nitrate and ammonium availability on uptake kinetics of stream biofilms. Freshw Sci 32:1155–1167. https://doi.org/10.1899/12-209.1

Riemann B, Carstensen J, Dahl K et al (2016) Recovery of Danish coastal ecosystems after reductions in nutrient loading: a holistic ecosystem approach. Estuaries Coasts 39:82–97. https://doi.org/10.1007/s12237-015-9980-0

Riis T, Dodds WK, Kristensen PB, Baisner AJ (2012) Nitrogen cycling and dynamics in a macrophyte-rich stream as determined by a release. Freshw Biol 57:1579–1591. https://doi.org/10.1111/j.1365-2427.2012.02819.x

Riis T, Levi PS, Baattrup-Pedersen A et al (2017) Experimental drought changes ecosystem structure and function in a macrophyte-rich stream. Aquat Sci 79:841–853. https://doi.org/10.1007/s00027-017-0536-1

Ritz S, Dähnke K, Fischer H (2018) Open-channel measurement of denitrification in a large lowland river. Aquat Sci 80: 11 doi: 10.1007/s00027-017-0560-1

Royer TV, Tank JL, David MB (2004) Transport and fate of nitrate in headwater agricultural streams in Illinois. J Environ Qual 33:1296–1304. https://doi.org/10.2134/jeq2004.1296

Runkel RL (2002) A new metric for determining the importance of transient storage. J North Am Benthol Soc 21:529–543. doi: 10.2307/1468428

Runkel RL, McKnight DM, Andrews ED (1998) Analysis of transient storage subject to unsteady flow: diel flow variation in an Antarctic stream. J North Am Benthol Soc 17:143–154. https://doi.org/10.2307/1467958

Russell MJ, Weller DE, Jordan TE et al (2008) Net anthropogenic phosphorus inputs: spatial and temporal variability in the Chesapeake Bay region. Biogeochemistry 88:285–304. https://doi.org/10.1007/s10533-008-9212-9

Sabater F, Butturini A, Martí E et al (2000) Effects of riparian vegetation removal on nutrient retention in a Mediterranean stream. J North Am Benthol Soc 19:609–6290. https://doi.org/10.2307/1468120

Schade JD, Seybold EC, Drake T, et al (2016) Variation in summer nitrogen and phosphorus uptake among Siberian headwater streams. Polar Res 35. https://doi.org/10.3402/polar.v35.24571

Schaefer SC, Alber M (2007) Temperature controls a latitudinal gradient in the proportion of watershed nitrogen exported to coastal ecosystems. Biogeochemistry 85:333–346. https://doi.org/10.1007/s10533-007-9144-9

Schlesinger WH, Bernhardt ES (2013) Biogeochemistry: an analysis of global change, 3rd edn. Academic Press, Waltham, MA

Seitzinger S, Harrison JA, Bohlke JK, et al (2006) Denitrification across landscapes and waterscapes: a synthesis. Ecol Appl 16:2064–2090. https://doi.org/10.1890/1051-0761(2006)016%5b2064:dalawa%5d2.0.co;2

Seitzinger SP, Mayorga E, Bouwman AF, et al (2010) Global river nutrient export: a scenario analysis of past and future trends. Global Biogeochem Cycles 24:GB0A08. https://doi.org/10.1029/2009gb003587

Seitzinger SP, Styles R V, Boyer EW, et al (2002) Nitrogen retention in rivers: model development and application to watersheds in the northeastern U.S.A. Biogeochemistry 57/58:199–237

Sharpley AN (1995) Dependence of runoff phosphorus on extractable soil phosphorus. J Environ Qual 24:920–926. https://doi.org/10.2134/jeq1995.00472425002400050020x

Sharpley AN, Gburek WJ, Folmar G, Pionke HB (1999) Sources of phosphorus exported from an agricultural watershed in Pennsylvania. Agric Water Manag 41:77–89. https://doi.org/10.1016/S0378-3774(99)00018-9

Sharpley AN, Jarvie HP, Buda A et al (2013) Phosphorus legacy: overcoming the effects of past management practices to mitigate future water quality impairment. J Environ Qual 42:1308–1326. https://doi.org/10.2134/jeq2013.03.0098

Simon KS, Townsend CR, Biggs BJF, Bowden WB (2005) Temporal variation of N and P uptake in 2 New Zealand streams. J North Am Benthol Soc 24:1–18. https://doi.org/10.1899/0887-3593(2005)024%3c0001:TVONAP%3e2.0.CO;2

Small GE, Ardón M, Duff JH et al (2016) Phosphorus retention in a lowland Neotropical stream following an eight-year enrichment experiment. Freshw Sci 35:1–11. https://doi.org/10.1086/684491

Smith RA, Alexander RB, Schwarz GE (2003) Natural background concentrations of nutrients in streams and rivers of the conterminous United States. Environ Sci Technol 37:3039–3047. https://doi.org/10.1021/es020663b

Smith VH, Joye SB, Howarth RW (2006) Eutrophication of freshwater and marine ecosystems. Limnol Ocean 51:351–355. https://doi.org/10.4319/lo.2006.51.1_part_2.0351

Sterner R, Elser J (2002) Ecological stoichiometry: the biology of elements from molecules to the biosphere. Princeton University Press, Princeton

Stofleth JM, Shields FD, Fox GA (2008) Hyporheic and total transient storage in small, sand-bed streams. Hydrol Process 22:1885–1894. https://doi.org/10.1002/hyp.6773

Storey RG, Williams DD, Fulthorpe RR (2004) Nitrogen processing in the hyporheic zone of a pastoral stream. Biogeochemistry 69:285–313. https://doi.org/10.1023/B:BIOG.0000031049.95805.ec

Stream Solute Workshop (1990) Concepts and methods for assessing solute dynamics in stream ecosystems. J North Am Benthol Soc 9:95–119. https://doi.org/10.2307/1467445

Strokal M, Kroeze C, Lili L et al (2015) Increasing dissolved nitrogen and phosphorus export by the Pearl River (Zhujiang): a modeling approach at the sub-basin scale to assess effective nutrient management. Biogeochemistry 125:221–242. https://doi.org/10.1007/s10533-015-0124-1

Subalusky AL, Dutton CL, Rosi EJ, Post DM (2017) Annual mass drownings of the Serengeti wildebeest migration influence nutrient cycling and storage in the Mara River. Proc Natl Acad Sci 114:7647–7652. https://doi.org/10.1073/pnas.1614778114

Subalusky AL, Post DM (2019) Context dependency of animal resource subsidies. Biol Rev 94:517–538. https://doi.org/10.1111/brv.12465

Tank JL, Martí E, Riis T et al (2018) Partitioning assimilatory nitrogen uptake in streams: an analysis of stable isotope tracer additions across continents. Ecol Monogr 88:120–138. https://doi.org/10.1002/ecm.1280

Tank JL, Reisinger AJ, Rosi EJ (2017) Nutrient limitation and uptake. In: Lamberti GA, Hauer FR (eds) Methods in stream ecology, vol 2: ecosystem function, 3rd edn. Academic Press, pp 147–171

Tank JL, Rosi-Marshall EJ, Baker MA, Hall RO (2008) Are rivers just big streams? A pulse method to quantify nitrogen demand in a large river. Ecology 89:2935–2945. https://doi.org/10.1890/07-1315.1

Ti C, Yan X (2013) Spatial and temporal variations of river nitrogen exports from major basins in China. Environ Sci Pollut Res 20:6509–6520. https://doi.org/10.1007/s11356-013-1715-9

Trentman MT, Dodds WK, Fencl JS et al (2015) Quantifying ambient nitrogen uptake and functional relationships of uptake versus concentration in streams: a comparison of stable isotope, pulse, and plateau approaches. Biogeochemistry 125:65–79. https://doi.org/10.1007/s10533-015-0112-5

Triska FJ, Oremland RS (1981) Denitrification associated with periphyton communities. Appl Environ Microbiol 42:745 LP–748

Valett HM, Morrice JA, Dahm CN, Campana ME (1996) Parent lithology, surface-groundwater exchange, and nitrate retention in headwater streams. Limnol Oceanogr 41:333–345. https://doi.org/10.4319/lo.1996.41.2.0333

Vanni MJ (2002) Nutrient cycling by animals in freshwater ecosystems. Annu Rev Ecol Syst 33:341–370. https://doi.org/10.1146/annurev.ecolsys.33.010802.150519

Vanni MJ, Flecker AS, Hood JM, Headworth JL (2002) Stoichiometry of nutrient recycling by vertebrates in a tropical stream: linking species identity and ecosystem processes. Ecol Lett 5:285–293. https://doi.org/10.1046/j.1461-0248.2002.00314.x

Vanni MJ, Renwick WH, Headworth JL et al (2001) Dissolved and particulate nutrient flux from three adjacent agricultural watersheds: a five-year study. Biogeochemistry 54:85–114. https://doi.org/10.1023/A:1010681229460

Vilmin L, Mogollón JM, Beusen AHW, Bouwman AF (2018) Forms and subannual variability of nitrogen and phosphorus loading to global river networks over the 20th century. Glob Planet Change 163:67–85. https://doi.org/10.1016/j.gloplacha.2018.02.007

Webster JR, Mulholland PJ, Tank JL et al (2003) Factors affecting ammonium uptake in streams-an inter-biome perspective. Freshw Biol 48:1329–1352. https://doi.org/10.1046/j.1365-2427.2003.01094.x

Webster JR, Patten BC (1979) Effects of watershed perturbation on stream potassium and calcium dynamics. Ecol Monogr 49:51–72. https://doi.org/10.2307/1942572

Webster JR, Tank JL, Wallace JB et al (2000) Effects of litter exclusion and wood removal on phosphorus and nitrogen retention in a forest stream. Verhandlungen der Int Vereinigung für Theor und Angew Limnol 27:1337–1340. https://doi.org/10.1080/03680770.1998.11901453

Weld JL, Sharpley AN, Beegle DB, Gburek WJ (2001) Identifying critical sources of phosphorus export from agricultural watersheds. Nutr Cycl Agroecosystems 59:29–38. https://doi.org/10.1023/A:1009838927800

Wenger SJ, Subalusky AL, Freeman MC (2019) The missing dead: the lost role of animal remains in nutrient cycling in North American Rivers. Food Webs 18:e00106. https://doi.org/10.1016/j.fooweb.2018.e00106

Whiles MR, Hall RO, Dodds WK et al (2013) Disease-driven amphibian declines alter ecosystem processes in a tropical stream. Ecosystems 16:146–157

White DS, Elzinga CH, Hendricks SP (1987) Temperature patterns within the hyporheic zone of a northern Michigan river. J North Am Benthol Soc 6:85–91. https://doi.org/10.2307/1467218

Wigand C, Finn M, Findlay S, Fischer D (2001) Submersed macrophyte effects on nutrient exchanges in riverine sediments. Estuaries 24:398. https://doi.org/10.2307/1353241

Wilcke W, Leimer S, Peters T et al (2013) The nitrogen cycle of tropical montane forest in Ecuador turns inorganic under environmental change. Glob Biogeochem Cycles 27:1194–1204. https://doi.org/10.1002/2012GB004471

Wilson HF, Xenopoulos MA (2011) Nutrient recycling by fish in streams along a gradient of agricultural land use. Glob Chang Biol 17:130–139. https://doi.org/10.1111/j.1365-2486.2010.02284.x

Withers PJA, Jarvie HP (2008) Delivery and cycling of phosphorus in rivers: a review. Sci Total Environ 400:379–395. https://doi.org/10.1016/j.scitotenv.2008.08.002

Withers PJA, Jarvie HP, Hodgkinson RA et al (2009) Characterization of phosphorus sources in rural watersheds. J Environ Qual 38:1998. https://doi.org/10.2134/jeq2008.0096

Worrall F, Jarvie HP, Howden NJK, Burt TP (2016) The fluvial flux of total reactive and total phosphorus from the UK in the context of a national phosphorus budget: comparing UK river fluxes with phosphorus trade imports and exports. Biogeochemistry 130:31–51. https://doi.org/10.1007/s10533-016-0238-0

WWAP (United Nations World Water Assessment Programme) (2017) The United Nations world water development report 2017. The Untapped Resource. Paris, Wastewater

Yang L, Lei K (2018) Effects of land use on the concentration and emission of nitrous oxide in nitrogen-enriched rivers. Environ Pollut 238:379–388. https://doi.org/10.1016/j.envpol.2018.03.043

Yang X, Jomaa S, Büttner O, Rode M (2019) Autotrophic nitrate uptake in river networks: a modeling approach using continuous high-frequency data. Water Res 157:258–268. https://doi.org/10.1016/j.watres.2019.02.059

Ye S, Reisinger AJ, Tank JL et al (2017) Scaling dissolved nutrient removal in river networks: a comparative modeling investigation. Water Resour Res 53:9623–9641. https://doi.org/10.1002/2017WR020858

Zarnetske JP, Haggerty R, Wondzell SM, Baker MA (2011) Dynamics of nitrate production and removal as a function of residence time in the hyporheic zone. J Geophys Res Biogeosciences 116:G01025. https://doi.org/10.1029/2010JG001356

Zhang W, Swaney DP, Hong B et al (2015) Net anthropogenic phosphorus inputs and riverine phosphorus fluxes in highly populated headwater watersheds in China. Biogeochemistry 126:269–283. https://doi.org/10.1007/s10533-015-0145-9

Carbon Dynamics and Stream Ecosystem Metabolism

Ecologists have long been fascinated by the biological, chemical, and physical processes that regulate the flow of carbon, or energy, within ecosystems (Odum 1968). Carbon resources govern the structure of ecological communities and influence many other ecological processes. Therefore, understanding patterns in the spatial and temporal variation of the biological availability and quantity of organic resources moving within and through systems is an essential component in the study of stream ecology.

Energy sources in streams fall into two broad categories: autochthonous inputs derived from aquatic primary producers, and allochthonous inputs of organic matter from terrestrial ecosystems (Fig. 14.1). Heterotrophs—microorganisms, meiofauna, and macrofauna—decompose and consume supplies of organic carbon, ultimately mineralizing some fraction of the total as CO_2. Variable amounts of carbon may be stored within sediments, the riparian zone, and the floodplain (Fig. 14.2; Sutfin et al. 2016), and substantial quantities of allochthonous and autochthonous energy are exported downstream. This whole-ecosystem view brings into focus a series of topics that comprise the study of riverine carbon dynamics.

Lotic ecosystems are open, meaning that they receive energy from and supply energy to upstream and downstream habitats. Lateral energy exchanges between terrestrial and aquatic habitats are often substantial, especially when rivers are connected to a floodplain. Allochthonous organic matter is comprised of material derived from the tissues of plants and animals; it is a mixture of molecules that includes carbon and other elements. Allochthonous inputs of coarse, fine, and dissolved organic matter are substantial sources of carbon in many stream settings, especially in small streams shaded by a forested riparian zone, and in streams with high sediment load where algal primary production tends to be light-limited. In contrast, autochthonous production by algae and other primary producers is expected to make a greater contribution to the carbon pool in wider streams and rivers with reduced canopy cover, but less so in deeper and more turbid rivers when light becomes limiting. Shifts between autochthony and allochthony represent a conversion from reliance on internal to external energy sources, and the relative contribution of internal versus external sources is expected to vary through space and time with landscape setting and along the river continuum (e.g., Vannote et al. 1980; Minshall et al. 1985; Thorp et al. 2006; Winemiller et al. 2010).

In this chapter, we focus on the biological processes that influence carbon fluxes as revealed by tracing the sources and fates of allochthonous organic matter, quantified from carbon budgets and transport estimates, and by measuring primary production (carbon fixed through photosynthesis) and respiration (carbon released through cellular respiration), collectively referred to as stream metabolism. Comparison of primary production and respiration across streams and seasons provides the basis to explore how environmental setting influences stream metabolism, and to gain insight into the relative contributions of autochthonous and allochthonous carbon sources. Quantification of the inputs, rates of utilization, transport, and storage of carbon, including particulate and dissolved organic matter originating from terrestrial and upstream sources, makes it possible to construct carbon budgets and estimate carbon utilization, providing additional insight into ecosystem processing of energy supplies. Together these approaches provide a powerful foundation allowing stream ecologists to measure and model carbon dynamics, and better understand some of the biological, chemical, and physical factors that influence carbon cycling in rivers and streams.

14.1 Energy Flow in Lotic Systems

It is apparent that energy flow in lotic ecosystems is spatially and temporally complex, and is dominated by a longitudinal gradient that is frequently interrupted by lakes, dams, stepped changes due to tributary inputs, and discrete habitat

© Springer Nature Switzerland AG 2021
J. D. Allan et al., *Stream Ecology*,
https://doi.org/10.1007/978-3-030-61286-3_14

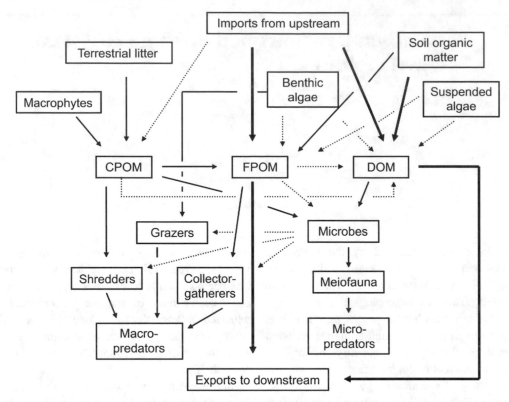

Fig. 14.1 Simplified model of principal carbon fluxes within a stream ecosystem (CPOM, coarse particulate organic matter; FPOM, fine particulate organic matter; DOM, dissolved organic matter). Heavier lines indicate dominant pathways of transport or metabolism of organic matter in a temperate woodland stream. Note that mineralization of organic carbon to CO_2 by respiration and storage are omitted. (Reproduced from Wetzel 1983)

Fig. 14.2 2 Organic carbon is stored within four primary reservoirs in river systems: (**a**) above- and below-ground standing biomass as riparian vegetation, (**b**) large in-stream and downed wood on the floodplain, (**c**) sediment on the floodplain surface and in the shallow subsurface, including soil organic carbon, litter, and humus, and (**d**) in-stream biomass including filamentous algae, periphyton, benthic invertebrates, fish, and particulate organic matter. Values indicate the estimated range of organic carbon per area (Mg C ha^{-1}). Note that in-stream biomass (**d**) accounts for a relatively small portion of carbon stored in river systems per area when compared to the other three reservoirs. (Reproduced from Sutfin et al. 2016)

types (Poole 2002; Webster 2007). The food webs of streams and rivers are fueled by a complex mixture of autochthonous and allochthonous energy sources, such that unraveling their relative contributions to higher trophic levels is a considerable challenge that stream ecologists have been studying for decades (Cummins and Klug 1979; Cummins et al. 1973; Minshall 1978).

Unlike terrestrial systems where large plants visible to the naked eye dominate autotrophic biomass, the primary producers of greatest significance in streams are the mostly microscopic benthic algae. These are found on stones, wood, and other surfaces and occur where light, nutrients, and other conditions are suitable for their growth. Organic matter that enters the stream from the surrounding land, such as leaf fall and other plant and animal detritus, is a significant energy source in most streams. Bacteria and fungi are the immediate consumers of organic substrates and in doing so create a microbe-rich and nutritious food supply for consumers, including biofilms on both inorganic and organic surfaces, and autumn-shed leaves riddled with fungal mycelia. These were the topics of Chaps. 6, 7 and 8.

The sources, processing, and fate of organic carbon are the determinants of energy flow in rivers and streams. Landscape constraints on the physical structure of a stream ecosystem determine which processes have local preeminence. For example, when lateral connectivity is high, the stream system will be strongly influenced by floodplain interactions; when vertical connectivity is high, the stream system will be strongly influenced by interactions with the hyporheos; and when both lateral and vertical connectivity are constrained, the stream will be most strongly impacted by upstream processes and by interruptions due to lakes and dams. Measuring the fluxes of primary production and respiration, and quantifying the flow of carbon through aquatic communities, are some of the principal methods to help researchers unravel the magnitude, range of variation, and response of carbon dynamics to both natural and anthropogenic factors.

Scientists have developed many conceptual frameworks to predict energy flow in rivers and streams. Collectively, these concepts provide a foundation for research in aquatic biogeochemistry and food web ecology. They also highlight some of the important, site-specific aspects of rivers to consider when examining carbon dynamics, including network connectivity and the ease of assimilation of diverse carbon sources. Some of the most important concepts pertain to changes along a river's length, the river's interactions with its floodplain, and the relative importance of different energy sources in supporting higher consumers in the food web.

The river continuum concept (RCC) has been widely applied to examine trophic interactions, productivity, and respiration in rivers and streams throughout the globe (Vannote et al. 1980). First introduced in Chap. 1, we include this figure again here to portray the integration of stream order, energy sources, food webs, and to a lesser degree nutrients, into a longitudinal model of energy flow in streams (Fig. 14.3). Originally conceived for river systems flowing through forested regions in the temperate zone, the RCC asserts that headwaters (stream order 1–3) should be heavily shaded and receive abundant leaf litter, but algal growth often will be light-limited. Streams of order four through six are expected to support more algae and aquatic plant life because they are wider and less shaded, and also should be fueled by organic particles from upstream. According to the RCC, headwaters should have relatively more allochthonous inputs, indicated by a ratio of primary production to respiration well below one, whereas the mid reaches should have more autochthonous production and a higher ratio of primary production to respiration. Higher-order rivers are thought to be too wide to have energy supplies dominated by riparian leaf fall, and too deep for energy to be primarily derived from algal production on the bed. Instead, organic inputs from upstream and the floodplain, along with river plankton, should play a greater role.

The river continuum concept has proven to be a resilient summary of the relative roles played by different basal resources along an idealized river system. Furthermore, as previously described (Sect. 9.1.5), the longitudinal distribution of functional feeding groups often, although not invariably, can be shown to be at least approximately in accord with expectations from the RCC. Nonetheless, the applicability of this model to running waters worldwide has been questioned (e.g., Winterbourn et al. 1981; Lake et al. 1985), and research has demonstrated that factors including climate, dominant land cover, and altitude also are known to influence resource gradients and the functional groups of aquatic consumers in streams (Tomanova et al. 2007).

To re-examine the predicted changes in dominant carbon resources along a river continuum, researchers compared gut contents from macroinvertebrate specimens archived from the original RCC study in 1976 with specimens collected from the same study sites in 2009 (Rosi-Marshall et al. 2016). Macroinvertebrate diets remained similar through time, and as predicted by the RCC, there was a longitudinal pattern in the dominance of allochthonous resources in macroinvertebrate diets. However, in contrast to expectations, autochthonous resources were exceptionally important (~35–75% of diets) at all of the sites (Rosi-Marshall et al. 2016), suggesting that the hypothesized pattern of changing energy pathways from headwaters to river mouth idealized in the RCC is only a first approximation of a more complete understanding of how energy is acquired within lotic ecosystems.

Fig. 14.3 The river continuum concept summarizes expected longitudinal changes in energy inputs and consumers as one proceeds from a first-order stream to a large river. A low ratio of primary production to respiration (P/R) indicates that the majority of the energy supplied to the food web derives from organic matter and microbial activity, and mostly originates as terrestrial production outside the stream channel. A P/R approaching one indicates that much more energy to the food web is supplied by primary production within the stream channel. An important upstream-downstream linkage is the export of fine particulate organic matter (FPOM) from the headwaters to locations downstream. (Reproduced from Vannote et al. 1980)

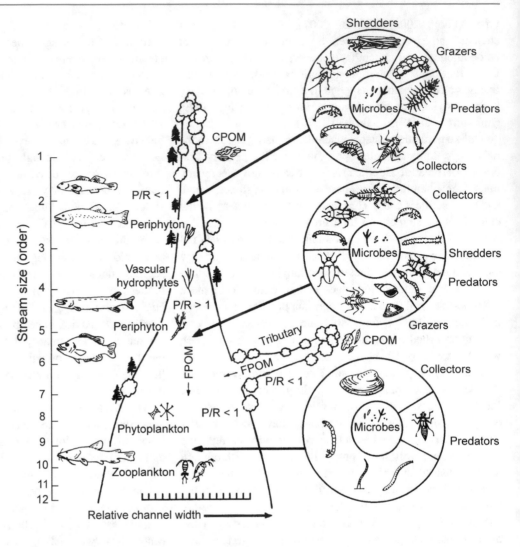

Though rivers are often conceptualized as continua of flowing waters, in many regions rivers are periodically interrupted by naturally occurring or anthropogenically-derived lentic waters. Lake Tonlé Sap in the Mekong basin in Cambodia and Lake Saint-Pierre in the Saint Lawrence basin in Canada are good examples of large, natural lentic habitats within river networks. In contrast, the impoundments caused by damming of rivers and streams have well-documented impacts on upstream-downstream linkages, including habitat fragmentation, changes to flow and thermal regimes, and altered transport of sediments, nutrients, and organic matter. The effects of a dam eventually dissipate, although often not for many tens of kilometers. Because many rivers have multiple dams, they can experience repeated breaks in the river continuum, referred to as serial discontinuity (Ward and Stanford 1983). Recovery of the river downstream of each dam depends on dam size, its position on the river network, tributary inputs, and other factors. In the case of rivers that historically were connected to extensive floodplains, dams and levees may permanently

sever lateral connectivity (Ward and Stanford 1995), resulting in the loss of critical ecosystem functions.

For many lowland rivers, energy inputs derive primarily from upstream sources, including tributaries and any production that occurs within the main channel, but lateral inputs can be substantial in rivers that inundate their floodplains (Junk et al. 1989). Often, larger rivers in both temperate and tropical regions are characterized by seasonal floods that redefine both terrestrial and aquatic habitats, and shift conditions from lotic to lentic predictably on an annual basis (Fig. 14.4). During the annual flood pulse, organic matter from the floodplain as well as algae and organic matter from fringing channels and floodplain lakes make substantial contributions to the secondary production of the river-floodplain biota (Tockner et al. 2000). These ideas form the basis of the flood-pulse concept (FPC; Junk et al. 1989; Junk and Wantzen 2004).

Authors of the FPC assert that in large, undammed rivers, riverine animal biomass is primarily supported by organic matter inputs from the floodplain, rather than by carbon sources transported from upstream (Fig. 14.5). Indeed, the

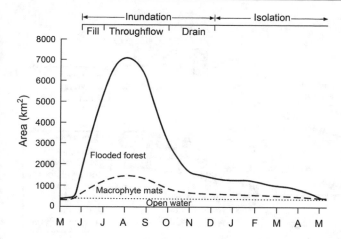

Fig. 14.4 Contributions of flooded forest, macrophyte mat, and open water to the Orinoco floodplain area throughout the hydrological cycle (Reproduced from Lewis et al. 2001)

most productive freshwater fisheries are located in large rivers with extensive floodplains, where the recruitment of young fishes correlates with interannual variation in the strength of flooding and thus determines the size of the catch when those juveniles mature into harvested size classes (Welcomme 1979). In the Rio Solimões, the growth of omnivore fishes was clearly linked to hydrological seasonality (Bayley 1988), as was also true in the lower Mississippi river provided that flooding coincided with temperatures above 15 °C (Schramm and Eggleton 2006).

Recent work has documented nuanced responses of fish populations and communities to the flood-pulse, suggesting that behavior, ontogeny, and species identity are important factors mediating consumer response to the flood-pulse. In a study examining the influence of degree of flooding on fish abundance, Castello and others (2019) documented a positive relationship between flooding and the abundance of pacu (*Colossoma macropomum*), a long-lived, over-harvested fish in the Brazilian Amazon, but these effects were only realized by fishes during their early stages of development. Tropical floodplain fishes in Tonlé Sap Lake in the Mekong River network are subject to annual flooding that increases freshwater habitat by approximately 500%. Research by McMeans et al. (2019) suggests that fishes in this system have diverse trophic responses to rising waters. Some species—especially small piscivores—shifted their trophic position by increasing their consumption of invertebrates and plant material during flooding. In Lake Saint-Pierre in the St. Lawrence River of North America, researchers used stable isotope techniques to demonstrate that fishes using floodplain habitat in the early portion of the growing season benefitted most from resources associated with flooding. Notably, the contribution of floodplain resources to fish diets declined through time, indicating that the quality and quantity of resources associated with the

flood-pulse may change throughout the period of inundation (Farly et al. 2019). Similar to findings from tropical systems in the Amazon and Mekong basins, the fish community in the St. Lawrence also demonstrated ontogenetic and species-specific responses to the flood pulse (Farly et al. 2019), underscoring the point that flood-pulse dynamics are not limited to tropical river networks. Understanding how aquatic communities respond physiologically, behaviorally, and trophically to seasonal changes in resource availability is an emerging frontier in stream ecology (McMeans et al. 2019).

The effects of the flood pulse are not realized in all rivers. As important as the floodplain may be to secondary production in large rivers, at least one-fourth of the fish species from a number of large temperate rivers can complete their life cycle in the main channel (Galat and Zweimüller 2001). Fishes including larvae and juveniles were abundant in the main channel of the Illinois and Mississippi Rivers and appeared to be supported by in-channel production based on the presence of zooplankton and invertebrates in their diet (Dettmers et al. 2001). These apparently contrasting findings may reflect differences in the role of floodplain inundation in tropical versus temperate settings, or between more pristine rivers versus rivers with more developed floodplains. In rivers characterized by high rates of primary production, regulated flows, or where the floodplain is not as productive as the main channel due to the timing of the flood, fish production may be more dependent on in-channel production (Junk and Wantzen 2004). In rivers with extensive flooding driven by an annual flood pulse, the original model may apply.

There is no question that detrital energy inputs are important sources of organic carbon in virtually all lotic ecosystems; however, budgetary accounts of inputs and exports may fail to provide an accurate view of the energy supplies that fuel higher trophic levels. Even in larger rivers, there is evidence that instream productivity and energy contributions from instream carbon dynamics can be important in supporting secondary production (Rosi-Marshall et al. 2016; Brett et al. 2017). By analyzing the signature of certain isotopes in animal consumers, it is possible to identify their primary food supplies, and in a number of instances where the energy sources were assumed to be allochthonous, a surprising dependency on autochthonous production was revealed. Isotopic signatures of fishes and invertebrates indicated that transported organic matter, including living and detrital algal components, was the main source of carbon for primary consumers in both constricted and floodplain reaches of the Ohio River (Thorp et al. 1998). In the Orinoco floodplain, macrophytes and leaf litter from the flooded forest represented 98% of the total carbon available, but isotope analysis showed that phytoplankton and periphyton were the major carbon sources for fish and macroinvertebrates. In addition, isotope data did not

Fig. 14.5 Depiction of the flood-pulse concept. During the annual hydrological cycle in a floodplain river, the littoral boundary of the river moves laterally with the rise and fall of the flood pulse, influencing fish recruitment and exchanges of nutrients and organic matter. The right-hand column indicates typical life-history traits of fish. DO refers to dissolved oxygen. (Reproduced from Bayley 1995)

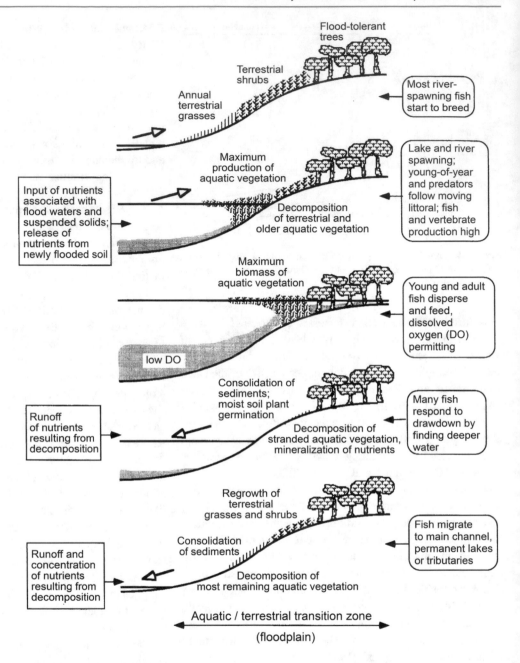

indicate that vascular plant carbon reached invertebrates through the microbial loop, suggesting instead that virtually all detrital carbon entered a "microbial dead end" and thus did not contribute to animal secondary production in the Orinoco floodplain (Lewis et al. 2001).

From the perspective of ecosystem metabolism, large lowland rivers are highly heterotrophic, reflecting high microbial respiration supported by high concentrations of dissolved and particulate organic matter. Secondary production by macroconsumers, however, may be based to a much greater extent on autochthonous production that occurs within the channel or in side-channels and floodplain lakes (Fig. 14.6). The riverine productivity model (RPM; Thorp and Delong 1994) emphasizes that carbon resources derived from autochthonous production may be important because they can be assimilated easily and are stored for longer periods of time near the bank where benthic macroinvertebrates tend to aggregate. According to the RPM, autochthonous carbon fuels much of secondary production in rivers, especially those with constricted channels, and can be an important energy supply to macroconsumers in rivers with floodplains (Thorp and Delong 1994, 2002).

Vertical connectivity is the third important spatial dimension of rivers that may influence energy flow. Groundwater enters stream channels along multiple flow paths, both deep and shallow, which vary with rainfall, soil

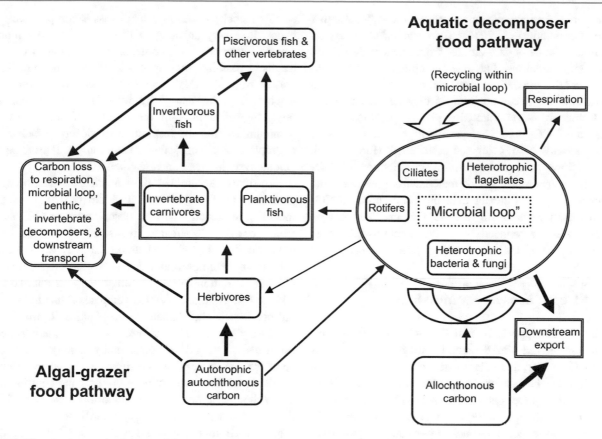

Fig. 14.6 The riverine productivity model proposes that secondary production by macroinvertebrates and fishes depends on autochthonous organic matter produced in the river channel and in the riparian zone, which are more labile but less abundant than organic matter of allochthonous origin transported from upstream reaches. The latter dominates the total amount of organic matter transported by rivers and contributes to high rates of microbial respiration but contributes little to the higher food web. (Reproduced from Thorp and Delong 2002)

moisture, and season, and result in distinctive signatures in their chemical constituents, including nutrients, dissolved organic carbon, and carbon dioxide. Conditions within the sediments can be particularly important to nutrient and carbon cycling because sediment characteristics affect abiotic uptake and storage, and because patches that differ in the availability of oxygen and organic matter strongly influence other biogeochemical cycles (Stegen et al. 2018). Connectivity between the water column and hyporheic zone can be mediated by discharge; therefore, changes in flow may alter carbon dynamics in the hyporheic zone (Fasching et al. 2016). For example, Romeijn et al. (2019) used an incubation experiment to demonstrate that the quality and quantity of organic matter in stream sediments can significantly influence CO_2 production from the hyporheic zone. They estimated that, under certain conditions, stream sediments can account for 35% of total stream evasion documented in other studies, a rate of CO_2 production that is much greater than typically reported.

Until relatively recently, rivers were often considered to be pipes, transporting carbon from terrestrial to marine systems. While still a research frontier, the incorporation of river systems into global carbon models generates a macro-scale view of the relative contribution of rivers as storage sites, active processors, or exporters of carbon downstream. Recent work has shown that the quantity of terrestrially derived carbon entering streams and rivers substantially exceeds the amount of riverine carbon delivered to the oceans, indicating that streams may play important but understudied roles in carbon storage and processing via respiration and subsequent release of CO_2 to the atmosphere (Cole et al. 2007; Wohl et al. 2017; Battin et al. 2008; Battin et al. 2009). This idea is borne out by recent estimates by Raymond et al. (2013) that indicate the amount of carbon emitted from rivers by outgassing of CO_2 exceeds the amount of carbon exported by rivers to oceans.

Estimates of global CO_2 evasion from inland waters are much higher for rivers than for lakes and reservoirs, but these numbers are constantly updated as technology improves and measurements are collected in more and more systems (Rocher-Ros et al. 2019). Though the assessments are tentative due to lack of data from some regions, it appears that a great majority of stream CO_2 evasion ($\sim 70\%$) originates from tropical and sub-tropical waters—systems that comprise a relatively small proportion of the Earth's surface (Raymond et al. 2013; Allen and Pavelsky 2018). The primary source of CO_2 emitted from inland waters is not known with certainty, but lateral inputs of CO_2 from groundwater that are derived from carbon fixation in forested systems, and the decomposition of organic matter within streams and rivers, undoubtedly play large roles in the generation of CO_2 and changes along a stream network (Hotchkiss et al. 2015; Campeau et al. 2019; Horgby et al. 2019). In lower order streams, the contribution of CO_2 derived from decomposition of material from the terrestrial environment is much greater than the contribution of CO_2 from in-stream metabolism, a ratio that changes moving downstream as the connectivity of the stream to the terrestrial systems declines with increasing water volume.

14.2 Stream Ecosystem Metabolism

The energy that supports the majority of Earth's ecosystems ultimately is derived from the sun. Photosynthetic organisms convert CO_2, water, and solar energy into reduced forms of carbon that can be consumed by heterotrophic organisms. This conversion process is termed gross primary production (GPP), and the rate of GPP often is measured from the amount of oxygen generated as a by-product of photosynthesis. Both autotrophic and heterotrophic organisms consume oxygen to use the energy contained in organic carbon compounds through the process of cellular respiration. Net primary production (NPP) is the difference between carbon fixed by autotrophs through GPP and autotrophic respiration (R_A), or the fraction of carbon used to meet their own metabolic demands. Net primary production is represented by Eq. 14.1, where R_A is a negative term as it is often measured as the amount of oxygen consumed:

$$NPP = GPP + R_A \qquad (14.1)$$

Ecosystem respiration (ER) is an aggregate estimate of R_A and heterotrophic respiration (R_H) in a system. Net ecosystem productivity (NEP), also referred to as net ecosystem metabolism, is the sum of GPP and ER, where ER is a negative term, and is described by the following relationship:

$$NEP = GPP + ER \qquad (14.2)$$

Estimating the contribution of respiration by autotrophs to ER is essential to describe carbon dynamics. In terrestrial ecosystems, estimates of metabolism may include estimates of both NEP and NPP, as scientists can obtain relatively robust estimates of R_A. In contrast, estimates of net production in rivers are typically restricted to NEP because the turnover of autotrophic biomass is very high and the

standing stock of autotrophic biomass is comparatively low; hence, it is very challenging to quantify R_A. Additionally, stream autotrophs form complex communities with heterotrophs in biofilms, making it infeasible to measure R_A separately from R_H (Hall and Hotchkiss 2017). Autotrophic respiration is influenced by many factors, including the physiological activity of the algae, self-shading by algal communities, and the respiration of closely associated heterotrophic organisms; thus, variation in R_A is expected among systems (Hall and Beaulieu 2013).

To address the problem of estimating the fraction of GPP consumed by R_A, Hall and Beaulieu (2013) developed a modeling approach based on observations of GPP and ER from more than 20 streams. In systems where GPP and ER did not covary (i.e., sites where productivity was not driving patterns in respiration) and that were characterized by large temporal variation in GPP, average R_A was approximately 44% of ER. Though the authors emphasize that their method did not address the challenge of separating R_A from the R_H of closely associated heterotrophs, and acknowledge that estimates of R_A varied substantially among streams, this approach can be applied to estimate R_A in other systems.

Net ecosystem production represents the contribution of autochthonous production in supporting heterotrophic production, and hence to whole system metabolism. Thus, NEP can be used to evaluate internal versus external organic carbon inputs to a stream reach—estimates that can be scaled up empirically or through modelling to examine metabolic transitions along the length of a river through time. At one extreme, NEP can be much less than zero, indicating that GPP contributes relatively little energy to the system. At the other extreme, NEP values greater than zero implies that primary production contributions to total heterotrophic respiration exceed the reliance on allochthonous materials, producing excess carbon that may be exported downstream.

Stream metabolism is a metric that integrates physicochemical characteristics of a stream (i.e., estimates of gas exchange with the environment) with estimates of biological activity (i.e., photosynthetic activity and aerobic respiration). Metabolism is measured in a river as a function of oxygen concentrations (Odum 1956) using the following equation:

$$\frac{dO}{dt} = GPP + ER + K\left(O_{def}\right) \qquad (14.3)$$

where $\frac{dO}{dt}$ is the change in oxygen concentration through time. In this relationship, GPP is the rate of O_2 produced through photosynthesis and is a positive flux; ER is the rate of O_2 consumed through respiration, and is a negative flux. The net exchange of O_2 between water and air is the product of a gas exchange rate K and the oxygen deficit (O_{def}). The O_{def} is the difference between the O_2 concentration at saturation in water at a given temperature and atmospheric pressure, and the measured O_2 concentration in the water. Carbon production can be estimated from metabolism estimates using the formula g C = 0.286 $\times$ g O_2 and reported in g C m^{-2} day^{-1}.

14.2.1 Factors Controlling Autochthonous Production

The photosynthetic activities of benthic algae, macrophytes, and phytoplankton constitute the principal autochthonous inputs to lotic ecosystems. Compared to terrestrial systems where vascular plants dominate autotrophic biomass, the periphyton, consisting mainly of benthic algae and cyanobacteria, are the most important autotrophs in most rivers and streams. As discussed in detail in Chap. 6, patterns in benthic primary production are governed by many factors including, but not limited to, light and nutrient availability, grazing, discharge, and disturbance.

Productivity in streams is very heterogeneous, with highly productive systems producing an average of approximately 13 g O_2 m^{-2} d^{-1} (Hoellein et al. 2013). Low rates of GPP are often found in turbid systems, in streams characterized by high levels of hydrologic disturbance, and in streams with dense riparian shading. In contrast, streams and rivers with exceptionally high rates of GPP often have elevated concentrations of nutrients, higher temperatures, and greater light availability when compared to relatively undisturbed systems (Hall 2016).

Instream productivity and metabolism are strongly related to light availability. For instance, daily and seasonal variation in GPP were strongly related to light availability in a small stream in Tennessee, US, with peaks during early spring prior to leaf out (Fig. 14.7; Roberts et al. 2007). In a study of streams from various biomes across North America, Mulholland and others (2001) highlighted the strong relationship between GPP and light availability. Similarly, in their review of over 60 estimates from stream sites in eastern North America, Webster et al. (1995) found that primary production in forested streams was about half that of open streams, although results were highly variable. Predictions of in-stream light availability are complex, as in-stream light regimes are strongly influenced by channel features such as canyon walls (Hall et al. 2015) or incised channels (Blaszczak et al. 2018), sediment and organic matter load affecting turbidity, and the phenology and timing of leaf-out of riparian plants.

Estimates of the contribution of macrophytes to ecosystem primary production are too few to generalize, but at least in some circumstances they can be significant. In the New River, Virginia, US, short-term production by *Podostemum ceratophyllum* was about equivalent to periphyton production (Hill and Webster 1982, 1983). Short-term estimates

also suggested that macrophytes contributed about 9% of the annual primary production in the Fort River, Massachusetts (Fisher and Carpenter 1976), and about 15% in the Red River, Michigan (King and Ball 1967). In a study of Brazilian streams, Tromboni and others (2017) documented that where present, macrophytes can generate a large fraction of reach-scale GPP.

14.2.2 Factors Controlling Ecosystem Respiration

Ecosystem respiration is the integrative measure of the utilization of organic carbon from all sources and by all organisms within the stream channel. It includes respiration by primary producers, microbial heterotrophs, and animals, which conceptually can be separated into the R_A of autotrophs and R_H of heterotrophs. As previously mentioned, it is very challenging to quantify R_A in flowing waters, as the turnover of autotrophic biomass is very high, and the standing stock of autotrophic biomass is relatively low. Therefore, respiration estimates in streams are often limited to ER (Hall and Hotchkiss 2017).

Though direct contributions by microorganisms have yet to be measured and compared within streams, we assume their respiration is the largest component of R_H, reflecting the roles of bacteria and fungi in the breakdown of organic matter and their ability to use labile DOM from stream water. Because metabolic processes can be strongly temperature dependent (Demars et al. 2011), respiration is expected to vary with temperature and season. Total respiration should also increase with increasing amounts of benthic organic matter (BOM), but its biological availability is at least as important as its quantity (Findlay et al. 1986). In the woodland stream mentioned earlier (Fig. 14.7), respiration was highest in early spring due to high GPP, and again in autumn after leaf fall, which are periods of moderate temperature, and respiration was low during the warmer mid-summer period because of low organic matter supplies. A downstream increase in benthic respiration might be expected if total carbon inputs increase, because warmer temperatures stimulate higher rates, or because larger rivers receive greater inputs from domestic sewage or agricultural runoff. Due to the relative scarcity of data for large rivers, relationships between ER and longitudinal position are still poorly documented. However, downstream increases in ER have been reported in systems as disparate as the highly autotrophic Salmon River (Minshall et al. 1992) and highly heterotrophic blackwater rivers in Georgia (Meyer and Edwards 1990).

Respiration rates within a site can also vary with discharge. In a study of two streams within the Glensaugh Research Station in north-east Scotland, Demars (2019) demonstrated that rates of ER following peak flows can be exceptionally

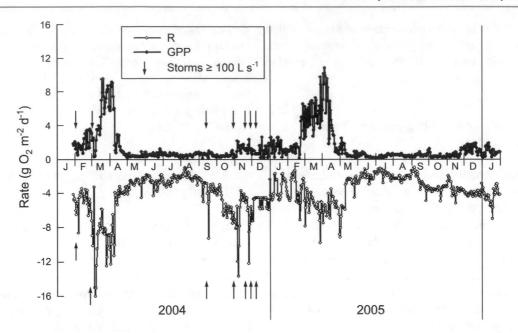

Fig. 14.7 Daily rates of gross primary production (GPP: positive values, black line) and ecosystem respiration (R: negative values, gray line) measured in Walker Branch in Tennessee in the eastern US from 28 January 2004 through 31 January 2006. Vertical lines separate years. Arrows indicate storms during which maximum instantaneous discharge was greater than or equal to 100 L s^{-1}. Variance in GPP correlates with seasonal and day-to-day variation in light levels. Variance in ecosystem R correlates with seasonal and day-to-day variation in GPP and autumn leaf inputs. (Reproduced from Roberts et al. 2007)

high. Storm-associated pulses of respiration have been observed previously, but this was one of the first investigations to document how hydrological connectivity between riparian and stream habitats during storm events can enhance the instream supply of DOC and stimulate respiration.

In a cross-biome comparison of 22 streams, Sinsabaugh (1997) summarized stream benthic respiration rates in relation to benthic organic matter, temperature, primary production, and other system variables (Fig. 14.8). Benthic respiration was directly proportional to stream temperature and, presumably due to high rates of utilization, the standing stock of benthic organic matter was inversely related to stream temperature. Owing to these offsetting trends, respiration per gram of benthic organic carbon was strongly related to temperature. Because the coefficient of this relationship was too high for a simple metabolic response, Sinsabaugh inferred that other factors also must be operating, such as higher quality BOM or greater nutrient availability in streams of warmer climates.

14.2.3 Factors Controlling Gas Exchange

Stream metabolism relies on the exchange of oxygen between the atmosphere and water to maintain adequate amounts of oxygen in solution. This flux is a function of the transfer velocity of the gas at the air-water interface, the solubility coefficient of the gas, and the difference in gas concentrations between the air and the water. In streams, the turbulence generated by water flowing over benthic substrates is the dominant driver of oxygen exchange, producing spatial and temporal heterogeneity within and among systems (Fig. 14.9; Ulseth et al. 2019). Unlike lacustrine systems, in some streams, gas exchange rates rather than biological processes can effectively control oxygen concentrations and make it difficult to estimate gas exchange (Hall 2016). An analysis by Ulseth et al. (2019) suggests that gas exchange in streams exists in two different states. In low-energy streams that are characterized by channels with shallow slopes, turbulent diffusion of gasses—the transfer of gasses at the air-water interface due to irregular or chaotic motion—is the primary factor influencing the exchange of gas with the atmosphere. In contrast, in high-energy systems characterized by steep slopes, turbulence generates air bubbles in the water column that dominate gas exchange processes. The ability to accurately estimate the factors mediating gas-exchange in streams is essential in estimating stream metabolism and in quantifying the global contribution of streams to fluxes of greenhouse gases.

14.2.4 Methods to Estimate Stream Metabolism

Until recently, advances in our knowledge of stream metabolism were limited mainly by technological capabilities and expense. Initial estimates of metabolism were made by

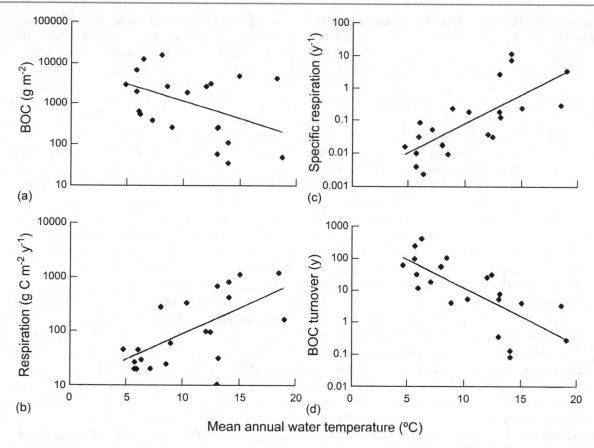

Fig. 14.8 Relationships of respiration rate and standing stock of benthic organic carbon (BOC) with stream temperature for 22 streams. (**a**) BOC decreases and (**b**) respiration rate increases with mean annual water temperature. (**c**) Specific respiration increases and (**d**) the turnover rate of BOC decreases with temperature. See text for further explanation. (Reproduced from Sinsabaugh 1997)

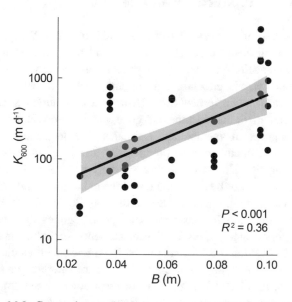

Fig. 14.9 Gas exchange (k_{600}) increased with median streambed roughness (B) across eight of the Swiss alpine stream reaches studied. The black line is the fit from log linear regression and the grey band represents 95% CI of the predicted k_{600}. (Reproduced from Ulseth et al. 2019)

collecting water samples every few hours throughout the day and laboriously titrating each sample to determine fluctuations in dissolved oxygen levels (e.g., Odum 1956). Subsequently, large, expensive, and finicky sensors were developed at the end of the last century to generate continuous oxygen and temperature data; however, the size, cost, and maintenance of these sensors prevented most researchers from deploying them in all but the most easily monitored systems during periods of stable in-stream conditions. Therefore, our initial understanding of within-stream variation in metabolism was primarily limited to smaller streams in the temperate zone during low flows on sunny days. In the past decade, the arrival of smaller, cost-effective sensors has allowed ecologists to collect large amounts of data from a wide range of rivers and streams throughout the world. These sensors can be deployed continuously (collecting data in intervals of seconds or minutes) and for longer periods of time (months or years), documenting diel patterns in oxygen concentrations. Local habitat heterogeneity can strongly influence estimates of stream metabolism and should be considered when deploying sensors (Siders et al. 2017; Dodds et al. 2018). Coupled with new computational tools, data generated by these sensors are providing insights into the metabolic regimes of rivers (Hall 2016), and an improved understanding of human influence on diel and seasonal patterns in stream ecosystem function (Arroita et al. 2019).

In addition to open water measurement of oxygen flux to estimate whole stream metabolism, increasingly by continuous monitoring using sensors, researchers have employed closed chambers to estimate metabolism. Many of the initial estimates of stream metabolism were made using enclosed benthic chambers, where oxygen change in the light measures NEP, and oxygen change at night or in darkened chambers provides an estimate of ecosystem respiration (Bott et al. 2006). At least in small streams, all primary production and virtually all respiration can reasonably be assumed to occur at the streambed. Using benthic chambers, GPP is estimated by adding respiration measured during the night to net oxygen change in the light, and NEP is calculated as the difference between GPP and 24-hour ecosystem respiration. Benthic chambers are especially useful for measuring local-scale heterogeneity and testing of environmental variables, but are difficult to scale up to the entire ecosystem, unless extensively replicated. It is important also to note that in larger systems, metabolism can be dominated by planktonic photosynthesis and respiration, which would require suspended bottles containing the biota of the water column to estimate planktonic contributions to system-wide metabolism (e.g., Reisinger et al. 2015), as is also done in lakes.

Gas exchange rates often are measured by injecting tracer gases, such as sulfur hexafluoride, propane, or argon, into streams and measuring the decline in tracer concentration over the study reach (Raymond et al. 2012; Hall and Ulseth 2020). Velocities can also be estimated using equations derived from channel geomorphology and hydraulics or ecosystem metabolism models. In large, slow-moving systems, floating chambers may also be an effective way to estimate gas exchange (Beaulieu et al. 2012). Methods to estimate gas exchange present challenges to stream ecologists. First, with gas injection, the sampling effort and cost of analysis can be significant hurdles to overcome if researchers are working on limited budgets, in remote sites, or are comparing numerous sites. Additionally, if the stream bed is relatively uniform or the flow is low, gas exchange rates can be exceptionally low and decreases in tracer concentrations difficult to measure within a given reach (Hall 2016). There is also some uncertainty in scaling gas exchange rates across systems with empirical equations, because they cannot be generalized for all streams and rivers. This is especially true in streams with steep slopes and great hydrologic energy, such as many streams in mountainous regions (Ulseth et al. 2019). Global efforts to collect more DO and gas exchange data are constantly enhancing the power of the models needed to successfully estimate reaeration, and future work will be improved by this effort.

The power of whole stream metabolism estimation using near-continuous water column monitoring is the integration of all GPP and ER for a stream reach, which can be expanded to a much greater temporal and spatial coverage by deploying multiple sensors. In addition, production and respiration in the benthic and hyporheic zones can influence system-wide metabolism in rivers, processes that are not robustly estimated using benthic chambers (Mulholland et al. 2001; Webster et al. 1995). Thus, many recent estimates of metabolism have been made using open-water methods. Spatial and temporal variability remain as challenges, however. Large, within-system variability in metabolism is most likely one of the reasons why researchers have had difficulty identifying the factors governing metabolic rates within and among river networks (Rodríguez-Castillo et al. 2019; Koenig et al. 2019).

Readers interested in learning more about the specific methods associated with measuring, modeling, and interpreting metabolism data should consider contributing to, and reading work by researchers associated with the StreamPULSE Project (http://streampulse.org/; e.g., Appling et al. 2018b; Hall 2016; Hall and Hotchkiss 2017; Hall et al. 2016; Appling et al. 2018a). Tradeoffs associated with specific methods used to measure metabolism have been discussed by many authors, including Hall et al. (2007), Staehr et al. (2012), Song et al. (2016), and Dodds et al. (2018).

14.2.5 Interpretation of Relationships Between Productivity and Respiration

The ratio of GPP to ER (often referred to as the P/R ratio), has long been used as a simple index of the relative importance of energy fixed by primary producers within the stream, versus allochthonous organic matter derived from terrestrial plant production. However, continuous sensor measurement has documented great variation in within- and among-stream metabolism estimates due to spatial and temporal fluctuations in productivity and respiration, suggesting that short-term and/or spatially-restricted measurements may not be appropriate metrics to compare metabolic activity among streams (Bernhardt et al. 2018; Hall 2016). Therefore, conclusions about metabolic regimes that were supported using short-term measurements, including many studies estimating P/R, should be interpreted with caution.

To better understand the difficulties of interpreting P/R ratios, recall that R is the sum of respiration by autotrophs (R_A) and by heterotrophs (R_H). Heterotrophic respiration can be further broken down into respiration supported by autotrophic production and respiration supported by allochthonous sources. Whichever is the larger fraction of R_H is the true measure of autotrophy versus heterotrophy. As a further caveat, predicting the fraction of ecosystem respiration that is supported by autochthonous versus allochthonous sources, and estimating how much respiration is generated by microorganisms, as compared to other organisms, remains a substantial challenge. For example, microorganisms may derive their energy from both autochthonous and allochthonous sources, and metazoans primarily from autochthonous sources (Thorp and Delong 1994, 2002).

This sort of reasoning may help to explain a striking discrepancy between Bayley's (1989a) analysis of carbon flux in the Rio Solimões, the whitewater branch of the Amazon River, which showed most carbon originating as detritus from aquatic and floodplain macrophytes, and other investigations that focused their studies on carbon flow within the food web of the river. For instance, an analysis of the stable isotope of carbon, ^{13}C, in fish tissue and in various plants found that the food chain supporting an abundant group of detritivorous fishes, the Characiformes, begins with phytoplankton and not macrophyte detritus as might be expected (Araujo-Lima et al. 1986). More recent work has reached similar conclusions. Mortillaro et al. (2015) demonstrated that detritivores were positioned at the base of the food chain as expected, but fatty acid analysis pointed to inclusion of autochthonous food sources, such as microalgae, in their diets. In the Orinoco floodplain, phytoplankton and attached microalgae again are the main source of carbon for fishes and aquatic invertebrates, despite the greater abundance of macrophytes and terrestrial litter (Hamilton et al. 1992; Lewis et al. 2001).

14.2.6 Patterns in Stream Metabolism

With the advent of continuous monitoring of oxygen flux at multiple sites, it is now feasible to search for broad patterns in the net ecosystem productivity of streams and rivers (Bernhardt et al. 2018). Studies of terrestrial and lake ecosystems have documented predictable patterns related to seasonal variation in environmental variables such as light, temperature, and nutrients; however, the same cannot be said for river ecosystem metabolism. In flowing waters, the seasonality of light and temperature often are not synchronous, and autotrophic biomass can be quickly reduced or eliminated by scouring flows associated with seasonal or aseasonal patterns in precipitation. Pulses of allochthonous resources, such as leaf-litter in the fall in the temperate zone, can decouple relationships between GPP and ER. Allochthonous inputs of carbon can exceed autochthonous production, differentially influencing productivity and respiration. Increased sediment load, due to surface runoff, reduces the amount of light reaching the benthos and influences GPP. Fluctuating discharge throughout the year can also produce scouring, burial,

and drying events that influence autotrophic biomass and community composition, and subsequently alter patterns in productivity (Bernhardt et al. 2018).

The two years of daily measurement of GPP and ER from a small stream in the southeastern United States clearly shows how factors controlling stream metabolism change seasonally (Fig. 14.7). This first-order, deciduous forest stream was heterotrophic throughout the year except during the open-canopy spring, when GPP and ER were equal. Leaf phenology was the main control of seasonal variation, day-to-day weather variation influenced light availability and GPP, and storms suppressed GPP in spring by scouring algae but stimulated GPP in fall by removing leaf litter and increasing light availability. Daily ER was controlled by autotrophic activity in the spring and allochthonous organic matter inputs from leaf litter in autumn. After an initial decrease following storms, labile organic matter inputs from the surrounding terrestrial system led to a multi-day stimulation of ER. Thus, variability in ecosystem metabolism was evident on all time scales, and attributable to daily and seasonal influence of light interacting with vegetation, and episodic high flows.

The decoupling of GPP and ER is frequently driven by large fluctuations in ER in streams characterized by relatively low GPP. However, in streams characterized by high GPP, algae and their associated bacteria generate a large proportion of system-wide ER. Under these circumstances GPP and ER do frequently covary. An excellent example of such covariation was documented in a study of NEP by Huryn et al. (2014) in a spring-fed stream in Alaska in the northern United States (Fig. 14.10). Peak summer rates of GPP and ER were comparable to those of productive streams at temperate latitudes. In contrast, winter rates were low. They suggested that light availability was responsible for patterns in GPP, whereas carbon limitation of heterotrophs, due to low GPP, limited ER.

In both temperate and tropical rivers, seasonal changes in allochthonous inputs can dominate energy flow and influence metabolic regimes, but the factors controlling leaf senescence differ between regions. In the temperate zone, the phenology of deciduous vegetation is largely controlled by temperature and photoperiod (Piao et al. 2019). In autumn, as temperatures cool and day length shortens, leaf fall provides a pulse of allochthonous matter entering streams. In contrast, precipitation, rather than temperature, plays a dominant role in leaf senescence in the tropics. Shedding of leaves during the driest months may help tropical plants reduce water stress (Reich and Borchert 1984), and generate seasonal inputs of riparian leaf litter in some tropical biomes (Tonin et al. 2017). Because anthropogenic climate change is expected to both increase temperatures and alter

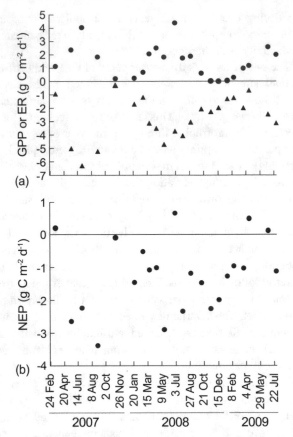

Fig. 14.10 Seasonal patterns of (**a**) gross primary production (GPP, circles) and ecosystem respiration (ER triangles, top panel) and (**b**) net ecosystem production (NEP) for Ivishak Spring, Alaska, US. All values were estimated semi-monthly from March 2007 to August 2009. (Reproduced from Huryn et al. 2014)

precipitation regimes, climate change has the potential to influence the quantity, biological availability, and timing of allochthonous inputs, and alter the metabolic regimes of rivers throughout the world (Larsen et al. 2016).

Stream size may also affect patterns in GPP and ER. Reach-scale metabolism measurements have only recently been collected in larger systems, now providing insights into how the size and position of a river within a watershed influence patterns in metabolism. In a study of 14 mid-sized rivers in the western and midwestern United States, variation in GPP among rivers spanned much of the range of GPP that has been documented in smaller streams (Hall et al. 2016). However, the rivers included in the study had lower rates of heterotrophic respiration relative to GPP, and both GPP and ER peaked in rivers of medium size (Fig. 14.11). In a study of Spanish rivers by Rodríguez-Castillo et al. (2019), NEP was lowest and the difference between GPP and ER was the greatest in the smallest tributaries; however, there was no distinct pattern in NEP that was associated with stream size.

Much of what we have learned about patterns in river energy dynamics has been gleaned from studies conducted

in systems with highly predictable flow regimes (e.g., regulated or spring fed systems; Bernhardt et al. 2018). Yet, stream metabolism is likely to vary with hydrologic disturbance and stream channel retentiveness because these factors directly influence organic matter storage, and thus may alter patterns in benthic respiration (Demars 2019). Even when exposed to intense light, streams with relatively frequent bed-moving flows have relatively low productivity. Frequent hydrologic disturbance can influence stream metabolism by scouring periphyton and biofilms from stone surfaces, and in more extreme cases, through bed transport and up-ending of stones. Measurement of ecosystem metabolism in a sixth-order, gravel-bed Swiss River for 447 days showed strong effects due to bed-moving spates (Uehlinger and Naegeli 1998). Immediately after spates, primary production and ecosystem respiration both declined. Primary production recovered more rapidly in summer than in winter, whereas recovery of respiration showed less seasonal dependency. Spates may have less effect on respiration than primary production because heterotrophic processing of organic matter within the streambed is likely to be less affected by disturbance than autotrophic activity on the bed surface. Thus, depth of scouring, amount of organic matter storage within the streambed, and magnitude of the disturbance will determine the extent to which ecosystem metabolism is altered.

Recent work offers preliminary evidence that it may be possible to classify river systems by their rates of total and net productivity and by the seasonal patterns of photosynthesis and respiration. In an analysis of long-term metabolism datasets from 47 rivers in the United States, Savoy et al. (2019) documented two dominant riverine productivity regimes characterized by the timing of peak productivity. Summer-peak rivers had the mean date of peak productivity in midsummer and an extended period of high GPP. In contrast, spring-peak rivers were characterized by a discrete peak in GPP earlier in the year, followed by very unproductive summer months. A set of environmental variables, including watershed area, water temperature, and discharge, placed most rivers into one of these two productivity regimes. The analysis also hinted at the existence of two additional productivity clusters: aseasonal systems with relatively constant, low rates of GPP year-round, and summer decline rivers with early productivity peaks that declined gradually throughout the summer. This work suggests that a classification system for streams based on patterns in net productivity may be an additional tool to estimate the impact of human activities on stream ecosystem function. Figure 14.12 depicts a conceptualization of the different patterns in climate, light, and hydrology that may occur along the river continuum. Acting together, these drivers of metabolic rates can produce a wide range of values for both GPP and ER, and their sum, NEP. However, additional

Fig. 14.11 (**a**) Gross primary production (GPP), (**b**) ecosystem respiration (ER), (**c**) heterotrophic respiration (|HR|), and (**d**) GPP/ER as a function of river discharge. All ER values are absolute values. Black points are the 14 rivers from this study; gray points are data from other studies. Axes are log scaled. The point far to the right is from the Mississippi River and represents the largest possible size for a North American river. Because of the zero density in points between the Mississippi River and the second largest river in the dataset, the regression line was not fit to include the Mississippi River. (Reproduced from Hall et al. 2016)

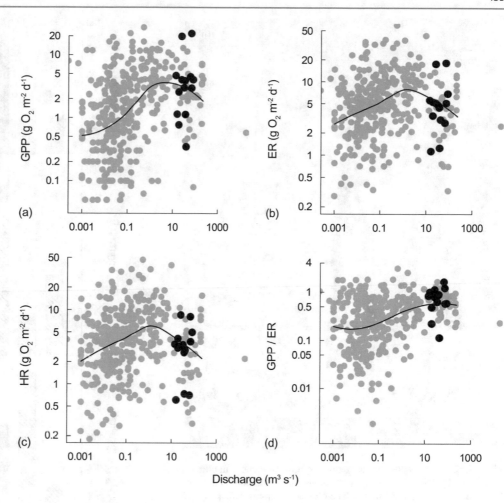

long-term, continuous data are needed from a greater number and greater diversity of rivers and streams to create a robust classification system.

Predicting how energy flows from headwaters to a river delta is made even more complicated because streams and rivers are often part of complex freshwater networks made up of both lentic and lotic habitats that are impacted by human activities (Hotchkiss et al. 2018). Because water residence times differ markedly between rivers and lakes, the opportunity for metabolic processing and for export along the river system vary as well; this may be especially true in river systems that have been dammed. Modelling carbon dynamics in complex river networks is an emerging challenge for stream ecologists, and essential for developing global carbon budgets.

14.2.7 Additional Factors Influencing Metabolic Processes

The distribution of plants and animals may also influence metabolic processes in streams and rivers. An evaluation of substrate-specific GPP and ER in Atlantic Rainforest streams

in Brazil by Tromboni et al. (2017) documented strong contributions to GPP from substrate covered by epilithon and macrophytes, and large contributions to system-wide respiration estimates from substrate covered by leaf litter (Fig. 14.13). Animals may also influence patterns in GPP and ER, as demonstrated for aggregations of mussels. Though significant effects were not observed for ER or NEP, reaches with mussel beds had much greater rates of GPP than reaches without beds (Atkinson et al. 2018). Plants and animals that enter rivers and streams from other habitats also have the potential to influence stream metabolism. Though not often considered, respiring roots from riparian tree species may influence in-stream oxygen dynamics, especially in smaller systems (Dodds et al. 2017). Migrating and senescing salmon can have substantial effects on instream GPP and ER (Levi et al. 2013). In the cobble-bottom streams of southeast Alaska, US, GPP doubled during the salmon run (Fig. 14.14). However, GPP responded inconsistently to the presence of salmon in sand-bottom streams in Michigan in the north-central US, possibly because salmon-derived nutrients enriched autotrophic and heterotrophic communities in all streams, but the changes in nitrogen and phosphorus in Michigan were not as dramatic as were the changes in Alaskan streams (Levi et al. 2013).

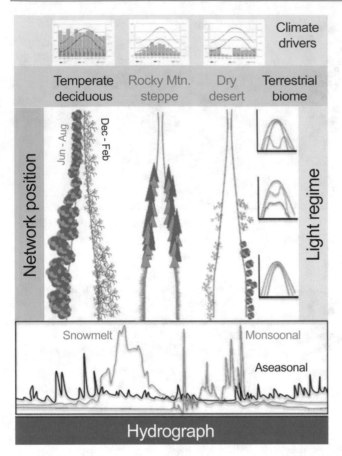

Fig. 14.12 A conceptual model depicting how differences in climate, light, and hydrologic regimes vary along the river continuum and between three terrestrial biomes. The climate diagrams across the top show average monthly precipitation in blue bars with daily air temperatures shown as blue (minimum) and red (maximum) lines. (Reproduced from Bernhardt et al. 2018)

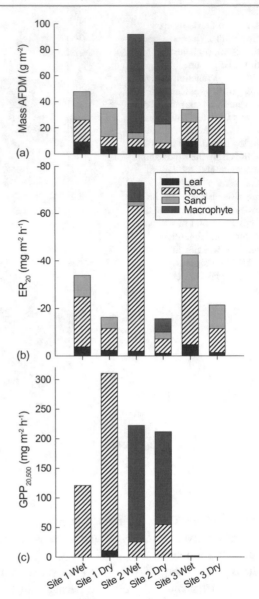

Fig. 14.13 (a) Standing stock of ash-free mass per unit stream area by stream and by season, (b) ecosystem respiration (ER), and (c) gross primary production (GPP) rates per unit area for three streams in wet and dry seasons and on different substrates. (Reproduced from Tromboni et al. 2017)

Land use change can alter metabolic processes in rivers and streams, often by increases in light and nutrient concentrations associated with land conversion (Masese et al. 2017; Griffiths et al. 2013; Tank et al. 2010). Pooling data from periodic daily measurements of whole-stream metabolism from six nutrient-rich streams draining row-crop agriculture in the midwestern United States, Griffiths et al. (2013) documented the influence of variation in light, water temperature, and nutrient concentrations associated with agricultural development (Fig. 14.15). Primary production varied with light level, which was influenced by stream incision and aspect despite the lack of riparian canopy (Fig. 14.15a). Higher water temperatures and greater concentrations of soluble reactive phosphorus were linked with greater rates of respiration (Fig. 14.15b). Measured only during baseflow, both productivity and respiration were high relative to more pristine systems, and one-fourth of all daily measurements had a P/R > 1. Interestingly, the range of metabolic rates was similar across the six streams, possibly because of imposed homogeneity due to agriculture.

At the global scale, access to wastewater treatment can be quite limited. Even in regions with wastewater infrastructure, large volumes of untreated waste may be discharged into surrounding rivers and streams (Connor et al. 2017), influencing stream metabolism. Using a 20-year dissolved oxygen record following the construction of a wastewater treatment plant on the Oria River in northern Spain, Arroita et al. (2019) demonstrated that respiration considerably exceeded GPP in the sewage-impacted river. Wastewater treatment reduced the summer peaks of productivity (Fig. 14.16a) and had an even greater dampening effect on annual rates of respiration (annual rates of respiration by

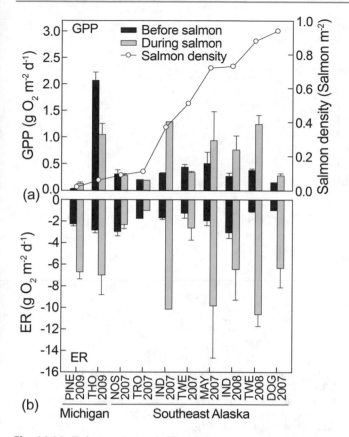

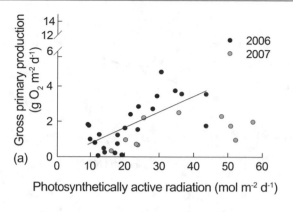

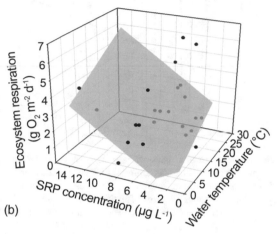

Fig. 14.14 Estimates (mean ± SE) of daily (**a**) gross primary production (GPP) and (**b**) ecosystem respiration (ER) before and during the salmon run for streams in Alaska and Michigan. ER, representing the consumption of oxygen, is displayed as negative values to provide contrast with concurrent GPP. Streams are ordered from left to right according to increasing peak salmon density (line connecting open circles) along the horizontal axis. (Reproduced from Levi et al. 2014)

Fig. 14.15 Relationships between (**a**) gross primary production (GPP) and photosynthetically active radiation (PAR), and (**b**) ecosystem respiration (ER) with streamwater temperature and soluble reactive phosphorus concentration (SRP). (Reproduced from Griffiths et al. 2014)

autotrophs and heterotrophs combined, Fig. 14.16b), resulting in an increase in net ecosystem productivity in the river (Fig. 14.16c). Consequently, river metabolism shifted from strongly heterotrophic to near equilibrium between primary production and ecosystem respiration, indicating that autotrophs were the main drivers of metabolism, and resulting in conditions that facilitated the recovery of aquatic macroinvertebrates and fishes.

Climate change may also induce changes in the processes underpinning stream metabolism (Harjung et al. 2019; Song et al. 2018), and alpine systems may be especially at risk. In 12 study reaches of the Ybbs River network in Austria, researchers documented peaks in productivity in ten of the sites during spring snowmelt that were linked to patterns in light reaching the stream surface and catchment area. As winter precipitation shifted from snow to rain in the spring of a low-snow year, the streams experienced increases in respiration, which converted net ecosystem production in the spring from autotrophy to heterotrophy (Ulseth et al. 2018).

These finding suggest that climate-induced changes in temperature and precipitation regimes may also transform the source-sink dynamics of carbon in streams and rivers. Warming temperatures and reduced snow pack may influence food web structure and ecosystem processes throughout river networks if streams in alpine regions begin emitting more within-stream respiratory CO_2 and supplying less autochthonous energy to reaches downstream. Research in other cold regions of the globe also has demonstrated that increasing water temperatures can influence patterns in instream productivity. In an experimental manipulation of streams in Iceland, Hood et al. (2018) artificially increased water temperatures but retained seasonal changes in light. Irrespective of light seasonality, primary production was greater under warmer temperatures. This change was linked to a shift in the autotroph community, suggesting that altered thermal regimes associated with climate change may influence aquatic community structure in ways that change important ecosystem processes in streams (Fig. 14.17).

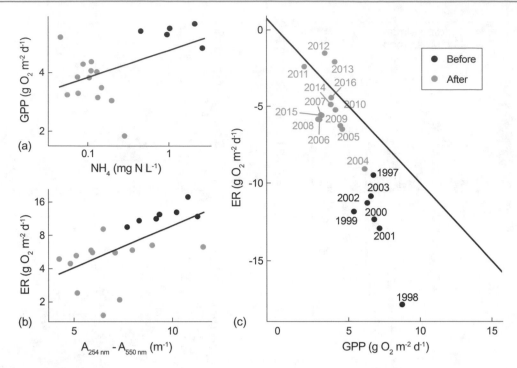

Fig. 14.16 Relationships between (**a**) the annual peak of gross primary production (GPP) and ammonium concentration and (**b**) the annual peak in ecosystem respiration (ER) and the absorbance of dissolved organic matter (a proxy for the concentration of dissolved organic matter) before and after wastewater treatment plant installation in the Oria River in the northern Iberian peninsula. (**c**) The decrease in ER exceeded that for GPP, and the river went from heterotrophy to equilibrium between GPP and ER (i.e., closer to the 1:1 line represented by the black line). All values correspond to summer means. (Reproduced from Arroita et al. 2019)

The intensification of extreme weather events, such as drought and hurricanes, is associated with anthropogenic climate change (Stott 2016; Ornes 2018). Though work is relatively limited, studies have begun to reveal some of the implications of these events on stream metabolism. In an investigation of the response of urban streams to Superstorm Sandy on the east coast of the United States, Reisinger et al. (2017) found that both productivity and respiration declined precipitously following floods associated with storm events; however, the impact was greater on primary production. Both processes recovered quickly (4–18 days) after the disturbance event, and did not differ significantly in recovery rate. This suggests that metabolic processes in urban streams may be more susceptible to change in response to an event, but may recover more quickly after intense storms compared to streams in less disturbed watersheds, because urban streams are often characterized by a flashy hydrograph (Reisinger et al. 2017).

Extreme weather can create conditions that catalyze other environmental events that subsequently influence ecosystem process in rivers and streams. For example, a drought-induced defoliation event by larval gypsy moths (*Lymantria dispar*) in Rhode Island in the northeastern US reduced canopy cover by over 50%. Relative to the prior year of data, water temperatures were warmer, light availability was greater, and autotrophic activity was enhanced following defoliation (Addy et al. 2018). In addition, both caterpillar frass (feces) and leaf detritus associated with the event added particulate carbon and organic nutrients to the stream that may have enhanced respiration. During the defoliation event, the stream experienced lower mean daily levels and wider diel cycles of dissolved oxygen. Though both instream productivity and respiration were significantly higher during the defoliation event, the impact on respiration was greater.

Although flow extremes are initially disruptive, organic matter may be deposited within the sediments as the flood subsides, and so any decline ecosystem respiration may be short-lived. Light, nutrients, and other factors favoring algal growth will of course influence how rapidly the autotrophic community recovers. In comparison to sites with frequent disturbance events, rates of GPP often are maximized in clear streams with high light availability and low flows. In these systems, productivity rates can be as high as those recorded for temperate forests (Bernhardt et al. 2018). Globally, the pressure on freshwater resources is increasing, and more and more streams are subject to low flows and drying events. Desiccation, like flooding, can influence the timing and magnitude of riverine productivity. Hence, changes in flow are predicted to alter patterns in stream metabolism.

As our understanding of patterns in stream metabolism expands, it is likely that rivers and streams will be characterized by their "metabolic regimes" and serve as the basis

Fig. 14.17 An ecosystem-level temperature manipulation undertaken to quantify how coupling of stream ecosystem metabolism and nutrient uptake responded to a realistic warming scenario. Water temperature and gross primary production (GPP) are shown before (**a**) and after (**b**) warming. Water temperatures were higher during the warming manipulation, but retained the same seasonality. Gross primary production (GPP) was higher during the warming manipulation, and the autotroph community shifted toward dominance of *Ulva*, a macroscopic green alga, during June and July. During the warming manipulation, GPP peaked in April–May and June–July. The second peak in production was associated with an *Ulva* bloom as shown in the photographs of the experimental stream in July before (**c**, 2011) and during (**d**, 2013) the warming manipulation. (Reproduced from Hood et al 2018)

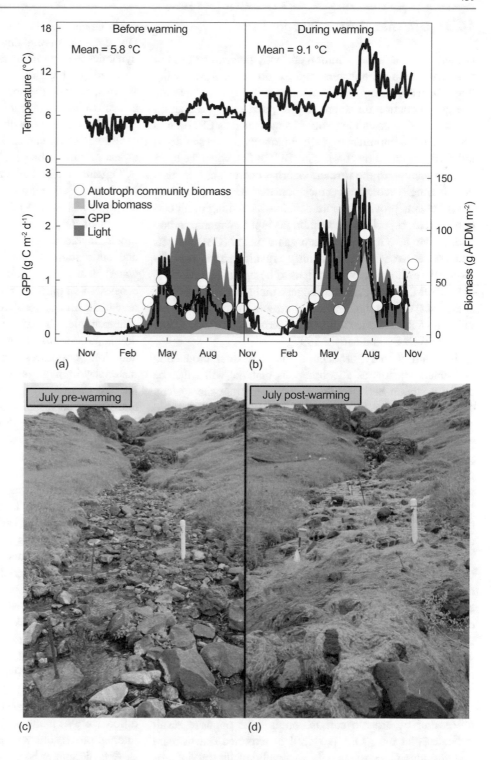

for a new functional classification system for rivers and streams (Ulseth et al. 2019; Bernhardt et al. 2018). Anthropogenic activities, including but not limited to, land use change, water infrastructure, and increasing temperatures and altered precipitation regimes associated with climate change, are all expected to influence spatial and temporal variation in metabolic processes. Systems characterized by short periods of peak metabolism are expected to experience the most severe effects of human disturbance, as small shifts in the magnitude and timing of productivity could result in large changes in energy dynamics in the system (Bernhardt et al. 2018).

14.3 Organic Matter Budgets

Organic carbon is the common currency that can be used to quantify all inputs, transfers, and exports of energy flowing through an ecosystem. The previous section explored internal energy production measured as GPP, and energy consumption by autotrophs and heterotrophs measured as ER. As was discussed, the importance of allochthonous carbon sources is indirectly captured by comparing GPP to ER; when the latter is large relative to the former, external energy sources must be fueling ecosystem metabolism. Organic carbon (OC) budgets provide a more detailed accounting of all carbon sources, especially allochthonous inputs. Organic carbon budgets are also referred to as organic matter (OM) budgets because researchers often quantify organic matter, and subsequently estimate carbon content to be approximately 50% of total OM. Organic matter budgets, including energy inputs and losses, can provide insightful cross-system comparisons of the overall efficiency with which ecosystems use available energy (Webster and Meyer 1997).

Instream primary production and terrestrial production that enters streams as allochthonous material will either be used by stream heterotrophs or exported from the system, either as CO_2 that is respired and outgasses to the atmosphere, or via downstream transport, potentially to the oceans. Organic matter can accumulate over relatively short periods, on the timescale of weeks to months, and storage on or within the streambed and on banks and floodplains can occur on the timescale of years to decades, and perhaps even longer. Storage of organic matter depends on flow variation, as material tends to accumulate during low flows and be exported by high flows. Averaged over long periods, storage was originally thought to be negligible, at least for streams of low order. However, more recent work suggests that mountain streams can store large amounts of carbon (Wohl et al. 2017). Thus, factors that influence the relative rates of conversion of organic carbon to CO_2 versus transport largely determine what fraction of organic matter is mineralized within stream ecosystems, and this is expected to differ among OM compartments. A high rate of utilization relative to transport indicates that OM is contributing to stream metabolism and the stream ecosystem is efficient in its processing of organic carbon inputs. The opposite result indicates that most OM is stored or exported downstream and the stream ecosystem is relatively inefficient in processing carbon resources.

It should be noted that dissolved inorganic carbon (DIC) can comprise a large amount of the total carbon budget in streams (Argerich et al. 2016; Campeau et al. 2017). In-stream DIC is derived from biological and geological sources, originating in both terrestrial and aquatic

environments. Soil respiration is often a primary source of DIC entering streams, but its relative importance to carbon dynamics is often regulated by underlying geology (e.g., the weathering of carbonate minerals). Evasion of CO_2 from streams, stream metabolism, and anaerobic processes are also often key components in DIC cycling (Campeau et al. 2018). As our focus is on organic matter budgets we will not explore DIC dynamics in depth, but this is an important aspect of carbon biogeochemistry.

Organic matter budgets are constructed for some delimited area of an ecosystem. This can be a stream or river reach, or in the case of small headwater streams, the entire catchment. Organic matter budgets attempt to measure all inputs, including primary production, POM from leaf litter and other sources, and DOM from upstream and groundwater; all standing stocks of CBOM, FBOM, and wood; and ecosystem outputs as respiration and export. Budgets can reveal transformations that occur within the study system (for example, CPOM might dominate inputs while FPOM dominates outputs), thereby lending insight into the physical and biological processes that alter the quantity and quality of material within the stream. Coupled with measurement of internal fluxes and the processes that are responsible, the budget approach can provide considerable insight into the flow of material through ecosystems.

In their landmark study of a 1,700 m reach of Bear Brook, a small woodland stream in New Hampshire, Fisher and Likens (1973) pioneered the use of organic matter budgets in running waters. OM inputs from litter, throughfall, and surface and subsurface water were quantified. Because impermeable bedrock underlies this drainage basin, all hydrologic outputs could be estimated from streamflow and organic matter concentrations measured at a weir. The amount of stored material in Bear Brook was assumed to be constant, and on this basis, respiration was estimated from the excess of imports over exports. From the annual energy budget for Bear Brook (Table 14.1), it appears that greater than 99% of the energy inputs were due to allochthonous material (with particulates contributing more than dissolved matter), and about 65% of this was exported downstream (Webster and Meyer 1997). More POM was exported from the study segment than entered it from upstream, and this difference was made up by inputs of litter fall. Virtually all internal processing was attributed to microorganisms.

Organic matter budgets have since been constructed for a number of river ecosystems spanning a range of conditions. A predominance of allochthonous inputs seems to be the rule wherever there is ample riparian vegetation. In a first-order blackwater stream in Virginia with a tree canopy along its entire length, litterfall represented 100% of total inputs (Smock 1997). Similarly, the carbon budget of the Kuparuk

Table 14.1 Organic matter budget for Bear Brook, New Hampshire, in the Hubbard Brook Experimental Forest. Bear Brook is a second order stream, with a catchment area of 132 ha and a streambed area of 6,377 m^2. Based on a compilation of studies by Findlay et al. (1997)

Organic matter parameters	
Inputs (g AFDM m^{-2} y^{-1})	
Gross primary production	3.5
Litterfall and lateral movement	594
Groundwater DOM	95
Standing crops (g m^{-2})	
Wood > 1 mm	530
CBOM > 1 mm (not including wood)	610
FPOM < 1 mm	53
Outputs	
Autotrophic respiration (g m^{-2} y^{-1})	1.75
Heterotrophic respiration (g m^{-2} y^{-1})	101
Particulate transport (kg y^{-1})	1700
Dissolved transport (kg y^{-1})	514

River, originating in the Brooks Range of Alaska, US, and flowing northwards into the Arctic Ocean, is almost totally dominated by allochthonous inputs (Peterson et al. 1986). In this tundra stream meandering through peatland, allochthonous inputs of peat and tundra plant litter exceeded benthic algae primary production by almost an order of magnitude. Although the Kuparuk River is unshaded, cold temperatures and low phosphorus concentrations limit periphyton production. Subsequent estimates showed that net primary production by mosses is similar in magnitude to benthic algal production, increasing the total contribution of autochthonous carbon to this river but not altering the main finding that primary production is modest (Harvey et al. 1997).

The autochthonous component of organic matter budgets is expected to increase downstream as rivers increase in width and the effects of shading and allochthonous inputs from riparian vegetation diminish. In subarctic streams in Quebec, Canada, allochthonous material contributed over 75% of total inputs in streams of low order (Naiman and Link 1997). In contrast, allochthonous inputs contributed only 6–18% the total in larger streams of order five and six. The contribution of autochthonous organic matter to total inputs was positively related to stream order in a synthesis of organic matter budgets from 35 streams located in North America, the Caribbean, Europe, and Antarctica (Webster and Meyer 1997). Arid-land streams were an exception because they are open to the sun and receive few litter inputs.

Instream primary production typically dominates desert stream organic matter budgets (Bunn et al. 2006; Fisher et al. 1982) and high latitude streams (McKnight and Tate 1997; Huryn and Benstead 2019). Primary production in Sycamore Creek, Arizona, was sufficiently high that it substantially exceeded community respiration (Table 14.2); the excess was accounted for by accrual of algal biomass and by downstream export (Grimm 1988). In a meltwater stream in the McMurdo Dry Valleys of Antarctica, primary production by algal mats, composed primarily of filamentous cyanobacteria, was the only carbon source; unsurprisingly, in a land without terrestrial vegetation, allochthonous inputs were zero. Although autochthonous production may be low in many stream types, many researchers have argued that the role of instream primary production has been under-appreciated (Brett et al. 2017). Primary production exceeds litter inputs in a number of examples (Table 14.3), and there is a fairly obvious alternation in their relative importance depending upon forest canopy development.

Seasonality and land use can also drive organic matter dynamics in lotic systems (Tank et al. 2010). In temperate forested streams, peak litterfall occurs in the autumn. Large litter inputs also can result from water stress, as seen in many tropical systems during the dry season (Tonin et al. 2017). In streams draining landscapes with limited riparian vegetation, such as desert streams, seasonal pulses can be less pronounced or nonexistent (Schade and Fisher 1997). Streams in forested watersheds have significantly higher POM inputs when compared to streams draining non-forested watersheds (Golladay 1997). Furthermore, in the temperate zone, streams draining undisturbed watersheds typically have greater leaf litter inputs than do streams draining watersheds that have been logged (Webster et al. 1990).

Fewer organic matter budgets have been constructed for segments of large rivers. Bayley (1989b) approximated a carbon budget for a 187 km stretch with a maximum inundated area of 5,330 km^2 of the Solimões River (the Amazon above Manaus, Brazil). Only a small fraction of the total carbon supply originated with transport of material from upstream (<1%), or as primary production by river phytoplankton (5.4%) and periphyton attached to macrophytes (1.5%). Production by aquatic and terrestrial macrophytes in the littoral regions and floodplain, and litter inputs from the flooded forest, collectively accounted for approximately 90% of carbon production, and so river-floodplain interactions were of far greater consequence than events within the channel. Findings from the Orinoco floodplain of Venezuela were similar: forest litter represented 27% and macrophytes 68% of total carbon sources, and inputs from phytoplankton and periphyton production together contributed only 2% (Lewis et al. 2001). In a sixth-order blackwater river in Georgia, river channel gross primary production accounted for only about one-fifth of total inputs, which were dominated by floodplain organic matter originating in extensive riparian swamps of up to 1–2 km in width (Meyer and

Table 14.2 Organic matter budget for Sycamore Creek. Arizona. Sycamore Creek is a 5th-order stream with a catchment area of 50,500 ha and a streambed area of 33.1 m². Budget is based on a compilation of studies by Jones et al. (1997)

Organic matter parameters	
Inputs (g AFDM m^{-2} y^{-1})	
Gross primary production	1,888
Litterfall	16.5
Lateral movement	3.1
Standing crops (g m^{-2})	
CBOM > 1 mm (not including wood)	5.2
BOM (not including leaves and wood)	104
Hyporheic FPOM	39
Outputs	
Autotrophic respiration (g m^{-2} y^{-1})	944
Heterotrophic respiration (g m^{-2} y^{-1})	372
Hyporheic respiration (g m^{-2} y^{-1})	3259
Particulate transport-baseflow (kg y^{-1})	11,900
Dissolved transport (kg y^{-1})	506,000

A synthesis of 36 organic matter budgets from six different biomes reveals distinct trends related to landscape controls of inputs to streams (Webster and Meyer 1997). A principal components analysis of major budget components categorized streams along a first axis that was positively correlated with litterfall and BOM, and negatively correlated with primary production; and a second axis that was strongly correlated with POM and DOM concentrations in transport (Fig. 14.18). Small mountain streams cluster in the lower right of Fig. 14.18, sharing the characteristics of high litterfall and BOM, and low GPP. Lowland streams have much higher organic matter concentrations and thus greater transport, and arid-land streams fall at the opposite end of the first axis with high GPP and low litterfall and BOM. Thus climate, terrestrial biome, and position along the elevational gradient can be seen to be important underlying controls on stream organic matter budgets.

The budget approach to organic matter dynamics has been highly informative, but its limitations must be acknowledged. Missing terms are common, particularly DOM sources, POM inputs from floodplains, and storm

Table 14.3 Comparison of energy inputs from net primary production (NPP) versus litter fall for a number of spring and running water studies. Additional inputs (e.g., groundwater, transport from upstream) are not considered here. From Peterson et al. (1986) after Minshall (1978). See Petersen (1986) for citations to individual studies

River	Energy input (g C m^{-2}y^{-1})		
	Autochthonous NPP	Allochthonous litter inputs	Reference
Bear Brook, NH	0.6	251	Fisher and Likens (1973)
Kuparuk River, AK	13	100–300	Peterson et al. (1986)
Root Spring, MA	73	261	Teal (1957)
New Hope Creek, NC	73	238	Hall (1972)
Fort River, MA	169	213	Fisher (1977)
Cone Spring, IA	119	70	Tilly (1968)
Deep Creek, ID 1	206	0.2	Minshall (1978)
Deep Creek, ID 2	368	7	Minshall (1978)
Deep Creek, ID 3	761	1.1	Minshall (1978)
Thames River, U.K.	667	16	Mann et al. (1970)
Silver Springs, FL	981	54	Odum (1957)
Tecopa Bore, CA[a]	1229	0	Naiman (1976)

[a]Thermal spring

Edwards 1990). However, as discussed earlier in the chapter, food webs can be fueled largely by autochthonous sources even in systems dominated by allochthonous inputs. Hence, when considered in conjunction with data indicating that many aquatic organisms derive their energy from autochthonous resources, these studies collectively highlight that OM budgets can be used to track carbon through ecosystems, but do not necessarily identify the carbon sources that support secondary production.

transport of POM. Dissolved organic matter inputs are influenced by stream size, precipitation, dominant land use, the presence of wetlands in a watershed, hydraulic conductivity, and hydrologic flow paths (Tank et al. 2010). Stream DOM is predominantly derived from riparian soils and terrestrial leaf litter and accounts for a large percentage of total organic matter inputs (Tank et al. 2018). Additionally, DOM can be derived from instream primary production, suggesting that some DOM is a byproduct of photosynthesis. Lower concentrations of DOM tend to be found in watersheds where soils have high adsorption capacities (e.g., soils rich in clay; Tank et al. 2010).

Inputs, outputs, and storage can vary substantially among years, but because of the effort involved, organic matter budgets often are estimated for just one year, or pieced together with data from multiple years. In a comparison of 23 organic matter budgets from rivers of various sizes, located in different biomes, only one was in steady state (Cummins et al. 1983). Substantial accrual of stored organic matter occurred in 14 budgets, while exports exceeded imports in the remaining eight. The input-output balance for 17 streams reported by Webster and Meyer (1997) included several cases where outputs exceeded inputs by a large margin. These authors argue that it is unlikely that exports will be higher than inputs in annual budgets, and suggest that underestimation of inputs from groundwater and floodplains may be responsible for the imbalance. However, interannual variation in disturbances such as fire, storms, and logging that occur infrequently are important to ecosystem dynamics,

and they are unlikely to be incorporated in a one-year "snap-shot". For example, FPOM export varied fourfold over a seven-year study in a stream in the southeastern United States (Wallace et al. 1997). Ideally, any ecosystem budget should be placed in a historical context in order to capture among-year variation in processing, storage and export.

A more recent effort to document multi-year variation in the carbon budget of a headwater stream by Argerich et al. (2016) estimated that 159 kg C ha^{-1} yr^{-1} was exported from the system. This was notable because the stream comprised only 0.4% of the watershed area, but was exporting carbon at a rate similar to published estimates for much larger systems. Through time, carbon export was dominated by the transport of DIC downstream ($\sim$40% of the total) and by the evasion of CO_2 to the atmosphere ($\sim$26% of the total). Dissolved ($\sim$11%) and particulate ($\sim$21%) organic carbon comprised a much smaller fraction of the total carbon exported from the system. Carbon export was seasonally variable, as 90% of total carbon export occurred between November and April,

Fig. 14.18 A principal components analysis of 25 stream energy budgets from six biomes shows that arid streams, small mountain streams, and lowland streams separate along axes determined by energy inputs, benthic organic matter, and transport rates of organic matter. See text for further explanation. (Reproduced from Webster and Meyer 1997)

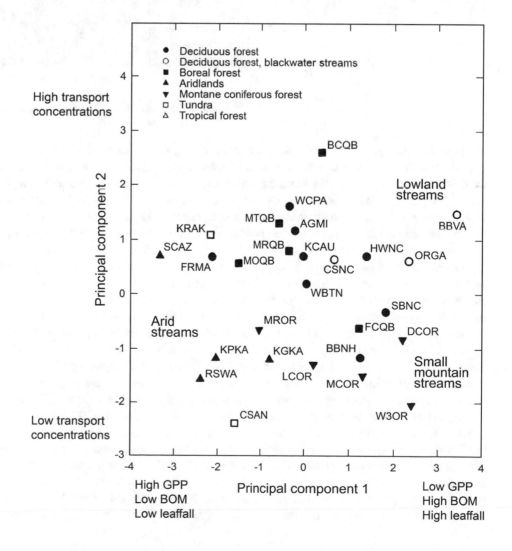

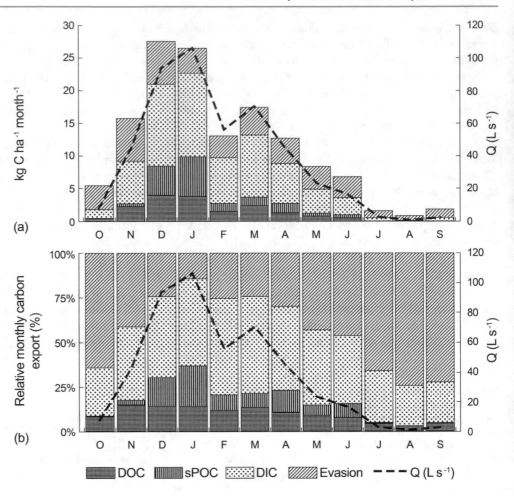

Fig. 14.19 The total (**a**) and relative monthly (**b**) contribution of different sources of carbon to stream carbon export. Mean monthly discharge (Q) is represented by the dotted line in both figures. The carbon values were calculated from weekly composite samples (n = 3) that were collected between 2004–2013. (Reproduced from Argerich et al. 2016)

the period of the year associated with the greatest flows (Fig. 14.19). Interannual variation was also observed, but the drivers of variation differed among the different forms of carbon. Flow was a good predictor of the export of DIC and DOC, but not POC. In contrast, annual variation in stream metabolism was related to in-stream temperatures and photosynthetic active radiation.

Quantification of the amount or organic carbon stored within river systems makes clear that vastly more carbon is stored within riparian vegetation, downed wood, and sediments than is contained within all of the instream biotic compartments (Fig. 14.2). Organic matter in riparian systems is primarily stored in above-ground standing biomass, large woody debris, and sediment on and beneath the floodplain surface (Sutfin et al. 2016). The residence time of retained organic material ranges from days (e.g., labile sugars) to hundreds of years (e.g., woody debris; Tank et al. 2010). The mechanisms promoting retention differ among the fractions of organic matter because of their varying physical and chemical characteristics. For example, DOM is biologically retained, whereas POM is first physically retained and then processed biologically (Tank et al. 2010). Though many components of the carbon budget are

positively correlated with increasing drainage area, carbon standing stocks are not, indicating that other factors must account for variability in this term (Fig. 14.20; Wohl et al. 2017). For instance, the size of carbon pools varies with environmental variables, such as riparian vegetation, soil type, and microbial activity that are influenced by climate, flow regime, valley geometry and underlying geology, making it difficult to generalize about patterns in OC standing crop. However, rivers in cool, wet regions with complex channel geometry within unconfined valleys are optimal conditions for the retention and storage of organic matter in riparian habitats (Fig. 14.21; Sutfin et al. 2016).

Human activities likely have increased riverine carbon flux, especially since the mid-20th Century. From the world's longest record of DOC concentrations, some 130 years for the Thames Basin, UK, Noacco et al. (2017) found that 90% of the long-term rise in fluvial DOC is explained by increased urbanization and is linked to rising population and increased sewage effluent. Land disturbance also has increased carbon export, related to the conversion of grasslands to agriculture and the mobilization of carbon stored in soils. Recent studies analyzing the ^{14}C content of transported organic matter have found that a major

Fig. 14.20 Scaling of organic carbon fluxes and standing stock with river drainage area. If the slope of the line is significantly less than 1, the flux or stock decreases more slowly relative to the increase in drainage area. If slope of the line is significantly greater than 1, the flux or stock decreases at a faster rate than drainage area. A slope of 0 indicates no relationship between flux or stock and drainage area. (**a**) The regression between dissolved organic carbon (DOC) flux and drainage area. (**b**) The regression between particulate organic carbon (POC) flux and drainage area. (**c**) The regression between total organic carbon (TOC = DOC + POC) and drainage area. (**d**) The regression between sedimentation rate of OC within the riparian zone and drainage area. (**e**) No relationship was seen between the stock of OC and the drainage area. (Reproduced from Wohl et al. 2017)

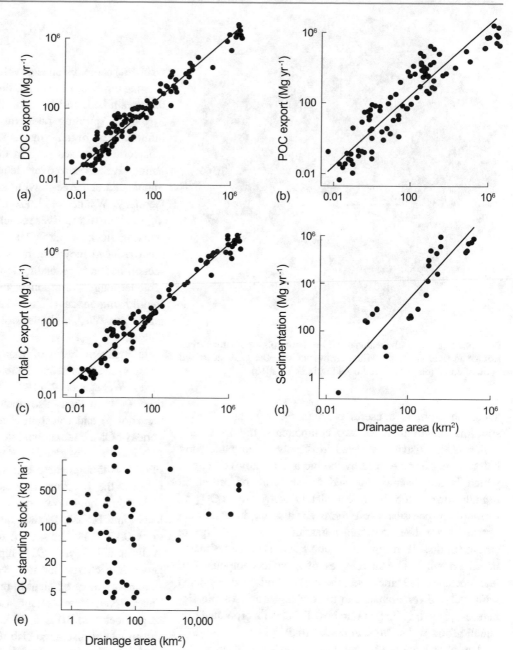

component in most rivers is highly aged material >1,000 years in age, further evidence of the mobilization of stored carbon by human disturbance. Analysis of a global data set of radiocarbon ages of riverine dissolved organic carbon found that the age of dissolved organic carbon in rivers increases with population density and the proportion of human-dominated landscapes within a watershed, and decreases with annual precipitation (Butman et al. 2015). Although one might expect that organic material that has withstood decomposition for thousands of years would be a poor food source compared to recently produced material, studies have found surprisingly significant incorporation of

aged carbon into planktonic food webs in the Hudson River, New York, US (Caraco et al. 2010).

14.4 Carbon Spiraling

Carbon spiraling, a measure of the distance traveled by an atom of carbon in organic form until it is mineralized to CO_2, serves as a comparative measure of an ecosystem's efficiency in processing organic material (Newbold et al. 1982). Because organic matter transport is such a dominant process in streams, estimates of the travel time or distance of

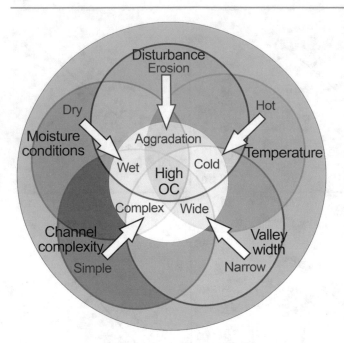

Fig. 14.21 Some of the regional and local controls on organic carbon storage in river corridors. White arrows indicate the gradient toward optimal conditions. (Modified from Sutfin et al. 2016)

a carbon atom is a useful comparative measure. Carbon spiraling length includes two components: uptake length, which is the distance traveled in dissolved inorganic form before being immobilized by the biota; and turnover length, which is the distance traveled by an atom of carbon in organic form before being completely converted to CO_2 by metabolic processes. These terms are also used for nutrient uptake and subsequent mineralization or release. Uptake length for dissolved organic carbon (DOC) can be estimated based on whole-stream releases of organic compounds or leaf leachates. In some cases, the material released is labeled with ^{13}C and researchers can trace changes in isotopic signatures, providing insight into how DOC is incorporated into aquatic food webs (Mineau et al. 2016).

Turnover length, a measure of ecosystem efficiency, can be estimated from downstream carbon flux divided by ecosystem respiration (Newbold et al. 1982). The turnover length (S_p) of different types of POM is estimated from the average particle velocity (V_p) divided by breakdown rate (k):

$$Sp = \frac{Vp}{k} \quad (14.3)$$

This represents the distance a particle travels before entering the next pool of organic matter (for CPOM to become FPOM, or for FPOM to become DOM). The breakdown rate k for FPOM is estimated from its respiration rate in the laboratory. V_p can be calculated as:

$$Vp = \frac{Sw}{Tt + \frac{Sw}{Vw}} \quad (14.4)$$

where S_w is the distance traveled by the particle in the water, T_t is the turnover time or the time that the particle remains on the stream bed, and V_w is water velocity.

Carbon spiraling rates vary among rivers and streams, influenced by organic matter inputs, retention capacity, and metabolic processes such as GPP and ER. In a study comparing carbon turnover length and turnover times in impounded and free-flowing sections of the Spree River in Germany, Wanner et al. (2002) documented that free flowing sections of the river recycled approximately 50% of the standing stock of particulate organic carbon, but impoundment reduced recycling rates to just 25%. The impounded section had larger standing stocks of carbon, shorter carbon turnover lengths, and longer turnover times compared to the free-flowing reach (Fig. 14.22), suggesting that impoundments can alter multiple stocks and flows of the carbon budget.

In their synthesis of many studies of breakdown and transport in forested small streams in the southeastern United States, Webster et al. (1999) compared biological turnover time (a term that also includes physical and chemical breakdown) and transport distance for the four main categories of OM. Breakdown rates ranged from nearly six years for sticks to a few months for leaves, and exceeded a year for FPOM. Although these estimates are provisional for many reasons, the outcome is reasonable: transport rates were higher than breakdown rates for sticks in comparison with leaves and FPOM. Particle turnover lengths were estimated to be 0.15, 0.11 and 42 km for sticks, leaves, and FPOM. Webster and Meyer (1997) reported a significant correlation between discharge and turnover length, implying that small streams are more efficient in the use of organic matter. In the Taieri River, New Zealand, organic carbon turnover length ranged between 10 and 98 km, with higher values downstream where discharge also was higher (Young and Huryn 1997). In the Snake River, Idaho, turnover lengths were between 11 and 108 km and were related to patterns in current velocity (Thomas et al. 2005).

Spiraling length also varies over time and is influenced by land use (Lisboa et al. 2016). In small streams draining agricultural landscapes in Indiana in the central US, spiraling length changed seasonally, from 7.7–54.4 km in winter to 0.2–9.0 km in summer (Griffiths et al. 2012). Unsurprisingly, the authors suggested seasonality was primarily driven by differences in discharge, suggesting that hydrology tightly controls the fate of organic carbon in these streams. This work also provided evidence that relative to forested streams, agricultural streams tended to be less retentive of

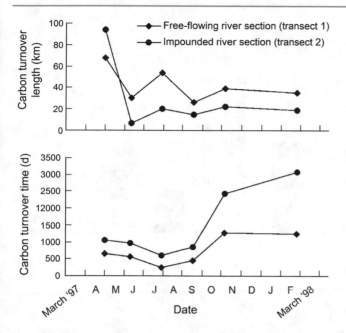

Fig. 14.22 Temporal variation in turnover lengths (**a**) and turnover times (**b**) of organic carbon for free-flowing and impounded sections of the River Spree, Germany. (Reproduced from Wanner et al. 2002)

organic carbon. Griffiths et al. (2012) inferred that small streams draining agricultural areas primarily function as conduits transporting organic carbon downstream, except during low, stable-flow periods when they can be as retentive of organic carbon as forested headwaters (Fig. 14.23).

14.5 Summary

In a prescient essay published in 1975, Noel Hynes wrote that "in every respect the valley rules the stream". Geology determines the availability of ions and the supply of sediments, topography determines slope and degree of containment, climate and soils determine vegetation and hence the availability of autochthonous and allochthonous organic matter, and so on. Decades of research support this view. The river continuum concept describes how basal resources and thus consumer assemblages and stream metabolism change along a river's length owing to changes in river size and terrestrial influences. The flood pulse model reminds us that carbon dynamics in rivers is strongly influenced by hydrology and connectivity with the surrounding terrestrial environment. In addition, our perspective on rivers within landscapes has expanded to encompass more explicit consideration of the physical template and spatial hierarchy provided by the river network. Intriguingly, Hynes (1975) also opined that every stream "is likely to be individual and

thus not really very easily classifiable". Yet, decades of effort to place the individuality of streams within the frameworks of scale and landscape have significantly advanced our understanding of the causes of that individuality. New advances, supported by the increasing availability of sensors, adoption of open-water methods to measure whole stream metabolism, and new statistical tools will support emerging efforts to classify streams by their metabolic processes.

Sources of organic carbon in lotic ecosystems include autochthonous production by algae and aquatic plants, and allochthonous inputs of dead organic matter from terrestrial primary production. Studies of stream ecosystem metabolism address two central questions: the relative magnitude of internal versus external energy sources, including their variation along a river's length and with landscape setting; and the efficiency of the stream ecosystem in metabolizing those energy supplies versus loss of carbon to downstream ecosystems, the atmosphere, and the oceans. Principal approaches include the comparison of gross primary production to ecosystem respiration, mass balance estimation of all inputs and exports, and measures of the efficiency with which organic carbon is utilized.

Stream metabolism is measured by accounting for the oxygen produced through primary production, lost through ecosystem respiration, and exchanged with the atmosphere. The relationship between gross primary production and ecosystem respiration can indicate whether an ecosystem is reliant mainly on internal production, or requires organic matter subsidies to sustain respiration. The net flux of oxygen, measured as the sum of gross primary productivity (a positive value) and ecosystem respiration (a negative value) is called net ecosystem productivity. Macroscale patterns in stream metabolism and the biological, physical, and chemical factors that regulate productivity, respiration, and riverine gas exchange are becoming better understood as longer-term data are collected in a broader diversity of lotic systems. Ecologists and water resource managers may soon be able to use the metabolic regimes of streams to assess the influence of anthropogenic activities on the function of flowing waters.

Organic carbon budgets are based on the estimation of all inputs, standing stocks, and losses within a stream reach or, ideally, a catchment, although the latter is practical only for headwater steams. Budget studies demonstrate how the climate and the terrestrial biome influence the relative magnitude of allochthonous versus autochthonous inputs. Inputs of coarse, fine, and dissolved organic matter from terrestrial primary production typically dominate the energy supply in small, forested streams where algal primary production tends

Fig. 14.23 Left Panel: Carbon transported as (**a**) dissolved organic carbon (DOC), (**b**) fine particulate organic carbon (FPOC), and (**c**) coarse particulate organic carbon (CPOC). Right Panel: Organic carbon spiraling metrics: (**d**) organic carbon velocity (VOC), (**e**) biotic turnover rate of organic carbon (KOC), and (**f**) organic carbon turnover length (SOC). Mean values are reported (±standard error) for each biologically important time period (autumn, winter, early summer, and late summer). Letters represent significant differences between seasons based on results from Tukey's HSD post-hoc tests (Reproduced from Griffiths et al. 2012)

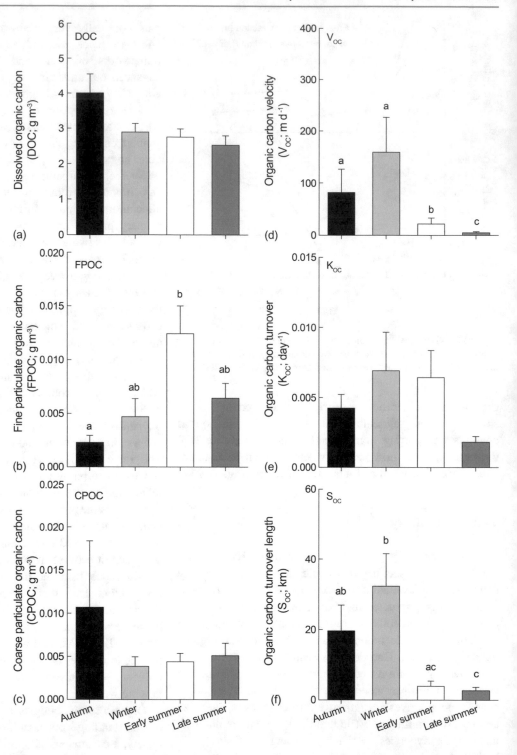

to be light-limited, but primary production is of greater importance in open locations that receive sufficient light. Thus, longitudinal position and landscape setting determine the relative magnitude of sources of organic carbon to stream ecosystems. In general, arid-land, meadow, and prairie streams have high primary production relative to detrital inputs, temperate forested streams are the opposite and highly dependent upon external energy inputs, and lowland streams have large quantities of DOC and POC in transport.

Organic matter that enters the channels of streams and rivers can be stored for some time on streambanks and by burial within the channel, but ultimately it is exported to downstream ecosystems or mineralized to CO_2 by the biota and lost from the system by outgassing. Export is the fate of a great deal of organic matter. Organic matter export is determined by the interaction of material available on the stream bottom, retentive capacity of the system, and hydrologic variability.

Globally, rivers respire significant quantities of carbon to the atmosphere and export significant quantities of organic carbon from terrestrial primary production to downstream locations and the oceans. Streams of low order are frequently inefficient, exporting large quantities of FPOM and DOM to downstream reaches. Our understanding of the role of smaller systems in watershed-level CO_2 evasion is just emerging. Large rivers transport substantial amounts of POM and DOM, but declines in DOC concentrations in lower reaches of large rivers and their super-saturation with CO_2 provide evidence of substantial metabolic activity, indicating that significant mineralization takes place near the lower terminus of rivers.

References

Addy K, Gold AJ, Loffredo JA et al (2018) Stream response to an extreme drought-induced defoliation event. Biogeochemistry 140:199–215

Allen GH, Pavelsky TM (2018) Global extent of rivers and streams. Science 361:585–588

Appling AP, Hall RO, Yackulic CB et al (2018a) Overcoming equifinality: leveraging long time series for stream metabolism estimation. J Geophys Res Biogeo 123:624–645

Appling AP, Read JS, Winslow LA et al (2018b) The metabolic regimes of 356 rivers in the United States. Sci data 5:180292

Araujo-Lima CA, Forsberg BR, Victoria R et al (1986) Energy sources for detritivorous fishes in the Amazon. Science 234:1256–1258

Argerich A, Haggerty R, Johnson SL et al (2016) Comprehensive multiyear carbon budget of a temperate headwater stream. J Geophys Res Biogeo 121:1306–1315

Arroita M, Elosegi A, Hall RO Jr (2019) Twenty years of daily metabolism show riverine recovery following sewage abatement. Limnol Oceanogr 64:S77–S92

Atkinson CL, Sansom BJ, Vaughn CC et al. (2018) Consumer aggregations drive nutrient dynamics and ecosystem metabolism in nutrient-limited systems. Ecosystems:1–15

Battin TJ, Kaplan LA, Findlay S et al (2008) Biophysical controls on organic carbon fluxes in fluvial networks. Nat Geosci 1:95–100. https://doi.org/10.1038/ngeo101

Battin TJ, Luyssaert S, Kaplan LA et al (2009) The boundless carbon cycle. Nat Geosci 2:598

Bayley PB (1988) Factors affecting growth rates of young tropical floodplain fishes: seasonality and density-dependence. Environ Biol Fishes 21:127–142

Bayley PB (1989a) Aquatic environments in the Amazon Basin, with an analysis of carbon sources, fish production, and yield. Can J Fish Aquat Sci 106:399–08

Bayley PB (1989b) Understanding large river-floodplain ecosystems. BioScience 45:153–158

Beaulieu JJ, Shuster WD, Rebholz JA (2012) Controls on gas transfer velocities in a large river. J Geophys Res Biogeo 117. https://doi.org/10.1029/2011jg001794

Bernhardt ES, Heffernan JB, Grimm NB et al (2018) The metabolic regimes of flowing waters. Limnol Oceanogr 63:S99–S118

Blaszczak JR, Delesantro JM, Urban DL et al (2018) Scoured or suffocated: Urban stream ecosystems oscillate between hydrologic and dissolved oxygen extremes. Limnol Oceanogr 64:877–894

Bott TL, Newbold JD, Arscott DB (2006) Ecosystem metabolism in Piedmont streams: reach geomorphology modulates the influence of riparian vegetation. Ecosystems 9:398–421

Brett MT, Bunn SE, Chandra S et al (2017) How important are terrestrial organic carbon inputs for secondary production in freshwater ecosystems? Freshw Biol 62:833–853

Bunn SE, Balcombe SR, Davies PM et al. (2006) Aquatic productivity and food webs of desert river ecosystems. Ecol Desert Rivers:76–99

Butman DE, Wilson HF, Barnes RT et al (2015) Increased mobilization of aged carbon to rivers by human disturbance. Nat Geosci 8:112

Campeau A, Bishop K, Amvrosiadi N et al (2019) Current forest carbon fixation fuels stream CO_2 emissions. Nat Commun 10:1–9

Campeau A, Bishop K, Nilsson MB et al (2018) Stable carbon isotopes reveal soil-stream DIC linkages in contrasting headwater catchments. J Geophys Res-Biogeosci 123:149–167. https://doi.org/10.1002/2017jg004083

Campeau A, Wallin MB, Giesler R et al (2017) Multiple sources and sinks of dissolved inorganic carbon across Swedish streams, refocusing the lens of stable C isotopes. Sci Rep 7:9158

Caraco N, Bauer JE, Cole JJ et al (2010) Millennial-aged organic carbon subsidies to a modern river food web. Ecology 91:2385–2393

Castello L, Bayley PB, Fabré NN et al (2019) Flooding effects on abundance of an exploited, long-lived fish population in river-floodplains of the Amazon. Rev Fish Biol Fish 29:487–500

Cole JJ, Prairie YT, Caraco NF et al (2007) Plumbing the global carbon cycle: integrating inland waters into the terrestrial carbon budget. Ecosystems 10:171–184. https://doi.org/10.1007/s10021-006-9013-8

Connor R, Renata A, Ortigara C et al. (2017) The United Nations World Water Development Report 2017. Wastewater: The untapped resource. The United Nations World Water Development Report. United Nations Educational, Scientific and Cultural Organization, Paris

Cummins K, Sedell J, Swanson F et al (1983) Organic matter budgets for stream ecosystems: problems in their evaluation. In: Barnes JR, Minshall GW (eds) Stream ecology: application and testing of general ecological theory. Springer, pp 299–353

Cummins KW, Klug MT (1979) Feeding ecology of stream invertebrates. Annu Rev Ecol Syst 10:147–172

Cummins KW, Petersen RC, Howard FO et al (1973) Utilization of leaf litter by stream detritivores. Ecology 54:336–345

Demars BO (2019) Hydrological pulses and burning of dissolved organic carbon by stream respiration. Limnol Oceanogr 64:406–421

Demars BO, Russell Manson J, Olafsson JS et al (2011) Temperature and the metabolic balance of streams. Freshw Biol 56:1106–1121

Dettmers JM, Gutreuter S, Wahl DH et al (2001) Patterns in abundance of fishes in main channels of the upper Mississippi River system. Can J Fish Aquat Sci 58:933–942

Dodds WK, Higgs SA, Spangler MJ et al (2018) Spatial heterogeneity and controls of ecosystem metabolism in a Great Plains river network. Hydrobiologia 813:85–102. https://doi.org/10.1007/s10750-018-3516-0

Dodds WK, Tromboni F, Saltarelli WA et al (2017) The root of the problem: direct influence of riparian vegetation on estimation of stream ecosystem metabolic rates. Limnol Oceanogr Lett 2:9–17

Farly L, Hudon C, Cattaneo A et al (2019) Seasonality of a floodplain subsidy to the fish community of a large temperate river. Ecosystems 22:1823–1837

Fasching C, Ulseth AJ, Schelker J et al (2016) Hydrology controls dissolved organic matter export and composition in an Alpine stream and its hyporheic zone. Limnol Oceanogr 61:558–571

Findlay SEG, Likens GE, Hedin L, Fisher SG, McDowell WH (1997) Organic matter dynamics in Bear Brook, Hubbard Brook Experimental Forest, new Hampshire, USA. J N Am Benthol Soc 16:43–46

Findlay S, Smith PJ, Meyer JL (1986) Effect of detritus addition on metabolism of river sediment. Hydrobiologia 137: 257-263

Fisher SG, Carpenter SR (1976) Ecosystem and macrophyte primary production of the Fort River, Massachusetts. Hydrobiologia 49:175-187

Fisher SG (1977) Organic matter processing by a stream-segment ecosystem: Fort River, Massachusetts, USA. Int Rev Ges Hydrobiol 62:701-727

Fisher SG, Gray LJ, Grimm NB et al (1982) Temporal succession in a desert stream ecosystem following flash flooding. Ecol Monogr 52:93–110

Fisher SG, Likens GE (1973) Energy flow in Bear Brook, New Hampshire–integrative approach to stream ecosystem metabolism. Ecol Monogr 43:421–439. https://doi.org/10.2307/1942301

Galat DL, Zweimüller I (2001) Conserving large-river fishes: is the highway analogy an appropriate paradigm? J N Am Benthol Soc 20:266–279

Golladay SW (1997) Suspended particulate organic matter concentration and export in streams. J N Am Benthol Soc 16:122–131

Griffiths NA, Tank JL, Royer TV et al (2013) Agricultural land use alters the seasonality and magnitude of stream metabolism. Limnol Oceanogr 58:1513–1529

Griffiths NA, Tank JL, Royer TV et al (2012) Temporal variation in organic carbon spiraling in Midwestern agricultural streams. Biogeochemistry 108:149–169. https://doi.org/10.1007/s10533-011-9585-z

Grimm NB (1988) Role of macroinvertebrates in nitrogen dynamics of a desert stream. Ecology 69:1884–1893

Hall R, Thomas S, Gaiser EE (2007) Measuring freshwater primary production. In: Fahey TJ, Knapp AK (eds) Principles and standards for measuring primary production. Oxford University Press, Oxford, pp 175–203

Hall RO (2016) Chapter 4 - Metabolism of streams and rivers: Estimation, controls, and application. In: Jones JB, Stanley EH (eds) Stream ecosystems in a changing environment. Academic Press, Boston, pp 151–180. https://doi.org/10.1016/B978-0-12-405890-3.00004-X

Hall RO, Beaulieu JJ (2013) Estimating autotrophic respiration in streams using daily metabolism data. Freshw Sci 32:507–516

Hall RO, Hotchkiss ER (2017) Stream metabolism. In: Lamberti GA, Hauer FR (eds) Methods in stream ecology (Third Edition). Academic Press, pp 219–233. https://doi.org/10.1016/B978-0-12-813047-6.00012-7

Hall RO, Tank JL, Baker MA et al (2016) Metabolism, gas exchange, and carbon spiraling in rivers. Ecosystems 19:73–86. https://doi.org/10.1007/s10021-015-9918-1

Hall RO, Ulseth AJ (2020) Gas exchange in streams and rivers. Wiley Interdiscip Rev Water 7:e1391

Hall RO, Yackulic CB, Kennedy TA et al (2015) Turbidity, light, temperature, and hydropeaking control primary productivity in the Colorado River, Grand Canyon. Limnol Oceanogr 60:512–526

Hamilton S, Lewis W, Sippel S (1992) Energy sources for aquatic animals in the Orinoco River floodplain: evidence from stable isotopes. Oecologia 89:324–330

Harjung A, Ejarque E, Battin T et al (2019) Experimental evidence reveals impact of drought periods on dissolved organic matter quality and ecosystem metabolism in subalpine streams. Limnol Oceanogr 64:46–60

Harvey CJ, Peterson BJ, Bowden WB et al (1997) Organic matter dynamics in the Kuparuk River, a tundra river in Alaska, USA. J N Am Benthol Soc 16:18–23

Hill BH, Webster JR (1982) Aquatic macrophyte breakdown in an Appalachian river. Hydrobiologia 89:53-59

Hill BH, Webster JR (1983) Aquatic macrophyte contribution to the New River organic matter budget. Dynamics of Lotic Systems, Ann Arbor Science, Ann Arbor MI, pp 273-282

Hoellein TJ, Bruesewitz DA, Richardson DC (2013) Revisiting Odum (1956): a synthesis of aquatic ecosystem metabolism. Limnol Oceanogr 58:2089–2100

Hood JM, Benstead JP, Cross WF et al (2018) Increased resource use efficiency amplifies positive response of aquatic primary production to experimental warming. Glob Change Biol 24:1069–1084

Horgby Å, Boix Canadell M, Ulseth AJ et al (2019) High-resolution spatial sampling identifies groundwater as driver of CO2 dynamics in an alpine stream network. J Geophys Res Biogeo 124:1961–1976

Hotchkiss E, Hall R Jr, Sponseller R et al (2015) Sources of and processes controlling CO_2 emissions change with the size of streams and rivers. Nat Geosci 8:696

Hotchkiss E, Sadro S, Hanson P (2018) Toward a more integrative perspective on carbon metabolism across lentic and lotic inland waters. Limnol Oceanogr Lett 3:57–63

Huryn AD, Benstead JP (2019) Seasonal changes in light availability modify the temperature dependence of secondary production in an arctic stream. Ecology 100:e02690

Huryn AD, Benstead JP, Parker SM (2014) Seasonal changes in light availability modify the temperature dependence of ecosystem metabolism in an arctic stream. Ecology 95:2826–2839

Hynes HBN (1975) The stream and its valley. Verhandlungen der Internationalen Vereinigung fur Limnologie 19:1–15

Jones JB, Schade JD, Fisher SG, Grimm NB (1997) Organic matter dynamics in Sycamore Creek, a desert stream in Arizona, USA. J N Am Benthol Soc 16:78–82

Junk W, Bayley P, Sparks R (1989) The flood pulse concept in river-floodplain systems. Can J Fish Aquat Sci 106:110–127

Junk WJ, Wantzen KM (2004) (2004) The flood pulse concept: new aspects, approaches and applications-an update. Second international symposium on the management of large rivers for fisheries. Food and Agriculture Organization and Mekong River Commission, FAO pp, pp 117–149

King D Ball RC (1967) Comparative energetics of a polluted stream. Limnol Oceanogr 12: 27–33

Koenig LE, Helton AM, Savoy P et al (2019) Emergent productivity regimes of river networks. Limnol Oceanogr Lett 4:173–181

Lake P, Barmuta L, Boulton A et al (1985) Australian streams and Northern Hemisphere stream ecology: comparisons and problems. Proc Ecol Soc Aust 1985:61–82

Larsen S, Muehlbauer JD, Marti E (2016) Resource subsidies between stream and terrestrial ecosystems under global change. Glob Change Biol 22:2489–2504. https://doi.org/10.1111/gcb.13182

Levi PS, Tank JL, Rüegg J et al (2013) Whole-stream metabolism responds to spawning Pacific salmon in their native and introduced ranges. Ecosystems 16:269–283. https://doi.org/10.1007/s10021-012-9613-4

Lewis WM, Hamilton SK, Rodríguez MA et al (2001) Foodweb analysis of the Orinoco floodplain based on production estimates and stable isotope data. J N Am Benthol Soc 20:241–254

Lisboa LK, Thomas S, Moulton TP (2016) Reviewing carbon spiraling approach to understand organic matter movement and transformation in lotic ecosystems. Acta Limnologica Brasiliensia 28

Mann KH (1969) The dynamics of aquatic ecosystems. Advanc Ecol Res 6:1-81

Masese FO, Salcedo-Borda JS, Gettel GM et al (2017) Influence of catchment land use and seasonality on dissolved organic matter composition and ecosystem metabolism in headwater streams of a Kenyan river. Biogeochemistry 132:1–22

McKnight DM, Tate C (1997) Canada stream: a glacial meltwater stream in Taylor Valley, south Victoria Land, Antarctica. J N Am Benthol Soc 16:14–17

McMeans BC, Kadoya T, Pool TK et al (2019) Consumer trophic positions respond variably to seasonally fluctuating environments. Ecology 100:e02570

Meyer JL, Edwards RT (1990) Ecosystem metabolism and turnover of organic carbon along a blackwater river continuum. Ecology 71:668–677

Mineau MM, Wollheim WM, Buffam I et al (2016) Dissolved organic carbon uptake in streams: a review and assessment of reach-scale measurements. J Geophys Res Biogeo 121:2019–2029

Minshall GW (1978) Autotrophy in stream ecosystems. Bioscience 28:767–771

Minshall GW, Cummins KW, Petersen RC et al (1985) Developments in stream ecosystem theory. Can J Fish Aquat Sci 42:1045–1055

Mortillaro J-M, Pouilly M, Wach M et al (2015) Trophic opportunism of central Amazon floodplain fish. Freshw Biol 60:1659–1670

Mulholland P, Fellows C, Tank J et al (2001) Inter-biome comparison of factors controlling stream metabolism. Freshw Biol 46:1503–1517

Naiman RJ (1976) Primary production, standing stock, and export of organic matter in a Mohave Desert thermal stream 1. Limnol Oceanogr 21: 60-73

Naiman RJ, Link GL (1997) Organic matter dynamics in 5 subarctic streams, Quebec, Canada. J N Am Benthol Soc 16:33–39

Newbold J, Mulholland P, Elwood J et al (1982) Organic carbon spiralling in stream ecosystems. Oikos:266–272

Noacco V, Wagener T, Worrall F et al (2017) Human impact on long-term organic carbon export to rivers. J Geophys Res Biogeo 122:947–965

Odum EP (1968) Energy flow in ecosystems: a historical review. Am Zool 8:11–18

Odum HT (1956) Primary production in flowing waters. Limnol Oceanogr 2:85–97

Odum HT (1957) Trophic structure and productivity of Silver Springs, Florida. Ecol Monogr 27:55-112

Ornes S (2018) Core concept: how does climate change influence extreme weather? Impact attribution research seeks answers. PNAS 115:8232–8235. https://doi.org/10.1073/pnas.1811393115

Peterson BJ, Hobbie JE, Corliss TL (1986) Carbon flow in a tundra stream ecosystem. Can J Fish 43:1259–1270

Piao S, Liu Q, Chen A et al (2019) Plant phenology and global climate change: current progresses and challenges. Glob Change Biol. https://doi.org/10.1111/gcb.14619

Poole GC (2002) Fluvial landscape ecology: addressing uniqueness within the river discontinuum. Freshw Biol 47:641–660

Raymond PA, Hartmann J, Lauerwald R et al (2013) Global carbon dioxide emissions from inland waters. Nature 503:355

Raymond PA, Zappa CJ, Butman D et al (2012) Scaling the gas transfer velocity and hydraulic geometry in streams and small rivers. Limnol Oceanogr Fluids Environ 2:41–53

Reich PB, Borchert R (1984) Water stress and tree phenology in a tropical dry forest in the lowlands of Costa Rica. J of Eco 72:61–74. https://doi.org/10.2307/2260006

Reisinger AJ, Rosi EJ, Bechtold HA et al (2017) Recovery and resilience of urban stream metabolism following Superstorm Sandy and other floods. Ecosphere 8:e01776

Reisinger AJ, Tank JL, Rosi-Marshall EJ et al (2015) The varying role of water column nutrient uptake along river continua in contrasting landscapes. Biogeochemistry 125:115–131. https://doi.org/10.1007/s10533-015-0118-z

Roberts BJ, Mulholland PJ, Hill WR (2007) Multiple scales of temporal variability in ecosystem metabolism rates: results from 2 years of continuous monitoring in a forested headwater stream. Ecosystems 10:588–606

Rocher-Ros G, Sponseller RA, Lidberg W et al (2019) Landscape process domains drive patterns of CO_2 evasion from river networks. Limnol Oceanogr Lett 4:87–95. https://doi.org/10.1002/lol2.10108

Rodríguez-Castillo T, Estévez E, González-Ferreras AM et al (2019) Estimating ecosystem metabolism to entire river networks. Ecosystems 22:892–911. https://doi.org/10.1007/s10021-018-0311-8

Romeijn P, Comer-Warner SA, Ullah S et al (2019) Streambed organic matter controls on carbon dioxide and methane emissions from streams. Environ Sci Technol 53:2364–2374

Rosi-Marshall EJ, Vallis KL, Baxter CV et al (2016) Retesting a prediction of the river continuum concept: autochthonous versus allochthonous resources in the diets of invertebrates. Freshw Sci 35:534–543. https://doi.org/10.1086/686302

Savoy P, Appling AP, Heffernan JB, et al. (2019) Metabolic rhythms in flowing waters: An approach for classifying river productivity regimes. Limnol Oceanogr 64:1835-1851

Schade JD, Fisher SG (1997) Leaf litter in a Sonoran Desert stream ecosystem. J N Am Benthol Soc 16:612–626

Schramm HL, Eggleton MA (2006) Applicability of the flood-pulse concept in a temperate floodplain river ecosystem: thermal and temporal components. River Res Appl 22:543–553

Siders AC, Larson DM, Rüegg J et al (2017) Probing whole-stream metabolism: influence of spatial heterogeneity on rate estimates. Freshw Biol 62:711–723

Sinsabaugh RL (1997) Large-scale trends for stream benthic respiration. J N Am Benthol Soc 16:119–122

Smock LA (1997) Organic matter dynamics in Buzzards Branch, a blackwater stream in Virginia, USA. J N Am Benthol Soc 16:54–58

Song C, Dodds WK, Rüegg J et al (2018) Continental-scale decrease in net primary productivity in streams due to climate warming. Nat Geosci 11:415

Song C, Dodds WK, Trentman MT et al (2016) Methods of approximation influence aquatic ecosystem metabolism estimates. Limnol Oceanogr Methods 14:557–569

Staehr PA, Testa JM, Kemp WM et al (2012) The metabolism of aquatic ecosystems: history, applications, and future challenges. Aquat Sci 74:15–29

Stegen JC, Johnson T, Fredrickson JK et al (2018) Influences of organic carbon speciation on hyporheic corridor biogeochemistry and microbial ecology. Nat Commun 9:585

Stott P (2016) How climate change affects extreme weather events. Science 352:1517–1518. https://doi.org/10.1126/science.aaf7271

Sutfin NA, Wohl EE, Dwire KA (2016) Banking carbon: a review of organic carbon storage and physical factors influencing retention in floodplains and riparian ecosystems. Earth Surf Process Landf 41:38–60

Tank JL, Rosi-Marshall EJ, Griffiths NA et al (2010) A review of allochthonous organic matter dynamics and metabolism in streams. J N Am Benthol Soc 29:118–146

Tank SE, Fellman JB, Hood E et al (2018) Beyond respiration: controls on lateral carbon fluxes across the terrestrial-aquatic interface. Limnol Oceanog Lett 3:76–88

Thomas SA, Royer TV, Snyder EB et al (2005) Organic carbon spiraling in an Idaho river. Aquat Sci 67:424–433

Thorp JH, Alexander J, James E, Bukaveckas BL et al (1998) Responses of Ohio River and Lake Erie dreissenid molluscs to changes in temperature and turbidity. Can J Fish Aquat Sci 55:220–229

Thorp JH, Delong MD (1994) The riverine productivity model: an heuristic view of carbon sources and organic processing in large river ecosystems. Oikos 94:305–308

Thorp JH, Delong MD (2002) Dominance of autochthonous autotrophic carbon in food webs of heterotrophic rivers. Oikos 96:543–
550

Thorp JH, Thoms MC, Delong MD (2006) The riverine ecosystem
synthesis: biocomplexity in river networks across space and time.
River Res Appl 22:123–147

Tilly, L. J. (1968) The structure and dynamics of Cone Spring. Eco
Mono 38: 169-197

Tockner K, Malard F, Ward J (2000) An extension of the flood pulse
concept. Hydrol Process 14:2861–2883

Tomanova S, Tedesco PA, Campero M et al (2007) Longitudinal and
altitudinal changes of macroinvertebrate functional feeding groups
in neotropical streams: a test of the River Continuum Concept.
Archiv für Hydrobiologie 170:233–241

Tonin AM, Gonçalves JF, Bambi P et al (2017) Plant litter dynamics in
the forest-stream interface: precipitation is a major control across
tropical biomes. Sci Rep 7:10799. https://doi.org/10.1038/s41598-
017-10576-8

Tromboni F, Dodds WK, Neres Lima V et al. (2017) Heterogeneity and
scaling of photosynthesis, respiration, and nitrogen uptake in three
Atlantic Rainforest streams. Ecosphere 8

Uehlinger U, Naegeli MW (1998) Ecosystem metabolism, disturbance,
and stability in a prealpine gravel bed river. J N Am Benthol Soc
17:165-178

Ulseth AJ, Bertuzzo E, Singer GA et al (2018) Climate-induced
changes in spring snowmelt impact ecosystem metabolism and
carbon fluxes in an alpine stream network. Ecosystems 21:373–390

Ulseth AJ, Hall RO, Canadell MB et al (2019) Distinct air–water gas
exchange regimes in low-and high-energy streams. Nat Geosci:1

Vannote RL, Minshall GW, Cummins KW et al (1980) River
continuum concept. Can J Fish Aquat Sci 37:130–137. https://doi.
org/10.1139/f80-017

Wallace JB, Cuffney T, Eggert S et al (1997) Stream organic matter
inputs, storage, and export for Satellite Branch at Coweeta
Hydrologic Laboratory, North Carolina, USA. J N Am Benthol
Soc 16:67–74

Wanner S, Ockenfeld K, Brunke M et al (2002) The distribution and
turnover of benthic organic matter in a lowland river: influence of

hydrology, seston load and impoundment. River Res Appl 18:107–
122

Ward J, Stanford J (1983) The serial discontinuity concept of lotic
ecosystems. In: Fontaine TD and Bartell SM (eds) Dynamics of
lotic ecosystems. Ann Arbor Science Publishers, Ann. Arbor,
pp 29–42

Ward J, Stanford J (1995) The serial discontinuity concept: extending
the model to floodplain rivers. Reg Rivers Resear Manag 10:159–
168

Webster J, Golladay S, Benfield E et al (1990) Effects of forest
disturbance on particulate organic matter budgets of small streams.
J N Am Benthol Soc 9:120–140

Webster J, Meyer JL (1997) Organic matter budgets for streams: a
synthesis. J N Am Benthol Soc 16:141–161

Webster J, Wallace J, Benfield E (1995) Organic processes in streams
of the eastern United States. In: Cushing CE, Cummins KW,
Minshall GW (eds) River and stream ecosystems-ecosystems of the
world. University of California Press, Berkeley, pp 117–187

Webster JR (2007) Spiraling down the river continuum: stream ecology
and the U-shaped curve. J N Am Benthol Soc 26:375–389. https://
doi.org/10.1899/06-095.1

Webster JR, Benfield EF, Ehrman TP et al (1999) What happens to
allochthonous material that falls into streams? A synthesis of new
and published information from Coweeta. Freshw Biol 41:687–705

Welcomme RL (1979) Fisheries ecology of floodplain rivers [tropics].
Longman, Rome

Wetzel RG (1983) Limnology, 2end edn. Harcourt Brace, Fort Worth

Winemiller KO, Flecker AS, Hoeinghaus DJ (2010) Patch dynamics
and environmental heterogeneity in lotic ecosystems. J N Am
Benthol Soc 29:84–99. https://doi.org/10.1899/08-048.1

Winterbourn MJ, Rounick J, Cowie B (1981) Are New Zealand stream
ecosystems really different? N Z J Mar Freshw Res 15:321–328

Wohl E, Hall RO, Lininger KB et al (2017) Carbon dynamics of river
corridors and the effects of human alterations. Ecol Monogr 87:379–
409

Young R, Huryn AD (1997) Longitudinal patterns of organic matter
transport and turnover along a New Zealand grassland river. Freshw
Biol 38:93–107

The world's ecosystems have been extensively altered throughout the age of the Anthropocene, river ecosystems perhaps most of all. With nearly half of global river volume moderately to severely impacted by dams and other waterworks (Grill et al. 2015; Lange et al. 2018), over half of available freshwater runoff captured for human use (Jackson et al. 2001), one-fourth of the global sediment load trapped behind dams before it reaches the oceans (Vörösmarty and Sahagian 2000), and several of the world's great rivers no longer flowing to the sea during dry periods (Postel 2000), the physical evidence is overwhelming (Best 2019). Although these are amongst the most dramatic examples of anthropogenic impact, other pressures resulting from human actions are of major importance. Broad categories of threats described earlier include pollution, flow modification, habitat degradation, over-exploitation, species invasions, and climate change (Table 1.1).

Sadly, there is no shortage of evidence that biological assemblages and ecosystem processes have been profoundly altered by these pressures acting alone or, often, in combination. Freshwater ecosystems are among the most threatened and their communities among the most imperiled on Earth. By area, freshwater ecosystems occupy less than 1% of the Earth's surface yet contain 10% of all known species, including about one third of all vertebrates (Strayer and Dudgeon 2010). A compilation of geographical range data for 7,083 freshwater species of mammals, amphibians, reptiles, fishes, crabs, and crayfish found that almost one in three is threatened with extinction world-wide (Fig. 15.1). In addition, all six groups exhibited a higher risk of extinction than their terrestrial counterparts, and extinction risk was estimated to be higher in lotic habitats than wetlands and lakes (Collen et al. 2014). Urgent and concerted global action is needed to stem the loss of freshwater biodiversity while there is still time (Tickner et al. 2020).

The primary purpose of this chapter is to describe what can be done to reverse this harm. Specific actions will depend on many variables associated with local circumstances. Is a particular location and stream or river best viewed within the framework of repair, restore, or protect (Fig. 1.10)? What direct and indirect economic values are at issue, and what aesthetic and natural values? What institutions, and what policy and legal frameworks will influence the process, and what is the level of community engagement with the resource? While many factors will influence how best to manage a riverine ecosystem to improve its condition, we believe the starting point should be at a higher level. Why should we endeavor to repair, restore, and protect rivers, and what should be our over-arching goal?

Most of us have an intuitive idea that streams and rivers benefit humans. They are a source of drinking water and harvestable fish, of hydropower and irrigation water when harnessed by dams and canals, useful for navigation, and have functioned as a defensive barrier for ancient cities. Their floodplains absorb flood waters, slowing downstream passage while capturing sediments and nutrients that enrich the floodplain's agricultural potential. In addition to these tangible benefits, running waters have aesthetic values that include the pleasures people experience from fishing, paddling, or strolling along a riverbank, but can extend much further into the spiritual realm. Beyond the sciences of hydrology, geomorphology, ecology, and other disciplines that contribute to our understanding of rivers, running waters have served as muse and metaphor for philosophers, poets, and humanist writings about people and nature. Before concluding this book with an exploration of how scientists, citizens, managers, and decision-makers can most effectively work to improve the status of rivers, we begin with the most important question: why should we do so?

© Springer Nature Switzerland AG 2021
J. D. Allan et al., *Stream Ecology*,
https://doi.org/10.1007/978-3-030-61286-3_15

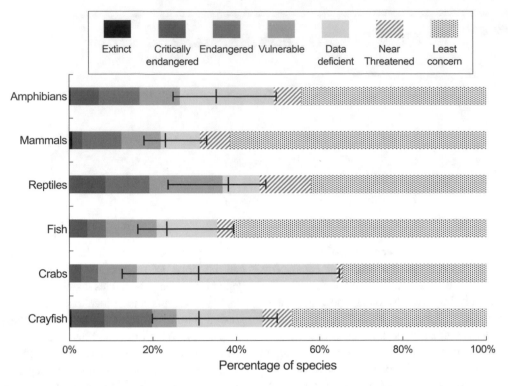

Fig. 15.1 Extinction risk of global freshwater fauna by taxonomic group. Central vertical lines represent the best estimate of the proportion of species threatened with extinction, with whiskers showing confidence limits. Data for fish and reptiles are samples from the respective group; all other data are comprehensive assessments of all species (n = 568 crayfish, 1191 crabs, 630 fish, 57 reptiles, 490 mammals and 4147 amphibians). Solid colors are threatened species, from left to right: black, extinct; darkest grey, critically endangered; mid-grey, endangered; light grey, vulnerable; lightest grey, data deficient. Patterned bars are non-threatened species: hatched, near threatened; dotted, least concern. (Reproduced from Collen et al. 2014)

15.1 Benefits from Intact Rivers

15.1.1 Ecosystem Services

Recent years have seen the development of conceptual frameworks for assessing the value of an ecosystem, as well as increased application of economic tools in an effort to monetize different benefits that an ecosystem provides. The supply of ecosystem services, defined variously as the goods and services that an ecosystem provides free of charge, and as the benefits that people receive from ecosystems (Millennium Ecosystem Assessment 2005), is widely recognized as a potent framework to justify management and restoration actions. Ecosystem services (ES) often are grouped into four broad categories. Provisioning services include the production of directly consumed resources, such as fish, drinking water, and hydropower. Regulating services are the benefits obtained from regulating processes, including waste decomposition and water purification, flood control, and pest suppression. Supporting services (sometimes combined with habitat services) include basal resources, nutrient and other biogeochemical cycles, degradation of organic wastes, and species' habitat. Cultural services include educational, recreational, aesthetic, and spiritual benefits.

Recent work favors what is known as the ES cascade, a useful framework for operationalizing ES quantification by breaking the concept into measurable entities (Boerema et al. 2017). The cascade framework links natural systems to aspects of human well-being, following a pattern similar to a production chain: from ecological structures and processes generated by ecosystems, to the services and benefits eventually derived by humans (Haines-Young and Potschin 2010). For example, the existence of floodplains and riparian wetlands in a catchment may dissipate the energy and slow the passage of a flood. This function of the ecosystem connotes its capacity to do something that is potentially useful to people, and so can be considered an ecosystem service. The human benefit of this ecosystem service will depend on the extent of harm that may occur due to flooding, and its value can be estimated by methods described below, such as peoples' willingness to pay to maintain this service. A central point is that ecosystem services exist (are realized) when some benefit accrues to people. That benefit can be experienced directly at the location where the service is realized, but benefits also can be experienced at a distance, for example when clean drinking water is the result of land management in headwaters, or when someone in North America gains satisfaction from the existence of river dolphins in the Amazon River.

Two important ideas flow from explicit consideration of ecosystem services. First, human actions that degrade ecosystem condition and function may compromise the ecosystem's ability to deliver ES. An ES framework makes explicit that benefits may be lost due to environmental stressors. It also provides a way to message and quantify the benefits gained under ecological restoration. Second, the recognition that ecosystems provide not one but many benefits raises the question of the inter-relationships among multiple services. Ecosystem properties, processes, and components that are the basis of service provision will be affected, usually adversely, by myriad human activities and, hopefully more positively, by management and restoration aimed at reversing environmental degradation. As a consequence, individual services as well as the complete bundle of services provided by that ecosystem will change. In some instances, management intended to improve some ecosystem service may benefit other services as well, in what is known as co-benefits or a "win-win" outcome. Protecting a wetland provides both wildlife habitat and improved water quality for human uses. In contrast, management intended to improve one ecosystem service may adversely affect another, indicating tradeoffs in choices and outcomes. Providing for a spring flood pulse may conflict with ensuring that enough water remains in the river for year-round navigation. Similarly, an intense storm event may replenish drinking water supplies for a given jurisdiction, but flood neighborhoods downstream. Conflicting outcomes can be common, especially in urbanizing watersheds, and the distribution of services and disservices experienced by a community are often influenced by income and demographics (Keeler et al. 2019).

The total value of a river is the sum of its direct and indirect uses, and its non-use values, which require different methods for their estimation. Direct-use value refers to the value of some ecosystem product as a commodity that can be sold in a market at a known exchange price. Harvested fish and electricity generated from hydropower are examples. Indirect use value stems from benefits to human society from indirect utilization of ecosystem services. Flood protection afforded by floodplains, natural water filtration, and carbon sequestration are examples, as are recreational uses and activities. Additional values that have been recognized include preserving the option to utilize ecosystem services in the future (option value), satisfaction that an ecosystem exists (existence value), recognition of the welfare the ecosystem may give other people (altruistic value), and preserving the ecosystem for future generations (bequest value).

When an economic value cannot be derived from existing markets, methods to monetize indirect and non-use services include stated preference, revealed preference, and benefit transfer. Stated preference methods such as contingent valuation ask people for their willingness to pay for a certain ecosystem or service, typically with a survey presenting choices or alternative scenarios. Revealed preference methods relate peoples' willingness-to-pay for a service to their actual expenditures or the value of some market good or service. Hedonic pricing estimates a value for some ecosystem service such as water quality from its statistical relationship with the price of a good for which a market actually exists, such as waterfront housing. A number of studies have shown that water views and proximity to shoreline is highly desirable in residential housing markets. Sales data on land parcels adjacent to the Neuse River in North Carolina, US, established that a riparian property generally commands a significant premium compared to an otherwise equivalent property (Bin et al. 2009). Travel cost methods use gas mileage costs, entry fees, on-site expenditures, and outlays on recreational equipment as substitutes for the market price of some environmental good or service. Recreational fishing in the rivers of the Pantanal region of South America draws several tens of thousands of Brazilian anglers for typically week-long trips (Shrestha et al. 2002). Travel costs determined by angler survey of $86–$140 per day are high relative to similar estimate for the US of around $33 per person day in 1996 dollars. Aggregate value of recreational fishing in the Brazilian Pantanal ranges from $35 to $56 million.

Stated and revealed preference methods are both widely used. Revealed preference has the advantage that it is based on estimates of actual dollars spent. Stated preference methods have the disadvantage that it is unclear whether peoples' expressed willingness-to-pay translates into actual dollars, but have the advantage that they can be applied to non-use values such as existence values of fish and wildlife (Bergstrom and Loomis 2017).

The transfer of estimated benefits from one site to another, known as benefit transfers, is widely used to argue for site protection in environmental decision-making (Plummer 2009; Richardson et al. 2015). In benefit transfers, a single value from an empirical study of a site, or the mean from multiple study sites, is used to provide a value estimate for that ecosystem service at similar sites over a large region. Often used to reduce effort and expense in analyzing options, its greatest potential pitfall is the assumption of correspondence among locations (Plummer 2009). A preferred alternative is a benefit function that relates an estimated willingness-to-pay to a set of site characteristics, including its socio-economic setting. When a benefit function is based on multiple sites exhibiting a range of conditions, a more robust benefit transfer can be accomplished by measuring the function's variables at a new site and evaluating the function at those values.

Stated willingness-to-pay methods have shown that people value ecosystem protection for some non-use amenity, even from a distance. Using contingent valuation to assess respondents' willingness-to-pay for removing two dams to restore the ecosystem and its anadromous fishery of the Elwha River in Washington State, US, Loomis (1996) estimated aggregate benefits to residents of the state at $138 million annually for 10 years. Reasoning that restoration of a river in a national park and increases in salmonid populations are public goods available to all, the survey was also sent to residents throughout the US. Results indicated that the general public would be willing to pay between $3 and $6 billion, revealing the substantial nonmarket value of removing old dams to restore salmon and steelhead runs in the Pacific Northwest. Asked their willingness to pay for five ecosystems services that depended on a trade-off between instream flows versus off-stream uses, residents along the Platte River, Colorado, US, found that, on average, individuals would pay an additional $250 annually via a higher water bill (Loomis et al. 2000). When extrapolated to the population living along the river, ES values exceeded costs of alternative conservation efforts. Using geo-tagged photographs as a proxy for recreational visits to lakes in Minnesota and Iowa, US, Keeler et al. (2015) showed that number of visits increased with improved water clarity. Recreational lake users were willing to travel farther and incur increased travel costs to visit lakes of greater clarity, a finding consistent with stated preference studies.

Despite the potential for estimates of non-market value to benefit environmental decision-making, actual monetization studies still are relatively few. A literature search for studies of river restoration that quantified and valued one or more ecosystem goods or services found 32 examples, including 24 in the United States, six in Europe, one in Mexico, and one in China (Bergstrom and Loomis 2017). Restoration of fisheries accounted for two-thirds of the examples, and included studies focused on threatened species, native species, and recreational fishing. Estimation methods included stated preference by contingent valuation and choice experiments, revealed preference including hedonic price and travel cost estimates, and benefit transfer. Stated preference estimates were most common. Willingness-to-pay estimates for river restoration increased with length of river restored and number of goods and services valued. The authors inferred that these valuation estimates were used primarily as background information, although in some cases they entered more directly into decisions.

While advances in the valuation of ecosystem services have provided new tools for capturing the worth of ecosystems and communicating this to an audience more familiar with monetary valuation, such approaches have yet to capture all dimensions of value. Chan et al. (2012) propose a typology that recognizes eight dimensions of value, including self- vs other-oriented, physical vs metaphysical, and anthropocentric vs eco-centric, among others. Reliance on an ES perspective raises questions about the judgments we make in assigning value to nature, and poorly resolved ethical concerns concerning the relationships between humans and non-human nature (Jax et al. 2013). By emphasizing monetary valuation, the result is that other, less tangible non-use values are marginalized in decision-making. Some experiences of nature are especially intangible, such as a love of nature or explicit, spiritual connection with some natural feature. Few would advocate monetizing a sacred site in order to negotiate a trade-off with resource extraction. In a similar vein, all ecosystems, including rivers and streams, have intrinsic value that many people will feel uncomfortable expressing in currency. This segues to a second, certainly co-equal answer to why we protect rivers, which we will call rheophilia.

15.1.2   Rheophilia

Are rivers an amenity, that we may utilize as needed, or even replace with manufactured alternatives such as de-salinized drinking water, fish production by aquaculture, and designed streams flowing through designed landscapes? Or do streams and rivers play a deeper role in supporting human well-being, and if so is that role enhanced by, or does it even require, the opportunity to experience diverse river settings in as near-natural a state as we can achieve? And if the latter is closer to what we humans desire, how can we best make the case for the required effort to protect rivers? One line of argument, born of the need to demonstrate human benefits in terms of economic value, is described in the above discussion of ecosystem services. A second line of argument, only partially captured by cultural ES, is expressed beautifully by the title of a conference address by Luna Leopold in 1977, *A Reverence for Rivers* (Leopold 1977), and of a book by Kurt Fausch, *For the Love of Rivers* (Fausch 2015). These are in the tradition of scholarly explorations of the human basis for the love of nature and its restorative benefits, developed in the writings of E.O. Wilson (*Biophilia*, (Wilson 1984) and Rachel and Stephen Kaplan (*The Experience of Nature: A Psychological Perspective*, (Kaplan and Kaplan 1989). Nor should this perspective require academic scholarship, one might argue, for it also is captured in the frequent mention of running waters in literature, art, and song. Perhaps restoring rivers is not only to improve the condition of the river ecosystem, but is equally or even more about enhancing the river's restorative capacity for human well-being. Cultural ecosystem services recognize this, but to date have had limited success in capturing these values, which to us seem best viewed as a separate and equal domain.

Perception of the attractiveness of river and riparian scenes can be determined from preferences expressed for photographs that depict different settings. Using a questionnaire to assess reactions to 20 pictures of rivers from over 2,000 students across ten countries, in terms of naturalness, danger, aesthetics, and need for improvement, Le Lay et al. (2008) found that a preference for mountain streams with turbulent flow and boulders was common to all. Whitewater and scattered large boulders characterized the most aesthetic and natural riverscapes, whereas rivers characterized by extensive gravel bars, narrow bands of water, and large amounts of wood were considered the least attractive. Surveys of the public's perception of river corridors in southern England and Wales found strong preferences for mature, sinuous rivers with natural channels and vegetated banks (House and Sangster 1991).

Interestingly, despite much scientific evidence of the benefits of wood in streams, several studies have found negative public perception of wood in rivers. River channels with woody debris can be considered less aesthetically pleasing, more dangerous, and needing more improvement than those without wood (Chin et al. 2008). Cross-cultural comparisons indicate that perceptions of riverscapes are influenced by cultural setting. Respondents from some European countries and the US had a negative perception of regulated rivers, and a positive perception of wood within streams; participants from India, China, and Russia were the opposite (Le Lay et al. 2008). Such differences may stem from many elements of lived experiences, culture, and local environmental history, as well as from education and its communication.

Perception of the attractiveness of river sounds can be assessed using audio-recordings that compare urban to natural or park-like settings. A survey of preferred sounds in two squares in the city center of Sheffield, England, found a preference for natural sounds and especially for the sound of water in park fountains, making soundscape an important element of the design of urban spaces (Yang and Kang 2005). Subjects exposed to stress (given three seconds to determine if an equation was correct or false) and monitored for physiological response exhibited faster mood recovery when experiencing nature sounds (a fountain and bird calls) in comparison with urban noise such as traffic (Alvarsson et al. 2010). When presented with sound and image combinations representing a stream, a village, a quiet park, a busy park, and a residential neighborhood, subjects expressed an overall preference for natural and rural over urban and man-made scenes. The most highly rated combination was the sound and image of a stream (Carles et al. 1992).

Beyond an expression of preference, studies show that visual images of natural environments facilitate attention restoration, improve mood, and can more generally enhance health. There is evidence that interacting with nature has cognitive benefits, in part because of the attention-capturing distractions of navigating an urban environment in comparison with a more natural setting. Participants assigned to walk for 50 min in a large urban park performed better on a cognitive task than others who walked in the downtown area of a city (Berman et al. 2008). A second experiment found that participants performed better at more complex attentional functions when viewing photographs that depicted scenes from nature in comparison with city scenes.

We cannot do justice here to a humanist perspective on flowing waters, revealed in art, poetry, and great works of literature, but we would be remiss not to mention it. The Hudson River School was a mid-19th century group of American landscape painters whose work drew inspiration from the Hudson River and the surrounding area. Known for their realistic, detailed, and sometimes idealized portrayal of nature, their paintings often juxtaposed peaceful agriculture and the remaining wilderness, or portrayed an idyllic scene of still-pure nature. Celebrated landscape artists Frederic Edwin Church and Albert Bierstadt were a second generation of this school, and running waters were central to some of their most famous paintings, including Church's *Niagara*, *Morning in the Tropics*, and *Heart of the Andes*. Looking to a different culture and a different time, *Along the River During the Qingming Festival* (the *Qingming Shanghe Tu*) painted by the Song dynasty artist Zhang Zeduan (1085–1145) depicts the daily life of people and the landscape during a period of the Song Dynasty. Said to celebrate the festive spirit and worldly commotion at the Qingming Festival, this is considered to be the most renowned work among all Chinese paintings.

Many fine works of literature, both fiction and non-fiction, draw inspiration from rivers. Mark Twain's great novels, *The Adventures of Tom Sawyer* (1876) and *The Adventures of Huckleberry Finn* (1884) surely drew their river settings from Twain's years as a river boat pilot, which also formed the basis for his *Life on the Mississippi* (1884). Henry David Thoreau, an American philosopher of nature best known for *Walden, Or Life in the Woods* (1854), earlier published *A Week on the Concord and Merrimack Rivers* (1849), describing his 1839 hiking and boating trip with his brother through parts of Massachusetts and New Hampshire. And in one of the finest short stories about fishing every written, *The Big Two-Hearted River* (1925), Earnest Hemingway describes not just the dedicated chase after a large trout, but the healing and restorative power of nature following the devastation of the First World War. For more contemporary American writings, an admittedly selective short list would include *A River Runs Through It*, Norman Maclean semi-autobiographical account of coming of age in an early 20th-century Montana family in which "there was no clear line between religion and fly fishing"; *River Horse*, William Least Heat-Moon's account of travelling across America not by road but by water; and *Desert Solitaire*, Edward Abbey's vignettes of river running and explorations in the American southwest. There is no shortage of mention of rivers in poetry and song, and here we highlight just one example. Any who has travelled a river by canoe cannot help but be moved by "The Song My Paddle Sings" by the Canadian poet Pauline Johnson (1862–1913), also known as Tekahionwake, the daughter of a Mohawk Chief and a woman of English parentage.

In the end, the answer to the question, "Why protect rivers?" is straightforward. River ecosystems provide many benefits to humans, and both our understanding and our ability to quantify these benefits are advancing rapidly. Arguably, however, this is the lesser of two rationales. As the Senegalese forest engineer, Baba Dioum, said in a 1968 presentation to the International Union for the Conservation of Nature, "In the end we will conserve only what we love, we will love only what we understand". The ocean explorer Jacques Cousteau summed it up more succinctly: "people will only protect what they love". Shōzō Tanaka, considered to be Japan's first conservationist, said: "The care of rivers is not a question of rivers but of the human heart". Protecting rivers ultimately is the responsibility to protect what we love.

Understanding the why gives motivation and urgency. The how of repairing, restoring, and protecting rivers blends river science, human perceptions and beliefs, socio-economics, politics, and much more. The following sections attempts to inform readers of some of the major approaches and challenges.

15.2   Goals in River Management

Setting realistic goals for river management must, as a beginning point, be guided by some appraisal of threats and opportunities. There must also be some level of societal support, or a plan to garner support. Individual streams and rivers span an enormous range of settings and challenges, from highly compromised systems, to those where restorative actions hold great promise, to still others where protection against future threats may suffice to preserve them in a near-natural state. Specific objectives likely will depend very much on both the condition of the river system and how societies view its uses and values.

Over time, perspectives on river management have shifted from a more limited focus on meeting human needs while attempting to mitigate environmental costs, to one that emphasizes sustained human benefits, including water for direct human use and water to support other services supplied by healthy ecosystems. The terminology of river management can be distracting, as management actions can be described as restoration, rehabilitation, and improvement;

and integrated river basin management (IRBM) is interchangeable with integrated watershed and catchment management. More generally, these approaches fall under the rubric of ecosystem-based management and adaptive management, ideas whose ascendency dates to the 1990s. Ecosystem-based management advocates a holistic approach that recognizes the full array of interactions within an ecosystem, including people and their activities, and the need for cooperative management over large jurisdictional areas (Slocombe 1993). Adaptive management is an integrated, interdisciplinary approach that emphasizes on-going cycles of learning through management interventions, whether they succeed or fail, and the harmonizing of environmental and societal goals as the guiding framework (Walters and Holling 1990).

Management actions aimed at improving rivers increasingly emphasize a holistic approach that attempts to create or maintain some aspect of river form and function that aligns with hydrologic, geomorphic, and ecological processes (Wohl et al. 2015). This can be accomplished by relieving pressures that degrade and harm a river system, thereby promoting natural recovery; and by active measures to assist recovery that may include dam removal, addition of habitat elements, control of an invasive species, ensuring environmentally beneficial flows, and many more such actions. More generally, it includes diverse management activities intended to improve the hydrologic, geomorphic, and ecological processes within a degraded watershed and replace lost, damaged, or compromised elements of the natural system. Other rivers of the same region and approximately the same environmental setting, that are relatively undisturbed, often serve as the benchmark, and the aspiration of achieving healthy ecosystems is at the forefront. Unfortunately, however, coordination of multiple projects throughout a catchment, and consideration of pressures arising at large spatial scales, too often are ignored (Bernhardt et al. 2007; Feld et al. 2011; Friberg et al. 2016).

From the 1980s onward, ecological restoration came into wide use to describe management activities intended to restore damaged ecosystems to a more natural, undisturbed state. Characterized by a more explicit pairing of science and practice, and by goals focused more strongly on recovering historic form and function, this perspective rapidly took hold in river management, resulting in a dramatic rise of projects characterized as river restoration (Bernhardt et al. 2005). This has resulted in a rapidly expanding literature that describes individual projects as restoration work. In addition, there has been much discussion concerning the feasibility of attempting a return to pre-disturbance condition, as well as much analysis of their success or lack thereof, both discussed below. In this chapter we prefer to lump all such activities under the label of management, but where researchers describe their work as restoration, we do as well.

River restoration projects have a wide range of objectives. Based on over 37,000 projects compiled from governmental databases, gray literature, and contacts from seven regions of the coterminous US, Bernhardt et al. (2005) identified 13 categories of river restoration, together with median cost and typical activities or measures taken (Table 15.1). From a database of 813 hydromorphological river restoration projects mostly from Europe, Friberg et al. (2016) identified 53 specific measures grouped within 8 categories: water quantity, sediment quantity, flow dynamics, longitudinal connectivity, in-channel habitat, riparian zone, river planform, and floodplain. Habitat improvements were common, including the removal of artificial embankments, the addition of large wood, and the provision of spawning gravel. A compilation of information on 644 restoration projects from 149 published studies that provided quantitative information on the effectiveness of restoration projects found the most common objectives to be related to increasing biodiversity, stabilizing channels, improving riparian and in-stream habitat, and improving water quality (Palmer et al. 2014, Fig. 15.2a). Methods used were dominated by physical manipulations, such as moving channels laterally, adding sinuosity, or raising/lowering the bed or floodplain for reconnection; and addition of in-stream structures, such as boulders, logs, and gravel (Fig. 15.2b). From these and other reviews of restoration activities it is

Table 15.1 Common river restoration goals and activities, following Bernhardt et al. (2005) and Wohl et al. (2015). Although these categories are not fully independent, they are common rationales for most restoration projects. Projects aimed at improving water quality, riparian management, and habitat improvements were amongst the least expensive and most frequently carried out in the analysis of Bernhardt et al. (2005). Stormwater management, floodplain restoration, and dam removal were more expensive and less common

Goal	Description of activities
Esthetics/recreation/education	Activities that increase community value: use, appearance, access, safety, and knowledge
Bank stabilization	Practices designed to reduce or eliminate erosion or slumping of bank material into the river channel; this category does not include stormwater management
Channel reconfiguration	Alteration of channel geometry, planform, and/or longitudinal profile and/or daylighting (converting pipes or culverts to open channels); includes meander restoration and in-channel structures that alter the thalweg
Dam removal/retrofit	Removal of dams and weirs or modifications/retrofits to existing dams to reduce negative impacts; excludes dam modifications that are simply for improving fish passage
Fish passage	Removal of barriers to upstream/downstream migration of fishes; includes the physical removal of barriers, construction of alternative pathways, and migration barriers placed at strategic locations along streams to prevent undesirable species from accessing upstream areas
Floodplain reconnection	Practices that increase the inundation frequency, magnitude, or duration of floodplain areas and/or promote fluxes of organisms and materials between channels and floodplain areas
Flow modification	Practices that alter the timing and delivery of water quantity (does not include stormwater management); typically but not necessarily associated with releases from impoundments and constructed flow regulators
Instream habitat improvement	Altering structural complexity to increase habitat availability and diversity for target organisms and provision of breeding habitat and refugia from disturbance and predation
Instream species management	Practices that directly alter aquatic native species distribution and abundance through the addition (stocking) or translocation of animal and plant species and/or removal of exotic species; excludes physical manipulations of habitat/breeding territory
Land acquisition	Practices that obtain lease/title/easements for streamside land for the explicit purpose of preservation or removal of impacting agents and/or to facilitate future restoration projects
Riparian management	Practices that improve riparian and bank condition including riparian buffer creation and maintenance, revegetation, eradication of weeds and nonnative plants, livestock exclusion
Stormwater management	Practices intended to reduce stormwater runoff at source and reduce hydrologic scouring by means of rain gardens, pervious pavers, holding/retention ponds; constructed wetlands lower in watershed to filter sediments and nutrients
Water quality management	Similar to above but including water treatment infrastructure and regulatory control of pollutants, as well as landscape-scale best management practices to reduce runoff and capture sediments and nutrients

Restoration goal

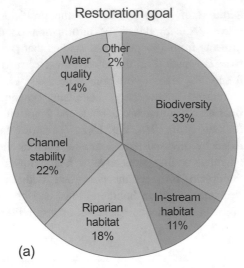

(a)

Restoration method

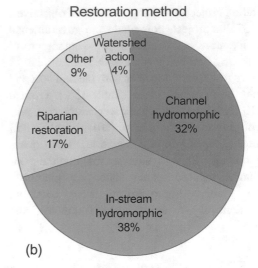

(b)

Fig. 15.2 Summary of the most common restoration goals and implementation methods for 644 river or stream restoration projects; values are percentages of projects using a goal or method. (**a**) The primary goal of each restoration project, (**b**) restoration methods depending on how the project was implemented. Channel hydromorphic projects involved reconfiguring the channel, such as moving it laterally, adding sinuosity, or raising/lowering the bed or floodplain for reconnection, and they often included addition of in-stream structures, such as boulders, logs, and gravel; in-stream hydromorphic projects were less intensive projects that involved only manipulating in-stream structures, adding large woody debris, armoring the bank, or creating

artificial riffles without major channel excavation or reconfiguration; riparian restoration projects were those projects implemented by planting of riparian vegetation or removal of nonnative vegetation as the primary or sole restoration method; watershed action projects were those in which the project was implemented up in the watershed without manipulation of the channel, and they included, for example, addition of stormwater management, creation of wetlands, or use of cover crops; and "other" projects were varied, including, for example, treatment of acid mine drainage, dam removal, changes in reservoir releases to restore natural flow regime, or creation of an in-stream or riparian wetland. (Reproduced from Palmer et al. 2014)

apparent that the most common management activities involve channels, sediments, and flows, sometimes combined as hydrogeomorphic; and addition of wood, boulders, and gravel, all measures to improve habitat for the biota. Much river restoration aims to improve the physical environment, relying on expertise in hydrology, geomorphology, and the habitat requirements of organisms.

Clearly, management activities intended to improve stream condition are many and diverse. Are they successful? If a project calls for addition of boulders and wood to a one-km river reach to improve habitat for invertebrates and fishes, the proximate measure is whether indeed the habitat elements remain in place following flood events, perhaps as observed after one or more years. But the desired outcome is self-sustaining animal populations, which may not be evaluated. The reason for emphasizing measures of success in river restoration is that much time and money has been invested, with little knowledge of what has been gained, little opportunity to learn from experience, and insufficient sharing of what works and what does not. The average costs were summed for 37,000 projects to obtain the conservative estimate that, from 1990 thru 2003, more than 1 billion dollars a year were being invested in efforts to restore US rivers (Bernhardt et al. 2005). However, only 10% mentioned any form of monitoring. A similar inventory in the UK as of 2016

contained over 2800 completed projects with only 21% stating some degree of monitoring (England et al. 2019).

When projects have been deemed successful, the criteria used may not be rigorous. Interviews with managers of over 300 US projects considered successful revealed that post-project appearance and positive public opinion were the main measures of success (Bernhardt et al. 2007). An evaluation of 44 French river restoration projects found that the quality of evaluation strategies often was inadequate for understanding the link between a project and ecological outcomes, and projects with the poorest evaluation strategies generally reached the most positive conclusions about the effects of restoration (Morandi et al. 2014). Of 848 Swiss restoration projects, success was evaluated for 232, with methods ranging from very comprehensive ecological assessments to counting the number of fish through a fish pass (Kurth and Schirmer 2014). The authors commented that comparison of results among projects and with projects elsewhere was difficult because individual projects varied in aim and method of evaluation.

The above makes clear that many restoration projects have been undertaken at considerable expense and with minimal systematic accounting. And while the intent is laudatory, efforts to re-shape rivers are significant interventions, often involving heavy equipment (Fig. 15.3). Adding to the

Fig. 15.3 A restoration crew works to add large wood to a Pacific Northwest stream where this material was aggressively removed for decades. Mckenzie River, Oregon. Photo by David Herasimtshshchuk, Freshwaters Illustrated

concern, restoration projects have been observed to fail physically, or not achieve desired ecological outcomes. Explosive growth in the field of river science has drawn valid criticism for inappropriate project design (Kondolf 2006), an emphasis on restoring structure rather than function (Palmer et al. 2014), inadequate monitoring (Bernhardt et al. 2005; England et al. 2019), and lack of demonstrable improvement in the biota (Palmer et al. 2010; Feld et al. 2011). A synthesis of findings from over 800 recent European Union (EU) river restoration projects identified four ways in which projects are prone to failure: inadequate consideration of large-scale pressures related to land use in the catchment, absence of source populations to re-colonize a site, inadequate habitat and microhabitat, and failure to rejuvenate processes related to flow and sediment regime (Friberg et al. 2016).

Despite these legitimate criticisms, river restoration has helped to drive fundamental research to address knowledge gaps that limit successful restoration (Wohl et al. 2015). Individual projects have unquestionably restored some elements of river function, allowing the river's natural dynamism to reconfigure the channel and associated habitat features (Fig. 15.4). A recent evaluation of existing river restorations across Europe concluded that outcomes have been highly variable with, on balance, more positives than negatives (Friberg et al. 2016). Modest success is attributed to the local scale at which much restoration is carried out, largely ignoring the larger-scale pressures related to catchment land use or the lack of source populations to recolonize restored habitats. What one sees at the local scale in a natural river ecosystem is determined not only by micro-habitat features of the site, but by processes nested within and occurring at larger spatial scales in a hierarchy of controls. Restoration activities likewise will be scale dependent and

linked to the spatial and temporal heterogeneity provided by natural stream reaches, as illustrated in Fig. 15.5. One very important implication is that efforts undertaken at the local scale may fail to produce the desired outcome, if stressors operating at a large scale remain unaddressed.

We should not under-estimate the scientific and technological knowledge needed to provide general guidance for management actions intended to restore stream ecosystems towards a good ecological status. This complexity is illustrated here by a conceptual model depicting the hierarchical relationships between catchment land-use, catchment pressures, riparian buffer management, instream abiotic states and instream biological states (Fig. 15.6). Establishment of riparian buffers has been shown to be an effective form of river restoration, acting to stabilize streambanks, mitigate diffuse pollution by agriculture, and moderate stream temperatures (Feld et al. 2011). However, assessing the outcomes from riparian planting is difficult because of the many features that characterize riparian buffers, such as buffer length, width, and density, or the species planted, as well as the multiple pathways by which buffers affect stream processes and potentially mitigate human influences. Based on a structured literature review of a large number of studies published between 1990 and 2017, Feld et al. (2018) concluded that riparian management has beneficial effects on the supply of coarse particulate organic matter, large woody debris, and shade (and thus thermal damping) that are largely independent of conditions further upstream in the catchment. In contrast, expected benefits in retention of nutrients and fine sediments from riparian management are more likely to be affected by conditions upstream of the restored section, thus requiring catchment-scale as well as local interventions. Given the substantial number of restoration activities that may be considered, the multiple response variables, and the need to take into account upstream and catchment-scale pressures, effective restoration design is a non-trivial exercise in the application of best scientific knowledge.

The shortcomings of existing river restoration practice are increasingly well understood. As Wohl et al. (2015) point out, criticisms fall into three main categories. Monitoring commonly is inadequate to quantitatively and objectively determine whether restoration goals are achieved. Many restoration projects fail to achieve significant improvements as shown by measures of water quality or biological communities. Finally, and perhaps most importantly, inclusion of the nonscientific community in river restoration planning and implementation often is inadequate. Even so, the growing practice of river restoration provides a testing ground for scientific understanding of rivers, and a context in which societal attitudes toward rivers and humanity's ability to sustain river ecosystems can advance.

Fig. 15.4 Views of the Mareta River, Italy, before (left, 2005) and after (right, 2010) river restoration that removed grade-control structures. *Source* Archivio fotografico dell'Agenzia per la Protezione civile/Luca Messina–Provincia Autonoma di Bolzano

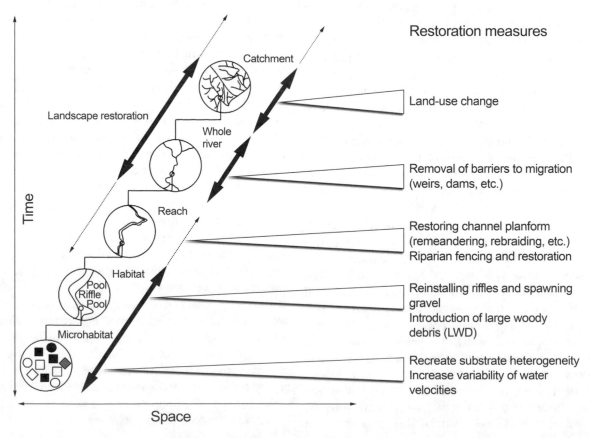

Fig. 15.5 Restoration challenges often are present across a range of spatial and temporal scales, such that impairments at the catchment scale may limit success of restoration efforts undertaken at the local scale. (Reproduced from Friberg et al. 2016)

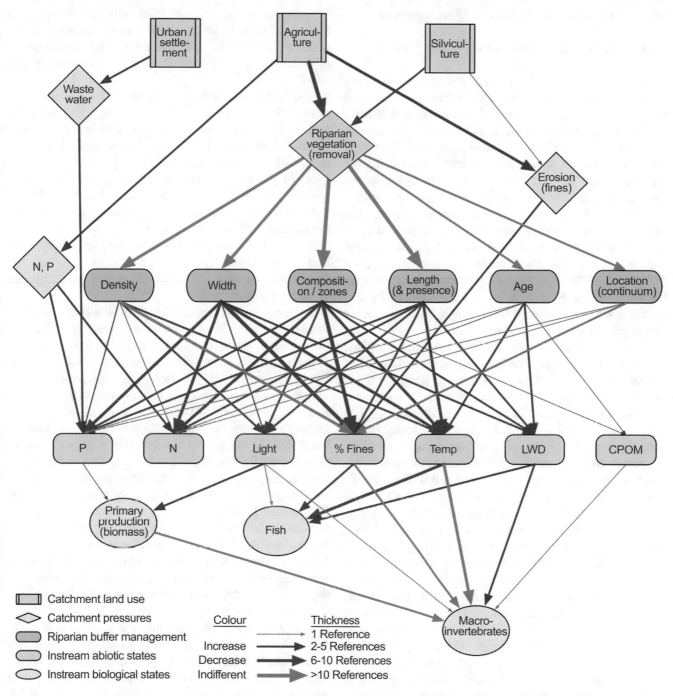

Fig. 15.6 A conceptual model based on extensive literature synthesis depicting the hierarchical relationships between catchment land-use, catchment pressures, riparian buffer management, instream abiotic states, and instream biological states. Arrows represent consistent evidence of negative (blue) and positive (red) relationships, or unclear evidence (grey) with both positive and negative effects reported in the literature. Arrow thickness is proportional to the number of studies supporting a significant relationship between two elements of the model. (Reproduced from Feld et al. 2018)

15.3　Frameworks for River Management

15.3.1　Integrated River Basin Management

Policy prescriptions for the management of water resources have long agreed on two fundamental principles. First, the river basin (catchment, watershed) is the appropriate scale for organizing water management, because water sources and uses in a watershed are interrelated. Second, because political boundaries rarely correspond with watersheds, and watershed-scale decision making structures typically are lacking (although that is changing), they should be created. Watershed-scale organizations are needed to bring together all "stakeholders" and produce integrated watershed management, avoiding a fragmented approach to management and decision-making. The authority for integrated actions likely rests with cooperative coordination among existing agencies, facilitated through some basin-wide entity that promotes a common agenda by serving as the advocate for shared priorities and an integrated approach.

Despite the attractiveness and the consistency of this message, integrated river basin management or IRBM has met with criticism for failing to achieve in practice what is intended, with shortcomings often blamed on social and political obstacles. More specifically, such efforts struggle to resolve fundamental political questions about where boundaries should be drawn, how participation should be structured, and how and to whom decision makers within a watershed are accountable ((Blomquist and Schlager 2005). Decisions typically involve trade-offs, and it is difficult to imagine all of the local impacts of different choices. Small-scale local users may feel a loss of control over resource allocation decisions as their region is subsumed under decisions made from the perspective of the larger watershed.

Integrated river basin management as a multi-faceted planning process has been re-invigorated and formalized under the Water Framework Directive (WFD) of the European Union. The Water Framework Directive, in force since 2000, is considered to be the most significant piece of European water legislation in decades, modernizing much of earlier EU water legislation and extending the concepts of river basin management to the whole of Europe (Griffiths 2002). Its aim is to take a holistic approach to water management and achieve "good water status". Key elements included water management at hydrological scales, the involvement of non-state actors in water planning, and various economic principles such as cost-benefit analyses, as well as a common strategy to support the 28 EU member states (Boeuf and Fritsch 2016). Ecological status is determined from biological parameters referenced to what would

be expected in the absence of significant anthropogenic influence. The need for some exceptions is recognized; for example, for bodies of water that are artificial in construction or where the physical structure has been irrevocably and heavily modified.

The WFD has stimulated an enormous amount of activity across the European Union, leading to numerous project-specific publications and several reviews of progress (Hering et al. 2010; Boeuf and Fritsch 2016; Voulvoulis et al. 2017). Perhaps unsurprisingly, given the ambitious intent of the WFD, difficulties with implementation have received a good deal of attention, and attainment of good status for many waters remains to be demonstrated. Among the challenges, Boeuf and Fritsch (2016) point to the mismatch between ecological (river basins) and political (political and administrative institutions) scales, lack of attention to synergetic ecological effects, and low acceptance by target groups.

As river restoration matures both as science and practice, there is increasing recognition of the need for a planning framework to guide practitioners and place project-specific restoration within a river basin context. A critical review of 663 published studies of European restoration projects identified poor or improper project planning as the most frequent shortcoming (Angelopoulos et al. 2017). One recently proposed planning framework is depicted in Fig. 15.7a. At the project identification stage, clear objectives are set for ecological condition at local scale, while keeping the project in a river basin/catchment context. Benchmarks are measurable targets for restoring degraded river sections by comparison with sites that have the required ecological status in that river system or elsewhere. Endpoints are feasible targets for river restoration and so are the basis for determining success or failure. Monitoring can include a wide range of physical, chemical, and biological variables, depending on what concerns motivated the project, but should relate to outcomes. In other words, if habitat elements are added to benefit fish populations, it is useful to quantify the habitat created, but important to assess whether fish populations benefited. Monitoring should include before and after sampling, and will be even more insightful if paired with a control site not receiving restorative measures. A full evaluation of project success requires clear objectives, endpoints, and measurable indicators that are sensitive to gradual improvements (Fig. 15.7b).

Regardless of mixed progress to date, the WFD has accomplished a great deal. Implementation of the WFD is greatly increasing knowledge on the ecology of European surface waters. Rather than relying mainly on chemical quality of surface waters, condition is assessed using a wide range of biological measures referenced against best

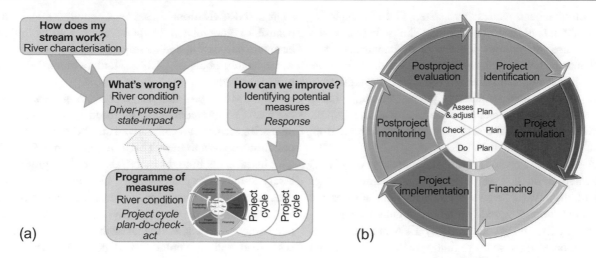

Fig. 15.7 (**a**) Project planning cycle at a catchment scale using a five step approach starting at the top left text box: (1) River characterization —at a catchment scale to identify river styles and understand their processes. (2) River status—understand the current condition of the aquatic biota or biological quality elements. (3) River restoration potential—to understand the level of restoration a river can achieve. (4) Project identification—to identify specific restoration projects at a reach scale and identify suitable rehabilitation measures and project objectives. (5) The project cycle—planning, formulation and implement of projects at a local scale. (**b**) Six stages for restoration project planning: (1) project formulation; (2) financing; (3) project implementation; (4) post-project monitoring; (5) post-project evaluation and (6) adjustment or maintenance. (Reproduced from Friberg et al. 2016)

attainable condition for a water body type, thereby placing aquatic ecology in the center of water management. Monitoring methods are being improved and standardized. Much attention is given to difficulties encountered in the planning process cycle, and with engagement of the public, areas where further improvements are needed. Mechanisms are in place for river basin management across national borders, including international commissions for transboundary basins such as the Rhine and the Danube. Much can be learned from this ambitious experiment to improve water quality and ecological status throughout the European Union, which cannot be recounted briefly. Interested readers may consult EEA (2018) and Carvalho et al. (2019).

15.3.2   The US Clean Water Act and TMDLs

Protection of the waters of the United States is largely accomplished through the Clean Water Act (CWA), which provides the basic structure for regulating discharges of pollutants, giving the Environmental Protection Agency (EPA) authority to implement pollution controls. Other Acts of Congress, including the Endangered Species Act, National Environmental Policy Act, Wild and Scenic Rivers Acts, and Surface Mining control and Reclamation Act, among others, provide further means to protect freshwater ecosystems. Numerous federal and state agencies have roles in making decisions and implementing regulations, and nongovernmental organizations (NGOs) such as The Nature Conservancy, the Sierra Club, and American Rivers can bring attention and pressure when regulatory enforcement is perceived as lax. However, it is the CWA, and its main regulatory tool, the total maximum daily load (TMDL), that is the principal mechanism for managing freshwater ecosystems. Many of the US restoration projects described earlier likely were initiated in response to a TMDL finding.

The objective of the Clean Water Act (CWA) of the United States, in effect since 1972, is to restore and maintain the chemical, physical, and biological integrity of the nation's waters. It requires states to compile lists of water bodies that do not fully support beneficial uses such as aquatic life, fisheries, drinking water, recreation, industry, or agriculture. These inventories are known as 303(d) Lists, and characterize waters as fully supporting, impaired, or in some cases, threatened for beneficial uses. Water quality standards set by a state, territory, or authorized tribe provide a narrative and numeric criteria for determining whether a waterbody is attaining or not attaining its designated uses; waters designated as not attaining then require the establishment of total maximum daily loads (TMDLs) for all pollutants identified as causing impairment. The US EPA assists states (this term includes territories and authorized tribes) in listing impaired waters and developing TMDLs for these waterbodies (Fig. 15.8). A TMDL is the maximum amount of a pollutant allowed to enter a waterbody so that the waterbody will meet and continue to meet water quality standards for that particular pollutant. It is determined by the sum of all point and nonpoint sources entering the waterbody, plus a margin of safety. The EPA's regulations require public involvement in developing TMDLs, although the level of citizen involvement varies by state. Once completed, TMDLs should clearly identify the links between the waterbody use impairment, the causes of impairment, and the pollutant load reductions needed to meet the applicable water quality standards. States are not explicitly required to develop TMDL implementation plans, although many include some type of implementation plan with the TMDL. The TMDL alone is sufficient to remove the waterbody from the state's list of impaired waters.

Management of point-source pollutants generally is implemented through a permitting process under another section of the CWA. Reductions in non-point sources are implemented through a wide variety of regulatory and voluntary programs, and incentivized by EPA program funds that grant money to the states to fund specific projects aimed at reducing the nonpoint source pollution. Substantial progress has been made in improving water quality through regulatory and permitting processes for wastewater treatment plants and industrial dischargers, two prominent point

sources (NRC (National Research Council) 2001). However, control of unregulated nonpoint sources of pollution has been less successful, and largely for that reason, the nation's water quality goals of "fishable and swimmable" have not been achieved.

The TMDL program has been controversial, in part because of requirements and costs faced by states, as well as by industries, farmers, and others who may be required to use new pollution controls to meet TMDL requirements. Often, prodding by lawsuits brought by NGOs or citizen groups is necessary to move the process along. Development of a TMDL does not guarantee that improvements will follow. Success at achieving targets is considerably higher when addressing point sources, which are managed through a permitting process, than with nonpoint source pollution, which usually requires the coordination of a suite of voluntary activities. Most TMDLs have a non-point source component, and common barriers to success include inadequate funding, incomplete knowledge of the effectiveness of various management practices such as riparian buffer strips, cover crops, and other land management measures, and inability to demonstrate causal connections is system response.

The TMDL process increasingly is being used to address water quality impairments in large systems where nonpoint pollution is the main driver of impairment and the most effective combination of best management practices (BMPs) and their spatial deployment is largely uncharted territory. BMPs typically are intended to retain stormwater and enhance infiltration so that nutrient processing and sediment storage can reduce loads to downstream systems. Their efficiencies vary and depend on design, maintenance and placement within the watershed.

The TMDL process has been used to implement BMPs over a large region in the case of the Chesapeake Bay of the eastern US. The Bay TMDL (www.epa.gov/chesapeake-bay-tmdl), completed in 2010, identified reductions in total nitrogen, phosphorus, and suspended solids needed to meet water quality standards, and specified that 60% of its goals be implemented in 2017, with total implementation by 2025. The need to reduce nutrient and sediment loads to meet these requirements has led to the implementation of novel stream corridor restoration designs, including the conversion of eroded channels to stream—wetland complexes that enhance a channel's capacity to trap and retain suspended materials delivered from upstream. Unfortunately, their effectiveness has been less than desired. Input-output budgets of total suspended solids at Bay tributaries where stream-wetland complexes had been constructed showed insignificant changes (Filoso et al. 2015). Rather than attempt to trap nutrients near the stream's juncture with the Bay, this research suggested that BMPs might better be placed in upstream locations where most inputs originate. Williams et al. (2017) also reported better nutrient and sediment

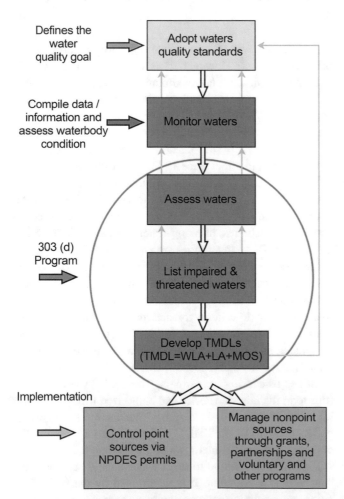

Fig. 15.8 A depiction of the TMDL process under the Clean Water Act. WLA, wasteload allocation; LA, load allocation; MOS, margin of safety (Reproduced from https://www.epa.gov/tmdl)

retention from a headwater stormwater retention project than a downstream constructed wetland. Based on their findings, the authors argue that headwater restoration projects and urban BMPs are likely to be better investments than large-scale stream-wetland complexes constructed in the lower watershed.

The development of Total Maximum Daily Loads (TMDLs) has increased markedly in recent years and as of spring 2014, over 70,000 TMDLs had been completed (https://www.epa.gov/tmdl/impaired-waters-restoration-process). As staggering as this number is, the additional need is huge. The most recently reported National Rivers and Stream Assessment (USEPA 2016) gives a rating of poor to 46% of assessed stream based on biological condition. An additional 25% were rated as fair, and only 28% were found in good condition. Nutrients, riparian condition, and sediments were identified as leading causes. Implementation of TMDLs takes time, and demonstration of success likely requires even more time, making it difficult to quantify. Published research on TMDL projects does not appear to be common in the scientific literature, and broad, comparative assessments appear to be lacking. There likely is overlap between projects initiated in response to a TMDL and many of the restoration projects described earlier, where success often is not adequately assessed. The fact that implementation of TMDLs is not required for de-listing adds another layer of uncertainty in evaluating success of this program.

15.3.3 Freshwater Conservation Planning

The rationale for conservation planning is clear: to provide adequate protection to the full complement of species and ecosystem types in freshwaters. Surveys find that as much as one in three freshwater species are threatened with extinction world-wide, as described earlier in this chapter (Fig. 15.1), and freshwater taxa often exhibit a higher risk of extinction than their terrestrial counterparts (Collen et al. 2014). Regrettably, freshwater conservation planning has lagged behind terrestrial and marine efforts. Despite increasing efforts to establish protected areas (PAs), their effectiveness for freshwater conservation is uncertain and freshwater biodiversity continues to decline (Hermoso et al. 2016). Among the principal reasons are lack of consideration of freshwater needs when designing protected areas, fewer resources devoted to freshwater conservation management, and poor understanding of complex management problems beyond the limits of the protected area. Limited information on the geographic distribution of species and ecosystem types has also hampered freshwater conservation planning. Fortunately, however, efforts to develop global syntheses of freshwater biogeography and threats are making good progress (Abell et al. 2008; Vörösmarty et al. 2010). A map of

freshwater ecoregions of the world, based on the distribution and composition of freshwater fish species, provides a useful tool to support global and regional conservation planning efforts, and to identify outstanding and imperiled freshwater systems. Presently including 830 ecoregional units, an interactive map can be viewed at https://www.feow.org/. Combining high-resolution hydrographic and land-use data, Grill et al. (2019) have generated a global map of free-flowing rivers, thereby identifying least-impacted areas that can be viewed as conservation opportunities.

At this time, freshwater protection relies heavily on protected areas designed largely around terrestrial features, with often limited consideration of their effectiveness in representing freshwater features of conservation concern (Hermoso et al. 2016). Rivers located within parks have experienced contaminant spills and invasive species, and often are affected by dams even within park boundaries. Most protected areas are not designed with biodiversity protection as a goal, and so whether their boundaries include species of concern may be accidental. Existing freshwater protected areas often are situated downstream from disturbed lands (Abell et al. 2007), in some cases rivers form the border of a reserve and so receive protection on only one side (Roux et al. 2008), and many are small, fragmented areas that lack sufficient connectivity to a broad suite of habitats (Pringle 2001). In France, all mainland national parks are located at high elevations, whereas most imperiled fishes occur downstream (Keith 2000). Using a database of conservation and recreational lands in the state of Michigan, US, Herbert et al. (2010) found uneven representation of key freshwater features. Wetlands were well represented, but riparian zones were not, particularly for headwater streams and large rivers, and terrestrial rare species received better coverage than their aquatic counterparts.

At present, nearly 15% of the world's land is in some form of protected area, close to the goal of the Convention on Biological Diversity to conserve 17% of inland waters by 2020 (CBD 2010). (Visit the CBD website for most recent statements of goals and progress at https://www.cbd.int/). For freshwater conservation to be effective, however, we require both better knowledge of the diversity and distribution of taxa and ecosystem types, and conservation planning based on an understanding of freshwater ecosystems.

When knowledge of the diversity and distribution of many taxa is inadequate, it often is assumed that better-known groups will act as surrogates for conservation planning purposes (Rodrigues and Brooks 2007). Yet, comparisons of the geographic distributions of terrestrial and aquatic taxa generally find that the former are inadequate surrogates for patterns of both richness and threat for many freshwater groups. Utilizing a comprehensive assessment of freshwater biodiversity for the entire continent of Africa, Darwall et al. (2011) examined patterns of richness and threat for all known species of freshwater fish, crabs, mollusks, dragonflies, and damselflies, and compared patterns for these aquatic groups with those of birds, mammals, and amphibians. In general, they found that groups that have been the focus of most conservation research are poor surrogates for patterns of both richness and threat for many freshwater groups, and the existing protected area network underrepresented freshwater species. In addition, freshwater groups had significantly lower surrogacy values for each other than did birds, mammals, and amphibians. In short, conservation efforts targeted at the better-known taxonomic groups may not confer adequate benefits for other species. In their global survey six freshwater taxonomic groups, Collen et al. (2014) also found little congruence among these groups in species richness, threatened-species richness, and endemism.

There also is recognition that, due to their linear, branching, and hierarchical nature, aquatic systems are not well suited to terrestrial PA planning approaches. Authors agree that the catchment scale is appropriate for freshwater conservation (Saunders et al. 2002; Dudgeon et al. 2006), but problematic in practice because the area required can be impracticably large and the exclusion of people rarely is feasible. When one considers the need to protect the entire upstream drainage network, the riparian zone, and much of the surrounding landscape, and to avoid dams, pollution, or other activities that might prevent passage of migratory species, the challenges of whole-catchment conservation are apparent. Abell et al. (2007) argue that the solution requires looking beyond the protection of individual sites, and instead developing a spatially distributed set of conservation strategies intended to protect specific populations or target areas.

The literature addressing effective design of freshwater PAs has largely combined core principles of freshwater ecology with common sense (Hermoso et al. 2016). These principles emphasize the importance of preserving upstream–downstream and lateral linkages both for biophysical functioning and species movements; that abating threats requires a catchment-wide approach, and that maintaining or restoring hydrological regimes is critical to a PA's success. Further, PAs should be representative of different types of freshwater ecosystems, large enough to provide adequate habitat, sufficiently connected via upstream-downstream linkages to allow the movements of biota and transport of materials, and cognizant of external threats. Various approaches to freshwater conservation planning have been explored, including manual exercises and others automated by software, but the end product is the same: a map of locations targeted for conservation action, superimposed on a river network, showing levels of human disturbance, extent of protection, and gaps in the protected network. From this one can inventory the number of river segments by level of disturbance, level of protection, and priority for conservation action.

Development of a network of PAs for the Upper Mississippi River basin illustrates a sequential process that combines a coarse-and fine-filter approach. The coarse filter relies on a hierarchical spatial classification based on broader scale zoogeographic and hydrologic units, and the fine filter uses detailed species distribution data where available (Higgins et al. 2005). Planning for the Upper Mississippi River basin benefited from a relatively large amount of data on species occurrences available for the fine filter (Khoury et al. 2011). The coarse filter identified 238 unique types of aquatic ecosystems which, combined with the 129 species that were elements of the fine filter, resulted in 606 areas of biodiversity significance, primarily small rivers, headwaters, and creeks (Fig. 15.9). If implemented, this network would ensure some representation for 78% of the 129 fine-filter species. Working in a data-scarce region where biological and physical data were almost completely lacking, Thieme et al. (2007) used remote sensing to map basins/sub-basins for a large river system in the southwest Amazon and classify ecosystem types based on physical features and vegetation. The resulting network of conservation areas was

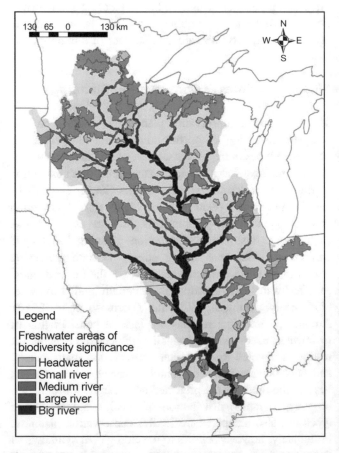

Fig. 15.9 Freshwater areas of biodiversity significance in the Upper Mississippi River. (Reproduced from Khoury et al. 2011)

partially encompassed within protected lands and indigenous areas, but their analysis identified 84 currently unprotected sub-basins necessary to fulfill representation and connectivity goals.

Further evolution of freshwater PA planning lies in the direction of formalizing the necessary steps, and increasingly is computationally intensive, often relying on specialized software to establish a network of protected areas representing the full variety of species or ecosystems. Referred to as systematic conservation planning, it emphasizes three overarching principles—representation, persistence, and quantitative conservation target setting (Nel et al. 2009). Representation refers to the need to include the full variety of biodiversity and ecosystem types in the planning region. Persistence requires maintenance of ecological condition to support the natural processes that maintain species and ecosystem integrity; in effect, assessing and addressing threats. Setting quantitative conservation targets requires defining the number of occurrences of a particular river type or the number of occurrences of a species that are desired. Together these allow computation of various conservation portfolios by calculating the contribution each site makes to conservation targets not yet achieved in the existing set of conservation areas. Many other factors must be taken into consideration, of course. Vulnerability of a site, essentially a threat assessment, can be assessed using information on planned or potential development within a planning unit. A number of societal considerations are critical to project implementation and success, including cooperation among different actors, building capacity in conservation agencies, raising awareness of the need for conservation, and developing an appropriate monitoring and evaluation system.

An advantage of computer-driven planning algorithms is the ability to examine alternative scenarios, vary conservation targets, and factor in costs. Esselman and Allan (2011) used Marxan conservation planning software to design a network of river sites intended to capture 15% of the range of each of 63 fish species in Belize, Central America. Upstream risk intensity was modeled from a GIS of landscape-based sources of stress, and solutions were constrained to account for river basin divides. The proposed reserve network encompassed 11% of the study area, of which half was within existing protected areas, and remaining areas were identified as gaps in protection. Addition of critical management zones, as defined by Abell et al. (2007), including riparian buffers and fish migration corridors, expanded the network area by one-fifth.

The principles of freshwater conservation planning are increasingly understood and embraced by agencies and NGOs. However, many completed plans are research undertakings whose impact on actions in the real world is not obvious. Some, such as the Upper Mississippi Basin plan, are the work of influential NGOs such as The Nature

Conservancy, with presumably a higher likelihood of implementation. Expanded monitoring and evaluation of freshwater PAs is of critical importance to learn what influences their effectiveness and, hopefully, to demonstrate their benefits. As with restoration, implementation, monitoring, and assessment of effectiveness are challenges still to be met.

15.4 Three Pillars of River Management

Successful management actions to repair, restore, and protect rivers will require expertise from many sectors, confidence that actions taken are likely to produce desired results, and support from the public and institutions. Here we put forth what might be called the three pillars of river management: fundamental science, measurement of progress, and societal support.

15.4.1 Understanding the Fundamentals of River Systems

Current scientific knowledge provides a sound underpinning for river management. Recent years have seen significant advances in methods of hydrologic analysis, flow metrics that have ecological relevance, and our understanding of the relationships between flows and ecological processes. Using data on the relationship between flow and ecology over a wide range of flows and species, including life cycle stages and seasonal timing, the science of environmental flows, or e-flows, offers a holistic approach with the potential to recommend a hydrologic regime that can achieve desired outcomes linked to explicit quantitative or qualitative ecological, geomorphological, and perhaps also social and economic responses. The ecological limits of hydrologic analysis (ELOHA) provides a framework for determining environmental flows needed to meet ecological and societal needs (Fig. 2.20). Beginning with hydrologic analysis and streamflow classification, this multi-step framework assigns flow-altered streams to a presumed pre-impact stream type, using flow alteration-ecological response relationships drawn from existing data and knowledge or new studies. More than a science framework, ELOHA seeks to incorporate expert and traditional knowledge and differing priorities and social perspectives to provide a decision-making framework to aid planning and address water conflicts.

Research in fluvial geomorphology (Chap. 3) provides a process-based understanding of the balance between river discharge and sediment supply, how together they determine many aspects of channel form and habitat, and how alterations to either can be major drivers of river degradation. This is the knowledge base to understand the influence of altered flows and changes in sediment supply, especially reductions in sediment supply due to trapping behind dams and mining of river beds for gravel. Stream restoration requires an understanding of the interactions of discharge and sediment supply in determining channel form and physical habitat within the channel. Most restoration work begins with these considerations, often referred to as hydrogeomorphic because of the intersection of hydrologic and geomorphologic processes. Recognition of the dynamic nature of river systems allows us to make realistic choices between when to do nothing, when to undertake management interventions to assist system recovery, and when the best solution is engineered design, necessary to accommodate human presence in the landscape (Fig. 15.10).

How organisms respond to different features of their abiotic environment, including dissolved oxygen, current, temperature, and physical habitat elements, has received extensive study for decades (Chap. 5). The roles played by variability of flows, substrate heterogeneity, wood, alternating pools and riffles, gravel for spawning by fishes, habitat for invertebrates, and surfaces for attachment by algae and biofilms, are well understood. It has long been recognized that structurally simple or extreme environments tend to support fewer species, whereas more moderate and heterogeneous habitats support more species. Thus, the knowledge exists to reverse the anthropogenic degradation and homogenization of habitat resulting from human disturbance.

To be sure, our knowledge is incomplete, and complex interactions among the main drivers of environmental degradation make it difficult to predict with certainty how a system will respond to management intervention. The conceptual model of linked pressures and response variables associated with riparian buffer management (Fig. 15.6) illustrates the complexity of our current understanding of cause-and-effect relationships, and the amount of research

examining each linkage. Expectations of outcomes from management intervention must always be paired with acceptance of uncertainty, and thus the need for evaluating system response and possibly for the employment of new management measures.

15.4.2   Measuring Progress

Monitoring and assessment are key aspects of any management program seeking to improve ecosystem condition. Ideally, this takes place within a larger framework in which objectives are clearly established, management is targeted at probable causal factors, and lessons learned from a rigorous monitoring program feed back into future actions in a cycle of adaptive management. Monitoring of water quality and sensitive species, especially of species responsive to organic enrichment and low oxygen levels, has been practiced for at least a century. Methods began to change after the 1970s, at least partly because 1972 amendments to the Clean Water Act called for maintaining and restoring the biological integrity of fresh waters. Today, monitoring to assess river condition is increasingly sophisticated and standardized, employing integrative ecological indices based on the biota and on aspects of habitat. The goal of these indices is to measure river condition, and increasingly this is referred to as 'river health', in the very broad sense that a healthy river is one in good condition (Karr and Chu 1999).

Development of the Index of Biological Integrity (IBI, Karr et al. 1986), was a significant milestone, as it provided the link between the goal of maintaining and restoring the biological integrity of fresh waters, and the means to measure its attainment. The IBI is a multi-metric index, meaning that it is the sum of ten or more individual metrics, including species richness and composition, local indicator species, trophic composition, fish abundance, and fish condition. Because it is based on multiple metrics, which are expected to be sensitive to different levels and types of environmental stress, the IBI is considered a useful integrator of multiple stressors affecting biological assemblages. By the late 1990s, a multitude of rapid bioassessment protocols were in use by various state agencies in the US (Barbour et al. 1999), with subsequent extensive updates (USEPA 2013, 2016). This provided technical guidance to protocols, although the choice of metrics often differed among states. In Europe, implementation of the WFD has led to the development of many different approaches, which, although diverse, are sufficiently intercalibrated to provide robust comparisons across countries and regions (Birk et al. 2012). Multimetric indices based on benthic invertebrates are most common, but other methods, including those based on traits, biomarkers, and functional feeding groups, also are used (Bonada et al. 2006).

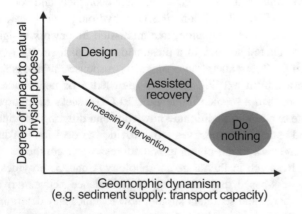

Fig. 15.10 The need for restoration intervention is a function of degree of impact and geomorphic dynamism of the system at relevant scales. (Reproduced from Friberg et al. 2016)

The need for indices suitable for monitoring river condition based on the biota has led to much innovation and a diversity of approaches. A comparison of 13 protocols for monitoring streams from different regions and countries found that methods of sampling and level of taxonomic resolution were similar (Buss et al. 2015). However, translation of data into an index has followed different paths, resulting in at least three main approaches. Multi-metric indices, described above, are in wide use. In addition, biotic indices based on the presumed sensitivity of individual taxa to impairment assign a sensitivity or tolerance score to each family or genus of aquatic invertebrates, and then aggregate these into a single score based on the assemblage of invertebrates at a site. The Saprobian system, long used in Europe to assess organic pollution (Bonada et al. 2006), and the Family Biotic Index (Hilsenhoff 1988), developed in North America, are examples. "Predictive modeling" is another approach that uses statistical models to predict the expected set of species from environmental site characteristics based on a multivariate model developed using undisturbed reference sites. When a test site is to be evaluated, its environmental conditions are used to predict the expected assemblage assuming the site is unaltered, and then the observed assemblage is compared to the expected (O/E). Predictive modeling has been pursued in England (Clarke et al. 2003), Australia (Simpson and Norris 2000), and the US (Hawkins et al. 2000). The US National Rivers and Streams Assessment uses both multi-metric and O/E indices for assessing wadeable streams (USEPA 2016). In the future, methods in the early stages of development, including genetic methods that rely on DNA sampled from organisms or directly from the water ("environmental DNA") may play a more prominent role (Hering et al. 2018).

Each approach requires the establishment of reference conditions for comparison with test or impaired sites. While the need for benchmark or reference conditions is widely recognized, locating a suite of undisturbed sites often is challenging. Stoddard et al. (2006) advocate use of terms such as minimally disturbed and least disturbed to acknowledge that reference condition may not reflect the historical, undisturbed ecological condition of streams. Use of reference condition encounters another challenge when bioassessment covers large geographic areas with differing climate, geology, vegetation, etc. Biological assessment is most efficient when reference condition is regionalized, and so comparisons across Europe or the United States rely on indices that are referenced to different regional expectations. To aid in implementation of the WFD, European workers have developed a river typology based on altitude, size, and geology, which captures almost 80% of all European rivers for purposes of intercalibration of monitoring results (Lyche Solheim et al. 2019).

The development of biological indices spurred the transition away from reliance mainly on chemical water quality standards, such as dissolved oxygen and nutrient concentrations, by providing a mechanism to determine the ecological condition of a waterbody. Indices translate a matrix of species names and abundances decipherable only by another biologist into a single metric that can be used to classify a water body as poor, fair, good, or excellent. They are a powerful tool for communicating the ecological condition of a region's or a nation's streams and rivers to decision-makers and the public. By combining sampled locations from a national assessment of US streams rated in good, fair, or poor biological condition, with a model relating stream and landscape condition, Hill et al. (2017) mapped the condition of every stream in the US. Such comprehensive information is extremely useful in identifying priorities for conservation and restoration.

A freshwater monitoring program in South East Queensland, Australia, illustrates how such information can be used to garner public support and influence decision-makers. The Queensland project took an important step beyond monitoring by converting its findings into scorecards that communicate easily and effectively with the public. Each indicator was standardized from 0 to 1, where 1 is the reference condition for a particular stream type (i.e. 'best case'), and 0 is the 90th percentile recorded or the theoretical minimum (i.e. 'worst case'). These report cards (Fig. 15.11) are well publicized, and presented each year to politicians and senior policy makers in a televised ceremony (Bunn et al. 2010).

15.4.3 Societal Support

Existing scientific knowledge and practical expertise are not the barriers to repairing, restoring, and protecting river ecosystems. The necessary technical expertise to improve the ecological condition of rivers exists. Expertise in hydrology, geomorphology, and ecology is sufficiently advanced to give confidence that the underlying principles of river science are well understood. The science and practice of monitoring ecological condition also is on sound footing, ensuring that measurement of condition and tracking of trends can be reliably accomplished. Practitioners of river restoration in the public and private sectors have a wealth of experience to draw upon, although some controversies exist regarding approaches, and whether the sharing of lessons learned is as effective as it could be. With the caveat that the response of any complex ecosystem to management intervention includes a measure of uncertainty, hence compels an adaptive management perspective, the necessary science and technical expertise is sufficient to undertake a wide range of actions to rehabilitate and restore river ecosystems.

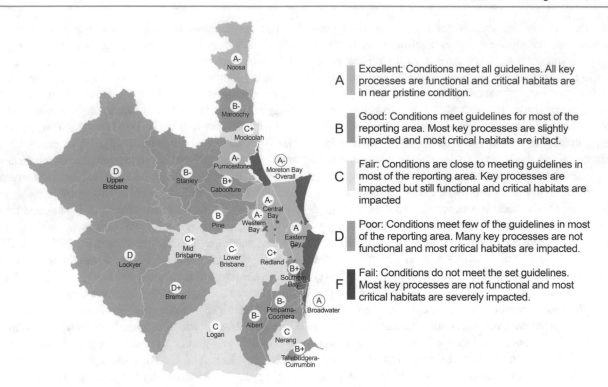

Fig. 15.11 A report card showing environmental conditions across multiple catchments in South East Queensland, Australia. This reporting system has been in place since 2000, and the report itself is presented annually by the non-governmental organization, Healthy Land and Water, to politicians and senior policy makers in a public, televised ceremony. Image courtesy of the Healthy Land and Water, Queensland, Australia

Successful river restoration requires more than this, of course. It requires the commitment of resources, both human and monetary, and the support of responsible agencies and institutions. This in turn requires more effective transmission of knowledge to policy and decision-makers, and better communication with the public. Going further, it requires recognition that ecosystems are complex, interactive socio-ecological systems (Young et al. 2006). At the same time that human economic uses of rivers and their landscapes are ultimate drivers of the many direct causes of river degradation, society desires that river resources be conserved and restored to functional states. Therein lies the great challenge—how to use riverine resources in a manner that is socially and ecologically acceptable (Naiman 2013). To fully appreciate river ecosystems as complex socio-ecological systems, and to manage them to sustainably meet societal and ecological goals, will require new approaches and further maturation. Encouragingly, however, efforts to communicate with, learn from, and actively engage the public are increasingly woven into existing efforts through the participation of stakeholders.

Stakeholders include all who have some association with the river ecosystem and may be affected by management actions. This may include scientists, representatives of interest groups and businesses, members of the community, employees of government agencies and non-governmental organizations (NGOs), and activists, representing a diversity of perspectives. Where their participation is embraced, stakeholders play an increasingly important role in public policy and are frequently used as a source to better inform public decision making. River restoration centers and environmental NGOs can have the resources to exert a strong role in shaping dialogues and influencing outcomes. Community groups can participate in the collection of data and monitoring of projects within a citizen science framework, providing the opportunity for much more extensive post-project appraisal and a more productive relationship between scientists and nonscientists. The involvement of local communities in monitoring has the potential to stimulate learning, promote engagement, and provide useful information to inform management.

Stakeholder participation goes beyond simply receiving input from the public, requiring that stakeholders have some influence over decisions made and actions taken. It also brings a mix of risks and advantages (Luyet et al. 2012). For example, stakeholder participation builds trust and local knowledge can improve project design; however, the process can be time-consuming, expensive, and not fully representative of all interests. Who participates, how they participate, and when within the project timeline they participate, are all important concerns addressed by a social science literature that discusses participation techniques. Differences in interests and objectives can complicate decision-making. One can imagine a situation where local stakeholders prioritize aesthetic goals such as a riverside path and an uncluttered river by the removal of wood, while science professionals advocate for an undisturbed riparian and in-channel wood. In a survey of local stakeholders with respect to the construction of a dam on a Dutch river, Verbrugge et al. (2017) found that local residents, recreational users, and shipping professionals differed in their level of trust, attachment to the river landscape, and evaluation of the effects of dams.

Involving stakeholders in scenario planning can be a powerful tool for envisioning alternative outcomes of river management. Based on detailed input from local stakeholders, Baker et al. (2004) developed three alternative future landscapes for the Willamette River Basin in the year 2050, and evaluated the likely effects of these landscape changes on four endpoints: water availability, Willamette River physical and biological condition, ecological condition of tributary streams, and terrestrial wildlife. Stakeholder input was extensive, with monthly meeting held for over two years to develop very detailed assumptions in designing each alternative future (Hulse et al. 2004). The process led to greater stakeholder understanding and a feeling of ownership in the final product, and strongly reflected stakeholder values, assumptions, and visions. Researchers concluded that expert-based scenarios outside the experience of current stakeholder experience may be less readily accepted, but perhaps should be blended with stakeholder-based perspectives to broaden the range of envisioned futures.

Equitably engaging stakeholders in the governance of water resources is challenging (Butler and Adamowski 2015). If not actively engaged in advocacy groups, community members may not be aware of their legal rights or their ability to have their voice heard in watershed planning activities. Historical relationships between decision-making bodies and groups of varying demographics may influence the power and perceived legitimacy of stakeholder groups. The ability to attend meetings and other events may be restricted to stakeholders who have transportation, childcare, and flexible work hours. Education and outreach materials may only be presented in a single language, inhibiting stakeholders who speak and read different languages from fully participating in planning and governance. These are some of the reasons why communities continue to struggle with environmental justice issues associated with water resource management. The USEPA defines environmental justice as "the fair treatment and meaningful involvement of all people regardless of race, color, national origin, or income with respect to the development, implementation, and enforcement of environmental laws, regulations, and policies." As evidenced by events including, but not limited to the Flint water crisis in Michigan in the US (Hanna-Attisha et al. 2016), the consequences of the inequitable distribution and management of water resources can be profound.

Participation by members of local communities also can be especially important in relatively undeveloped areas where resource conflicts affect the livelihoods of indigenous populations. In remote regions, scientific study may be very limited, making local knowledge all the more valuable. In areas where indigenous people or local communities of long-standing have strong cultural associations with and livelihood dependency on natural resources and the state of the environment, these individuals and communities are not only key stakeholders, but also valuable sources of knowledge. Traditional ecological knowledge (TEK), also called indigenous ecological knowledge, increasingly is seen as a complement to scientific knowledge, especially in areas where scientific data are scant. Although capturing all facets of TEK in a short definition is difficult, TEK is typically defined as a cumulative body of knowledge, practices, and beliefs, handed down through generations by cultural transmission, concerning the relationship of living beings (including humans) with one another and with their environment (Berkes et al. 2000; Failing et al. 2007). In some ways TEK is similar to the local knowledge that can be provided by a number of identifiable groups, including long-time community residents, indigenous people, and resource users with specialized knowledge such as fishers, farmers, or hunters. Usher (2000) suggests that the knowledge of indigenous people is likely to be richer, however, covering a larger region and accumulated over a longer time, and consequently the breadth of aboriginal environmental knowledge and the scope for drawing connections among phenomena may be greater.

Traditional ecological knowledge has been incorporated into environmental flow assessments in remote locations with a strong indigenous presence. The Patuca River, Honduras, Central America, flows through three national protected areas that include the roadless territory of two indigenous groups, the Miskito and the Tawahka, who depend on riverine and riparian ecosystems for navigation, agriculture, artisanal fisheries, bush meat, edible and useful plants, and drinking water. To develop environmental flow recommendations to mitigate effects of hydropower development in a data-poor region, Esselman and Opperman (2010) held workshops with representatives of the indigenous communities where participants annotated photographs and used hand-drawn maps to show water levels associated with different river conditions, how different flood levels affected crops and communities, and the most challenging passage points for boat traffic. The ability of indigenous fishermen to recognize taxa, behavioral traits, and spatiotemporal changes in fish assemblage composition across seasons was consistent with findings from the fish biology literature. Flow prescriptions for this data-poor region, based on hydrologic analysis, research published in the scientific literature, and local knowledge included low flows for each month, high-flow pulses, and floods, in dry, normal, and wet years (Fig. 15.12).

The threats posed to indigenous people's belief systems and livelihoods can lead to very strong opposition to a development project. In the Altai region of Siberia, a combination of TEK and cultural beliefs stopped a large dam and changed the course of development in a region. The Katun River in the Altai region of Siberia contains large numbers of important cultural sites dating from the Neolithic, and is considered central to the culture of the indigenous Altaians. Studies by cultural anthropologists describe spiritual beliefs related to the river, including curative sacred springs, special words said while crossing the river, and avoidance of taking water at night, which may upset the spirit of the river (Klubnikin et al. 2000). Proposed construction of a large dam would have significantly altered the ecology of the region, and the Altaian people would have lost much of their sacred and cultural landscape. Opposition to this project united indigenous people, well-known Siberian writers, and scientists in a protest that successfully defeated plans to build this 80-m high dam.

Today, all of the major rivers of the Altai Republic remain free of dams, new buildings stress energy efficient construction, and energy is supplied by renewables including solar, wind, and small-scale hydro (https://www.altaiproject.org/).

Finally, it is important to note that the advances of recent decades have occurred mainly in rich countries. A global analysis comparing human water security threat with biodiversity threat found stark contrasts between rich and poor nations (Vörösmarty et al. 2010). Much of the developed world faces the challenge of protecting biodiversity while maintaining established water services. In contrast, the developing world faces threats to both human water security and biodiversity. In many parts of the world, environmental protection is less robust because resources are fewer and improvements in livelihoods are more urgent. The world's large rivers and their floodplains are home to some 2.7 billion people. For many of these river systems, human pressures are intense, and economic dependence on these river basins for water, power, and food is high (Best 2019). In such settings, river basin management will need to focus on co-benefits of simultaneously supporting economic development while protecting biodiversity and key ecosystem functions (Poff and Matthews 2013).

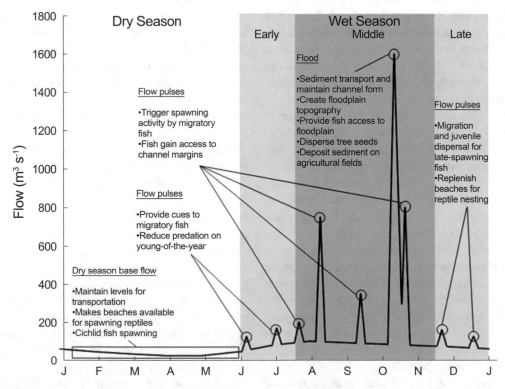

Fig. 15.12 A graphical summary of environmental flow recommendation (dark black line) for the Patuca River, Honduras, in response to a proposed hydropower dam. Dry and wet seasons are indicated by white and gray shaded areas, respectively. Environmental flow components for a "normal" hydrological year. (base flow, high flow pulses, and floods) are labeled with some details of the important ecological and social values that they support. Ideally, the timing of flow pulses and floods will be adjusted as a function of reservoir inflow, rather than having the static shape presented here as an example. (Reproduced from Esselman and Opperman 2010)

15.5   Progress Made, Progress Needed

The urgent need to repair, restore, and conserve river ecosystems around the globe is beyond question. Rivers face an abundance of threats, and many of these threats appear to be on an upward trajectory. Biological diversity is severely imperiled. Surveys of river condition find that most are in poor or fair condition, especially in developed countries. The success of restoration activities is poorly known and at best, mixed. Land-based protected area design may be inadequate for freshwater ecosystems. Change is desperately needed.

It is important, however, not to lose sight of what has been accomplished. As a thought exercise, consider the progress of the past two decades, using the year 2000 as an inflection point. Threats to rivers and the imperiled status of freshwater biodiversity were highlighted in papers by Allan and Flecker (1993), Malmqvist and Rundle (2002), and Strayer (2006), among many others. Ecological restoration became the primary objective of a great deal of stream management, resulting in dramatic growth of published research and projects described as restoration activities (Bernhardt et al. 2005). Emphasis has shifted from restoring structure to restoring function, and there is wide recognition that local actions will be most successful if embedded in coordinated actions at catchment or river basin scale. Increasingly, these are carried out within a formalized planning process (Fig. 15.7). At about the turn of the century, the United States and countries of the European Union underwent a major change from an emphasis on effluent water quality standards based on pollutants, to recognition that the condition of waterbodies should be assessed using biological and ecological measures. Although biological assessment has a long history, and some major methodological advances date from earlier decades, widespread adoption and standardization of biological assessment has taken place within the past two decades (Barbour et al. 1999). Scholarly recognition that fresh waters had received inadequate attention in conservation planning can be seen in papers published around the year 2000 (Saunders et al. 2002). At about this time, a focus on ecosystem services gained prominence and sparked advances in defining and quantifying human benefits (Millennium Ecosystem Assessment 2005). Public engagement and stakeholder participation have become more widely recognized as critical to project success. Finally, as described throughout this third edition of *Stream Ecology*, myriad research advances have greatly improved out understanding of the basic principles that govern the structure and function of stream and river ecosystems.

No doubt, for every advance just described over the past two or so decades, one can find antecedents in excellent work performed over many decades of the 20th century. But that should not obscure the main point of this thought exercise: repairing, restoring, and conserving river ecosystems is a long, complicated process, and not much time has elapsed since the urgency has been widely acknowledged. This is important to keep in mind when confronted with the legitimate and important argument that progress to date has been disappointing. Looking forward, we can hope to see continued progress in every area, but three seem especially important to call out.

First, better understanding of the ecological success of management interventions is urgently needed. This is widely recognized (Palmer et al. 2014; Wohl et al. 2015; Friberg et al. 2016), and it is sobering to realize how little information exists to assess the overall success of restoration projects, the European Union's WFD, USEPA's TMDL program, or freshwater conservation planning. One explanation is simply the time needed for system recovery. Ecosystems may recover slowly, and improvement may become more evident in future years. Another is the challenging task of assembling reports of widely varying detail from thousands of projects carried out by different agencies and states. These issues can be addressed through improvements in monitoring, reporting, and the establishment of common data banks, and likely would be especially helpful for lessons learned about actions at the local or reach scale. Although the science and practice of monitoring have undergone major advances in the past several decades, the need remains to develop more targeted methods able to detect gradual improvements and to improve diagnosis of the cause of deterioration (Carvalho et al. 2019). A greater challenge lies in the need to understand how local projects integrate at the catchment or river basin scale, and how these collective actions may be at the mercy of unaddressed threats within and beyond the catchment. The emphasis on integrated river basin planning within the WFD is a welcome signal. However, determining how best to design a suite of complementary projects at the catchment scale, and then assessing their collective success, is uncharted territory. Conservation planning approaches may help, but are intended to develop the optimum suite of targets, rather than the optimum suite of threat-ameliorating actions. Because so much of the real work of river management is carried out by agencies, there is a need for cross-jurisdictional collaboration, aided by institutional arrangements that ensure common goals, methods, and programmatic frameworks informed by best available science. Finally, resources are limited, and so it is important to be able to identify a suite of actions that is sufficient to achieve a desired level of river condition, justifying the return on investment.

Second, the influence of a changing climate will manifest in many ways that are poorly understood. Our ability to forecast future river flows is affected by what has been called the end of stationary, meaning our ability to predict future flows, floods, and droughts based on the historical record. Flow prescriptions intended to benefit the biota become more uncertain, possibly requiring a shift towards prescribing a range of flows that can sustain resilient, socially valued ecological characteristics in a changing world. Climate change will influence temperature regimes as well as overall runoff and flow regimes, with consequences for extreme events, seasonal timing, and all of the metrics describing flow magnitude, frequency, duration, and so on. In turn, the composition of riparian vegetation, rates of productivity, organic matter decay, and biogeochemical cycling will be altered. The full complement of changes can at best be speculated upon. Climate change also is expected to impact societies by changes to water runoff and supply, increasing water stress in some areas and hazard risk from flooding in others (Palmer et al. 2008). These impacts are likely to be greatest in river systems already affected by dams, relative to free-flowing rivers, and require a range of management interventions that may or may not benefit the ecosystem. One can hope that the global community will act to slow climate change. Managers of river ecosystems must prepare to adapt to its consequences through a deeper understanding of its impacts and appropriate actions that enhance system resiliency.

Third, greater appreciation is needed that what is being managed is not just an ecosystem, but a socioecological system. Humans are an integral part of virtually every ecosystem, directly and indirectly contributing to their degradation, benefiting from and enjoying their services, and serving as the source of the human capital to study and advocate for their protection. Promising steps have been made in public participation, stakeholder engagement, and incorporation of traditional ecological knowledge in planning and decision-making. A change in emphasis is underway, from a limited focus on meeting human needs while attempting to mitigate environmental costs, to an emphasis on sustained human benefits, including water for direct human use and water to support healthy ecosystems and the services they provide. It will be necessary to go one important step further. Protection of rivers is served by a shared societal vision of what a healthy river provides, aesthetically as well as functionally. Appreciation for the views and sounds of nature, and of rivers, is universal. To return to an earlier theme, people will protect what they love. The greatest challenge in protecting rivers is to vastly increase the numbers who care about protecting rivers. The future of river conservation lies in a full accommodation of human needs and values into a shared and equitable vision of the healthy rivers and streams that we wish to bequeath to future generations.

15.6 Summary

In the age of the Anthropocene, the world's rivers are amongst the planet's most highly altered ecosystems. Freshwater biological diversity is highly threatened, as much or more than terrestrial counterparts for taxa where comparative data exist. Yet, healthy streams and rivers benefit humans in myriad ways. They provide drinking water and harvestable fish, generate hydropower and supply irrigation water when harnessed by dams and canals, are useful for navigation, and absorb flood waters. In addition to these tangible benefits, running waters have aesthetic values that include the pleasures people experience from fishing, paddling, or strolling along a riverbank, but extend much further into the spiritual realm.

There are many reasons why society should repair, restore, and protect river systems for present and future generations. One important rationale centers on ecosystem services, the goods and services that an ecosystem provides free of charge, and the benefits that people receive from ecosystems. These can be grouped into provisioning services, such as fish, drinking water, and hydropower; regulating services such as waste decomposition and water purification; supporting services, including basal resources, nutrient cycling, and habitat provisioning; and cultural services, including educational, recreational, aesthetic, and spiritual benefits. An ecosystem services perspective makes explicit that benefits may be lost due to environmental stressors, and also provides a way to message and quantify the benefits gained under ecological restoration. Quantification of ecosystem services relies on a mix of methods ranging from direct market valuation to indirect methods assessing peoples' willingness-to-pay for a benefit and proxy estimates.

A second rationale, arguably more fundamental than service valuation, emphasizes the importance of nature to human well-being. This includes the spiritual and psychological, the restorative experience that derives from encounters with natural environments, and much that is difficult to put into words. This is the love of nature broadly, and of rivers specifically, which we call rheophilia. Support for this perspective can be found in scholarly analysis of human responses to landscapes and sounds, from measurements of mood, and in the art, literature, poetry, and music that draws inspiration from and celebrates natural settings. It should be noted that such thinking is not restricted to the most wild and remote images of nature, but also includes rural and urban landscapes in which people live their lives. The restorative power of nature is not limited to a natural world with no human presence, but is enhanced by experience with the sights and sounds found in more natural settings.

These two rationales explain the "why" of repairing, restoring, and protecting rivers, providing motivation and urgency. The "how" blends river science, human perceptions and beliefs, socio-economics, politics, and much more. Management actions aimed at improving rivers increasingly emphasize a holistic approach that attempts to create or maintain some aspect of river form and function that aligns with hydrologic, geomorphic, and ecological processes. This can be accomplished by relieving pressures that degrade and harm a river system, thereby promoting natural recovery; and by active measures to assist recovery that may include dam removal, addition of habitat elements, control of an invasive species, ensuring environmentally beneficial flows, and many more such actions. In recent decades, such actions increasingly are referred to using the term river restoration, and can be characterized by a more explicit pairing of science and practice, and by goals focused more strongly on recovering historic form and function to the extent feasible. There has been lively debate over the appropriateness of benchmarking against historical, undisturbed condition, whether form or function is the more suitable perspective, and the success or lack thereof to date.

Preceding chapters in this book describe the fundamental science that provides the understanding and the toolkit for managing rivers. Here we emphasize the frameworks for implementing management actions, under three headings. Integrated river basin management (IRBM) recognizes that the river basin (catchment, watershed) is the appropriate scale for organizing water management, because water sources and uses in a watershed are interrelated. Because political boundaries rarely correspond with watersheds, watershed-scale decision making requires institutional arrangements that provide for cross-jurisdictional cooperation. Enactment of the Water Framework Directive with the goal to attain good ecological status for waters of the 28 member states of the European Union, in force since 2000, has led to re-invigoration of IRBM and significant improvements in science, environmental monitoring, and formalization of the planning process in river management. In the United States, the Clean Water Act requires states, territories, and tribes to set water quality standards, which, if not met, require development of a Total Minimum Daily Load (TMDL) to address the cause of failure to meet designated uses. Rather distinct from these agency-driven approaches to remedy degraded waters, the field of conservation planning attempts to identify rivers systems, portions of river systems, or locations important for their habitat elements and species, to be designated as protected areas. Often driven by conservation organizations and supported by governments and international conventions, these efforts seek to ensure representation of the most biologically important representatives of ecosystems and habitats. While freshwater conservation planning has lagged behind similar efforts in the terrestrial realm, this field is advancing rapidly, developing frameworks that take into account the longitudinal and lateral connectivity that characterize rivers, as well as their vulnerability to external threats.

Successful management actions to repair, restore, and protect rivers will require expertise from many sectors, confidence that actions taken are likely to produce desired results, and support from the public and institutions. These are the three pillars of river management: fundamental science, measurement of progress, and societal support. There is realistic fear that river ecosystems, so highly threatened by human impacts, will continue to decline. There also is justifiable concern that efforts to date have failed to deliver the hoped-for gains. But the acceleration of knowledge, concern, and effort is relatively recent. Ecosystems are complex entities with many interacting parts, such that system responses to management action generally contain some element of uncertainty. There is still much to learn about the timeline and pathway followed by recovering ecosystems. Continued efforts are called for.

Further maturation of river management should build on lessons learned, through improvements in monitoring, reporting, and the establishment of common data banks. More work is needed to understand how local projects integrate at the catchment or river basin scale, and how these collective actions may be at the mercy of unaddressed threats within and beyond the catchment. Lastly and most importantly, greater appreciation is needed that what is being managed is not just an ecosystem, but a socioecological system. Humans are an integral part of virtually every ecosystem, directly and indirectly contributing to their degradation, benefiting from and enjoying their services, and providing the human capital to study these ecosystems and advocate for their protection. A change in emphasis is underway, from a limited focus on meeting human needs while attempting to mitigate environmental costs, to an emphasis on sustained human benefits, including water for direct human use and water to support healthy ecosystems and the services they provide. Protection of rivers is best served by a shared societal vision of what a healthy river provides, aesthetically as well as functionally.

References

Abell R, Allan JD, Lehner B (2007) Unlocking the potential of protected areas for freshwaters. Biol Conserv 134:48–63. https://doi.org/10.1016/j.biocon.2006.08.017

Abell R, Thieme ML, Revenga C et al (2008) Freshwater ecosystems of the World: a new map of biogeographic units for freshwater biodiversity conservation. Bioscience 58:403–414

Allan JD, Flecker AS (1993) Biodiversity conservation in running waters. Bioscience 43:32–43. https://doi.org/10.2307/1312104

Alvarsson JJ, Wiens S, Nilsson ME (2010) Stress recovery during exposure to nature sound and environmental noise. Int J Environ Res Public Health 7:1036–1046. https://doi.org/10.3390/ijerph7031036

Angelopoulos NV, Cowx IG, Buijse AD (2017) Integrated planning framework for successful river restoration projects: upscaling lessons learnt from European case studies. Environ Sci Policy 76:12–22. https://doi.org/10.1016/j.envsci.2017.06.005

Baker JP, Hulse DW, Gregory SV et al (2004) Alternative futures for the Willamette River Basin, Oregon. Ecol Appl 14:313–324. https://doi.org/10.1890/02-5011

Barbour M, Gerritsen J, Snyder BD, Stribling JB (1999) Rapid Bioassessment Protocols for Use in Streams and Wadeable Rivers: Periphyton, Benthic Macroinvertebrates, and Fish-Second Edition. EPA 841-B-99-002. U.S. Enviromental Protection Agency; Office of water, Washington, D.C

Bergstrom JC, Loomis JB (2017) Economic valuation of river restoration: an analysis of the valuation literature and its uses in decision-making. Water Resour Econ 17:9–19. https://doi.org/10.1016/j.wre.2016.12.001

Berkes F, Colding J, Folke C (2000) Rediscovery of traditional ecological knowledge as adaptive management. Ecol Appl 10:1251–1262

Berman MG, Jonides J, Kaplan S (2008) The cognitive benefits of interacting with nature. Psychol Sci 19:1207–1212. https://doi.org/10.1111/j.1467-9280.2008.02225.x

Bernhardt ES, Palmer MA, Allan JD, et al (2005) Synthesizing U.S. river restoration efforts. Science (80-) 308. https://doi.org/10.1126/science.1109769

Bernhardt ES, Sudduth EB, Palmer MA, et al (2007) Restoring rivers one reach at a time: results from a survey of U.S. river restoration practitioners. Restor Ecol 15. https://doi.org/10.1111/j.1526-100x.2007.00244.x

Best J (2019) Anthropogenic stresses on the world's big rivers. Nat Geosci 12:7–21. https://doi.org/10.1038/s41561-018-0262-x

Bin O, Landry CE, Meyer GF (2009) Riparian buffers and hedonic prices: a quasi-experimental analysis of residential property values in the neuse river basin. Am J Agric Econ 91:1067–1079. https://doi.org/10.1111/j.1467-8276.2009.01316.x

Birk S, Bonne W, Borja A et al (2012) Three hundred ways to assess Europe's surface waters: an almost complete overview of biological methods to implement the Water Framework Directive. Ecol Indic 18:31–41. https://doi.org/10.1016/j.ecolind.2011.10.009

Blomquist W, Schlager E (2005) Political pitfalls of integrated watershed management. Soc Nat Resour 18:101–117. https://doi.org/10.1080/08941920590894435

Boerema A, Rebelo AJ, Bodi MB et al (2017) Are ecosystem services adequately quantified? J Appl Ecol 54:358–370. https://doi.org/10.1111/1365-2664.12696

Boeuf B, Fritsch O (2016) Studying the implementation of the water framework directive in Europe: A meta-analysis of 89 journal articles. Ecol Soc 21:19. https://doi.org/10.5751/ES-08411-210219

Bonada N, Prat N, Resh VH, Statzner B (2006) Developments in aquatic insect biomonitoring: a comparative analysis of recent approaches. Annu Rev Entomol 51:495–523. https://doi.org/10.1146/annurev.ento.51.110104.151124

Bunn SE, Abal EG, Smith MJ et al (2010) Integration of science and monitoring of river ecosystem health to guide investments in catchment protection and rehabilitation. Freshw Biol 55:223–240. https://doi.org/10.1111/j.1365-2427.2009.02375.x

Buss DF, Carlisle DM, Chon TS, et al (2015) Stream biomonitoring using macroinvertebrates around the globe: a comparison of large-scale programs. Environ Monit Assess 187. https://doi.org/10.1007/s10661-014-4132-8

Butler C, Adamowski J (2015) Empowering marginalized communities in water resources management: addressing inequitable practices in participatory model building. J Environ Manage 153:153-162

Carles J, Bernáldez F, De Lucio J (1992) Audiovisual interactions and soundscape preferences. Landsc Res 17:52–56. https://doi.org/10.1080/01426399208706361

Carvalho L, Mackay EB, Cardoso AC et al (2019) Protecting and restoring Europe's waters: an analysis of the future development needs of the water framework directive. Sci Total Environ 658:1228–1238. https://doi.org/10.1016/j.scitotenv.2018.12.255

CBD (2010) Convention on biological diversity-Aichi biodiversity targets. http://www.cbd.int/sp/targets/

Chan KMA, Satterfield T, Goldstein J (2012) Rethinking ecosystem services to better address and navigate cultural values. Ecol Econ 74:8–18. https://doi.org/10.1016/j.ecolecon.2011.11.011

Chin A, Daniels MD, Urban MA et al (2008) Perceptions of wood in rivers and challenges for stream restoration in the United States. Environ Manage 41:893–903. https://doi.org/10.1007/s00267-008-9075-9

Clarke RT, Wright JF, Furse MT (2003) RIVPACS models for predicting the expected macroinvertebrate fauna and assessing the ecological quality of rivers. Ecol Model 160:219–233

Collen B, Whitton F, Dyer EE et al (2014) Global patterns of freshwater species diversity, threat and endemism. Glob Ecol Biogeogr 23:40–51. https://doi.org/10.1111/geb.12096

Darwall W, Holland R, Smith K et al (2011) Investment for freshwater species. Conserv Lett 4:474–482

Dudgeon D, Arthington AH, Gessner MO et al (2006) Freshwater biodiversity: importance, threats, status and conservation challenges. Biol Rev 81:163–182. https://www.eea.europa.eu/publications/state-of-water

EEA (2018) European waters. Assessment of status and pressures

England J, Naura M, Mant J, Skinner K (2019) Seeking river restoration appraisal best practice: supporting wider national and international environmental goals. Water Environ J 0:1–9. https://doi.org/10.1111/wej.12517

Esselman PC, Allan JD (2011) Application of species distribution models and conservation planning software to the design of a reserve network for the riverine fishes of northeastern Mesoamerica. Freshw Biol 56:71–88. https://doi.org/10.1111/j.1365-2427.2010.02417.x

Esselman PC, Opperman JJ (2010) Overcoming information limitations for the prescription of an environmental flow regime for a central american river. Ecol Soc 15:6. https://doi.org/10.5751/ES-03058-150106

Failing L, Gregory R, Harstone M (2007) Integrating science and local knowledge in environmental risk management: a decision-focused approach. Ecol Econ 64:47–60. https://doi.org/10.1016/j.ecolecon.2007.03.010

Fausch KD (2015) For the love of rivers. Oregon State University Press, Corvallis, OR

Feld CK, Birk S, Bradley DC et al (2011) From natural to degraded rivers and back again. A test of restoration ecology theory and practice. Adv Ecol Res 44:119–209. https://doi.org/10.1016/B978-0-12-374794-5.00003-1

Feld CK, Fernandes MR, Ferreira MT et al (2018) Evaluating riparian solutions to multiple stressor problems in river ecosystems—a conceptual study. Water Res 139:381–394. https://doi.org/10.1016/j.watres.2018.04.014

Filoso S, Smith SMC, Williams MR, Palmer MA (2015) The efficacy of constructed stream-wetland complexes at reducing the flux of suspended solids to Chesapeake Bay. Environ Sci Technol 49:8986–8994. https://doi.org/10.1021/acs.est.5b00063

Friberg N, Angelopoulos NV, Buijse AD et al (2016) Effective river restoration in the 21st Century: from trial and error to novel evidence-based approaches. Adv Ecol Res 55:535–611. https://doi.org/10.1016/bs.aecr.2016.08.010

Griffiths M (2002) The European water framework directive: an approach to integrated river basin management. Eur Water Manag Online 1–15

Grill G, Lehner B, Lumsdon AE et al (2015) An index-based framework for assessing patterns and trends in river fragmentation and flow regulation by global dams at multiple scales. Environ Res Lett 10:1–15. https://doi.org/10.1088/1748-9326/10/1/015001

Grill G, Lehner B, Thieme M et al (2019) Mapping the world's free-flowing rivers. Nature 569:215–221. https://doi.org/10.1038/s41586-019-1111-9

Haines-Young R, Potschin M (2010) The links between biodiversity, ecosystem services and human well-being. In: Raffaelli D, Frid C (eds) Ecosystem ecology: a new synthesis. BES Ecological Reviews, CUP, Cambridge, UK

Hanna-Attisha M, LaChance J, Sadler RC, Others (2016) Elevated blood lead levels in children associated with the Flint drinking water crisis: a spatial analysis of risk and public health response. Am J Public Health 196:283–290

Hawkins CP, Norris RH, Hogue JN, Feminella JW (2000) Development and evaluation of predictive models for measuring the biological integrity of streams. Ecol Appl 10:1456–1477

Herbert ME, McIntyre PB, Doran PJ, et al (2010) Terrestrial reserve networks do not adequately represent aquatic ecosystems. Conserv Biol 24:1002–1011. https://doi.org/10.1111/j.1523-1739.2010.01460.x

Hering D, Borja A, Carstensen J, et al (2010) The European water framework directive at the age of 10 : a critical review of the achievements with recommendations for the future. Sci Total Environ 408:4007-4019. https://doi.org/10.1016/j.scitotenv.2010.05.031

Hering D, Borja A, Jones JI et al (2018) Implementation options for DNA-based identification into ecological status assessment under the European water framework directive. Water Res 138:192–205. https://doi.org/10.1016/j.watres.2018.03.003

Hermoso V, Abell R, Linke S, Boon P (2016) The role of protected areas for freshwater biodiversity conservation: challenges and opportunities in a rapidly changing world. Aquat Conserv Mar Freshw Ecosyst 26:3–11. https://doi.org/10.1002/aqc.2681

Higgins JV, Bryer MT, Khoury ML, Fitzhugh TW (2005) A freshwater classification approach for biodiversity conservation planning. Conserv Biol 19:432–445

Hill RA, Fox EW, Leibowitz SG et al (2017) Predictive mapping of the biotic condition of conterminous U.S. rivers and streams. Ecol Appl 27:2397–2415. https://doi.org/10.1002/eap.1617

Hilsenhoff WL (1988) Rapid field assessment of organic pollution with a family-level biotic index. J North Am Benthol Soc 7:65–68. https://doi.org/10.2307/1467832

House MA, Sangster EK (1991) Public perception of river-corridor management. JIWEM 5:312–317

Hulse DW, Branscomb A, Payne SG (2004) Envisioning alternatives: using citizen guidance to map future land and water use. Ecol Appl 14:325–341

Jackson RB, Carpenter SR, Dahm CN, et al (2001) Water in a changing world. Ecol Appl 11:1027–1045. https://doi.org/10.1890/0012-9623(2008)89%5b341:iie%5d2.0.co;2

Jax K, Barton DN, Chan KMA et al (2013) Ecosystem services and ethics. Ecol Econ 93:260–268. https://doi.org/10.1016/j.ecolecon.2013.06.008

Kaplan R, Kaplan S (1989) The experience of nature: a psychological perspective. Cambridge University Press, Cambridge, UK

Karr JR, Chu EW (1999) Restoring life in running waters: better biological monitoring. Island Press, Washington DC

Karr JR, Fausch KD, Angermeier PL, et al (1986) Assessing biological integrity in running waters a method and its rationale. Illinois Natural History Survey Special Publication No. 5

Keeler BL, Hamel P, McPhearson T, Others (2019) Social-ecological and technological factors moderate the value of urban nature. Nat Sustain 2:29–38

Keeler BL, Wood SA, Polasky S et al (2015) Recreational demand for clean water: evidence from geotagged photographs by visitors to lakes. Front Ecol Environ 13:76–81. https://doi.org/10.1890/140124

Keith P (2000) The part played by protected areas in the conservation of threatened French freshwater fish. Biol Conserv 92:265–273. https://doi.org/10.1016/S0006-3207(99)00041-5

Khoury M, Higgins J, Weitzell R (2011) A freshwater conservation assessment of the Upper Mississippi River basin using a coarse-and fine-filter approach. Freshw Biol 56:162–179. https://doi.org/10.1111/j.1365-2427.2010.02468.x

Klubnikin K, Annett C, Cherkasova M, et al (2000) The sacred and the scientific: traditional ecological knowledge in Siberian river conservation. Ecol Appl 10:1296–1306. https://doi.org/10.1890/1051-0761(2000)010%5b1296:tsatst%5d2.0.co;2

Kondolf GM (2006) River restoration and meanders. Ecol Soc 11:42. https://doi.org/10.5751/ES-01795-110242

Kurth AM, Schirmer M (2014) Thirty years of river restoration in Switzerland: implemented measures and lessons learned. Environ Earth Sci 72:2065–2079. https://doi.org/10.1007/s12665-014-3115-y

Lange K, Meier P, Trautwein C et al (2018) Basin-scale effects of small hydropower on biodiversity dynamics. Front Ecol Environ 16:397–404. https://doi.org/10.1002/fee.1823

Le Lay YF, Piégay H, Gregory K et al (2008) Variations in cross-cultural perception of riverscapes in relation to in-channel wood. Trans Inst Br Geogr 33:268–287. https://doi.org/10.1111/j.1475-5661.2008.00297.x

Leopold LB (1977) A reverence for rivers. Geology 5:429–430

Loomis J, Kent P, Strange L et al (2000) Measuring the total economic value of restoring ecosystem services in an impaired river basin: results from a contingent valuation survey. Ecol Econ 33:103–117. https://doi.org/10.1016/S0921-8009(99)00131-7

Loomis JB (1996) Measuring the economic benefits of removing dams and restoring the Elwha River: results of a contingent valuation survey. Water Resour Res 32:441–447. https://doi.org/10.1029/95WR03243

Luyet V, Schlaepfer R, Parlange MB, Buttler A (2012) A framework to implement stakeholder participation in environmental projects. J Environ Manage 111:213–219. https://doi.org/10.1016/j.jenvman.2012.06.026

Lyche Solheim A, Globevnik L, Austnes K et al (2019) A new broad typology for rivers and lakes in Europe: development and application for large-scale environmental assessments. Sci Total Environ 697:134043. https://doi.org/10.1016/j.scitotenv.2019.134043

Malmqvist B, Rundle S (2002) Threats to the running water ecosystems of the world. Environ Conserv 29:134–153. https://doi.org/10.1017/S0376892902000097

Millennium Ecosystem Assessment (2005) Ecosystems and human well-being: synthesis. Island Press, Washington DC

Morandi B, Piégay H, Lamouroux N, Vaudor L (2014) How is success or failure in river restoration projects evaluated? Feedback from French restoration projects. J Environ Manage 137:178–188. https://doi.org/10.1016/j.jenvman.2014.02.010

Naiman RJ (2013) Socio-ecological complexity and the restoration of river ecosystems. Inl Waters 3:391–410. https://doi.org/10.5268/IW-3.4.667

Nel JL, Roux DJ, Abell R et al (2009) Progress and challenges in freshwater conservation planning. Aquat Conserv Mar Freshw Ecosyst 19:474–485. https://doi.org/10.1002/aqc.1010

NRC (National Research Council) (2001) Assessing the TMDL approach to water quality management. The National Academies Press, Washington DC

Palmer MA, Hondula KL, Koch BJ (2014) Ecological restoration of streams and rivers: shifting strategies and shifting goals. Annu Rev Ecol Evol Syst 45:247–269. https://doi.org/10.1146/annurev-ecolsys-120213-091935

Palmer MA, Menninger HL, Bernhardt E (2010) River restoration, habitat heterogeneity and biodiversity: a failure of theory or practice? Freshw Biol 55:205–222. https://doi.org/10.1111/j.1365-2427.2009.02372.x

Palmer MA, Reidy Liermann CA, Nilsson C et al (2008) Climate change and the world's river basins: anticipating management options. Front Ecol Environ 6:81–89. https://doi.org/10.1890/060148

Plummer ML (2009) Assessing benefit transfer for the valuation of ecosystem services. Front Ecol Environ 7:38–45. https://doi.org/10.1890/080091

Poff NLR, Matthews JH (2013) Environmental flows in the Anthropocence: past progress and future prospects. Curr Opin Environ Sustain 5:667–675. https://doi.org/10.1016/j.cosust.2013.11.006

Postel SL (2000) Entering an era of water scarcity: the challenges ahead. Ecol Appl 10:941–948. https://doi.org/10.1890/1051-0761(2000)010%5b0941:eaeows%5d2.0.co;2

Pringle CM (2001) Hydrologic connectivity and the management of biological reserves: a global perspective. Ecol Appl 11:981–998. https://doi.org/10.1890/1051-0761(2001)011%5b0981:hcatmo%5d2.0.co;2

Richardson L, Loomis J, Kroeger T, Casey F (2015) The role of benefit transfer in ecosystem service valuation. Ecol Econ 115:51–58. https://doi.org/10.1016/j.ecolecon.2014.02.018

Rodrigues ASL, Brooks TM (2007) Shortcuts for biodiversity conservation planning: the effectiveness of surrogates. Annu Rev Ecol Evol Syst 38:713–737. https://doi.org/10.1146/annurev.ecolsys.38.091206.095737

Roux DJ, Nel JL, Ashton PJ et al (2008) Designing protected areas to conserve riverine biodiversity: lessons from a hypothetical redesign of the Kruger National Park. Biol Conserv 141:100–117. https://doi.org/10.1016/j.biocon.2007.09.002

Saunders DL, Meeuwig JJ, Vincent ACJ (2002) Freshwater protected areas: strategies for conservation. Conserv Biol 16:30–41. https://doi.org/10.1046/j.1523-1739.2002.99562.x

Shrestha RK, Seidl AF, Moraes AS (2002) Value of recreational fishing in the Brazilian Pantanal: A travel cost analysis using count data models. Ecol Econ 42:289–299. https://doi.org/10.1016/S0921-8009(02)00106-4

Simpson JC, Norris RH (2000) Biological assessment of river quality: development of AUSRIVAS models and outputs. In: Wright JF, Sutcliffe DW, Furse MT (eds) Freshwater biomonitoring and benthic macroinvertebrates. Freshwater Biological Association, Ambleside, Cumbria, UK, pp 125–142

Slocombe DS (1993) Implementing ecosystem-based management managing a region. Bioscience 43:612–622

Stoddard JL, Larsen DP, Hawkins CP et al (2006) Setting expectations for the ecological condition of streams: the concept of reference condition. Ecol Appl 16:1267–1276. https://doi.org/10.1890/1051-0761(2006)

Strayer DL (2006) Challenges for freshwater invertebrate conservation. J North Am Benthol Soc 25:271–287. https://doi.org/10.1899/0887-3593(2006)25%5b271:cffic%5d2.0.co;2

Strayer DL, Dudgeon D (2010) Freshwater biodiversity conservation: recent progress and future challenges. J North Am Benthol Soc 29:344–358. https://doi.org/10.1899/08-171.1

Thieme M, Lehner B, Abell R et al (2007) Freshwater conservation planning in data-poor areas: an example from a remote Amazonian basin (Madre de Dios River, Peru and Bolivia). Biol Conserv 135:484–501. https://doi.org/10.1016/j.biocon.2006.10.054

Tickner D, Opperman J, Abell R, et al (2020) Bending the curve of global freshwater biodiversity loss-an emergency recovery plan. BioScience 70:330-342

USEPA (2016) National rivers and streams assessment 2008-2009: a collaborative survey. Washington, DC

USEPA (2013) Biological assessment program review: assesing level of technical rigor to support water quality management. U.S. Environmental Protection Agency, Office of Science and Technology, Office of Water EPA 820-R-13-001, Washington DC

Usher PJ (2000) Communicating about contaminants in country food: the experience in aboriginal communities. Arctic 53:183–193. ISBN 0-9699774-0-9

Verbrugge LNH, Ganzevoort W, Fliervoet JM et al (2017) Implementing participatory monitoring in river management: the role of stakeholders' perspectives and incentives. J Environ Manage 195:62–69. https://doi.org/10.1016/j.jenvman.2016.11.035

Vörösmarty CJ, McIntyre PB, Gessner MO et al (2010) Global threats to human water security and river biodiversity. Nature 467:555–561

Vörösmarty CJ, Sahagian D (2000) Anthropogenic disturbance of the terrestrial water cycle. Bioscience 50:753. https://doi.org/10.1641/0006-3568(2000)050%5b0753:adottw%5d2.0.co;2

Voulvoulis N, Arpon KD, Giakoumis T (2017) The EU water framework directive: from great expectations to problems with implementation. Sci Total Environ 575:358–366. https://doi.org/10.1016/j.scitotenv.2016.09.228

Walters CJ, Holling CS (1990) Large-scale management experiments and learning by doing. Ecology 71:2060–2068

Williams MR, Bhatt G, Filoso S, Yactayo G (2017) Stream restoration performance and its contribution to the Chesapeake Bay TMDL: challenges posed by climate change in urban areas. Estuaries Coasts 40:1227–1246. https://doi.org/10.1007/s12237-017-0226-1

Wilson EO (1984) Biophilia. Harvard College, Cambridge, MA

Wohl E, Lane SN, Wilcox AC (2015) The science and practice of river restoration. Water Resour Res 51:5974–5997. https://doi.org/10.1002/2014WR016874

Yang W, Kang J (2005) Soundscape and sound preferences in urban squares: a case study in Sheffield. J Urban Des 10:61–80. https://doi.org/10.1080/13574800500062395

Young OR, Berkhout F, Gallopin GC et al (2006) The globalization of socio-ecological systems: an agenda for scientific research. Glob Environ Chang 16:304–316. https://doi.org/10.1016/j.gloenvcha.2006.03.004

Index

© Springer Nature Switzerland AG 2021
J. D. Allan et al., *Stream Ecology*,
https://doi.org/10.1007/978-3-030-61286-3

Printed in the United States
by Baker & Taylor Publisher Services